Progress in Scientific Computing
Vol. 4

Edited by
S. Abarbanel
R. Glowinski
G. Golub
P. Henrici
H.-O. Kreiss

Birkhäuser
Boston · Basel · Stuttgart

Jane K. Cullum
Ralph A. Willoughby

Lanczos Algorithms for Large Symmetric Eigenvalue Computations Vol. II Programs

1985

Birkhäuser
Boston · Basel · Stuttgart

Authors:
Jane K. Cullum
Ralph A. Willoughby
IBM T. J. Watson Research Center
Yorktown Heights, NY 10598 (USA)

Library of Congress Cataloging in Publication Data

(Revised for volume 2)
Cullum, Jan K., 1938–
Lanczos algorithms for large symmetric eigenvalue computations.
(Progress in Scientific computing ; v. 3–4)
Includes bibliographies and indexes.
Contents: v. 1. Theory – – v. 2. Programs.
1. Symmetric matrices – – Data processing.
2. Eigenvalues – – Data processing. I. Willoughby, Ralph A. II. Title. III. Series.
QA193.C84 512.9'434 81-38450
ISBN 3-7643-3058-9 (v. 1)

CIP-Kurztitelaufnahme der Deutschen Bibliothek

Cullum, Jane K.:
Lanczos algorithms for large symmetric eigenvalue computations / Jane K. Cullum ; Ralph A. Willoughby.
– Boston ; Basel ; Stuttgart : Birkhäuser
ISBN 3-7643-3295-6 (Basel . . .);
ISBN 0-8176-3295-6 (Boston)
NE: Willoughby, Ralph A.:
Vol. 2. Programs. – 1985.
(Progress in Scientific computing ; Vol. 4)
ISBN 3-7643-3294-8 (Basel . . .)
ISBN 0-8176-3294-8 (Boston)
NE: GT

Printed in Germany
ISBN 0-8176-3292-6 (complete set)
ISBN 0-8176-3058-9 (volume 1)
ISBN 0-8176-3294-8 (volume 2)
ISBN 3-7643-3292-6 (complete set)
ISBN 3-7643-3058-9 (volume 1)
ISBN 3-7643-3294-8 (volume 2)

Table of Contents

CHAPTER 1
LANCZOS PROCEDURES

SECTION 1.1 INTRODUCTION

The FORTRAN codes contained in this volume are designed for computing eigenvalues and eigenvectors or singular values and singular vectors of large, sparse matrices. Large means of order several hundred to perhaps 10,000. The largest matrix which we tested was real symmetric and had order 4900. This book is divided into 9 chapters. In this first chapter we give a brief description of Lanczos eigenelement procedures and then make some comments about what the Lanczos codes in this book can and cannot be expected to compute. Detailed analyses of the ideas used in these procedures are contained in Volume 1 of this book.

Chapters 2 through 7 contain procedures which are based upon the single-vector Lanczos recursion with no reorthogonalization of any kind. Six different classes of problems are addressed in these 6 chapters: Eigenelement computations for real symmetric matrices (Chapter 2); for Hermitian matrices (Chapter 3); for factored inverses of real symmetric matrices (Chapter 4); for real symmetric, generalized problems (Chapter 5) and for nondefective, complex symmetric matrices (Chapter 7); and Singular value and vector computations for real, rectangular matrices (Chapter 6). Chapters 8 and 9 contain Lanczos procedures which are based upon a 'block' version of the Lanczos recursions. These iterative block procedures include some reorthogonalization within each iteration, but this reorthogonalization is limited to reorthogonalizations w.r.t. certain vectors in each first Lanczos block.

The single-vector procedures can be used to compute anywhere from a very few to very many eigenvalues (singular values). These eigenvalues (singular values) need not be at the extremes of the spectrum. For some matrices it is even possible to compute all of the eigenvalues. The iterative block procedures can only be used to compute a few extreme eigenvalues of the specified matrix. The single vector codes consist of two phases. First eigenvalues or singular values are computed and then corresponding eigenvectors or singular vectors are computed. The iterative 'block' codes compute eigenvalues and corresponding eigenvector approximations simultaneously. Block codes for computing singular values are not included in this book. See for example, Golub, Luk, and Overton [1981] for an example of such a block algorithm.

With three exceptions which are given below, each Chapter 2 - 9 contains the following types of information for the particular class of problems considered in that chapter: documentation; main program(s); LANCZS subroutine for computing Lanczos matrices; sample matrix-vector multiply and/or solve subroutines; other subroutines needed by the codes in that chapter; and definitions of the files used by the programs together with sample input files. Because of the similarities between the variables, flags, etc., the documentation for the codes contained in Chapters 2,3,4, and 5 was combined and is contained in Section 2.2 of Chapter 2. The codes in Chapters 2, 3, 4 and 5 use essentially (with 2 exceptions) the same set of 'other or additional subroutines' so these subroutines were combined and are given only in Chapter 2, Section 2.6. Similarly, the block codes in Chapters 8 and 9 use the same set of additional subroutines and these are given only in Section 8.5. Some additional optional, preprocessing codes are also provided, and again each of these is included in only one of the chapters and not in each of the ones where it might be useful.

Each set of codes contains many write statements. These write statements serve two major functions: to provide consistency checks on the information supplied by the user, and to provide running commentary on the progress of the computations. Much of the code has been modularized to help make the program logic more transparent to the user. These codes are not designed as efficiently as they could be. Many internal comments have been included. Numerous consistency checks have been used to verify that the user has set up the procedure properly. Basically, we have compromised some efficiency for safety and robustness.

Each LANCZS subroutine together with the corresponding sample matrix-vector multiply and solve subroutines are in files labelled as *MULT. For example in Chapter 2 where real symmetric matrices are discussed this file is labelled LEMULT. The user should note that within a given *MULT file, each sample USPEC* and *MATV subroutine has been given two names so that these subroutines can co-exist with similar subroutines for other test matrices. However, two different *MULT files cannot co-exist because subroutine names are reused in going from one category of matrices to another category. In particular for the codes in Chapters 2,3, and 7, the matrix-vector multiply subroutine is called CMATV. Moreover, in all of the chapters, the matrix specification subroutines are called USPEC*. This reuse of names makes it easier for the user to pass from one set of codes to another. Furthermore, from category to category, subroutines with similar function were typically given the same name. For example, all of the subroutines which generate families of Lanczos matrices are named LANCZS. There are two BISEC bisection subroutines for computing eigenvalues of real symmetric tridiagonal matrices, one for Chapters 2,3,4, and 5 and the other one is for Chapter 6. If these sets of codes had to co-exist in one computer file, then it would be necessary for the user to devise a scheme for renaming those subroutines which have the same names.

With respect to portability, each of these programs and subroutines has been individually checked for portability by the PFORT Verifier [1981], but the communications between these subroutines have not been checked. Obvious problems with portability like non-Fortran items in the format statements have all been removed. However, certain nonportable constructions have been retained because they make the programs somewhat easier to use. The header of each of the programs contains a list of those constructions in that program which were identified by the PFORT verifier as being nonportable. These headers can be used to locate the nonportable items so that if necessary they can be modified. A list of most of the nonportable items and the reasons for retaining them are given in Table 1.1.1.

TABLE 1.1.1 Nonportable Constructions Used in the Codes

Nonportable Construction	Where Used	Why Used
Entry	Passes storage locations of arrays and parameters needed to define user-specified matrix from subroutine USPEC where arrays are dimensioned and initialized to the corresponding matrix-vector multiply or solve subroutine.	Codes do not need to 'see' the user-specified matrix. Codes need only output from matrix-vector multiply or solve subroutines for the matrix being used. User does not have to alter the calling sequences to these subroutines every time the number or kind of arrays needed to define the given matrix is changed.
Formats (20A4) and (4Z20)	(20A4) is used to read and write explanatory comments within the main programs and in sample USPEC subroutines.	Allows the user to easily modify headers describing the matrix and code being used.
	Machine format (4Z20) is used to read in and write out the Lanczos tridiagonal matrices generated and other quantities for which conversion errors could cause numerical problems.	Prevents format conversion errors incurred in input/output conversions.
Free Format Read (5, *)	Used in main program and in sample USPEC subroutines on read-ins of user-specified parameters from input file 5.	Ease of input. User does not have to have the input values properly aligned in the input file.
Complex*16 Variables	Used only in the Hermitian and in the complex symmetric Lanczos codes.	Computations require double precision complex arithmetic.
Specification of Machine Epsilon	Used in main programs	Required to define tolerances used at various points in the computations.

The single vector Lanczos codes in Chapters 2 - 7 are essentially self-contained. The user must provide the matrix-vector multiply and/or solve subroutines which are required by these codes, together with a matrix specification subroutine which defines, dimensions and initializes the matrix which will be used by the Lanczos procedure. The sample matrix-specification subroutines and sample matrix-vector multiply and solve subroutines contained with these codes can be modified and used if appropriate or they can be replaced completely. All of these procedures require a random number generator subroutine, inner product subroutines, and a subroutine to mask underflow. These procedures assume that each time the random number generator is called that the seed for this generator is automatically reset to a different value.

The iterative 'block' Lanczos codes in Chapters 8 and 9 require matrix specification and matrix-vector multiply and solve subroutines very similar to those used in the single vector codes, plus the same type of random number generating subroutine, inner product subroutine, and mask subroutine. However, as implemented here the block codes are not self-contained. These codes call two subroutines from the EISPACK Library [1976,1977], TRED2 and IMTQL2, which are used repeatedly to compute the eigenvalues and eigenvectors of the small Lanczos matrices generated on each iteration of the block procedures. The user can of course replace these calls by calls to subroutines which perform similar functions, if the EISPACK Library is not available.

The optional preprocessing programs in Sections 2.7, 4.5, 6.7, and 7.7 are stand-alone (if one includes the programs which must be supplied by the user), except for the subroutine PERMUT given in Section 4.5. PERMUT can be used in conjunction with the procedures in Chapters 4, 5 and 9. It calls the SPARSPAK Library [1979] (A. George, J. Liu, E. Ng, U. Waterloo) to try to determine a reordering of the given sparse matrix for which the sparsity of the given matrix translates into a sparse factorization of the reordered matrix.

SECTION 1.2 WHAT ARE LANCZOS PROCEDURES?

Lanczos procedures for computing eigenvalues and eigenvectors of real symmetric matrices are based upon one or more variants of the basic single-vector Lanczos recursion for tridiagonalizing a real symmetric matrix A. Given a starting vector v_1 which is typically-generated randomly, the Lanczos recursion implements a Gram-Schmidt orthogonalization of the matrix-vector products Av_i corresponding to the Lanczos vectors v_i generated by the recursion. See for example Bjorck [1967]. Specifically, we have that for i = 2,...,m,

$$\beta_{i+1}v_{i+1} = Av_i - \alpha_i v_i - \beta_i v_{i-1} \tag{1.2.1}$$

where $\alpha_i \equiv v_i^T A v_i$ and $\beta_{i+1} \equiv v_{i+1}^T A v_i$. By definition $\alpha_i v_i$ and $\beta_i v_{i-1}$ are the projections of Av_i onto the two most recently-generated Lanczos vectors v_i and v_{i-1}. In practice to improve the numerical stability of this recursion, the above formulas are replaced by the following ones.

$$\alpha_i \equiv v_i^T(Av_i - \beta_i v_{i-1}) \quad \text{and} \quad \beta_{i+1} \equiv \| Av_i - \alpha_i v_i - \beta_i v_{i-1} \| . \tag{1.2.2}$$

The α_i as defined in Eqn(1.2.2) correspond to a modified Gram-Schmidt orthogonalization procedure. The formula for β_{i+1} given in Eqn(1.2.2) is theoretically equivalent to the one given in Eqn(1.2.1). However, it is superior numerically because this choice directly controls the sizes of the Lanczos vectors. See Paige [1976].

Rewriting Eqn(1.2.1) in matrix form, we obtain

$$AV_j = V_jT_j + \beta_{j+1}v_{j+1}e_j^T \tag{1.2.3}$$

where T_j denotes the real symmetric tridiagonal Lanczos matrix of order j whose diagonal entries are the scalars α_i, $1 \leq i \leq j$, and whose subdiagonal (superdiagonal) entries are the scalars β_{i+1}, $1 \leq i \leq j-1$, generated by the Lanczos recursion. In Eqn(1.2.3), $V_j = \{v_1, v_2, ..., v_j\}$, the matrix whose columns are the Lanczos vectors generated by the recursion, and e_j is the coordinate vector whose jth component is 1 and whose other components are 0.

It is easy to demonstrate by induction that in exact arithmetic each set of vectors V_j generated by the recursion in Eqns(1.2.1) - (1.2.2) is an orthonormal set. Therefore for any A-matrix with n distinct eigenvalues and any starting vector v_1 which has a projection on every eigenspace of A, we have that for each $j \leq n$,

$$T_j = V_j^T A V_j. \tag{1.2.4}$$

Thus the symmetric tridiagonal matrices T_j are representations of the projections of the given matrix A onto the subspaces spanned by the corresponding sets of Lanczos vectors V_j. The eigenvalues of these matrices are the eigenvalues of the A-matrix restricted to these subspaces. Since the Lanczos vectors are obtained by orthogonalizing vectors of the form $\{v_1, Av_1, A^2v_1, ...,\}$, we expect the eigenvalues of the T_j to provide good approximations to some of the eigenvalues of A, if j is sufficiently large. Clearly, at least theoretically, if we extend the recursion to $j = n$, then the eigenvalues of T_n will be the eigenvalues of A. T_n is simply an orthogonal transformation of A and must therefore have the same eigenvalues as A. Moreover, any Ritz vector V_ju obtained from an eigenvector u of some T_j is an approximation to a corresponding eigenvector of A.

Basic steps in any Lanczos procedure for computing eigenvalues and eigenvectors of 'symmetric' matrices are the following. (1) Use a variant of the Lanczos recursion to transform the given 'symmetric' matrix A into a family of 'symmetric' tridiagonal matrices of varying sizes. (2) Compute eigenvalues and eigenvectors of certain members of this family. Because of the real symmetric tridiagonal structure this is a much simpler problem than computing the eigenvalues and eigenvectors of A directly. (3) Take some or all of these eigenvalues as approximations to eigenvalues of A and map the corresponding eigenvectors of the tridiagonal matrix into Ritz vectors for the matrix A. (4) Use these Ritz vectors as approximations to the eigenvectors of A.

The Lanczos recursion in Eqn(1.2.1) has several properties which make it particularly attractive for dealing with large but sparse matrices. First the given matrix enters the recursion only through the matrix-vector multiply terms Av_i. Thus contrary to what is done in the standard methods for solving small or medium size eigenvalue problems, see for example EISPACK [1976,1977], the given matrix is not explicitly modified. The user must provide only a subroutine which computes Ax for any given vector x. If the matrix A is sparse, this computation can be done using an amount of storage that is only linear in the size of the matrix instead of quadratic. Second, the recursion uses only the two most recently-generated Lanczos vectors. The Gram-Schmidt orthogonalization of an arbitrary set of vectors would require that at any given stage in the process that all of the vectors which have already been orthogonalized be available for orthogonalizing each additional vector as it is considered. Thus, the storage requirements for implementing the basic Lanczos recursion are minimal. If we use Eqns(1.2.1) - (1.2.2) then only 2 n-vectors are needed for the two most recently-generated Lanczos vectors plus storage for the α and β arrays.

There are however numerical problems if only a simple direct implementation of this recursion is programmed. In general such an implementation yields Lanczos matrices which have extra eigenvalues in addition to the 'good' eigenvalues which are approximations to eigenvalues of A. These extraneous or 'spurious' eigenvalues are caused by the losses in the orthogonality of the Lanczos vectors which in turn are caused by the combination of the roundoff errors resulting from the finite computer arithmetic and the convergence (as j is increased) of eigenvalues of the Lanczos matrices to eigenvalues of the original matrix A. This interaction between the computer arithmetic and the convergence of eigenvalues is discussed in Paige [1971,1980].

During the past 5-10 years many different types of Lanczos eigenelement algorithms have been proposed. See Volume 1, Chapter 2 of this book for a brief survey of the literature. Most of these procedures incorporate modifications to the basic Lanczos recursion in Eqns(1.2.1) - (1.2.2) which force the Lanczos vectors to stay nearly orthonormal. These approaches require either the repeated computation of Ritz vectors or the repeated reorthogonalization of the Lanczos vectors as they are generated or some combination of these two computations. In either case as the size of the Lanczos matrix generated is increased to be able to compute more eigenvalues, the associated Ritz vectors or the Lanczos vectors needed for the reorthogonalizations require more and more storage. These modifications often work well but destroy much of the simplicity of the basic procedure, and because of the added storage requirements resulting from the reorthogonalizations they limit the number of eigenelements which can be computed.

The approach which we have chosen and which is implemented in the enclosed FORTRAN programs in Chapters 2 - 7 is not to force the orthogonality of the Lanczos vectors by reorthogonalizing, but to work directly with the basic Lanczos recursion, accepting the losses in orthogonality, and then unraveling the effects of these losses. This approach allows us to retain the basic simplicity of the Lanczos recursion, to minimize the storage requirements, and to therefore maximize the number of eigenvalues of A which can be computed. In our approach in the single-vector algorithms in Chapters 2 - 7, Ritz vectors are not computed until after the eigenvalues have been computed accurately. Consequently, the basic storage requirements for our eigenvalue (singular value) algorithms are only a small multiple of the size of the largest Lanczos matrix used in the computations. Thus, we can compute many eigenvalues of very large but sparse matrices. Depending upon what is to be computed and upon the eigenvalue distribution in the given matrix A, the sizes of the Lanczos matrices used in these computations may be much smaller or considerably larger than the original A-matrix. However the Lanczos matrices generated by the procedures in Chapters 2 - 6 are real symmetric and tridiagonal so that these matrices can be very large and still not present insurmountable computational problems. Eigenvalue and eigenvector computations for such matrices require minimal amounts of storage and fairly reasonable numbers of arithmetic operations.

The computational problems which arise from not maintaining near orthogonality of the Lanczos vectors and which we must address in our single-vector codes are of two types. First and most importantly, we must deal with the question of sorting the eigenvalues of the Lanczos matrices into 2 classes, one corresponding to the 'good' eigenvalues which are approximations to the eigenvalues of A and the other corresponding to the extra or 'spurious' eigenvalues caused by the losses in orthogonality. The identification test used for doing this is discussed in Volume 1, Chapter 4, Section 4.5 of this book. For the procedures discussed in Chapters 2 - 6, this identification test is an integral and inexpensive part of the eigenvalue (singular value) computations. For the complex symmetric procedure discussed in Chapter 7 this test is handled in a considerably less eloquent manner and is expensive.

The second but much less serious difficulty we must address is the question of false multiplicities. The multiplicity of a particular 'good' eigenvalue as an eigenvalue of the Lanczos matrices is not related to the multiplicity of that eigenvalue as an eigenvalue of the A-matrix. 'Good' eigenvalues may replicate many times as eigenvalues of a Lanczos matrix, but be only simple eigenvalues of the original A-matrix. Thus, these single-vector procedures cannot directly determine the true multiplicities of the computed 'good' eigenvalues. Of course, this latter comment is also applicable to any single-vector Lanczos procedure not just to our procedures. Theoretically, at most one eigenvector for each distinct eigenvalue of the A-matrix can be obtained using the single-vector Lanczos recursion given in Eqns(1.2.1) and (1.2.2). (This of course is not true for iterative block Lanczos procedures.) It is interesting to note however that if the Lanczos recursion is used without any reorthogonalization, then it can yield sets of linearly independent eigenvectors for eigenvalues which are multiple in the A-matrix. The amount of work required to compute these additional eigenvectors depends upon the particular matrix in question and upon the particular eigenvalue. The codes provided in Chapters 2 - 7 of this book do not however incorporate this capability.

The iterative 'block' Lanczos procedures for real symmetric matrices given in Chapters 8 -

9 are based upon a block version of the Lanczos recursion

$$Q_{j+1}B_{j+1} = AQ_j - Q_jA_j - Q_{j-1}B_j^T \qquad (1.2.5)$$

for j=1,2,...,s where Q_1 is nxq and the coefficient matrices A_j and B_{j+1} are block analogs of the scalar coefficients in the single-vector Lanczos recursion in Eqns(1.2.1) and (1.2.2). The number of blocks s used on each iteration is chosen such that $qs << n$, where n is the order of the given A-matrix and q is chosen such that $q \geq q'$, the number of eigenvalues and eigenvectors desired. The Lanczos matrices are real symmetric, block tridiagonal matrices. In Eqn(1.2.5) we used Q_j instead of V_j because in our block Lanczos procedures we maintain near-orthogonality of the blocks generated within each iteration by incorporating reorthogonalization of the blocks of Lanczos vectors with respect to certain vectors in the first Lanczos block.

The 'block' procedures provided in Chapters 8 and 9 are really hybrid algorithms, something between a true block Lanczos procedure, see for example Cullum and Donath [1974], and the single-vector Lanczos procedures given in Chapters 2 - 7. The sequence of 'blocks' generated on each iteration of this hybrid method has the property that the first Q-block contains at least as many vectors as the user is trying to compute, but the second and succeeding blocks each contain only one vector. The corresponding resulting Lanczos matrices are not block tridiagonal. Each Lanczos matrix has a border of blocks in the first q rows and columns and is tridiagonal below this border.

At the beginning of each chapter, a brief description is given of the particular variant of the Lanczos recursion used in the Lanczos codes included in that chapter, along with some additional comments relevant to the particular types of problems being considered in that chapter.

SECTION 1.3 COMMENTS AND DISCLAIMERS

The single-vector Lanczos procedures contained in Chapters 2 - 7 do not behave like standard eigenelement procedures. Their behavior is both non-classical and somewhat unorthodox. If one of these codes were run on two different kinds of computers but with the same original matrix and the same initial specifications, the computed results could be quite different. A primary cause for such differences can of course be a difference in the starting vector caused by a difference in the random number generators. However even if the same starting vector were read in, the results would almost surely differ due to the differences in the computer arithmetic. In practice, the Lanczos matrices generated on two different kinds of computers may agree for a certain number of Lanczos steps but will begin to diverge upon the convergence of one or more of the eigenvalues of these Lanczos matrices to eigenvalues of the A-matrix. If after a reasonable number of steps in the Lanczos recursion we were to compare the entries in the Lanczos matrices generated by the two different computers, the values would probably be very different.

Furthermore, if we were to compute the eigenvalues of the two sets of Lanczos matrices for various sizes and 'spurious' eigenvalues were present, then these spurious eigenvalues would be different and even appear in different portions of the spectrum. In fact, prior to the convergence of a particular 'good' eigenvalue, the values of that good eigenvalue, in terms of how accurate it is at any given stage in the computations, may differ. However once a 'good' eigenvalue in either set has converged, that 'good' eigenvalue will agree with a true eigenvalue of the original user-specified matrix to as many digits as can be expected.

Therefore, if the user carries out the sample eigenvalue computation provided in Chapter 2, he/she should not be alarmed or surprised if the output from the computer being used does not agree with what is shown in the sample, as long as the converged 'good' eigenvalues agree. Actually one may observe different rates of convergence on different kinds of computers, depending upon the computer arithmetic. With increased arithmetic precision in all of the computations, these procedures may converge more rapidly. With decreased precision, they will converge less rapidly. All of our codes use double precision arithmetic (for an IBM 3083) and any precision less than that is not recommended.

Each of these procedures requires the user to supply either a matrix-vector multiply subroutine or a matrix-vector solve subroutine. (Both types of subroutines are required for the codes in Chapter 5.) Such subroutines should perform the required computations rapidly and accurately, taking advantage of any special properties or structure in the given matrix. Our Lanczos programs see the original matrix as the outputs of these subroutines. The codes provided include sample matrix-vector multiply subroutines for a general sparse 'symmetric' matrix given in a particular sparse format. These are available for the user to use or modify as desired. Note that similar programs are also provided for the singular value/vector computations. Accuracy is important in these subroutines because consistency must be maintained in the information being provided to the LANCZS subroutine which is generating the Lanczos matrices. There is no built-in mechanism for preserving symmetry. Therefore, the matrix-vector multiply and solve subroutines must be coded with care. Without such consistency the Lanczos codes will not function properly.

The convergence characteristics of the two types of Lanczos procedures considered are quite different. These differences are discussed in Chapters 4 and 7 of Volume 1 of this book. However, in both cases, the degree of difficulty in computing the desired eigenvalues depends upon the eigenvalue gaps. For the single-vector procedures the primary factor in determining whether or not it is feasible to compute either large numbers of eigenvalues or the eigenvalues with the smallest gaps, is the gap ratio, the ratio of the largest gap between two neighboring eigenvalues to the smallest such gap. The smaller this ratio, the easier it is to compute all of the eigenvalues of the given matrix. The larger this ratio, the harder it is to compute those eigenvalues with the smallest gaps. The locations of the desired eigenvalues in the spectrum of the given matrix also play a significant role in the rate of convergence of individual eigenvalues.

Both types of Lanczos procedures favor extreme eigenvalues. The iterative block codes, in fact, can only compute a few extreme eigenvalues. However for the single-vector codes, it is possible for interior eigenvalues which have gaps which are significantly larger than the gaps for some of the extreme eigenvalues to converge prior to the convergence of those extreme eigenvalues. Examples of the convergence achievable are given in Volume 1, Chapter 4 of this book.

The convergence of the iterative block procedures depends primarily upon the gaps between the eigenvalues being computed and the closest eigenvalue not being approximated, the spread of the matrix, and the overall eigenvalue distribution. The block procedures discussed in Chapters 8 and 9 are iterative and the codes track the rate of convergence. If the observed rate is too slow (as specified by the user), these block procedures will terminate without achieving convergence. The user then has the option of restarting the block procedure with a different choice of parameters and using the current approximation to the basis for the desired eigenspace as the starting vectors.

Thus, the amount of work required for a particular eigenelement computation for a given matrix using a particular method depends directly upon the eigenvalue distribution in that matrix and upon which portion of the spectrum is being computed. Some problems are 'easy', others are hard. Therefore failure can occur, in the sense that these procedures may not be able to compute the information desired by the user within the computational bounds specified by the user. However the single-vector Lanczos procedures, even in 'failure', provide a great deal of information about the eigenvalue spectrum of the given matrix.

In deciding which procedure to use on a given problem, our preference is a single-vector procedure, although the iterative block procedures can often quickly provide simultaneously the desired eigenvalues and eigenvectors. If the user wants extreme eigenvalues and the user knows or suspects that one or more of these is multiple, then the block procedure is probably preferable. More details about the Lanczos procedures contained in this book can be found in Volume 1. Any questions about these programs including the question of obtaining copies of these codes or of problems with these codes, should be addressed directly to the authors. We hope that these codes will prove useful in many different applications in the engineering and scientific community.

CHAPTER 2
REAL SYMMETRIC MATRICES

SECTION 2.1 INTRODUCTION

The FORTRAN codes in this chapter address the question of computing distinct eigenvalues and corresponding eigenvectors of real symmetric matrices, using a single-vector Lanczos procedure. For a given real symmetric matrix A, these codes compute real scalars λ and corresponding real vectors $x \neq 0$, such that

$$Ax = \lambda x. \tag{2.1.1}$$

DEFINITION 2.1.1 A real nxn matrix $A \equiv (a_{ij}),\ 1 \leq i,j \leq n,$ is a real symmetric matrix if and only if for every i and j, $a_{ij} = a_{ji}$.

Real symmetric matrices are discussed in detail in Stewart [1973]. Properties which we use are: (1) Real symmetric matrices have complete eigensystems. That is, the dimension of the eigenspace corresponding to any given eigenvalue of the given matrix A is the same as the multiplicity of that eigenvalue as a root of the characteristic polynomial of A. (2) For any two distinct eigenvalues of A, λ and μ, and corresponding eigenvectors x and y, $x^T y = 0$. Thus, eigenvectors corresponding to different eigenvalues are orthogonal, and we can construct an eigenvector basis which is orthonormal. Vectors are orthonormal if they are orthogonal and each has a Euclidean norm of 1. (The Euclidean norm of a vector is just the square root of the sum of the squares of its components.) (3) Small perturbations in the matrix cause only small perturbations in the eigenvalues. Of the classes of matrices which we consider, the class of real symmetric matrices is the most well-behaved and thus the 'easiest'.

The Lanczos codes contained in this chapter correspond to the most straight-forward implementation of the Lanczos recursion included in this book. These codes can be used to compute either a very few or very many of the distinct eigenvalues of the given real symmetric matrix. As the documentation in the next section indicates, the A-multiplicity of a given computed 'good' Lanczos eigenvalue can be obtained only with additional computation, and the modifications required to do this additional computation are not included in these versions of the codes. This implementation uses the basic Lanczos recursion given in Eqns(1.1.2) - (1.2.1) in Section 1.2 of Chapter 1 to generate a family of real symmetric, tridiagonal matrices (T-matrices) whose sizes are specified by the user. There is no reorthogonalization of the Lanczos vectors at any stage in any of the computations.

LEVAL, the main program for the real symmetric eigenvalue computations, calls the subroutine BISEC to compute eigenvalues of the user-specified Lanczos tridiagonal matrices on the user-specified intervals. BISEC simultaneously computes these T-eigenvalues with their

T-multiplicities and sorts the computed T-eigenvalues into two classes, the 'good' T-eigenvalues and the 'spurious' T-eigenvalues. The 'good' T-eigenvalues are accepted as approximations to eigenvalues of the user-specified matrix A. The accuracy of these 'good' T-eigenvalues as eigenvalues of A is then estimated using error estimates computed by subroutine INVERR. Error estimates are computed only for isolated 'good' T-eigenvalues. All other 'good' T-eigenvalues are assumed to have converged. If convergence has not yet occurred and a larger Lanczos matrix has been specified by the user, the program will continue on to a larger Lanczos matrix, repeating the above procedure on this larger matrix.

Once the eigenvalues have been computed accurately enough, the user can select a subset of the 'converged' eigenvalues for which eigenvectors are to be computed. The main program LEVEC, for computing eigenvectors of real symmetric matrices, is then used to compute these desired eigenvectors.

All computations are in double precision real arithmetic. The user must supply a subroutine USPEC which defines and initializes the user-specified matrix A and a subroutine CMATV which computes matrix-vector multiplies Ax for any given vector x. These subroutines must be constructed in such a way as to take advantage of the sparsity (and/or structure) of the user-supplied A-matrix and such that these computations are done accurately. More details about these real symmetric single-vector Lanczos procedures are given in Chapter 4 of Volume 1 of this book.

SECTION 2.2 DOCUMENTATION FOR PROGRAMS IN CHAPTERS 2 - 5

```
C-----DOCUMENTATION FOR SINGLE-VECTOR------------------------------------
C     LANCZOS EIGENVALUE/EIGENVECTOR PROGRAMS FOR
C     (1) REAL SYMMETRIC MATRICES
C     (2) HERMITIAN MATRICES
C     (3) FACTORED INVERSES OF REAL SYMMETRIC MATRICES
C     (4) REAL SYMMETRIC, GENERALIZED PROBLEMS WHERE ONE OF THE
C         MATRICES IS POSITIVE DEFINITE AND ITS CHOLESKY FACTORS ARE
C         AVAILABLE
C
C
C     AUTHORS:  JANE CULLUM AND RALPH A. WILLOUGHBY, IBM RESEARCH,
C               YORKTOWN HEIGHTS, NY 10598.  PHONE: 914-945-2227
C
C
C     REAL SYMMETRIC MATRICES:
C
C     GIVEN A REAL SYMMETRIC MATRIX A OF ORDER N THE THREE SETS OF
C     FORTRAN FILES LABELLED LEVAL, LESUB, AND LEMULT CAN BE USED TO
C     COMPUTE DISTINCT EIGENVALUES OF THE USER-SPECIFIED MATRIX
C     IN USER-SPECIFIED INTERVALS.
C
C     CORRESPONDING EIGENVECTORS FOR SELECTED, COMPUTED EIGENVALUES CAN
C     BE COMPUTED USING THE SETS OF FILES LABELLED LEVEC, LESUB, AND
C     LEMULT.
C
C
C     HERMITIAN MATRICES:
C
C     GIVEN A HERMITIAN MATRIX A OF ORDER N THE THREE SETS OF
C     FORTRAN FILES LABELLED HLEVAL, LESUB, AND HLEMULT CAN BE USED
C     TO COMPUTE DISTINCT EIGENVALUES IN USER-SPECIFIED INTERVALS.
C
C     CORRESPONDING EIGENVECTORS FOR SELECTED, COMPUTED EIGENVALUES
C     CAN BE COMPUTED USING THE SETS OF PROGRAMS LABELLED HLEVEC,
C     LESUB, AND HLEMULT.
C
C
C     FACTORED INVERSES OF REAL SYMMETRIC MATRICES:
C
C     GIVEN A REAL SYMMETRIC MATRIX A, THE LANCZOS RECURSION IS
C     APPLIED TO THE INVERSE OF A, USING A FACTORIZATION
C     OF A.  THE SETS OF FILES LIVAL, LESUB, AND LIMULT
C     CAN BE USED TO COMPUTE THE DISTINCT EIGENVALUES OF THE
C     INVERSE OF THE A-MATRIX AND OF A IN  USER-SPECIFIED
C     INTERVALS.  THE PROGRAMS ACTUALLY ALLOW ONE TO WORK WITH
C     ANY MATRIX B = PCP' WHERE  C = SO*A + SHIFT*I, WHERE
C     SO AND SHIFT ARE SCALARS CHOSEN BY THE USER AND P IS A
C     PERMUTATION MATRIX CHOSEN SUCH THAT THE FACTORIZATION
C     OF THE B-MATRIX RETAINS SPARSITY.  IN THE
C     SAMPLE LIMULT SUBROUTINES PROVIDED, SO AND SHIFT MUST BE
C     CHOSEN SO THAT THE RESULTING B-MATRIX IS POSITIVE DEFINITE,
C     AND THE CHOLESKY FACTORS ARE USED TO SOLVE B*U = V.
C     HOWEVER, THE USER CAN EASILY REPLACE THE SAMPLE USPEC AND
C     BSOLV SUBROUTINES PROVIDED BY SUBROUTINES THAT ALLOW THE
C     GENERAL FACTORIZATION L*D*(L-TRANSPOSE).  THESE LANCZOS
C     PROGRAMS APPLY THE LANCZOS RECURSION TO B-INVERSE, USING
C     THE FACTORIZATION PROVIDED.  OPTIONAL PREPROCESSING PROGRAMS
C     PERMUT, LORDER, LFACT, AND LTEST ARE PROVIDED FOR SET-UP PURPOSES.
C     PERMUT USES THE SPARSPAK PACKAGE OF A. GEORGE, J. LIU AND
C     E. NG TO OBTAIN A REORDERING OF THE GIVEN MATRIX THAT
C     PRESERVES SPARSENESS ON SUBEQUENT FACTORIZATION.  LORDER
C     CAN BE USED TO REORDER A GIVEN MATRIX, USING A GIVEN
C     PERMUTATION.  LFACT CAN BE USED TO COMPUTE THE CHOLESKY
C     FACTORS OF A GIVEN POSITIVE DEFINITE B-MATRIX.  LTEST CAN
```

```
C     BE USED TO ESTIMATE THE NUMERICAL CONDITION OF THE
C     B-MATRIX.
C
C     CORRESPONDING EIGENVECTORS FOR SELECTED, COMPUTED
C     EIGENVALUES CAN BE COMPUTED USING THE SETS OF FILES
C     LABELLED LIVEC, LESUB, AND LIMULT.
C
C     GENERALIZED REAL SYMMETRIC PROBLEMS:
C
C     GIVEN 2 REAL SYMMETRIC MATRICES A AND B WHERE IN ADDITION B IS
C     POSITIVE DEFINITE AND ITS CHOLESKY FACTORS ARE AVAILABLE,
C     THE SETS OF FILES LGVAL, LGMULT, AND LESUB CAN BE USED
C     TO COMPUTE THE DISTINCT EIGENVALUES OF THE GENERALIZED
C     PROBLEM A*X = EVAL*B*X.
C
C     CORRESPONDING EIGENVECTORS CAN BE COMPUTED USING THE PROGRAMS
C     LGVEC, LGMULT, AND LESUB.  NOTE THAT THE PREPROCESSING PROGRAMS
C     AVAILABLE FOR USE IN CASE (3) (PERMUT, LORDER, LFACT, AND LTEST)
C     CAN ALSO BE USED IN THIS CASE TO OBTAIN A SUITABLE PERMUTATION,
C     AND A FACTORIZATION OF THE RESULTING B-MATRIX.  THE A-MATRIX
C     CAN THEN BE PERMUTED USING LORDER.
C
C
C     THESE PROGRAMS ALL USE LANCZOS TRIDIAGONALIZATION WITHOUT
C     REORTHOGONALIZATION TO GENERATE REAL SYMMETRIC TRIDIAGONAL
C     MATRICES, T(1,MEV), OF ORDER MEV.  SUBSETS OF THE EIGENVALUES OF
C     THESE T-MATRICES, LABELLED AS THE 'GOOD EIGENVALUES', YIELD
C     APPROXIMATIONS TO THE DESIRED EIGENVALUES.  CORRESPONDING
C     RITZ VECTORS ARE APPROXIMATIONS TO THE DESIRED EIGENVECTORS.
C     NOTE THAT FOR CASE (4) THE GENERALIZED LANCZOS RECURSION
C     B*V(I+1)*BETA(I+1) = A*V(I) - B*V(I)*ALPHA(I) - B*V(I-1)*BETA(I)
C     IS USED, ALONG WITH THE B-NORM.
C
C     THE IDEAS USED IN THESE PROGRAMS ARE DISCUSSED IN THE FOLLOWING
C     REFERENCES.
C
C     1.  JANE CULLUM AND RALPH A. WILLOUGHBY, LANCZOS ALGORITHMS
C         FOR LARGE SYMMETRIC MATRICES, VOLUME ?, PROGRESS IN
C         SCIENTIFIC COMPUTING, EDITORS, G. GOLUB, H.O. KREISS,
C         S. ARBARBANEL, AND R. GLOWINSKI,  BIRKHAUSER BOSTON INC.,
C         CAMBRIDGE, MASSACHUSETTS, 1983.
C
C     2.  JANE CULLUM AND RALPH A. WILLOUGHBY, COMPUTING EIGENVECTORS
C         (AND EIGENVALUES) OF LARGE, SYMMETRIC MATRICES USING
C         LANCZOS TRIDIAGONALIZATION, LECTURE NOTES IN MATHEMATICS,
C         773, NUMERICAL ANALYSIS PROCEEDINGS, DUNDEE 1979, EDITED BY
C         G. A. WATSON, SPRINGER-VERLAG, (1980), BERLIN, PP.46-63.
C
C     3. IBID, LANCZOS AND THE COMPUTATION IN SPECIFIED INTERVALS OF
C        THE SPECTRUM OF LARGE SPARSE, REAL SYMMETRIC MATRICES, SPARSE
C        MATRIX PROCEEDINGS 1978, ED. I.S. DUFF AND G. W. STEWART,
C        SIAM, PHILADELPHIA, PP.220-255, 1979.
C
C     4. IBID, COMPUTING EIGENVALUES OF VERY LARGE SYMMETRIC MATRICES-
C        AN IMPLEMENTATION OF A LANCZOS ALGORITHM WITHOUT
C        REORTHOGONALIZATION, J. COMPUT. PHYS. 44(1981), 329-358.
C
C
C-----PORTABILITY-----------------------------------------------------
C
C
C     PROGRAMS WERE TESTED FOR PORTABILITY USING THE PFORT VERIFIER.
C     FOR DETAILS OF THE VERIFIER SEE FOR EXAMPLE, B. G. RYDER AND
C     A. D. HALL, "THE PFORT VERIFIER", COMPUTING SCIENCE TECHNICAL
C     REPORT 12, BELL LABORATORIES, MURRAY HILL, NEW JERSEY 07974,
C     (REVISED), JANUARY 1981.
C
```

```
C      WITH THE EXCEPTION OF THE PROGRAMS FOR HERMITIAN MATRICES WHICH   LEV01320
C      ARE NOT PORTABLE BECAUSE OF THEIR USE OF COMPLEX*16 VARIABLES,    LEV01330
C      THE OTHER PROGRAMS INCLUDED ARE PORTABLE EXCEPT FOR A FEW         LEV01340
C      CONSTRUCTIONS WHICH, IF NECESSARY, WILL HAVE TO BE MODIFIED       LEV01350
C      BY THE USER FOR THE PARTICULAR COMPUTER BEING USED.               LEV01360
C                                                                        LEV01370
C      NONPORTABLE CONSTRUCTIONS:                                        LEV01380
C                                                                        LEV01390
C      REAL SYMMETRIC MATRICES:                                          LEV01400
C      IN LEVAL AND IN LEVEC                                             LEV01410
C          1.  DATA/MACHEP STATEMENT                                     LEV01420
C          2.  ALL READ(5,*) STATEMENTS (FREE FORMAT)                    LEV01430
C          3.  FORMAT(20A4) USED FOR THE EXPLANATORY HEADER ARRAY, EXPLANLEV01440
C          4.  FORMAT(4Z20) USED TO READ AND WRITE ALPHA/BETA FILES.     LEV01450
C      IN LEMULT                                                         LEV01460
C          1.  IN CMATV AND  USPEC THE ENTRY THAT PASSES THE STORAGE     LEV01470
C              LOCATIONS OF THE ARRAYS DEFINING THE USER-SPECIFIED       LEV01480
C              MATRIX.                                                   LEV01490
C          2.  IN THE SAMPLE USPEC PROVIDED:  FREE FORMAT (8,*),         LEV01500
C              THE FORMAT (20A4), AND DATA/MACHEP STATEMENT.             LEV01510
C                                                                        LEV01520
C      HERMITIAN MATRICES:                                               LEV01530
C      IN HLEVAL AND IN HLEVEC                                           LEV01540
C          1.  DATA/MACHEP STATEMENT                                     LEV01550
C          2.  ALL READ(5,*) STATEMENTS (FREE FORMAT)                    LEV01560
C          3.  FORMAT(20A4) USED FOR THE EXPLANATORY HEADER ARRAY, EXPLANLEV01570
C          4.  COMPLEX*16 VARIABLES AND FUNCTIONS SUCH AS DCMPLX.        LEV01580
C          5.  FORMAT (4Z20) USED TO READ AND WRITE ALPHA/BETA FILES.    LEV01590
C      IN HLEMULT                                                        LEV01600
C          1.  IN CMATV AND  USPEC THE ENTRY THAT PASSES THE STORAGE     LEV01610
C              LOCATIONS OF THE ARRAYS DEFINING THE USER-SPECIFIED       LEV01620
C              MATRIX.                                                   LEV01630
C          2.  COMPLEX*16 VARIABLES AND FUNCTIONS SUCH AS DCMPLX.        LEV01640
C          3.  IN THE SAMPLE USPEC PROVIDED:  FREE FORMAT (8,*),         LEV01650
C              THE FORMAT (20A4), AND DATA/MACHEP STATEMENT.             LEV01660
C                                                                        LEV01670
C      FACTORED INVERSES OF REAL SYMMETRIC MATRICES:                     LEV01680
C      IN LIVAL AND IN LIVEC                                             LEV01690
C          1.  DATA/MACHEP STATEMENT                                     LEV01700
C          2.  ALL READ(5,*) STATEMENTS (FREE FORMAT)                    LEV01710
C          3.  FORMAT(20A4) USED FOR THE EXPLANATORY HEADER ARRAY, EXPLANLEV01720
C          4.  FORMAT(4Z20) USED TO READ AND WRITE ALPHA/BETA FILES.     LEV01730
C      IN LIMULT                                                         LEV01740
C          1.  IN USPEC AND BSOLV, THE ENTRIES THAT PASS                 LEV01750
C              THE STORAGE LOCATIONS OF THE ARRAYS DEFINING THE          LEV01760
C              USER-SPECIFIED MATRIX.                                    LEV01770
C          2.  IN THE SAMPLE USPEC SUBROUTINES PROVIDED:                 LEV01780
C              FORMATS (20A4) AND (4Z20), FREE FORMAT (8,*), AND         LEV01790
C              DATA/MACHEP STATEMENTS.                                   LEV01800
C                                                                        LEV01810
C                                                                        LEV01820
C      GENERALIZED SYMMETRIC PROBLEM, B-MATRIX POSITIVE                  LEV01830
C      DEFINITE AND CHOLESKY FACTORS AVAILABLE:                          LEV01840
C      IN LGVAL AND IN LGVEC                                             LEV01850
C          1.  DATA/MACHEP STATEMENT                                     LEV01860
C          2.  ALL READ(5,*) STATEMENTS (FREE FORMAT)                    LEV01870
C          3.  FORMAT(20A4) USED FOR THE EXPLANATORY HEADER ARRAY, EXPLANLEV01880
C          4.  FORMAT(4Z20) USED TO READ AND WRITE ALPHA/BETA FILES.     LEV01890
C      IN LGMULT                                                         LEV01900
C          1.  IN USPECA, USPECB, AMATV AND LSOLV THE ENTRIES            LEV01910
C              THAT PASS THE STORAGE LOCATIONS OF THE ARRAYS DEFINING    LEV01920
C              THE USER-SPECIFIED MATRICES.                              LEV01930
C          2.  IN THE SAMPLE USPECA AND USPECB SUBROUTINES PROVIDED:     LEV01940
C              FORMATS (20A4) AND (4Z20), FREE FORMAT (8,*), AND         LEV01950
C              DATA/MACHEP STATEMENTS.                                   LEV01960
C                                                                        LEV01970
C      ALL 4 CASES USE THE FORTRAN FILE LESUB:                           LEV01980
```

```
C     IN LESUB ALL STATEMENTS ARE PORTABLE EXCEPT FOR:
C        (1)  THE ENTRY IN SUBROUTINE LPERM THAT PASSES THE
C             PERMUTATION FROM THE USPEC SUBROUTINE TO LPERM.
C             (THIS IS USED ONLY IN CASES (3) AND (4)).
C        (2)  THE COMPLEX*16 VARIABLES AND FUNCTIONS USED IN
C             SUBROUTINE CINPRD. (THIS IS USED ONLY IN CASE (2)).
C
C     IN THE COMMENTS BELOW:
C
C     COMPLEX*16 = COMPLEX VARIABLE, 16 BYTES OF STORAGE
C     REAL*8 = REAL VARIABLE, 8 BYTES OF STORAGE
C     REAL*4 = REAL VARIABLE, 4 BYTES OF STORAGE
C     INTEGER*4 = INTEGER VARIABLE, 4 BYTES
C
C
C-----MATRIX SPECIFICATION-------------------------------------------
C
C
C     IN CASES (1) AND (2), SUBROUTINE USPEC IS USED TO SPECIFY THE
C     USER-SUPPLIED A-MATRIX.  SIMILARLY, IN CASE (4) SUBROUTINES
C     USPECA AND USPECB DEFINE THE USER-SUPPLIED A-MATRIX AND B-MATRIX.
C     IN CASE (3) ((4)), SUBROUTINE USPECB DEFINES THE FACTORIZATION
C     OF THE MATRIX (B-MATRIX) USED BY THE LANCZOS PROCEDURE.
C     (IN CASE (3) THE A-MATRIX IS NOT USED DIRECTLY.)
C
C     IN CASES (1) AND (2), SUBROUTINE CMATV IS A CORRESPONDING
C     MATRIX-VECTOR MULTIPLY SUBROUTINE WHICH SHOULD BE DESIGNED
C     TO TAKE ADVANTAGE OF ANY SPECIAL PROPERTIES OF THE GIVEN
C     MATRIX.  IN CASE (4) THIS SUBROUTINE IS NEEDED FOR THE
C     A-MATRIX AND THUS IS CALLED AMATV.  IN CASES (3) AND (4)
C     SUBROUTINES THAT CAN SOLVE B*U = V, USING A SPARSE
C     FACTORIZATION OF B ARE NEEDED.  THESE SUBROUTINES ARE
C     CALLED RESPECTIVELY, BSOLV AND LSOLV.  IN ALL CASES,
C     ANY MATRIX-VECTOR MULTIPLY AND SOLVE SUBROUTINES USED
C     MUST BE DESIGNED TO COMPUTE RAPIDLY AND ACCURATELY.
C
C     IN ALL CASES:
C     SUBROUTINE USPEC(A OR B) HAS THE CALLING SEQUENCE
C
C           CALL USPEC(N,MATNO)
C
C     WHERE N IS THE ORDER OF THE USER-SUPPLIED MATRIX A, AND
C     MATNO IS A <= 8 DIGIT INTEGER USED AS A MATRIX AND
C     TEST IDENTIFICATION NUMBER.  IN ALL CASES THIS (THESE)
C     SUBROUTINE(S) DEFINES (DIMENSIONS) THE ARRAYS REQUIRED
C     TO SPECIFY THE MATRIX (MATRICES IN CASE (4)) THAT WILL BE
C     USED BY THE LANCZS SUBROUTINE.  IN CASES (1) AND (2)
C     THIS IS THE A-MATRIX; IN CASE (3) THIS IS THE FACTORIZATION
C     OF A SCALED, SHIFTED AND PERMUTED VERSION OF THE
C     USER-SPECIFIED A-MATRIX.  IN CASE (4) THE A-MATRIX
C     IS SPECIFIED AS WELL AS THE FACTORIZATION OF THE
C     B-MATRIX.  THIS SUBROUTINE ALSO INITIALIZES THE ARRAYS
C     AND ANY OTHER PARAMETERS NEEDED TO DEFINE THE MATRIX
C     (MATRICES).  THE STORAGE LOCATIONS OF THESE PARAMETERS
C     AND ARRAYS ARE THEN PASSED TO THE MATRIX-VECTOR MULTIPLY
C     SUBROUTINE CMATV IN CASES (1) AND (2), TO THE SUBROUTINE
C     BSOLV IN CASE (3), AND TO THE SUBROUTINES AMATV
C     AND LSOLV IN CASE (4) VIA ENTRY CALLS.  IN CASES (3) AND (4)
C     WHENEVER A MATRIX HAS BEEN PERMUTED, THERE IS ALSO AN
C     ENTRY INTO THE SUBROUTINE LPERM TO PASS THE LOCATIONS OF
C     THE PERMUTATIONS IPR AND IPRT USED.  SAMPLE USPECS, CMATV,
C     AMATV, BSOLV AND LSOLV SUBROUTINES ARE INCLUDED
C     IN THE RELEVANT FILES.  THESE SAMPLE PROGRAMS ASSUME THAT
C     THE USER-SUPPLIED A-MATRIX IS STORED ON FILE 8 IN CASES (1),
C     (2), AND (4), AND THAT THE FACTORIZATION OF THE B-MATRIX
C     IS ON FILE 7 IN CASES (3) AND (4).  THE USER SHOULD SEE
C     THE INDIVIDUAL SAMPLE SUBROUTINES FOR MORE DETAILS.
```

```
C
C     IN CASES (1) AND (2):
C     SUBROUTINE CMATV HAS THE CALLING SEQUENCE
C
C            CALL CMATV(W,U,SUM)
C
C     IN THE REAL SYMMETRIC CASE, U AND W ARE REAL*8 VECTORS
C     AND SUM IS A REAL*8 SCALAR.  IN THE HERMITIAN CASE, U
C     AND W ARE COMPLEX*16 VECTORS AND SUM IS A REAL*8 SCALAR.
C     CMATV CALCULATES U = A*W - SUM*U FOR THE USER-SPECIFIED
C     MATRIX A. ONE OF THE SAMPLE CMATV SUBROUTINES INCLUDED
C     COMPUTES MATRIX-VECTOR MULTIPLIES FOR AN ARBITRARY SPARSE,
C     SYMMETRIC MATRIX STORED IN THE SPARSE FORMAT SPECIFIED IN THE
C     CORRESPONDING SAMPLE USPEC SUBROUTINE.  FOR CASES (1) AND
C     (2) CMATV IS THE SUBROUTINE USED BY THE LANCZS SUBROUTINE
C     THAT GENERATES THE T-MATRICES.  IN CASE (4) SUBROUTINE
C     AMATV HAS THE SAME CALLING SEQUENCE AS CMATV IN CASE (1).
C
C     IN CASES (3) AND (4):
C     ALPHA/BETA HISTORY IS GENERATED USING SPARSE MATRIX INVERSION.
C     IN CASE (3), AT EACH ITERATION OF THE LANCZOS RECURSION
C     GIVEN A FACTORIZATION OF THE MATRIX BEING USED, THE
C     SUBROUTINE BSOLV FOR A GIVEN V, COMPUTES U SUCH THAT B*U = V.
C     THE CALLING SEQUENCE OF BSOLV IS
C
C                  CALL BSOLV(V,U,IBSOLV)
C
C     WHEN IBSOLV = 2,  U = (B-INVERSE)*V IS RETURNED.  IN CASE (4),
C     AT EACH ITERATION OF THE GENERALIZED LANCZOS RECURSION BOTH THE
C     SUBROUTINE AMATV AND THE SUBROUTINE LSOLV ARE USED.  THE
C     CALLING SEQUENCE OF LSOLV IS
C
C                  CALL LSOLV(V,U,ISOLV)
C
C     WHERE U AND V ARE REAL*8 VECTORS.  LSOLV PERFORMS 4 FUNCTIONS.
C     LET L DENOTE THE CHOLESKY FACTOR OF THE B-MATRIX USED IN LANCZS.
C     WHEN ISOLV = 1, LSOLV COMPUTES U = L*V.  WHEN ISOLV = 2,
C     LSOLV COMPUTES U = (L-TRANSPOSE)*V.  WHEN ISOLV = 3, LSOLV
C     COMPUTES U = (L-INVERSE)*V.  WHEN ISOLV = 4, LSOLV
C     COMPUTES  U = ((L-TRANSPOSE)-INVERSE)*V.
C
C     SAMPLE PROGRAMS ASSUME THAT THE A-MATRIX (CASES (1),(2),(4))
C     IS ON FILE 8 AND STORED IN THE FOLLOWING SPARSE FORMAT:
C     ICOL(K), K = 1,NZL, NUMBER OF SUBDIAGONAL NONZEROS IN COLUMN K.
C     IROW(K), K = 1,NZS, ROW INDEX OF ASD(K).
C     AD(K), K=1,N, CONTAINS THE DIAGONAL ELEMENTS OF THE A-MATRIX.
C     ASD(K), K=1,NZS  CONTAINS THE SUBDIAGONAL ELEMENTS OF A BY COLUMN.
C     NZS = NUMBER OF NONZERO ELEMENTS BELOW THE DIAGONAL OF A
C     NZL = INDEX OF LAST COLUMN WITH NONZERO SUBDIAGONAL ENTRIES
C     N = ORDER OF THE A-MATRIX.
C
C     NOTE THAT THE OPTIONAL PREPROCESSING PROGRAMS PERMUT AND
C     LORDER ASSUME THAT THE GIVEN MATRIX IS ON FILE 8.  CASES (3)
C     AND (4) ASSUME THAT THE SPARSE FACTORIZATION OF B IS STORED ON
C     FILE 7.  THE SAMPLE BSOLV SUBROUTINE SUPPLIED ASSUMES
C     THAT THE B-MATRIX IS POSITIVE DEFINITE AND THAT ITS CHOLESKY
C     FACTOR IS PROVIDED ON FILE 7, STORED IN SPARSE FORMAT IN
C     ARRAYS BD AND BSD.  THE USER CAN EASILY REPLACE THIS SAMPLE
C     BSOLV SUBROUTINE AND THE CORRESPONDING SAMPLE USPEC
C     SUBROUTINE BY SUBROUTINES THAT DEFINE AND USE A GENERAL
C     FACTORIZATION L*D*(L TRANSPOSE).
C
C     THE SAMPLE USPEC, CMATV (CASES (1) AND (2)), AMATV (CASE (4)),
C     BSOLV (CASE (3)), AND LSOLV (CASE(4)) MUST BE MODIFIED BY
C     THE USER TO ACCOMODATE THE USER-SPECIFIED MATRIX OR MATRICES.
C
C
```

```
C-----MACHEP-----------------------------------------------------------
C
C
C     MACHEP IS A MACHINE DEPENDENT PARAMETER SPECIFYING THE RELATIVE
C     PRECISION OF THE FLOATING POINT ARITHMETIC USED.
C     MACHEP = 2.2 * 10**-16 FOR DOUBLE PRECISION ARITHMETIC ON
C     IBM 370-3081.
C
C     THE USER WILL HAVE TO RESET THIS PARAMETER TO
C     THE CORRESPONDING VALUE FOR THE MACHINE BEING USED.  NOTE THAT
C     IF A MACHINE WITH A MACHINE EPSILON THAT IS MUCH LARGER THAN THE
C     VALUE GIVEN HERE IS BEING USED, THEN THERE COULD BE
C     PROBLEMS WITH THE TOLERANCES.
C
C
C-----SUBROUTINES AND FUNCTIONS USER MUST SUPPLY------------------------
C
C
C     GENRAN,  FINPRO,  MASK,  USPEC,  AND
C     CASES (1) AND (2),  CMATV:  CASE (3), BSOLV:
C     CASE (4),  AMATV AND LSOLV.
C
C     GENRAN = COMPUTES K PSEUDO-RANDOM NUMBERS AND STORES THEM IN
C              THE REAL*4 ARRAY, G.  THIS SUBROUTINE IS USED TO
C              GENERATE A STARTING VECTOR FOR THE LANCZOS PROCEDURE
C              IN THE SUBROUTINE LANCZS AND A STARTING RIGHT-HAND SIDE
C              FOR INVERSE ITERATION IN THE SUBROUTINE INVERR.
C
C              TESTS REPORTED IN THE REFERENCES USED EITHER GGL1 OR
C              GGL2 FROM THE IBM LIBRARY  SLMATH.
C              THE EXISTING CALLING SEQUENCE IS:
C
C                      CALL GENRAN(IIX,G,K).
C
C              WHERE IIX =INTEGER SEED, G = REAL*4 ARRAY WHOSE
C              DIMENSION MUST BE >= K.  K RANDOM NUMBERS ARE GENERATED
C              AND PLACED IN G.
C
C     FINPRO = DOUBLE PRECISION FUNCTION WHICH COMPUTES THE INNER
C              PRODUCT OF 2 DOUBLE PRECISION VECTORS OF DIMENSION  N.
C              TESTS REPORTED IN THE REFERENCES USED THE HARWELL
C              LIBRARY SUBROUTINE FM02AD.
C              EXISTING CALLING SEQUENCE IS
C
C                      CALL FINPRO(N,V,J,W,K).
C
C              COMPUTES THE INNER PRODUCT OF DIMENSION N OF THE VECTORS
C              V AND W.  SUCCESSIVE COMPONENTS OF V AND OF W ARE STORED
C              AT LOCATIONS THAT ARE ,RESPECTIVELY, J AND K UNITS APART.
C
C     MASK = MASKS OVERFLOW AND UNDERFLOW.
C            USER MUST SUPPLY OR COMMENT OUT CALL.
C
C    USPEC = DIMENSIONS AND INITIALIZES ARRAYS NEEDED TO SPECIFY
C            MATRIX THAT WILL BE USED BY LANCZS SUBROUTINE.
C            IN CASE (4) A-MATRIX AND B-MATRIX MUST BOTH BE SPECIFIED.
C
C    CMATV = MATRIX-VECTOR MULTIPLY FOR USER-SUPPLIED MATRIX.
C            CASES (1) AND (2).  SEE MATRIX SPECIFICATION SECTION.
C
C    AMATV = MATRIX-VECTOR MULTIPLY FOR USER-SUPPLIED A-MATRIX.
C            CASES (4) ONLY.  SEE MATRIX SPECIFICATION SECTION.
C
C    BSOLV = GIVEN A VECTOR V COMPUTES U SUCH THAT B*U = V,
C            USING THE FACTORIZATION OF B.  USED IN CASE (3) ONLY.
C            SEE MATRIX SPECIFICATION SECTION.
C
```

```
C     LSOLV = PERFORMS 4 FUNCTIONS.  GIVEN A VECTOR V COMPUTES
C             U = L*V, U = (L-TRANSPOSE)*V, U = (L-INVERSE)*V OR
C             U = (L-TRANSPOSE)-INVERSE*V, USING THE CHOLESKY
C             FACTORS OF B.  USED ONLY IN CASE (4).  SEE MATRIX
C             SPECIFICATION SECTION.
C
C
C-----------------------------------------------------------------------
C
C      COMMENTS FOR EIGENVALUE COMPUTATIONS
C
C-----------------------------------------------------------------------
C
C
C-----PARAMETER CONTROLS FOR EIGENVALUE PROGRAMS------------------------
C
C
C      PARAMETER CONTROLS ARE INTRODUCED TO ALLOW SEGMENTATION OF THE
C      EIGENVALUE COMPUTATIONS AND TO ALLOW VARIOUS COMBINATIONS OF
C      READ/WRITES.
C
C      THE FLAG ISTART CONTROLS THE T-MATRIX (ALPHA/BETA HISTORY)
C      GENERATION.
C
C      ISTART = (0,1)  MEANS
C
C              (0) THERE IS NO EXISTING ALPHA/BETA HISTORY AND ONE
C                  MUST BE GENERATED.
C
C              (1) THERE IS AN EXISTING ALPHA/BETA HISTORY AND IT IS
C                  TO BE READ IN FROM FILE 2 AND EXTENDED IF NECESSARY.
C
C      THE FLAG ISTOP CAN BE USED IN CONJUNCTION WITH THE FLAG ISTART TO
C      ALLOW SEGMENTATION OF THE EIGENVALUE COMPUTATIONS.
C
C      ISTOP  = (0,1)  MEANS
C
C              (0) PROGRAM COMPUTES ONLY THE REQUESTED ALPHAS/BETAS,
C                  STORES THEM AND THE LAST 2 LANCZOS VECTORS GENERATED
C                  IN FILE 1 AND THEN TERMINATES.  IN CASE (4) THERE
C                  ARE ACTUALLY 3 VECTORS TO BE SAVED.
C
C              (1) PROGRAM COMPUTES REQUESTED ALPHAS/BETAS AND THEN
C                  USES THE BISEC SUBROUTINE TO CALCULATE EIGENVALUES
C                  OF THE TRIDIAGONAL MATRICES GENERATED FOR THE ORDERS
C                  SPECIFIED BY THE USER AND ON THE USER-SPECIFIED
C                  INTERVALS.  PROGRAM THEN USES THE SUBROUTINE INVERR
C                  TO COMPUTE ERROR ESTIMATES FOR THE ISOLATED GOOD
C                  T-EIGENVALUES WHICH ARE USED TO CHECK THE
C                  CONVERGENCE OF THESE T-EIGENVALUES.
C
C      CONTROL PARAMETERS FOR WRITES
C
C      IHIS   = (0,1)  MEANS
C
C              (0) IF ISTOP .GT. 0 THEN ALPHA/BETAS ARE NOT SAVED ON
C                  FILE 1.
C
C              (1) PROGRAM WRITES ALPHAS/BETAS AND LAST 2 LANCZOS
C                  VECTORS TO FILE 1 SO THAT THE T-MATRIX GENERATION
C                  MAY BE REUSED OR CONTINUED LATER IF NECESSARY.
C                  TYPICALLY ONE WOULD ALWAYS DO THIS ON ANY RUN WHERE
C                  A HISTORY FILE IS BEING GENERATED.  HISTORY MUST BE
C                  SAVED IN MACHINE FORMAT ((4Z20) FOR IBM 3081) SO
C                  THAT NO ERRORS ARE INTRODUCED BY FORMAT CONVERSIONS.
C
C      IDIST  = (0,1)  MEANS
```

```
C
C             (0) DISTINCT EIGENVALUES OF T-MATRICES ARE NOT SAVED.
C
C             (1) PROGRAM WRITES COMPUTED DISTINCT EIGENVALUES OF
C                 T-MATRICES ALONG WITH THEIR T-MULTIPLICITIES
C                 TO FILE 11.
C
C     IWRITE = (0,1)   MEANS
C
C             (0) NO EXTENDED OUTPUT FROM SUBROUTINES BISEC AND INVERR
C                 IS SENT TO FILE 6.
C
C             (1) INDIVIDUAL COMPUTED T-EIGENVALUES AND CORRESPONDING
C                 ERROR ESTIMATES FROM THE SUBROUTINES BISEC AND INVERR
C                 ARE PRINTED OUT TO FILE 6 AS THEY ARE COMPUTED.
C
C     THE PROGRAM ALWAYS MAKES A SEPARATE LIST OF THE COMPUTED GOOD
C     T-EIGENVALUES ALONG WITH THEIR MINIMAL GAPS AND WRITES THEM OUT
C     TO FILE 3.  CORRESPONDING ERROR ESTIMATES FOR ANY ISOLATED
C     GOOD T-EIGENVALUES ARE ALWAYS WRITTEN TO FILE 4.
C
C
C-----INPUT/OUTPUT FILES FOR EIGENVALUE PROGRAMS------------------------
C
C     ANY INPUT DATA OTHER THAN THE ALPHA/BETA HISTORY SHOULD BE STORED
C     ON FILE 5. SEE SAMPLE INPUT/OUTPUT FROM TYPICAL RUN.
C     THE READ STATEMENTS IN THE GIVEN FORTRAN PROGRAM ASSUME THAT
C     THE DATA STORED ON FILE 5 IS IN FREE FORMAT.  USER SHOULD NOTE
C     THAT 'FREE FORMAT' IS NOT CLASSIFIED AS PORTABLE BY PFORT SO THAT
C     THE USER MAY HAVE TO MODIFY THE READ STATEMENTS FROM FILE 5 TO
C     CONFORM TO WHAT IS PERMISSIBLE ON THE MACHINE BEING USED.
C
C     FILE 6 WAS USED AS THE INTERACTIVE TERMINAL OUTPUT FILE.
C     THIS FILE PROVIDES A RUNNING ACCOUNT OF THE PROGRESS OF THE
C     COMPUTATIONS.  THE AMOUNT OF INFORMATION PRINTED OUT IS
C     CONTROLLED BY THE PARAMETER IWRITE.
C
C DESCRIPTION OF OTHER I/O FILES
C
C FILE (K)     CONTAINS:
C
C      (1)     OUTPUT FILE:
C              HISTORY FILE OF NEWLY-GENERATED T-MATRIX (ALPHA AND
C              BETA VECTORS) AND LAST 2 LANCZOS VECTORS USED
C              IN THE T-MATRIX GENERATION.  NOTE THAT IN CASE (4)
C              THREE 'LANCZOS' VECTORS ARE WRITTEN TO FILE 1.
C              IF IHIS = 0 AND ISTOP = 1, FILE 1 IS NOT WRITTEN.
C
C      (2)     INPUT FILE:
C              SAME AS FILE 1 EXCEPT THAT IT CONTAINS A
C              PREVIOUSLY-GENERATED T-MATRIX (IF ANY).  IF ISTART = 1,
C              PROGRAM ASSUMES THAT THERE IS A HISTORY FILE OF ALPHAS
C              AND BETAS ON FILE 2.  THESE ALPHAS AND BETAS ARE
C              READ IN ALONG WITH THE LAST 2 LANCZOS VECTORS THAT
C              WERE GENERATED.  IN CASE (4) THREE 'LANCZOS' VECTORS
C              ARE READ IN FROM FILE 2.
C
C      (3)     OUTPUT FILE:
C              COMPUTED GOOD EIGENVALUES OF THE T-MATRICES USED. ALSO
C              CONTAINS T-MULTIPLICITIES OF THESE EIGENVALUES AS
C              EIGENVALUES OF THE T-MATRIX, AND THEIR GAPS AS
C              EIGENVALUES IN THE A-MATRIX AND IN THE T-MATRIX.
C              FILE 3 IS ALWAYS WRITTEN.  IN CASE (3) THIS OUTPUT
C              CONTAINS THE EIGENVALUES OF THE B-INVERSE MATRIX
C              SINCE IN THIS CASE THE T-MATRICES CORRESPOND TO
C              THE B-INVERSE MATRIX AND NOT TO THE A-MATRIX.  IN
C              THIS CASE, 3 SETS OF GAPS ARE GIVEN, THOSE IN
```

```
C            THE T-MATRIX, IN THE B-INVERSE MATRIX AND THOSE
C            FOR THE CORRESPONDING EIGENVALUES IN THE A-MATRIX.
C
C      (4)   OUTPUT FILE:
C            ERROR ESTIMATES FOR THE ISOLATED GOOD T-EIGENVALUES
C            WHICH ARE OBTAINED USING THE SUBROUTINE INVERR. THESE
C            ESITMATES USE THE LAST COMPONENTS OF THE ASSOCIATED
C            T-EIGENVECTORS WHICH ARE COMPUTED USING INVERSE
C            ITERATION.  FILE 4 IS ALWAYS WRITTEN.
C
C
C      (7)   INPUT FILE:
C            USED ONLY IN CASES (3) AND (4), FACTORED INVERSES
C            OF REAL SYMMETRIC MATRICES AND GENERALIZED EIGENVALUE
C            PROBLEM.  CONTAINS THE REQUIRED FACTORIZATION OF THE
C            B-MATRIX.
C
C      (8)   INPUT FILE:
C            SAMPLE USPEC SUBROUTINE ASSUMES THAT THE ARRAYS
C            REQUIRED TO SPECIFY THE USER'S-MATRIX ARE STORED ON
C            FILE 8.  USERS MUST MAKE WHATEVER DEFINITIONS ARE
C            APPROPRIATE FOR THEIR MATRICES.  NOTE THAT IN CASE
C            (3) THE LANCZS SUBROUTINE DOES NOT USE THE MATRIX
C            ON FILE 8 IN THE T-MATRIX GENERATION, RATHER IT
C            USES THE FACTORIZATION OF AN ASSOCIATED
C            B-MATRIX WHICH IS STORED ON FILE 7.  IN CASE (4),
C            THE INFORMATION STORED ON BOTH FILES 7 AND 8 IS USED.
C
C      (9)   INPUT AND OUTPUT FILE:
C            CAN BE USED TO STORE THE TRUE EIGENVALUES OF THE
C            GIVEN PROBLEM,  WHEN THE PROCEDURE
C            IS BEING EXERCISED ON A TEST MATRIX.
C
C     (11)   OUTPUT FILE:
C            COMPUTED DISTINCT EIGENVALUES OF T-MATRICES USED.
C            ALSO CONTAINS THEIR T-MULTIPLICITIES AND T-GAPS TO
C            NEAREST DISTINCT EIGENVALUES, AND THE T-MULTIPLICITY
C            PATTERN OF THE GOOD AND THE SPURIOUS T-EIGENVALUES.
C            FILE 11 IS WRITTEN ONLY IF IDIST = 1.
C
C
C-----PARAMETERS SET BY THE EIGENVALUE PROGRAMS------------------------
C
C
C     THESE PARAMETERS ARE SET INTERNALLY IN THE PROGRAM
C
C     SCALEK     K = 1,2,3,4
C
C               THE SCALING FACTORS SCALEK HAVE BEEN INTRODUCED IN AN
C               ATTEMPT TO MAKE THE TOLERANCES USED IN THE
C               T-MULTIPLICITY, SPURIOUS, ISOLATION AND PRTESTS ADJUST
C               TO THE SCALE OF THE GIVEN MATRIX.  THESE FACTORS MUST
C               NOT BE MODIFIED.  THESE TOLERANCES OCCUR IN LUMP,
C               ISOEV, AND PRTEST.
C
C     NOTE:    THE USER SHOULD NOTE THAT IF THE MATRIX BEING
C     PROCESSED IS VERY STIFF, THAT IS THE RATIO OF THE LARGEST
C     EIGENVALUE IN MAGNITUDE TO THE SMALLEST IN MAGNITUDE IS VERY
C     LARGE, THEN THE TOLERANCES BEING USED IN BISEC, LUMP, ISOEV
C     AND PRTEST MAY NOT TREAT THE SMALL END (SMALL IN MAGNITUDE)
C     VERY WELL.  IN SOME SUCH CASES A USER-INTRODUCED REDUCTION
C     IN THE SIZE OF TKMAX AND THE SUBSEQUENT RECOMPUTATION OF
C     THE T-MATRIX EIGENVALUES IN (ONLY) THE LOWER END OF THE
C     SPECTRUM WITH THIS TKMAX MAY RESULT IN IMPROVED COMPUTATIONS
C     AT THE LOW END.
C
C     THE LUMP, ISOEV, AND PRTEST TOLERANCES THAT WERE USED
```

```
C      MOST IN  THE TESTING OF THESE ALGORITHMS WERE NOT
C      SCALE INVARIANT BUT SEEMED TO WORK WELL ON MATRICES THAT
C      HAD EIGENVALUES WITH MAGNITUDES BOTH GREATER THAN AND LESS
C      THAN 1.  THESE TOLERANCES ARE ALSO INCLUDED IN THESE THREE
C      SUBROUTINES BUT AS COMMENTED OUT STATEMENTS.  THEY CAN BE
C      REVIVED BY COMMENTING OUT THE CORRESPONDING TOLERANCES
C      SPECIFIED IN THE STATEMENT ABOVE EACH OF THESE.
C
C      IMPORTANT TOLERANCES OR SCALES THAT ARE USED REPEATEDLY
C      THROUGHOUT THE LANCZOS EIGENVALUE PROGRAMS ARE THE FOLLOWING:
C      SCALED MACHINE EPSILON:  TTOL = TKMAX*EPSM WHERE
C      EPSM = 2*MACHINE EPSILON AND
C      TKMAX = MAX(|ALPHA(J)|,BETA(J), J = 1,MEV)
C      BISEC CONVERGENCE TOLERANCE:  BISTOL = DSQRT(1000+MEV)*TTOL
C      BISEC T-MULTIPLICITY TOLERANCE:  MULTOL = (1000+MEV)*TTOL
C      LANCZOS CONVERGENCE TOLERANCE:   CONTOL = BETA(MEV+1)*1.D-10
C
C
C      BTOL = RELATIVE TOLERANCE USED TO ESTIMATE ANY LOSS OF LOCAL
C             ORTHOGONALITY OF THE LANCZOS VECTORS AFTER THE T-MATRIX
C             HAS BEEN GENERATED.  THE LANCZOS PROCEDURE WORKS WELL
C             ONLY IF LOCAL ORTHOGONALITY BETWEEN SUCCESSIVE LANCZOS
C             VECTORS IS MAINTAINED.  THE TNORM SUBROUTINE TESTS
C             WHETHER OR NOT
C
C                  MINIMUM  |BETA(I)|/||A|| > BTOL.
C                  I=2,KMAX
C
C             IF THIS TEST IS VIOLATED BY SOME BETA AND A T-MATRIX THAT
C             WOULD INCLUDE SUCH A BETA IS REQUESTED, THEN THE LANCZOS
C             PROCEDURE WILL TERMINATE FOR THE USER TO DECIDE WHAT TO
C             DO.  THE USER CAN OVER-RIDE THIS TEST BY SIMPLY DECREASING
C             THE SIZE OF BTOL, BUT THEN CONVERGENCE IS NOT AS CERTAIN.
C             THE PROGRAM SETS BTOL = 1.D-8 WHICH IS A VERY CONSERVATIVE
C             CHOICE. THE || A || IS ESTIMATED BY USING AN ESTIMATE
C             OF THE NORM OF THE T-MATRIX, T(1,KMAX).
C
C      GAPTOL = RELATIVE TOLERANCE USED IN THE SUBROUTINE ISOEV
C               TO DETERMINE WHICH OF THE GOOD T-EIGENVALUES NEED
C               ERROR ESTIMATES.  THE PROGRAM SETS GAPTOL = 1.D-8.
C               IF FOR A GIVEN 'GOOD' T-EIGENVALUE THE COMPUTED GAP
C               IS TOO SMALL AND IS DUE TO A 'SPURIOUS' T-EIGENVALUE
C               THEN THE 'GOOD' T-EIGENVALUE IS ASSUMED TO HAVE CONVERGED
C               AND NO ERROR ESTIMATES ARE COMPUTED.
C
C
C-----USER-SPECIFIED PARAMETERS FOR EIGENVALUE PROGRAMS-----------------
C
C
C      RELTOL = RELATIVE TOLERANCE USED IN 'COMBINING' COMPUTED
C               EIGENVALUES OF T(1,MEV) PRIOR TO COMPUTING ERROR
C               ESTIMATES.
C
C      THE LUMPING OF T-EIGENVALUES OCCURS IN SUBROUTINE LUMP.
C      LUMPING IS NECESSARY BECAUSE IT IS IMPOSSIBLE TO ACCURATELY
C      PREDICT THE ACCURACY OF THE BISEC SUBROUTINE.  LUMP 'COMBINES'
C      T-EIGENVALUES THAT HAVE SLIPPED BY THE TOLERANCE THAT WAS USED
C      IN THE T-MULTIPLICITY TESTS.  IN PARTICULAR IF FOR SOME J,
C
C    |EVALUE(J)-EVALUE(J-1)| < DMAX1(RELTOL*|EVALUE(J)|,SCALE2*MULTOL)
C
C      THEN THESE T-EIGENVALUES ARE 'COMBINED'.  MULTOL IS THE TOLERANCE
C      THAT WAS USED IN THE T-MULTIPLICITY TEST IN BISEC.  SEE THE HEADER
C      ON THE LUMP SUBROUTINE FOR MORE DETAILS.
C
C      RELTOL IS SET TO 1.D-10.
C
```

```
C      MXINIT = MAXIMUM NUMBER OF INVERSE ITERATIONS ALLOWED IN
C               SUBROUTINE INVERR FOR EACH ISOLATED GOOD T-EIGENVALUE.
C               TYPICALLY ONLY ONE ITERATION IS REQUIRED.
C
C      SEEDS FOR RANDOM NUMBER GENERATORS = INTEGER*4 SCALARS.
C
C                 (1) SVSEED = SEED FOR STARTING VECTOR USED IN
C                     T-MATRIX GENERATION IN LANCZS SUBROUTINE
C
C                 (2) RHSEED = SEED FOR RIGHT-HAND SIDE USED IN
C                     INVERSE ITERATION COMPUTATIONS IN INVERR.
C
C      BISEC DATA
C
C      (1) NINT  =  NUMBER OF SUBINTERVALS ON WHICH EIGENVALUES ARE
C                   TO BE COMPUTED.
C
C      (2) LB(J) = (J = 1,NINT)  =  LEFT END POINTS OF THESE INTERVALS.
C                  MUST BE PROVIDED IN INCREASING ORDER.  THAT IS,
C                  LB(J) < LB(J+1) FOR J = 1,NINT.
C
C      (3) UB(J) = (J = 1,NINT) =  RIGHT END POINTS OF THESE INTERVALS.
C                  MUST BE PROVIDED IN INCREASING ORDER.  THAT IS,
C                  UB(J) < UB(J+1) FOR J = 1,NINT.
C
C      (4) MXSTUR  =  MAXIMUM NUMBER OF STURM ITERATIONS ALLOWED FOR
C                     ENTIRE SET OF EIGENVALUE CALCULATIONS OVER ALL
C                     SPECIFIED SIZE T-MATRICES.  PROGRAM WILL
C                     TERMINATE IF THIS LIMIT IS EXCEEDED.
C
C      T-MATRICES
C
C      SIZES OF T-MATRICES
C
C               (1) KMAX= MAXIMUM ORDER FOR T-MATRIX THAT USER IS WILLING
C                         TO CONSIDER.
C
C               (2) NMEVS = MAXIMUM NUMBER OF T-MATRICES THAT WILL BE
C                           CONSIDERED.
C
C               (3) NMEV(J)  (J=1,NMEVS)  = SIZES OF T-MATRIX TO BE
C                                          CONSIDERED SEQUENTIALLY.
C
C      T-MATRIX-GENERATION
C
C      USER SHOULD NOTE THAT THIS PROGRAM FIRST COMPUTES A T-MATRIX
C      OF ORDER KMAX AND THEN CYCLES THROUGH THE T-MATRICES SPECIFIED
C      A PRIORI BY THE USER, USING THE SUBROUTINE BISEC TO COMPUTE THE
C      EIGENVALUES OF THE T-MATRICES ON THE INTERVALS SPECIFIED BY
C      THE USER.
C
C      IDEALLY, ONE WOULD COMPUTE THE EIGENVALUE APPROXIMATIONS AT A
C      REASONABLE SIZE T-MATRIX, LOOK AT THE ACCURACY OF THE COMPUTED
C      RESULTS AND USE THAT TO DETERMINE AN APPROPRIATE
C      INCREMENT FOR THE SIZE OF THE T-MATRIX BASED UPON WHAT
C      HAS ALREADY CONVERGED AND UPON THE SIZES OF THE ERROR ESTIMATES
C      ON THOSE EIGENVALUES THAT ARE DESIRED BUT THAT HAVE NOT YET
C      CONVERGED. HOWEVER, IN THE INTERESTS OF GENERALITY AND
C      SIMPLICITY WE DID NOT DO THAT HERE.
C
C
C-----CONVERGENCE TESTS FOR THE EIGENVALUE PROGRAMS--------------------
C
C
C      THE CONVERGENCE TEST INCORPORATED IN THIS PROGRAM IS
C      BASED UPON THE ASSUMPTION THAT THOSE T-EIGENVALUES AND THEIR
C      ASSOCIATED T-EIGENVECTORS WHICH CORRESPOND TO THE
```

```
C     EIGENVALUES AND RITZVECTORS WHICH WE WISH TO COMPUTE
C     CONVERGE AS THE T-SIZE IS INCREASED.
C
C     AS CURRENTLY PROGRAMMED, CONVERGENCE IS CHECKED BY EXAMINING
C     THE SIZES OF ALL OF THE COMPUTED ERROR ESTIMATES ON ALL OF THE
C     INTERVALS SPECIFIED BY THE USER.  IDEALLY CONVERGENCE SHOULD
C     BE CHECKED ONLY ON THOSE EIGENVALUES OF INTEREST AND
C     ONCE THE EIGENVALUES ON SUB-INTERVALS OF THESE INTERVALS HAVE
C     CONVERGED, ANY SUBSEQUENT EIGENVALUE COMPUTATIONS SHOULD BE
C     MADE ONLY ON THE UNCONVERGED PORTIONS.  OBVIOUSLY, IT WOULD BE
C     DIFFICULT TO INCORPORATE CODE TO DO THE ABOVE WITHOUT KNOWING
C     A PRIORI PRECISELY WHAT THE USER IS TRYING TO COMPUTE.
C     THEREFORE, WE DID NOT ATTEMPT TO DO THIS.  IF ONE WISHES TO
C     MAKE SUCH A MODIFICATION THEN ONE MUST ALSO MODIFY THE PROGRAM
C     SO THAT IT CREATES AN OVERALL LIST OF THE CONVERGED 'GOOD'
C     T-EIGENVALUES AS THEY ARE COMPUTED, SINCE CONVERGED 'GOOD'
C     T-EIGENVALUES WHICH WERE COMPUTED AT A PARTICULAR VALUE OF MEV
C     WOULD NO LONGER BE RECOMPUTED AT LARGER VALUES OF MEV.
C
C     IF ONLY A FEW EIGENVALUES ARE TO BE COMPUTED THEN SUCH CHANGES
C     WOULD NOT MAKE MUCH DIFFERENCE IN THE RUNNING TIME.
C
C
C-----ARRAYS REQUIRED BY THE EIGENVALUE PROGRAMS-----------------------
C
C
C     ALL 4 CASES
C
C     ALPHA(J) = REAL*8 ARRAY. ITS DIMENSION MUST BE AT LEAST KMAX,
C                THE LENGTH OF THE LARGEST T-MATRIX ALLOWED.  THIS
C                ARRAY CONTAINS THE DIAGONAL ENTRIES OF THE T-MATRICES.
C
C     BETA(J) = REAL*8 ARRAY.  ITS DIMENSION MUST BE AT LEAST KMAX+1.
C               THIS ARRAY CONTAINS THE SUBDIAGONAL ENTRIES OF THE
C               T-MATRICES.
C
C               THE ALPHA AND BETA VECTORS ARE NOT ALTERED
C               DURING THE CALCULATIONS.
C
C     V1(J),V2(J),VS(J) = REAL*8 ARRAYS IN REAL SYMMETRIC CASES.
C                         V1 AND V2 ARE COMPLEX*16 IN HERMITIAN CASE.
C                         IN CASES (1) AND (2) VS MUST BE OF
C                         DIMENSION AT LEAST KMAX.  IN CASES (3) AND
C                         (4) VS MUST BE OF DIMENSION AT LEAST
C                         MAX(N,KMAX).  IN REAL SYMMETRIC CASES
C                         V1 MUST BE OF DIMENSION AT LEAST
C                         MAX(KMAX+1,N) AND V2 MUST BE OF DIMENSION
C                         MAX(KMAX,N).  IN HERMITIAN CASES V1
C                         MUST BE OF DIMENSION MAX(N,(KMAX+1)/2)
C                         AND V2 OF DIMENSION AT LEAST MAX(N,KMAX/2).
C                         IN ALL CASES HOWEVER, THE ABOVE DIMENSIONS
C                         FOR V2 ARE VALID ONLY IF NO MORE
C                         THAN KMAX/2 EIGENVALUES OF THE GIVEN
C                         T-MATRICES ARE TO BE COMPUTED IN ANY GIVEN
C                         SUBINTERVAL. V2 IS USED IN THE SUBROUTINE
C                         BISEC TO HOLD THE UPPER AND LOWER
C                         ENDPOINTS OF THE SUBINTERVALS GENERATED
C                         DURING THE BISECTIONS.  THEREFORE, ITS
C                         REAL*8 DIMENSION MUST ALWAYS BE AT LEAST
C                         2*Q WHERE Q IS THE MAXIMUM NUMBER OF
C                         EIGENVALUES OF THE SPECIFIED T-MATRIX IN ANY
C                         ONE OF THE SPECIFIED INTERVALS.
C                         NOTE THAT IN THE HERMITIAN CASE, V1 AND V2
C                         ARE DEFINED AS COMPLEX*16 IN THE MAIN PROGRAM
C                         AND IN THE LANCZS SUBROUTINE BUT ARE
C                         REDEFINED AS REAL*8 IN OTHER SUBROUTINES.
C
```

```
C     LB(J),UB(J) = REAL*8 ARRAYS. EACH MUST BE OF DIMENSION AT LEAST
C                   NINT, THE NUMBER OF SUBINTERVALS TO BE CONSIDERED.
C                   LB CONTAINS THE LEFT-END POINTS OF THE INTERVALS
C                   ON WHICH EIGENVALUES ARE TO BE COMPUTED.  UB
C                   CONTAINS THE RIGHT-END POINTS.
C
C     EXPLAN(J) = REAL*4 ARRAY.  ITS DIMENSION IS 20.  THIS ARRAY IS
C                 USED TO ALLOW EXPLANATORY COMMENTS IN THE INPUT FILES.
C
C     G(J) = REAL*4 ARRAY.  ITS DIMENSION MUST BE >= MAX(2*KMAX,N)
C            IT IS USED FOR HOLDING THE RANDOM VECTORS GENERATED,
C            HOLDING THE COMPUTED ERROR ESTIMATES AND THE COMPUTED
C            MINIMAL GAPS FOR THE GOOD T-EIGENVALUES.
C
C     MP(J) = INTEGER*4 ARRAY.  ITS DIMENSION MUST BE AT LEAST KMAX,
C             THE MAXIMUM SIZE OF THE T-MATRICES ALLOWED.  IT CONTAINS
C             THE T-MULTIPLICITIES OF THE COMPUTED EIGENVALUES. NOTE
C             THAT 'SPURIOUS' T-EIGENVALUES ARE DENOTED BY A
C             T-MULTIPLICITY OF 0.  T-EIGENVALUES THAT THE SUBROUTINE
C             PRTEST HAS IDENTIFIED AS 'GOOD' BUT HIDDEN ARE IDENTIFIED
C             BY A T-MULTIPLICITY OF -10.
C
C     NMEV(J) = INTEGER*4 ARRAY.  ITS DIMENSION MUST BE AT LEAST THE
C               NUMBER OF T-MATRICES ALLOWED.  IT CONTAINS THE ORDERS
C               OF THE T-MATRICES TO BE CONSIDERED.
C
C
C     FOR CASE (3) ONLY:
C     GR(J),GC(J) = REAL*8 ARRAYS. USED ONLY IN THE HERMITIAN CASE.
C                   GR AND GC MUST EACH BE OF DIMENSION AT LEAST N.
C                   BOTH ARE USED IN THE RANDOM VECTOR GENERATION.
C                   GR IS ALSO USED TO STORE MINIMAL GAPS BETWEEN
C                   'GOOD' T-EIGENVALUES.
C
C     FOR CASES (3) AND (4) FOR THE PERMUTATION:
C
C     IPR(J), IPT(J) = INTEGER*4 ARRAYS.  EACH OF DIMENSION AT LEAST N.
C                      USED TO STORE THE REORDERING OF THE GIVEN MATRIX
C                      OR MATRICES.
C
C
C     OTHER ARRAYS
C
C     THE USER MUST SPECIFY IN THE SUBROUTINE USPEC (A OR B) WHATEVER
C     ARRAYS ARE REQUIRED TO DEFINE THE MATRIX OR MATRICES BEING USED.
C     ALSO IN CASES (3) AND (4) ONLY, WHEN WORKING WITH INVERSES
C     OF SPARSE SYMMETRIC MATRICES, IN THE OPTIONAL, PREPROCESSING
C     PROGRAMS PERMUT, LFACT, LORDER, AND LTEST IT IS NECESSARY TO
C     SPECIFY ADDITIONAL ARRAYS JUST FOR THESE COMPUTATIONS.  THE USER
C     IS REFERRED TO THOSE PROGRAMS FOR DETAILS.
C
C
C-----SUBROUTINES INCLUDED----------------------------------------------
C
C
C     LANCZS = COMPUTES THE ALPHA/BETA HISTORY. IN CASES (1) AND (2)
C              REAL SYMMETRIC AND HERMITIAN MATRICES, USES SUBROUTINES
C              FINPRO, GENRAN AND CMATV.  IN CASE (3), INVERSES OF
C              REAL SYMMETRIC MATRICES, USES SUBROUTINES FINPRO,
C              GENRAN AND BSOLV.  IN CASE (4), GENERALIZED EIGENVALUE
C              PROBLEM, USES SUBROUTINES FINPRO, GENRAN, AMATV AND
C              LSOLV.
C
C     BISEC = COMPUTES EIGENVALUES OF THE SPECIFIED T-MATRIX
C             USING STURM SEQUENCING, ON SEQUENCE OF INTERVALS
C             SPECIFIED BY THE USER.  EACH SUBINTERVAL IS TREATED
C             AS OPEN ON THE LEFT AND CLOSED ON THE RIGHT.
```

```
C            EIGENVALUES ARE COMPUTED WITH SIMULTANEOUS DETERMINATION
C            OF THE T-MULTIPLICITIES AND OF SPURIOUS T-EIGENVALUES.
C
C      INVERR = USES INVERSE ITERATION ON T-MATRICES TO COMPUTE ERROR
C             ESTIMATES ON COMPUTED GOOD T-EIGENVALUES. (USES GENRAN)
C
C      LUMP = 'COMBINES' EIGENVALUES OF T-MATRIX USING THE RELATIVE
C            TOLERANCE RELTOL.
C
C      ISOEV = CALCULATES GAPS BETWEEN DISTINCT EIGENVALUES OF T-MATRIX
C            AND THEN USES THESE GAPS TO LABEL THOSE 'GOOD'
C            T-EIGENVALUES FOR WHICH ERROR ESTIMATES ARE NOT COMPUTED.
C
C      TNORM = COMPUTES THE SCALE TKMAX USED IN DETERMINING THE
C            TOLERANCES FOR THE SPURIOUS, T-MULTIPLICITY AND PRTESTS.
C            IT ALSO CHECKS FOR LOCAL ORTHOGONALITY OF THE LANCZOS
C            VECTORS BY TESTING THE RELATIVE SIZE OF THE BETAS USING
C            THE RELATIVE TOLERANCE BTOL.
C
C      PRTEST = LOOKS FOR GOOD T-EIGENVALUES THAT HAVE BEEN MISLABELLED
C             BY THE SPURIOUS TEST BECAUSE THEY HAD 'TOO SMALL' A
C             PROJECTION ON THE STARTING LANCZOS VECTOR.
C             (LESS THAN SINGLE PRECISION)
C             TESTS INDICATE THAT SUCH EIGENVALUES ARE RARE.
C             PRTEST SHOULD BE CALLED ONLY AFTER CONVERGENCE
C             HAS BEEN ESTABLISHED.
C
C      INVERM = USED TO COMPUTE ERROR ESTIMATES FOR ANY T-EIGENVALUES
C             WHICH PRTEST INDICATES MAY HAVE BEEN MISLABELLED.
C             SUCH T-EIGENVALUES ARE RELABELLED ONLY IF THEIR ERROR
C             ESTIMATES ARE SUFFICIENTLY SMALL.  PRIMARY USE OF
C             INVERM IS IN THE CORRESPONDING EIGENVECTOR COMPUTATIONS.
C
C      CASES (3) AND (4) ONLY, FACTORED INVERSES:
C
C      FOR OPTIONAL, PRELIMINARY PROCESSING:
C      PERMUT  (PROGRAM CALLS SPARSPAK PACKAGE) :
C      USES THE NONZERO STRUCTURE OF A GIVEN MATRIX A.
C      CAN BE USED TO OBTAIN A REORDERING OF A THAT PRESERVES
C      SPARSITY UNDER FACTORIZATION.  PERMUT CALLS
C      THE SPARSPAK PROGRAMS, (A. GEORGE, J. LIU, E.NG,
C      U. WATERLOO).  PERMUT ALSO TAKES THE USER-SPECIFIED MATRIX,
C      APPLIES THE SCALE SO AND THE SHIFT TO IT, AND THEN WRITES
C      OUT THE CORRESPONDING SPARSE MATRIX DATA FILE FOR THE
C      RESULTING MATRIX C = SO*A + SHIFT*I.  SEE THE PERMUT FORTRAN
C      CODE FOR DETAILS.
C
C      LORDER (STAND-ALONE PROGRAM):
C      GIVEN A MATRIX C IN SPARSE FORMAT AND A PERMUTATION P,
C      COMPUTES THE REORDERED MATRIX B = P*C*P' AND WRITES IT
C      TO FILE 9 IN SPARSE FORMAT.  SEE THE LORDER FORTRAN CODE
C      FOR DETAILS.
C
C      LFACT (STAND-ALONE PROGRAM) :
C      GIVEN A POSITIVE DEFINITE MATRIX B IN SPARSE FORMAT
C      COMPUTES THE SPARSE CHOLESKY FACTOR L OF B AND WRITES IT
C      TO FILE 7 IN SPARSE FORMAT.  THUS, B = L*L'.
C      SEE THE LFACT FORTRAN CODE FOR DETAILS.
C
C      LTEST (CALLS 3 USER-SUPPLIED PROGRAMS CMATV, CMATS, AND BSOLV):
C      GIVEN THE FACTORIZATION OF A MATRIX B, LTEST COMPUTES
C      THE SOLUTION OF THE EQUATION B*U = B*V1 FOR A SPECIFIC RANDOMLY-
C      GENERATED V1, WITH AND WITHOUT ITERATIVE REFINEMENT, TO
C      OBTAIN A ROUGH CHECK ON THE NUMERICAL CONDITION OF THE MATRIX B.
C      THIS PROGRAM USES 3 SUBROUTINES CMATV, CMATS, AND BLSOLV.
C      SEE THE LTEST FORTRAN PROGRAM FOR DETAILS.
C
```

```
C
C-----OTHER PROGRAMS PROVIDED--------------------------------------------
C
C     LECOMPAC (STAND ALONE PROGRAM):
C                 TRANSLATES A REAL SYMMETRIC MATRIX PROVIDED IN THE
C                 FORMAT  I, J, A(I,J) INTO THE SPARSE MATRIX
C                 FORMAT USED IN THE SAMPLE USPEC, CMATV, BSOLV AND
C                 LSOLV SUBROUTINES PROVIDED. IT ASSUMES THAT THE
C                 MATRIX ENTRIES ARE GIVEN EITHER COLUMN BY COLUMN OR
C                 ROW BY ROW.  THE DATA SET CREATED IS WRITTEN TO
C                 FILE 8.
C
C
C-----COMMENTS ON THE STORAGE REQUIRED FOR EIGENVALUE PROGRAMS----------
C
C
C     CASE (1), REAL SYMMETRIC MATRICES:
C
C     THE ARRAYS IN THE REAL SYMMETRIC EIGENVALUE PROGRAM REQUIRE
C     APPROXIMATELY THE EQUIVALENT OF ONE REAL*8 ARRAY OF DIMENSION
C
C        3.5*KMAX + 2*MAX(KMAX,N) + .5* MAX(2*KMAX,N)
C
C     PLUS WHATEVER IS NEEDED TO GENERATE A*X FOR THE GIVEN MATRIX A.
C     THE ARRAYS ALPHA, BETA, VS AND MP CONSUME 3.5*KMAX*8 BYTES.
C     THE ARRAYS V1 AND V2 CONSUME 2*MAXIMUM(KMAX,N)*8 BYTES, WITH THE
C     QUALIFICATION STATED ABOVE WHERE V2 IS DEFINED.  THE G-ARRAY
C     CONSUMES .5*MAX(2*KMAX,N)*8 BYTES.
C
C     CASE (2), HERMITIAN MATRICES:
C
C     THE ARRAYS IN THE HERMITIAN EIGENVALUE PROGRAMS REQUIRE
C     THE EQUIVALENT OF ONE REAL*8 ARRAY OF DIMENSION
C
C        3.5*KMAX + 4*MAX(KMAX/2,N) + .5*MAX(2*KMAX,N) + 2*N
C
C     PLUS WHATEVER IS NEEDED TO GENERATE A*X FOR THE GIVEN MATRIX A.
C     THE ARRAYS ALPHA, BETA, VS, AND MP CONSUME 3.5*KMAX*8 BYTES.
C     THE ARRAYS V1 AND V2 CONSUME 4*MAXIMUM(KMAX/2,N)*8 BYTES, WITH THE
C     QUALIFICATION STATED ABOVE WHERE V2 IS DEFINED.  THE G-ARRAY
C     CONSUMES .5*MAX(2*KMAX,N)*8 BYTES.  GR REQUIRES
C     AND GC REQUIRE 2*N*8BYTES.
C
C
C     CASE (3), INVERSES OF REAL SYMMETRIC MATRICES:
C
C     THE ARRAYS IN THE EIGENVALUE PROGRAMS DESIGNED FOR
C     CASE (3), INVERSES OF REAL SYMMETRIC MATRICES USING
C     REORDERING AND FACTORIZATION, REQUIRE
C     THE EQUIVALENT OF ONE REAL*8 ARRAY OF DIMENSION
C
C        3*KMAX + 3*MAX(KMAX,N) + .5*MAX(2*KMAX,N)
C
C     PLUS WHATEVER IS NEEDED TO GENERATE B(INVERSE)*X  FOR THE
C     SCALED, SHIFTED AND PERMUTED VERSION OF A WHICH WE DENOTE
C     BY B.  THE ARRAYS ALPHA, BETA, MP, AND MP2 CONSUME 3*KMAX*8
C     BYTES.  THE ARRAYS V1, V2, AND VS CONSUME 3*MAX(KMAX,N)*8 BYTES,
C     WITH THE QUALIFICATION STATED ABOVE WHERE V2 IS DEFINED.
C     THE G ARRAY CONSUMES .5*MAX(2*KMAX,N)*8 BYTES. THESE NUMBERS
C     DO NOT INCLUDE THE STORAGE REQUIRED BY THE PREPROCESSING PROGRAMS
C     PERMUT, LORDER, LFACT, AND LTEST.
C
C
C     A SYMMETRIC, SPARSE MATRIX OF ORDER N WITH NZS NONZERO ELEMENTS
C     BELOW THE MAIN DIAGONAL WOULD REQUIRE THE EQUIVALENT OF ONE
C     REAL*8 ARRAY OF DIMENSION 1.5*(NZS + N) IF THE POINTERS USED
C     ARE INTEGER*4.
```

```
C
C     SOME OF THE ARRAY STORAGE IS NOT ESSENTIAL AND COULD BE
C     ELIMINATED IF STORAGE IS A PROBLEM.
C     THE FOLLOWING COMMENTS APPLY DIRECTLY ONLY TO CASE (1),
C     THE PROGRAMS FOR REAL SYMMETRIC MATRICES, HOWEVER, SIMILAR
C     STATEMENTS COULD BE MADE ABOUT THE OTHER CASES.
C
C     CASE (1), REAL SYMMETRIC PROGRAMS:
C     THE G ARRAY COULD BE REMOVED IF THE USER IS WILLING TO
C
C            (1) REGENERATE THE RANDOM STARTING VECTOR IN INVERR
C                FOR EACH ERROR ESTIMATE
C            (2) WRITE OUT THE ERROR ESTIMATES AND VARIOUS GAPS AS
C                THEY ARE GENERATED RATHER THAN STORING THEM IN G FOR
C                LATER PRINTOUT
C            (3) CHECK CONVERGENCE WITHIN INVERR
C
C     CLEARLY THE INDEX VECTOR MP COULD BE AN INTEGER*2 ARRAY AS COULD
C     THE POINTERS USED TO DEFINE THE USER'S MATRIX.
C
C     THE USER SHOULD NOTE THAT WITH AN EIGENVALUE SUBROUTINE THAT
C     USES BISECTION (LIKE BISEC) IF MORE THAN 25% OF THE
C     EIGENVALUES ARE TO BE COMPUTED, THEN IT MAY BE MORE
C     ECONOMICAL TO USE THE EISPACK SUBROUTINE IMTQL1.
C     (SEE MATRIX EIGENSYTEM ROUTINES-EISPACK GUIDE (2ND EDITION)
C     B.T. SMITH ET AL, SPRINGER-VERLAG, NEW YORK, 1976, P213.).
C     HOWEVER, IF THE SUBROUTINE IMTQL1 IS TO BE USED IN PLACE
C     OF BISEC, THEN NONTRIVIAL CHANGES IN THE LANCZOS CODE MUST BE
C     MADE. FOR DETAILS OF ONE SUCH IMPLEMENTATION SEE
C     IBM RESEARCH REPORT 8298, COMPUTING
C     EIGENVALUES OF LARGE SYMMETRIC MATRICES - AN IMPLEMENTATION OF A
C     LANCZOS ALGORITHM WITH NO REORTHOGONALIZATION.  PART II. COMPUTER
C     PROGRAMS., DECEMBER 1980, WHICH CONTAINS A GENERAL
C     LANCZOS CODE WHICH INCLUDES AN IMTQL1 OPTION OR
C     PREFERABLY CONTACT THE AUTHORS.
C
C     THE BISEC SUBROUTINE WHICH IS INCLUDED IS A MODIFIED FORM OF
C     THE BISECT SUBROUTINE IN EISPACK. BISEC ASSUMES THAT THE
C     VECTOR V2 IS LONG ENOUGH TO HOLD BOTH THE UPPER AND THE
C     LOWER BOUNDS ON THE BISECTION INTERVALS USED TO COMPUTE
C     THE EIGENVALUES OF THE T-MATRICES.  THEREFORE, IF THE
C     LENGTH OF V2 IS ONLY KMAX, BISEC CAN COMPUTE ONLY AT MOST
C     KMAX/2 EIGENVALUES OF THE GIVEN T-MATRIX IN ANY GIVEN
C     SUBINTERVAL.
C
C     AS PROGRAMMED BISEC USES THE ARRAYS ALPHA,BETA,V1,V2,VS AND MP.
C     HOWEVER, V1 IS USED ONLY TO STORE BETA(J)**2 SO THAT THEY DO NOT
C     HAVE TO BE REGENERATED ON EACH STURM.  IF THE USER IS WILLING TO
C     COMPUTE THE BETA(J)**2 AS NEEDED, THEN V1 COULD BE ELIMINATED
C     FROM BISEC. BISEC STORAGE THEN BECOMES A REAL*8 ARRAY OF DIMENSION
C     4.25*KMAX IF WE ALSO REDUCE MP TO INTEGER*2.  FURTHERMORE,
C     IF ONE KNEW THAT ONLY Q*MEV EIGENVALUES OF T(1,MEV) WERE TO BE
C     COMPUTED AT EACH STAGE FOR SOME Q<.5 THEN FURTHER REDUCTIONS IN
C     STORAGE COULD BE MADE IN BISEC.
C
C     AS PROGRAMMED INVERR USES ALPHA, BETA,V1,V2,VS,G AND MP.
C     VS CONTAINS THE COMPUTED EIGENVALUES OF T(1,MEV). MP GIVES
C     THEIR T-MULTIPLICITIES AND FLAGS WHICH EIGENVALUES ARE TO HAVE
C     ERROR ESTIMATES COMPUTED.  V2 IS USED FOR THE SOLUTION
C     VECTOR IN THE INVERSE ITERATION AND V1 FOR THE FACTORIZATION.
C     G CONTAINS THE RANDOMLY-GENERATED STARTING VECTOR FOR THE
C     INVERSE ITERATION. THE BASIC STORAGE FOR INVERR IS THEREFORE
C     A REAL*8 ARRAY OF DIMENSION 4*KMAX PLUS THE STORAGE NEEDED FOR
C     THE COMPUTED T-EIGENVALUES AND THEIR T-MULTIPLICITIES.
C
C     VS COULD BE USED TO STORE ONLY THOSE COMPUTED EIGENVALUES OF
C     T(1,MEV) THAT ARE OF INTEREST. IN THAT CASE THE DIMENSIONS OF VS
```

```
C     AND OF MP NEED NOT BE ANY LONGER THAN THE NUMBER OF SUCH
C     EIGENVALUES. AS PROGRAMMED, ALL THE COMPUTED DISTINCT EIGENVALUES
C     OF T(1,MEV) ARE STORED IN VS.  THEREFORE TO TAKE ADVANTAGE OF
C     SUCH A REDUCTION IN STORAGE THE USER WOULD HAVE TO MODIFY THAT
C     PART OF THE PROGRAM AND ALSO COMMENT OUT THE CALL TO THE
C     SUBROUTINE PRTEST.
C
C
C-----------------------------------------------------------------------
C
C     COMMENTS FOR EIGENVECTOR COMPUTATIONS
C
C-----------------------------------------------------------------------
C
C
C     THE EIGENVALUES WHOSE EIGENVECTORS ARE TO BE COMPUTED MUST
C     HAVE BEEN COMPUTED USING THE CORRESPONDING LANCZOS EIGENVALUE
C     PROGRAMS BECAUSE THE EIGENVECTOR PROGRAMS WILL USE THE SAME
C     FAMILY OF LANCZOS TRIDIAGONAL MATRICES THAT WAS USED IN THE
C     CORRESPONDING EIGENVALUE COMPUTATIONS.
C
C     THESE PROGRAMS ASSUME THAT THE EIGENVALUES SUPPLIED TO IT
C     HAVE BEEN COMPUTED ACCURATELY, AS MEASURED BY THE
C     ERROR ESTIMATES COMPUTED IN THE CORRESPONDING LANCZOS
C     EIGENVALUE COMPUTATIONS, ALTHOUGH THESE ESTIMATES ARE
C     TYPICALLY CONSERVATIVE.  IN CASES (1), (2) AND (4), THE
C     EIGENVALUES OF INTEREST ARE STORED IN THE ARRAY GOODEV(J),
C     J=1,NGOOD.  IN CASE (3) THE PROGRAM WORKS WITH THE
C     EIGENVALUES OF B(INVERSE) WHICH ARE STORED IN THE ARRAY
C     GOODBI(J), J=1,NGOOD.  THE CORRESPONDING EIGENVALUES
C     OF A ARE STORED IN GOODA(J), J=1,NGOOD.
C
C     FOR EACH GOODEV(J), THE SUBROUTINE STURMI COMPUTES THE
C     SMALLEST SIZE LANCZOS TRIDIAGONAL MATRIX, T(1,M1(J)), FOR
C     WHICH  GOODEV(J) IS AN EIGENVALUE TO WITHIN A SPECIFIED
C     TOLERANCE.  IT ALSO ATTEMPTS TO COMPUTE THE SIZE, M2(J),
C     BY WHICH THE GIVEN EIGENVALUE BECOMES A DOUBLE EIGENVALUE
C     TO WITHIN THE GIVEN TOLERANCE.  THESE VALUES ARE USED
C     TO DETERMINE 1ST GUESSES AT SIZES FOR THE T-EIGENVECTORS
C     THAT WILL BE USED IN THE RITZ VECTOR COMPUTATIONS.
C     SUBROUTINE INVERM SUCCESSIVELY COMPUTES CORRESPONDING
C     EIGENVECTORS OF ENLARGED T-MATRICES UNTIL A SUITABLE
C     SIZE T-MATRIX IS DETERMINED FOR EACH J.  UP TO 10 SUCH
C     EIGENVECTOR COMPUTATIONS ARE ALLOWED FOR EACH EIGENVALUE.
C
C     AFTER APPROPRIATE T-EIGENVECTORS HAVE BEEN COMPUTED,
C     RITZ VECTOR CORRESPONDING TO THESE T-EIGENVECTORS ARE THEN
C     COMPUTED AND TAKEN AS APPROXIMATE EIGENVECTORS OF A FOR THE
C     GIVEN EIGENVALUES, GOODEV(J), J = 1, ..., NGOOD.
C
C     THIS IMPLEMENTATION FIRST COMPUTES ALL OF THE RELEVANT
C     EIGENVECTORS OF THE SYMMETRIC TRIDIAGONAL MATRICES
C     IN THE VECTOR, TVEC.
C
C     THEN, AS EACH OF THE LANCZOS VECTORS IS REGENERATED, ALL
C     OF THE RITZ VECTORS CORRESPONDING TO THESE
C     T-EIGENVECTORS ARE UPDATED USING THE CURRENTLY-GENERATED
C     LANCZOS VECTOR.  LANCZOS VECTORS ARE GENERATED (NOTE
C     THAT THEY ARE NOT BEING KEPT), UNTIL ENOUGH HAVE
C     BEEN GENERATED TO MAP THE LONGEST T-EIGENVECTOR INTO ITS
C     CORRESPONDING RITZ VECTOR.  THE ARRAY RITVEC CONTAINS THE
C     SUCCESSIVE RITZ VECTORS WHICH ARE THE APPROXIMATE
C     EIGENVECTORS OF A.
C
C
C-----PARAMETER CONTROLS FOR EIGENVECTOR PROGRAMS-----------------------
C
```

```
C
C     IN CASES (3) AND (4) WHERE A SPARSE FACTORIZATION OF A
C     SPECIFIED MATRIX IS USED, THE USER SPECIFIES USING THE FLAG
C     JPERM WHETHER OR NOT THE FACTORIZATION SUPPLIED CORRESPONDS
C     TO THE ORIGINAL MATRIX OR TO A PERMUTATION OF THE ORIGINAL
C     MATRIX.
C
C     JPERM = (0,1) MEANS
C            0     NO PERMUTATION
C            1     MATRIX HAS BEEN PERMUTED.  NOTE THAT IN
C                  CASE (4) THE PROGRAM WILL ASSUME THAT THE
C                  DATA SUPPLIED FOR THE A-MATRIX CORRESPONDS TO THE
C                  CORRESPONDING PERMUTATION OF THE ORIGINAL A-MATRIX.
C                  IN BOTH CASES THE LANCZS CODES WILL WORK WITH THE
C                  PERMUTED MATRICES AND THE PERMUTATION WILL BE
C                  UNDONE ONLY IN THE EIGENVECTOR PROGRAM AFTER
C                  THE RITZ VECTORS FOR THE PERMUTED PROBLEM HAVE
C                  BEEN COMPUTED.
C
C     OTHER PARAMETER CONTROLS ARE INTRODUCED TO ALLOW SEGMENTATION
C     OF THE EIGENVECTOR COMPUTATIONS AND TO ALLOW VARIOUS COMBINATIONS
C     OF READ/WRITES.
C
C     THE FLAG MBOUND ALLOWS THE USER TO DETERMINE A FIRST GUESS ON THE
C     STORAGE THAT WILL BE REQUIRED BY THE T-EIGENVECTORS FOR THE
C     EIGENVALUES WHOSE EIGENVECTORS ARE TO BE COMPUTED.
C     THIS CAN BE USED TO ESTIMATE THE REQUIRED SIZE OF THE TVEC ARRAY.
C
C     MBOUND = (0,1) MEANS
C
C                (0)  PROGRAM COMPUTES FIRST GUESSES AT THE SIZES
C                     OF THE T-MATRICES REQUIRED BY EACH OF THE
C                     EIGENVALUES SUPPLIED AND THEN CONTINUES WITH
C                     THE CORRESPONDING T-EIGENVECTOR COMPUTATIONS.
C
C                (1)  PROGRAM COMPUTES FIRST GUESSES AT THE SIZES
C                     OF THE T-MATRICES REQUIRED BY EACH OF THE
C                     EIGENVALUES SUPPLIED, STORES THESE IN FILE 10
C                     AND THEN TERMINATES.  THE USER CAN USE THESE
C                     SIZES TO ESTIMATE THE SIZE TVEC ARRAY NEEDED
C                     FOR THE DESIRED T-EIGENVECTOR COMPUTATIONS.
C
C     THE FLAGS NTVCON, TVSTOP, LVCONT, AND ERCONT CONTROL THE STOPPING
C     CRITERIA FOR INTERMEDIATE POINTS IN THE LANCZOS PROCEDURE.  THEY
C     TERMINATE THE PROCEDURE IF VARIOUS QUANTITIES COULD NOT BE
C     COMPUTED AS DESIRED.
C
C     NTVCON = (0,1) MEANS
C
C                 (0)  IF THE ESTIMATED STORAGE FOR THE T-EIGENVECTORS
C                      EXCEEDS THE USER-SPECIFIED DIMENSION OF THE
C                      TVEC ARRAY PROGRAM DOES NOT CONTINUE WITH THE
C                      T-EIGENVECTOR COMPUTATIONS.  TERMINATION OCCURS.
C
C                 (1)  CONTINUE WITH THE T-EIGENVECTOR COMPUTATIONS
C                      EVEN IF THE ESTIMATED STORAGE FOR TVEC EXCEEDS
C                      THE USER-SPECIFIED DIMENSION OF THE TVEC ARRAY.
C                      IN THIS SITUATION THE PROGRAM COMPUTES AS MANY
C                      T-EIGENVECTORS AS IT HAS ROOM FOR, IN THE SAME
C                      ORDER IN WHICH THE EIGENVALUES ARE PROVIDED.
C
C      SVTVEC = (0,1) MEANS
C
C                 (0)  DO NOT STORE THE COMPUTED T-EIGENVECTORS ON
C                      FILE 11 UNLESS ALSO HAVE THE FLAG TVSTOP = 1,
C                      IN WHICH CASE THE T-EIGENVECTORS ARE ALWAYS
C                      WRITTEN TO FILE 11.
```

```
C
C             (1)  STORE THE COMPUTED T-EIGENVECTORS ON FILE 11.
C
C      TVSTOP = (0,1) MEANS
C
C             (0)  ATTEMPT TO CONTINUE ON TO THE COMPUTATION
C                  OF THE RITZVECTORS AFTER COMPLETING THE
C                  COMPUTATION OF THE T-EIGENVECTORS.
C
C             (1)  TERMINATE AFTER COMPUTING THE
C                  T-EIGENVECTORS AND STORING THEM ON FILE 11.
C
C      LVCONT = (0,1) MEANS
C
C             (0)  IF SOME OF THE T-EIGENVECTORS THAT WERE
C                  REQUESTED WERE NOT COMPUTED, EXIT
C                  FROM THE PROGRAM WITHOUT COMPUTING THE
C                  CORRESPONDING RITZ VECTORS.
C
C             (1)  CONTINUE ON TO THE RITZ VECTOR COMPUTATIONS
C                  EVEN IF NOT ALL OF THE T-EIGENVECTORS
C                  REQUESTED WERE COMPUTED.
C
C      ERCONT = (0,1) MEANS
C
C             (0)  PROCEDURE WILL NOT COMPUTE A RITZ VECTOR FOR
C                  ANY EIGENVALUE FOR WHICH NO T-EIGENVECTOR WHICH
C                  SATISFIES THE ERROR ESTIMATE TEST (ERTOL) HAS
C                  BEEN IDENTIFIED.
C
C             (1)  A RITZ VECTOR WILL BE COMPUTED FOR EVERY
C                  EIGENVALUE FOR WHICH A T-EIGENVECTOR HAS BEEN
C                  COMPUTED REGARDLESS OF WHETHER OR NOT THAT
C                  T-EIGENVECTOR SATISFIED THE ERROR ESTIMATE TEST.
C
C
C-----INPUT/OUTPUT FILES FOR THE EIGENVECTOR COMPUTATIONS---------------
C
C
C     ANY INPUT DATA OTHER THAN THE T-MATRIX HISTORY FILE AND THE
C     PREVIOUSLY COMPUTED EIGENVALUES AND CORRESPONDING ERROR
C     ESTIMATES SHOULD BE STORED ON FILE 5 IN FREE FORMAT.
C     SEE SAMPLE INPUT/OUTPUT FOR TYPICAL INPUT FILE.
C
C     FILE 6 WAS USED AS THE INTERACTIVE TERMINAL OUTPUT FILE.
C     THIS FILE PROVIDES A RUNNING ACCOUNT OF THE PROGRESS OF THE
C     COMPUTATIONS.  ADDITIONAL PRINTOUT IS GENERATED WHEN
C     THE FLAG IWRITE = 1.
C
C
C DESCRIPTION OF OTHER I/O FILES
C
C FILE (K)     CONTAINS:
C
C     (2)      INPUT FILE:
C              PREVIOUSLY-GENERATED T-MATRICES (ALPHA/BETA ARRAYS)
C              AND THE FINAL TWO LANCZOS VECTORS USED ON THAT
C              COMPUTATION.  THIS PROGRAM ALLOWS ENLARGEMENT
C              OF ANY T-MATRICES PROVIDED ON FILE 2.  NOTE THAT
C              IN CASE (4), THREE 'LANCZOS' VECTORS ARE ON FILE 2.
C
C     (3)      INPUT FILE:
C              THE GOOD EIGENVALUES OF THE T-MATRIX  T(1,MEV)
C              FOR WHICH RITZ VECTORS ARE REQUESTED.
C              FILE 3 ALSO CONTAINS THE T-MULTIPLICITIES OF THESE
C              EIGENVALUES AND THEIR COMPUTED GAPS IN THE
C              T-MATRICES AND IN THE USER-SUPPLIED MATRIX. IN
```

```
C            CASE (3) FILE 3 CONTAINS THE EIGENVALUES OF THE
C            B-INVERSE MATRIX AND 3 SETS OF CORRESPONDING GAPS.
C            THIS FILE IS CREATED IN THE LANCZOS EIGENVALUE
C            COMPUTATIONS.
C
C      (4)   INPUT FILE:
C            ERROR ESTIMATES FOR THE ISOLATED GOOD T-EIGENVALUES
C            ON FILE 3.  THIS FILE IS CREATED DURING THE LANCZOS
C            EIGENVALUE COMPUTATIONS.
C
C      (7)   INPUT FILE:
C            IN CASE (3) ((4)),
C            CONTAINS SPARSE MATRIX REPRESENTATION OF FACTORIZATION
C            OF MATRIX (B-MATRIX) USED BY LANCZS SUBROUTINE.
C
C      (8)   INPUT FILE:
C            IN CASES (1),(2) AND (4), USPEC SUBROUTINE ASSUMES
C            THAT USER-SUPPLIED A-MATRIX IS ON FILE 8.  IN CASE (3)
C            A-MATRIX CAN BE STORED ON FILE 8, BUT IT IS NOT
C            USED BY THE LANCZOS PROGRAMS.
C
C      (9)   OUTPUT FILE:
C            ERROR ESTIMATES FOR THE COMPUTED RITZ VECTORS CONSIDERED
C            AS EIGENVECTORS OF THE MATRIX USED BY THE LANCZS
C            SUBROUTINE.  THESE ESTIMATES ARE OF THE FORM
C                  AERROR = || A*RITVEC - EVAL*RITVEC ||
C            WHERE A DENOTES THE MATRIX USED BY LANCZS, EVAL DENOTES
C            THE EIGENVALUE BEING CONSIDERED AND RITVEC DENOTES
C            THE COMPUTED RITZ VECTOR.
C
C      (10)  OUTPUT FILE:
C            GUESSES AT APPROPRIATE SIZE T-MATRICES FOR THE
C            T-EIGENVECTORS FOR EACH SUPPLIED EIGENVALUE, GOODEV(J).
C
C      (11)  OUTPUT FILE:
C            COMPUTED T-EIGENVECTORS CORRESPONDING TO EIGENVALUES
C            IN THE GOODEV ARRAY.  NOTE THAT IT IS POSSIBLE IN
C            CERTAIN SITUATIONS THAT FOR SOME EIGENVALUES IN THE
C            GOODEV ARRAY A T-EIGENVECTOR WILL NOT BE COMPUTED.
C
C      (12)  OUTPUT FILE:
C            CONTAINS COMPUTED RITZ VECTORS CORRESPONDING TO
C            THE T-EIGENVECTORS ON FILE 11. NOTE THAT IN
C            SOME SITUATIONS THAT FOR SOME EIGENVALUES IN
C            THE GOODEV ARRAY FOR WHICH T-EIGENVECTORS HAVE
C            BEEN COMPUTED NO RITZ VECTOR WILL HAVE BEEN
C            COMPUTED.
C
C      (13)  OUTPUT FILE:
C            ADDITIONAL INFORMATION ABOUT THE BOUNDS AND ERROR
C            ESTIMATES OBTAINED.
C
C
C-----SEEDS FOR EIGENVECTOR PROGRAMS-----------------------------------
C
C      SEEDS FOR RANDOM NUMBER GENERATOR GENRAN
C               (1) SVSEED = INTEGER*4 SCALAR USED IN THE SUBROUTINE
C                            GENRAN TO GENERATE THE STARTING VECTOR FOR
C                            THE REGENERATION OF THE LANCZOS VECTORS.
C
C               (2) RHSEED = INTEGER*4 SCALAR USED IN THE SUBROUTINE
C                            GENRAN TO GENERATE A RANDOM VECTOR FOR
C                            USE IN SUBROUTINE INVERM.
C
C      USER SHOULD NOTE THAT SVSEED MUST BE THE SAME SEED THAT
C      WAS USED TO GENERATE THE T-MATRICES THAT WERE USED TO
C      COMPUTE THE EIGENVALUES WHOSE EIGENVECTORS ARE TO BE COMPUTED.
```

```
C       SVSEED IS READ IN FROM FILE 3.
C
C
C-----USER-SPECIFIED PARAMETERS FOR THE EIGENVECTOR PROGRAMS-----------
C
C
C       NGOOD    = NUMBER OF EIGENVALUES READ INTO THE GOODEV ARRAY
C                  READ FROM FILE 3.
C
C       N        = SIZE OF THE USER-SUPPLIED MATRIX.
C
C       MEV      = SIZE OF THE T-MATRIX THAT WAS USED TO COMPUTE
C                  THE EIGENVALUES WHOSE EIGENVECTORS ARE REQUESTED.
C                  MEV IS READ IN FROM FILE 3.
C
C       KMAX     =  SIZE OF THE T-MATRIX PROVIDED ON FILE 2.
C
C       MDIMTV   = MAXIMUM CUMULATIVE SIZE OF THE TVEC ARRAY ALLOWED
C                  FOR ALL OF THE T-EIGENVECTORS REQUIRED.  MDIMTV
C                  MUST NOT EXCEED THE USER-SPECIFIED DIMENSION OF
C                  THE TVEC ARRAY.  PROGRAM CAN BE RUN WITH THE FLAG
C                  MBOUND = 1 TO DETERMINE AN EDUCATED GUESS ON AN
C                  APPROPRIATE DIMENSION FOR THE TVEC ARRAY.
C
C       MDIMRV = MAXIMUM CUMULATIVE SIZE OF THE RITVEC ARRAY ALLOWED
C                FOR ALL OF THE RITZ VECTORS TO BE COMPUTED. MDIMRV
C                MUST NOT EXCEED THE USER-SPECIFIED DIMENSION OF
C                THE RITVEC ARRAY.  MUST BE SELECTED SO THAT
C                THERE IS ENOUGH ROOM FOR A RITZ VECTOR FOR EVERY
C                GOODEV(J) READ INTO PROGRAM.  (>= NGOOD*N)
C
C
C-----ARRAYS REQUIRED BY THE EIGENVECTOR PROGRAMS--------------------
C
C
C       ALL 4 CASES
C
C       ALPHA(J) = REAL*8 ARRAY.  ITS DIMENSION MUST BE AT LEAST
C                  KMAXN, THE LARGEST SIZE T-MATRIX CONSIDERED BY
C                  THE PROGRAM.  NOTE THAT KMAXN IS THE LARGER OF
C                  THE SIZE OF THE ALPHA, BETA HISTORY PROVIDED
C                  ON FILE 2 (IF ANY ) AND THE SIZE WHICH THE PROGRAM
C                  SPECIFIES INTERNALLY, THIS LATTER IS ALWAYS
C                  < = 11*MEV / 8  +  12, WHERE MEV IS THE SIZE
C                  T-MATRIX THAT WAS USED IN THE CORRESPONDING EIGENVALUE
C                  COMPUTATIONS.  ALPHA CONTAINS THE DIAGONAL ENTRIES
C                  OF THE LANCZOS T-MATRICES.  ALPHA IS NOT DESTROYED
C                  IN THE COMPUTATIONS.
C
C       BETA(J) = REAL*8 ARRAY.  ITS DIMENSION MUST BE AT LEAST 1
C                 MORE THAN THAT OF ALPHA.  DIMENSION COMMENTS ABOVE
C                 ABOUT ALPHA APPLY ALSO TO THE BETA ARRAY.  BETA
C                 CONTAINS THE SUBDIAGONAL ENTRIES OF THE T-MATRICES.
C                 BETA IS NOT DESTROYED IN THE COMPUTATIONS.
C
C       RITVEC(J) = REAL*8 ARRAY IN REAL SYMMETRIC AND INVERSE OF
C                   REAL SYMMETRIC CASES.  COMPLEX*16 IN CASE (2),
C                   HERMITIAN MATRICES.  IN EACH CASE ITS DIMENSION >=
C                   NGOOD*N WHERE N IS THE ORDER OF THE USER-SUPPLIED
C                   MATRIX AND NGOOD IS THE NUMBER OF EIGENVALUES WHOSE
C                   EIGENVECTORS ARE TO BE COMPUTED.  IT CONTAINS THE
C                   COMPUTED APPROXIMATE EIGENVECTORS OF A.  THESE
C                   COMPUTED RITZ VECTORS ARE STORED ON FILE 12.
C
C       TVEC(J)   = REAL*8 ARRAY.  ITS DIMENSION MUST BE AT LEAST
C                   MTOL = |MA(1)| + |MA(2)| + ... + |MA(NGOOD)|
C                   WHERE NGOOD IS THE NUMBER OF EIGENVALUES BEING
```

```
C                 CONSIDERED AND |MA(J)| IS THE SIZE OF THE
C                 T-MATRIX BEING USED FOR THE RITZ VECTOR
C                 COMPUTATION FOR GOODEV(J).  THESE SIZES
C                 ARE COMPUTED BY THE PROGRAM.  AN ESTIMATE OF
C                 MTOL CAN BE OBTAINED BY SETTING MBOUND = 1,
C                 RUNNING THE PROGRAM, AND MULTIPLYING THE RESULTING
C                 TOTAL OF THE T-SIZES SPECIFIED BY 5/4.  THE ARRAY
C                 TVEC CONTAINS THE COMPUTED T-EIGENVECTORS.  IF THE
C                 FLAG SVTVEC = 1 OR THE FLAG TVSTOP = 1, THEN
C                 THESE VECTORS ARE SAVED ON FILE 11.
C
C     V1(J)     = REAL*8 ARRAY IN REAL SYMMETRIC AND INVERSE OF
C                 REAL SYMMETRIC CASES.  COMPLEX*16 IN CASE (2),
C                 HERMITIAN MATRICES.  IN THE REAL CASES ITS
C                 DIMENSION MUST BE THE MAXIMUM OF KMAX AND N.
C                 IN THE HERMITIAN CASE ITS DIMENSION MUST BE
C                 THE MAXIMUM OF KMAX/2 AND N WHERE KMAX IS THE
C                 LARGEST SIZE T-MATRIX THAT IS TO BE CONSIDERED
C                 IN THE T-EIGENVECTOR COMPUTATIONS.  V1 IS USED
C                 IN THE SUBROUTINE INVERM AND IN THE REGENERATION
C                 OF THE LANCZOS VECTORS.
C
C     V2(J)     = REAL*8 ARRAY IN THE REAL SYMMETRIC AND INVERSE
C                 OF REAL SYMMETRIC CASES.  COMPLEX*16 IN CASE (2),
C                 HERMITIAN MATRICES.  IN CASES (1),(3) AND (4), ITS
C                 DIMENSION MUST BE > = MAX(KMAX,N); IN CASE (2)
C                 > = MAX(KMAX/2,N).  IT IS USED IN THE REGENERATION
C                 OF THE LANCZOS VECTORS AND IN SUBROUTINE INVERM.
C
C     GOODEV(J),  = REAL*8 ARRAYS EACH OF DIMENSION AT LEAST NGOOD.
C     EVNEW(J)      CONTAIN THE EIGENVALUES FOR WHICH EIGENVECTORS
C                   ARE REQUESTED.  EIGENVALUES IN GOODEV ARE READ
C                   IN FROM FILE 3.  IN CASE (3) GOODEV IS REPLACED
C                   BY GOODA AND GOODBI ARRAYS, SEE BELOW.
C
C     AMINGP(J), = REAL*4 ARRAYS OF DIMENSION AT LEAST NGOOD.
C     TMINGP(J)    CONTAIN, RESPECTIVELY, THE MINIMAL GAPS FOR
C                  CORRESPONDING EIGENVALUES IN GOODEV ARRAY IN
C                  A-MATRIX AND IN T-MATRIX.
C
C     TERR(J), ERR(J),      = REAL*4 ARRAYS (EXCEPT TLAST WHICH IS
C     ERRDGP(J), TLAST(J)     REAL*8).  EACH MUST BE OF DIMENSION
C     RNORM(J), TBETA(J)      AT LEAST NGOOD.  USED TO STORE QUANTITIES
C                             GENERATED DURING THE COMPUTATIONS FOR
C                             LATER PRINTOUT.
C
C     G(J)     = REAL*4 ARRAY WHOSE DIMENSION MUST BE AT LEAST
C                MAX(KMAX,N).  USED IN SUBROUTINE GENRAN TO HOLD
C                RANDOM NUMBERS NEEDED FOR THE LANCZOS VECTOR
C                REGENERATION AND FOR THE INVERSE ITERATION
C                COMPUTATIONS IN THE SUBROUTINE INVERM.
C
C     MP(J) = INTEGER*4 ARRAY WHOSE DIMENSION IS AT LEAST NGOOD.
C             INITIALLY CONTAINS THE MULTIPLICITY OF THE EIGENVALUE
C             GOODEV(J) AS AN EIGENVALUE OF THE T-MATRIX T(1,MEV).
C             USED TO FLAG EIGENVALUES FOR WHICH NO T-EIGENVECTOR
C             OR NO RITZ VECTOR IS TO BE COMPUTED.
C
C     MA(J)    = INTEGER*4 ARRAYS EACH OF WHOSE DIMENSIONS
C                IS AT LEAST NGOOD.  USED IN DETERMINING
C                AN APPROPRIATE T-MATRIX FOR EACH EIGENVALUE
C                IN GOODEV ARRAY.
C
C     MINT(J),MFIN(J) = INTEGER*4 ARRAYS WHOSE DIMENSIONS MUST BE AT
C                       LEAST NGOOD.  USED TO POINT TO THE BEGINNINGS
C                       AND THE ENDS OF THE COMPUTED EIGENVECTOR
C                       OF THE T-MATRIX, T(1,|MA(J)|).
```

```
C
C     IDELTA(J)  = INTEGER*4 ARRAY WHOSE DIMENSION MUST BE AT
C                  LEAST NGOOD.  CONTAINS INCREMENTS USED IN LOOPS
C                  ON APPROPRIATE SIZE T-MATRIX FOR THE T-EIGENVECTOR
C                  COMPUTATIONS.
C
C
C     CASE (2) ONLY, HERMITIAN MATRICES:
C
C     GR(J),GC(J)      = REAL*8 ARRAYS USED ONLY IN CASE (2),
C                        HERMITIAN MATRICES.  EACH MUST BE AT
C                        LEAST MAX(N,KMAX).  USED TO HOLD
C                        STARTING VECTORS FOR LANCZS
C                        COMPUTATIONS AND FOR INVERM SUBROUTINES.
C
C     CASES (3) AND (4) ONLY, FACTORED INVERSES OF REAL SYMMETRIC
C     MATRICES AND GENERALIZED EIGENVALUE PROBLEMS:
C
C     VS(J) =  REAL*8 ARRAY WHOSE DIMENSION MUST BE AT LEAST N.
C              USED IN REGENERATION OF THE LANCZOS VECTORS.
C
C     IPR(J), IPT(J) = INTEGER*4 ARRAYS.  EACH MUST BE OF DIMENSION
C                      AT LEAST N, THE ORDER OF A. USED TO STORE
C                      THE REORDERING OF THE GIVEN MATRIX.
C
C     CASE (3) ONLY, INVERSES OF REAL SYMMETRIC MATRICES:
C
C     GOODA(J), GOODBI(J) = REAL*8 ARRAYS.  EACH MUST BE OF DIMENSION
C                           AT LEAST NGOOD, THE NUMBER OF EIGENVALUES
C                           BEING CONSIDERED.  GOODA CONTAINS THE
C                           EIGENVALUES OF A AND GOODBI CONTAINS THE
C                           EIGENVALUES OF B(INVERSE).  THE PROGRAM
C                           WORKS DIRECTLY WITH THE GOODBI ARRAY.
C
C
C-----SUBROUTINES INCLUDED FOR THE EIGENVECTOR COMPUTATIONS------------
C
C
C     STURMI = FOR EACH GIVEN EIGENVALUE GOODEV(J) DETERMINES
C              THE SMALLEST SIZE T-MATRIX FOR WHICH GOODEV(J) IS
C              AN EIGENVALUE (TO WITHIN A GIVEN TOLERANCE) AND IF
C              POSSIBLE THE SMALLEST SIZE T-MATRIX FOR WHICH
C              IT IS A DOUBLE EIGENVALUE (TO WITHIN THE SAME
C              TOLERANCE).  THE SIZE T-MATRIX USED IN THE
C              EIGENVECTOR COMPUTATIONS IS THEN DETERMINED BY
C              STARTING WITH AN INITIAL GUESS BASED UPON THE
C              INFORMATION FROM STURMI, AND LOOPING ON THE SIZE
C              OF THE T-EIGENVECTOR COMPUTATIONS.
C
C     LBISEC = RECOMPUTES THE VALUE OF THE GIVEN EIGENVALUE AT THE
C              T-SIZE SPECIFIED FOR THE T-EIGENVECTOR COMPUTATION.
C              LBISEC IS A SIMPLIFICATION OF THE BISEC SUBROUTINE
C              USED IN THE LANCZOS EIGENVALUE COMPUTATIONS.
C
C     INVERM = FOR THE T-SIZES CONSIDERED BY THE PROGRAM COMPUTES
C              THE CORRESPONDING EIGENVECTORS OF THESE T-MATRICES
C              CORRESPONDING TO THE USER-SUPPLIED EIGENVALUES IN
C              THE GOODEV ARRAY.
C
C     THE LANCZS, TNORM , AND CINPRD (CASE (2) ONLY) SUBROUTINES
C     ARE USED HERE AS WELL AS IN THE EIGENVALUE COMPUTATIONS.
C
C     IN CASES (3) AND (4) ONLY AND THEN ONLY IF THE ORIGINAL MATRIX
C     (MATRICES) HAS (HAVE) BEEN PERMUTED:
C
C     LPERM  = (USED IN CASE (3) AND (4) ONLY). GIVEN A B-MATRIX AND
C              A PERMUTATION P DEFINED IN THE VECTORS IPR AND IPT,
```

```
C             AND A VECTOR X COMPUTE EITHER (P-TRANSPOSE)*X OR PX.      LEV15390
C                                                                       LEV15400
C-----------------------------------------------------------------------LEV15410
```

SECTION 2.3 MAIN PROGRAM, EIGENVALUE CALCULATIONS

```
C-----LEVAL  (EIGENVALUES OF REAL SYMMETRIC MATRICES)-------------------
C
C     CONTAINS MAIN PROGRAM FOR COMPUTING DISTINCT EIGENVALUES OF
C     A REAL SYMMETRIC MATRIX USING LANCZOS TRIDIAGONALIZATION
C     WITHOUT REORTHOGONALIZATION.
C
C     PFORT VERIFIER IDENTIFIED THE FOLLOWING NONPORTABLE
C     CONSTRUCTIONS
C
C     1.  DATA/MACHEP/ STATEMENT
C     2.  ALL READ(5,*) STATEMENTS (FREE FORMAT)
C     3.  FORMAT(20A4) USED WITH EXPLANATORY HEADER EXPLAN.
C     4.  HEXADECIMAL FORMAT (4Z20) USED IN ALPHA/BETA FILES 1 AND 2.
C
C-----------------------------------------------------------------------
C
      DOUBLE PRECISION  ALPHA(5000),BETA(5001)
      DOUBLE PRECISION  V1(5000),V2(5000),VS(5000)
      DOUBLE PRECISION  LB(20),UB(20)
      DOUBLE PRECISION  BTOL,GAPTOL,TTOL,MACHEP,EPSM,RELTOL
      DOUBLE PRECISION  SCALE1,SCALE2,SCALE3,SCALE4,BISTOL,CONTOL,MULTOL
      DOUBLE PRECISION  ONE,ZERO,TEMP,TKMAX,BETAM,BKMIN,TO,T1
      REAL  G(5000),EXPLAN(20)
      INTEGER  MP(5000),NMEV(20)
      INTEGER  SVSEED,RHSEED,SVSOLD
      INTEGER IABS
      REAL  ABS
      DOUBLE PRECISION  DABS, DSQRT, DFLOAT
      EXTERNAL CMATV
C
C-----------------------------------------------------------------------
      DATA MACHEP/Z3410000000000000/
      EPSM = 2.0D0*MACHEP
C-----------------------------------------------------------------------
C
C     ARRAYS MUST BE DIMENSIONED AS FOLLOWS:
C     DIMENSION OF V2 ASSUMES THAT NO MORE THAN KMAX/2 EIGENVALUES
C     OF THE T-MATRICES ARE BEING COMPUTED IN ANY ONE OF THE
C     SUB-INTERVALS BEING CONSIDERED.  V2 CONTAINS THE UPPER AND LOWER
C     BOUNDS FOR EACH T-EIGENVALUE BEING COMPUTED BY BISEC IN ANY ONE
C     GIVEN INTERVAL.
C
C     1.  ALPHA: >= KMAX,    BETA: >= (KMAX+1)
C     2.  V1:  >= MAX(N,KMAX+1)
C     3.  V2:    >= MAX(N,KMAX)
C     4.  VS:  >= KMAX
C     5.  G:  >= MAX(N,2*KMAX)
C     6.  MP:     >= KMAX
C     7.  LB,UB:  >= NUMBER OF SUBINTERVALS SUPPLIED TO BISEC.
C     8.  NMEV:  >= NUMBER OF T-MATRICES ALLOWED.
C     9.  EXPLAN:  DIMENSION IS 20.
C
C
C     IMPORTANT TOLERANCES OR SCALES THAT ARE USED REPEATEDLY
C     THROUGHOUT THIS PROGRAM ARE THE FOLLOWING:
C     SCALED MACHINE EPSILON:  TTOL = TKMAX*EPSM WHERE
C     EPSM = 2*MACHINE EPSILON AND
C     TKMAX = MAX(|ALPHA(J)|,BETA(J), J = 1,MEV)
C     BISEC CONVERGENCE TOLERANCE:  BISTOL = DSQRT(1000+MEV)*TTOL
C     BISEC T-MULTIPLICITY TOLERANCE:  MULTOL = (1000+MEV)*TTOL
C     LANCZOS CONVERGENCE TOLERANCE:   CONTOL = BETA(MEV+1)*1.D-10
C-----------------------------------------------------------------------
C     OUTPUT HEADER
      WRITE(6,10)
```

```
   10 FORMAT(/' LANCZOS PROCEDURE FOR REAL SYMMETRIC MATRICES'/)
C
C     SET PROGRAM PARAMETERS
C     SCALEK ARE USED IN TOLERANCES NEEDED IN SUBROUTINES LUMP,
C     ISOEV AND PRTEST.  USER MUST NOT MODIFY THESE SCALES.
      SCALE1 = 5.0D2
      SCALE2 = 5.0D0
      SCALE3 = 5.0D0
      SCALE4 = 1.0D4
      ONE  = 1.0D0
      ZERO = 0.0D0
      BTOL = 1.0D-8
C     BTOL = EPSM
      GAPTOL = 1.0D-8
      ICONV = 0
      MOLD = 0
      MOLD1 = 1
      ICT = 0
      MMB = 0
      IPROJ = 0
C-----------------------------------------------------------------------
C     READ USER-SPECIFIED PARAMETERS FROM INPUT FILE 5 (FREE FORMAT)
C
C     READ USER-PROVIDED HEADER FOR RUN
      READ(5,20) EXPLAN
      WRITE(6,20) EXPLAN
      READ(5,20) EXPLAN
      WRITE(6,20) EXPLAN
   20 FORMAT(20A4)
C
C     READ ORDER OF MATRICES (N) , MAXIMUM ORDER OF T-MATRIX (KMAX),
C     NUMBER OF T-MATRICES ALLOWED (NMEVS), AND MATRIX IDENTIFICATION
C     NUMBERS (MATNO)
      READ(5,20) EXPLAN
      READ(5,*) N,KMAX,NMEVS,MATNO
C
C     READ SEEDS FOR LANCZS AND INVERR SUBROUTINES (SVSEED AND RHSEED)
C     READ MAXIMUM NUMBER OF ITERATIONS ALLOWED FOR EACH INVERSE
C     ITERATION (MXINIT) AND MAXIMUM NUMBER OF STURM SEQUENCES
C     ALLOWED (MXSTUR)
      READ(5,20) EXPLAN
      READ(5,*) SVSEED,RHSEED,MXINIT,MXSTUR
C
C     ISTART = (0,1):  ISTART = 0 MEANS ALPHA/BETA FILE IS NOT
C     AVAILABLE.  ISTART = 1 MEANS ALPHA/BETA FILE IS AVAILABLE ON
C     FILE 2.
C     ISTOP = (0,1):  ISTOP = 0 MEANS PROCEDURE GENERATES ALPHA/BETA
C     FILE AND THEN TERMINATES.  ISTOP = 1 MEANS PROCEDURE GENERATES
C     ALPHAS/BETAS IF NEEDED AND THEN COMPUTES EIGENVALUES AND ERROR
C     ESTIMATES AND THEN TERMINATES.
      READ(5,20) EXPLAN
      READ(5,*) ISTART,ISTOP
C
C     IHIS = (0,1):  IHIS = 0 MEANS ALPHA/BETA FILE IS NOT WRITTEN
C     TO FILE 1.  IHIS = 1 MEANS ALPHA/BETA FILE IS WRITTEN TO FILE 1.
C     IDIST = (0,1):  IDIST = 0 MEANS DISTINCT T-EIGENVALUES
C     ARE NOT WRITTEN TO FILE 11.  IDIST = 1 MEANS DISTINCT
C     T-EIGENVALUES ARE WRITTEN TO FILE 11.
C     IWRITE = (0,1):  IWRITE = 0 MEANS NO INTERMEDIATE OUTPUT
C     FROM THE COMPUTATIONS IS WRITTEN TO FILE 6.  IWRITE = 1 MEANS
C     T-EIGENVALUES AND ERROR ESTIMATES ARE WRITTEN TO FILE 6
C     AS THEY ARE COMPUTED.
      READ(5,20) EXPLAN
      READ(5,*) IHIS,IDIST,IWRITE
C
C     READ IN THE RELATIVE TOLERANCE (RELTOL) FOR USE IN THE
C     SPURIOUS, T-MULTIPLICITY, AND PRTESTS.
```

```
      READ(5,20) EXPLAN
      READ(5,*) RELTOL
C
C     READ IN THE SIZES OF THE T-MATRICES TO BE CONSIDERED.
      READ(5,20) EXPLAN
      READ(5,*) (NMEV(J), J=1,NMEVS)
C
C     READ IN THE NUMBER OF SUBINTERVALS TO BE CONSIDERED.
      READ(5,20) EXPLAN
      READ(5,*) NINT
C
C     READ IN THE LEFT-END POINTS OF THE SUBINTERVALS TO BE CONSIDERED.
C     THESE MUST BE IN ALGEBRAICALLY-INCREASING ORDER
      READ(5,20) EXPLAN
      READ(5,*) (LB(J), J=1,NINT)
C
C     READ IN THE RIGHT-END POINTS OF THE SUBINTERVALS TO BE CONSIDERED.
C     THESE MUST BE IN ALGEBRAICALLY-INCREASING ORDER
      READ(5,20) EXPLAN
      READ(5,*) (UB(J), J=1,NINT)
C
C-----------------------------------------------------------------------
C     INITIALIZE THE ARRAYS FOR THE USER-SPECIFIED MATRIX
C     AND PASS THE STORAGE LOCATIONS OF THESE ARRAYS TO THE
C     MATRIX-VECTOR MULTIPLY SUBROUTINE CMATV.
C
      CALL USPEC(N,MATNO)
C
C-----------------------------------------------------------------------
C     MASK UNDERFLOW AND OVERFLOW
C
      CALL MASK
C
C-----------------------------------------------------------------------
C
C     WRITE TO FILE 6, A SUMMARY OF THE PARAMETERS FOR THIS RUN
C
      WRITE(6,30) MATNO,N,KMAX
   30 FORMAT(/3X,'MATRIX ID',4X,'ORDER OF A',4X,'MAX ORDER OF T'/
     1 I12,I14,I18/)
C
      WRITE(6,40) ISTART,ISTOP
   40 FORMAT(/2X,'ISTART',3X,'ISTOP'/2I8/)
C
      WRITE(6,50) IHIS,IDIST,IWRITE
   50 FORMAT(/4X,'IHIS',3X,'IDIST',2X,'IWRITE'/3I8/)
C
      WRITE(6,60) SVSEED,RHSEED
   60 FORMAT(/' SEEDS FOR RANDOM NUMBER GENERATOR'//
     1 4X,'LANCZS SEED',4X,'INVERR SEED'/2I15/)
C
      WRITE(6,70) (NMEV(J), J=1,NMEVS)
   70 FORMAT(/' SIZES OF T-MATRICES TO BE CONSIDERED'/(6I12))
C
      WRITE(6,80) RELTOL,GAPTOL,BTOL
   80 FORMAT(/' RELATIVE TOLERANCE USED TO COMBINE COMPUTED T-EIGENVALUE
     1S'/E15.3/' RELATIVE GAP TOLERANCES USED IN INVERSE ITERATION'/
     1E15.3/' RELATIVE TOLERANCE FOR CHECK ON SIZE OF BETAS'/E15.3/)
C
      WRITE(6,90) (J,LB(J),UB(J), J=1,NINT)
   90 FORMAT(/' BISEC WILL BE USED ON THE FOLLOWING INTERVALS'/
     1 (I6,2E20.6))
C
       IF (ISTART.EQ.0) GO TO 140
C
C     READ IN ALPHA BETA HISTORY
C
```

```
      READ(2,100)MOLD,NOLD,SVSOLD,MATOLD
  100 FORMAT(2I6,I12,I8)
C
      IF (KMAX.LT.MOLD) KMAX = MOLD
      KMAX1 = KMAX + 1
C
C     CHECK THAT ORDER N, MATRIX ID MATNO, AND RANDOM SEED SVSEED
C     AGREE WITH THOSE IN THE HISTORY FILE.  IF NOT PROCEDURE STOPS.
C
      ITEMP = (NOLD-N)**2+(MATNO-MATOLD)**2+(SVSEED-SVSOLD)**2
C
      IF (ITEMP.EQ.0) GO TO 120
C
      WRITE(6,110)
  110 FORMAT(' PROGRAM TERMINATES'/    ' READ FROM FILE 2 CORRESPONDS TO
     1 DIFFERENT MATRIX THAN MATRIX SPECIFIED'/)
      GO TO 640
C
  120 CONTINUE
      MOLD1 = MOLD+1
C
      READ(2,130)(ALPHA(J), J=1,MOLD)
      READ(2,130)(BETA(J), J=1,MOLD1)
  130 FORMAT(4Z20)
C
      IF (KMAX.EQ.MOLD) GO TO 160
C
      READ(2,130)(V1(J), J=1,N)
      READ(2,130)(V2(J), J=1,N)
C
  140 CONTINUE
      IIX = SVSEED
C
C-----------------------------------------------------------------------
C
      CALL LANCZS(CMATV,ALPHA,BETA,V1,V2,G,KMAX,MOLD1,N,IIX)
C
C-----------------------------------------------------------------------
C
      KMAX1 = KMAX + 1
C
      IF (IHIS.EQ.0.AND.ISTOP.GT.0) GO TO 160
C
      WRITE(1,150) KMAX,N,SVSEED,MATNO
  150 FORMAT(2I6,I12,I8,' = KMAX,N,SVSEED,MATNO')
C
      WRITE(1,130)(ALPHA(I), I=1,KMAX)
      WRITE(1,130)(BETA(I), I=1,KMAX1)
C
      WRITE(1,130)(V1(I), I=1,N)
      WRITE(1,130)(V2(I), I=1,N)
C
      IF (ISTOP.EQ.0) GO TO 540
C
  160 CONTINUE
      BKMIN = BTOL
      WRITE(6,170)
  170 FORMAT(/' T-MATRICES (ALPHA AND BETA) ARE NOW AVAILABLE'/)
C
C-----------------------------------------------------------------------
C     SUBROUTINE TNORM CHECKS MIN(BETA)/(ESTIMATED NORM(A)) > BTOL .
C     IF THIS IS VIOLATED IB IS SET EQUAL TO THE NEGATIVE OF THE INDEX
C     OF THE MINIMAL BETA.  IF(IB < 0) THEN SUBROUTINE TNORM IS
C     CALLED FOR EACH VALUE OF MEV TO DETERMINE WHETHER OR NOT THERE
C     IS A BETA IN THE T-MATRIX SPECIFIED THAT VIOLATES THIS TEST.
C     IF THERE IS SUCH A BETA THE PROGRAM TERMINATES FOR THE USER
C     TO DECIDE WHAT TO DO.  THIS TEST CAN BE OVER-RIDDEN BY
```

```
C     SIMPLY MAKING BTOL SMALLER, BUT THEN THERE IS THE POSSIBILITY
C     THAT LOSSES IN THE LOCAL ORTHOGONALITY MAY HURT THE COMPUTATIONS.
C     BTOL = 1.D-8 IS HOWEVER A CONSERVATIVE CHOICE FOR BTOL.
C
C     TNORM ALSO COMPUTES TKMAX = MAX(|ALPHA(K)|,BETA(K), K=1,KMAX).
C     TKMAX IS USED TO SCALE THE TOLERANCES USED IN THE
C     T-MULTIPLICITY AND SPURIOUS TESTS IN BISEC. TKMAX IS ALSO USED IN
C     THE PROJECTION TEST FOR HIDDEN EIGENVALUES THAT HAD 'TOO SMALL'
C     A PROJECTION ON THE STARTING VECTOR.
C
      CALL TNORM(ALPHA,BETA,BKMIN,TKMAX,KMAX,IB)
C
C-----------------------------------------------------------------------
C
      TTOL = EPSM*TKMAX
C
C     LOOP ON THE SIZE OF THE T-MATRIX
C
  180 CONTINUE
      MMB = MMB + 1
      MEV = NMEV(MMB)
C     IS MEV TOO LARGE ?
      IF(MEV.LE.KMAX) GO TO 200
      WRITE(6,190) MMB, MEV, KMAX
  190 FORMAT(/' TERMINATE PRIOR TO CONSIDERING THE',I6,'TH T-MATRIX'/
     1' BECAUSE THE SIZE REQUESTED',I6,' IS GREATER THAN THE MAXIMUM SIZ
     1E ALLOWED',I6/)
      GO TO 540
C
  200 MP1 = MEV + 1
      BETAM = BETA(MP1)
C
      IF (IB.GE.0) GO TO 210
C
      T0 = BTOL
C
C-----------------------------------------------------------------------
C
      CALL TNORM(ALPHA,BETA,T0,T1,MEV,IBMEV)
C
C-----------------------------------------------------------------------
C
      TEMP = T0/TKMAX
      IBMEV = IABS(IBMEV)
      IF (TEMP.GE.BTOL) GO TO 210
      IBMEV = -IBMEV
      GO TO 600
C
  210 CONTINUE
      IC = MXSTUR-ICT
C
C-----------------------------------------------------------------------
C     BISEC LOOP. THE SUBROUTINE BISEC INCORPORATES DIRECTLY THE
C     T-MULTIPLICITY AND SPURIOUS TESTS. T-EIGENVALUES WILL BE
C     CALCULATED BY BISEC SEQUENTIALLY ON INTERVALS
C     (LB(J),UB(J)), J = 1,NINT).
C
C     ON RETURN FROM BISEC
C     NDIS = NUMBER OF DISTINCT EIGENVALUES OF T(1,MEV) ON UNION
C            OF THE (LB,UB) INTERVALS
C     VS = DISTINCT T-EIGENVALUES IN ALGEBRAICALLY INCREASING ORDER
C     MP = MULTIPLICITIES OF THE T-EIGENVALUES IN VS
C     MP(I) = (0,1,MI), MI>1, I=1,NDIS  MEANS:
C        (0)  VS(I) IS SPURIOUS
C        (1)  VS(I) IS T-SIMPLE AND GOOD
C        (MI) VS(I) IS MULTIPLE AND IS THEREFORE NOT ONLY GOOD BUT
C              ALSO A CONVERGED GOOD T-EIGENVALUE.
```

```
C
C
      CALL BISEC(ALPHA,BETA,V1,V2,VS,LB,UB,EPSM,TTOL,MP,NINT,
     1 MEV,NDIS,IC,IWRITE)
C
C-----------------------------------------------------------------------
C
      IF (NDIS.EQ.0) GO TO 620
C
C     COMPUTE THE TOTAL NUMBER OF STURM SEQUENCES USED TO DATE
C     COMPUTE THE BISEC CONVERGENCE AND T-MULTIPLICITY TOLERANCES USED.
C     COMPUTE THE CONVERGENCE TOLERANCE FOR EIGENVALUES OF A.
      ICT = ICT + IC
      TEMP = DFLOAT(MEV+1000)
      MULTOL = TEMP*TTOL
      TEMP = DSQRT(TEMP)
      BISTOL = TTOL*TEMP
      CONTOL = BETAM*1.D-10
C
C-----------------------------------------------------------------------
C    SUBROUTINE LUMP 'COMBINES' T-EIGENVALUES THAT ARE 'TOO CLOSE'.
C    NOTE HOWEVER THAT CLOSE SPURIOUS T-EIGENVALUES ARE NOT AVERAGED
C    WITH GOOD ONES. HOWEVER, THEY MAY BE USED TO INCREASE THE
C    MULTIPLICITY OF A GOOD T-EIGENVALUE.
C
      LOOP = NDIS
      CALL LUMP(VS,RELTOL,MULTOL,SCALE2,MP,LOOP)
C
C-----------------------------------------------------------------------
C
      IF(NDIS.EQ.LOOP) GO TO 230
C
      WRITE(6,220) NDIS, MEV, LOOP
  220 FORMAT(/I6,' DISTINCT T-EIGENVALUES WERE COMPUTED IN BISEC AT MEV
     1',I6/ 2X,' LUMP SUBROUTINE REDUCES NUMBER OF DISTINCT EIGENVALUES
     1O',I6)
C
  230 CONTINUE
      NDIS = LOOP
      BETA(MP1) = BETAM
C
C-----------------------------------------------------------------------
C     THE SUBROUTINE ISOEV LABELS THOSE SIMPLE EIGENVALUES OF T(1,MEV)
C     WITH VERY SMALL GAPS BETWEEN NEIGHBORING EIGENVALUES OF T(1,MEV)
C     TO AVOID COMPUTING ERROR ESTIMATES FOR ANY SIMPLE GOOD
C     T-EIGENVALUE THAT IS TOO CLOSE TO A SPURIOUS EIGENVALUE.
C     ON RETURN FROM ISOEV, G CONTAINS CODED MINIMAL GAPS
C     BETWEEN THE DISTINCT EIGENVALUES OF T(1,MEV). (G IS REAL).
C     G(I) < 0 MEANS MINGAP IS DUE TO LEFT GAP G(I) > 0 MEANS DUE TO
C     RIGHT GAP. MP(I) = -1 MEANS THAT THE GOOD T-EIGENVALUE IS SIMPLE
C     AND HAS A VERY SMALL MINGAP IN T(1,MEV) DUE TO A SPURIOUS
C     EIGENVALUE.  NG = NUMBER OF GOOD T-EIGENVALUES.
C     NISO = NUMBER OF ISOLATED GOOD T-EIGENVALUES.
C
      CALL ISOEV(VS,GAPTOL,MULTOL,SCALE1,G,MP,NDIS,NG,NISO)
C
C-----------------------------------------------------------------------
C
      WRITE(6,240)NG,NISO,NDIS
  240 FORMAT(/I6,' GOOD T-EIGENVALUES HAVE BEEN COMPUTED'/
     1 I6,' OF THESE ARE T-ISOLATED'/
     2 I6,' = NUMBER OF DISTINCT T-EIGENVALUES COMPUTED'/)
C
C     DO WE WRITE DISTINCT EIGENVALUES OF T-MATRIX TO FILE 11?
      IF (IDIST.EQ.0) GO TO 280
C
      WRITE(11,250) NDIS,NISO,MEV,N,SVSEED,MATNO
```

```
  250 FORMAT(/4I6,I12,I8,' = NDIS,NISO,MEV,N,SVSEED,MATNO'/)
C
      WRITE(11,260) (MP(I),VS(I),G(I), I=1,NDIS)
  260 FORMAT(2(I3,E25.16,E12.3))
C
      WRITE(11,270) NDIS, (MP(I), I=1,NDIS)
  270 FORMAT(/I6,' = NDIS, T-MULTIPLICITIES (0 MEANS  SPURIOUS)'/(20I4))
C
  280 CONTINUE
C
      IF (NISO.NE.0) GO TO 310
C
      WRITE(4,290) MEV
  290 FORMAT(/' AT MEV = ',I6,' THERE ARE NO ISOLATED T-EIGENVALUES'/
     1' SO NO ERROR ESTIMATES WERE COMPUTED/')
C
      WRITE(6,300)
  300 FORMAT(/' ALL COMPUTED GOOD T-EIGENVALUES ARE MULTIPLE'/
     1 ' THEREFORE ALL SUCH EIGENVALUES ARE ASSUMED TO HAVE CONVERGED')
C
      ICONV = 1
      GO TO 350
C
  310 CONTINUE
C
C-----------------------------------------------------------------------
C     SUBROUTINE INVERR COMPUTES ERROR ESTIMATES FOR ISOLATED GOOD
C     T-EIGENVALUES USING INVERSE ITERATION ON T(1,MEV). ON RETURN
C     G(J) = MINIMUM GAP IN T(1,MEV) FOR EACH VS(J), J=1,NDIS
C     G(MEV+I) = BETAM*|U(MEV)| = ERROR ESTIMATE FOR ISOLATED GOOD
C              T-EIGENVALUES, WHERE I = 1, NISO AND  BETAM = BETA(MEV+1)
C              U(MEV) IS MEVTH COMPONENT OF THE UNIT EIGENVECTOR OF T
C              CORRESPONDING TO THE ITH ISOLATED GOOD T-EIGENVALUE.
C     A NEGATIVE ERROR ESTIMATE MEANS THAT FOR THAT PARTICULAR
C     EIGENVALUE THE INVERSE ITERATION DID NOT CONVERGE IN <= MXINIT
C     STEPS AND THAT THE CORRESPONDING ERROR ESTIMATE IS QUESTIONABLE.
C
C     V2 CONTAINS THE ISOLATED GOOD T-EIGENVALUES
C     V1 CONTAINS THE MINGAPS TO THE NEAREST DISTINCT  EIGENVALUE
C        OF T(1,MEV) FOR EACH ISOLATED GOOD T-EIGENVALUE IN V2.
C     VS CONTAINS THE NDIS DISTINCT EIGENVALUES OF T(1,MEV)
C     MP CONTAINS THE CORRESPONDING CODED T-MULTIPLICITIES
C
      IT = MXINIT
      CALL INVERR(ALPHA,BETA,V1,V2,VS,EPSM,G,MP,MEV,MMB,NDIS,NISO,N,
     1    RHSEED,IT,IWRITE)
C
C-----------------------------------------------------------------------
C
C     SIMPLE CHECK FOR CONVERGENCE. CHECKS TO SEE IF ALL OF THE ERROR
C     ESTIMATES ARE SMALLER THAN CONTOL = BETAM*1.D-10.
C     IF THIS TEST IS SATISFIED, THEN CONVERGENCE FLAG, ICONV IS SET
C     TO 1.  TYPICALLY ERROR ESTIMATES ARE VERY CONSERVATIVE.
C
      WRITE(6,320) CONTOL
  320 FORMAT(/' CONVERGENCE IS TESTED USING THE CONVERGENCE TOLERANCE',
     1E13.4/)
C
      II = MEV +1
      IF = MEV+NISO
      DO 330 I = II,IF
      IF (ABS(G(I)).GT.CONTOL) GO TO 350
  330 CONTINUE
      ICONV = 1
      MMB = NMEVS
C
      WRITE(6,340) CONTOL
```

```
  340 FORMAT(' ALL COMPUTED ERROR ESTIMATES WERE LESS THAN',E15.4/
     1 ' THEREFORE PROCEDURE TERMINATES'/)
C
  350 CONTINUE
C
C     IF CONVERGENCE IS INDICATED, THAT IS ICONV = 1 ,THEN
C     THE SUBROUTINE PRTEST IS CALLED TO CHECK FOR ANY CONVERGED
C     T-EIGENVALUES THAT HAVE BEEN MISLABELLED AS SPURIOUS BECAUSE
C     THE PROJECTION OF THEIR EIGENVECTOR(S) ON THE STARTING
C     VECTOR WERE TOO SMALL.
C     NUMERICAL TESTS INDICATE THAT SUCH EIGENVALUES ARE RARE.
C     IF FOR SOME REASON MANY OF THESE HIDDEN EIGENVALUES APPEAR
C     ON SOME RUN, YOU CAN BE CERTAIN THAT SOMETHING IS FOULED UP.
C
      IF (ICONV.EQ.0) GO TO 480
C
C-----------------------------------------------------------------------
C
      CALL PRTEST (ALPHA,BETA,VS,TKMAX,EPSM,RELTOL,SCALE3,SCALE4,
     1 MP,NDIS,MEV,IPROJ)
C
C-----------------------------------------------------------------------
C
      IF(IPROJ.EQ.0) GO TO 470
C
      IF(IDIST.EQ.1)  WRITE(11,360) IPROJ
  360 FORMAT(' SUBROUTINE PRTEST WANTS TO RELABEL',I6,' SPURIOUS T-EIGEN
     1VALUES'/' WE ACCEPT RELABELLING ONLY IF LAST COMPONENT OF T-EIGENV
     1ECTOR IS L.T. 1.D-10'/)
C
      IIX = RHSEED
C
C-----------------------------------------------------------------------
C
      CALL GENRAN(IIX,G,MEV)
C
C-----------------------------------------------------------------------
C
      ITEN = -10
      NISOM = NISO + MEV
      IWRITO = IWRITE
      IWRITE = 0
C
      DO 390 J = 1,NDIS
      IF(MP(J).NE.ITEN) GO TO 390
      TO = VS(J)
C
C-----------------------------------------------------------------------
C
      IT = MXINIT
      CALL INVERM(ALPHA,BETA,V1,V2,TO,TEMP,T1,EPSM,G,MEV,IT,IWRITE)
C
C-----------------------------------------------------------------------
C
      IF(TEMP.LE.1.D-10) GO TO 380
C     ERROR ESTIMATE WAS NOT SMALL REJECT RELABELLING OF THIS EIGENVALUE
      IF(IDIST.EQ.1)  WRITE(11,370) J,TO,TEMP
  370 FORMAT(/' LAST COMPONENT FOR',I6,'TH EIGENVALUE',E20.12/' IS TOO L
     1ARGE = ',E15.6,' SO DO NOT ACCEPT PRTEST RELABELLING'/)
      MP(J) = 0
      IPROJ = IPROJ - 1
      GO TO 390
C     RELABELLING ACCEPTED
  380 NISOM = NISOM + 1
      G(NISOM) = BETAM*TEMP
  390 CONTINUE
```

```
      IWRITE = IWRITO
C
      IF(IPROJ.EQ.0) GO TO 430
      WRITE(6,400) IPROJ
  400 FORMAT(/I6,' T-EIGENVALUES WERE RECLASSIFIED AS GOOD.'/
     1' THESE ARE IDENTIFIED IN FILE 3 BY A T-MULTIPLICITY OF -10'/' USE
     2R SHOULD INSPECT EACH TO MAKE SURE NEIGHBORS HAVE CONVERGED'/)
C
      IF(IDIST.EQ.1)  WRITE(11,410) IPROJ
  410 FORMAT(/I6,' T-EIGENVALUES WERE RELABELLED AS GOOD'/
     1' BELOW IS CORRECTED T-MULTIPLICITY PATTERN'/)
C
      WRITE(6,420) NDIS, (MP(I), I=1,NDIS)
      IF(IDIST.EQ.1)  WRITE(11,420) NDIS, (MP(I), I=1,NDIS)
  420 FORMAT(/I6,' = NDIS, T-MULTIPLICITIES (0 MEANS  SPURIOUS)'/
     1 6X, ' (-10) MEANS SPURIOUS T-EIGENVALUE RELABELLED AS GOOD'/(20I4
     1))
C
C     RECALCULATE MINGAPS FOR DISTINCT T(1,MEV) EIGENVALUES.
  430 NM1 = NDIS - 1
      G(NDIS) = VS(NM1)-VS(NDIS)
      G(1) = VS(2)-VS(1)
C
      DO 440 J = 2,NM1
      T0 = VS(J)-VS(J-1)
      T1 = VS(J+1)-VS(J)
      G(J) = T1
      IF (T0.LT.T1) G(J) = -T0
  440 CONTINUE
      IF(IPROJ.EQ.0) GO TO 470
C     WRITE TO FILE 4 ERROR ESTIMATES FOR THOSE T-EIGENVALUES RELABELLED
      NGOOD = 0
      DO 450 J = 1,NDIS
      IF(MP(J).EQ.0) GO TO 450
      NGOOD = NGOOD + 1
      IF(MP(J).NE.ITEN) GO TO 450
      T0 = VS(J)
      NISO = NISO + 1
      NISOM = MEV + NISO
      WRITE(4,460) NGOOD,T0,G(NISOM),G(J)
  450 CONTINUE
  460 FORMAT(I10,E25.16,2E14.3)
C
  470 CONTINUE
C
C     WRITE THE GOOD T-EIGENVALUES TO FILE 3.  FIRST TRANSFER THEM
C     TO V2 AND THEIR T-MULTIPLICITIES TO THE CORRESPONDING POSITIONS
C     IN MP AND COMPUTE THE A-MINGAPS, THE MINIMAL GAPS BETWEEN THE
C     GOOD T-EIGENVALUES.  THESE GAPS WILL BE PUT IN THE ARRAY G.
C     SINCE G CURRENTLY CONTAINS THE MINIMAL GAPS BETWEEN THE DISTINCT
C     EIGENVALUES OF THE T-MATRIX, THESE GAPS WILL FIRST BE
C     TRANSFERRED TO V1.  NOTE THAT V1<0 MEANS THAT THAT MINIMAL GAP
C     IN THE T-MATRIX IS DUE TO A SPURIOUS T-EIGENVALUE.
C     ALL THIS INFORMATION IS PRINTED TO FILE 3
C
  480 CONTINUE
C
      NG = 0
      DO 490 I = 1,NDIS
      IF (MP(I).EQ.0) GO TO 490
      NG = NG+1
      MP(NG) = MP(I)
      V2(NG) = VS(I)
      TEMP = G(I)
      TEMP = DABS(TEMP)
      J = I+1
      IF (G(I).LT.ZERO) J = I-1
```

```
      IF (MP(J).EQ.0) TEMP = -TEMP                                      LEV06000
      V1(NG) = TEMP                                                     LEV06010
  490 CONTINUE                                                          LEV06020
C                                                                       LEV06030
      WRITE(6,500)MEV                                                   LEV06040
  500 FORMAT(//' T-EIGENVALUE CALCULATION AT MEV = ',I6,'    IS COMPLETELEV06050
     1')                                                                LEV06060
C                                                                       LEV06070
C     NG = NUMBER OF COMPUTED DISTINCT GOOD T-EIGENVALUES.  NEXT        LEV06080
C     GENERATE GAPS BETWEEN GOOD T-EIGENVALUES (AMINGAPS) AND PUT THEM  LEV06090
C     IN G.  G(J) < 0 MEANS THE AMINGAP IS DUE TO THE LEFT-HAND GAP.    LEV06100
C                                                                       LEV06110
      NGM1 = NG - 1                                                     LEV06120
      G(NG) = V2(NGM1)-V2(NG)                                           LEV06130
      G(1) = V2(2)-V2(1)                                                LEV06140
C                                                                       LEV06150
      DO 510 J = 2,NGM1                                                 LEV06160
      T0 = V2(J)-V2(J-1)                                                LEV06170
      T1 = V2(J+1)-V2(J)                                                LEV06180
      G(J) = T1                                                         LEV06190
      IF (T0.LT.T1) G(J) = -T0                                          LEV06200
  510 CONTINUE                                                          LEV06210
C                                                                       LEV06220
C     WRITE GOOD T-EIGENVALUES OUT TO FILE 3.                           LEV06230
C                                                                       LEV06240
      WRITE(3,520)NG,NDIS,MEV,N,SVSEED,MATNO,MULTOL,IB,BTOL             LEV06250
  520 FORMAT(4I6,I12,I8,' = NG,NDIS,MEV,N,SVEED,MATNO'/                 LEV06260
     1 E20.12,I6,E13.4,' = MUTOL,INDEX MINIMAL BETA,BTOL'/              LEV06270
     1' EV NO',1X,'TMULT',10X,'GOOD EIGENVALUE',7X,'TMINGAP',7X,'AMINGAPLEV06280
     1')                                                                LEV06290
C                                                                       LEV06300
      WRITE(3,530)(I,MP(I),V2(I),V1(I),G(I), I=1,NG)                    LEV06310
  530 FORMAT(2I6,E25.16,2E14.3)                                         LEV06320
C                                                                       LEV06330
C     IF CONVERGENCE FLAG ICONV.NE.1 AND NUMBER OF T-MATRICES           LEV06340
C     CONSIDERED TO DATE IS LESS THAN NUMBER ALLOWED, INCREMENT MEV.    LEV06350
C     AND LOOP BACK TO 210 TO REPEAT COMPUTATIONS.  RESTORE BETA(MEV+1).LEV06360
C                                                                       LEV06370
      BETA(MP1) = BETAM                                                 LEV06380
C                                                                       LEV06390
      IF (MMB.LT.NMEVS.AND.ICONV.NE.1) GO TO 180                        LEV06400
C                                                                       LEV06410
C     END OF LOOP ON DIFFERENT SIZE T-MATRICES ALLOWED.                 LEV06420
C                                                                       LEV06430
  540 CONTINUE                                                          LEV06440
C                                                                       LEV06450
      IF(ISTOP.EQ.0)  WRITE(6,550)                                      LEV06460
  550 FORMAT(/' T-MATRICES (ALPHA AND BETA) ARE NOW AVAILABLE, TERMINATELEV06470
     1')                                                                LEV06480
      IF (IHIS.EQ.1.AND.KMAX.NE.MOLD) WRITE(1,560)                      LEV06490
  560 FORMAT(/' ABOVE ARE THE FOLLOWING VECTORS '/                      LEV06500
     1 '  ALPHA(I), I = 1,KMAX'/                                        LEV06510
     2 '  BETA(I), I = 1,KMAX+1'/                                       LEV06520
     3 ' FINAL TWO LANCZOS VECTORS OF ORDER N FOR I = KMAX,KMAX+1'/     LEV06530
     4 ' ALL VECTORS IN THIS FILE HAVE HEX FORMAT 4Z20 '/               LEV06540
     5 ' ----- END OF FILE 1 NEW ALPHA, BETA HISTORY---------------'///)LEV06550
C                                                                       LEV06560
      IF (ISTOP.EQ.0) GO TO 640                                         LEV06570
C                                                                       LEV06580
      WRITE(3,570)                                                      LEV06590
  570 FORMAT(/' ABOVE ARE COMPUTED GOOD T-EIGENVALUES'/                 LEV06600
     1 ' NG = NUMBER OF GOOD T-EIGENVALUES COMPUTED'/                   LEV06610
     2 ' NDIS = NUMBER OF COMPUTED DISTINCT EIGENVALUES OF T(1,MEV)'/   LEV06620
     3 ' N = ORDER OF A,  MATNO = MATRIX IDENT'/                        LEV06630
     4 ' MULTOL = T-MULTIPLICITY TOLERANCE FOR T-EIGENVALUES IN BISEC'/ LEV06640
     4 ' TMULT IS THE T-MULTIPLICITY OF GOOD T-EIGENVALUE'/             LEV06650
     5 ' TMULT = -1 MEANS SPURIOUS T-EIGENVALUE TOO CLOSE'/             LEV06660
```

```
     6 ' DO NOT COMPUTE ERROR ESTIMATES FOR SUCH EIGENVALUES'/
     7 ' AMINGAP = MINIMAL GAP BETWEEN THE COMPUTED A-EIGENVALUES'/
     8 ' AMINGAP .LT. 0. MEANS MINIMAL GAP IS DUE TO LEFT-HAND GAP'/
     9 ' TMINGAP= MINIMAL GAP W.R.T.  DISTINCT EIGENVALUES IN T(1,MEV)'/
     1 ' TMINGAP .LT. 0. MEANS MINGAP IS DUE TO SPURIOUS T-EIGENVALUE'/
     2 ' ----- END OF FILE 3 GOODEIGENVALUES-----------------------'///)
C
      IF (IDIST.EQ.1) WRITE(11,580)
  580 FORMAT(/' ABOVE ARE THE DISTINCT EIGENVALUES OF T(1,MEV).'/
     2 ' THE FORMAT IS       T-MULTIPLICITY    T-EIGENVALUE   TMINGAP'/
     3 '          THIS FORMAT IS REPEATED TWICE ON EACH LINE.'/
     4 ' T-MULTIPLICITY = -1 MEANS THAT THE SUBROUTINE ISOEV HAS TAGGED'
     5/'   THIS SIMPLE T-EIGENVALUE AS HAVING A VERY CLOSE SPURIOUS'/
     6 '   T-EIGENVALUE SO THAT NO ERROR ESTIMATE WILL BE COMPUTED'/
     7 '   FOR THAT EIGENVALUE IN SUBROUTINE INVERR.'/
     8 ' TMINGAP .LT. 0, TMINGAP IS DUE TO LEFT GAP .GT. 0, RIGHT GAP.'/
     9 ' EACH OF THE DISTINCT T-EIGENVALUE TABLES IS FOLLOWED'/
     9 ' BY THE T-MULTIPLICITY PATTERN.'/
     1 ' NDIS = NUMBER OF COMPUTED DISTINCT EIGENVALUES OF T(1,MEV).'/
     2 ' NG = NUMBER OF GOOD T-EIGENVALUES. '/
     3 ' NISO = NUMBER OF ISOLATED GOOD T-EIGENVALUES. '/
     4 ' NISO ALSO IS THE COUNT OF +1 ENTRIES IN T-MULTIPLICITY PATTERN.
     5 '/' ----- END OF FILE 11 DISTINCT T-EIGENVALUES--------------'///
     6 )
C
      IF(NISO.NE.0)  WRITE(4,590)
  590 FORMAT(/' ABOVE ARE THE ERROR ESTIMATES OBTAINED FOR THE ISOLATED
     1GOOD T-EIGENVALUES'/
     1' OBTAINED VIA INVERSE ITERATION IN THE SUBROUTINE INVERR.'/
     1' ALL OTHER GOOD T-EIGENVALUES HAVE CONVERGED.'/
     2' ERROR ESTIMATE = BETAM*ABS(UM)'/
     2' WHERE BETAM = BETA(MEV+1) AND UM = U(MEV).'/
     3' U = UNIT EIGENVECTOR OF T WHERE T*U = EV*U AND EV = ISOLATED GOO
     3D T-EIGENVALUE.'/
     4' TMINGAP = GAP TO NEAREST DISTINCT EIGENVALUE OF T(1,MEV).'/
     5' TMINGAP .LT. 0. MEANS MINGAP IS DUE TO A LEFT NEIGHBOR.'/
     6' ERROR ESTIMATE L.T. 0 MEANS INVERSE ITERATION DID NOT CONVERGE'/
     7' ------ END OF FILE 4 ERRINV -------------------------------'//)
      GO TO 640
C
  600 CONTINUE
C
      IBB = IABS(IBMEV)
      IF (IBMEV.LT.0) WRITE(6,610) MEV,IBB,BETA(IBB)
  610 FORMAT(/' PROGRAM TERMINATES BECAUSE MEV REQUESTED = ',I6,' IS .GT
     1',I6/' AT WHICH AN ABNORMALLY SMALL BETA = ' , E13.4,' OCCURRED'/)
      GO TO 640
C
  620 IF (NDIS.EQ.0.AND.ISTOP.GT.0) WRITE(6,630)
  630 FORMAT(/' INTERVALS SPECIFIED FOR BISECT DID NOT CONTAIN ANY T-EIG
     1ENVALUES'/' PROGRAM TERMINATES')
C
  640 CONTINUE
C
      STOP
C-----END OF MAIN PROGRAM FOR LANCZOS EIGENVALUE COMPUTATIONS-----------
      END
```

SECTION 2.4 MAIN PROGRAM, EIGENVECTOR COMPUTATIONS

```
C-----LEVEC (EIGENVECTORS OF REAL SYMMETRIC MATRICES)-------------------
C
C     CONTAINS MAIN PROGRAM FOR COMPUTING AN EIGENVECTOR CORRESPONDING
C     TO EACH OF A SET OF EIGENVALUES WHICH HAVE BEEN COMPUTED
C     ACCURATELY BY THE CORRESPONDING LANCZOS EIGENVALUE PROGRAM
C     (LEVAL) FOR REAL SYMMETRIC MATRICES.  THIS PROGRAM COULD BE
C     MODIFIED TO COMPUTE ADDITIONAL EIGENVECTORS FOR ANY EIGENVALUE
C     WHICH IS A MULTIPLE EIGENVALUE OF THE GIVEN A-MATRIX.  THE
C     AMOUNT OF ADDITIONAL COMPUTATION REQUIRED WOULD DEPEND UPON
C     THE GIVEN A-MATRIX AND UPON WHAT PART OF THE SPECTRUM OF
C     A IS INVOLVED.
C
C     THE LANCZOS EIGENVECTOR COMPUTATIONS ASSUME THAT EACH
C     EIGENVALUE THAT IS BEING CONSIDERED HAS CONVERGED AS AN
C     EIGENVALUE OF THE LANCZOS TRIDIAGONAL MATRICES.
C
C     PFORT VERIFIER IDENTIFIED THE FOLLOWING NONPORTABLE
C     CONSTRUCTIONS
C
C     1.  DATA/MACHEP/ STATEMENT
C     2.  ALL READ(5,*) STATEMENTS (FREE FORMAT)
C     3.  FORMAT(20A4) USED WITH THE EXPLANATORY HEADER, EXPLAN
C     4.  HEXADECIMAL FORMAT (4Z20) USED IN ALPHA/BETA FILES 1 AND 2.
C
C     IMPORTANT NOTE:  THIS PROGRAM ALLOWS ENLARGEMENT OF THE ALPHA,
C     BETA ARRAYS.  IN PARTICULAR, IF ANY ONE OF THE EIGENVALUES
C     SUPPLIED IS T-SIMPLE AND NOT CLOSE TO A SPURIOUS EIGENVALUE,
C     THE PROGRAM REQUIRES THAT KMAX BE AT LEAST 11*MEV/8 + 12.  IF
C     KMAX IS NOT THIS LARGE, THEN THE PROGRAM RESETS KMAX TO THIS
C     SIZE AND EXTENDS THE ALPHA, BETA HISTORY IF REQUIRED.
C     THUS, THE DIMENSIONS OF THE ALPHA AND BETA ARRAYS MUST BE
C     LARGE ENOUGH TO ALLOW FOR THIS POSSIBILITY.
C     REMEMBER THAT THE BETA ARRAY, BETA(J), IS SUCH THAT
C     J = 1,..., KMAX+1.  SO IF THE KMAX USED BY THE PROGRAM
C     IS TO BE 3000, THEN BETA MUST BE OF LENGTH AT LEAST 3001.
C
C-----------------------------------------------------------------------
      DOUBLE PRECISION  ALPHA(5000),BETA(5001)
      DOUBLE PRECISION  V1(5000),V2(5000)
      DOUBLE PRECISION  RITVEC(30000),TVEC(30000),GOODEV(50),EVNEW(50)
      DOUBLE PRECISION  EVAL,EVALN,TOLN,TTOL,ERTOL,ALFA,BATA
      DOUBLE PRECISION  MULTOL,SCALE0,STUTOL,BTOL,LB,UB
      DOUBLE PRECISION  ONE,ZERO,MACHEP,EPSM,TEMP,SUM,ERRMIN,BKMIN
      DOUBLE PRECISION  RELTOL,ERROR,TERROR,TLAST(50)
      REAL  G(10000),AMINGP(50),TMINGP(50),EXPLAN(20)
      REAL  TERR(50),ERR(50),ERRDGP(50),RNORM(50),TBETA(50)
      INTEGER  MP(50),M1(50),M2(50),MA(50),ML(50),MINT(50),MFIN(50)
      INTEGER  SVSEED,SVSOLD,RHSEED,IDELTA(50)
      INTEGER  MBOUND,NTVCON,SVTVEC,TVSTOP,LVCONT,ERCONT,TFLAG
      DOUBLE PRECISION  FINPRO
      DOUBLE PRECISION  DABS, DMAX1, DSQRT, DFLOAT
      REAL ABS
      INTEGER  IABS
C-----------------------------------------------------------------------
      EXTERNAL CMATV
      DATA MACHEP/Z3410000000000000/
      EPSM = 2.D0*MACHEP
C-----------------------------------------------------------------------
C
C     ARRAYS MUST BE DIMENSIONED AS FOLLOWS:
C     1.  ALPHA:  >= KMAXN,  BETA: >= (KMAXN+1) WHERE KMAXN, THE
C                 LARGEST SIZE T-MATRIX CONSIDERED BY THE PROGRAM,
C                 IS THE LARGER OF THE SIZE OF THE ALPHA, BETA HISTORY
C                 PROVIDED ON FILE 2 (IF ANY ) AND THE SIZE WHICH THE
```

```
C              PROGRAM SPECIFIES INTERNALLY, THIS LATTER IS ALWAYS
C              < = 11*MEV / 8  +  12, WHERE MEV IS THE SIZE
C              T-MATRIX THAT WAS USED IN THE CORRESPONDING EIGENVALUE
C              COMPUTATIONS.
C     2.  V1:  >= MAX(N,KMAX)
C     3.  V2:   >= N
C     4.  G:  >= MAX(N,KMAX)
C     5.  RITVEC:  >= N*NGOOD, WHERE NGOOD IS NUMBER OF EIGENVALUES
C                SUPPLIED TO THIS PROGRAM.
C     6.  TVEC:  >= CUMULATIVE LENGTH OF ALL THE T-EIGENVECTORS
C         NEEDED TO GENERATE THE DESIRED RITZ VECTORS.  AN EDUCATED
C         GUESS AT AN APPROPRIATE LENGTH CAN BE OBTAINED BY RUNNING THE
C         PROGRAM WITH THE FLAG MBOUND = 1 AND MULTIPLYING THE
C         RESULTING SIZE BY 5/4.
C     7.  GOODEV, AMINGP, TMINGP, TERR, ERR, ERRGDP, RNORM, TBETA,
C         TLAST, EVNEW, MP, MA, M1, M2, MINT, MFIN AND IDELTA ALL MUST
C         BE >= NGOOD.
C
C-----------------------------------------------------------------------
C     OUTPUT HEADER
      WRITE(6,10)
   10 FORMAT(/' LANCZOS EIGENVECTOR PROCEDURE FOR REAL SYMMETRIC MATRICE
     1S'/)
C
C     SET PROGRAM PARAMETERS
C     USER MUST NOT MODIFY SCALEO
      SCALEO = 5.0D0
      ZERO = 0.0D0
      ONE = 1.0D0
      MPMIN = -1000
C     SET CONVERGENCE CRITERION FOR T-EIGENVECTORS.
      ERTOL = 1.D-10
C
C     READ USER-SPECIFIED PARAMETER FROM INPUT FILE 5 (FREE FORMAT)
C
C     READ USER-PROVIDED HEADER FOR RUN
      READ(5,20) EXPLAN
      WRITE(6,20) EXPLAN
   20 FORMAT(20A4)
C
C     READ IN THE MAXIMUM PERMISSIBLE DIMENSIONS FOR THE TVEC ARRAY
C     (MDIMTV), FOR THE RITVEC ARRAY (MDIMRV), AND FOR THE BETA
C     ARRAY (MBETA).
      READ(5,20) EXPLAN
      READ(5,*) MDIMTV, MDIMRV, MBETA
C
C     READ IN RELATIVE TOLERANCE (RELTOL) USED IN DETERMINING
C     APPROPRIATE SIZES FOR THE T-MATRICES TO BE USED IN THE RITZ
C     VECTOR COMPUTATIONS.
      READ(5,20) EXPLAN
      READ(5,*) RELTOL
C
C     SET FLAGS TO 0 OR 1:
C     MBOUND = 1:  PROGRAM TERMINATES AFTER COMPUTING 1ST GUESSES
C                  ON APPROPRIATE T-SIZES FOR USE IN THE RITZ VECTOR
C                  COMPUTATIONS
C     NTVCON = 0:  PROGRAM TERMINATES IF THE TVEC ARRAY IS NOT
C                  LARGE ENOUGH TO HOLD ALL THE T-EIGENVECTORS REQUIRED.
C     SVTVEC = 0:  THE T-EIGENVECTORS ARE NOT WRITTEN TO FILE 11
C                  UNLESS TVSTOP = 1
C     SVTVEC = 1:  WRITE THE T-EIGENVECTORS TO FILE 11.
C     TVSTOP = 1:  PROGRAM TERMINATES AFTER COMPUTING THE
C                  T-EIGENVECTORS
C     LVCONT = 0:  PROGRAM TERMINATES IF THE NUMBER OF T-EIGENVECTORS
C                  COMPUTED IS NOT EQUAL TO THE NUMBER OF RITZ
C                  VECTORS REQUESTED.
C     ERCONT = 0:  MEANS FOR ANY GIVEN EIGENVALUE, A RITZ VECTOR
```

```
C                 WILL NOT BE COMPUTED FOR THAT EIGENVALUE UNLESS
C                 A T-EIGENVECTOR HAS BEEN IDENTIFIED WITH A LAST
C                 COMPONENT WHICH SATISFIES THE SPECIFIED
C                 CONVERGENCE CRITERION.
C     ERCONT = 1: MEANS FOR ANY GIVEN EIGENVALUE, A RITZ VECTOR
C                 WILL BE COMPUTED.  IF A T-EIGENVECTOR CANNOT
C                 BE IDENTIFIED WHICH SATISFIES THE LAST
C                 COMPONENT CRITERION, THEN THE PROGRAM WILL
C                 USE THE T-VECTOR THAT CAME CLOSEST TO
C                 SATISFYING THE CRITERION.
C     IWRITE = 1: EXTENDED OUTPUT OF INTERMEDIATE COMPUTATIONS
C                 IS WRITTEN TO FILE 6
C     IREAD = 0:  ALPHA/BETA FILE IS REGENERATED.
C     IREAD = 1:  ALPHA/BETA FILE USED IN EIGENVALUE COMPUTATIONS
C                 IS READ IN AND EXTENDED IF NECESSARY.  IN BOTH
C                 CASES IREAD = 0 OR 1, THE LANCZOS VECTORS ARE
C                 ALWAYS REGENERATED FOR THE RITZ VECTOR
C                 COMPUTATIONS
C
      READ(5,20) EXPLAN
      READ(5,*) MBOUND,NTVCON,SVTVEC,IREAD
C
      READ(5,20) EXPLAN
      READ(5,*) TVSTOP,LVCONT,ERCONT,IWRITE
      IF (TVSTOP.EQ.1) SVTVEC = 1
C
C     READ IN SEED (RHSEED) FOR GENERATING RANDOM STARTING VECTOR
C     FOR INVERSE ITERATION ON THE T-MATRICES.
      READ(5,20) EXPLAN
      READ(5,*) RHSEED
C
C     READ IN MATNO = MATRIX/RUN IDENTIFICATION NUMBER AND
C     N = ORDER OF A-MATRIX
      READ(5,20) EXPLAN
      READ(5,*) MATNO,N
C
C-----------------------------------------------------------------------
C     INITIALIZE THE ARRAYS FOR THE USER-SPECIFIED MATRIX
C     AND PASS THE STORAGE LOCATIONS OF THESE ARRAYS TO THE
C     MATRIX-VECTOR MULTIPLY SUBROUTINE CMATV.
C
      CALL USPEC(N,MATNO)
C
C-----------------------------------------------------------------------
C     MASK UNDERFLOW AND OVERFLOW
      CALL MASK
C
C-----------------------------------------------------------------------
C     WRITE RUN PARAMETERS OUT TO FILE 6
C
      WRITE(6,30) MATNO,N
   30 FORMAT(/' MATRIX IDENTIFICATION NO. = ',I10,' ORDER OF A = ',I5)
C
      WRITE(6,40) MBOUND,NTVCON,SVTVEC,IREAD
   40 FORMAT(/3X,'MBOUND',3X,'NTVCON',3X,'SVTVEC',3X,'IREAD'/3I9,I8)
C
      WRITE(6,50) TVSTOP,LVCONT,ERCONT,IWRITE
   50 FORMAT(/3X,'TVSTOP',3X,'LVCONT',3X,'ERCONT',3X,'IWRITE'/4I9)
C
      WRITE(6,60) MDIMTV,MDIMRV,MBETA
   60 FORMAT(/3X,'MDIMTV',3X,'MDIMRV',3X,'MBETA'/2I9,I8)
C
      WRITE(6,70) RELTOL,RHSEED
   70 FORMAT(/7X,'RELTOL',3X,'RHSEED'/E13.4,I9)
C
C
C     FROM FILE 3 READ IN THE NUMBER OF EIGENVALUES (NGOOD) FOR WHICH
```

```
C     EIGENVECTORS ARE REQUESTED, THE ORDER (MEV) OF THE LANCZOS
C     TRIDIAGONAL MATRIX USED IN COMPUTING THESE EIGENVALUES, THE
C     ORDER (NOLD) OF THE USER-SPECIFIED MATRIX USED IN THE EIGENVALUE
C     COMPUTATIONS, THE SEED (SVSEED) USED FOR GENERATING THE STARTING
C     VECTOR THAT WAS USED IN THOSE LANCZOS EIGENVALUE COMPUTATIONS,
C     AND THE MATRIX/RUN IDENTIFICATION NUMBER (MATOLD) USED IN THOSE
C     COMPUTATIONS.  ALSO READ IN THE NUMBER (NDIS) OF DISTINCT
C     EIGENVALUES OF T(1,MEV) THAT WERE COMPUTED BUT THIS VALUE IS
C     NOT USED IN THE EIGENVECTOR COMPUTATIONS.
C
      READ(3,80) NGOOD,NDIS,MEV,NOLD,SVSEED,MATOLD
   80 FORMAT(4I6,I12,I8)
C
C     READ IN THE T-MULTIPLICITY TOLERANCE USED IN THE BISEC SUBROUTINE
C     DURING THE COMPUTATION OF THE GIVEN EIGENVALUES.
C     ALSO READ IN THE FLAG IB.  IF IB < 0, THEN SOME BETA(I) IN THE
C     T-MATRIX FILE PROVIDED ON FILE 2 FAILED THE ORTHOGONALITY
C     TEST IN THE TNORM SUBROUTINE.  USER SHOULD NOTE THAT THIS VECTOR
C     PROGRAM PROCEEDS INDEPENDENTLY OF THE SIZE OF THE BETA USED.
C
      READ(3,90) MULTOL,IB,BTOL
   90 FORMAT(E20.12,I6,E13.4)
C
      TEMP = DFLOAT(MEV+1000)
      TTOL = MULTOL/TEMP
      WRITE(6,100) MULTOL,TTOL
  100 FORMAT(/' T-MULTIPLICITY TOLERANCE USED IN THE EIGENVALUE COMPUTAT
     1IONS WAS',E13.4/' SCALED MACHINE EPSILON IS',E13.4)
C
C     CONTINUE WRITE TO FILE 6 OF THE PARAMETERS FOR THIS RUN
C
      WRITE(6,110)NGOOD,NDIS,MEV,NOLD,MATOLD,SVSEED,MULTOL,IB,BTOL
  110 FORMAT(/' EIGENVALUES SUPPLIED ARE READ IN FROM FILE 3'/' FILE 3
     1HEADER IS'/4X,'NG',2X,'NDIS',3X,'MEV',2X,'NOLD',2X,'MATOLD',4X,
     1'SVSEED',6X,'MULTOL',6X,'IB',9X,'BTOL'/4I6,I8,I10,E12.3,I8,E13.4/)
C
C     IS THE ARRAY RITVEC LONG ENOUGH TO HOLD ALL OF THE DESIRED
C     RITZ VECTORS (APPROXIMATE EIGENVECTORS)?
      NMAX = NGOOD*N
      IF(MBOUND.NE.0) GO TO 120
      IF(TVSTOP.NE.1.AND.NMAX.GT.MDIMRV) GO TO 1310
C
C     CHECK THAT THE ORDER N AND THE MATRIX IDENTIFICATION NUMBER
C     MATNO SPECIFIED BY THE USER AGREE WITH THOSE READ IN FROM
C     FILE 3.
  120 ITEMP = (NOLD-N)**2+(MATOLD-MATNO)**2
      IF (ITEMP.NE.0) GO TO 1330
C
C     READ IN FROM FILE 3, THE T-MULTIPLICITIES OF THE EIGENVALUES
C     WHOSE EIGENVECTORS ARE TO BE COMPUTED, THE VALUES OF THESE
C     EIGENVALUES AND THEIR MINIMAL GAPS AS EIGENVALUES OF THE
C     USER-SPECIFIED MATRIX AND AS EIGENVALUES OF THE T-MATRIX.
C
      READ(3,20) EXPLAN
      READ(3,130) (MP(J),GOODEV(J),TMINGP(J),AMINGP(J), J=1,NGOOD)
  130 FORMAT(6X,I6,E25.16,2E14.3)
C
      WRITE(6,140) (J,GOODEV(J),MP(J),TMINGP(J),AMINGP(J), J=1,NGOOD)
  140 FORMAT(/' EIGENVALUES READ IN, T-MULTIPLICITIES, T-GAPS AND A-GAPS
     1 '/4X,' J ',5X,'GOOD EIGENVALUE',5X,'MULT',4X,'  TMINGAP ',4X,
     1'  AMINGAP '/(I6,E25.16,I4,2E15.4))
C
C     READ IN ERROR ESTIMATES
      WRITE(6,150) MEV,SVSEED
  150 FORMAT(/' THESE EIGENVALUES WERE COMPUTED USING A T-MATRIX OF
```

```
     1ORDER ',I5/' AND SEED FOR RANDOM NUMBER GENERATOR =',I12)
C     CHECK WHETHER OR NOT THERE ARE ANY T-ISOLATED EIGENVALUES IN
C     THE EIGENVALUES PROVIDED
      DO 160 J=1,NGOOD
      IF(MP(J).EQ.1) GO TO 170
  160 CONTINUE
      GO TO 190
  170 READ(4,20) EXPLAN
      READ(4,20) EXPLAN
      READ(4,20) EXPLAN
      READ(4,180) NISO
  180 FORMAT(18X,I6)
      READ(4,20) EXPLAN
      READ(4,20) EXPLAN
      READ(4,20) EXPLAN
  190 DO 220 J=1,NGOOD
      ERR(J) = 0.D0
      IF(MP(J).NE.1) GO TO 220
      READ(4,200) EVAL, ERR(J)
  200 FORMAT(10X,E25.16,E14.3)
      IF(DABS(EVAL - GOODEV(J)).LT.1.D-10) GO TO 220
      WRITE(6,210) EVAL,GOODEV(J)
  210 FORMAT(' PROBLEM WITH READ IN OF ERROR ESTIMATES'/' EIGENVALUE REAL
     1D IN',E20.12,' DOES NOT MATCH GOODEV(J) ='/E20.12)
      GO TO 1550
C
  220 CONTINUE
C
      WRITE(6,230) (J,GOODEV(J),ERR(J), J=1,NGOOD)
  230 FORMAT(' ERROR ESTMATES ='/4X,' J',5X,'EIGENVALUE',10X,' ESTIMATE
     1'/(I6,E20.12,E14.3))
C
      IF(IREAD.EQ.0)  GO TO 330
C
C     READ IN THE SIZE OF THE T-MATRIX PROVIDED ON FILE 2.  READ IN
C     THE ORDER OF THE USER-SPECIFIED MATRIX , THE SEED FOR THE
C     RANDOM NUMBER GENERATOR, AND THE MATRIX/TEST IDENTIFICATION
C     NUMBER THAT WERE USED IN THE LANCZOS EIGENVALUE COMPUTATIONS.
C     THESE ARE USED IN A CONSISTENCY CHECK
C     IF FLAG IREAD = 0 REGENERATE ALPHA, BETA
C
      READ(2,240) KMAX,NOLD,SVSOLD,MATOLD
  240 FORMAT(2I6,I12,I8)
C
      WRITE(6,250) KMAX,NOLD,SVSOLD,MATOLD
  250 FORMAT(/' READ IN THE T-MATRICES STORED ON FILE 2'/' FILE 2 HEADER
     1 IS'/2X,'KMAX',2X,'NOLD',6X,'SVSOLD',2X,'MATOLD'/2I6,I12,I8/)
C
C     CHECK THAT THE ORDER, THE MATRIX/TEST IDENTIFICATION NUMBER
C     AND THE SEED FOR THE RANDOM NUMBER GENERATOR USED IN THE
C     LANCZOS COMPUTATIONS THAT GENERATED THE ALPHA,BETA FILE
C     BEING USED AGREE WITH WHAT THE USER HAS SPECIFIED.
      IF (NOLD.NE.N.OR.MATOLD.NE.MATNO.OR.SVSOLD.NE.SVSEED) GO TO 1350
C
      KMAX1 = KMAX + 1
C
C     READ IN THE T-MATRICES FROM FILE 2.  THESE ARE USED TO GENERATE
C     THE T-EIGENVECTORS THAT WILL BE USED IN THE RITZ VECTOR
C     COMPUTATIONS.  HISTORY MUST BE STORED IN MACHINE FORMAT
C     ((4Z20) FOR IBM/3081)
C
      READ(2,260) (ALPHA(J), J=1,KMAX)
      READ(2,260) (BETA(J), J=1,KMAX1)
  260 FORMAT(4Z20)
C
      READ(2,260) (V1(J), J=1,N)
```

```
      READ(2,260) (V2(J), J=1,N)
C
C     KMAX MAY BE ENLARGED IF THE SIZE AT WHICH THE EIGENVALUE
C     COMPUTATIONS WERE PERFORMED IS ESSENTIALLY KMAX AND
C     THERE IS AT LEAST ONE EIGENVALUE THAT IS T-SIMPLE AND
C     T-ISOLATED, IN THE SENSE THAT IF ITS NEAREST NEIGHBOR IS TOO
C     CLOSE THAT NEIGHBOR IS A 'GOOD' T-EIGENVALUE.
      DO 270 J = 1,NGOOD
      IF(MP(J).EQ.1) GO TO 290
  270 CONTINUE
      WRITE(6,280)
  280 FORMAT(/' ALL EIGENVALUES USED ARE T-MULTIPLE OR CLOSE TO SPURIOUS
     1 T-EIGENVALUES'/' SO KMAX IS NOT INCREASED')
      IF(KMAX.LT.MEV) GO TO 1370
      GO TO 310
C
  290 KMAXN= 11*MEV/8 + 12
      IF(MBETA.LE.KMAXN) GO TO 1530
      IF(KMAX.GE.KMAXN )  GO TO 310
      WRITE(6,300) KMAX, KMAXN
  300 FORMAT(' ENLARGE KMAX FROM ',I6,' TO ',I6)
      MOLD1 = KMAX + 1
      KMAX = KMAXN
      GO TO 380
C
  310 WRITE(6,320) KMAX
  320 FORMAT(/' T-MATRICES HAVE BEEN READ IN FROM FILE 2'/' THE LARGEST
     1SIZE T-MATRIX ALLOWED IS',I6/)
C
      IF(IREAD.EQ.1) GO TO 400
C
C     REGENERATE THE ALPHA AND BETA
C
  330 MOLD1 = 1
C
      DO 340 J = 1,NGOOD
      IF(MP(J).EQ.1) GO TO 360
  340 CONTINUE
      KMAX = MEV + 12
      WRITE(6,350) KMAX
  350 FORMAT(/' ALL EIGENVALUES FOR WHICH EIGENVECTORS ARE TO BE COMPUTE
     1D ARE EITHER T-MULTIPLE OR CLOSE TO'/' A SPURIOUS T-EIGENVALUE. TH
     1EREFORE SET KMAX = MEV + 12 = ',I7)
      GO TO 380
C
  360 KMAXN = 11*MEV/8 + 12
      IF(MBETA.LE.KMAXN) GO TO 1530
      WRITE(6,370) KMAXN
  370 FORMAT(' SET KMAX EQUAL TO ',I6)
      KMAX = KMAXN
C
  380 WRITE(6,390) MOLD1,KMAX
  390 FORMAT(/' LANCZS SUBROUTINE GENERATES ALPHA(J), BETA(J+1), J =',
     1 I6,' TO ', I6/)
C
C-----------------------------------------------------------------------
C
      IIX = SVSEED
      CALL LANCZS(CMATV,ALPHA,BETA,V1,V2,G,KMAX,MOLD1,N,IIX)
C
C-----------------------------------------------------------------------
C
  400 CONTINUE
C
C     THE SUBROUTINE STURMI DETERMINES THE SMALLEST SIZE T-MATRIX FOR
C     WHICH THE EIGENVALUE IN QUESTION IS A T-EIGENVALUE (TO WITHIN A
C     GIVEN TOLERANCE) AND IF POSSIBLE THE SMALLEST SIZE T-MATRIX
```

```
C     FOR WHICH IT IS A DOUBLE T-EIGENVALUE (TO WITHIN THE SAME
C     TOLERANCE).  THE SIZE T-MATRIX USED IN THE RITZ VECTOR
C     COMPUTATIONS IS THEN DETERMINED BY LOOPING ON SIZE OF THE
C     T-EIGENVECTORS STARTING WITH A T-SIZE DETERMINED FROM THE
C     OUTPUT FROM STURMI.
C
C
      STUTOL = SCALE0*MULTOL
      IF(IWRITE.EQ.1) WRITE(6,410)
  410 FORMAT(' FROM STURMI')
      DO 450 J = 1,NGOOD
      EVAL = GOODEV(J)
C     COMPUTE THE TOLERANCES USED BY STURMI TO DETERMINE AN INTERVAL
C     CONTAINING THE EIGENVALUE EVAL.
      TEMP = DABS(EVAL)*RELTOL
      TOLN = DMAX1(TEMP,STUTOL)
C
C-----------------------------------------------------------------------
C
      CALL STURMI(ALPHA,BETA,EVAL,TOLN,EPSM,KMAX,MK1,MK2,IC,IWRITE)
C
C-----------------------------------------------------------------------
C
C     STORE THE COMPUTED ORDERS OF T-MATRICES FOR LATER PRINTOUT
      M1(J) = MK1
      M2(J) = MK2
      ML(J) = (MK1 + 3*MK2)/4
      IF(MK2.EQ.KMAX)  ML(J) = KMAX
C
      IF(IC.GT.0) GO TO 430
C     IC = 0 MEANS THERE WAS NO EIGENVALUE IN THE DESIGNATED INTERVAL
C     BY T-SIZE KMAX.  THIS MEANS THAT THE EIGENVALUE PROVIDED HAS
C     NOT YET CONVERGED SO ITS EIGENVECTOR WILL NOT BE COMPUTED.
      WRITE(6,420) J,GOODEV(J),MK1,MK2
  420 FORMAT(I6,'TH EIGENVALUE',E20.12,' HAS NOT CONVERGED '/
     1' SO DO NOT COMPUTE ANY T-EIGENVECTOR OR RITZ VECTOR FOR IT'
     1/' MK1 AND MK2 FOR THIS EIGENVALUE WERE',2I6)
      MP(J) = MPMIN
      MA(J) = -2*KMAX
      GO TO 450
C     COMPUTE AN APPROPRIATE SIZE T-MATRIX FOR THE GIVEN EIGENVALUE.
  430 IF(M2(J).EQ.KMAX) GO TO 440
C     M1 AND M2 WERE BOTH DETERMINED
      MA(J) = (3*M1(J) + M2(J))/4  + 1
      GO TO 450
C     M2 NOT DETERMINED
  440 MA(J) = (5*M1(J))/4  + 1
C
  450 CONTINUE
C
      IF (IWRITE.EQ.1) WRITE(6,460) (MA(JJ), JJ=1,NGOOD)
  460 FORMAT(/' 1ST GUESS AT APPROPRIATE SIZE T-MATRICES'/
     1 ' ACTUAL VALUES WILL PROBABLY BE 1/4 AGAIN AS MUCH'/(13I6))
C
C     PRINT OUT TO FILE 10 1ST GUESSES AT SIZES OF THE T-MATRICES TO
C     BE USED IN THE EIGENVECTOR COMPUTATIONS.
C     PROGRAM LOOPS ON T-SIZE TO DETERMINE APPROPRIATE SIZE T-MATRIX.
      WRITE(10,470) N,KMAX
  470 FORMAT(2I8,' = ORDER OF USER MATRIX AND MAX ORDER OF T(1,MEV)')
C
      WRITE(10,480)
  480 FORMAT(/' 1ST GUESS AT APPROPRIATE SIZE T-MATRICES'/
     1 ' ACTUAL VALUES WILL PROBABLY BE 1/4 AGAIN AS MUCH'/)
C
      WRITE(10,490)
  490 FORMAT(4X,'J',4X,'A-EIGENVALUE',4X,'M1(J)',1X,'M2(J)',1X,'MA(J)')
C
```

```
      WRITE(10,500) (J,GOODEV(J),M1(J),M2(J), MA(J), J=1,NGOOD)
  500 FORMAT(I5,E19.12,3I6)
C
      IF(MBOUND.EQ.1) WRITE(10,510)
  510 FORMAT(/'  EV = GOODEV(J) IS A GOOD EIGENVALUE OF T(1,MEV)'/
     1' M1 = SMALLEST VALUE OF M SUCH THAT T(1,M) HAS AT LEAST'/
     1 '      ONE EIGENVALUE IN THE INTERVAL (EV-TOLN,EV+TOLN)'/
     1 '      TOLN(J) = DMAX1(GOODEV(J)*RELTOL, SCALEO*MULTOL)'/
     1 ' M2 = SMALLEST M (IF ANY) SUCH THAT IN THE ABOVE INTERVAL'/
     1 '       T(1,M) HAS AT LEAST TWO EIGENVALUES '/
     1 ' IABS(MA(J)) = APPROPRIATE SIZE T-MATRIX FOR GOODEV(J)'/
     1 ' INITIAL VALUE OF MA(J) IS CHOSEN HEURISTICALLY'/
     1 ' PROGRAM LOOPS ON SIZE OF T-MATRIX TO GET BETTER SIZE'/
     1 ' END OF SIZES OF T-MATRICES FILE 10'///)
C
C     TERMINATE AFTER COMPUTING 1ST GUESSES AT SIZES OF THE
C     T-MATRICES REQUIRED FOR THE GIVEN EIGENVALUES?
      IF(MBOUND.EQ.1) GO TO 1390
C
C     IS THERE ROOM FOR ALL OF THE REQUESTED T-EIGENVECTORS?
      MTOL = 0
      DO 520 J = 1,NGOOD
      IF(MP(J).EQ.MPMIN) GO TO 520
      MTOL = MTOL + IABS(MA(J))
  520 CONTINUE
      MTOL = (5*MTOL)/4
      IF(MTOL.GT.MDIMTV.AND.NTVCON.EQ.0) GO TO 1410
C
C-----------------------------------------------------------------------
C     GENERATE A RANDOM VECTOR TO BE USED REPEATEDLY BY
C     SUBROUTINE INVERM
C
      CALL GENRAN(RHSEED,G,KMAX)
C
C-----------------------------------------------------------------------
C
C     LOOP ON GIVEN EIGENVALUES TO COMPUTE THE CORRESPONDING
C     T-EIGENVECTOR.
C
      MTOL = 0
      NTVEC = 0
      ILBIS = 0
      DO 710 J = 1,NGOOD
      ICOUNT = 0
      ERRMIN = 10.D0
      MABEST = MPMIN
      IF(MP(J).EQ.MPMIN) GO TO 710
      TFLAG = 0
      EVAL = GOODEV(J)
      TEMP = DABS(EVAL)*RELTOL
      UB = EVAL + DMAX1(STUTOL,TEMP)
      LB = EVAL - DMAX1(STUTOL,TEMP)
  530 KMAXU = IABS(MA(J))
C
C     SELECT A SUITABLE INCREMENT FOR THE ORDERS OF THE T-MATRICES
C     TO BE CONSIDERED IN DETERMINING APPROPRIATE SIZES FOR THE RITZ
C     VECTOR COMPUTATIONS.
      IF(ICOUNT.GT.0) GO TO 550
C     SELECT IDELTA(J) BASED UPON THE T-MULTIPLICITY OBTAINED
      IF(M2(J).EQ.KMAX) GO TO 540
C     M2 DETERMINED
      IDELTA(J) = ((3*M1(J) + 5*M2(J))/8  + 1 - IABS(MA(J)))/10  +  1
      GO TO 550
C     M2 NOT DETERMINED
  540 MAMAX = MIN0((11*MEV)/8 + 12, (13*M1(J))/8 + 1)
      IDELTA(J) = (MAMAX - IABS(MA(J)))/10  +  1
```

```
  550 ICOUNT = ICOUNT + 1
C
C-----------------------------------------------------------------------
C     TO MIMIMIZE THE EFFECT OF THE ONE-SIDED ACCEPTANCE TEST FOR
C     EIGENVALUES IN THE BISEC SUBROUTINE, RECOMPUTE THE GIVEN
C     EIGENVALUE AT THE SPECIFIED KMAXU
C
      CALL LBISEC(ALPHA,BETA,EPSM,EVAL,EVALN,LB,UB,TTOL,KMAXU,NEVT)
C
C-----------------------------------------------------------------------
C
C     CHECK WHETHER OR NOT GIVEN T-MATRIX HAS AN EIGENVALUE IN THE
C     SPECIFIED INTERVAL AND IF SO WHAT ITS T-MULTIPLICITY IS.
C
      IF(NEVT.EQ.1) GO TO 590
      IF(NEVT.NE.0) GO TO 570
      ILBIS = 1
      WRITE(6,560) EVAL,KMAXU
  560 FORMAT(/' PROBLEM ENCOUNTERED IN RECOMPUTATION OF USER-SUPPLIED EI
     1GENVALUE',E20.12/' THE SIZE T-MATRIX SPECIFIED',I6,' DOES NOT
     1HAVE AN EIGENVALUE IN THE INTERVAL SPECIFIED'/' THEREFORE NO EIGEN
     1VECTOR WILL BE COMPUTED FOR THIS PARTICULAR EIGENVALUE'/)
      GO TO 610
C
  570 IF(NEVT.GT.1)  WRITE(6,580) EVAL,KMAXU
  580 FORMAT(/' PROBLEM ENCOUNTERED IN RECOMPUTATION OF USER-SUPPLIED
     1EIGENVALUE',E20.12/' FOR THE SIZE T-MATRIX SPECIFIED =',I6,' THE
     1GIVEN EIGENVALUE IS T-MULTIPLE IN THE INTERVAL SPECIFIED'/' SOMETH
     1ING IS WRONG, THEREFORE NO EIGENVECTOR WILL BE COMPUTED FOR THIS E
     1IGENVALUE'/)
C
      MP(J) = MPMIN
      MA(J) = -2*KMAX
      GO TO 710
C
  590 CONTINUE
      ILBIS = 0
C
      EVNEW(J) = EVALN
      EVAL = EVALN
      MTOL = MTOL+KMAXU
C
C     IS THERE ROOM IN TVEC ARRAY FOR THE NEXT T-EIGENVECTOR?
C     IF NOT, SKIP TO RITZ VECTOR COMPUTATIONS.
      IF (MTOL.GT.MDIMTV) GO TO 720
C
      IT = 3
      KINT = MTOL - KMAXU +1
C
C     RECORD THE BEGINNING AND END OF THE T-EIGENVECTOR BEING COMPUTED
      MINT(J) = KINT
      MFIN(J) = MTOL
C
C-----------------------------------------------------------------------
C     SUBROUTINE INVERM DOES INVERSE ITERATION, I.E. SOLVES
C     (T(1,KMAXU) - EVAL)*U = RHS  FOR EACH EIGENVALUE TO OBTAIN THE
C     DESIRED T-EIGENVECTOR.
C
      IF(IWRITE.EQ.1)  WRITE(6,600) J
  600 FORMAT(/I6,'TH EIGENVALUE')
C
      CALL INVERM(ALPHA,BETA,V1,TVEC(KINT),EVAL,ERROR,TERROR,EPSM,
     1 G,KMAXU,IT,IWRITE)
C
C-----------------------------------------------------------------------
C
      TERR(J) = TERROR
```

```
      TLAST(J) = ERROR
      KMAXU1 = KMAXU + 1
      TBETA(J) = BETA(KMAXU1)*ERROR
C
C     AFTER EACH OF THE T-EIGENVECTORS IS COMPUTED, THE
C     SIZE OF THE ERROR ESTIMATE, ERROR IS CHECKED.
C     IF THIS ESTIMATE IS NOT AS SMALL AS DESIRED AND
C     |MA(J)| < ML(J), PROGRAM ATTEMPTS TO INCREASE THE SIZE OF |MA(J)|
C     AND REPEAT THE T-EIGENVECTOR COMPUTATIONS.
C
      IF(ERROR.LT.ERTOL.OR.TFLAG.EQ.1)  GO TO 700
C
      IF(ERROR.GE.ERRMIN) GO TO 610
C     LAST COMPONENT IS LESS THAN MINIMAL TO DATE
      ERRMIN = ERROR
      MABEST = MA(J)
  610 CONTINUE
C
      IF(MA(J).GT.0)  ITEST = MA(J) + IDELTA(J)
      IF(MA(J).LT.0)  ITEST = -(IABS(MA(J)) + IDELTA(J))
      IF(IABS(ITEST).LE.ML(J).AND.ICOUNT.LE.10) GO TO 630
C     NEW MA(J) IS GREATER THAN MAXIMUM ALLOWED.
      IF(ERCONT.EQ.0.OR.MABEST.EQ.MPMIN) GO TO 650
      TFLAG = 1
      MA(J) = MABEST
      IF(ILBIS.EQ.0)  MTOL = MTOL - KMAXU
      WRITE(6,620) MA(J)
  620 FORMAT(' 10 ORDERS WERE CONSIDERED.  NONE SATISFIED THE ERROR TEST
     1'/' THEREFORE USE THE BEST ORDER OBTAINED FOR THE EIGENVECTORS'
     1,I6)
      GO TO 530
C
  630 MA(J) = ITEST
C
      MT = IABS(MA(J))
      IF(IWRITE.EQ.1) WRITE(6,640)  MT
  640 FORMAT(/' CHANGE SIZE OF T-MATRIX TO ',I6,' RECOMPUTE T-EIGENVECTO
     1R')
C
      IF(ILBIS.EQ.0)  MTOL = MTOL - KMAXU
C
      GO TO 530
C
C     APPROPRIATE SIZE T-MATRIX WAS NOT OBTAINED
  650 CONTINUE
      WRITE(10,660) J,EVAL,MP(J)
  660 FORMAT(/' ON 10 INCREMENTS NOT ABLE TO IDENTIFY APPROPRIATE SIZE
     1T-MATRIX FOR'/
     1' EIGENVALUE(',I4,') = ',E20.12,' T-MULTIPLICITY =',I4/)
      IF(M2(J).EQ.KMAX) WRITE(10,670)
      IF(M2(J).LT.KMAX) WRITE(10,680)
  670 FORMAT(/' ORDERS TESTED RANGED FROM 5*M1(J)/4 TO APPROXIMATELY
     1 '/'  MIN(11*MEV/8,13*M1(J)/8)'/)
  680 FORMAT(/' ORDERS TESTED RANGED FROM (3*M1(J)+M2(J))/4 TO APPROXIMA
     1TELY'/'  (3*M1(J) + 5*M2(J))/8.'/)
      WRITE(10,690)
  690 FORMAT(' ALLOWING LARGER ORDERS FOR THE T-MATRICES MAY RESULT IN
     1 SUCCESS'/' BUT PROBABLY WILL NOT.  PROBLEM IS PROBABLY DUE TO'
     1 /' LACK OF CONVERGENCE OF GIVEN EIGENVALUE, CHECK THE ERROR ESTIM
     1ATE'/)
      MP(J) = MPMIN
      IF(ILBIS.EQ.0)  MTOL = MTOL - KMAXU
      GO TO 710
  700 NTVEC = NTVEC + 1
C
  710 CONTINUE
      NGOODC = NGOOD
```

```
      GO TO 740
C
C     COME HERE IF THERE IS NOT ENOUGH ROOM FOR ALL OF T-EIGENVECTORS
  720 NGOODC = J-1
      WRITE(6,730)  J, MTOL, MDIMTV
  730 FORMAT(/'  NOT ENOUGH ROOM IN TVEC FOR ',I4,'TH T-VECTOR'/'  T-DIM
     1ENSION REQUESTED = ',I6,' BUT TVEC HAS DIMENSION = ',I6/)
      IF(NGOODC.EQ.0)  GO TO 1430
      MTOL = MTOL-KMAXU
C
  740 CONTINUE
C
C     THE LOOP ON T-EIGENVECTOR COMPUTATIONS IS COMPLETE.
C     WRITE OUT THE SIZE T-MATRICES THAT WILL BE USED FOR
C     THE RITZ VECTOR COMPUTATIONS.
C
      WRITE(10,750)
  750 FORMAT(/' SIZES OF T-MATRICES THAT WILL BE USED IN THE RITZ COMPUT
     1ATIONS'/5X,'J',16X,'GOODEV(J)',1X,'MA(J)')
C
      WRITE(10,760)   (J,GOODEV(J),MA(J), J=1,NGOOD)
  760 FORMAT(I6,E25.14,I6)
      WRITE(10,510)
C
      WRITE(6,770) MTOL
  770 FORMAT(/' THE CUMULATIVE LENGTH OF THE T-EIGENVECTORS IS',I18)
C
      WRITE(6,780) NTVEC,NGOOD
  780 FORMAT(/I6,' T-EIGENVECTORS OUT OF',I6,' REQUESTED WERE COMPUTED')
C
C     SAVE THE T-EIGENVECTORS ON FILE 11?
      IF(TVSTOP.NE.1.AND.SVTVEC.EQ.0) GO TO 840
C
      WRITE(11,790) NTVEC,MTOL,MATNO,SVSEED
  790 FORMAT(I6,3I12,' = NTVEC,MTOL,MATNO,SVSEED')
C
      DO 820 J=1,NGOODC
C     IF MP(J) = MPMIN THEN NO SUITABLE T-EIGENVECTOR IS AVAILABLE
C     FOR THAT EIGENVALUE.
      IF(MP(J).EQ.MPMIN) WRITE(11,800) J,MA(J),GOODEV(J),MP(J)
  800 FORMAT(2I6,E20.12,I6/' TH EIGVAL,T-SIZE,EVALUE,FLAG,NO EIGVEC')
      IF(MP(J).NE.MPMIN) WRITE(11,810) J,MA(J),GOODEV(J),MP(J)
  810 FORMAT(I6,I6,E20.12,I6/' T-EIGVEC,SIZE T,EVALUE OF A,MP(J)')
      IF(MP(J).EQ.MPMIN) GO TO 820
      KI = MINT(J)
      KF = MFIN(J)
C
      WRITE(11,260) (TVEC(K), K=KI,KF)
C
  820 CONTINUE
C
      IF(TVSTOP.NE.1) GO TO 840
C
      WRITE(6,830) TVSTOP, NTVEC,NGOOD
  830 FORMAT(/' USER SET TVSTOP = ',I1/
     1' THEREFORE PROGRAM TERMINATES AFTER T-EIGENVECTOR COMPUTATIONS'/
     1' T-EIGENVECTORS THAT WERE COMPUTED ARE SAVED ON FILE 11'/
     1I8,' T-EIGENVECTORS WERE COMPUTED OUT OF',I7,' REQUESTED'/)
C
      GO TO 1550
C
  840 CONTINUE
C     IF NOT ABLE TO COMPUTE ALL THE REQUESTED T-EIGENVECTORS,
C     CONTINUE WITH THE LANCZOS VECTOR COMPUTATIONS ANYWAY?
      IF(NTVEC.NE.NGOOD.AND.LVCONT.EQ.0) GO TO 1450
C
C     COMPUTE THE MAXIMUM SIZE OF THE T-MATRIX USED FOR THOSE
```

```
C     EIGENVALUES WITH GOOD ERROR ESTIMATES.
C
      KMAXU = 0
      DO 850 J = 1,NGOODC
      MT = IABS(MA(J))
      IF(MT.LT.KMAXU.OR.MP(J).EQ.MPMIN) GO TO 850
      KMAXU = MT
  850 CONTINUE
C
      IF(KMAXU.EQ.0) GO TO 1490
C
      WRITE(6,860) KMAXU
  860 FORMAT(/I6,' = LARGEST SIZE T-MATRIX TO BE USED IN THE RITZ VECTOR
     1 COMPUTATIONS')
C
C     COUNT THE NUMBER OF RITZ VECTORS NOT BEING COMPUTED
      MREJEC = 0
      DO 870 J=1,NGOODC
  870 IF(MP(J).EQ.MPMIN)  MREJEC = MREJEC + 1
      MREJET = MREJEC + (NGOOD-NGOODC)
      IF(MREJET.NE.0) WRITE(6,880) MREJET
  880 FORMAT(/' RITZ VECTORS ARE NOT COMPUTED FOR',I6,' OF THE EIGENVALU
     1ES'/)
      NACT = NGOODC - MREJEC
      WRITE(6,890) NGOOD,NTVEC,NACT
  890 FORMAT(/I6,' RITZ VECTORS WERE REQUESTED'/I6,' T-EIGENVECTORS WERE
     1 COMPUTED'/I6,' RITZ VECTORS WILL BE COMPUTED'/)
C     CHECK IF THERE ARE ANY RITZ VECTORS TO COMPUTE
      IF(MREJEC.EQ.NGOODC) GO TO 1470
C
C     CONTINUE WITH THE LANCZOS VECTOR COMPUTATIONS?
      IF(LVCONT.EQ.0.AND.MREJEC.NE.0) GO TO 1450
C
C     NOW COMPUTE THE RITZ VECTORS.  REGENERATE THE
C     LANCZOS VECTORS.
C
      DO 900 I = 1,NMAX
  900 RITVEC(I) = ZERO
C
C-----------------------------------------------------------------------
C     REGENERATE THE STARTING VECTOR. THIS MUST BE GENERATED AND
C     NORMALIZED PRECISELY THE WAY IT WAS DONE IN THE EIGENVALUE
C     COMPUTATIONS, OTHERWISE THERE WILL BE A MISMATCH BETWEEN
C     THE T-EIGENVECTORS THAT HAVE BEEN COMPUTED FROM THE T-MATRICES
C     READ IN FROM FILE 2 AND THE LANCZOS VECTORS THAT ARE
C     BEING REGENERATED.
C
      IIL = SVSEED
      CALL GENRAN(IIL,G,N)
C
C-----------------------------------------------------------------------
C
      DO 910 J = 1,N
  910 V2(J) = G(J)
C
      SUM = FINPRO(N,V2(1),1,V2(1),1)
      SUM = ONE/DSQRT(SUM)
C
      DO 920 J = 1,N
      V1(J) = ZERO
  920 V2(J) = V2(J)*SUM
C
C     LOOP FOR GENERATING RITZ VECTORS  (IVEC = 1,KMAXU)
      IVEC = 1
      BATA = ZERO
C
```

```
      GO TO 980
C
  930 CONTINUE
C
C     COMPUTE V1 = A*V2 - BATA*V1
C
C-----------------------------------------------------------------------
C
      CALL CMATV(V2,V1,BATA)
C
C-----------------------------------------------------------------------
C
      ALFA = FINPRO(N,V1(1),1,V2(1),1)
C
      DO 940 J = 1,N
  940 V1(J) = V1(J)-ALFA*V2(J)
C
      BATA = FINPRO(N,V1(1),1,V1(1),1)
      BATA = DSQRT(BATA)
      SUM = ONE/BATA
C
      TEMP = BETA(IVEC)
      TEMP = DABS(BATA - TEMP)/TEMP
      IF (TEMP.LT.1.0D-10)GO TO 960
C
C     THE BETA BEING REGENERATED DO NOT MATCH THE BETA IN FILE 2.
C     SOMETHING IS WRONG IN THE LANCZOS VECTOR GENERATION.
C     PROGRAM TERMINATES FOR USER TO CORRECT THE PROBLEM
C     WHICH MUST BE IN THE STARTING VECTOR GENERATION OR IN
C     THE MATRIX-VECTOR MULTIPLY SUBROUTINE CMATV SUPPLIED.
C     THIS SUBROUTINE MUST BE THE SAME ONE USED IN THE
C     EIGENVALUE COMPUTATIONS OR A MISMATCH WILL ENSUE.
C
      WRITE(6,950) IVEC,BATA,BETA(IVEC),TEMP
  950 FORMAT(/2X,'IVEC',16X,'BATA',10X,'BETA(IVEC)',14X,'RELDIF'/I6,
     13E20.12/' IN LANCZOS VECTOR REGENERATION THE ENTRIES OF THE TRIDIA
     1GONAL MATRICES BEING'/' GENERATED ARE NOT THE SAME AS THOSE IN THE
     1 MATRIX SUPPLIED ON FILE 2.'/' THEREFORE SOMETHING IS BEING INITIA
     1LIZED OR COMPUTED DIFFERENTLY FROM THE WAY'/' IT WAS COMPUTED IN T
     1HE EIGENVALUE COMPUTATIONS'/' THE PROGRAM TERMINATES FOR THE USER 
     1TO DETERMINE WHAT THE PROBLEM IS'/)
      GO TO 1550
C
C
  960 CONTINUE
      DO 970 J = 1,N
      TEMP = SUM*V1(J)
      V1(J) = V2(J)
  970 V2(J) = TEMP
C
  980 CONTINUE
C
      LFIN = 0
      DO 1000 J = 1,NGOODC
      LL = LFIN
      LFIN = LFIN + N
C
      IF(IABS(MA(J)).LT.IVEC.OR.MP(J).EQ.MPMIN) GO TO 1000
      II = IVEC + MINT(J) - 1
      TEMP = TVEC(II)
C     II IS THE (IVEC)TH COMPONENT OF THE T-EIGENVECTOR CONTAINED
C     IN TVEC(MINT(J)).
C
      DO 990 K = 1,N
      LL = LL + 1
  990 RITVEC(LL) = TEMP*V2(K) + RITVEC(LL)
C
```

```
 1000 CONTINUE
C
      IVEC = IVEC + 1
      IF (IVEC.LE.KMAXU) GO TO 930
C
C
C     RITZVECTOR GENERATION IS COMPLETE. NORMALIZE EACH RITZVECTOR.
C     NOTE THAT IF CERTAIN RITZ VECTORS WERE NOT COMPUTED THEN THAT
C     PORTION OF THE RITVEC ARRAY WAS NOT UTILIZED.
C
      LFIN = 0
      DO 1050 J = 1,NGOODC
C
      KK = LFIN
      LFIN = LFIN + N
      IF(MP(J).EQ.MPMIN) GO TO 1050
C
      DO 1010 K = 1,N
      KK = KK + 1
 1010 V2(K) = RITVEC(KK)
C
      SUM = FINPRO(N,V2(1),1,V2(1),1)
      SUM = DSQRT(SUM)
      RNORM(J) = SUM
      TEMP = DABS(ONE-SUM)
      SUM = ONE/SUM
C
      KK = LFIN - N
      DO 1020 K = 1,N
      KK = KK + 1
      V2(K) = SUM*V2(K)
 1020 RITVEC(KK) = V2(K)
C
      IF (IWRITE.NE.0) WRITE(6,1030) J,GOODEV(J)
 1030 FORMAT(/I5,' TH EIGENVALUE CONSIDERED = ',E20.12/)
C
      IF (IWRITE.NE.0) WRITE(6,1040) TERR(J),TBETA(J),TEMP
 1040 FORMAT(' NORM OF ERROR IN T-EIGENVECTOR = ',E14.3/
     1 ' BETA(MA(J)+1)*U(MA(J)) = ',E14.3/
     1 ' ABS(NORM(RITVEC) - 1.0)  = ',E14.3/)
C
      LINT = LFIN - N + 1
      EVAL = EVNEW(J)
C
C-----------------------------------------------------------------------
C
      CALL CMATV(RITVEC(LINT),V2,EVAL)
C
C-----------------------------------------------------------------------
C
C     COMPUTE ERROR IN RITZ VECTOR CONSIDERED AS A EIGENVECTOR OF A.
C     V2 = A*RITVEC - EVAL*RITVEC
C
      SUM = FINPRO(N,V2(1),1,V2(1),1)
      SUM = DSQRT(SUM)
      ERR(J) = SUM
      GAP = ABS(AMINGP(J))
      ERRDGP(J) = SUM/GAP
C
 1050 CONTINUE
C
C
C     RITZVECTORS ARE NORMALIZED AND ERROR ESTIMATES ARE IN ERR ARRAY
C     AND IN ERRDGP ARRAY. STORE EVERYTHING
C
C
      WRITE(9,1060)
```

```
 1060 FORMAT(3X,'A-EIGENVALUE',2X,'MA(J)',3X,'A-MINGAP',6X,'AERROR',2X,
     1 'AERROR/GAP',6X,'TERROR')
C
      WRITE(13,1070)
 1070 FORMAT(16X,'GOODEV(J)',5X,'RITZNORM',6X,'AMINGAP',5X,
     1 'TBETA(J)',5X,'TLAST(J)')
C
      DO 1100 J=1,NGOODC
C
      IF(MP(J).EQ.MPMIN) GO TO 1100
C
      WRITE(9,1080)EVNEW(J),MA(J),AMINGP(J),ERR(J),ERRDGP(J),TERR(J)
 1080 FORMAT(E15.8,I6,4E12.4)
C
      WRITE(13,1090) EVNEW(J),RNORM(J),AMINGP(J),TBETA(J),TLAST(J)
 1090 FORMAT(E25.14,4E13.5)
C
 1100 CONTINUE
C
      IF(MREJEC.EQ.0) GO TO 1180
      WRITE(9,1110)
 1110 FORMAT(/' RITZ VECTORS WERE NOT COMPUTED FOR THE FOLLOWING EIGENVA
     1LUES'/' EITHER BECAUSE THEY HAD NOT CONVERGED OR BECAUSE THE ERROR
     1 ESTIMATE'/'  WAS NOT AS SMALL AS DESIRED'/)
C
      DO 1170 J = 1,NGOODC
      IF(MP(J).NE.MPMIN) GO TO 1170
C     WRITE OUT MESSAGE FOR EACH EIGENVALUE FOR WHICH NO EIGENVECTOR
C     WAS COMPUTED.
C
      WRITE(9,1120)
 1120 FORMAT(6X,'GOODEV(J)',3X,'MA(J)',5X,'AMINGP(J)',6X,'TLAST(J)',3X,
     1'MP(J)')
      WRITE(9,1130) GOODEV(J),MA(J),AMINGP(J),TBETA(J),MP(J)
 1130 FORMAT(E15.8,I8,2E14.4,I8)
C
      WRITE(13,1140)
 1140 FORMAT(/' RITZ VECTORS WERE NOT COMPUTED FOR THE FOLLOWING EIGENVA
     1LUES'/' BECAUSE THEY HAD NOT CONVERGED'/)
C
      WRITE(13,1150)
 1150 FORMAT(6X,'GOODEV(J)',3X,'MA(J)',3X,'M1(J)',3X,'M2(J)',3X,'MP(J)'
     1/)
      WRITE(13,1160) GOODEV(J),MA(J),M1(J),M2(J),MP(J)
 1160 FORMAT(E15.8,4I8)
C
 1170 CONTINUE
 1180 CONTINUE
C
      WRITE(9,1190)
 1190 FORMAT(/' ABOVE ARE ERROR ESTIMATES FOR THE A AND T EIGENVECTORS'/
     1 ' ASSOCIATED WITH THE GOODEV LISTED IN COLUMN 1'/
     1 ' AERROR = NORM(A*X - EV*X)  TERROR = NORM(T*Y - EV*Y) '/
     1 ' WHERE T = T(1,MA(J))    X = RITZ VECTOR = V*Y  V = SUCCESSIVE'/
     1 ' LANCZOS VECTORS. AMINGAP = GAP TO NEAREST A-EIGENVALUE'//)
C
      WRITE(13,1200)
 1200 FORMAT(/' ABOVE ARE ERROR ESTIMATES ASSOCIATED WITH THE GOODEV'/
     1 ' RITZNORM = NORM(COMPUTED RITZ VECTOR)'/
     1 ' TBETA(J) = BETA(MA(J)+1)*Y(MA(J)),   T*Y = EVAL*Y'/
     1 ' TLAST(J) = Y(MA(J))'/
     1 ' AMINGAP = GAP TO NEAREST A-EIGENVALUE'/)
C
C     NUMBER OF RITZ VECTORS COMPUTED
      NCOMPU = NGOODC - MREJEC
      WRITE(12,1210) N,NCOMPU,NGOODC,MATNO
```

```
 1210 FORMAT(3I6,I12,' SIZE A, NO.RITZVECS, NO.EVALUES,MATNO')
C
      LFIN = 0
      DO 1270 J = 1,NGOODC
      LINT = LFIN + 1
      LFIN = LFIN + N
C
      IF(MP(J).EQ.MPMIN) GO TO 1250
C     RITZ VECTOR WAS COMPUTED
      WRITE(12,1220) J, GOODEV(J), MP(J)
 1220 FORMAT(I6,4X,E20.12,I6,' J, EIGENVAL, MP(J)')
C
      WRITE(12,1230) ERR(J),ERRDGP(J)
 1230 FORMAT(2E15.5,' = NORM(A*Z-EVAL*Z) AND  NORM(A*Z-EVAL*Z)/MINGAP')
C
      WRITE(12,1240) (RITVEC(LL), LL=LINT,LFIN)
 1240 FORMAT(4E20.12)
      GO TO 1270
C     NO RITZ VECTOR WAS COMPUTED FOR THIS EIGENVALUE
 1250 WRITE(12,1260) J,GOODEV(J),MP(J)
 1260 FORMAT(I6,4X,E20.12,I6,' J,EIGVALUE,NO RITZ VECTOR COMPUTED')
C
 1270 CONTINUE
C
C     DID ANY T-MATRICES INCLUDE OFF-DIAGONAL ENTRIES SMALLER THAN
C     DESIRED, AS SPECIFIED BY BTOL?
C
      IF(IB.GT.0) GO TO 1300
C
      WRITE(6,1280) KMAXU
 1280 FORMAT(/' FOR LARGEST T-MATRIX CONSIDERED',I7,' CHECK THE SIZE OF
     1BETAS')
C
C-----------------------------------------------------------------------
C
      CALL TNORM(ALPHA,BETA,BKMIN,TEMP,KMAXU,IBMT)
C
C-----------------------------------------------------------------------
C
      IF(IBMT.LT.0) WRITE (6,1290)
 1290 FORMAT(/' WARNING THE T-MATRICES FOR ONE OR MORE OF THE EIGENVALUE
     1S CONSIDERED'/' HAD AN OFF-DIAGONAL ENTRY THAT WAS SMALLER THAN TH
     1E BETA TOLERANCE THAT WAS SPECIFIED'/)
 1300 CONTINUE
C
       GO TO 1550
C
 1310 WRITE(6,1320) NGOOD,NMAX,MDIMRV
 1320 FORMAT(/I4,' RITZ VECTORS WERE REQUESTED BUT THE REQUIRED DIMENSIO
     1N',I6/' IS LARGER THAN THE USER-SPECIFIED DIMENSION OF RITVEC',I6
     1/' THEREFORE, THE EIGENVECTOR PROCEDURE TERMINATES FOR THE USER TO
     1 INTERVENE')
C
      GO TO 1550
C
 1330 WRITE(6,1340) NOLD,N,MATOLD,MATNO
 1340 FORMAT(//' PARAMETERS READ FROM FILE 3 DO NOT AGREE WITH THOSE SPE
     1CIFIED BY THE USER'/' N,NOLD,MATOLD,MATNO = ',2I6,2I12/' PROGRAM T
     1ERMINATES FOR USER TO RESOLVE PROBLEM'/)
C
      GO TO 1550
C
 1350 WRITE(6,1360)
 1360 FORMAT(//' PARAMETERS IN THE ALPHA,BETA FILE HEADER DO NOT AGREE W
     1ITH PARAMTERS'/' SPECIFIED BY THE USER.  THEREFORE THE PROGRAM TER
     1MINATES FOR THE USER'/' TO RESOLVE THE PROBLEM'/)
C
```

```
      GO TO 1550
C
 1370 WRITE(6,1380) KMAX,MEV
 1380 FORMAT(/' ALPHA,BETA FILE HEADER GIVES KMAX =',I6/
     1' BUT EIGENVALUES WERE COMPUTED AT MEV = ',I6,' PROGRAM STOPS'/)
C
      GO TO 1550
C
 1390 WRITE(6,1400)
 1400 FORMAT(//' PROGRAM COMPUTED 1ST GUESSES AT T-MATRIX SIZES AND READ
     1 THEM TO'/' FILE 10, THEN TERMINATED AS REQUESTED.')
      GO TO 1550
C
 1410 WRITE(6,1420) MTOL, MDIMTV
 1420 FORMAT(/' PROGRAM TERMINATES BECAUSE THE TVEC DIMENSION ANTICIPATE
     1D',I7/'  IS LARGER THAN THE TVEC DIMENSION',I7,' SPECIFIED BY THE
     1USER.'/'  USER MAY RESET THE TVEC DIMENSION AND RESTART THE PROGRA
     1M')
      GO TO 1550
C
 1430 WRITE(6,1440)
 1440 FORMAT(/' PROGRAM TERMINATES BECAUSE NO SUITABLE T-EIGENVECTORS WE
     1RE IDENTIFIED'/'  FOR ANY OF THE EIGENVALUES SUPPLIED.  PROBLEM CO
     1ULD BE CAUSED'/'  BY TOO SMALL A TVEC DIMENSION OR SIMPLY THAT SUI
     1TABLE T-VECTORS COULD'/'  NOT BE IDENTIFIED.  USER SHOULD CHECK OU
     1TPUT'/)
      GO TO 1550
C
 1450 WRITE(6,1460) LVCONT,NTVEC,NGOOD
 1460 FORMAT(/' LVCONT FLAG =',I2,' AND NUMBER ',I5,' OF T-EIGENVECTORS
     1 COMPUTED N.E.'/' NUMBER',I5,' REQUESTED SO PROGRAM TERMINATES'/)
      GO TO 1550
C
 1470 WRITE(6,1480)
 1480 FORMAT(//' PROGRAM TERMINATES WITHOUT COMPUTING RITZ VECTORS'/
     1' BECAUSE ALL T-EIGENVECTORS WERE REJECTED AS NOT SUITABLE FOR THE
     1 RITZ VECTOR'/' COMPUTATIONS.  PROBABLE CAUSE IS LACK OF CONVERGEN
     1CE OF THE EIGENVALUES SUPPLIED'/)
      GO TO 1550
C
 1490 WRITE(6,1500)
 1500 FORMAT(/' PROGRAM INDICATES THAT IT IS NOT POSSIBLE TO COMPUTE ANY
     1 OF THE'/' REQUESTED EIGENVECTORS. THEREFORE PROGRAM TERMINATES')
      DO 1510 J=1,NGOODC
 1510 WRITE(6,1520)  J,GOODEV(J),MP(J)
 1520 FORMAT(/4X,' J',11X,'GOODEV(J)',4X,'MP(J)'/I6,E20.12,I9)
      GO TO 1550
C
 1530 WRITE(6,1540) MBETA,KMAXN
 1540 FORMAT(/' PROGRAM TERMINATES BECAUSE THE STORAGE ALLOTTED FOR THE
     1BETA ARRAY',I8/' IS NOT SUFFICIENT FOR THE ENLARGED KMAX =',I8,' T
     1HAT THE PROGRAM WANTS'/' USER CAN ENLARGE THE DIMENSIONS OF THE AL
     1PHA AND BETA ARRAYS'/' AND RERUN THE PROGRAM'/)
C
 1550 CONTINUE
C
      STOP
CC-----END OF MAIN PROGRAM FOR LANCZOS EIGENVECTORS---------------------
      END
```

SECTION 2.5 LANCZS AND SAMPLE MATRIX-VECTOR MULTIPLY SUBROUTINES

```
C-----LEMULT-------------------------------------------------------------
C
C     CONTAINS SUBROUTINES  LANCZS, USPEC, AND CMATV
C     TO BE USED WITH THE REAL SYMMETRIC VERSION OF THE LANCZOS
C     EIGENVALUE/EIGENVECTOR PROCEDURES.
C     ALSO CONTAINS SUBROUTINES FOR POISSON TEST MATRICES THAT ALLOW
C     COMPUTATION OF TRUE ERRORS IN COMPUTED EIGENVALUES AND
C     IN CORRESPONDING EIGENVECTORS.
C
C     NONPORTABLE CONSTRUCTIONS:
C     1.  THE ENTRY MECHANISM USED TO PASS THE STORAGE
C         LOCATIONS OF THE USER-SPECIFIED MATRIX FROM THE
C         SUBROUTINE USPEC TO THE MATRIX-VECTOR SUBROUTINE
C         CMATV.
C     2.  IN THE SAMPLE USPEC AND CMATV FOR DIAGONAL TEST MATRICES:
C         FREE FORMAT (8,*) AND THE FORMAT (20A4).
C     3.  IN THE POISSON SUBROUTINES PROVIDED, THE DATA MACHEP
C         DEFINITION AND MANY OF THE INDICES FOR ARRAYS ARE NOT
C         IN A PORTABLE CONSTRUCTION.  THESE PROGRAMS SHOULD BE
C         REMOVED FROM THE LEMULT FILE IF THE USER IS NOT USING THEM.
C
C-----LANCZS-COMPUTE THE LANCZOS TRIDIAGONAL MATRICES-------------------
C
      SUBROUTINE LANCZS(MATVEC,ALPHA,BETA,V1,V2,G,KMAX,MOLD1,N,IIX)
C
C-----------------------------------------------------------------------
      DOUBLE PRECISION  ALPHA(1),BETA(1),V1(1),V2(1),SUM,TEMP,ONE,ZERO
      REAL  G(1)
      DOUBLE PRECISION FINPRO, DSQRT
      EXTERNAL MATVEC
C-----------------------------------------------------------------------
C
      ZERO = 0.D0
      ONE = 1.D0
C
      IF(MOLD1.GT.1)GO TO 30
C
C    ALPHA/BETA GENERATION STARTS AT I = 1
C    MOLD1 = 1 SET V1 = 0. AND V2 = RANDOM UNIT VECTOR
      BETA(1) = ZERO
      IIL=IIX
C
C-----------------------------------------------------------------------
      CALL GENRAN(IIL,G,N)
C-----------------------------------------------------------------------
C
      DO 10 I = 1,N
   10 V2(I) = G(I)
C
C-----------------------------------------------------------------------
      SUM = FINPRO(N,V2(1),1,V2(1),1)
C-----------------------------------------------------------------------
C
      SUM = ONE/DSQRT(SUM)
      DO 20 I = 1,N
      V1(I) = ZERO
   20 V2(I) = V2(I)*SUM
C
C    ALPHA BETA GENERATION LOOP
   30 CONTINUE
C
      DO 60 I=MOLD1,KMAX
      SUM = BETA(I)
C    MATVEC(V2,V1,SUM) CALCULATES  V1 = A*V2 - SUM*V1
```

```
C
C-----------------------------------------------------------------------
      CALL MATVEC(V2,V1,SUM)
C-----------------------------------------------------------------------
C
C-----------------------------------------------------------------------
      SUM = FINPRO(N,V1(1),1,V2(1),1)
C-----------------------------------------------------------------------
C
      ALPHA(I) = SUM
      DO 40 J=1,N
   40 V1(J) = V1(J)-SUM*V2(J)
C
C-----------------------------------------------------------------------
      SUM = FINPRO(N,V1(1),1,V1(1),1)
C-----------------------------------------------------------------------
C
      IN = I+1
      BETA(IN) = DSQRT(SUM)
      SUM = ONE/BETA(IN)
      DO 50 J=1,N
      TEMP = SUM*V1(J)
      V1(J) = V2(J)
   50 V2(J) = TEMP
   60 CONTINUE
C
C     END ALPHA, BETA GENERATION LOOP
C
      RETURN
C-----END OF LANCZS-----------------------------------------------------
      END
C
C-----USPEC (GENERAL SYMMETRIC SPARSE MATRICES)-------------------------
C
C     SUBROUTINE USPEC(N,MATNO)
      SUBROUTINE GUSPEC(N,MATNO)
C
C-----------------------------------------------------------------------
      DOUBLE PRECISION  A(10000),AD(5010)
      INTEGER  IROW(10000),ICOL(5010)
C-----------------------------------------------------------------------
C    USPEC DIMENSIONS AND INITIALIZES THE ARRAYS NEEDED TO DEFINE
C    THE USER-SPECIFIED MATRIX AND THEN PASSES THE STORAGE LOCATIONS
C    OF THESE ARRAYS TO THE MULTIPLY SUBROUTINE CMATV.
C
C    MATRIX IS STORED IN FOLLOWING SPARSE MATRIX FORMAT:
C    N = ORDER OF A-MATRIX,
C    NZS = NUMBER OF NONZERO SUBDIAGONAL ENTRIES,
C    NZL = INDEX OF LAST COLUMN CONTAINING NONZERO SUBDIAGONAL ENTRIES,
C    ICOL(J), J=1,NZL IS THE NUMBER OF NONZERO SUBDIAGONAL ELEMENTS
C             IN COLUMN J.
C    IROW(K), K = 1,NZS IS THE CORRESPONDING ROW INDEX FOR A(K).
C    AD(I), I=1,N CONTAINS DIAGONAL ENTRIES (INCLUDING ANY 0
C          DIAGONAL ENTRIES).
C    A(K), K=1,NZS CONTAINS NONZERO SUBDIAGONAL ENTRIES, BY COLUMN
C    FOR J > NZL THERE ARE NO NONZERO SUBDIAGONAL ELEMENTS IN COLUMN J.
C    ICOL(J) = 0 IS ALLOWED
C
C-----------------------------------------------------------------------
C     ARRAYS THAT DEFINE THE MATRIX ARE READ IN FROM FILE 8
C
      READ(8,10) NZS,NOLD,NZL,MATOLD
   10 FORMAT(I10,2I6,I8)
C
      WRITE(6,20) NZS,NOLD,NZL,MATOLD
```

```
   20 FORMAT(I10,2I6,I8,' = NZS,NOLD,NZL,MATOLD'/)
C
C     TEST OF PARAMETER CORRECTNESS
      ITEMP = (NOLD-N)**2 + (MATNO-MATOLD)**2
C
      IF(ITEMP.EQ.0) GO TO 40
C
      WRITE(6,30)
   30 FORMAT(' PROGRAM TERMINATES BECAUSE EITHER ORDERS OF OR LABELS FOR
     1 MATRIX DISAGREE')
      GO TO 70
C
   40 CONTINUE
C
C     NUMBER OF NONZERO SUBDIAGONAL ENTRIES IN EACH COLUMN IS READ
C     THEN THE CORRESPONDING ROW INDEX FOR EACH SUCH ENTRY IS READ
      READ(8,50) (ICOL(K), K=1,NZL)
      READ(8,50) (IROW(K), K=1,NZS)
   50 FORMAT(13I6)
C
C     DIAGONAL IS READ FIRST, THEN NONZERO BELOW DIAGONAL ENTRIES
      READ(8,60) (AD(K), K=1,N)
      READ(8,60) (A(K), K=1,NZS)
   60 FORMAT(4E19.10)
C
C-----------------------------------------------------------------------
C     PASS STORAGE LOCATIONS OF ARRAYS THAT DEFINE THE MATRIX TO
C     THE MATRIX-VECTOR MULTIPLY SUBROUTINE CMATV
C
      CALL CMATVE(A,AD,ICOL,IROW,N,NZL)
C-----------------------------------------------------------------------
C
      RETURN
   70 STOP
C-----END OF USPEC------------------------------------------------------
      END
C
C-----MATRIX-VECTOR MULTIPLY FOR REAL SPARSE SYMMETRIC MATRICES---------
C
C     SUBROUTINE CMATV(W,U,SUM)
      SUBROUTINE GCMATV(W,U,SUM)
C
C-----------------------------------------------------------------------
      DOUBLE PRECISION   U(1),W(1),A(1),AD(1),SUM
      INTEGER  IROW(1),ICOL(1)
C-----------------------------------------------------------------------
C    SPARSE MATRIX-VECTOR MULTIPLY FOR LANCZS  U = A*W - SUM*U
C    SEE USPEC SUBROUTINE FOR DESCRIPTION OF THE ARRAYS THAT DEFINE
C    THE MATRIX
C-----------------------------------------------------------------------
C
C     COMPUTE THE DIAGONAL TERMS
      DO 10 I = 1,N
   10 U(I) = AD(I)*W(I)-SUM*U(I)
C
C     COMPUTE BY COLUMN
      LLAST = 0
      DO 30 J = 1,NZL
C
      IF (ICOL(J).EQ.0) GO TO 30
      LFIRST = LLAST + 1
      LLAST = LLAST + ICOL(J)
C
      DO 20 L = LFIRST,LLAST
      I = IROW(L)
C
      U(I) = U(I) + A(L)*W(J)
```

```
      U(J) = U(J) + A(L)*W(I)
C
   20 CONTINUE
C
   30 CONTINUE
C
      RETURN
C
C-----------------------------------------------------------------------
C     STORAGE LOCATIONS OF ARRAYS ARE PASSED TO CMATV FROM USPEC
C
      ENTRY CMATVE(A,AD,ICOL,IROW,N,NZL)
C-----------------------------------------------------------------------
C
      RETURN
C-----END OF CMATV------------------------------------------------------
      END
C
C-----MATRIX-VECTOR MULTIPLY FOR DIAGONAL TEST MATRICES-----------------
C
      SUBROUTINE CMATV(W,U,SUM)
C     SUBROUTINE DCMATV(W,U,SUM)
C
C     CMATV COMPUTES  U = (DIAGONAL MATRIX) * W - SUM * U
C-----------------------------------------------------------------------
      DOUBLE PRECISION  W(1),U(1),SUM
      DOUBLE PRECISION  D(1)
C-----------------------------------------------------------------------
C
      DO 10 I=1,N
   10 U(I)= D(I)*W(I) - SUM*U(I)
      RETURN
C
C-----------------------------------------------------------------------
      ENTRY MVDIAE(D,N)
C-----------------------------------------------------------------------
C
      RETURN
C-----END OF DIAGONAL TEST MATRIX  MULTIPLY-----------------------------
      END
C
C
C-----START OF USPEC FOR DIAGONAL TEST MATRIX---------------------------
C
      SUBROUTINE USPEC(N,MATNO)
C     SUBROUTINE DUSPEC(N,MATNO)
C
C-----------------------------------------------------------------------
      DOUBLE PRECISION  D(1000), SHIFT, SPACE
      DOUBLE PRECISION  DABS, DFLOAT
      REAL  EXPLAN(20)
C-----------------------------------------------------------------------
C
      READ(8,10) EXPLAN
   10 FORMAT(20A4)
      READ(8,*) NOLD,NUNIF,SPACE,D(1),SHIFT
      NNUNIF = NOLD - NUNIF
      WRITE(6,20) NOLD,SPACE,NNUNIF,D(1),SHIFT
   20 FORMAT(/' DIAGONAL TEST MATRIX, SIZE = ',I4/' MOST ENTRIES ARE ',
     1E10.3,' UNITS APART.',I3,' ENTRIES'/' ARE IRREGULARLY SPACED. FIRS
     1T ENTRY IS ',E10.3,' SHIFT = ',E10.3/)
C
      IF(N.NE.NOLD) GO TO 90
C     COMPUTE THE UNIFORM PORTION OF THE SPECTRUM
      DO 30 J=2,NUNIF
   30 D(J) = D(1) - DFLOAT(J-1)*SPACE
      NUNIF1=NUNIF + 1
```

```
      READ(8,10)  EXPLAN
      DO 40 J=NUNIF1,N
   40 READ(8,*) D(J)
      NB = NUNIF - 2
C
      IF(SHIFT.EQ.0.) GO TO 60
      DO 50 J=1,N
   50 D(J) = D(J) + SHIFT
C
C     PRINT OUT THE EIGENVALUES OF INTEREST
   60 WRITE(6,70) (D(I), I=1,10 )
      WRITE(6,80) (D(I), I = NB,N)
   70 FORMAT(/' REAL SYMMETRIC LANCZOS TEST, 1ST 10 ENTRIES OF DIAGONAL
     1TEST MATRIX = '/(3E22.14))
   80 FORMAT(/' MIDDLE UNIFORM PORTION OF MATRIX IS NOT PRINTED OUT'/
     1' END OF UNIFORM PLUS NONUNIFORM SECTION = '/(3E25.16))
C
C     DIAGONAL GENERATION COMPLETE
C
C-----------------------------------------------------------------------
C    CALL ENTRY TO MATRIX-VECTOR MULTIPLY SUBROUTINE  TO PASS
C    STORAGE LOCATION OF D-ARRAY AND ORDER OF A-MATRIX.
       CALL MVDIAE(D,N)
C-----------------------------------------------------------------------
C
      RETURN
   90 WRITE(6,100) NOLD,N
  100 FORMAT(' PROGRAM TERMINATES BECAUSE NOLD = ',I5,'DOES NOT EQUAL N
     1 =',I5)
C-----END OF USPEC SUBROUTINE FOR DIAGONAL TEST MATRICES----------------
      STOP
      END
C
C-----POISSON TEST MATRICES---------------------------------------------
C
C     CONTAINS SUBROUTINES USPEC, CMATV, EXEVG, EXERR AND  EXVEC
C
C-----START OF USPEC----------------------------------------------------
C
C     SUBROUTINE USPEC(N,MATNO)
      SUBROUTINE PUSPEC(N,MATNO)
C
C-----------------------------------------------------------------------
      DOUBLE PRECISION  C0,C1,C2,HALF,ONE
      REAL  EXPLAN(20)
C---------------------------------------------------------------------
      HALF = 0.5D0
      ONE  = 1.0D0
C
C     READ USER-SPECIFIED PARAMETERS FROM INPUT FILE 8 (FREE FORMAT)
C
      READ(8,10) EXPLAN
      WRITE(6,10) EXPLAN
   10 FORMAT(20A4)
C
      READ(8,10) EXPLAN
      READ(8,*) KX,KY,C0
      N = KX*KY
      C1 = HALF-C0
      C2 = ONE
C
      WRITE(6,20) N,KX,KY,C2,C0,C1
   20 FORMAT(/5X,'N',4X,'KX',4X,'KY',7X,'DIAGONAL',3X,'X-CODIAGONAL',
     1 3X,'Y-CODIAGONAL'/3I6,3E15.8/)
C
C-----------------------------------------------------------------------
      CALL PMATVE(C0,C1,C2,KX,KY)
```

```
      CALL EXEVE(C0,C1,C2,KX,KY)
      CALL EXERRP(C0,C1,C2,KX,KY)
      CALL EXVECP(C0,C1,C2,KX,KY)
C-----------------------------------------------------------------------
C
      RETURN
C-----END OF USPEC------------------------------------------------------
      END
C
C-----START OF CMATV----------------------------------------------------
C     CALCULATE U = A*W - SUM*U FOR REAL POISSON MATRICES
C
C     SUBROUTINE CMATV(W,U,SUM)
      SUBROUTINE PMATV(W,U,SUM)
C
C-----------------------------------------------------------------------
      DOUBLE PRECISION  U(1),W(1)
      DOUBLE PRECISION  C0,C1,C2,CC0,CC1,SUM
C-----------------------------------------------------------------------
C
      N = KX*KY
      KX1 = KX-1
      KY1 = KY-1
C
      KK = 1
      U(KK) = (C2*W(KK)+C0*W(KK+1)+C1*W(KK+KX)) - SUM*U(KK)
      KK = KX
      U(KK) = (C2*W(KK)+C0*W(KK-1)+C1*W(KK+KX)) - SUM*U(KK)
      KK = N - KX + 1
      U(KK) = (C2*W(KK)+C0*W(KK+1)+C1*W(KK-KX)) - SUM*U(KK)
      KK = N
      U(KK) = (C2*W(KK)+C0*W(KK-1)+C1*W(KK-KX)) - SUM*U(KK)
C
      DO 10 J = 2,KX1
      KK = J
      U(KK) = (C2*W(KK)+C0*W(KK-1)+C0*W(KK+1)+C1*W(KK+KX)) - SUM*U(KK)
      KK = J+N-KX
      U(KK) = (C2*W(KK)+C0*W(KK-1)+C0*W(KK+1)+C1*W(KK-KX))-SUM*U(KK)
   10 CONTINUE
C
      DO 30 J = 2,KY1
      KK = (J-1)*KX + 1
      U(KK) = (C2*W(KK)+C0*W(KK+1)+C1*W(KK-KX)+C1*W(KK+KX)) - SUM*U(KK)
      DO 20 I = 2,KX1
      KK = KK + 1
      U(KK) = (C2*W(KK)+C0*W(KK-1)+C0*W(KK+1)+C1*W(KK-KX)
     1 +C1*W(KK+KX)) - SUM*U(KK)
   20 CONTINUE
      KK = KK + 1
      U(KK) = (C2*W(KK)+C0*W(KK-1)+C1*W(KK-KX)+C1*W(KK+KX)) - SUM*U(KK)
   30 CONTINUE
C
      RETURN
C
C-----------------------------------------------------------------------
      ENTRY  PMATVE(CC0,CC1,C2,KX,KY)
C-----------------------------------------------------------------------
C
      C0 = -CC0
      C1 = -CC1
C
      RETURN
C-----END OF CMATV-------------------------- ----------------------------
      END
C
C-----START OF EXEVG----------------------------------------------------
C
```

```
C     COMPUTES TRUE EIGENVALUES OF POISSON MATRIX, GAPS BETWEEN
C     TRUE EIGENVALUES, AND MULTIPLICITIES OF TRUE EIGENVALUES
C     AND STORE THESE VALUES, RESPECTIVELY, IN U, G, AND MP.
C     THESE QUANTITIES ARE WRITTEN OUT TO FILE 9
C
      SUBROUTINE EXEVG(U,G,MP)
C
C-----------------------------------------------------------------------
      DOUBLE PRECISION  U(1)
      DOUBLE PRECISION MACHEP,EPSM,C0,C1,C2,T0,T1,PIK,PIL,ONE,TWO
      DOUBLE PRECISION ATOLN,EE
      REAL G(1)
      INTEGER MP(1)
C-----------------------------------------------------------------------
      DATA MACHEP/Z3410000000000000/
      EPSM = 2.0D0*MACHEP
C-----------------------------------------------------------------------
      N = KX*KY
      ONE  = 1.0D0
      TWO  = 2.0D0
      T0 = DARCOS(-ONE)
      T1 = DFLOAT(KX+1)
      PIK = T0/T1
      T1 = DFLOAT(KY+1)
      PIL = T0/T1
C     GENERATE TRUE EIGENVALUES
      KP = 0
      DO 20 J = 1,KY
      T1 = PIL*DFLOAT(J)
      T0 = C2 - TWO*C1*DCOS(T1)
      DO 10 I = 1,KX
      KP = KP+1
      T1 = PIK*DFLOAT(I)
   10 U(KP) = T0 - TWO*C0*DCOS(T1)
   20 CONTINUE
C
C     ORDER U VECTOR BY INCREASING ALGEBRAIC SIZE
      DO 40 K = 2,N
      KM1 = K-1
      DO 30 L = 1,KM1
      JJ = K-L
      IF (U(JJ+1).GE.U(JJ)) GO TO 40
      T0 = U(JJ)
      U(JJ) = U(JJ+1)
   30 U(JJ+1) = T0
   40 CONTINUE
      ATOLN = DMAX1(DABS(U(1)),DABS(U(N)))*EPSM
C
      WRITE(9,50)
   50 FORMAT(' TRUE EIGENVALUES FOR POISSON'/)
C
      WRITE(9,60)N,KX,KY,C2,C0,C1,ATOLN
      WRITE(6,60) N,KX,KY,C2,C0,C1,ATOLN
   60 FORMAT(1X,'A-SIZE',2X,'X-DIM',2X,'Y-DIM'/3I7/
     1 5X,'A-DIAGONAL',3X,'X-CODIAGONAL',3X,'Y-CODIAGONAL',10X,'ATOLN'/
     2 4E15.8)
C
C     DETERMINE MULTIPLICITIES FOR TRUE EIGENVALUES
      I = 1
      IDEX = 1
      J = 1
      NEXACT = 0
   70 J = J+1
      IF (J.GT.N) GO TO 80
      EE = DABS(U(J)-U(I))
      IF (EE.GT.ATOLN) GO TO 80
      IDEX = IDEX+1
```

```
      GO TO 70
   80 NEXACT = NEXACT+1
      U(NEXACT) = U(I)
      MP(NEXACT) = IDEX
C     MP(K) = MULTIPLICITY OF KTH EIGENVALUE CLUSTER FOR A
      IDEX = 1
      I = J
      IF (I.GT.N) GO TO 90
      GO TO 70
   90 CONTINUE
C
C     MULTIPLICITIES HAVE BEEN DETERMINED
C     NEXACT = NUMBER OF DISTINCT A-EIGENVALUES
C
      WRITE(9,100)NEXACT
      WRITE(6,100)NEXACT
  100 FORMAT(I6,' = NUMBER OF TRUE A-EIGENVALUES WHICH ARE DISTINCT'/)
C
C     MINGAP CALCULATION FOR DISTINCT A-EIGENVALUES
      NM1 = NEXACT - 1
      G(NEXACT) = U(NM1)-U(NEXACT)
      G(1) = U(2)-U(1)
C
      DO 110 J = 2,NM1
      T0 = U(J)-U(J-1)
      T1 = U(J+1)-U(J)
      G(J) = T1
      IF (T0.LT.T1) G(J) = -T0
  110 CONTINUE
C
C     NEXACT DISTINCT A-EIGENVALUES ARE IN U IN ASCENDING ORDER
C     MP = MULTIPLICITIES OF THE DISTINCT EIGENVALUES OF A
C     G = TRUE MINIMUM GAP IN A FOR EACH OF THESE EIGENVALUES
C     G < 0 INDICATES THE LEFT-HAND GAP WAS MINIMAL.
C     OUTPUT MULTIPLICITIES, DISTINCT EVS, AND MINGAPS TO FILE 9
C
      WRITE(9,120)
  120 FORMAT(5X,'I',1X,'AMULT',5X,'TRUE A-EIGENVALUE(I)',
     1 3X,'A-MINGAP(I)')
C
      WRITE(9,130)(J,MP(J),U(J),G(J), J=1,NEXACT)
  130 FORMAT(2I6,E25.16,E14.3)
C
      WRITE(9,140)
  140 FORMAT(' NEXACT DISTINCT A-EIGENVALUES ARE IN ASCENDING ORDER'/
     1 ' AMULT = MULTIPLICITIES OF THE DISTINCT EIGENVALUES OF A.'/
     2 ' A-MINGAP(I) = TRUE MINIMUM GAP IN A FOR EACH EIGENVALUE.'/
     3 ' A-MINGAP(I) LT 0 INDICATES THE LEFT-HAND GAP WAS MINIMAL.'//)
C
C     WE ORDER U VECTOR BY INCREASING SIZE OF THE GAPS
C
      DO 150 K = 1,NEXACT
  150 MP(K) = K
C
      DO 170 K = 2,NEXACT
      KM1 = K-1
C
      DO 160 L = 1,KM1
      JJ = K - L
      IF (ABS(G(JJ+1)).GE.ABS(G(JJ))) GO TO 170
      EE = U(JJ)
      U(JJ) = U(JJ+1)
      U(JJ+1) = EE
      GG = G(JJ)
      G(JJ) = G(JJ+1)
      G(JJ+1) = GG
      IEE = MP(JJ)
```

```
      MP(JJ) = MP(JJ+1)
  160 MP(JJ+1) = IEE
C
  170 CONTINUE
C
      WRITE(9,180)
  180 FORMAT(5X,'K',6X,'A-MINGAP',5X,'TRUE A-EIGENVALUE(I)',2X,'A-EVNO')
C
      WRITE(9,190)(J,G(J),U(J),MP(J), J=1,NEXACT)
  190 FORMAT(I6,E14.3,E25.16,I8)
C
      WRITE(9,200)
  200 FORMAT(' NEXACT DISTINCT A-EIGENVALUES. GAPS IN ASCENDING ORDER'/
     2 ' A-MINGAP(I) = TRUE MINIMUM GAP IN A FOR EACH EIGENVALUE.'/
     3 ' A-MINGAP(I) LT 0 INDICATES THE LEFT-HAND GAP WAS MINIMAL.'/
     3 ' A-MATRIX IS BLOCK TRIDIAGONAL AND EACH DIAGONAL BLOCK IS OF ORD
     3ER NX.'/
     4 ' NX = NUMBER OF POINTS ON EACH X-LINE. THERE ARE NY DIAGONAL BLO
     4CKS.'/
     5 ' NY = NUMBER OF POINTS ON EACH Y-LINE.'/
     5 ' A-DIAGONAL    = A(K,K)'/
     6 ' X-CODIAGONAL = A(I,I+1)'/
     7 ' Y-CODIAGONAL = A(I,I+NX)'/
     8 ' ----- END OF FILE 9 TRUEEV----------------------------'//)
C
      RETURN
C
C-----------------------------------------------------------------------
      ENTRY EXEVE(C0,C1,C2,KX,KY)
C-----------------------------------------------------------------------
C
      RETURN
C-----END OF EXEVG------------------------------------------------------
      END
C
C-----START OF EXERR----------------------------------------------------
C
C     FOR GIVEN COMPUTED EIGENVALUES, V(I), I=1,2,...,NG
C     COMPUTES THE CLOSEST TRUE EIGENVALUES AND THE ERROR IN THE
C     COMPUTED EIGENVALUES, AND STORES THESE RESPECTIVELY
C     IN U(I) AND IN G(MEV+I).  THESE QUANTITIES ARE WRITTEN
C     TO FILE 10.
C
      SUBROUTINE EXERR(V,U,G,MP,MEV,NG,NEXACT,IWRITE)
C
C-----------------------------------------------------------------------
      DOUBLE PRECISION U(1),V(1)
      DOUBLE PRECISION EV,EE,T0,T1,C0,C1,C2,PIK,PIL
      DOUBLE PRECISION ATOLN,EPSM,MACHEP,ZERO,ONE,TWO
      REAL G(1)
      INTEGER MP(1)
C-----------------------------------------------------------------------
      DATA MACHEP/Z3410000000000000/
      EPSM = 2.0D0*MACHEP
C-----------------------------------------------------------------------
C     ON ENTRY V CONTAINS NG GOOD EIGENVALUES OF T(1,MEV)
C     MP CONTAINS THE MULTIPLICITIES OF THESE EIGENVALUES.
C     U(I) = GAP TO NEAREST DISTINCT TMEV I=1,NG
C     U < 0. MEANS GAP IS DUE TO SPURIOUS EV
C
C     ON EXIT G(MEV+I) = ERROR FOR V(I) < 0. IF MULT EV > 1
C     K = MP(I) MEANS |V(I) - U(K)| = MIN
C     MP < 0 MEANS MORE THAN ONE I USES SAME K
C
C     T0 = C2 - 2*C1*COS(PIL*J)
C     U(KP) = T0 - 2*C0*COS(PIK*I)
C     KP = (J-1)*KX + I
```

```
C     C2 = ONE
C     C0 = X-CODIAGONAL = INPUT
C     C1 = Y-CODIAGONAL = HALF - C0
C-----------------------------------------------------------------------
      N = KX*KY
      ZERO = 0.0D0
      ONE  = 1.0D0
      TWO  = 2.0D0
C
C     SET G(I) = GAP FROM GOOD T(MEV) TO NEAREST DISTINCT TMEV I=1,NG
      DO 10 I = 1,NG
      G(I) = U(I)
   10 CONTINUE
C
C     REGENERATE A-EIGENVALUES
      T0 = DARCOS(-ONE)
      T1 = DFLOAT(KX+1)
      PIK = T0/T1
      T1 = DFLOAT(KY+1)
      PIL = T0/T1
      KP = 0
C
      DO 30 J = 1,KY
      T1 = PIL*DFLOAT(J)
      T0 = C2 - TWO*C1*DCOS(T1)
      DO 20 I = 1,KX
      KP = KP+1
      T1 = PIK*DFLOAT(I)
   20  U(KP) = T0 - TWO*C0*DCOS(T1)
   30 CONTINUE
C
C     ORDER U VECTOR BY INCREASING ALGEBRAIC SIZE
      DO 50 K = 2,N
      KM1 = K-1
      DO 40 L = 1,KM1
      JJ = K-L
      IF (U(JJ+1).GE.U(JJ)) GO TO 50
      T0 = U(JJ)
      U(JJ) = U(JJ+1)
   40 U(JJ+1) = T0
   50 CONTINUE
C
      ATOLN = DMAX1(DABS(U(1)),DABS(U(N)))*EPSM
C
C     DETERMINE MULTIPLICITIES FOR TRUE EIGENVALUES
      I = 1
      J = 1
      NEXACT = 0
   60 J = J+1
      IF (J.GT.N) GO TO 70
      EE = DABS(U(J)-U(I))
      IF (EE.GT.ATOLN) GO TO 70
      IDEX = IDEX+1
      GO TO 60
   70 NEXACT = NEXACT+1
      U(NEXACT) = U(I)
      I = J
      IF (I.GT.N) GO TO 80
      GO TO 60
   80 CONTINUE
C
C     NEXACT = NUMBER OF DISTINCT A-EIGENVALUES
C     U CONTAINS TRUE DISTINCT A-EV ORDERED BY INCREASING SIZE
C
      IF ( IWRITE.EQ.1) WRITE(6,90)MEV,NG,NEXACT
   90 FORMAT(/3I6,' = MEV, NG, NEXACT, POISZ CASE'/
```

```
     1 ' TRUE ERRORS FOR GOOD EIGENVALUES'/)                            LEM06650
C                                                                       LEM06660
C      WRITE(6,61) (K,U(K), K=1,NEXACT)                                 LEM06670
C  61 FORMAT(4(I5,E15.8))                                               LEM06680
C                                                                       LEM06690
C     CALCULATION OF THE TRUE ERRORS.                                   LEM06700
      KL = 1                                                            LEM06710
      DO 110 ITEV = 1,NG                                                LEM06720
      EV = V(ITEV)                                                      LEM06730
      K = KL                                                            LEM06740
      T1 = DABS(EV - U(KL))                                             LEM06750
C                                                                       LEM06760
      DO 100 KP = KL,NEXACT                                             LEM06770
      T0 = DABS(EV - U(KP))                                             LEM06780
      IF (T0.GE.T1) GO TO 100                                           LEM06790
      K = KP                                                            LEM06800
      T1 = T0                                                           LEM06810
  100 CONTINUE                                                          LEM06820
C                                                                       LEM06830
      IF (K.EQ.KL.AND.ITEV.GT.1) T1 = -T1                               LEM06840
      KL = K                                                            LEM06850
      MP(ITEV) = K                                                      LEM06860
      G(MEV+ITEV) = T1                                                  LEM06870
  110 CONTINUE                                                          LEM06880
C                                                                       LEM06890
C     TRUE ERRORS HAVE BEEN COMPUTED OUTPUT THEM TO FILE 10             LEM06900
C     FORM HEADER FOR ERREXACT FILE 10                                  LEM06910
      WRITE(10,120)N,KX,KY,C2,C0,C1                                     LEM06920
  120 FORMAT(' POISSONZ TRUE ERROR FOR GOOD EIGENVALUES'/               LEM06930
     1 5X,'N',4X,'NX',4X,'NY'/3I6//                                     LEM06940
     2 5X,'A-DIAGONAL',3X,'X-CODIAGONAL',3X,'Y-CODIAGONAL'/3E15.8//)    LEM06950
C                                                                       LEM06960
      WRITE(10,130)MEV,NG,NEXACT                                        LEM06970
  130 FORMAT(/3I6,' = MEV,NG,NEXACT'/1X,'T-EV NO',1X,'A-EV NO',         LEM06980
     1 10X,'GOOD EIGENVALUE',5X,'TRUEERROR',7X,'TMINGAP')               LEM06990
C                                                                       LEM07000
      WRITE(10,140)(I,MP(I),V(I),G(MEV+I),G(I), I=1,NG)                 LEM07010
  140 FORMAT(2I8,E25.16,2E14.3)                                         LEM07020
C                                                                       LEM07030
      WRITE(10,150)                                                     LEM07040
  150 FORMAT(' ABOVE ARE THE TRUE ERRORS FOR POISSON GOODEV'/           LEM07050
     1 ' IF A-EV NO LT 0 THEN GOODEV HAS MULTIPLICITY GT 1'/            LEM07060
     1 ' IF TRUE ERROR LT 0 THEN MORE THAN ONE GOODEV APPROXIMATES'/    LEM07070
     1 ' THE SAME TRUE POISSON EIGENVALUE'/                             LEM07080
     1 ' IF TMINGAP LT 0 THE MINGAP IS DUE TO SPURIOUS EIGENVALUE'//)   LEM07090
C                                                                       LEM07100
      RETURN                                                            LEM07110
C                                                                       LEM07120
C-----------------------------------------------------------------------LEM07130
      ENTRY EXERRP(C0,C1,C2,KX,KY)                                      LEM07140
C-----------------------------------------------------------------------LEM07150
C                                                                       LEM07160
C-----END OF EXERR------------------------------------------------------LEM07170
      RETURN                                                            LEM07180
      END                                                               LEM07190
C                                                                       LEM07200
C-----START OF EXVEC----------------------------------------------------LEM07210
C                                                                       LEM07220
C     (JVEC = 1): FOR A GIVEN RITZ VECTOR V AND EIGENVALUE X1, COMPUTES LEM07230
C     THE CLOSEST EIGENVALUE Y1 AND CORRESPONDING TRUE EIGENVECTOR U,   LEM07240
C     AND THEN CALCULATES THE NORM OF THE DIFFERENCE BETWEEN            LEM07250
C     V AND U AND THE MAXIMAL DIFFERENCE BETWEEN THE COMPONENTS.        LEM07260
C     THESE QUANTITIES ARE WRITTEN TO FILE 6.                           LEM07270
C                                                                       LEM07280
C     (JVEC = 2):  COMPUTES THE PROJECTION OF EACH                      LEM07290
C     OF THE TRUE EIGENVECTORS ON THE LANCZOS STARTING VECTOR           LEM07300
```

```
C     USED BY THE LANCZS SUBROUTINE AND WRITES THEM TO FILE 12.
C
      SUBROUTINE EXVEC(U,V,X1,Y1,G,MP,IIX,JVEC,ICOUNT)
C
C-----------------------------------------------------------------------
      DOUBLE PRECISION U(1),V(1)
      DOUBLE PRECISION WI(110),WJ(110),WII(110)
      DOUBLE PRECISION X1,Y1,EV,EE,WS,PIK,PIL,SUM,PROJ,TEMP,S
      DOUBLE PRECISION ATOLN,EPSM,MACHEP,ZERO,HALF,ONE,TWO
      DOUBLE PRECISION C0,C1,C2,T0,T1,T2
      REAL G(1),GG
      INTEGER MP(1)
      DOUBLE PRECISION FINPRO
C-----------------------------------------------------------------------
C     THIS PROGRAM CALCULATES THE TRUE EIGENVALUES AND EIGENVECTORS
C     OF THE POISSON MATRIX A OF ORDER  N = KX*KY
C     A CONSISTS OF KY TRIDIAGONAL BLOCKS OF ORDER KX
C     KX = X-DIMENSION     KY = Y-DIMENSION.
C
C     IIX = SEED FOR RANDOM NUMBER GENERATOR USED TO CALCULATE
C           STARTING LANCZOS VECTOR IN LANCZS
C     V = RANDOM UNIT STARTING VECTOR FOR LANCZS
C     A*U = EV*U    ||U|| = ONE
C
C     C2 = DIAGONAL OF KX BY KX MATRIX
C    -C0 = C0-DIAGONAL OF THE KX BY KX MATRIX.
C    -C1 = Y-CODIAGONAL.
C
C     NOTE THAT THE VECTORS WI,WJ,WII ARE DIMENSIONED INTERNALLY
C     THEY ARE USED JUST TO KEEP FROM REGENERATING INFORMATION.
C     WI,WII = REAL*8 ARRAYS OF DIMENSION AT LEAST KX
C     WJ     = REAL*8 ARRAY  OF DIMENSION AT LEAST KY.
C
C     NOTATION USED IN PROGRAM
C
C     PIK = ARCOS(-1)/(KX+1)     PIL = ARCOS(-1)/(KY+1)
C     WI(I) = PIK*I           WJ(J) = PIL*J
C
C     U(K) IS A-EV ORDERED BY INCREASING SIZE, K = 1,N
C     LATER U IS USED TO STORE THE TRUE EIGENVECTOR
C     T0 = C2 - 2*C1*COS(PIL*J)     EV(I,J) = T0 - 2*C0*COS(PIK*I)
C     I = 1,KX     J = 1,KY     KP = (J-1)*KX + I
C
C     U(KV) = SIN(PIK*I*IK)*SIN(PIL*J*JK)
C     IK = 1,KX     JK = 1,KY     KV = (JK-1)*KX + IK
C     U IS UNSCALED EIGENVECTOR FOR EV(I,J) = Y1
C     WS = 1/||U||: ||U|| = .5*DSQRT(T2*T3)  T2 = KX+1  T3 = KY+1
C
C    JVEC  = (1,2) FLAGS COMPUTATIONS TO BE PERFORMED.
C
C           = (1) MEANS GIVEN X1 FIND Y1 AND KVEC SUCH THAT
C                       Y1 = EV(KVEC) AND |X1-Y1| = MIN
C                 ALSO GIVEN UNIT RITZ VECTOR ASSOCIATED WITH X1
C                      CALCULATE UNIT EIGENVECTOR U, A*U = Y1*U
C                      T2 = ||V-U||  T1 = MAX(|V(K)-U(K)|, K= 1,N)
C                      MAX OCCURS FIRST AT K = KK
C
C           = (2) MEANS CALCULATION OF THE PROJECTION OF THE STARTING
C                      LANCZOS VECTOR ON EACH EIGENVECTOR OF A.
C
C-----------------------------------------------------------------------
      DATA MACHEP/Z3410000000000000/
      EPSM = 2.0D0*MACHEP
C-----------------------------------------------------------------------
C     SPECIFY PARAMETERS
      N = KX*KY
      ZERO = 0.0D0
```

```
      HALF = 0.5D0
      ONE  = 1.0D0
      TWO  = 2.0D0
      T0 = DARCOS(-ONE)
      T1 = DFLOAT(KX+1)
      PIK = T0/T1
      T2 = DFLOAT(KY+1)
      PIL = T0/T2
      WS = TWO/DSQRT(T1*T2)
C
C     GENERATE WI WJ VECTORS
      KP = 0
      DO 20 J = 1,KY
      T1 = PIL*DFLOAT(J)
      WJ(J) = T1
      T0 = C2 - TWO*C1*DCOS(T1)
      DO 10 I = 1,KX
      KP = KP+1
      T1 = PIK*DFLOAT(I)
      WI(I) = T1
   10 U(KP) = T0 - TWO*C0*DCOS(T1)
   20 CONTINUE
C     U(KP) = EV(I,J) = C2 - 2*C1*COS(PIL*J) - 2*C0*COS(PIK*I)
C
C     INITIALIZE MP VECTOR
      DO 30 K = 1,N
   30 MP(K) = K
C
C     WE ORDER U VECTOR BY INCREASING SIZE OF THE EVS
      DO 50 K = 2,N
      KM1 = K-1
C
      DO 40 L = 1,KM1
      JJ = K - L
      IF (U(JJ+1).GE.U(JJ)) GO TO 50
      EE = U(JJ)
      U(JJ) = U(JJ+1)
      U(JJ+1) = EE
      IEE = MP(JJ)
      MP(JJ) = MP(JJ+1)
   40 MP(JJ+1) = IEE
C
   50 CONTINUE
C
      ATOLN = DMAX1(DABS(U(1)),DABS(U(N)))*EPSM
C
      IF (ICOUNT.EQ.1) WRITE(6,60) N,KX,KY,JVEC,C2,C0,C1,ATOLN
   60 FORMAT(/' TRUE ERRORS FOR CONVERGED GOODEV'/
     1 4I6,' = N KX KY JVEC'//
     1 4E12.5,' = C2 C0 C1 ATOLN'//)
C
C     KP = MP(K) MEANS EIGENVALUE U(K) CORRESPONDS TO EIGENVECTOR W(KP)
C     COMPUTE TOLERANCE USED IN COMPUTING TRUE MULTIPLICITIES
C
      IF (JVEC.EQ.1) GO TO 180
C
C     JVEC = 2 SO CALCULATE PROJECTIONS AND WRITE IN FILE 12
C
      WRITE(12,70)
   70 FORMAT(' PROJECTIONS OF LANCZOS STARTING VECTOR ON A-EIGENVECS')
C
      WRITE(12,80)N,KX,KY,IIX,C2,C0,C1,ATOLN
   80 FORMAT(1X,'A-SIZE',2X,'X-DIM',2X,'Y-DIM',6X,'SVSEED'/3I7,I12/
     1 5X,'A-DIAGONAL',3X,'X-CODIAGONAL',3X,'Y-CODIAGONAL',5X,'ATOLN'/
     2 3E15.8,E10.3)
C
      WRITE(12,90)
```

```
   90 FORMAT(5X,'PROJECTION',8X,'TRUE A-EIGENVALUE',1X,'EV NO'
     1,2X,'VEC NO')
C
C     GENERATE SAME RANDOM UNIT VECTOR USED IN THE LANCZS RECURSIONS.
      IIL=IIX
C
C-----------------------------------------------------------------------
      CALL GENRAN(IIL,G,N)
C-----------------------------------------------------------------------
C
      DO 100 I = 1,N
  100 V(I) = G(I)
C
C-----------------------------------------------------------------------
      SUM = FINPRO(N,V(1),1,V(1),1)
C-----------------------------------------------------------------------
C
      SUM = 1.DO/DSQRT(SUM)
C
      DO 110 I = 1,N
  110 V(I) = V(I)*SUM
C
C     DETERMINE UNIT EIGENVECTOR W ASSOCIATED WITH EACH EV(I,J) = Y1
C     AND CALCULATE THE PROJECTION G(K) OF U ON THE STARTING VECTOR V
C     A*U = EV*U        WS = 1/||WU||: WU = UNSCALED EIGENVECTOR
C
      DO 160 K =1,N
C     DETERMINE I J FROM K: MP(K) = KP = (J-1)*KX+I
      KP = MP(K)
      I = MOD(KP,KX)
      IF (I.EQ.0) I = KX
      T1 = WI(I)
      J = 1 + (KP-1)/KX
      T0 = WJ(J)
      T0 = WJ(J)
C
      Y1 = C2 - TWO*C1*DCOS(WJ(J)) - TWO*CO*DCOS(WI(I))
C     Y1 = EV(I,J)
C
      DO 120 II = 1,KX
      T2 = T1*DFLOAT(II)
  120 WII(II) = WS*DSIN(T2)
C
      KV = 0
      DO 140 JJ = 1,KY
      T2 = T0*DFLOAT(JJ)
      T2 = DSIN(T2)
C
      DO 130 II = 1,KX
      KV = KV + 1
  130 U(KV) = T2*WII(II)
C
  140 CONTINUE
C
C     U IS UNIT EIGENVECTOR OF A ASSOCIATED WITH EV(I,J) = Y1
C     G(K) IS THE PROJECTION OF U ON V  FOR Y1
C
C-----------------------------------------------------------------------
      PROJ = FINPRO(N,U(1),1,V(1),1)
C-----------------------------------------------------------------------
C
      TEMP = DABS(PROJ)
      G(K) = TEMP
C
C     DESIRED PROJECTION HAS BEEN COMPUTED OUTPUT IT TO FILE 12.
      WRITE(12,150) G(K),Y1,K,MP(K)
```

```
  150 FORMAT(E15.8,E25.16,I6,I8)
C
  160 CONTINUE
C
      WRITE(12,170)
  170 FORMAT(' ----- END OF FILE 12 PROJECT -----------------------'//)
C
      GO TO 310
C
C     JVEC = 1
C
C     X1 IS AN INPUT PARAMETER. WE CALCULATE TRUE
C     A-EIGENVALUE WHICH IS CLOSEST TO X1, LABEL IT Y1 AND CALCULATE
C     UNIT EIGENVECTOR OF A ASSOCIATED WITH Y1. A*U = Y1*U, ||U|| = 1.
C     Y1 = EV(I,J). EIGENVALUES OF A ARE ORDERED BY INCREASING SIZE.
C     V = RITZ VECTOR ASSOCIATED WITH GOODEV X1
C
  180 CONTINUE
      KX1 = 0
      IF (X1.LE.U(1)) KX1 = 1
      IF (X1.GE.U(N)) KX1 = N
      NM1 = N-1
      IF (KX1.NE.0) GO TO 200
C
      DO 190 KVEC = 2,N
      IF (X1.GE.U(KVEC)) GO TO 190
C     U(KVEC-1).LE.X1.LT.U(KVEC)
      T1 = X1 - U(KVEC-1)
      T2 = U(KVEC) - X1
      KX1 = KVEC - 1
      IF (T1.GT.T2) KX1 = KVEC
      GO TO 200
  190 CONTINUE
C
  200 Y1 = U(KX1)
C
      IF (KX1.EQ.1) EE = U(2) - U(1)
      IF (KX1.EQ.N) EE = U(N) - U(NM1)
      IF (KX1.EQ.1.OR.KX1.EQ.N) GO TO 210
      EE = DMIN1(U(KX1+1)-U(KX1),U(KX1)-U(KX1-1))
  210 CONTINUE
C
      TO = DABS(ONE - X1/Y1)
C
      WRITE(6,220) N,KX1,ICOUNT,Y1,X1,TO,EE
  220 FORMAT(3I8,' = N, A-EV NUMBER,RITZ NUMBER'//
     1 18X,' TRUEEV',19X,'GOODEV',4X,'RELERROR',4X,'A-MINGAP'/
     1 2E25.16,2E12.3/)
C
      IF (EE.GT.ATOLN) GO TO 240
C
      WRITE(6,230)
  230 FORMAT(' Y1 IS A MULTIPLE EIGENVALUE OF A SO WE EXIT'/)
C
      GO TO 310
C
C     Y1 IS TOEPLITZ EIGENVALUE CLOSEST TO X1.
C     CALCULATION OF EIGENVECTOR ASSOCIATED WITH EIGENVALUE Y1
C     A*U = Y1*U
C     DETERMINE I J FROM K: MP(K) = KP = (J-1)*KX+I
  240 CONTINUE
      K = KX1
      KP = MP(K)
      I = MOD(KP,KX)
      IF (I.EQ.0) I = KX
      T1 = WI(I)
      J = 1 + (KP-1)/KX
```

```
      T2 = WJ(J)
C
      DO 250 II = 1,KX
      T0 = T1*DFLOAT(II)
  250 WII(II) = WS*DSIN(T0)
C
      KV = 0
      DO 270 JJ = 1,KY
      T0 = T2*DFLOAT(JJ)
      T0 = DSIN(T0)
C
      DO 260 II = 1,KX
      KV = KV + 1
  260 U(KV) = T0*WII(II)
C
  270 CONTINUE
C
C     U IS UNIT TRUE EIGENVECTOR OF A ASSOCIATED WITH Y1
C     V IS UNIT RITZVECTOR OF A ASSOCIATED WITH X1
C
      KK = 0
      S = ONE
      T1 = ZERO
C
      DO 280 K = 1,N
      IF (DABS(U(K)).LE.T1) GO TO 280
      T1 = DABS(U(K))
      KK = K
  280 CONTINUE
      IF (U(KK)*V(KK).LT.ZERO) S = - ONE
C
      KK = 0
      T1 = ZERO
      T2 = ZERO
      DO 290 K = 1,N
      TEMP = DABS(S*U(K) - V(K))
      T2 = T2 + TEMP**2
      IF (TEMP.LE.T1) GO TO 290
      KK = K
      T1 = TEMP
  290 CONTINUE
C
      T2 = DSQRT(T2)
      WRITE(6,300) KK,T1,T2
  300 FORMAT(' EIGENVECTOR ERROR. MAX ERROR AT COMPONENT = ',I6/
     1 ' MAX DABS(TRUEVEC(K)-RITZVEC(K)) = ',E12.5/
     1 ' NORM(TRUEVEC-RITZVEC)  = ',E12.5/)
C
  310 CONTINUE
C
      RETURN
C
C-----------------------------------------------------------------------
      ENTRY EXVECP(C0,C1,C2,KX,KY)
C-----------------------------------------------------------------------
C
      RETURN
C-----END OF EXVEC------------------------------------------------------
      END
```

SECTION 2.6 OTHER SUBROUTINES USED BY THE PROGRAMS IN CHAPTERS 2,3,4,5

```
C-----LESUB  (1) REAL SYMMETRIC-----------------------------------------
C            (2) HERMITIAN MATRICES
C            (3) FACTORED INVERSES OF REAL SYMMETRIC MATRICES AND
C            (4) REAL SYMMETRIC GENERALIZED, A*X = EVAL*B*X WHERE
C                B IS POSITIVE DEFINITE, CHOLESKY FACTOR AVAILABLE
C
C     ACCORDING TO PFORT THESE SUBROUTINES ARE PORTABLE EXCEPT FOR:
C     (1) THE COMPLEX*16 VARIABLES AND THE CORRESPONDING FUNCTIONS
C         FOR COMPLEX VARIABLES, DCMPLX, DREAL AND DCONJG USED IN
C         THE SUBROUTINE CINPRD (USED ONLY IN CASE (2), HERMITIAN)
C     (2) THE ENTRY IN THE SUBROUTINE LPERM USED TO PASS THE
C         PERMUTATION FROM THE UPSEC SUBROUTINE TO LPERM. (USED
C         ONLY IN CASES (3) AND (4), INVERSE AND GENERALIZED).
C
C     SUBROUTINES    BISEC, INVERR, TNORM, LUMP, ISOEV, PRTEST, AND
C                    INVERM ARE USED WITH THE LANCZOS EIGENVALUE
C                    PROGRAMS LEVAL, HLEVAL, LIVAL AND LGVAL. STURMI,
C                    INVERM, LBISEC, AND TNORM ARE USED WITH THE
C                    EIGENVECTOR PROGRAMS LEVEC, HLEVEC, LIVEC AND
C                    LGVEC.  LPERM IS USED WITH LIVEC AND LGVEC.
C                    IN THE HERMITIAN CASE, THE SUBROUTINE CINPRD
C                    IS ALSO USED.
C
C-----COMPUTE T-EIGENVALUES BY BISECTION--------------------------------
C
      SUBROUTINE BISEC(ALPHA,BETA,BETA2,VB,VS,LBD,UBD,EPS,TTOL,MP,
     1 NINT,MEV,NDIS,IC,IWRITE)
C
C-----------------------------------------------------------------------
      DOUBLE PRECISION  ALPHA(1),BETA(1),BETA2(1),VB(1),VS(1)
      DOUBLE PRECISION  LBD(1),UBD(1),EPS,EPT,EP0,EP1,TEMP,TTOL
      DOUBLE PRECISION  ZERO,ONE,HALF,YU,YV,LB,UB,XL,XU,X1,X0,XS,BETAM
      INTEGER  MP(1),IDEF(10)
      DOUBLE PRECISION  DABS, DSQRT, DMAX1, DMIN1, DFLOAT
C-----------------------------------------------------------------------
C     COMPUTES EIGENVALUES OF T(1,MEV) BY LOOPING INTERNALLY ON THE
C     USER-SPECIFIED INTERVALS,  (LB(J),UB(J)), J = 1,NINT.  INTERVALS
C     ARE TREATED AS OPEN ON THE LEFT AND CLOSED ON THE RIGHT.
C     THE BISEC SUBROUTINE SIMULTANEOUSLY LABELS SPURIOUS T-EIGENVALUES
C     AND DETERMINES THE T-MULTIPLICITIES OF EACH GOOD T-EIGENVALUE.
C     SPURIOUS T-EIGENVALUES ARE LABELLED BY A T-MULTIPLICITY = 0.
C     ANY T-EIGENVALUE WITH A T-MULTIPLICITY >= 1 IS 'GOOD'.
C
C     IF IWRITE = 0 THEN MOST OF THE WRITES TO FILE 6 ARE NOT
C     ACTIVATED.
C
C     NOTE THAT PROGRAM ASSUMES THAT NO MORE THAN MMAX/2 EIGENVALUES
C     OF T(1,MEV) ARE TO BE COMPUTED IN ANY ONE OF THE SUBINTERVALS
C     CONSIDERED, WHERE MMAX = DIMENSION OF VB SPECIFIED BY THE USER
C     IN THE MAIN PROGRAM LEVAL.
C
C     ON ENTRY
C     BETA2(J) IS SET = BETA(J)*BETA(J).  THE STORAGE FOR BETA2 COULD
C     BE ELIMINATED BY RECOMPUTING THE BETA(J)**2 FOR EACH STURM
C     SEQUENCE.
C
C     EPS = 2*MACHEP =  4.4 * 10**-16 ON IBM 3081.
C     TTOL = EPS*TKMAX WHERE
C     TKMAX - MAX(|ALPHA(K)|,BETA(K), K-1,KMAX)
C
C     ON EXIT
C     NDIS = TOTAL NUMBER OF COMPUTED DISTINCT EIGENVALUES OF
C            T(1,MEV) ON THE UNION OF THE (LB,UB) INTERVALS.
C     VS = COMPUTED DISTINCT EIGENVALUES OF T(1,MEV) IN ALGEBRAICALLY-
```

```
C          INCREASING ORDER
C     MP = CORRESPONDING T-MULTIPLICITIES OF THESE EIGENVALUES
C     MP(I) = (0,1,MI), MI>1, I=1,NDIS  MEANS:
C        (0)  V(I) IS SPURIOUS
C        (1)  V(I) IS ISOLATED AND GOOD
C        (MI) V(I) IS MULTIPLE AND HENCE A CONVERGED GOOD T-EIGENVALUE.
C     IC = TOTAL NUMBER OF STURMS USED
C
C     DEFAULTS
C     ISKIP = 0 INITIALLY. IF DEFAULT OCCURS ON J-TH SUB-INTERVAL, SET
C             ISKIP=ISKIP+1 AND IDEF(ISKIP) = J
C             DEFAULTS OCCUR IF THERE ARE NO T-EIGENVALUES  IN THE
C             SUBINTERVAL SPECIFIED OR IF THE NUMBER
C             OF STURMS SEQUENCES REQUIRED EXCEEDS MXSTUR.
C             WHEN A DEFAULT OCCURS THE PROGRAM
C             SKIPS THE INTERVAL INVOLVED AND GOES ON TO THE NEXT
C             INTERVAL.
C
C-----------------------------------------------------------------------
C     SPECIFY PARAMETERS
      ZERO = 0.0D0
      ONE  = 1.0D0
      HALF = 0.5D0
      MXSTUR = IC
      NDIS = 0
      IC = 0
      ISKIP = 0
      MP1 = MEV+1
C     SAVE THEN SET BETA(MEV+1) = 0. GENERATE BETA**2
      BETAM = BETA(MP1)
      BETA(MP1) = ZERO
C
      DO 10 I = 1,MP1
   10 BETA2(I) = BETA(I)*BETA(I)
C
C     EPO IS USED IN T-MULTIPLICITY AND SPURIOUS TESTS
C     EP1 AND EPS ARE USED IN THE BISEC CONVERGENCE TEST
C
      TEMP = DFLOAT(MEV+1000)
      EP0  = TEMP*TTOL
      EP1  = DSQRT(TEMP)*TTOL
C
      WRITE(6,20)MEV,NINT
   20 FORMAT(/' BISEC CALCULATION'/' ORDER OF T IS',I6/
     1' NUMBER OF INTERVALS IS',I6/)
C
      WRITE(6,30) EP0,EP1
   30 FORMAT(/' MULTOL, TOLERANCE USED IN T-MULTIPLICITY AND SPURIOUS TE
     1STS = ',E10.3/' BISTOL, TOLERANCE USED IN BISEC CONVERGENCE TEST =
     1 ',E10.3/)
C
C     LOOP ON THE NINT INTERVALS  (LB(J),UB(J)), J=1,NINT
      DO 430 JIND = 1,NINT
      LB = LBD(JIND)
      UB = UBD(JIND)
C
      WRITE(6,40)JIND,LB,UB
   40 FORMAT(//1X,'BISEC INTERVAL NO',2X,'LOWER BOUND',2X,'UPPER BOUND'/
     1I18,2E13.5/)
C
C     INITIALIZATION AND PARAMETER SPECIFICATION
C     ICT IS TOTAL STURM COUNT ON (LB,UB)
C
      NA = 0
      MD = 0
      NG = 0
```

```
      ICT = 0
C
C     START OF T-EIGENVALUE CALCULATIONS
      X1 = UB
      ISTURM = 1
      GO TO 330
C     FORWARD STURM CALCULATION TO DETERMINE NA = NO. T-EIGENVALUES > UB
   50 NA = NEV
C
      X1 = LB
      ISTURM = 2
      GO TO 330
C     FORWARD STURM CALC TO DETERMINE MT = NO. T-EIGENVALUES ON (LB,UB)
   60 CONTINUE
      MT=NEV
      ICT = ICT +2
C
      WRITE(6,70)MT,NA
   70 FORMAT(/2I6,' = NO. TMEV ON (LB,UB) AND NO. .GT. UB'/)
C
C     DEFAULT TEST: IS ESTIMATED NUMBER OF STURMS > MXSTUR?
      IEST = 30*MT
      IF (IEST.LT.MXSTUR) GO TO 90
C
      WRITE(6,80)
   80 FORMAT(//' ESTIMATED NUMBER OF STURMS REQUIRED EXCEEDS USER LIMIT'
     1/' SKIP THIS SUBINTERVAL')
      GO TO 110
C
   90 CONTINUE
C
      IF (MT.GE.1) GO TO 120
C
      WRITE(6,100)
  100 FORMAT(//' THERE ARE NO T-EIGENVALUES ON THIS INTERVAL)'/)
C
  110 ISKIP = ISKIP+1
      IDEF(ISKIP) = JIND
      GO TO 430
C
C     REGULAR CASE.
  120 CONTINUE
C
      IF (IWRITE.NE.0) WRITE(6,130)
  130 FORMAT(/' DISTINCT T-EIGENVALUES COMPUTED USING BISEC'/
     1 13X,'T-EIGENVALUE',2X,'TMULT',3X,'MD',4X,'NG')
C
C     SET UP INITIAL UPPER AND LOWER BOUNDS FOR T-EIGENVALUES
      DO 140 I=1,MT
      VB(I) = LB
      MTI = MT + I
  140 VB(MTI) = UB
C
C     CALCULATE T-EIGENVALUES FROM LB UP TO UB  K = MT,...,1
C     MAIN LOOP FOR FINDING KTH T-EIGENVALUE
C
      K = MT
  150 CONTINUE
      ICO = 0
      XL = VB(K)
      MTK = MT+K
      XU = VB(MTK)
C
      ISTURM = 3
      X1 = XU
      ICO = ICO + 1
```

```
      GO TO 330
C     FORWARD STURM CALCULATION AT XU
  160 NU=NEV
C
C     BISECTION LOOP FOR KTH T-EIGENVALUE. TEST  X1=MIDPOINT OF (XL,XU)
      ISTURM = 4
  170 CONTINUE
      X1 = (XL+XU)*HALF
      XS = DABS(XL)+DABS(XU)
      X0 = XU-XL
      EPT = EPS*XS+EP1
C
C     EPT IS CONVERGENCE TOLERANCE FOR KTH T-EIGENVALUE
C
      IF (X0.LE.EPT) GO TO 230
C
C     T-EIGENVALUE HAS NOT YET CONVERGED
C
      ICO = ICO + 1
      GO TO 330
C     FORWARD STURM CALCULATION AT CURRENT T-EIGENVALUE APPROXIMATION.
  180 CONTINUE
C
C     UPDATE T-EIGENVALUE INTERVAL (XL,XU)
C
      IF (NEV.LT.K) GO TO 190
C
C     NUMBER OF T-EIGENVALUES NEV = K
      XL = X1
      GO TO 170
  190 CONTINUE
C     NUMBER OF T-EIGENVALUES NEV<K
      XU = X1
      NU = NEV
C
C     UPDATE OF T-EIGENVALUE BOUNDS
C
      IF (NEV.EQ.0) GO TO 210
C
      DO 200 I = 1,NEV
  200 VB(I) = DMAX1(X1,VB(I))
C
  210 NEV1 = NEV+1
C
      DO 220 II = NEV1,K
      I = MT+II
  220 VB(I) = DMIN1(X1,VB(I))
C
      GO TO 170
C
C     END (XL,XU) BISECTION LOOP FOR KTH T-EIGENVALUE ON (LB,UB)
C     TEST FOR T-MULTIPLICITY AND IF SIMPLE THEN TEST FOR SPURIOUSNESS
C
  230 CONTINUE
      NDIS = NDIS+1
      MD = MD+1
      VS(NDIS) = X1
C
      JSTURM = 1
      X1 = XL-EP0
      GO TO 370
C     BACKWARD STURM CALCULATION
  240 KL = KEV
      JL = JEV
C
      JSTURM = 2
      ICO = ICO + 2
```

```
      X1 = XU+EPO
      GO TO 370
C     BACKWARD STURM CALCULATION
  250 JU = JEV
      KU = KEV
C
C     FOR T(1,MEV)
C     NU - KU = NO. T-EIGENVALUES ON (XU, XU + EPO)
C     KL - KU = NO. T-EIGENVALUES ON (XL - EPO, XU + EPO)
C
C     FOR T(2,MEV)
C     JL -JU = NO. T-EIGENVALUES ON (XL - EPO, XU + EPO)
C
C     IS THIS A SIMPLE T-EIGENVALUE?
C
      IF (KL-KU-1.EQ.0) GO TO 290
C
C     VS(NDIS) = KTH-T-EIGENVALUE OF (LB,UB) IS MULTIPLE AND HENCE GOOD
      IF (KU.EQ.NU) GO TO 280
C     CONTINUE TO CHECK FOR T-MULTIPLICITY
  260 CONTINUE
      ISTURM = 5
      X1 = X1+EPO
      ICO = ICO + 1
      GO TO 330
C     FORWARD STURM CALCULATION
  270 KNE = KU-NEV
      KU = NEV
      IF (KNE.NE.0) GO TO 260
C     SPECIFY T-MULTIPLICITY = MP(NDIS)
  280 MPEV = KL-KU
      KNEW = KU
      GO TO 300
C     END MULTIPLE CASE
C
C     T-EIGENVALUE IS SIMPLE   CHECK IF IT IS SPURIOUS
  290 CONTINUE
      MPEV = 1
      IF (JU.LT.JL) MPEV=0
      KNEW = K-1
C
C     X1 >= XU+EPO
C     SPURIOUS TEST AND T-SIMPLE CASE COMPLETED
C     START OF NEXT T-EIGENVALUE COMPUTATION
C
  300 K = KNEW
      MP(NDIS) = MPEV
      IF (MPEV.GE.1) NG = NG + 1
C
      IF (IWRITE.NE.0) WRITE(6,310) VS(NDIS),MPEV,MD,NG
  310 FORMAT(E25.16,3I6)
C
C     UPDATE STURM COUNT. ICO = STURM COUNT FOR KTH T-EIGENVALUE
      ICT = ICT + ICO
C
C     EXIT TEST FOR K DO LOOP
C
      IF (K.LE.0) GO TO 410
C
C     UPDATE LOWER BOUNDS
      DO 320 I=1,KNEW
  320 VB(I) = DMAX1(X1,VB(I))
C
      GO TO 150
C     END OF BISECTION LOOP FOR KTH T-EIGENVALUE
C
C     FORWARD STURM CALCULATION
```

```
  330 NEV = -NA
      YU  = ONE
C
      DO 360 I = 1,MEV
      IF (YU.NE.ZERO) GO TO 340
      YV = BETA(I)/EPS
      GO TO 350
  340 YV = BETA2(I)/YU
  350 YU = X1 - ALPHA(I) - YV
      IF (YU.GE.ZERO) GO TO 360
      NEV = NEV + 1
  360 CONTINUE
C     NEV = NUMBER OF T-EIGENVALUES ON (X1,UB)
C
      GO TO (50,60,160,180,270), ISTURM
C
C     BACKWARD STURM CALCULATION FOR T(1,MEV) AND T(2,MEV)
  370 KEV = -NA
      YU  = ONE
C
      DO 400 II = 1,MEV
      I = MP1-II
      IF (YU.NE.ZERO) GO TO 380
      YV = BETA(I+1)/EPS
      GO TO 390
  380 YV = BETA2(I+1)/YU
  390 YU = X1-ALPHA(I)-YV
      JEV = 0
      IF (YU.GE.ZERO) GO TO 400
      KEV = KEV+1
      JEV = 1
  400 CONTINUE
      JEV = KEV-JEV
C
      GO TO (240,250), JSTURM
C
C     KEV = -NA + (NUMBER OF T(1,MEV) EIGENVALUES) > X1
C     JEV = -NA + (NUMBER OF T(2,MEV) EIGENVALUES) > X1
C     SET PARAMETERS FOR NEXT INTERVAL
  410 CONTINUE
      IC = ICT+IC
      MXSTUR = MXSTUR-ICT
C
      WRITE(6,420) JIND,NG,MD
  420 FORMAT(/' T-EIGENVALUE CALCULATION ON INTERVAL',I6,'  IS COMPLETE'
     1 /3X,'NO. GOOD',3X,'NO. DISTINCT'/I10,I13)
C
  430 CONTINUE
C
C     END LOOP ON THE SUBINTERVALS (LB(J),UB(J)), J=1,NINT
C     ISKIP OUTPUT
C
      IF (ISKIP.GT.0) WRITE(6,440)ISKIP
  440 FORMAT(' BISEC DEFAULTED ON',I3,3X,'INTERVALS'/
     1 ' DEFAULTS OCCUR IF AN INTERVAL HAS NO T-EIGENVALUES'/
     2 ' OR THE STURM ESTIMATE EXCEEDS THE USER-SPECIFIED LIMIT'/)
C
      IF (ISKIP.GT.0) WRITE(6,450)(IDEF(I), I=1,ISKIP)
  450 FORMAT(' BISEC DEFAULTED ON INTERVALS'/(10I8))
C
C     RESET BETA AT I = MP1
      BETA(MP1) = BETAM
C-----END OF BISEC--------------------------------------------------
      RETURN
      END
C
```

```
C-----INVERSE ITERATION ON T(1,MEV)------------------------------------
C
      SUBROUTINE INVERR(ALPHA,BETA,V1,V2,VS,EPS,G,MP,MEV,MMB,NDIS,NISO,
     1 N,IKL,IT,IWRITE)
C
C-----------------------------------------------------------------------
      DOUBLE PRECISION ALPHA(1),BETA(1),V1(1),V2(1),VS(1)
      DOUBLE PRECISION X1,U,Z,EST,TEMP,TO,T1,RATIO,SUM,XU,NORM,TSUM
      DOUBLE PRECISION  BETAM,EPS,EPS3,EPS4,ZERO,ONE
      REAL G(1)
      INTEGER  MP(1)
      DOUBLE PRECISION FINPRO
      REAL  ABS
      DOUBLE PRECISION  DABS, DMIN1, DSQRT, DFLOAT
C-----------------------------------------------------------------------
C     COMPUTES ERROR ESTIMATES FOR COMPUTED ISOLATED GOOD T-EIGENVALUES
C     IN VS AND WRITES THESE T-EIGENVALUES AND ESTIMATES TO FILE 4.
C     BY DEFINITION A GOOD T-EIGENVALUE IS ISOLATED IF ITS
C     CLOSEST T-NEIGHBOR IS ALSO GOOD, OR ITS CLOSEST NEIGHBOR IS
C     SPURIOUS, BUT THAT NEIGHBOR IS FAR ENOUGH AWAY.  SO
C     IN PARTICULAR, WE COMPUTE ESTIMATES FOR GOOD T-EIGENVALUES
C     THAT ARE IN CLUSTERS OF GOOD T-EIGENVALUES.
C
C     USES INVERSE ITERATION ON T(1,MEV) SOLVING THE EQUATION
C     (T - X1*J)V2 = RIGHT-HAND SIDE (RANDOMLY-GENERATED)
C     FOR EACH SUCH GOOD T-EIGENVALUE X1.
C
C     PROGRAM REFACTORS T-X1*I ON EACH ITERATION OF INVERSE ITERATION.
C     TYPICALLY ONLY ONE ITERATION IS NEEDED PER EIGENVALUE X1.
C
C     POSSIBLE STORAGE COMPRESSION
C     G STORAGE COULD BE ELIMINATED BY REGENERATING THE RANDOM
C     RIGHT-HAND SIDE ON EACH ITERATION AND PRINTING OUT THE
C     ERROR ESTIMATES AS THEY ARE GENERATED.
C
C     ON ENTRY AND EXIT
C     MEV = ORDER OF T
C     ALPHA, BETA CONTAIN THE NONZERO ENTRIES OF THE T-MATRIX
C     VS = COMPUTED DISTINCT EIGENVALUES OF T(1,MEV)
C     MP = T-MULTIPLICITY OF EACH T-EIGENVALUE IN VS. MP(I) = -1 MEANS
C          VS(I) IS A GOOD T-EIGENVALUE BUT THAT IT IS SITTING CLOSE TO
C          A SPURIOUS T-EIGENVALUE.  MP(I) = 0 MEANS VS(I) IS SPURIOUS.
C          ESTIMATES ARE COMPUTED ONLY FOR THOSE T-EIGENVALUES
C          WITH MP(I) = 1. FLAGGING WAS DONE IN SUBROUTINE ISOEV
C          PRIOR TO ENTERING INVERR.
C     NISO = NUMBER OF ISOLATED GOOD T-EIGENVALUES CONTAINED IN VS
C     NDIS =  NUMBER OF DISTINCT T-EIGENVALUES IN VS
C     IKL = SEED FOR RANDOM NUMBER GENERATOR
C     EPS = 2. * MACHINE EPSILON
C
C     IN PROGRAM:
C     ITER = MAXIMUM NUMBER OF INVERSE ITERATION STEPS ALLOWED FOR EACH
C            X1.  ITER = IT ON ENTRY.
C     G = ARRAY OF DIMENSION AT LEAST MEV + NISO.  USED TO STORE
C         RANDOMLY-GENERATED RIGHT-HAND SIDE.  THIS IS NOT
C         REGENERATED FOR EACH X1. G IS ALSO USED TO STORE ERROR
C         ESTIMATES AS THEY ARE COMPUTED FOR LATER PRINTOUT.
C     V1,V2 = WORK SPACES USED IN THE FACTORIZATION OF T(1,MEV).
C     AT THE END OF THE INVERSE ITERATION COMPUTATION FOR X1, V2
C     CONTAINS THE UNIT EIGENVECTOR OF T(1,MEV) CORRESPONDING TO X1.
C     V1 AND V2 MUST BE OF DIMENSION AT LEAST MEV.
C
C     ON EXIT
C     G(J) = MINIMUM GAP IN T(1,MEV) FOR EACH VS(J), J=1,NDIS
C     G(MEV+I) = BETAM*|V2(MEV)| = ERROR ESTIMATE FOR ISOLATED GOOD
C               T-EIGENVALUES, WHERE I = 1,NISO  AND  BETAM = BETA(MEV+1)
C               V2(MEV) IS LAST COMPONENT OF THE UNIT EIGENVECTOR OF
```

```
C              T(1,MEV) CORRESPONDING TO ITH ISOLATED GOOD T-EIGENVALUE.
C
C     IF FOR SOME X1 IT.GT.ITER THEN THE ERROR ESTIMATE IN G IS MARKED
C     WITH A - SIGN.
C
C     V2 = ISOLATED GOOD T-EIGENVALUES
C     V1 = MINIMAL T-GAPS FOR THE T-EIGENVALUES IN V2.
C     THESE ARE CONSTRUCTED FOR WRITE-OUT PURPOSES ONLY AND NOT
C     NEEDED ELSEWHERE IN THE PROGRAM.
C-----------------------------------------------------------------------
C
C     LABEL OUTPUT FILE  4
      IF (MMB.EQ.1) WRITE(4,10)
   10 FORMAT(' INVERSE ITERATION ERROR ESTIMATES'/)
C
C     FILE 6 (TERMINAL) OUTPUT OF ERROR ESTIMATES
      IF (IWRITE.NE.0.AND.NISO.NE.0) WRITE(6,20)
   20 FORMAT(/' INVERSE ITERATION ERROR ESTIMATES'/'  JISO',' JDIST',8X
     1,'GOOD T-EIGENVALUE',4X,'BETAM*UM',5X,'TMINGAP')
C
C     INITIALIZATION AND PARAMETER SPECIFICATION
      ZERO = 0.0D0
      ONE = 1.0D0
      NG = 0
      NISO = 0
      ITER = IT
      MP1 = MEV+1
      MM1 = MEV-1
      BETAM = BETA(MP1)
      BETA(MP1) = ZERO
C
C     CALCULATE SCALE AND TOLERANCES
      TSUM = DABS(ALPHA(1))
      DO 30 I = 2,MEV
   30 TSUM = TSUM + DABS(ALPHA(I)) + BETA(I)
C
      EPS3 = EPS*TSUM
      EPS4 = DFLOAT(MEV)*EPS3
C
C     GENERATE SCALED RANDOM RIGHT-HAND SIDE
      ILL = IKL
C
C-----------------------------------------------------------------------
      CALL GENRAN(ILL,G,MEV)
C-----------------------------------------------------------------------
C
      GSUM = ZERO
      DO 40 I = 1,MEV
   40 GSUM = GSUM+ABS(G(I))
      GSUM = EPS4/GSUM
C
      DO 50 I = 1,MEV
   50 G(I) = GSUM*G(I)
C
C     LOOP ON ISOLATED GOOD T-EIGENVALUES IN VS (MP(I) = 1) TO
C     CALCULATE CORRESPONDING UNIT EIGENVECTOR OF T(1,MEV)
C
      DO 180 JEV = 1,NDIS
C
      IF (MP(JEV).EQ.0) GO TO 180
      NG = NG + 1
      IF (MP(JEV).NE.1) GO TO 180
C
      IT = 1
      NISO = NISO + 1
```

```
      X1 = VS(JEV)
C
C     INITIALIZE RIGHT HAND SIDE FOR INVERSE ITERATION
      DO 60 I = 1,MEV
   60 V2(I) = G(I)
C
C     TRIANGULAR FACTORIZATION WITH NEAREST NEIGHBOR PIVOT
C     STRATEGY. INTERCHANGES ARE LABELLED BY SETTING BETA < 0.
C
   70 CONTINUE
      U = ALPHA(1)-X1
      Z = BETA(2)
C
      DO 90 I = 2,MEV
      IF (BETA(I).GT.DABS(U)) GO TO 80
C     NO INTERCHANGE
      V1(I-1) = Z/U
      V2(I-1) = V2(I-1)/U
      V2(I) = V2(I)-BETA(I)*V2(I-1)
      RATIO = BETA(I)/U
      U = ALPHA(I)-X1-Z*RATIO
      Z = BETA(I+1)
      GO TO 90
   80 CONTINUE
C     INTERCHANGE CASE
      RATIO = U/BETA(I)
      BETA(I) = -BETA(I)
      V1(I-1) = ALPHA(I)-X1
      U = Z-RATIO*V1(I-1)
      Z = -RATIO*BETA(I+1)
      TEMP = V2(I-1)
      V2(I-1) = V2(I)
      V2(I) = TEMP-RATIO*V2(I)
   90 CONTINUE
      IF (U.EQ.ZERO) U = EPS3
C
C     SMALLNESS TEST AND DEFAULT VALUE FOR LAST COMPONENT
C     PIVOT(I-1) = |BETA(I)| FOR INTERCHANGE CASE
C     (I-1,I+1) ELEMENT IN RIGHT FACTOR = BETA(I+1)
C     END OF FACTORIZATION AND FORWARD SUBSTITUTION
C
C     BACK SUBSTITUTION
      V2(MEV) = V2(MEV)/U
      DO 110 II = 1,MM1
      I = MEV-II
      IF (BETA(I+1).LT.ZERO) GO TO 100
C     NO INTERCHANGE
      V2(I) = V2(I)-V1(I)*V2(I+1)
      GO TO 110
C     INTERCHANGE CASE
  100 BETA(I+1) = -BETA(I+1)
      V2(I) = (V2(I)-V1(I)*V2(I+1)-BETA(I+2)*V2(I+2))/BETA(I+1)
  110 CONTINUE
C
C     TESTS FOR CONVERGENCE OF INVERSE ITERATION
C     IF SUM |V2| COMPS. LE. 1 AND IT. LE. ITER DO ANOTHER INVIT STEP
C
      NORM = DABS(V2(MEV))
      DO 120 II = 1,MM1
      I = MEV-II
  120 NORM = NORM+DABS(V2(I))
C
      IF (NORM.GE.ONE) GO TO 140
      IT = IT+1
      IF (IT.GT.ITER) GO TO 140
      XU = EPS4/NORM
C
```

```
      DO 130 I = 1,MEV
  130 V2(I) = V2(I)*XU
C
      GO TO 70
C     ANOTHER INVERSE ITERATION STEP
C
C     INVERSE ITERATION FINISHED
C     NORMALIZE COMPUTED T-EIGENVECTOR : V2 = V2/||V2||
  140 CONTINUE
      SUM = FINPRO(MEV,V2(1),1,V2(1),1)
      SUM = ONE/DSQRT(SUM)
C
      DO 150 II = 1,MEV
  150 V2(II) = SUM*V2(II)
C
C     SAVE ERROR ESTIMATE FOR LATER OUTPUT
      EST = BETAM*DABS(V2(MEV))
      IF (IT.GT.ITER) EST = -EST
      MEVPNI = MEV + NISO
      G(MEVPNI) = EST
      IF (IWRITE.EQ.0) GO TO 180
C
C     FILE 6 (TERMINAL) OUTPUT OF ERROR ESTIMATES.
      IF (JEV.EQ.1) GAP = VS(2) - VS(1)
      IF (JEV.EQ.MEV) GAP = VS(MEV) - VS(MEV-1)
      IF (JEV.EQ.MEV.OR.JEV.EQ.1) GO TO 160
      TEMP = DMIN1(VS(JEV+1)-VS(JEV),VS(JEV)-VS(JEV-1))
      GAP = TEMP
  160 CONTINUE
C
      WRITE(6,170) NISO,JEV,X1,EST,GAP
  170 FORMAT(2I6,E25.16,2E12.3)
C
  180 CONTINUE
C
C     END ERROR ESTIMATE LOOP ON ISOLATED GOOD T-EIGENVALUES.
C     GENERATE DISTINCT MINGAPS FOR T(1,MEV).  THIS IS USEFUL AS AN
C     INDICATOR OF THE GOODNESS OF THE INVERSE ITERATION ESTIMATES.
C     TRANSFER ISOLATED GOOD T-EIGENVALUES AND CORRESPONDING TMINGAPS
C     TO V2 AND V1 FOR OUTPUT PURPOSES ONLY.
C
      NM1 = NDIS - 1
      G(NDIS) = VS(NM1)-VS(NDIS)
      G(1) = VS(2)-VS(1)
C
      DO 190 J = 2,NM1
      T0 = VS(J)-VS(J-1)
      T1 = VS(J+1)-VS(J)
      G(J) = T1
      IF (T0.LT.T1) G(J)=-T0
  190 CONTINUE
      ISO = 0
      DO 200 J = 1,NDIS
      IF (MP(J).NE.1) GO TO 200
      ISO = ISO+1
      V1(ISO) = G(J)
      V2(ISO) = VS(J)
  200 CONTINUE
C
      IF(NISO.EQ.0) GO TO 250
C
C     ERROR ESTIMATES ARE WRITTEN TO FILE 4
      WRITE(4,210)MEV,NDIS,NG,NISO,N,IKL,ITER,BETAM
  210 FORMAT(1X,'TSIZE',2X,'NDIS',1X,'NGOOD',2X,'NISO',1X,'ASIZE'/5I6/
     1 4X,'RHSEED',2X,'MXINIT',5X,'BETAM'/I10,I8,E10.3/
     2 2X,'GOODEVNO',8X,'GOOD T-EIGENVALUE',6X,'BETAM*UM',7X,'TMINGAP')
C
```

```
      ISPUR = 0
      I = 0
      DO 240 J = 1,NDIS
      IF(MP(J).NE.0) GO TO 220
      ISPUR = ISPUR + 1
      GO TO 240
  220 IF(MP(J).NE.1) GO TO 240
      I = I + 1
      MEVI = MEV + I
      IGOOD = J - ISPUR
      WRITE(4,230) IGOOD,V2(I),G(MEVI),V1(I)
  230 FORMAT(I10,E25.16,2E14.3)
  240 CONTINUE
      GO TO 270
C
  250 WRITE(4,260)
  260 FORMAT(/' THERE ARE NO ISOLATED T-EIGENVALUES SO NO ERROR ESTIMATE
     1S WERE COMPUTED')
C
C     RESTORE BETA(MEV+1) = BETAM
  270 BETA(MP1) = BETAM
C-----END OF INVERR------------------------------------------------------
      RETURN
      END
C
C-----START OF TNORM-----------------------------------------------------
C
      SUBROUTINE TNORM(ALPHA,BETA,BMIN,TMAX,MEV,IB)
C
C-----------------------------------------------------------------------
      DOUBLE PRECISION  ALPHA(1),BETA(1)
      DOUBLE PRECISION  TMAX,BMIN,BMAX,BSIZE,BTOL
      DOUBLE PRECISION  DABS, DMAX1
C-----------------------------------------------------------------------
C     COMPUTE SCALING FACTOR USED IN THE T-MULTIPLICITY, SPURIOUS AND
C     PRTESTS.   CHECK RELATIVE SIZE OF THE BETA(K), K=1,MEV
C     AS A TEST ON THE LOCAL ORTHOGONALITY OF THE LANCZOS VECTORS.
C
C              TMAX = MAX (|ALPHA(I)|, BETA(I),  I=1,MEV)
C              BMIN = MIN (BETA(I) I=2,MEV)
C              BSIZE = BMIN/TMAX
C              |IB| = INDEX OF MINIMAL(BETA)
C              IB < 0 IF BMIN/TMAX < BTOL
C-----------------------------------------------------------------------
C     SPECIFY PARAMETERS
      IB = 2
      BTOL = BMIN
      BMIN = BETA(2)
      BMAX = BETA(2)
      TMAX = DABS(ALPHA(1))
C
      DO 20 I = 2,MEV
      IF (BETA(I).GE.BMIN) GO TO 10
      IB = I
      BMIN = BETA(I)
   10 TMAX = DMAX1(TMAX,DABS(ALPHA(I)))
      BMAX = DMAX1(BETA(I),BMAX)
   20 CONTINUE
      TMAX = DMAX1(BMAX,TMAX)
C
C     TEST OF LOCAL ORTHOGONALITY USING SCALED BETAS
      BSIZE = BMIN/TMAX
      IF (BSIZE.GE.BTOL) GO TO 40
C
C     DEFAULT.  BSIZE IS SMALLER THAN TOLERANCE BTOL SPECIFIED IN MAIN
C     PROGRAM.  PROGRAM TERMINATES FOR USER TO DECIDE WHAT TO DO
C     BECAUSE LOCAL ORTHOGONALITY OF THE LANCZOS VECTORS COULD BE
```

```
C     LOST.
C
      IB = -IB
      WRITE(6,30) MEV
   30 FORMAT(/' BETA TEST INDICATES POSSIBLE LOSS OF LOCAL ORTHOGONALITY
     1OVER 1ST',I6,' LANCZOS VECTORS'/)
C
   40 CONTINUE
C
      WRITE(6,50) IB
   50 FORMAT(/' MINIMUM BETA RATIO OCCURS AT',I6,' TH BETA'/)
C
      WRITE(6,60) MEV,BMIN,TMAX,BSIZE
   60 FORMAT(/1X,'TSIZE',6X,'MIN BETA',5X,'TKMAX',6X,'MIN RATIO'/
     1 I6,E14.3,E10.3,E15.3/)
C
C-----END OF TNORM------------------------------------------------------
      RETURN
      END
C
C
C-----START OF LUMP-----------------------------------------------------
C
      SUBROUTINE LUMP(V1,RELTOL,MULTOL,SCALE2,LINDEX,LOOP)
C
C-----------------------------------------------------------------------
      DOUBLE PRECISION  V1(1),SUM,RELTOL,MULTOL,THOLD,ZERO,SCALE2
      INTEGER  LINDEX(1)
      DOUBLE PRECISION  DABS, DFLOAT, DMAX1
C-----------------------------------------------------------------------
C     LINDEX(J) = T-MULTIPLICITY OF JTH DISTINCT T-EIGENVALUE
C     LOOP = NUMBER OF DISTINCT T-EIGENVALUES
C     LUMP 'COMBINES' COMPUTED 'GOOD' T-EIGENVALUES THAT ARE
C     'TOO CLOSE'.
C     VALUE OF RELTOL IS 1.D-10.
C
C     IF IN A SET OF T-EIGENVALUES TO BE COMBINED THERE IS AN EIGENVALUE
C     WITH LINDEX=1, THEN THE VALUE OF THE COMBINED EIGENVALUES IS SET
C     EQUAL TO THE VALUE OF THAT EIGENVALUE.  NOTE THAT IF A SPURIOUS
C     T-EIGENVALUE IS TO BE 'COMBINED' WITH A GOOD T-EIGENVALUE, THEN
C     THIS IS DONE ONLY BY INCREASING THE INDEX, LINDEX, FOR THAT
C     T-EIGENVALUE.  NUMERICAL VALUES OF SPURIOUS EIGENVALUES ARE NEVER
C     COMBINE WITH THOSE OF GOOD T-EIGENVALUES.
C-----------------------------------------------------------------------
      ZERO = 0.0D0
      NLOOP = 0
      J = 0
      ICOUNT = 1
      JI = 1
      THOLD = DMAX1(RELTOL*DABS(V1(1)),SCALE2*MULTOL)
C     THOLD = DMAX1(RELTOL*DABS(V1(1)),RELTOL)
C
   10 J = J+1
      IF (J.EQ.LOOP) GO TO 20
      SUM = DABS(V1(J)-V1(J+1))
      IF (SUM.LT.THOLD) GO TO 60
   20 JF = JI + ICOUNT - 1
      INDSUM = 0
      ISPUR = 0
C
      DO 30 KK = JI,JF
      IF (LINDEX(KK).NE.0) GO TO 30
      ISPUR = ISPUR + 1
      INDSUM = INDSUM + 1
   30 INDSUM = INDSUM + LINDEX(KK)
C
C     IF (JF-JI.GE.1) WRITE(6,40) (V1(KKK), KKK=JI,JF)
```

```
   40 FORMAT(/' LUMP LUMPS THE T-EIGENVALUES'/(4E20.13))
C
C     COMPUTE THE 'COMBINED' T-EIGENVALUE AND THE RESULTING
C     T-MULTIPLICITY
      K = JI - 1
   50 K = K+1
      IF (K.GT.JF) GO TO 70
      IF (LINDEX(K) .NE.1) GO TO 50
      NLOOP = NLOOP + 1
      V1(NLOOP) = V1(K)
      GO TO 100
   60 ICOUNT = ICOUNT + 1
      GO TO 10
C
C     ALL INDICES WERE 0 OR >1
   70 NLOOP = NLOOP + 1
      IDIF = INDSUM - ISPUR
      IF (IDIF.EQ.0) GO TO 90
C
      SUM = ZERO
      DO 80 KK = JI,JF
   80 SUM = SUM + V1(KK) * DFLOAT(LINDEX(KK))
C
      V1(NLOOP) = SUM/DFLOAT(IDIF)
      GO TO 100
   90 V1(NLOOP) = V1(JI)
  100 LINDEX(NLOOP) = INDSUM
      IDIF = INDSUM - ISPUR
      IF (IDIF.EQ.0.AND.ISPUR.EQ.1) LINDEX(NLOOP) = 0
      IF (J.EQ.LOOP) GO TO 110
      ICOUNT = 1
      JI= J+1
      THOLD = DMAX1(RELTOL*DABS(V1(JI)),SCALE2*MULTOL)
C     THOLD = DMAX1(RELTOL*DABS(V1(JI)),RELTOL)
      IF (JI.LT.LOOP) GO TO 10
      NLOOP = NLOOP + 1
      V1(NLOOP)= V1(JI)
      LINDEX(NLOOP) = LINDEX(JI)
  110 CONTINUE
C
C     ON RETURN V1 CONTAINS THE DISTINCT T-EIGENVALUES
C     LINDEX CONTAINS THE CORRESPONDING T-MULTIPLICITIES
C
      LOOP = NLOOP
      RETURN
C-----END OF LUMP--------------------------------------------------------
      END
C
C
C-----START OF ISOEV-----------------------------------------------------
C
      SUBROUTINE ISOEV(VS,GAPTOL,MULTOL,SCALE1,G,MP,NDIS,NG,NISO)
C
C-----------------------------------------------------------------------
      DOUBLE PRECISION  VS(1),T0,T1,MULTOL,GAPTOL,SCALE1,TEMP
      REAL   G(1),GAP
      INTEGER  MP(1)
      REAL  ABS
      DOUBLE PRECISION  DABS, DMAX1
C-----------------------------------------------------------------------
C     GENERATE DISTINCT TMINGAPS AND USE THEM TO LABEL THE ISOLATED
C     GOOD T-EIGENVALUES THAT ARE VERY CLOSE TO SPURIOUS ONES.
C     ERROR ESTIMATES WILL NOT BE COMPUTED FOR THESE T-EIGENVALUES.
C
C     ON ENTRY AND EXIT
C     VS CONTAINS THE COMPUTED DISTINCT EIGENVALUES OF T(1,MEV)
C     MP CONTAINS THE CORRESPONDING T-MULTIPLICITIES
```

```
C     NDIS = NUMBER OF DISTINCT EIGENVALUES                             LES08640
C     GAPTOL = RELATIVE GAP TOLERANCE SET IN MAIN                       LES08650
C                                                                       LES08660
C     ON EXIT                                                           LES08670
C     G CONTAINS THE TMINGAPS.                                          LES08680
C     G(I) < 0 MEANS MINGAP IS DUE TO LEFT GAP                          LES08690
C     MP(I) IS NOT CHANGED EXCEPT THAT  MP(I)=-1, IF MP(I)=1,           LES08700
C     TMINGAP WAS TOO SMALL AND DUE TO A SPURIOUS T-EIGENVALUE.         LES08710
C                                                                       LES08720
C     IF MP(I)=-1 THAT SIMPLE GOOD T-EIGENVALUE WILL BE SKIPPED         LES08730
C     IN THE SUBSEQUENT ERROR ESTIMATE COMPUTATIONS IN INVERR           LES08740
C     THAT IS, WE COMPUTE ERROR ESTIMATES ONLY FOR THOSE GOOD           LES08750
C     T-EIGENVALUES WITH MP(I)=1.                                       LES08760
C-----------------------------------------------------------------------LES08770
C     CALCULATE MINGAPS FOR DISTINCT T(1,MEV) EIGENVALUES.              LES08780
      NM1 = NDIS - 1                                                    LES08790
      G(NDIS) = VS(NM1)-VS(NDIS)                                        LES08800
      G(1) = VS(2)-VS(1)                                                LES08810
C                                                                       LES08820
      DO 10 J = 2,NM1                                                   LES08830
      T0 = VS(J)-VS(J-1)                                                LES08840
      T1 = VS(J+1)-VS(J)                                                LES08850
      G(J) = T1                                                         LES08860
      IF (T0.LT.T1) G(J) = -T0                                          LES08870
   10 CONTINUE                                                          LES08880
C                                                                       LES08890
C     SET MP(I)=-1 FOR SIMPLE GOOD T-EIGENVALUES WHOSE MINGAPS  ARE     LES08900
C     'TOO SMALL' AND DUE TO SPURIOUS T-EIGENVALUES.                    LES08910
C                                                                       LES08920
      NISO = 0                                                          LES08930
      NG = 0                                                            LES08940
      DO 20 J = 1,NDIS                                                  LES08950
      IF (MP(J).EQ.0) GO TO 20                                          LES08960
      NG = NG+1                                                         LES08970
      IF (MP(J).NE.1) GO TO 20                                          LES08980
C     VS(J) IS NEXT SIMPLE GOOD T-EIGENVALUE                            LES08990
      NISO = NISO + 1                                                   LES09000
      I = J+1                                                           LES09010
      IF (G(J).LT.0.0) I = J-1                                          LES09020
      IF (MP(I).NE.0) GO TO 20                                          LES09030
      GAP = ABS(G(J))                                                   LES09040
      T0 = DMAX1(SCALE1*MULTOL,GAPTOL*DABS(VS(J)))                      LES09050
C     T0 = DMAX1(GAPTOL,GAPTOL*DABS(VS(J)))                             LES09060
      TEMP = T0                                                         LES09070
      IF (GAP.GT.TEMP) GO TO 20                                         LES09080
      MP(J) = -MP(J)                                                    LES09090
      NISO = NISO-1                                                     LES09100
   20 CONTINUE                                                          LES09110
C                                                                       LES09120
C-----END OF ISOEV-------------------------------------------------------LES09130
      RETURN                                                            LES09140
      END                                                               LES09150
C                                                                       LES09160
C-----START OF PRTEST----------------------------------------------------LES09170
C                                                                       LES09180
      SUBROUTINE PRTEST(ALPHA,BETA,TEIG,TKMAX,EPSM,RELTOL,SCALE3,SCALE4,LES09190
     1 TMULT,NDIST,MEV,IPROJ)                                           LES09200
C                                                                       LES09210
C-----------------------------------------------------------------------LES09220
      DOUBLE PRECISION  ALPHA(1), BETA(1),TEIG(1),SIGMA(10)             LES09230
      DOUBLE PRECISION  EPSM,RELTOL,PRTOL,TKMAX,LRATIO,URATIO           LES09240
      DOUBLE PRECISION  EPS,EPS1,BETAM,LBD,UBD,SIG,YU,YV,LRATS,URATS    LES09250
      DOUBLE PRECISION ZERO,ONE,TEN,BISTOL,SCALE3,SCALE4,AEV,TEMP       LES09260
      INTEGER  TMULT(1),ISIGMA(10)                                      LES09270
      DOUBLE PRECISION  DABS, DMAX1, DSQRT, DFLOAT                      LES09280
C-----------------------------------------------------------------------LES09290
C     AFTER CONVERGENCE HAS BEEN ESTABLISHED, SUBROUTINE PRTEST         LES09300
```

```
C     TESTS COMPUTED EIGENVALUES OF T(1,MEV) THAT HAVE BEEN LABELLED
C     SPURIOUS TO DETERMINE IF ANY EIGENVALUES OF A HAVE BEEN
C     MISSED BY LANCZOS PROCEDURE.  AN EIGENVALUE WITH A VERY SMALL
C     PROJECTION ON THE STARTING VECTOR (< SINGLE PRECISION)
C     CAN BE MISSED BECAUSE IT IS ALSO AN EIGENVALUE OF T(2,MEV) TO
C     WITHIN THE SQUARE OF THIS ORIGINAL PROJECTION.
C     OUR EXPERIENCE IS THAT SUCH SMALL PROJECTIONS OCCUR ONLY
C     VERY INFREQUENTLY.
C
C     THIS SUBROUTINE IS CALLED ONLY AFTER CONVERGENCE HAS BEEN
C     ESTABLISHED. ONCE CONVERGENCE HAS BEEN OBSERVED ON THE
C     OTHER EIGENVALUES THEN ONE CAN EXPECT TO ALSO HAVE CONVERGENCE
C     ON ANY SUCH HIDDEN EIGENVALUES.(IF THERE ARE ANY).  THIS
C     PROCEDURE CONSIDERS ONLY SPURIOUS T-EIGENVALUES AND ONLY THOSE
C     SPURIOUS T-EIGENVALUES THAT ARE ISOLATED FROM GOOD T-EIGENVALUES.
C     FOR EACH SUCH T-EIGENVALUE IT DOES 2 STURM SEQUENCES
C     AND A FEW SCALAR MULTIPLICATIONS.  UPON RETURN TO MAIN
C     PROGRAM ERROR ESTIMATES WILL BE COMPUTED FOR ANY EIGENVALUES
C     THAT HAVE BEEN LABELLED AS 'HIDDEN'.  SUCH T-EIGENVALUES
C     WILL BE RELABELLED AS 'GOOD' ONLY IF THESE ERROR ESTIMATES
C     ARE SUFFICIENTLY SMALL.
C-----------------------------------------------------------------------
      ZERO = 0.0D0
      ONE  = 1.0D0
      TEN  = 10.0D0
      PRTOL = 1.D-6
      TEMP = DFLOAT(MEV+1000)
      TEMP = DSQRT(TEMP)
      BISTOL = TKMAX*EPSM*TEMP
      NSIGMA = 4
      SIGMA(1) = TEN*TKMAX
C
      DO 10 J = 2,NSIGMA
   10 SIGMA(J) = TEN*SIGMA(J-1)
C
      IFIN = 0
      MF = 1
      ML = MEV
      BETAM = BETA(MF)
      BETA(MF) = ZERO
      IPROJ = 0
      J = 1
C
      IF (TMULT(1).NE.0) GO TO 110
C
      AEV = DABS(TEIG(1))
      TEMP = PRTOL*AEV
      EPS1 = DMAX1(TEMP,SCALE4*BISTOL)
C     EPS1 = DMAX1(TEMP,PRTOL)
      TEMP = RELTOL*AEV
      EPS  = DMAX1(TEMP,SCALE3*BISTOL)
C     EPS  = DMAX1(TEMP,RELTOL)
C
      IF (TEIG(2)-TEIG(1).LT.EPS1.AND.TMULT(2).NE.0) GO TO 110
C
   20 LBD = TEIG(J) - EPS
      UBD = TEIG(J) + EPS
      MEVL = 0
      IL = 0
      YU = ONE
C
      DO 50 I=MF,ML
      IF (YU.NE.ZERO) GO TO 30
      YV = BETA(I)/EPSM
      GO TO 40
   30 YV = BETA(I)*BETA(I)/YU
   40 YU = ALPHA(I)-LBD-YV
```

```
      IF (YU.GE.ZERO) GO TO 50
C     MEVL INCREMENTED
      MEVL = MEVL + 1
      IL = I
   50 CONTINUE
C
      LRATIO = YU
      MEV1L = MEVL
      IF (IL.EQ.ML) MEV1L=MEVL-1
C
C     MEVL = NUMBER OF EVS OF T(1,MEV) WHICH ARE < LBD
C     MEV1L = NUMBER OF EVS OF T(1,MEV-1) WHICH ARE < LBD
C     LRATIO = DET(T(1,MEV)-LBD)/DET(T(1,MEV-1)-LBD):
C
      MEVU = 0
      IL = 0
      YU = ONE
C
      DO 80 I=MF,ML
      IF (YU.NE.ZERO) GO TO 60
      YV = BETA(I)/EPSM
      GO TO 70
   60 YV = BETA(I)*BETA(I)/YU
   70 YU = ALPHA(I)-UBD-YV
      IF (YU.GE.ZERO) GO TO 80
C     MEVU INCREMENTED
      MEVU = MEVU + 1
      IL = I
   80 CONTINUE
C
      URATIO = YU
      MEV1U = MEVU
      IF (IL.EQ.ML) MEV1U=MEVU-1
C
C     MEVU = NUMBER OF EVS OF T(MEV) WHICH ARE < UBD
C     MEV1U = NUMBER OF EVS OF T(MEV-1) WHICH ARE < UBD
C     URATIO = DET(TM-UBD)/DET(T(M-1)-UBD): TM=T(MF,ML)
C
      NEV1 = MEV1U-MEV1L
C
      DO 90 K=1,NSIGMA
      SIG = SIGMA(K)
      LRATS = LRATIO-SIG
      URATS = URATIO-SIG
C     NOTE THE INCREMENT IS ON NUMBER OF EVALUES OF T(M-1)
      MEVLS = MEV1L
      IF (LRATS.LT.0.) MEVLS=MEV1L+1
      MEVUS = MEV1U
      IF (URATS.LT.0.) MEVUS=MEV1U+1
      ISIGMA(K) = MEVUS - MEVLS
   90 CONTINUE
C
      ICOUNT = 0
      DO 100 K=1,NSIGMA
  100 IF (ISIGMA(K).EQ.1) ICOUNT=ICOUNT + 1
C
      IF (ICOUNT.LT.2.OR.NEV1.EQ.0) GO TO 110
      TMULT(J) = -10
      IPROJ=IPROJ+1
C
  110 J=J+1
C
      IF (J.GE.NDIST) GO TO 120
      IF (TMULT(J).NE.0) GO TO 110
C
      AEV = DABS(TEIG(J))
      TEMP = PRTOL*AEV
```

```
      EPS1 = DMAX1(TEMP,SCALE4*BISTOL)
C     EPS1 = DMAX1(TEMP,PRTOL)
      TEMP = RELTOL*AEV
      EPS  = DMAX1(TEMP,SCALE3*BISTOL)
C     EPS  = DMAX1(TEMP,RELTOL)
C
      IF (TEIG(J)-TEIG(J-1).LT.EPS1.AND.TMULT(J-1).NE.0) GO TO 110
      IF (TEIG(J+1)-TEIG(J).LT.EPS1.AND.TMULT(J+1).NE.0) GO TO 110
C
      GO TO 20
C
  120 IF (IFIN.EQ.1) GO TO 130
      IF (TMULT(NDIST).NE.0) GO TO 130
C
      AEV = DABS(TEIG(NDIST))
      TEMP = PRTOL*AEV
      EPS1 = DMAX1(TEMP,SCALE4*BISTOL)
C     EPS1 = DMAX1(TEMP,PRTOL)
      TEMP = RELTOL*AEV
      EPS  = DMAX1(TEMP,SCALE3*BISTOL)
C     EPS  = DMAX1(TEMP,RELTOL)
C
      NDIST1=NDIST -1
      TEMP = TEIG(NDIST)-TEIG(NDIST1)
      IF (TEMP.LT.EPS1.AND.TMULT(NDIST1).NE.0) GO TO 130
      IFIN = 1
C
      GO TO 20
C
  130 BETA(MF) = BETAM
C
C-----END OF PRTEST----------------------------------------------------
      RETURN
      END
C
C------START OF STURMI-------------------------------------------------
C
      SUBROUTINE STURMI(ALPHA,BETA,X1,TOLN,EPSM,MMAX,MK1,MK2,IC,IWRITE)
C
C---------------------------------------------------------------------
      DOUBLE PRECISION  ALPHA(1),BETA(1)
      DOUBLE PRECISION  EPSM,X1,TOLN,EVL,EVU,BETA2
      DOUBLE PRECISION  U1,U2,V1,V2,ZERO,ONE
      INTEGER I,IC,ICD,ICO,IC1,IC2,MK1,MK2,MMAX
C---------------------------------------------------------------------
C
C     FOR ANY EIGENVALUE OF A THAT HAS CONVERGED AS AN EIGENVALUE
C     OF THE T-MATRICES THIS SUBROUTINE CALCULATES
C     THE SMALLEST SIZE OF THE T-MATRIX, T(1,MK1) DEFINED
C     BY THE ALPHA AND BETA ARRAYS SUCH THAT MK1.LE.MMAX
C     AND THE INTERVAL (X1-TOLN,X1+TOLN) CONTAINS AT LEAST ONE
C     EIGENVALUE OF T(1,MK1). IT ALSO CALCULATES MK2 <= MMAX
C     AS THE SMALLEST SIZE T-MATRIX (IF ANY) SUCH THAT THIS INTERVAL
C     CONTAINS AT LEAST TWO EIGENVALUES OF T(1,MK2).
C     IF NO T-MATRIX OF ORDER < MMAX SATISFIES THIS REQUIREMENT
C     THEN MK2 IS SET EQUAL TO MMAX.  THE EIGENVECTOR PROGRAM
C     USES THESE VALUES TO DETERMINE AN APPROPRIATE 1ST GUESS AT
C     AN APPROPRIATE SIZE T-MATRIX FOR THE EIGENVALUE X1.
C
C     ON EXIT IC = NUMBER OF EIGENVALUES OF T(1,MK2) IN THIS INTERVAL
C
C     STURMI REGENERATES THE QUANTITIES BETA(I)**2 EACH TIME IT IS
C     CALLED, OBVIOUSLY FOR THE PRICE OF ANOTHER VECTOR OF LENGTH
C     MMAX THIS GENERATION COULD BE DONE ONCE IN THE MAIN
C     PROGRAM BEFORE THE LOOP ON THE CALLS TO SUBROUTINE STURMI.
C
C     IF ANY OF THE EIGENVALUES BEING CONSIDERED WERE MULTIPLE
```

```
C     AS EIGENVALUES OF THE USER-SPECIFIED MATRIX, THEN
C     THIS SUBROUTINE COULD BE MODIFIED TO COMPUTE ADDITIONAL
C     SIZES MKJ, J = 3, ...  WHICH COULD THEN BE USED IN THE
C     MAIN LANCZOS EIGENVECTOR PROGRAM TO COMPUTE ADDITIONAL
C     EIGENVECTORS CORRESPONDING TO THESE MULTIPLE EIGENVALUES.
C     THE MAIN PROGRAM PROVIDED DOES NOT INCLUDE THIS OPTION.
C
C-----------------------------------------------------------------------
C     INITIALIZATION OF PARAMETERS
      MK1 = 0
      MK2 = 0
      ZERO = 0.0D0
      ONE  = 1.0D0
      BETA(1) = ZERO
      EVL = X1-TOLN
      EVU = X1+TOLN
      U1 = ONE
      U2 = ONE
      IC0 = 0
      IC1 = 0
      IC2 = 0
C
C     MAIN LOOP FOR CALCULATING THE SIZES MK1,MK2
      DO 60 I = 1,MMAX
      BETA2 = BETA(I)*BETA(I)
      IF (U1.NE.ZERO) GO TO 10
      V1 = BETA(I)/EPSM
      GO TO 20
   10 V1 = BETA2/U1
   20 U1 = EVL - ALPHA(I) - V1
      IF (U1.LT.ZERO) IC1 = IC1+1
      IF (U2.NE.ZERO) GO TO 30
      V2 = BETA(I)/EPSM
      GO TO 40
   30 V2 = BETA2/U2
   40 U2 = EVU - ALPHA(I) - V2
      IF (U2.LT.ZERO) IC2 = IC2+1
C     TEST FOR CHANGE IN NUMBER OF T-EIGENVALUES ON (EVL,EVU)
      ICD = IC1-IC2
      IC = ICD-IC0
      IF (IC.GE.1) GO TO 50
      GO TO 60
   50 CONTINUE
      IF (IC0.EQ.0) MK1 = I
      IC0 = IC0+1
      IF (IC0.GT.1) GO TO 70
   60 CONTINUE
C
      I = I-1
      IF (IC0.EQ.0) MK1 = MMAX
   70 MK2 = I
      IC = ICD
C
      IF (IWRITE.EQ.1) WRITE(6,80) X1,MK1,MK2,IC
   80 FORMAT(' EVAL =',E20.12,' MK1 =',I6,' MK2 =',I6,' IC =',I3/)
C
      RETURN
C-----END OF STURMI-----------------------------------------------------
      END
C
C
C-----START OF INVERM---------------------------------------------------
C
      SUBROUTINE INVERM(ALPHA,BETA,V1,V2,X1,ERROR,ERRORV,EPS,G,MEV,IT,
     1 IWRITE)
C
C-----------------------------------------------------------------------
```

```
      DOUBLE PRECISION  ALPHA(1),BETA(1),V1(1),V2(1)
      DOUBLE PRECISION  X1,U,Z,TEMP,RATIO,SUM,XU,NORM,TSUM,BETAM
      DOUBLE PRECISION  EPS,EPS3,EPS4,ERROR,ERRORV,ZERO,ONE
      REAL  G(1)
      DOUBLE PRECISION  DABS, DSQRT, DFLOAT
      DOUBLE PRECISION FINPRO
      REAL  ABS
C-----------------------------------------------------------------------
C
C     COMPUTES T-EIGENVECTORS FOR ISOLATED GOOD T-EIGENVALUES X1
C     USING INVERSE ITERATION ON T(1,MEV(X1)) SOLVING EQUATION
C     (T - X1*I)V2 = RIGHT-HAND SIDE (RANDOMLY-GENERATED) .
C     PROGRAM REFACTORS T- X1*I ON EACH ITERATION OF INVERSE ITERATION.
C     TYPICALLY ONLY ONE ITERATION IS NEEDED PER T-EIGENVALUE X1.
C
C     IF IWRITE = 1 THEN THERE ARE EXTENDED WRITES TO FILE 6 (TERMINAL)
C
C     ON ENTRY G CONTAINS A REAL*4 RANDOM VECTOR WHICH WAS GENERATED
C     IN MAIN PROGRAM.
C
C     ON ENTRY AND EXIT
C     MEV = ORDER OF T
C     ALPHA, BETA CONTAIN THE DIAGONAL AND OFFDIAGONAL ENTRIES OF T.
C     EPS = 2. * MACHINE EPSILON
C
C     IN PROGRAM:
C     ITER = MAXIMUM NUMBER STEPS ALLOWED FOR INVERSE ITERATION
C     ITER = IT ON ENTRY.
C     V1,V2 = WORK SPACES USED IN THE FACTORIZATION OF T(1,MEV).
C     V1 AND V2 MUST BE OF DIMENSION AT LEAST MEV.
C
C     ON EXIT
C     V2 = THE UNIT EIGENVECTOR OF T(1,MEV) CORRESPONDING TO X1.
C     ERROR = |V2(MEV)| = ERROR ESTIMATE FOR CORRESPONDING
C             RITZ VECTOR FOR X1.
C
C     ERRORV = || T*V2 - X1*V2 || = ERROR ESTIMATE ON T-EIGENVECTOR.
C     IF IT.GT.ITER THEN ERRORV = -ERRORV
C     IT = NUMBER OF ITERATIONS ACTUALLY REQUIRED
C-----------------------------------------------------------------------
C     INITIALIZATION AND PARAMETER SPECIFICATION
      ONE  = 1.0D0
      ZERO = 0.0D0
      ITER = IT
      MP1 = MEV+1
      MM1 = MEV-1
      BETAM = BETA(MP1)
      BETA(MP1) = ZERO
C
C     CALCULATE SCALE AND TOLERANCES
      TSUM = DABS(ALPHA(1))
      DO 10 I = 2,MEV
   10 TSUM = TSUM + DABS(ALPHA(I)) + BETA(I)
C
      EPS3 = EPS*TSUM
      EPS4 = DFLOAT(MEV)*EPS3
C
C     GENERATE SCALED RANDOM RIGHT-HAND SIDE
      GSUM = ZERO
      DO 20 I = 1,MEV
   20 GSUM = GSUM+ABS(G(I))
      GSUM = EPS4/GSUM
C
C     INITIALIZE RIGHT HAND SIDE FOR INVERSE ITERATION
      DO 30 I = 1,MEV
   30 V2(I) = GSUM*G(I)
```

```
      IT = 1
C
C     CALCULATE UNIT EIGENVECTOR OF T(1,MEV) FOR ISOLATED GOOD
C     T-EIGENVALUE X1.
C
C     TRIANGULAR FACTORIZATION WITH NEAREST NEIGHBOR PIVOT
C     STRATEGY. INTERCHANGES ARE LABELLED BY SETTING BETA < 0.
C
   40 CONTINUE
      U = ALPHA(1)-X1
      Z = BETA(2)
C
      DO 60 I=2,MEV
      IF (BETA(I).GT.DABS(U)) GO TO 50
C     NO PIVOT INTERCHANGE
      V1(I-1) = Z/U
      V2(I-1) = V2(I-1)/U
      V2(I) = V2(I)-BETA(I)*V2(I-1)
      RATIO = BETA(I)/U
      U = ALPHA(I)-X1-Z*RATIO
      Z = BETA(I+1)
      GO TO 60
C     PIVOT INTERCHANGE
   50 CONTINUE
      RATIO = U/BETA(I)
      BETA(I) = -BETA(I)
      V1(I-1) = ALPHA(I)-X1
      U = Z-RATIO*V1(I-1)
      Z = -RATIO*BETA(I+1)
      TEMP = V2(I-1)
      V2(I-1) = V2(I)
      V2(I) = TEMP-RATIO*V2(I)
   60 CONTINUE
C
      IF (U.EQ.ZERO) U=EPS3
C
C     SMALLNESS TEST AND DEFAULT VALUE FOR LAST COMPONENT
C     PIVOT(I-1) = |BETA(I)| FOR INTERCHANGE CASE
C     (I-1,I+1) ELEMENT IN RIGHT FACTOR = BETA(I+1)
C     END OF FACTORIZATION AND FORWARD SUBSTITUTION
C
C     BACK SUBSTITUTION
      V2(MEV) = V2(MEV)/U
      DO 80 II = 1,MM1
      I = MEV-II
      IF (BETA(I+1).LT.ZERO) GO TO 70
C     NO PIVOT INTERCHANGE
      V2(I) = V2(I)-V1(I)*V2(I+1)
      GO TO 80
C     PIVOT INTERCHANGE
   70 BETA(I+1) = -BETA(I+1)
      V2(I) = (V2(I)-V1(I)*V2(I+1)-BETA(I+2)*V2(I+2))/BETA(I+1)
   80 CONTINUE
C
C
C     TESTS FOR CONVERGENCE OF INVERSE ITERATION
C     IF SUM |V2| COMPS. LE. 1 AND IT. LE. ITER DO ANOTHER INVIT STEP
C
      NORM = DABS(V2(MEV))
      DO 90 II = 1,MM1
      I = MEV-II
   90 NORM = NORM+DABS(V2(I))
C
C     IS DESIRED GROWTH IN VECTOR ACHIEVED ?
C     IF NOT, DO ANOTHER INVERSE ITERATION STEP UNLESS NUMBER ALLOWED IS
C     EXCEEDED.
```

```
      IF (NORM.GE.ONE) GO TO 110
C
      IT=IT+1
      IF (IT.GT.ITER) GO TO 110
C
      XU = EPS4/NORM
      DO 100 I=1,MEV
  100 V2(I) = V2(I)*XU
C
      GO TO 40
C
C     NORMALIZE COMPUTED T-EIGENVECTOR : V2 = V2/||V2||
C
  110 CONTINUE
C
      SUM = FINPRO(MEV,V2(1),1,V2(1),1)
      SUM = ONE/DSQRT(SUM)
      DO 120 II = 1,MEV
  120 V2(II) = SUM*V2(II)
C
C     SAVE ERROR ESTIMATE FOR LATER OUTPUT
      ERROR = DABS(V2(MEV))
C
C     GENERATE ERRORV = ||T*V2 - X1*V2||.
      V1(MEV) = ALPHA(MEV)*V2(MEV)+BETA(MEV)*V2(MEV-1)-X1*V2(MEV)
      DO 130 J = 2,MM1
      JM = MP1 - J
      V1(JM) = ALPHA(JM)*V2(JM) + BETA(JM)*V2(JM-1) + BETA(JM+1)*V2(JM+1
     1) - X1*V2(JM)
  130 CONTINUE
C
      V1(1) = ALPHA(1)*V2(1) + BETA(2)*V2(2) - X1*V2(1)
      ERRORV = FINPRO(MEV,V1(1),1,V1(1),1)
      ERRORV = DSQRT(ERRORV)
      IF (IT.GT.ITER) ERRORV = -ERRORV
      IF (IWRITE.EQ.0) GO TO 150
C
C     FILE 6 (TERMINAL) OUTPUT OF ERROR ESTIMATES.
      WRITE(6,140) MEV,X1,ERROR,ERRORV
  140 FORMAT(2X,'TSIZE',15X,'EIGENVALUE',11X,'U(M)',9X,'ERRORV'/
     1 I6,E25.16,2E15.5)
C
C     RESTORE BETA(MEV+1) = BETAM
  150 CONTINUE
      BETA(MP1) = BETAM
C-----END OF INVERM-----------------------------------------------------
      RETURN
      END
C
C-----START OF LBISEC---------------------------------------------------
C
      SUBROUTINE LBISEC(ALPHA,BETA,EPSM,EVAL,EVALN,LB,UB,TTOL,M,NEVT)
C
C-----------------------------------------------------------------------
      DOUBLE PRECISION  ALPHA(1),BETA(1),X0,X1,XL,XU,YU,YV,LB,UB
      DOUBLE PRECISION  EPSM,EP1,EVAL,EVALN,EVD,EPT
      DOUBLE PRECISION  ZERO,ONE,HALF,TTOL,TEMP
      DOUBLE PRECISION  DABS,DSQRT,DFLOAT
C-----------------------------------------------------------------------
C     SPECIFY PARAMETERS
      ZERO = 0.0D0
      HALF = 0.5D0
      ONE  = 1.0D0
      XL = LB
      XU = UB
C
C     EP1 = DSQRT(1000+M)*TTOL     TTOL = EPSM*TKMAX
```

```
C     TKMAX = MAX(|ALPHA(K)|,BETA(K), K= 1,KMAX)
C
      TEMP = DFLOAT(1000+M)
      EP1 = DSQRT(TEMP)*TTOL
C
      NA = 0
      X1 = XU
      JSTURM = 1
      GO TO 60
C     FORWARD STURM CALCULATION
   10 NA = NEV
      X1 = XL
      JSTURM = 2
      GO TO 60
C     FORWARD STURM CALCULATION
   20 NEVT = NEV
C
C     WRITE(6,30) M,EVAL,NEVT,EP1
   30 FORMAT(/3X,'TSIZE',23X,'EV',9X/I8,E25.16/
     1 I6,' = NUMBER OF T(1,M) EIGENVALUES ON TEST INTERVAL'/
     1 E12.3,' =  CONVERGENCE TOLERANCE'/)
C
      IF (NEVT.NE.1) GO TO 120
C
C     BISECTION LOOP
      JSTURM = 3
   40 X1 = HALF*(XL+XU)
      XO = XU-XL
      EPT = EPSM*(DABS(XL) + DABS(XU)) + EP1
C     CONVERGENCE TEST
      IF (XO.LE.EP1) GO TO 100
      GO TO 60
C     FORWARD STURM CALCULATION
   50 CONTINUE
      IF(NEV.EQ.0) XU = X1
      IF(NEV.EQ.1) XL = X1
      GO TO 40
C     NEV = NUMBER OF T-EIGENVALUES OF T(1,M) ON (X1,XU)
C     THERE IS EXACTLY ONE T-EIGENVALUE OF T(1,M) ON (XL,XU)
C
C     FORWARD STURM CALCULATION
   60 NEV = -NA
      YU = ONE
      DO 90 I = 1,M
      IF (YU.NE.ZERO) GO TO 70
      YV = BETA(I)/EPSM
      GO TO 80
   70 YV = BETA(I)*BETA(I)/YU
   80 YU = X1 - ALPHA(I) - YV
      IF (YU.GE.ZERO) GO TO 90
      NEV = NEV+1
   90 CONTINUE
      GO TO (10,20,50), JSTURM
C
  100 CONTINUE
C
      EVALN = X1
      EVD = DABS(EVALN-EVAL)
C     WRITE(6,110) EVALN,EVAL,EVD
  110 FORMAT(/20X,'EVALN',21X,'EVAL',6X,'CHANGE'/2E25.16,E12.3/)
C
  120 CONTINUE
      RETURN
C-----END OF LBISEC--------------------------------------------------
      END
C-----START OF COMPLEX INNER PRODUCT-----------------------------------
C
```

```
C       COMPLEX INNER PRODUCT
C
        SUBROUTINE CINPRD(V2,V1,SUM,N)
C-----------------------------------------------------------------------
        DOUBLE PRECISION  ZERO,SUM
        COMPLEX*16 V2(1),V1(1),SUMC
C-----------------------------------------------------------------------
C
C       NOTE THAT THE ORDER MATTERS HERE
C       COMPUTES THE INNER PRODUCT OF THE CONJUGATE OF V2 WITH V1.
        ZERO = 0.D0
        SUMC = DCMPLX(ZERO,ZERO)
        DO 10 J=1,N
   10   SUMC = SUMC + DCONJG(V2(J))*V1(J)
        SUM = DREAL(SUMC)
C
        RETURN
C-----END OF COMPLEX INNER PRODUCT SUBROUTINE---------------------------
        END
C
C-----LPERM-PERMUTES VECTORS--------------------------------------------
C
        SUBROUTINE LPERM(W,U,IPERM)
C
C-----------------------------------------------------------------------
        DOUBLE PRECISION  U(1),W(1)
        INTEGER  IPR(1),IPT(1)
C-----------------------------------------------------------------------
C       SUBROUTINE HAS 2 BRANCHES:  IPERM = 1,  CALCULATES
C       U = P*W  WHERE  P IS THE PERMUTATION REPRESENTED BY IPR
C       LET J = IPR(K) THEN U(K) = W(J), K = 1,N. WE SET W(K)=U(K), K=1,N
C       IPERM = 2, USING THE PERMUTATION IPT (P-TRANSPOSE) U = P'*W, W=U
C       LET J = IPT(K) THEN U(K) = W(J), K=1,N. WE SET W(K) = U(K), K=1,N
C-----------------------------------------------------------------------
C
        IF(IPERM.EQ.2) GO TO 30
C       IPERM = 1
        DO 10 K = 1,N
        J = IPR(K)
   10   U(K) = W(J)
        DO 20 K = 1,N
   20   W(K) = U(K)
        GO TO 60
C       IPERM = 2
   30   DO 40 K = 1,N
        J = IPT(K)
   40   U(K) = W(J)
        DO 50 K = 1,N
   50   W(K) = U(K)
   60   CONTINUE
C
        RETURN
C
C-----------------------------------------------------------------------
        ENTRY LPERME(IPR,IPT,N)
C-----------------------------------------------------------------------
C
C-----END OF LPERM-------------------------------------------
        RETURN
        END
```

SECTION 2.7 OPTIONAL PREPROCESSING PROGRAM, FILE DEFINITIONS, SAMPLE INPUT/OUTPUT FILES

```
C-----LECOMPAC-(STAND-ALONE PROGRAM)-----------------------------------
C
C     THIS PROGRAM TRANSLATES A SPARSE SYMMETRIC N X N MATRIX A,
C     GIVEN AS I, J, A(I,J), INTO THE SPARSE MATRIX FORMAT
C     REQUIRED BY THE SAMPLE USPEC AND CMATV PROGRAMS PROVIDED
C     FOR USE WITH THE LANCZOS EIGENVALUE/EIGENVECTOR PROCEDURES.
C     THIS PROGRAM ASSUMES THAT THE MATRIX ENTRIES ARE PROVIDED
C     EITHER COLUMN BY COLUMN OR ROW BY ROW.
C     NOTE THAT THIS PROGRAM DOES NOT DIRECTLY APPLY TO THE
C     HERMITIAN CASE BECAUSE FOR HERMITIAN MATRICES THE DIAGONALS
C     ARE REAL AND THE OFF-DIAGONAL ENTRIES ARE COMPLEX VARIABLES.
C
C     NONPORTABLE STATEMENTS:  PFORT VERIFIER INDICATES THAT THIS
C                              IS PORTABLE.
C
C----------------------------------------------------------------------
      DOUBLE PRECISION A(15000), AD(2000)
      DOUBLE PRECISION ZERO
      INTEGER IROW(15000),ICOL(15000)
C----------------------------------------------------------------------
C    INPUT FILE 7 CONTAINS THE SPARSE SYMMETRIC NXN MATRIX STORED AS:
C
C              NZ,M,N,MATNO
C              I(K)  J(K)  A(K)  K = 1,NZ
C
C     WHERE NZ IS THE TOTAL NUMBER OF NONZEROS IN THE MATRIX A,
C     N IS THE ROW AND COLUMN DIMENSION OF A,
C     AND A(K) ARE THE NONZERO ENTRIES STORED ROW BY ROW OR
C     COLUMN BY COLUMN.  PROGRAM READS THIS IN AS IROW(K) = I(K),
C     ICOL(K) = J(K), AND A(K) = A(K).
C
C     OUTPUT FILE = 8 CONTAINS THE A-MATRIX IN SPARSE FORMAT
C
C            NZS,N,NZL,MATNO
C            ICOL(K)     K = 1,NZL
C            IROW(K)     K = 1,NZS
C              AD(K)     K = 1,N
C               A(K)     K = 1,NZS
C
C      WHERE N IS THE ORDER OF THE INPUT MATRIX A,
C      NZ EQUALS THE NUMBER OF NONZERO ELEMENTS IN A WHICH ARE ON
C      OR BELOW THE MAIN DIAGONAL.  NZL EQUALS THE NUMBER OF THE
C      LAST COLUMN HAVING NONZEROES BELOW THE DIAGONAL IN A.
C      NZS EQUALS THE NUMBER OF NONZERO ELEMENTS BELOW THE MAIN
C      DIAGONAL.   AD(K), K=1,N, CONTAINS THE DIAGONAL ELEMENTS OF A.
C      A(K), K=1,NZS, CONTAINS THE KTH NONZERO SUB-DIAGONAL ELEMENT
C      OF THE INPUT MATRIX.  A IS STORED COLUMN BY COLUMN.
C      IROW(K), K=1,NZS, CONTAINS THE ROW INDEX OF THE  NONZERO
C      STRICTLY LOWER TRIANGULAR ELEMENT A(K).
C      ICOL(K), K=1,NZL, EQUALS THE NUMBER OF STRICTLY LOWER
C      TRIANGULAR NONZEROES IN COLUMN K OF THE INPUT MATRIX.
C
C----------------------------------------------------------------------
      ZERO = 0.0D0
C
      READ(7,10) NZ,N,MATNO,IIROW
   10 FORMAT(2I6,I8,I4)
C
      WRITE(6,20) NZ,N,MATNO,IIROW
   20 FORMAT(I10,I6,I10,' = NO. NONZERO AIJ J.GE.I, ORDER OF A, MATNO'/
     1 I6,' = IIROW IF IIROW=0 ORDERING IS BY COLS IIROW=1 BY ROWS'/)
C
      DO 30 K = 1,N
```

```
   30 AD(K) = ZERO
C
      IF (IIROW.EQ.0) READ(7,40) (IROW(K),ICOL(K),A(K), K=1,NZ)
C
      IF (IIROW.EQ.1) READ(7,40) (ICOL(K),IROW(K),A(K), K=1,NZ)
   40 FORMAT(2I5,E14.7)
C
      LCOUNT = 0
      K = 1
C
C     START OF A NEW COLUMN
   50 CONTINUE
      J = ICOL(K)
      ICOL(J) = 0
   60 CONTINUE
C
      IF (J.NE.IROW(K)) GO TO 70
C
C     DIAGONAL CASE
      AD(J) = A(K)
      GO TO 80
C
C     SUB-DIAGONAL NONZERO
   70 CONTINUE
      NZL = J
      LCOUNT = LCOUNT + 1
      A(LCOUNT) = A(K)
      IROW(LCOUNT) = IROW(K)
      ICOL(J) = ICOL(J) + 1
C
   80 CONTINUE
      K = K+1
C
      IF(K.GT.NZ) GO TO 90
C
      IF(ICOL(K).GT.J) GO TO 50
C
      GO TO 60
C
   90 CONTINUE
      NZS = LCOUNT
C
      WRITE(8,100) NZS,N,NZL,MATNO
      WRITE(6,100) NZS,N,NZL,MATNO
  100 FORMAT(I10,2I6,I8,' = NZS N NZL MATNO')
C
      WRITE(8,110) (ICOL(I), I=1,NZL)
      WRITE(8,110) (IROW(K), K=1,NZS)
  110 FORMAT(13I6)
C
      WRITE(8,120) (AD(K), K=1,N)
      WRITE(8,120) (A(K), K=1,NZS)
  120 FORMAT(4E19.10)
C
C-----END LECOMPAC-----------------------------------------------------
      STOP
      END
```

-----INPUT/OUTPUT FILE DEFINITIONS FOR LEVAL AND LEVEC-----

Below is a listing of the input/output files which are accessed by the real symmetric Lanczos eigenvalue program, LEVAL. Included also is a sample of the input file which LEVAL requires on file 5. The parameters supplied in this file are in free format. LEVAL computes eigenvalues of real symmetric matrices on user-specified intervals.

SAMPLE DEFINITIONS OF THE INPUT/OUTPUT FILES FOR LEVAL

```
-----------------------------------------------------------------
*LEVAL EXEC LANCZOS EIGENVALUE CALCULATION REAL SYMMETRIC MATRICES
FI 06 TERM
FILEDEF  1 DISK &1        NHISTORY  A (RECFM F LRECL 80 BLOCK 80
FILEDEF  2 DISK &1        HISTORY   A (RECFM F LRECL 80 BLOCK 80
FILEDEF  3 DISK &1        GOODEV    A (RECFM F LRECL 80 BLOCK 80
FILEDEF  4 DISK &1        ERRINV    A (RECFM F LRECL 80 BLOCK 80
FILEDEF  5 DISK LEVAL     INPUT     A (RECFM F LRECL 80 BLOCK 80
FILEDEF  8 DISK &1        INPUT     A (RECFM F LRECL 80 BLOCK 80
FILEDEF 11 DISK &1        DISTINCT  A (RECFM F LRECL 80 BLOCK 80
LOAD   LEVAL   LESUB   LEMULT
-----------------------------------------------------------------
```

SAMPLE INPUT FILE FOR LEVAL

```
-----------------------------------------------------------------
 LANCZOS EIGENVALUE COMPUTATIONS, NO REORTHOGONALIZATION
 TEST MATRIX
LINE 1      N     KMAX     NMEVS      MATNO
          143     429          1     706830
LINE 2   SVSEED    RHSEED       MXINIT    MXSTUR
        7892713    147935            5    100000
LINE 3   ISTART      ISTOP
              0          1
LINE 4        IHIS      IDIST  IWRITE
                 1          0       1
LINE 5   RELTOL (RELATIVE TOLERANCE IN 'COMBINING' GOODEV)
    .0000000001
LINE 6   MB(1)   MB(2)   MB(3)   MB(4) (ORDERS OF T(1,MEV))
           190
LINE 7    NINT      (NUMBER OF SUB-INTERVALS FOR BISEC)
             1
LINE 8    LB(1)     LB(2)   LB(3)   (INTERVAL LOWER BOUNDS)
            0.0
LINE 9    UB(1)     UB(2)   UB(3)   (INTERVAL UPPER BOUNDS)
          1.001
-----------------------------------------------------------------
```

Below is a listing of the input/output files which are accessed by the real symmetric Lanczos eigenvector program, LEVEC. Included also is a sample of the input file which LEVEC requires on file 5. The parameters are supplied in free format. LEVEC computes eigenvectors for each of a user-specified subset of the eigenvalues computed by the companion program LEVAL.

```
SAMPLE DEFINITIONS OF THE INPUT/OUTPUT FILES FOR LEVEC
----------------------------------------------------------------
*LEVEC EXEC TO RUN LANCZOS EIGENVECTOR PROGRAM, REAL SYMMETRIC MATRICES
FI 06 TERM
FILEDEF  2 DISK &1      HISTORY   A (RECFM F LRECL 80 BLOCK 80
FILEDEF  3 DISK &1      GOODEV    A (RECFM F LRECL 80 BLOCK 80
FILEDEF  4 DISK &1      ERRINV    A (RECFM F LRECL 80 BLOCK 80
FILEDEF  5 DISK LEVEC   INPUT     A (RECFM F LRECL 80 BLOCK 80
FILEDEF  8 DISK &1      INPUT     A (RECFM F LRECL 80 BLOCK 80
FILEDEF  9 DISK &1      ERREST    A (RECFM F LRECL 80 BLOCK 80
FILEDEF 10 DISK &1      BOUNDS    A (RECFM F LRECL 80 BLOCK 80
FILEDEF 11 DISK &1      TEIGVECS  A (RECFM F LRECL 80 BLOCK 80
FILEDEF 12 DISK &1      RITZVECS  A (RECFM F LRECL 80 BLOCK 80
FILEDEF 13 DISK &1      PAIGE     A (RECFM F LRECL 80 BLOCK 80
LOAD  LEVEC  LESUB  LEMULT
----------------------------------------------------------------
```

```
SAMPLE INPUT FILE FOR LEVEC
----------------------------------------------------------------
 LEVEC REAL SYMMETRIC EIGENVECTOR COMPUTATIONS, NO REORTHOGONALIZATION
LINE 1  MDIMTV    MDIMRV  MBETA(MAX.DIMENSIONS, TVEC, RITVEC AND BETA
         10000     10000   2000
LINE 2       RELTOL
        .0000000001
LINE 3  MBOUND   NTVCON SVTVEC IREAD (FLAGS
             0        1      0     1
LINE 4  TVSTOP   LVCONT ERCONT  IWRITE (FLAGS
             0        1      1       1
LINE 5    RHSEED  (RANDOM GENERATOR SEED FOR STARTING VECTOR IN INVERM)
        45329517
LINE 6  MATNO       N
          100     100
----------------------------------------------------------------------
```

-----SAMPLE OUTPUT FOR REAL SYMMETRIC EIGENVALUE COMPUTATIONS-----

Below is a sample of the output from the LEVAL eigenvalue program. The output which is printed to file 6 is given first, followed by the outputs which are generated and written to other files. For this particular test matrix the true eigenvalues are known. Therefore, the true errors in the computed eigenvalue approximations are also listed below.

```
LANCZOS PROCEDURE FOR REAL SYMMETRIC MATRICES

LEVAL INPUT LANCZOS EIGENVALUE COMPUTATION, NO REORTHOGONALIZATION
POISSON MATRIX N = 528  KX =  33   KY = 16
     POISSON INPUT

    N     KX     KY       DIAGONAL    X-CODIAGONAL   Y-CODIAGONAL
  528     33     16 0.10000000D+01 0.25000000D+00 0.25000000D+00

A-SIZE  X-DIM  Y-DIM
   528     33     16
    A-DIAGONAL    X-CODIAGONAL   Y-CODIAGONAL           ATOLN
0.10000000D+01 0.25000000D+00 0.25000000D+00 0.883450490-15
  401 = NUMBER OF TRUE A-EIGENVALUES WHICH ARE DISTINCT


  MATRIX ID     ORDER OF A    MAX ORDER OF T
     607840            528              1320


 ISTART   ISTOP
     0       1


   IHIS   IDIST  IWRITE
      1       0       0


SEEDS FOR RANDOM NUMBER GENERATOR

   LANCZS SEED     INVERR SEED
       7892713          147935


SIZES OF T-MATRICES TO BE CONSIDERED
        792         1056

RELATIVE TOLERANCE USED TO COMBINE COMPUTED T-EIGENVALUES
     0.100D-09
RELATIVE GAP TOLERANCES USED IN INVERSE ITERATION
     0.100D-07
RELATIVE TOLERANCE FOR CHECK ON SIZE OF BETAS
     0.100D-07


BISEC WILL BE USED ON THE FOLLOWING INTERVALS
    1         0.900000D+00          0.100001D+01


T-MATRICES (ALPHA AND BETA) ARE NOW AVAILABLE


MINIMUM BETA RATIO OCCURS AT    616 TH BETA


TSIZE       MIN BETA      TKMAX       MIN RATIO
```

```
 1320      0.211D+00 0.153D+01       0.138D+00


BISEC CALCULATION
ORDER OF T IS   792
NUMBER OF INTERVALS IS      1


MULTOL, TOLERANCE USED IN T-MULTIPLICITY AND SPURIOUS TESTS =  0.122D-11
BISTOL, TOLERANCE USED IN BISEC CONVERGENCE TEST =  0.288D-13



BISEC INTERVAL NO  LOWER BOUND  UPPER BOUND
                1  0.90000D+00  0.10000D+01


   33   396 = NO. TMEV ON (LB,UB) AND NO. .GT. UB


T-EIGENVALUE CALCULATION ON INTERVAL      1  IS COMPLETE
  NO. GOOD   NO. DISTINCT
      33          33

   33 GOOD T-EIGENVALUES HAVE BEEN COMPUTED
   33 OF THESE ARE T-ISOLATED
   33 = NUMBER OF DISTINCT T-EIGENVALUES COMPUTED


CONVERGENCE IS TESTED USING THE CONVERGENCE TOLERANCE   0.4172D-10



T-EIGENVALUE CALCULATION AT MEV =    792    IS COMPLETE

BISEC CALCULATION
ORDER OF T IS  1056
NUMBER OF INTERVALS IS      1


MULTOL, TOLERANCE USED IN T-MULTIPLICITY AND SPURIOUS TESTS =  0.140D-11
BISTOL, TOLERANCE USED IN BISEC CONVERGENCE TEST =  0.308D-13



BISEC INTERVAL NO  LOWER BOUND  UPPER BOUND
                1  0.90000D+00  0.10000D+01


   40   528 = NO. TMEV ON (LB,UB) AND NO. .GT. UB


T-EIGENVALUE CALCULATION ON INTERVAL      1  IS COMPLETE
  NO. GOOD   NO. DISTINCT
      36          40

   36 GOOD T-EIGENVALUES HAVE BEEN COMPUTED
   36 OF THESE ARE T-ISOLATED
   40 = NUMBER OF DISTINCT T-EIGENVALUES COMPUTED


CONVERGENCE IS TESTED USING THE CONVERGENCE TOLERANCE   0.4755D-10
```

T-EIGENVALUE CALCULATION AT MEV = 1056 IS COMPLETE
RUNTIME = 1298 HNDRETHS OF A SECOND

33 33 792 528 7892713 607840 = NG,NDIS,MEV,N,SVEED,MATNO
0.121889836108D-11 616 0.1000D-07 = MUTOL,INDEX MINIMAL BETA,BTOL

EV NO	TMULT	GOOD EIGENVALUE	TMINGAP	AMINGAP
1	1	0.902308704549800 1D+00	0.960D-03	0.960E-03
2	1	0.9032683439082962D+00	0.960D-03	-0.960E-03
3	1	0.9075220662489024D+00	0.210D-03	0.210E-03
4	1	0.9077316472575748D+00	0.210D-03	-0.210E-03
5	1	0.9093026847308840D+00	0.157D-02	-0.157E-02
6	1	0.9117392539653555D+00	0.222D-02	0.222E-02
7	1	0.9139623171524736D+00	0.222D-02	-0.222E-02
8	1	0.9215518597033493D+00	0.371D-03	0.371E-03
9	1	0.9219228129503947D+00	0.371D-03	-0.371E-03
10	1	0.9272414797174763D+00	0.457D-02	0.457E-02
11	1	0.9318128595969843D+00	0.960D-03	0.960E-03
12	1	0.9327725002595535D+00	0.850D-03	0.850E-03
13	1	0.9336220180364069D+00	0.850D-03	-0.850E-03
14	1	0.9441957465130982D+00	0.200D-03	0.200E-03
15	1	0.9443958908053349D+00	0.200D-03	-0.200E-03
16	1	0.9538733953927414D+00	0.108D-02	0.108E-02
17	1	0.9549583603516555D+00	0.108D-02	-0.108E-02
18	1	0.9561549058563865D+00	0.120D-02	-0.120E-02
19	1	0.9582577195407838D+00	0.750D-03	0.750E-03
20	1	0.9590079809845515D+00	0.750D-03	-0.750E-03
21	1	0.9618498555646015D+00	0.259D-02	0.259E-02
22	1	0.9644445008375758D+00	0.259D-02	-0.259E-02
23	1	0.9672303320363761D+00	0.113D-02	0.113E-02
24	1	0.9683565650437763D+00	0.113D-02	-0.113E-02
25	1	0.9704879822455046D+00	0.213D-02	-0.213E-02
26	1	0.9738976059208907D+00	0.852D-03	0.852E-03
27	1	0.9747491588235540D+00	0.852D-03	-0.852E-03
28	1	0.9775269056442103D+00	0.278D-02	-0.278E-02
29	1	0.9813455306889490D+00	0.382D-02	-0.382E-02
30	1	0.9853232930778423D+00	0.398D-02	-0.398E-02
31	1	0.9894262717440089D+00	0.410D-02	-0.410E-02
32	1	0.9936194616944066D+00	0.419D-02	-0.419E-02
33	1	0.9999999999999889D+00	0.638D-02	-0.638E-02

36 40 1056 528 7892713 607840 = NG,NDIS,MEV,N,SVEED,MATNO
0.139846820891D-11 616 0.1000D-07 = MUTOL,INDEX MINIMAL BETA,BTOL

EV NO	TMULT	GOOD EIGENVALUE	TMINGAP	AMINGAP
1	1	0.9023087045495045D+00	0.960D-03	0.960E-03
2	1	0.9032683439081598D+00	0.960D-03	-0.960E-03
3	1	0.9075220637981604D+00	0.210D-03	0.210E-03
4	1	0.9077316405366920D+00	0.210D-03	-0.210E-03
5	1	0.9093026846956148D+00	-0.332D-03	-0.157E-02
6	1	0.9117392539584653D+00	0.210D-02	0.222E-02
7	1	0.9139623171477664D+00	0.222D-02	-0.222E-02
8	1	0.9215518596986422D+00	0.371D-03	0.371E-03
9	1	0.9219228129327942D+00	0.371D-03	-0.371E-03
10	1	0.9272414797172943D+00	-0.301D-04	0.457E-02
11	1	0.9318128595792929D+00	0.960D-03	0.960E-03
12	1	0.9327724989379482D+00	0.850D-03	0.850E-03
13	1	0.9336220180228539D+00	0.850D-03	-0.850E-03
14	1	0.9441957462784023D+00	0.200D-03	0.200E-03
15	1	0.9443958907455297D+00	0.200D-03	-0.200E-03
16	1	0.9538658202683405D+00	0.394D-03	0.394E-03
17	1	0.9542594208233769D+00	0.394D-03	-0.394E-03
18	1	0.9550432638722495D+00	0.784D-03	-0.784E-03
19	1	0.9560950958355893D+00	0.116D-03	0.116E-03
20	1	0.9562106619424611D+00	0.116D-03	-0.116E-03
21	1	0.9577516552053169D+00	0.112D-02	0.112E-02
22	1	0.9588724531626193D+00	0.781D-03	0.781E-03

```
  23     1  0.959653096449599 4D+00      0.781D-03     -0.781E-03
  24     1  0.961898763249054 9D+00      0.225D-02     -0.225E-02
  25     1  0.964469496366357 1D+00      0.257D-02     -0.257E-02
  26     1  0.967343363212947 6D+00      0.103D-02      0.103E-02
  27     1  0.968369026554663 9D+00      0.103D-02     -0.103E-02
  28     1  0.970495844970200 6D+00      0.213D-02     -0.213E-02
  29     1  0.973900045775317 9D+00      0.850D-03      0.850E-03
  30     1  0.974749564860223 9D+00      0.850D-03     -0.850E-03
  31     1  0.977526922187276 5D+00     -0.163D-02     -0.278E-02
  32     1  0.981345530975354 8D+00     -0.219D-02     -0.382E-02
  33     1  0.985323293115771 9D+00      0.398D-02     -0.398E-02
  34     1  0.989426271744463 6D+00      0.410D-02     -0.410E-02
  35     1  0.993619461694429 3D+00     -0.311D-04     -0.419E-02
  36     1  0.999999999999988 9D+00      0.638D-02     -0.638E-02

ABOVE ARE COMPUTED GOOD T-EIGENVALUES
NG = NUMBER OF GOOD T-EIGENVALUES COMPUTED
NDIS = NUMBER OF COMPUTED DISTINCT EIGENVALUES OF T(1,MEV)
N = ORDER OF A,  MATNO = MATRIX IDENT
MULTOL = T-MULTIPLICITY TOLERANCE FOR T-EIGENVALUES IN BISEC
TMULT IS THE T-MULTIPLICITY OF GOOD T-EIGENVALUE
TMULT = -1 MEANS SPURIOUS T-EIGENVALUE TOO CLOSE
DO NOT COMPUTE ERROR ESTIMATES FOR SUCH EIGENVALUES
AMINGAP = MINIMAL GAP BETWEEN THE COMPUTED A-EIGENVALUES
AMINGAP .LT. 0. MEANS MINIMAL GAP IS DUE TO LEFT-HAND GAP
TMINGAP= MINIMAL GAP W.R.T.  DISTINCT EIGENVALUES IN T(1,MEV)
TMINGAP .LT. 0. MEANS MINGAP IS DUE TO SPURIOUS T-EIGENVALUE
----- END OF FILE 3 GOODEIGENVALUES-----------------------



INVERSE ITERATION ERROR ESTIMATES

TSIZE  NDIS NGOOD  NISO ASIZE
  792    33    33    33   528
   RHSEED  MXINIT     BETAM
   147935       5 0.417D+00
 GOODEVNO        GOOD T-EIGENVALUE        BETAM*UM        TMINGAP
        1  0.902308704549800 1D+00      0.797E-06      0.960D-03
        2  0.903268343908296 2D+00      0.515E-06     -0.960D-03
        3  0.907522066248902 4D+00      0.590E-04      0.210D-03
        4  0.907731647257574 8D+00      0.970E-04     -0.210D-03
        5  0.909302684730884 0D+00      0.667E-05     -0.157D-02
        6  0.911739253965355 5D+00      0.273E-05      0.222D-02
        7  0.913962317152473 6D+00      0.212E-05     -0.222D-02
        8  0.921551859703349 3D+00      0.173E-05      0.371D-03
        9  0.921922812950394 7D+00      0.331E-05     -0.371D-03
       10  0.927241479717476 3D+00      0.297E-06      0.457D-02
       11  0.931812859596984 3D+00      0.255E-05      0.960D-03
       12  0.932772500259553 5D+00      0.215E-04      0.850D-03
       13  0.933622018036406 9D+00      0.212E-05     -0.850D-03
       14  0.944195746513098 2D+00      0.606E-05      0.200D-03
       15  0.944395890805334 9D+00      0.303E-05     -0.200D-03
       16  0.953873395392741 4D+00      0.282E-03      0.108D-02
       17  0.954958360351655 5D+00      0.149E-02     -0.108D-02
       18  0.956154905856386 5D+00      0.158E-02     -0.120D-02
       19  0.958257719540783 8D+00      0.360E-02      0.750D-03
       20  0.959007980984551 5D+00      0.169E-02     -0.750D-03
       21  0.961849855564601 5D+00      0.140E-02      0.259D-02
       22  0.964444500837575 8D+00      0.135E-02     -0.259D-02
       23  0.967230332036376 1D+00      0.341E-02      0.113D-02
       24  0.968356565043776 3D+00      0.113E-02     -0.113D-02
       25  0.970487982245504 6D+00      0.102E-02     -0.213D-02
       26  0.973897605920890 7D+00      0.647E-03      0.852D-03
       27  0.974749158823554 0D+00      0.271E-03     -0.852D-03
       28  0.977526905644210 3D+00      0.593E-04     -0.278D-02
```

```
       29   0.9813455306889490D+00      0.858E-05     -0.382D-02
       30   0.9853232930778423D+00      0.341E-05     -0.398D-02
       31   0.9894262717440089D+00      0.403E-06     -0.410D-02
       32   0.9936194616944066D+00      0.112E-06     -0.419D-02
       33   0.9999999999999889D+00      0.363E-07     -0.638D-02
TSIZE  NDIS NGOOD  NISO ASIZE
 1056    40    36    36   528
   RHSEED   MXINIT     BETAM
   147935        5 0.476D+00
  GOODEVNO         GOOD T-EIGENVALUE       BETAM*UM         TMINGAP
         1  0.9023087045495045D+00      0.391E-13      0.960D-03
         2  0.9032683439081598D+00      0.472E-13     -0.960D-03
         3  0.9075220637981604D+00      0.129E-11      0.210D-03
         4  0.9077316405366920D+00      0.297E-11     -0.210D-03
         5  0.9093026846956148D+00      0.219E-12      0.332D-03
         6  0.9117392539584653D+00      0.299E-14     -0.210D-02
         7  0.9139623171477664D+00      0.452E-13     -0.222D-02
         8  0.9215518596986422D+00      0.654E-14      0.371D-03
         9  0.9219228129327942D+00      0.356E-13     -0.371D-03
        10  0.9272414797172943D+00      0.198E-11     -0.301D-04
        11  0.9318128595792929D+00      0.386E-13      0.960D-03
        12  0.9327724989379482D+00      0.648E-13      0.850D-03
        13  0.9336220180228539D+00      0.488E-14     -0.850D-03
        14  0.9441957462784023D+00      0.164E-13      0.200D-03
        15  0.9443958907455297D+00      0.319E-13     -0.200D-03
        16  0.9538658202683405D+00      0.240E-09      0.394D-03
        17  0.9542594208233769D+00      0.232E-08     -0.394D-03
        18  0.9550432638722495D+00      0.201E-08     -0.784D-03
        19  0.9560950958355893D+00      0.189E-07      0.116D-03
        20  0.9562106619424611D+00      0.172E-07     -0.116D-03
        21  0.9577516552053169D+00      0.214E-08      0.112D-02
        22  0.9588724531626193D+00      0.474E-09      0.781D-03
        23  0.9596530964495994D+00      0.769E-09     -0.781D-03
        24  0.9618987632490549D+00      0.589E-10     -0.225D-02
        25  0.9644694963663571D+00      0.174E-10     -0.257D-02
        26  0.9673433632129476D+00      0.229E-10      0.103D-02
        27  0.9683690265546639D+00      0.727E-11     -0.103D-02
        28  0.9704958449702006D+00      0.514E-11     -0.213D-02
        29  0.9739000457753179D+00      0.301E-11      0.850D-03
        30  0.9747495648602239D+00      0.151E-11     -0.850D-03
        31  0.9775269221872765D+00      0.774E-12      0.163D-02
        32  0.9813455309753548D+00      0.765E-13     -0.219D-02
        33  0.9853232931157719D+00      0.137E-13     -0.398D-02
        34  0.9894262717444636D+00      0.436E-13     -0.410D-02
        35  0.9936194616944293D+00      0.114E-14     -0.311D-04
        36  0.9999999999999889D+00      0.155E-10     -0.638D-02

ABOVE ARE THE ERROR ESTIMATES OBTAINED FOR THE ISOLATED GOOD T-EIGENVALUES
OBTAINED VIA INVERSE ITERATION IN THE SUBROUTINE INVERR.
ALL OTHER GOOD T-EIGENVALUES HAVE CONVERGED.
ERROR ESTIMATE = BETAM*ABS(UM)
WHERE BETAM = BETA(MEV+1) AND UM = U(MEV).
U = UNIT EIGENVECTOR OF T WHERE T*U = EV*U AND EV = ISOLATED GOOD T-EIGENVALUE.
TMINGAP = GAP TO NEAREST DISTINCT EIGENVALUE OF T(1,MEV).
TMINGAP .LT. 0. MEANS MINGAP IS DUE TO A LEFT NEIGHBOR.
ERROR ESTIMATE L.T. 0 MEANS INVERSE ITERATION DID NOT CONVERGE
------ END OF FILE 4 ERRINV ---------------------------------


POISSONZ TRUE ERROR FOR GOOD EIGENVALUES
    N     NX     NY
  528     33     16

    A-DIAGONAL    X-CODIAGONAL    Y-CODIAGONAL
0.10000000D+01 0.25000000D+00 0.25000000D+00
```

792 33 401 = MEV,NG,NEXACT

T-EV NO	A-EV NO	GOOD EIGENVALUE	TRUEERROR	TMINGAP
1	166	0.9023087045498001D+00	0.292E-12	0.960E-03
2	167	0.9032683439082962D+00	0.145E-12	0.960E-03
3	168	0.9075220662489024D+00	0.245E-08	0.210E-03
4	169	0.9077316472575748D+00	0.672E-08	0.210E-03
5	170	0.9093026847308840D+00	0.353E-10	0.157E-02
6	171	0.9117392539653555D+00	0.688E-11	0.222E-02
7	172	0.9139623171524736D+00	0.470E-11	0.222E-02
8	173	0.9215518597033493D+00	0.471E-11	0.371E-03
9	174	0.9219228129503947D+00	0.176E-10	0.371E-03
10	175	0.9272414797174763D+00	0.187E-12	0.457E-02
11	176	0.9318128595969843D+00	0.177E-10	0.960E-03
12	177	0.9327725002595535D+00	0.132E-08	0.850E-03
13	178	0.9336220180364069D+00	0.136E-10	0.850E-03
14	179	0.9441957465130982D+00	0.235E-09	0.200E-03
15	180	0.9443958908053349D+00	0.598E-10	0.200E-03
16	181	0.9538733953927414D+00	0.758E-05	0.108E-02
17	183	0.9549583603516555D+00	0.849E-04	0.108E-02
18	185	0.9561549058563865D+00	0.558E-04	0.120E-02
19	186	0.9582577195407838D+00	0.506E-03	0.750E-03
20	187	0.9590079809845515D+00	0.136E-03	0.750E-03
21	189	0.9618498555646015D+00	0.489E-04	0.259E-02
22	190	0.9644445008375758D+00	0.250E-04	0.259E-02
23	191	0.9672303320363761D+00	0.113E-03	0.113E-02
24	192	0.9683565650437763D+00	0.125E-04	0.113E-02
25	193	0.9704879822455046D+00	0.786E-05	0.213E-02
26	194	0.9738976059208907D+00	0.244E-05	0.852E-03
27	195	0.9747491588235540D+00	0.406E-06	0.852E-03
28	196	0.9775269056442103D+00	0.165E-07	0.278E-02
29	197	0.9813455306889490D+00	0.286E-09	0.382E-02
30	198	0.9853232930778423D+00	0.379E-10	0.398E-02
31	199	0.9894262717440089D+00	0.450E-12	0.410E-02
32	200	0.9936194616944066D+00	0.268E-13	0.419E-02
33	201	0.9999999999999889D+00	0.105E-13	0.638E-02

ABOVE ARE THE TRUE ERRORS FOR POISSON GOODEV
IF A-EV NO LT 0 THEN GOODEV HAS MULTIPLICITY GT 1
IF TRUE ERROR LT 0 THEN MORE THAN ONE GOODEV APPROXIMATES
THE SAME TRUE POISSON EIGENVALUE
IF TMINGAP LT 0 THE MINGAP IS DUE TO SPURIOUS EIGENVALUE

POISSONZ TRUE ERROR FOR GOOD EIGENVALUES

N	NX	NY
528	33	16

A-DIAGONAL	X-CODIAGONAL	Y-CODIAGONAL
0.10000000D+01	0.25000000D+00	0.25000000D+00

1056 36 401 = MEV,NG,NEXACT

T-EV NO	A-EV NO	GOOD EIGENVALUE	TRUEERROR	TMINGAP
1	166	0.9023087045495045D+00	0.373E-14	0.960E-03
2	167	0.9032683439081598D+00	0.840E-14	0.960E-03
3	168	0.9075220637981604D+00	0.837E-14	0.210E-03
4	169	0.9077316405366920D+00	0.568E-14	0.210E-03
5	170	0.9093026846956148D+00	0.565E-14	-0.332E-03
6	171	0.9117392539584653D+00	0.598E-14	0.210E-02
7	172	0.9139623171477664D+00	0.543E-14	0.222E-02
8	173	0.9215518596986422D+00	0.172E-14	0.371E-03
9	174	0.9219228129327942D+00	0.393E-14	0.371E-03
10	175	0.9272414797172943D+00	0.480E-14	-0.301E-04
11	176	0.9318128595792929D+00	0.548E-14	0.960E-03

12	177	0.9327724989379482D+00	0.652E-14	0.850E-03
13	178	0.9336220180228539D+00	0.194E-14	0.850E-03
14	179	0.9441957462784023D+00	0.508E-14	0.200E-03
15	180	0.9443958907455297D+00	0.744E-14	0.200E-03
16	181	0.9538658202683405D+00	0.810E-14	0.394E-03
17	182	0.9542594208233769D+00	0.115E-13	0.394E-03
18	183	0.9550432638722495D+00	0.609E-14	0.784E-03
19	184	0.9560950958355893D+00	0.913E-14	0.116E-03
20	185	0.9562106619424611D+00	0.346E-14	0.116E-03
21	186	0.9577516552053169D+00	0.101E-13	0.112E-02
22	187	0.9588724531626193D+00	0.959E-14	0.781E-03
23	188	0.9596530964495994D+00	0.865E-14	0.781E-03
24	189	0.9618987632490549D+00	0.547E-14	0.225E-02
25	190	0.9644694963663571D+00	0.779E-14	0.257E-02
26	191	0.9673433632129476D+00	0.125E-14	0.103E-02
27	192	0.9683690265546639D+00	0.358E-14	0.103E-02
28	193	0.9704958449702006D+00	0.895E-14	0.213E-02
29	194	0.9739000457753179D+00	0.533E-14	0.850E-03
30	195	0.9747495648602239D+00	0.283E-14	0.850E-03
31	196	0.9775269221872765D+00	0.958E-15	-0.163E-02
32	197	0.9813455309753548D+00	0.171E-14	-0.219E-02
33	198	0.9853232931157719D+00	0.386E-14	0.398E-02
34	199	0.9894262717444636D+00	0.508E-14	0.410E-02
35	200	0.9936194616944293D+00	0.419E-14	-0.311E-04
36	201	0.9999999999999889D+00	0.105E-13	0.638E-02

ABOVE ARE THE TRUE ERRORS FOR POISSON GOODEV
IF A-EV NO LT 0 THEN GOODEV HAS MULTIPLICITY GT 1
IF TRUE ERROR LT 0 THEN MORE THAN ONE GOODEV APPROXIMATES
THE SAME TRUE POISSON EIGENVALUE
IF TMINGAP LT 0 THE MINGAP IS DUE TO SPURIOUS EIGENVALUE

-----SAMPLE OUTPUT FOR REAL SYMMETRIC EIGENVECTOR COMPUTATIONS-----

Below is a sample of the output from the LEVEC eigenvector program. The output which is printed to file 6 is given first, followed by the outputs which are generated and written to other files. Some of the middle portion of the printout to file 6 has been condensed.

```
LANCZOS EIGENVECTOR PROCEDURE FOR REAL SYMMETRIC MATRICES

LANCZOS EIGENVECTOR COMPUTATIONS, NO REORTHOGONALIZATION
     POISSON INPUT

    N    KX    KY        DIAGONAL    X-CODIAGONAL   Y-CODIAGONAL
  528    33    16 0.10000000D+01 0.25000000D+00 0.25000000D+00


MATRIX IDENTIFICATION NO. =      607840 ORDER OF A =    528

  MBOUND    NTVCON    SVTVEC    IREAD
       0         1         1        1

  TVSTOP    LVCONT    ERCONT    IWRITE
       0         1         0         1

  MDIMTV    MDIMRV    MBETA
   50000     20000     6000

     RELTOL   RHSEED
  0.1000D-09 45329517

T-MULTIPLICITY TOLERANCE USED IN THE EIGENVALUE COMPUTATIONS WAS    0.1398D-11
SCALED MACHINE EPSILON IS   0.6802D-15

EIGENVALUES SUPPLIED ARE READ IN FROM FILE 3
FILE 3  HEADER IS
   NG  NDIS    MEV   NOLD  MATOLD    SVSEED       MULTOL       IB          BTOL
   15    40   1056    528  607840    7892713   0.140D-11      616    0.1000D-07


EIGENVALUES READ IN, T-MULTIPLICITIES, T-GAPS AND A-GAPS
    J       GOOD EIGENVALUE      MULT       TMINGAP         AMINGAP
    1   0.9023087045495045D+00    1       0.9600E-03      0.9600E-03
    2   0.9032683439081598D+00    1       0.9600E-03     -0.9600E-03
    3   0.9075220637981604D+00    1       0.2100E-03      0.2100E-03
    4   0.9077316405366920D+00    1       0.2100E-03     -0.2100E-03
    5   0.9093026846956148D+00    1      -0.3320E-03     -0.1570E-02
    6   0.9443958907455297D+00    1       0.2000E-03     -0.2000E-03
    7   0.9538658202683405D+00    1       0.3940E-03      0.3940E-03
    8   0.9542594208233769D+00    1       0.3940E-03     -0.3940E-03
    9   0.9550432638722495D+00    1       0.7840E-03     -0.7840E-03
   10   0.9560950958355893D+00    1       0.1160E-03      0.1160E-03
   11   0.9813455309753548D+00    1      -0.2190E-02     -0.3820E-02
   12   0.9853232931157719D+00    1       0.3980E-02     -0.3980E-02
   13   0.9894262717444636D+00    1       0.4100E-02     -0.4100E-02
   14   0.9936194616944293D+00    1      -0.3110E-04     -0.4190E-02
   15   0.9999999999999889D+00    1       0.6380E-02     -0.6380E-02

THESE EIGENVALUES WERE COMPUTED USING A T-MATRIX OF      ORDER  1056
AND SEED FOR RANDOM NUMBER GENERATOR =      7892713
ERROR ESTMATES =
    J     EIGENVALUE            ESTIMATE
    1  0.902308704550D+00      0.391E-13
    2  0.903268343908D+00      0.472E-13
    3  0.907522063798D+00      0.129E-11
    4  0.907731640537D+00      0.297E-11
```

```
    5  0.909302684696D+00      0.219E-12
    6  0.944395890746D+00      0.319E-13
    7  0.953865820268D+00      0.240E-09
    8  0.954259420823D+00      0.232E-08
    9  0.955043263872D+00      0.201E-08
   10  0.956095095836D+00      0.189E-07
   11  0.981345530975D+00      0.765E-13
   12  0.985323293116D+00      0.137E-13
   13  0.989426271744D+00      0.436E-13
   14  0.993619461694D+00      0.114E-14
   15  0.100000000000D+01      0.155E-10

READ IN THE T-MATRICES STORED ON FILE 2
FILE 2 HEADER IS
 KMAX  NOLD      SVSOLD  MATOLD
 1320   528     7892713  607840

ENLARGE KMAX FROM    1320 TO    1464

LANCZS SUBROUTINE GENERATES ALPHA(J), BETA(J+1), J =  1321 TO   1464



1ST GUESS AT APPROPRIATE SIZE T-MATRICES
ACTUAL VALUES WILL PROBABLY BE 1/4 AGAIN AS MUCH
  943   934   994   999   969   968  1112  1132  1129  1151   972   931   901
  863   837

    9TH EIGENVALUE
 TSIZE                EIGENVALUE             U(M)           ERRORV
 1197   0.9550432638722377D+00    0.18352D-09    0.34361D-12

CHANGE SIZE OF T-MATRIX TO   1231 RECOMPUTE T-EIGENVECTOR

    9TH EIGENVALUE
 TSIZE                EIGENVALUE             U(M)           ERRORV
 1231   0.9550432638722377D+00    0.30709D-10    0.34845D-12

   11TH EIGENVALUE
 TSIZE                EIGENVALUE             U(M)           ERRORV
  972   0.9813455309753427D+00    0.38849D-11    0.13630D-10

THE CUMULATIVE LENGTH OF THE T-EIGENVECTORS IS             15314

   15 T-EIGENVECTORS OUT OF     15 REQUESTED WERE COMPUTED

 1343 = LARGEST SIZE T-MATRIX TO BE USED IN THE RITZ VECTOR COMPUTATIONS

   15 RITZ VECTORS WERE REQUESTED
   15 T-EIGENVECTORS WERE COMPUTED
   15 RITZ VECTORS WILL BE COMPUTED


   1 TH EIGENVALUE CONSIDERED =   0.902308704550D+00

NORM OF ERROR IN T-EIGENVECTOR =       0.300E-12
BETA(MA(J)+1)*U(MA(J)) =       0.617E-13
ABS(NORM(RITVEC) - 1.0)  =       0.660D-12

RUNTIME = 1653 HNDRETHS OF A SECOND


   A-EIGENVALUE  MA(J)   A-MINGAP      AERROR  AERROR/GAP      TERROR
 0.90230870D+00    943  0.9600E-03  0.4410E-12  0.4594E-09  0.3001E-12
 0.90326834D+00    934 -0.9600E-03  0.2019E-12  0.2103E-09  0.1340E-12
 0.90752206D+00    994  0.2100E-03  0.2028E-11  0.9659E-08  0.6610E-13
```

0.90773164D+00	999	-0.2100E-03	0.2570E-11	0.1224E-07	0.2588E-12
0.90930268D+00	969	-0.1570E-02	0.2734E-11	0.1742E-08	0.1902E-11
0.94439589D+00	968	-0.2000E-03	0.2073E-10	0.1036E-06	0.2815E-12
0.95386582D+00	1112	0.3940E-03	0.2563E-10	0.6505E-07	0.2056E-12
0.95425942D+00	1268	-0.3940E-03	0.1024E-10	0.2600E-07	0.1103E-12
0.95504326D+00	1231	-0.7840E-03	0.1489E-10	0.1899E-07	0.3485E-12
0.95609510D+00	1343	0.1160E-03	0.1062E-10	0.9155E-07	0.2351E-11
0.98134553D+00	972	-0.3820E-02	0.1957E-10	0.5124E-08	0.1363E-10
0.98532329D+00	931	-0.3980E-02	0.2979E-11	0.7484E-09	0.2031E-11
0.98942627D+00	901	-0.4100E-02	0.7486E-12	0.1826E-09	0.4603E-12
0.99361946D+00	888	-0.4190E-02	0.3438E-11	0.8206E-09	0.3755E-12
0.10000000D+01	861	-0.6380E-02	0.2566E-10	0.4022E-08	0.5213E-13

ABOVE ARE ERROR ESTIMATES FOR THE A AND T EIGENVECTORS
ASSOCIATED WITH THE GOODEV LISTED IN COLUMN 1
AERROR = NORM(A*X - EV*X)   TERROR = NORM(T*Y - EV*Y)
WHERE T = T(1,MA(J))   X = RITZ VECTOR = V*Y V = SUCCESSIVE
LANCZOS VECTORS. AMINGAP = GAP TO NEAREST A-EIGENVALUE

GOODEV(J)	RITZNORM	AMINGAP	TBETA(J)	TLAST(J)
0.90230870454952D+00	0.10000E+01	0.96000E-03	0.61736E-13	0.11215D-12
0.90326834390815D+00	0.10000E+01	-0.96000E-03	0.58491E-13	0.16274D-12
0.90752206379817D+00	0.10000E+01	0.21000E-03	0.20265E-11	0.37472D-11
0.90773164053670D+00	0.10000E+01	-0.21000E-03	0.25446E-11	0.49470D-11
0.90930268469560D+00	0.10000E+01	-0.15700E-02	0.12312E-12	0.32684D-12
0.94439589074552D+00	0.10000E+01	-0.20000E-03	0.20734E-10	0.44109D-10
0.95386582026835D+00	0.10000E+01	0.39400E-03	0.25604E-10	0.52969D-10
0.95425942082337D+00	0.10000E+01	-0.39400E-03	0.10235E-10	0.30085D-10
0.95504326387224D+00	0.10000E+01	-0.78400E-03	0.14891E-10	0.30709D-10
0.95609509583558D+00	0.10000E+01	0.11600E-03	0.97828E-11	0.17603D-10
0.98134553097534D+00	0.10000E+01	-0.38200E-02	0.20033E-11	0.38849D-11
0.98532329311576D+00	0.10000E+01	-0.39800E-02	0.42732E-12	0.81168D-12
0.98942627174445D+00	0.10000E+01	-0.41000E-02	0.31008E-12	0.11744D-11
0.99361946169444D+00	0.10000E+01	-0.41900E-02	0.33970E-11	0.57069D-11
0.10000000000000D+01	0.10000E+01	-0.63800E-02	0.25666E-10	0.76799D-10

ABOVE ARE ERROR ESTIMATES ASSOCIATED WITH THE GOODEV
RITZNORM = NORM(COMPUTED RITZ VECTOR)
TBETA(J) = BETA(MA(J)+1)*Y(MA(J)),   T*Y = EVAL*Y
TLAST(J) = Y(MA(J))
AMINGAP = GAP TO NEAREST A-EIGENVALUE

528   1464 = ORDER OF USER MATRIX AND MAX ORDER OF T(1,MEV)

1ST GUESS AT APPROPRIATE SIZE T-MATRICES
ACTUAL VALUES WILL PROBABLY BE 1/4 AGAIN AS MUCH

J	A-EIGENVALUE	M1(J)	M2(J)	MA(J)
1	0.902308704550D+00	754	1464	943
2	0.903268343908D+00	747	1464	934
3	0.907522063798D+00	795	1464	994
4	0.907731640537D+00	799	1464	999
5	0.909302684696D+00	775	1464	969
6	0.944395890746D+00	774	1464	968
7	0.953865820268D+00	889	1464	1112
8	0.954259420823D+00	905	1464	1132
9	0.955043263872D+00	903	1464	1129
10	0.956095095836D+00	920	1464	1151
11	0.981345530975D+00	777	1464	972
12	0.985323293116D+00	757	1452	931
13	0.989426271744D+00	725	1426	901
14	0.993619461694D+00	700	1350	863
15	0.100000000000D+01	679	1309	837

SIZES OF T-MATRICES THAT WILL BE USED IN THE RITZ COMPUTATIONS

J	GOODEV(J)	MA(J)
1	0.90230870454950D+00	943
2	0.90326834390816D+00	934
3	0.90752206379816D+00	994
4	0.90773164053669D+00	999
5	0.90930268469561D+00	969
6	0.94439589074553D+00	968
7	0.95386582026834D+00	1112
8	0.95425942082338D+00	1268
9	0.95504326387225D+00	1231
10	0.95609509583559D+00	1343
11	0.98134553097535D+00	972
12	0.98532329311577D+00	931
13	0.98942627174446D+00	901
14	0.99361946169443D+00	888
15	0.99999999999999D+00	861

```
EV = GOODEV(J) IS A GOOD EIGENVALUE OF T(1,MEV)
M1 = SMALLEST VALUE OF M SUCH THAT T(1,M) HAS AT LEAST
    ONE EIGENVALUE IN THE INTERVAL (EV-TOLN,EV+TOLN)
    TOLN(J) = DMAX1(GOODEV(J)*RELTOL, SCALEO*MULTOL)
M2 = SMALLEST M (IF ANY) SUCH THAT IN THE ABOVE INTERVAL
     T(1,M) HAS AT LEAST TWO EIGENVALUES
IABS(MA(J)) = APPROPRIATE SIZE T-MATRIX FOR GOODEV(J)
INITIAL VALUE OF MA(J) IS CHOSEN HEURISTICALLY
PROGRAM LOOPS ON SIZE OF T-MATRIX TO GET BETTER SIZE
END OF SIZES OF T-MATRICES FILE 10
```

CHAPTER 3

HERMITIAN MATRICES

SECTION 3.1 INTRODUCTION

The FORTRAN codes in this chapter address the question of computing distinct eigenvalues and corresponding eigenvectors of Hermitian matrices, using a single-vector Lanczos procedure. For a given Hermitian matrix A, these codes compute real scalars λ and corresponding complex vectors $x \neq 0$ such that

$$Ax = \lambda x . \qquad (3.1.1)$$

DEFINITION 3.1.1 A complex nxn matrix A, $A \equiv (a_{ij})$, $1 \leq i,j \leq n$, is a Hermitian matrix if and only if for every i and j, $a_{ij} = \bar{a}_{ji}$, where the overbar denotes the complex conjugate of the complex-valued entry a_{ij} .

It is straight-forward to demonstrate from Definition 3.1.1 that for any Hermitian matrix $A = B + Ci$, where B and C are real matrices and $i = \sqrt{-1}$, that B must be a real symmetric matrix and C must be a skew symmetric matrix. That is, $B^T = B$ and $C^T = -C$. Futhermore, it is not difficult to see that Hermitian matrices must have real diagonal entries and real eigenvalues. However, the eigenvectors are complex-valued. Any Hermitian matrix can be transformed into a real symmetric tridiagonal matrix for the purposes of computing the eigenvalues of the Hermitian matrix, Stewart [1973]. In fact, the Lanczos recursion which we use in the codes in this chapter transforms the given Hermitian matrix A into a family of real symmetric tridiagonal matrices rather than into a family of Hermitian tridiagonal matrices.

Hermitian matrices possess the 'same' properties as real symmetric matrices do, except that these properties are defined with respect to the complex or Hermitian norm, rather than with respect to the Euclidean norm, see Stewart [1973]. The Hermitian norm of a given complex-valued vector $x \equiv (x(i))$, $1 \leq i \leq n$, is defined as $\| x \|_{\mathscr{C}}^2 \equiv \sum_{i=1}^{n} \overline{x(i)} x(i)$. Three properties which we use are: (1) Hermitian matrices have complete eigensystems. That is, the dimension of the eigenspace corresponding to any given eigenvalue of a Hermitian matrix is the same as the multiplicity of that eigenvalue as a root of the characteristic polynomial of that matrix. (2) For any two distinct eigenvalues λ and μ, and corresponding eigenvectors x and y, $x^H y = 0$, where the superscript H denotes the complex conjugate transpose of the vector x. The complex conjugate transpose of a column vector x is the row vector whose ith component is $\overline{x(i)}$. Therefore there is a complete set of eigenvectors $X_n \equiv \{x_1,...,x_n\}$ such that X is a unitary matrix. (3) Small perturbations in a Hermitian matrix cause only small perturbations in the eigenvalues.

The single-vector Lanczos codes in this chapter can be used to compute either a very few or very many of the distinct eigenvalues of the given Hermitian matrix. The documentation for these codes is contained in Chapter 2, Section 2.2. As in the real symmetric case, the A-multiplicity of a given computed 'good' Lanczos eigenvalue can be obtained only with additional computation, and the modifications required to do this additional computation are not included in these versions of the codes. This implementation uses a Hermitian analog of the basic Lanczos recursion contained in Eqns(1.2.1) - (1.2.2) to generate a family of real symmetric tridiagonal matrices whose sizes are specified by the user. There is no reorthogonalization of the Lanczos vectors at any stage in any of the computations.

The Hermitian version of the Lanczos recursion which we use is given below. For $i=1,2,\ldots,m$ and a randomly-generated complex starting vector v_1 with $\|v_1\|_{\mathscr{C}} = 1$, generate Lanczos vectors v_i using the following recursion.

$$\beta_{i+1}v_{i+1} = Av_i - \alpha_i v_i - \beta_i v_{i-1}, \tag{3.1.2}$$

where

$$\alpha_i \equiv v_i^H A v_i \text{ and } \beta_{i+1} = \|Av_i - \alpha_i v_i - \beta_i v_{i-1}\|_{\mathscr{C}}. \tag{3.1.3}$$

We see from Eqns(3.1.3), that the Hermitian inner product is used. This is the 'natural' inner product for Hermitian matrices. Gram-Schmidt orthogonalization is used, unlike the real symmetric case where a modified Gram-Schmidt orthogonalization was used. This change in the local orthogonalization procedure increases the storage requirements for the implementation of the Lanczos recursion by one additional complex vector of length equal to the order of the original A-matrix. Modified Gram-Schmidt orthogonalization cannot be used in the Hermitian case because corrections to the α_i defined by this modification are complex-valued not real, and it would not be legitimate to accept the real portions of these corrections and simply ignore the complex portions.

It is easy to demonstrate that as we stated earlier, each Lanczos matrix (T-matrix) generated by this Hermitian recursion is a real symmetric tridiagonal matrix. In particular, we see from the formulas in Eqn(3.1.3) that the diagonal entries of each of these matrices are Rayleigh quotients of the given Hermitian matrix A, and therefore must all be real-valued. Furthermore by construction, the nonzero off-diagonal entries β_{i+1} are all real-valued. This use of real-valued β_i requires some justification. This justification is given in Section 4.9 of Chapter 4 of Volume 1 of this book.

HLEVAL, the main program for the Hermitian eigenvalue computations, calls the subroutine BISEC to compute eigenvalues of the specified tridiagonal Lanczos matrices on the user-specified intervals. BISEC simultaneously computes these T-eigenvalues with their T-multiplicities and sorts the computed T-eigenvalues into two classes, the 'good' T-eigenvalues and the 'spurious' T-eigenvalues. The 'good' T-eigenvalues are accepted as approximations to eigenvalues of the user-specified matrix A. The accuracy of these 'good' T-eigenvalues as eigenvalues of A is then estimated using error estimates computed by subroutine INVERR. Error estimates are computed only for isolated 'good' T-eigenvalues. All other 'good' T-eigenvalues are assumed to have converged. Convergence is then checked. If convergence has not yet occurred and a larger T-matrix has been specified by the user, the program will continue on to the larger T-matrix, repeating the above procedure on this larger matrix.

Once the eigenvalues have been computed accurately enough, the user can select a subset of the 'converged' eigenvalues for which eigenvectors are to be computed. The main program HLEVEC, for computing eigenvectors of Hermitian matrices, is then used to compute these desired eigenvectors.

The computations in the Lanczos recursion are a mixture of double precision real arithmetic and of double precision complex arithmetic. Once the Lanczos matrices have been computed, the remaining computations are all done in double precision real arithmetic, using the same subroutines that are used in the real symmetric case. In addition to the programs and subroutines provided here, the user must supply a subroutine USPEC which defines and initializes the user-specified matrix A and a subroutine CMATV which computes matrix-vector multiplies Ax for any given vector x. These subroutines must be constructed in such a way as to take advantage of the sparsity (and/or structure) of the user-supplied A-matrix and such that these computations are done accurately.

SECTION 3.2 MAIN PROGRAM, EIGENVALUE COMPUTATIONS

```
C-----HLEVAL  (EIGENVALUES OF HERMITIAN MATRICES)-----------------------
C
C     CONTAINS MAIN PROGRAM FOR COMPUTING DISTINCT EIGENVALUES OF
C     A HERMITIAN MATRIX USING LANCZOS TRIDIAGONALIZATION WITHOUT
C     REORTHOGONALIZATION
C
C     PORTABILITY:
C     THIS PROGRAM IS NOT PORTABLE DUE TO THE USE OF COMPLEX*16
C     VARIABLES.  MOREOVER, THE PFORT VERIFIER IDENTIFIED THE
C     FOLLOWING ADDITIONAL NONPORTABLE CONSTRUCTIONS:
C
C     1.  DATA/MACHEP/ STATEMENT
C     2.  ALL READ(5,*) STATEMENTS (FREE FORMAT)
C     3.  FORMAT(20A4) USED WITH EXPLANATORY HEADER EXPLAN.
C     4.  HEXADECIMAL FORMAT (4Z20) USED IN ALPHA/BETA FILES 1 AND 2.
C
C-----------------------------------------------------------------------
C
      DOUBLE PRECISION  ALPHA(5000),BETA(5001),VS(5000)
      COMPLEX*16  V1(5000),V2(5000)
      DOUBLE PRECISION  GR(1500),GC(1500),LB(20),UB(20)
      DOUBLE PRECISION  BTOL,GAPTOL,TTOL,MACHEP,EPSM,RELTOL
      DOUBLE PRECISION  SCALE1,SCALE2,SCALE3,SCALE4,BISTOL,CONTOL,MULTOL
      DOUBLE PRECISION  ONE,ZERO,TEMP,TKMAX,BETAM,BKMIN,T0,T1
      REAL  G(5000),EXPLAN(20)
      INTEGER  MP(5000),NMEV(20)
      INTEGER  SVSEED,RHSEED,SVSOLD
      INTEGER IABS
      REAL  ABS
      DOUBLE PRECISION  DABS, DSQRT, DFLOAT
      EXTERNAL CMATV
C
C-----------------------------------------------------------------------
      DATA MACHEP/Z3410000000000000/
      EPSM = 2.0D0*MACHEP
C-----------------------------------------------------------------------
C     THE ARRAYS V1 AND V2 ARE DEFINED AS COMPLEX*16 IN THE MAIN PROGRAM
C     AND IN THE SUBROUTINE LANCZS.  HOWEVER, IN THE OTHER SUBROUTINES
C     THEY ARE DECLARED AS DOUBLE PRECISION ARRAYS. NOTE THAT THE
C     DIMENSION OF V2 ASSUMES THAT NO MORE THAN KMAX/2 EIGENVALUES OF
C     THE T-MATRICES ARE BEING COMPUTED IN ANY ONE OF THE SUB-INTERVALS
C     BEING CONSIDERED.  V2 MUST CONTAIN UPPER AND LOWER BOUNDS
C     ON EACH T-EIGENVALUE COMPUTED BY BISEC IN ANY ONE GIVEN INTERVAL.
C
C     ARRAYS MUST BE DIMENSIONED AS FOLLOWS:
C          1. ALPHA:    >= KMAX.   BETA:  >= (KMAX+1)
C          2. V1:  >= MAX(N,(KMAX+1)/2).    V2:  >= MAX(N,KMAX/2)
C          3. VS:  >= MAX(2*N,KMAX).
C          4. GR,GC:  >=  N
C          5. G:  >= MAX(2*KMAX,N)
C          6. MP:    >= KMAX
C          7. LB,UB:  >= NUMBER OF SUB-INTERVALS SPECIFIED
C          8. NMEV:  >= NUMBER OF T-MATRICES SPECIFIED
C          9. EXPLAN:  DIMENSION IS 20.
C
C
C     IMPORTANT TOLERANCES OR SCALES THAT ARE USED REPEATEDLY
C     THROUGHOUT THE PROGRAM ARE THE FOLLOWING:
C     SCALED MACHINE EPSILON:  TTOL = TKMAX*EPSM WHERE
C     EPSM = 2*MACHINE EPSILON AND
C     TKMAX = MAX(|ALPHA(J)|,BETA(J), J = 1,MEV)
C     BISEC CONVERGENCE TOLERANCE:  BISTOL = DSQRT(1000+MEV)*TTOL
C     BISEC MULTIPLICITY TOLERANCE:  MULTOL = (1000+MEV)*TTOL
C     LANCZOS CONVERGENCE TOLERANCE:   CONTOL = BETA(MEV+1)*1.D-10
```

```
C-----------------------------------------------------------------------HLE00650
C     OUTPUT HEADER                                                     HLE00660
      WRITE(6,10)                                                       HLE00670
   10 FORMAT(/' LANCZOS PROCEDURE FOR HERMITIAN MATRICES'/)             HLE00680
C                                                                       HLE00690
C     SET PROGRAM PARAMETERS                                            HLE00700
C     SCALEK ARE USED IN TOLERANCES NEEDED IN SUBROUTINES LUMP,         HLE00710
C     ISOEV AND PRTEST.  USER MUST NOT MODIFY THESE SCALES.             HLE00720
      SCALE1 = 5.0D2                                                    HLE00730
      SCALE2 = 5.0D0                                                    HLE00740
      SCALE3 = 5.0D0                                                    HLE00750
      SCALE4 = 1.0D4                                                    HLE00760
      ONE  = 1.0D0                                                      HLE00770
      ZERO = 0.0D0                                                      HLE00780
C     BTOL = EPSM                                                       HLE00790
      BTOL = 1.0D-8                                                     HLE00800
      GAPTOL = 1.0D-8                                                   HLE00810
      ICONV = 0                                                         HLE00820
      MOLD = 0                                                          HLE00830
      MOLD1 = 1                                                         HLE00840
      ICT = 0                                                           HLE00850
      MMB = 0                                                           HLE00860
      IPROJ = 0                                                         HLE00870
C                                                                       HLE00880
C     READ USER-SPECIFIED PARAMETERS FROM INPUT FILE 5 (FREE FORMAT)    HLE00890
C                                                                       HLE00900
C     READ USER-PROVIDED HEADER FOR RUN                                 HLE00910
      READ(5,20) EXPLAN                                                 HLE00920
      WRITE(6,20) EXPLAN                                                HLE00930
      READ(5,20) EXPLAN                                                 HLE00940
      WRITE(6,20) EXPLAN                                                HLE00950
   20 FORMAT(20A4)                                                      HLE00960
C                                                                       HLE00970
C     READ ORDER OF MATRICES (N) , MAXIMUM ORDER OF T-MATRIX (KMAX),    HLE00980
C     NUMBER OF T-MATRICES ALLOWED (NMEVS), AND MATRIX IDENTIFICATION   HLE00990
C     NUMBERS (MATNO)                                                   HLE01000
      READ(5,20) EXPLAN                                                 HLE01010
      READ(5,*) N,KMAX,NMEVS,MATNO                                      HLE01020
C                                                                       HLE01030
C     READ SEEDS FOR LANCZS AND INVERR SUBROUTINES (SVSEED AND RHSEED)  HLE01040
C     READ MAXIMUM NUMBER OF ITERATIONS ALLOWED FOR EACH INVERSE        HLE01050
C     ITERATION (MXINIT) AND MAXIMUM NUMBER OF STURM SEQUENCES          HLE01060
C     ALLOWED (MXSTUR)                                                  HLE01070
      READ(5,20) EXPLAN                                                 HLE01080
      READ(5,*) SVSEED,RHSEED,MXINIT,MXSTUR                             HLE01090
C                                                                       HLE01100
C     ISTART = (0,1):  ISTART = 0 MEANS ALPHA/BETA FILE IS NOT          HLE01110
C     AVAILABLE.  ISTART = 1 MEANS ALPHA/BETA FILE IS AVAILABLE ON      HLE01120
C     FILE 2.                                                           HLE01130
C     ISTOP = (0,1):  ISTOP = 0 MEANS PROCEDURE GENERATES ALPHA/BETA    HLE01140
C     FILE AND THEN TERMINATES.  ISTOP = 1 MEANS PROCEDURE GENERATES    HLE01150
C     ALPHAS/BETAS IF NEEDED AND THEN COMPUTES EIGENVALUES AND ERROR    HLE01160
C     ESTIMATES AND THEN TERMINATES.                                    HLE01170
      READ(5,20) EXPLAN                                                 HLE01180
      READ(5,*) ISTART,ISTOP                                            HLE01190
C                                                                       HLE01200
C     IHIS = (0,1):  IHIS = 0 MEANS ALPHA/BETA FILE IS NOT WRITTEN      HLE01210
C     TO FILE 1.  IHIS = 1 MEANS ALPHA/BETA FILE IS WRITTEN TO FILE 1.  HLE01220
C     IDIST = (0,1):  IDIST = 0 MEANS DISTINCT T-EIGENVALUES            HLE01230
C     ARE NOT WRITTEN TO FILE 11.  IDIST = 1 MEANS DISTINCT             HLE01240
C     T-EIGENVALUES ARE WRITTEN TO FILE 11.                             HLE01250
C     IWRITE = (0,1):  IWRITF = 0 MEANS NO INTERMEDIATE OUTPUT          HLE01260
C     FROM THE COMPUTATIONS IS WRITTEN TO FILE 6.  IWRITE = 1 MEANS     HLE01270
C     T-EIGENVALUES AND ERROR ESTIMATES ARE WRITTEN TO FILE 6           HLE01280
C     AS THEY ARE COMPUTED.                                             HLE01290
      READ(5,20) EXPLAN                                                 HLE01300
```

```
      READ(5,*) IHIS,IDIST,IWRITE
C
C     READ IN THE RELATIVE TOLERANCE (RELTOL) FOR USE IN THE
C     SPURIOUS, T-MULTIPLICITY, AND PRTESTS.
      READ(5,20) EXPLAN
      READ(5,*) RELTOL
C
C     READ IN THE SIZES OF THE T-MATRICES TO BE CONSIDERED.
      READ(5,20) EXPLAN
      READ(5,*) (NMEV(J), J=1,NMEVS)
C
C     READ IN THE NUMBER OF SUBINTERVALS TO BE CONSIDERED.
      READ(5,20) EXPLAN
      READ(5,*) NINT
C
C     READ IN THE LEFT-END POINTS OF THE SUBINTERVALS TO BE CONSIDERED.
C     THESE MUST BE IN ALGEBRAICALLY-INCREASING ORDER
      READ(5,20) EXPLAN
      READ(5,*) (LB(J), J=1,NINT)
C
C     READ IN THE RIGHT-END POINTS OF THE SUBINTERVALS TO BE CONSIDERED.
C     THESE MUST BE IN ALGEBRAICALLY-INCREASING ORDER
      READ(5,20) EXPLAN
      READ(5,*) (UB(J), J=1,NINT)
C
C-----------------------------------------------------------------------
C
C     INITIALIZE THE ARRAYS FOR THE USER-SPECIFIED MATRIX
C     AND PASS THE STORAGE LOCATIONS OF THESE ARRAYS TO THE
C     MATRIX-VECTOR MULTIPLY SUBROUTINE CMATV.
C
      CALL USPEC(N,MATNO)
C
C-----------------------------------------------------------------------
C
C     MASK UNDERFLOW AND OVERFLOW
C
      CALL MASK
C
C-----------------------------------------------------------------------
C
C     WRITE TO FILE 6, A SUMMARY OF THE PARAMETERS FOR THIS RUN
C
      WRITE(6,30) MATNO,N,KMAX
   30 FORMAT(/3X,'MATRIX ID',4X,'ORDER OF A',4X,'MAX ORDER OF T'/
     1 I12,I14,I18/)
C
      WRITE(6,40) ISTART,ISTOP
   40 FORMAT(/2X,'ISTART',3X,'ISTOP'/2I8/)
C
      WRITE(6,50) IHIS,IDIST,IWRITE
   50 FORMAT(/4X,'IHIS',3X,'IDIST',2X,'IWRITE'/3I8/)
C
      WRITE(6,60) SVSEED,RHSEED
   60 FORMAT(/' SEEDS FOR RANDOM NUMBER GENERATOR'//
     1 4X,'LANCZS SEED',4X,'INVERR SEED'/2I15/)
C
      WRITE(6,70) (NMEV(J), J=1,NMEVS)
   70 FORMAT(/' SIZES OF T-MATRICES TO BE CONSIDERED'/(6I12))
C
      WRITE(6,80) RELTOL,GAPTOL,BTOL
   80 FORMAT(/' RELATIVE TOLERANCE USED TO COMBINE COMPUTED T-EIGENVALUE
     1S'/E15.3/' RELATIVE GAP TOLERANCES USED IN INVERSE ITERATION'/
     1E15.3/' RELATIVE TOLERANCE FOR CHECK ON SIZE OF BETAS'/E15.3/)
C
      WRITE(6,90) (J,LB(J),UB(J), J=1,NINT)
   90 FORMAT(/' BISEC WILL BE USED ON THE FOLLOWING INTERVALS'/
```

```
     1 (I6,2E20.6))
C
      IF (ISTART.EQ.0) GO TO 140
C
C     READ IN ALPHA BETA HISTORY
C
      READ(2,100)MOLD,NOLD,SVSOLD,MATOLD
  100 FORMAT(2I6,I12,I8)
C
      IF (KMAX.LT.MOLD) KMAX = MOLD
      KMAX1 = KMAX + 1
C
C     CHECK THAT ORDER N, MATRIX ID MATNO, AND RANDOM SEED SVSEED
C     AGREE WITH THOSE IN THE HISTORY FILE.  IF NOT PROCEDURE STOPS.
C
      ITEMP = (NOLD-N)**2+(MATNO-MATOLD)**2+(SVSEED-SVSOLD)**2
C
      IF (ITEMP.EQ.0) GO TO 120
C
      WRITE(6,110)
  110 FORMAT(' PROGRAM TERMINATES'/    ' READ FROM FILE 2 CORRESPONDS TO
     1 DIFFERENT MATRIX THAN MATRIX SPECIFIED'/)
      GO TO 640
C
  120 CONTINUE
      MOLD1 = MOLD+1
C
      READ(2,130)(ALPHA(J), J=1,MOLD)
      READ(2,130)(BETA(J), J=1,MOLD1)
  130 FORMAT(4Z20)
C
      IF (KMAX.EQ.MOLD) GO TO 160
C
      READ(2,130)(V1(J), J=1,N)
      READ(2,130)(V2(J), J=1,N)
C
  140 CONTINUE
      IIX = SVSEED
C
C-----------------------------------------------------------------------
C
      CALL LANCZS(CMATV,V1,V2,VS,ALPHA,BETA,GR,GC,G,KMAX,MOLD1,N,IIX)
C
C-----------------------------------------------------------------------
C
      KMAX1 = KMAX + 1
C
      IF (IHIS.EQ.0.AND.ISTOP.GT.0) GO TO 160
C
      WRITE(1,150) KMAX,N,SVSEED,MATNO
  150 FORMAT(2I6,I12,I8,' = KMAX,N,SVSEED,MATNO')
C
C     TO AVOID PERTURBATIONS CAUSED BY HEX TO DECIMAL AND DECIMAL TO HEX
C     CONVERSIONS, THE ALPHA AND BETA MUST BE WRITTEN OUT IN HEX.
      WRITE(1,130)(ALPHA(I), I=1,KMAX)
      WRITE(1,130)(BETA(I), I=1,KMAX1)
C
      WRITE(1,130)(V1(I), I=1,N)
      WRITE(1,130)(V2(I), I=1,N)
C
      IF (ISTOP.EQ.0) GO TO 540
C
  160 CONTINUE
      BKMIN = BTOL
      WRITE(6,170)
  170 FORMAT(/' T-MATRICES (ALPHA AND BETA) ARE NOW AVAILABLE'/)
C
```

```
C-----------------------------------------------------------------------
C     SUBROUTINE TNORM CHECKS MIN(BETA)/(ESTIMATED NORM(A)) > BTOL .
C     IF THIS IS VIOLATED IB IS SET EQUAL TO THE NEGATIVE OF THE INDEX
C     OF THE MINIMAL BETA.  IF(IB < 0) THEN SUBROUTINE TNORM IS
C     CALLED FOR EACH VALUE OF MEV TO DETERMINE WHETHER OR NOT THERE
C     IS A BETA IN THE T-MATRIX SPECIFIED THAT VIOLATES THIS TEST.
C     IF THERE IS SUCH A BETA THE PROGRAM TERMINATES FOR THE USER
C     TO DECIDE WHAT TO DO.  THIS TEST CAN BE OVER-RIDDEN BY
C     SIMPLY MAKING BTOL SMALLER, BUT THEN THERE IS THE POSSIBILITY
C     THAT LOSSES IN THE LOCAL ORTHOGONALITY MAY HURT THE COMPUTATIONS.
C     BTOL = 1.D-8 IS HOWEVER A CONSERVATIVE CHOICE FOR BTOL.
C
C     TNORM ALSO COMPUTES TKMAX = MAX(|ALPHA(K)|,BETA(K), K=1,KMAX).
C     TKMAX IS USED TO SCALE THE TOLERANCES USED IN THE
C     T-MULTIPLICITY AND SPURIOUS TESTS IN BISEC. TKMAX IS ALSO USED IN
C     THE PROJECTION TEST FOR HIDDEN EIGENVALUES THAT HAD 'TOO SMALL'
C     A PROJECTION ON THE STARTING VECTOR.
C
      CALL TNORM(ALPHA,BETA,BKMIN,TKMAX,KMAX,IB)
C
C-----------------------------------------------------------------------
C
      TTOL = EPSM*TKMAX
C
C     LOOP ON THE SIZE OF THE T-MATRIX
C
  180 CONTINUE
      MMB = MMB + 1
      MEV = NMEV(MMB)
C     IS MEV TOO LARGE ?
      IF(MEV.LE.KMAX) GO TO 200
      WRITE(6,190) MMB, MEV, KMAX
  190 FORMAT(/' TERMINATE PRIOR TO CONSIDERING THE',I6,'TH T-MATRIX'/
     1' BECAUSE THE SIZE REQUESTED',I6,' IS GREATER THAN THE MAXIMUM SIZ
     1E ALLOWED',I6/)
      GO TO 540
C
  200 MP1 = MEV + 1
      BETAM = BETA(MP1)
C
      IF (IB.GE.0) GO TO 210
C
      T0 = BTOL
C
C-----------------------------------------------------------------------
C
      CALL TNORM(ALPHA,BETA,T0,T1,MEV,IBMEV)
C
C-----------------------------------------------------------------------
C
      TEMP = T0/TKMAX
      IBMEV = IABS(IBMEV)
      IF (TEMP.GE.BTOL) GO TO 210
      IBMEV = -IBMEV
      GO TO 600
C
  210 CONTINUE
      IC = MXSTUR-ICT
C
C-----------------------------------------------------------------------
C     BISEC LOOP. THE SUBROUTINE BISEC INCORPORATES DIRECTLY THE
C     T-MULTIPLICITY AND SPURIOUS TESTS. T-EIGENVALUES WILL BE
C     CALCULATED BY BISEC SEQUENTIALLY ON INTERVALS
C     (LB(J),UB(J)), J = 1,NINT).
C
C     ON RETURN FROM BISEC
C     NDIS = NUMBER OF DISTINCT EIGENVALUES OF T(1,MEV) ON UNION
```

```
C            OF THE (LB,UB) INTERVALS
C      VS = DISTINCT T-EIGENVALUES IN ALGEBRAICALLY INCREASING ORDER
C      MP = MULTIPLICITIES OF THE T-EIGENVALUES IN VS
C      MP(I) = (0,1,MI), MI>1, I=1,NDIS  MEANS:
C         (0)  VS(I) IS SPURIOUS
C         (1)  VS(I) IS T-SIMPLE AND GOOD
C         (MI) VS(I) IS MULTIPLE AND IS THEREFORE NOT ONLY GOOD BUT
C              ALSO A CONVERGED GOOD T-EIGENVALUE.
C      WITHIN BISEC V1 AND V2 ARE DEFINED AS DOUBLE PRECISION ARRAYS
C
C
      CALL BISEC(ALPHA,BETA,V1,V2,VS,LB,UB,EPSM,TTOL,MP,NINT,
     1 MEV,NDIS,IC,IWRITE)
C
C-----------------------------------------------------------------------
C
      IF (NDIS.EQ.0) GO TO 620
C
C      COMPUTE THE TOTAL NUMBER OF STURM SEQUENCES USED TO DATE
C      COMPUTE THE BISEC CONVERGENCE AND T-MULTIPLICITY TOLERANCES USED.
C      COMPUTE THE CONVERGENCE TOLERANCE FOR EIGENVALUES OF A.
      ICT = ICT + IC
      TEMP = DFLOAT(MEV+1000)
      MULTOL = TEMP*TTOL
      TEMP = DSQRT(TEMP)
      BISTOL = TTOL*TEMP
      CONTOL = BETAM*1.D-10
C
C-----------------------------------------------------------------------
C     SUBROUTINE LUMP 'COMBINES' T-EIGENVALUES THAT ARE 'TOO CLOSE'.
C     NOTE HOWEVER THAT CLOSE SPURIOUS T-EIGENVALUES ARE NOT AVERAGED
C     WITH GOOD ONES. HOWEVER, THEY MAY BE USED TO INCREASE THE
C     MULTIPLICITY OF A GOOD T-EIGENVALUE.
C
      LOOP = NDIS
      CALL LUMP(VS,RELTOL,MULTOL,SCALE2,MP,LOOP)
C
C-----------------------------------------------------------------------
C
      IF(NDIS.EQ.LOOP) GO TO 230
C
      WRITE(6,220) NDIS, MEV, LOOP
  220 FORMAT(/I6,' DISTINCT T-EIGENVALUES WERE COMPUTED IN BISEC AT MEV
     1',I6/ 2X,' LUMP SUBROUTINE REDUCES NUMBER OF DISTINCT EIGENVALUES
     1TO',I6)
C
  230 CONTINUE
      NDIS = LOOP
      BETA(MP1) = BETAM
C
C-----------------------------------------------------------------------
C      THE SUBROUTINE ISOEV LABELS THOSE SIMPLE EIGENVALUES OF T(1,MEV)
C      WITH VERY SMALL GAPS BETWEEN NEIGHBORING EIGENVALUES OF T(1,MEV)
C      TO AVOID COMPUTING ERROR ESTIMATES FOR ANY SIMPLE GOOD
C      T-EIGENVALUE THAT IS TOO CLOSE TO A SPURIOUS T-EIGENVALUE.
C      ON RETURN FROM ISOEV, G CONTAINS CODED MINIMAL GAPS
C      BETWEEN THE DISTINCT EIGENVALUES OF T(1,MEV). (G IS REAL).
C      G(I) < 0 MEANS MINGAP IS DUE TO LEFT GAP G(I) > 0 MEANS DUE TO
C      RIGHT GAP. MP(I) = -1 MEANS THAT THE GOOD T-EIGENVALUE IS SIMPLE
C      AND HAS A VERY SMALL MINGAP IN T(1,MEV) DUE TO A SPURIOUS
C      T-EIGENVALUE.   NG = NUMBER OF GOOD EIGENVALUES.
C      NISO = NUMBER OF ISOLATED GOOD T-EIGENVALUES.
C
      CALL ISOEV(VS,GAPTOL,MULTOL,SCALE1,G,MP,NDIS,NG,NISO)
C
C-----------------------------------------------------------------------
C
```

```
      WRITE(6,240)NG,NISO,NDIS
  240 FORMAT(/I6,' GOOD T-EIGENVALUES HAVE BEEN COMPUTED'/
     1 I6,' OF THESE ARE T-ISOLATED'/
     2 I6,' = NUMBER OF DISTINCT T-EIGENVALUES COMPUTED'/)
C
C     DO WE WRITE DISTINCT EIGENVALUES OF T-MATRIX TO FILE 4?
      IF (IDIST.EQ.0) GO TO 280
C
      WRITE(11,250) NDIS,NISO,MEV,N,SVSEED,MATNO
  250 FORMAT(/4I6,I12,I8,' = NDIS,NISO,MEV,N,SVSEED,MATNO'/)
C
      WRITE(11,260) (MP(I),VS(I),G(I), I=1,NDIS)
  260 FORMAT(2(I3,E25.16,E12.3))
C
      WRITE(11,270) NDIS, (MP(I), I=1,NDIS)
  270 FORMAT(/I6,' = NDIS, T-MULTIPLICITIES (0 MEANS  SPURIOUS)'/(20I4))
C
  280 CONTINUE
C
      IF (NISO.NE.0) GO TO 310
C
      WRITE(4,290) MEV
  290 FORMAT(/' AT MEV = ',I6,' THERE ARE NO ISOLATED T-EIGENVALUES'/
     1' SO NO ERROR ESTIMATES WERE COMPUTED/')
C
      WRITE(6,300)
  300 FORMAT(/' ALL COMPUTED GOOD T-EIGENVALUES ARE MULTIPLE'/
     1 ' THEREFORE ALL SUCH EIGENVALUES ARE ASSUMED TO HAVE CONVERGED')
C
      ICONV = 1
      GO TO 350
C
  310 CONTINUE
C
C-----------------------------------------------------------------------
C     SUBROUTINE INVERR COMPUTES ERROR ESTIMATES FOR ISOLATED GOOD
C     T-EIGENVALUES USING INVERSE ITERATION ON T(1,MEV). ON RETURN
C     G(J) = MINIMUM GAP IN T(1,MEV) FOR EACH VS(J), J=1,NDIS
C     G(MEV+I) = BETAM*|U(MEV)| = ERROR ESTIMATE FOR ISOLATED GOOD
C              T-EIGENVALUES, WHERE I = 1, NISO AND  BETAM = BETA(MEV+1)
C              U(MEV) IS MEVTH COMPONENT OF THE UNIT EIGENVECTOR OF T
C              CORRESPONDING TO THE ITH ISOLATED GOOD T-EIGENVALUE.
C     A NEGATIVE ERROR ESTIMATE MEANS THAT FOR THAT PARTICULAR
C     EIGENVALUE THE INVERSE ITERATION DID NOT CONVERGE IN <= MXINIT
C     STEPS AND THAT THE CORRESPONDING ERROR ESTIMATE IS QUESTIONABLE.
C
C     V2 CONTAINS THE ISOLATED GOOD T-EIGENVALUES
C     V1 CONTAINS THE MINGAPS TO THE NEAREST DISTINCT  EIGENVALUE
C        OF T(1,MEV) FOR EACH ISOLATED GOOD EIGENVALUE IN V2.
C     VS CONTAINS THE NDIS DISTINCT EIGENVALUES OF T(1,MEV)
C     MP CONTAINS THE CORRESPONDING CODED T-MULTIPLICITIES
C     WITHIN INVERR V1 AND V2 ARE DOUBLE PRECISION ARRAYS
C
      IT = MXINIT
      CALL INVERR(ALPHA,BETA,V1,V2,VS,EPSM,G,MP,MEV,MMB,NDIS,NISO,N,
     1 RHSEED,IT,IWRITE)
C
C-----------------------------------------------------------------------
C
C     SIMPLE CHECK FOR CONVERGENCE. CHECKS TO SEE IF ALL OF THE ERROR
C     ESTIMATES ARE SMALLER THAN CONTOL.
C     IF THIS TEST IS SATISFIED, THEN CONVERGENCE FLAG, ICONV IS SET
C     TO 1.  TYPICALLY ERROR ESTIMATES ARE VERY CONSERVATIVE.
C
      WRITE(6,320) CONTOL
  320 FORMAT(/' CONVERGENCE IS TESTED USING THE CONVERGENCE TOLERANCE',
```

```
     1E13.4/)
C
      II = MEV +1
      IF = MEV+NISO
      DO 330 I = II,IF
      IF (ABS(G(I)).GT.CONTOL) GO TO 350
  330 CONTINUE
      ICONV = 1
      MMB = NMEVS
C
      WRITE(6,340) CONTOL
  340 FORMAT(' ALL COMPUTED ERROR ESTIMATES WERE LESS THAN',E15.4/
     1 ' THEREFORE PROCEDURE TERMINATES'/)
C
  350 CONTINUE
C
C     IF CONVERGENCE IS INDICATED, THAT IS ICONV = 1 ,THEN
C     THE SUBROUTINE PRTEST IS CALLED TO CHECK FOR ANY CONVERGED
C     EIGENVALUES THAT HAVE BEEN MISLABELLED AS SPURIOUS BECAUSE
C     THE PROJECTION OF THEIR EIGENVECTOR(S) ON THE STARTING
C     VECTOR WERE TOO SMALL.
C     NUMERICAL TESTS INDICATE THAT SUCH EIGENVALUES ARE RARE.
C     IF FOR SOME REASON MANY OF THESE HIDDEN EIGENVALUES APPEAR
C     ON SOME RUN, YOU CAN BE CERTAIN THAT SOMETHING IS FOULED UP.
C
      IF (ICONV.EQ.0) GO TO 480
C
C-----------------------------------------------------------------------
C
      CALL PRTEST (ALPHA,BETA,VS,TKMAX,EPSM,RELTOL,SCALE3,SCALE4,
     1 MP,NDIS,MEV,IPROJ)
C
C-----------------------------------------------------------------------
C
      IF(IPROJ.EQ.0) GO TO 470
C
      IF(IDIST.EQ.1)  WRITE(11,360) IPROJ
  360 FORMAT(' SUBROUTINE PRTEST WANTS TO RELABEL',I6,' SPURIOUS EIGENVA
     1LUES'/' WE ACCEPT RELABELLING ONLY IF LAST COMPONENT OF T-EIGENVEC
     1TOR IS L.T. 1.D-10'/)
C
      IIX = RHSEED
C
C-----------------------------------------------------------------------
C
      CALL GENRAN(IIX,G,MEV)
C
C-----------------------------------------------------------------------
C
      ITEN = -10
      NISOM = NISO + MEV
      IWRITO = IWRITE
      IWRITE = 0
C
      DO 390 J = 1,NDIS
      IF(MP(J).NE.ITEN) GO TO 390
      TO = VS(J)
C
C-----------------------------------------------------------------------
C
      IT = MXINIT
      CALL INVERM(ALPHA,BETA,V1,V2,TO,TEMP,T1,EPSM,G,MEV,IT,IWRITE)
C
C-----------------------------------------------------------------------
C
      IF(TEMP.LE.1.D-10) GO TO 380
C     ERROR ESTIMATE WAS NOT SMALL REJECT RELABELLING OF THIS EIGENVALUE
```

```
      IF(IDIST.EQ.1)  WRITE(11,370) J,TO,TEMP
  370 FORMAT(/' LAST COMPONENT FOR',I6,'TH EIGENVALUE',E20.12/' IS TOO L
     1ARGE = ',E15.6,' SO DO NOT ACCEPT PRTEST RELABELLING'/)
      MP(J) = 0
      IPROJ = IPROJ - 1
      GO TO 390
C     RELABELLING ACCEPTED
  380 NISOM = NISOM + 1
      G(NISOM) = BETAM*TEMP
  390 CONTINUE
      IWRITE = IWRITO
C
      IF(IPROJ.EQ.0) GO TO 430
      WRITE(6,400) IPROJ
  400 FORMAT(/I6,' T-EIGENVALUES WERE RECLASSIFIED AS GOOD.'/
     1' THESE ARE IDENTIFIED IN FILE 3 BY A T-MULTIPLICITY OF -10'/' USE
     2R SHOULD INSPECT EACH TO MAKE SURE NEIGHBORS HAVE CONVERGED'/)
C
      IF(IDIST.EQ.1)  WRITE(11,410) IPROJ
  410 FORMAT(/I6,' T-EIGENVALUES WERE RELABELLED AS GOOD'/
     1' BELOW IS CORRECTED T-MULTIPLICITY PATTERN'/)
C
      WRITE(6,420) NDIS, (MP(I), I=1,NDIS)
      IF(IDIST.EQ.1)  WRITE(11,420) NDIS, (MP(I), I=1,NDIS)
  420 FORMAT(/I6,' = NDIS, T-MULTIPLICITIES (0 MEANS  SPURIOUS)'/
     1 6X, ' (-10) MEANS SPURIOUS T-EIGENVALUE RELABELLED AS GOOD'/(20I4
     1))
C
C     RECALCULATE MINGAPS FOR DISTINCT T(1,MEV) EIGENVALUES.
  430 NM1 = NDIS - 1
      G(NDIS) = VS(NM1)-VS(NDIS)
      G(1) = VS(2)-VS(1)
C
      DO 440 J = 2,NM1
      TO = VS(J)-VS(J-1)
      T1 = VS(J+1)-VS(J)
      G(J) = T1
      IF (TO.LT.T1) G(J) = -TO
  440 CONTINUE
      IF(IPROJ.EQ.0) GO TO 470
C     WRITE TO FILE 4 ERROR ESTIMATES FOR THOSE T-EIGENVALUES RELABELLED
      NGOOD = 0
      DO 450 J = 1,NDIS
      IF(MP(J).EQ.0) GO TO 450
      NGOOD = NGOOD + 1
      IF(MP(J).NE.ITEN) GO TO 450
      TO = VS(J)
      NISO = NISO + 1
      NISOM = MEV + NISO
      WRITE(4,460) NGOOD,TO,G(NISOM),G(J)
  450 CONTINUE
  460 FORMAT(I10,E25.16,2E14.3)
C
  470 CONTINUE
C
C     WRITE THE GOOD T-EIGENVALUES TO FILE 3.  FIRST TRANSFER THEM
C     TO V2 AND THEIR T-MULTIPLICITIES TO THE CORRESPONDING POSITIONS
C     IN MP AND COMPUTE THE A-MINGAPS, THE MINIMAL GAPS BETWEEN THE
C     GOOD T-EIGENVALUES.  THESE GAPS WILL BE PUT IN THE ARRAY G.
C     SINCE G CURRENTLY CONTAINS THE MINIMAL GAPS BETWEEN THE DISTINCT
C     EIGENVALUES OF THE T-MATRIX, THESE GAPS WILL FIRST BE
C     TRANSFERRED TO GC.  NOTE THAT GC<0 MEANS THAT THAT MINIMAL GAP
C     IN THE T-MATRIX IS DUE TO A SPURIOUS T-EIGENVALUE.
C     ALL THIS INFORMATION IS PRINTED TO FILE 3
C
  480 CONTINUE
C
```

```
      NG = 0
      DO 490 I = 1,NDIS
      IF (MP(I).EQ.0) GO TO 490
      NG = NG+1
      MP(NG) = MP(I)
      GR(NG) = VS(I)
      TEMP = G(I)
      TEMP = DABS(TEMP)
      J = I+1
      IF (G(I).LT.ZERO) J = I-1
      IF (MP(J).EQ.0) TEMP = -TEMP
      GC(NG) = TEMP
  490 CONTINUE
C
      WRITE(6,500)MEV
  500 FORMAT(//' T-EIGENVALUE CALCULATION AT MEV = ',I6,'    IS COMPLETE
     1')
C
C     NG = NUMBER OF COMPUTED DISTINCT GOOD T-EIGENVALUES.  NEXT
C     GENERATE GAPS BETWEEN GOOD T-EIGENVALUES (AMINGAPS) AND PUT THEM
C     IN G.  G(J) < 0 MEANS THE AMINGAP IS DUE TO THE LEFT-HAND GAP.
C
      NGM1 = NG - 1
      G(NG) = GR(NGM1)-GR(NG)
      G(1) = GR(2)-GR(1)
C
      DO 510 J = 2,NGM1
      T0 = GR(J)-GR(J-1)
      T1 = GR(J+1)-GR(J)
      G(J) = T1
      IF (T0.LT.T1) G(J) = -T0
  510 CONTINUE
C
C     WRITE GOOD T-EIGENVALUES OUT TO FILE 3.
C
      WRITE(3,520)NG,NDIS,MEV,N,SVSEED,MATNO,MULTOL,IB,BTOL
  520 FORMAT(4I6,I12,I8,' = NG,NDIS,MEV,N,SVEED,MATNO'/
     1 E20.12,I6,E13.4,' = MUTOL,INDEX MINIMAL BETA,BTOL'/
     1' EV NO',2X,'TMULT',7X,'GOOD T-EIGENVALUE',7X,'TMINGAP',7X,'AMINGA
     1P')
C
      WRITE(3,530)(I,MP(I),GR(I),GC(I),G(I), I=1,NG)
  530 FORMAT(2I6,E25.16,2E14.3)
C
C     IF CONVERGENCE FLAG ICONV.NE.1 AND NUMBER OF T-MATRICES
C     CONSIDERED TO DATE IS LESS THAN NUMBER ALLOWED, INCREMENT MEV.
C     AND LOOP BACK TO 210 TO REPEAT COMPUTATIONS.  RESTORE BETA(MEV+1).
C
      BETA(MP1) = BETAM
C
      IF (MMB.LT.NMEVS.AND.ICONV.NE.1) GO TO 180
C
C     END OF LOOP ON DIFFERENT SIZE T-MATRICES ALLOWED.
C
  540 CONTINUE
C
      IF(ISTOP.EQ.0)  WRITE(6,550)
  550 FORMAT(/' T-MATRICES (ALPHA AND BETA) ARE NOW AVAILABLE, TERMINATE
     1')
      IF (IHIS.EQ.1.AND.KMAX.NE.MOLD) WRITE(1,560)
  560 FORMAT(/' ABOVE ARE THE FOLLOWING VECTORS '/
     1 '  ALPHA(I), I = 1,KMAX'/
     2 '  BETA(I), I = 1,KMAX+1'/
     3 ' FINAL TWO LANCZOS VECTORS OF ORDER N FOR I = KMAX,KMAX+1'/
     4 ' ALL ENTRIES IN THIS FILE HAVE FORMAT 4Z20 '/
     5 ' ----- END OF FILE 1 NEW ALPHA, BETA HISTORY---------------'///)
C
```

```
      IF (ISTOP.EQ.0) GO TO 640
C
      WRITE(3,570)
  570 FORMAT(/' ABOVE ARE COMPUTED GOOD T-EIGENVALUES'/
     1 ' NG = NUMBER OF GOOD T-EIGENVALUES COMPUTED'/
     2 ' NDIS = NUMBER OF COMPUTED DISTINCT EIGENVALUES OF T(1,MEV)'/
     3 ' N = ORDER OF A,  MATNO = MATRIX IDENT'/
     4 ' MULTOL = MULTIPLICITY TOLERANCE FOR T-EIGENVALUES IN BISEC'/
     4 ' TMULT IS THE T-MULTIPLICITY OF GOOD T-EIGENVALUE'/
     5 ' TMULT = -1 MEANS SPURIOUS T-EIGENVALUE TOO CLOSE'/
     6 ' DO NOT COMPUTE ERROR ESTIMATES FOR SUCH EIGENVALUES'/
     7 ' AMINGAP = MINIMAL GAP BETWEEN THE COMPUTED A-EIGENVALUES'/
     8 ' AMINGAP .LT. 0. MEANS MINIMAL GAP IS DUE TO LEFT-HAND GAP'/
     9 ' TMINGAP= MINIMAL GAP W.R.T.  DISTINCT EIGENVALUES IN T(1,MEV)'/
     1 ' TMINGAP .LT. 0. MEANS MINGAP IS DUE TO SPURIOUS EIGENVALUE'/
     2 ' ----- END OF FILE 3 GOODEIGENVALUES-----------------------'///)
C
      IF (IDIST.EQ.1) WRITE(11,580)
  580 FORMAT(/' ABOVE ARE THE DISTINCT EIGENVALUES OF T(1,MEV).'/
     2 ' THE FORMAT IS       T-MULTIPLICITY    EIGENVALUE   TMINGAP'/
     3 '        THIS FORMAT IS REPEATED TWICE ON EACH LINE.'/
     4 ' T-MULTIPLICITY = -1 MEANS THAT THE SUBROUTINE ISOEV HAS TAGGED'
     5 /'   THIS SIMPLE T-EIGENVALUE AS HAVING A VERY CLOSE SPURIOUS'/
     6 '   T-EIGENVALUE SO THAT NO ERROR ESTIMATE WILL BE COMPUTED'/
     7 '   FOR THAT EIGENVALUE IN SUBROUTINE INVERR.'/
     8 ' TMINGAP .LT. 0, TMINGAP IS DUE TO LEFT GAP .GT. 0, RIGHT GAP.'/
     9 ' EACH OF THE DISTINCT T-EIGENVALUE TABLES IS FOLLOWED'/
     9 ' BY THE T-MULTIPLICITY PATTERN.'/
     1 ' NDIS = NUMBER OF COMPUTED DISTINCT EIGENVALUES OF T(1,MEV).'/
     2 ' NG = NUMBER OF GOOD T-EIGENVALUES. '/
     3 ' NISO = NUMBER OF ISOLATED GOOD T-EIGENVALUES. '/
     4 ' NISO ALSO IS THE COUNT OF +1 ENTRIES IN T-MULTIPLICITY PATTERN.
     5 '/' ----- END OF FILE 4 DISTINCT T-EIGENVALUES----------------'//
     6 /)
C
      IF(NISO.NE.0)  WRITE(4,590)
  590 FORMAT(/' ABOVE ARE THE ERROR ESTIMATES OBTAINED FOR THE ISOLATED
     1GOOD EIGENVALUES'/
     1' OBTAINED VIA INVERSE ITERATION IN THE SUBROUTINE INVERR.'/
     1' ALL OTHER GOOD EIGENVALUES HAVE CONVERGED.'/
     2' ERROR ESTIMATE = BETAM*ABS(UM)'/
     2' WHERE BETAM = BETA(MEV+1) AND UM = U(MEV).'/
     3' U = UNIT EIGENVECTOR OF T WHERE T*U = EV*U AND EV = ISOLATED GOO
     3D EIGENVALUE.'/
     4' TMINGAP = GAP TO NEAREST DISTINCT EIGENVALUE OF T(1,MEV).'/
     5' TMINGAP .LT. 0. MEANS MINGAP IS DUE TO LEFT NEIGHBOR.'/
     6' ERROR ESTIMATE L.T. 0 MEANS INVERSE ITERATION DID NOT CONVERGE'/
     7' ------ END OF FILE 7 ERRINV -------------------------------'//)
      GO TO 640
C
  600 CONTINUE
C
      IBB = IABS(IBMEV)
      IF (IBMEV.LT.0) WRITE(6,610) MEV,IBB,BETA(IBB)
  610 FORMAT(/' PROGRAM TERMINATES BECAUSE MEV REQUESTED = ',I6,' IS .GT
     1',I6/' AT WHICH AN ABNORMALLY SMALL BETA = ' , E13.4,' OCCURRED'/)
      GO TO 640
C
  620 IF (NDIS.EQ.0.AND.ISTOP.GT.0) WRITE(6,630)
  630 FORMAT(/' INTERVALS SPECIFIED FOR BISECT DID NOT CONTAIN ANY EIGEN
     1VALUES'/' PROGRAM TERMINATES')
C
  640 CONTINUE
C
      STOP
C-----END OF MAIN PROGRAM FOR LANCZOS HERMITIAN EIGENVALUE COMPUTATIONS-
      END
```

SECTION 3.3 MAIN PROGRAM, EIGENVECTOR COMPUTATIONS

```
C-----HLEVEC (EIGENVECTORS OF HERMITIAN MATRICES)-----------------------
C
C     CONTAINS MAIN PROGRAM FOR COMPUTING AN EIGENVECTOR CORRESPONDING
C     TO EACH OF A SET OF EIGENVALUES THAT HAVE BEEN COMPUTED
C     ACCURATELY BY THE CORRESPONDING LANCZOS EIGENVALUE PROGRAM
C     (HLEVAL) FOR HERMITIAN MATRICES.  THIS PROGRAM COULD BE
C     MODIFIED TO COMPUTE ADDITIONAL EIGENVECTORS FOR ANY
C     MULTIPLE EIGENVALUE OF THE GIVEN A-MATRIX.  THE AMOUNT OF
C     ADDITIONAL COMPUTATION REQUIRED BY SUCH A MODIFICATION WOULD
C     DEPEND UPON THE GIVEN MATRIX AND UPON WHICH PART OF THE
C     SPECTRUM WAS INVOLVED.
C
C     THE LANCZOS EIGENVECTOR COMPUTATIONS ASSUME THAT EACH
C     EIGENVALUE THAT IS BEING CONSIDERED HAS CONVERGED AS AN
C     EIGENVALUE OF THE LANCZOS TRIDIAGONAL MATRICES.
C
C     PORTABILITY:
C     THIS PROGRAM IS NOT PORTABLE BECAUSE OF THE USE OF COMPLEX*16
C     VARIABLES.  MOREOVER, THE PFORT VERIFIER IDENTIFIED THE
C     FOLLOWING ADDITIONAL NONPORTABLE CONSTRUCTIONS:
C
C     1.  DATA/MACHEP/ STATEMENT
C     2.  ALL READ(5,*) STATEMENTS (FREE FORMAT)
C     3.  FORMAT(20A4) USED WITH THE EXPLANATORY HEADER, EXPLAN
C     4.  HEXADECIMAL FORMAT (4Z20) USED IN ALPHA/BETA FILES 1 AND 2.
C
C     IMPORTANT NOTE:  PROGRAM ALLOWS ENLARGEMENT OF THE ALPHA, BETA
C     ARRAYS.  IN PARTICULAR, IF ANY ONE OF THE EIGENVALUES SUPPLIED
C     IS T-SIMPLE AND NOT CLOSE TO A SPURIOUS T-EIGENVALUE, THE PROGRAM
C     REQUIRES THAT KMAX BE AT LEAST 11*MEV/8 + 12.  IF KMAX IS NOT
C     THIS LARGE, THEN THE PROGRAM WILL RESET KMAX TO THIS SIZE
C     AND EXTEND THE ALPHA, BETA HISTORY IF REQUIRED.
C     THUS, THE DIMENSIONS OF THE ALPHA AND BETA ARRAYS MUST BE
C     LARGE ENOUGH TO ALLOW FOR THIS POSSIBILITY.
C     REMEMBER THAT THE BETA ARRAY, BETA(J), IS SUCH THAT
C     J = 1,..., KMAX+1.  SO IF THE KMAX USED BY THE PROGRAM
C     IS TO BE 3000, THEN BETA MUST BE OF LENGTH AT LEAST 3001.
C
C
C-----------------------------------------------------------------------
      COMPLEX*16  V1(1000),V2(1000),VS(1000),RITVEC(10000),ZEROC,TEMPC
      DOUBLE PRECISION  ALPHA(1000),BETA(1001),GR(1000),GC(1000)
      DOUBLE PRECISION  TVEC(20000),GOODEV(50),EVNEW(50)
      DOUBLE PRECISION  EVAL,EVALN,TOLN,TTOL,ERTOL,ALFA,BATA
      DOUBLE PRECISION  MULTOL,SCALEO,STUTOL,BTOL,LB,UB
      DOUBLE PRECISION  ONE,ZERO,MACHEP,EPSM,TEMP,SUM,ERRMIN,BKMIN
      DOUBLE PRECISION  RELTOL,ERROR,TERROR,TLAST(50)
      REAL  G(1000),AMINGP(50),TMINGP(50),EXPLAN(20)
      REAL  TERR(50),ERR(50),ERRDGP(50),RNORM(50),TBETA(50)
      INTEGER  MP(50),M1(50),M2(50),MA(50),ML(50),MINT(50),MFIN(50)
      INTEGER  SVSEED,SVSOLD,RHSEED,IDELTA(50),MULEVA(50)
      INTEGER  MBOUND,NTVCON,SVTVEC,TVSTOP,LVCONT,ERCONT,TFLAG
      DOUBLE PRECISION  DABS, DMAX1, DSQRT, DFLOAT
      REAL ABS
      INTEGER  IABS
C-----------------------------------------------------------------------
      EXTERNAL CMATV
      DATA MACHEP/Z3410000000000000/
      EPSM = 2.D0*MACHEP
C-----------------------------------------------------------------------
C
C     ARRAYS MUST BE DIMENSIONED AS FOLLOWS:
C     1.  ALPHA:  >= KMAXN,  BETA: >= (KMAXN+1) WHERE KMAXN, THE
C                 LARGEST SIZE T-MATRIX CONSIDERED BY THE PROGRAM,
```

```
C               IS THE LARGER OF THE SIZE OF THE ALPHA, BETA HISTORY
C               PROVIDED ON FILE 2 (IF ANY ) AND THE SIZE WHICH THE
C               PROGRAM SPECIFIES INTERNALLY, THIS LATTER IS ALWAYS
C               < = 11*MEV / 8  +  12, WHERE MEV IS THE SIZE
C               T-MATRIX THAT WAS USED IN THE CORRESPONDING EIGENVALUE
C               COMPUTATIONS.
C     2.  V1:  >= MAX(N,KMAX/2)
C     3.  V2, VS:   >= N
C     4.  G:  >= MAX(N,KMAX).   GR, GC:  >= N
C     5.  RITVEC:  >= N*NGOOD, WHERE NGOOD IS NUMBER OF EIGENVALUES
C                 SUPPLIED TO THIS PROGRAM.
C     6.  TVEC:  >= CUMULATIVE LENGTH OF ALL THE T-EIGENVECTORS
C         NEEDED TO GENERATE THE DESIRED RITZ VECTORS.  AN EDUCATED
C         GUESS AT AN APPROPRIATE LENGTH CAN BE OBTAINED BY RUNNING THE
C         PROGRAM WITH THE FLAG MBOUND = 1 AND MULTIPLYING THE
C         RESULTING SIZE BY 5/4.
C     7.  GOODEV, EVNEW, AMINGP, TMINGP, TERR, ERR, ERRGDP, RNORM, TBETA
C         TLAST, MP, MA, M1, M2, MINT, MFIN, MULEVA, AND IDELTA ALL
C         MUST BE AT LEAST NGOOD.
C
C-----------------------------------------------------------------------
C     OUTPUT HEADER
      WRITE(6,10)
   10 FORMAT(/' LANCZOS EIGENVECTOR PROCEDURE FOR HERMITIAN MATRICES'/)
C
C     SET PROGRAM PARAMETERS
C     USER MUST NOT MODIFY SCALEO
      SCALEO = 5.0D0
      ZERO = 0.0D0
      ZEROC = DCMPLX(ZERO,ZERO)
      ONE = 1.0D0
      MPMIN = -1000
C     CONVERGENCE TOLERANCE FOR T-EIGENVECTORS FOR RITZ VECTORS
      ERTOL = 1.D-10
      ISREAL = 0
C
C     READ USER-SPECIFIED PARAMETER FROM INPUT FILE 5 (FREE FORMAT)
C
C     READ USER-PROVIDED HEADER FOR RUN
      READ(5,20) EXPLAN
      WRITE(6,20) EXPLAN
   20 FORMAT(20A4)
C
C     READ IN THE MAXIMUM PERMISSIBLE DIMENSIONS FOR THE TVEC ARRAY
C     (MDIMTV), FOR THE RITVEC ARRAY (MDIMRV), AND FOR THE BETA
C     ARRAY (MBETA).
C
      READ(5,20) EXPLAN
      READ(5,*) MDIMTV, MDIMRV, MBETA
C
C     READ IN RELATIVE TOLERANCE (RELTOL) USED IN DETERMINING
C     APPROPRIATE SIZES FOR THE T-MATRICES USED IN THE EIGENVECTOR
C     COMPUTATIONS
C
      READ(5,20) EXPLAN
      READ(5,*) RELTOL
C
C
C     SET FLAGS TO 0 OR 1:
C     MBOUND = 1:  PROGRAM TERMINATES AFTER COMPUTING 1ST GUESSES
C                  ON APPROPRIATE T-SIZES FOR USE IN THE RITZ VECTOR
C                  COMPUTATIONS
C     NTVCON = 0:  PROGRAM TERMINATES IF THE TVEC ARRAY IS NOT
C                  LARGE ENOUGH TO HOLD ALL THE T-EIGENVECTORS REQUIRED.
C     SVTVEC = 0:  THE T-EIGENVECTORS ARE NOT WRITTEN TO FILE 11
C                  UNLESS TVSTOP = 1
C     SVTVEC = 1:  WRITE THE T-EIGENVECTORS TO FILE 11.
```

```
C      TVSTOP = 1:  PROGRAM TERMINATES AFTER COMPUTING THE
C                   T-EIGENVECTORS
C      LVCONT = 0:  PROGRAM TERMINATES IF THE NUMBER OF T-EIGENVECTORS
C                   COMPUTED IS NOT EQUAL TO THE NUMBER OF RITZ
C                   VECTORS REQUESTED.
C      ERCONT = 0:  MEANS FOR ANY GIVEN EIGENVALUE, A RITZ VECTOR
C                   WILL NOT BE COMPUTED FOR THAT EIGENVALUE UNLESS
C                   A T-EIGENVECTOR HAS BEEN IDENTIFIED WITH A LAST
C                   COMPONENT WHICH SATISFIES THE SPECIFIED
C                   CONVERGENCE CRITERION.
C      ERCONT = 1:  MEANS FOR ANY GIVEN EIGENVALUE, A RITZ VECTOR
C                   WILL BE COMPUTED.  IF A T-EIGENVECTOR CANNOT
C                   BE IDENTIFIED WHICH SATISFIES THE LAST
C                   COMPONENT CRITERION, THEN THE PROGRAM WILL
C                   USE THE T-VECTOR THAT CAME CLOSEST TO
C                   SATISFYING THE CRITERION
C      IWRITE = 1:  EXTENDED OUTPUT OF INTERMEDIATE COMPUTATIONS
C                   IS WRITTEN TO FILE 6
C      IREAD = 0:   ALPHA/BETA FILE IS REGENERATED
C      IREAD = 1:   ALPHA/BETA FILE USED IN EIGENVALUE COMPUTATIONS
C                   IS READ IN AND EXTENDED IF NECESSARY.  IN BOTH
C                   CASES IREAD = 0 OR 1, THE LANCZOS VECTORS ARE
C                   ALWAYS REGENERATED FOR THE RITZ VECTOR
C                   COMPUTATIONS
C
      READ(5,20) EXPLAN
      READ(5,*) MBOUND,NTVCON,SVTVEC,IREAD
C
      READ(5,20) EXPLAN
      READ(5,*) TVSTOP,LVCONT,ERCONT,IWRITE
      IF (TVSTOP.EQ.1) SVTVEC = 1
C
C     READ IN SEED FOR GENERATING RANDOM STARTING VECTOR FOR THE
C     INVERSE ITERATION ON THE T-MATRICES.
C
      READ(5,20) EXPLAN
      READ(5,*) RHSEED
C
C     READ IN MATNO = MATRIX/RUN IDENTIFICATION NUMBER AND
C     N = ORDER OF A-MATRIX
C
      READ(5,20) EXPLAN
      READ(5,*) MATNO,N
C     IF MATNO < 0, THEN MATRIX SUPPLIED IS REAL AND USER WANTS TO
C     CHECK ON THE T-MULTIPLICITY OF THE EIGENVALUES OF GIVEN MATRIX
      IF(MATNO.GT.0) GO TO 30
      ISREAL = 1
      MATNO = - MATNO
   30 CONTINUE
C
C-----------------------------------------------------------------------
C     INITIALIZE THE ARRAYS FOR THE USER-SPECIFIED MATRIX
C     AND PASS THE STORAGE LOCATIONS OF THESE ARRAYS TO THE
C     MATRIX-VECTOR MULTIPLY SUBROUTINE CMATV.
C
      CALL USPEC(N,MATNO)
C
C-----------------------------------------------------------------------
C     MASK UNDERFLOW AND OVERFLOW
      CALL MASK
C
C-----------------------------------------------------------------------
C
C     WRITE RUN PARAMETERS OUT TO FILE 6
C
      WRITE(6,40) MATNO,N
```

```
   40 FORMAT(/' MATRIX IDENTIFICATION NO. = ',I10,' ORDER OF A = ',I5)
C
      WRITE(6,50) MBOUND,NTVCON,SVTVEC,IREAD
   50 FORMAT(/3X,'MBOUND',3X,'NTVCON',3X,'SVTVEC',3X,'IREAD'/3I9,I8)
C
      WRITE(6,60) TVSTOP,LVCONT,ERCONT,IWRITE
   60 FORMAT(/3X,'TVSTOP',3X,'LVCONT',3X,'ERCONT',3X,'IWRITE'/4I9)
C
      WRITE(6,70) MDIMTV,MDIMRV,MBETA
   70 FORMAT(/3X,'MDIMTV',3X,'MDIMRV',3X,'MBETA'/2I9,I8)
C
      WRITE(6,80) RELTOL,RHSEED
   80 FORMAT(/7X,'RELTOL',3X,'RHSEED'/E13.4,I9)
C
C
C     FROM FILE 3 READ IN THE NUMBER OF EIGENVALUES (NGOOD) FOR WHICH
C     EIGENVECTORS ARE REQUESTED, THE ORDER (MEV) OF THE LANCZOS
C     TRIDIAGONAL MATRIX USED IN COMPUTING THESE EIGENVALUES, THE
C     ORDER (NOLD) OF THE USER-SPECIFIED MATRIX USED IN THE EIGENVALUE
C     COMPUTATIONS, THE SEED (SVSEED) USED FOR GENERATING THE STARTING
C     VECTOR THAT WAS USED IN THOSE LANCZOS EIGENVALUE COMPUTATIONS,
C     AND THE MATRIX/RUN IDENTIFICATION NUMBER (MATOLD) USED IN THOSE
C     COMPUTATIONS.  ALSO READ IN THE NUMBER (NDIS) OF DISTINCT
C     EIGENVALUES OF T(1,MEV) THAT WERE COMPUTED BUT THIS VALUE IS
C     NOT USED IN THE EIGENVECTOR COMPUTATIONS.
C
      READ(3,90) NGOOD,NDIS,MEV,NOLD,SVSEED,MATOLD
   90 FORMAT(4I6,I12,I8)
C
C     READ IN THE T-MULTIPLICITY TOLERANCE USED IN THE BISEC SUBROUTINE
C     DURING THE COMPUTATION OF THE GIVEN EIGENVALUES.
C     ALSO READ IN THE FLAG IB.  IF IB < 0, THEN SOME BETA(I) IN THE
C     T-MATRIX FILE PROVIDED ON FILE 2 FAILED THE ORTHOGONALITY
C     TEST IN THE TNORM SUBROUTINE.  USER SHOULD NOTE THAT THIS
C     VECTOR PROGRAM PROCEEDS INDEPENDENTLY OF THE SIZE OF THE BETA.
C
      READ(3,100) MULTOL,IB,BTOL
  100 FORMAT(E20.12,I6,E13.4)
C
      TEMP = DFLOAT(MEV+1000)
      TTOL = MULTOL/TEMP
      WRITE(6,110) MULTOL,TTOL
  110 FORMAT(/' T-MULTIPLICITY TOLERANCE USED IN THE EIGENVALUE COMPUTAT
     1IONS WAS',E13.4/' SCALED MACHINE EPSILON IS',E13.4)
C
C     CONTINUE WRITE TO FILE 6 OF THE PARAMETERS FOR THIS RUN
C
      WRITE(6,120)NGOOD,NDIS,MEV,NOLD,MATOLD,SVSEED,MULTOL,IB,BTOL
  120 FORMAT(/' EIGENVALUES SUPPLIED ARE READ IN FROM FILE 3'/' FILE 3
     1HEADER IS'/4X,'NG',2X,'NDIS',3X,'MEV',2X,'NOLD',2X,'MATOLD',4X,
     1'SVSEED',6X,'MULTOL',6X,'IB',9X,'BTOL'/4I6,I8,I10,E12.3,I8,E13.4/)
C
C     IS THE ARRAY RITVEC LONG ENOUGH TO HOLD ALL OF THE DESIRED
C     RITZ VECTORS (APPROXIMATE EIGENVECTORS)?
      NMAX = NGOOD*N
      IF(MBOUND.NE.0) GO TO 130
      IF(TVSTOP.NE.1.AND.NMAX.GT.MDIMRV) GO TO 1390
C
C     CHECK THAT THE ORDER N AND THE MATRIX IDENTIFICATION NUMBER
C     MATNO SPECIFIED BY THE USER AGREE WITH THOSE READ IN FROM
C     FILE 3.
  130 ITEMP = (NOLD-N)**2+(MATOLD-MATNO)**2
      IF (ITEMP.NE.0) GO TO 1410
C
C
C     THE LANCZOS EIGENVECTOR COMPUTATIONS ASSUME THAT EACH
C     EIGENVALUE THAT IS BEING CONSIDERED HAS CONVERGED AS AN
```

```
C     EIGENVALUE OF THE LANCZOS TRIDIAGONAL MATRICES.
C
C     READ IN FROM FILE 3, THE T-MULTIPLICITIES OF THE EIGENVALUES
C     WHOSE EIGENVECTORS ARE TO BE COMPUTED, THE VALUES OF THESE
C     EIGENVALUES AND THEIR MINIMAL GAPS AS EIGENVALUES OF THE
C     USER-SPECIFIED MATRIX AND AS EIGENVALUES OF THE T-MATRIX.
C
      READ(3,20) EXPLAN
      READ(3,140) (MP(J),GOODEV(J),TMINGP(J),AMINGP(J), J=1,NGOOD)
  140 FORMAT(6X,I6,E25.16,2E14.3)
C
      WRITE(6,150) (J,GOODEV(J),MP(J),TMINGP(J),AMINGP(J), J=1,NGOOD)
  150 FORMAT(/' EIGENVALUES READ IN, T-MULTIPLICITIES, T-GAPS AND A-GAPS
     1 '/4X,' J ',5X,'GOOD EIGENVALUE',5X,'MULT',4X,'  TMINGAP ',4X,
     1'  AMINGAP '/(I6,E25.16,I4,2E15.4))
C
C     READ IN ERROR ESTIMATES
      WRITE(6,180) MEV,SVSEED
C     CHECK WHETHER OR NOT THERE ARE ANY ISOLATED T-EIGENVALUES IN
C     THE EIGENVALUES PROVIDED
      DO 160 J=1,NGOOD
      IF(MP(J).EQ.1) GO TO 170
  160 CONTINUE
      GO TO 200
  170 READ(4,20) EXPLAN
      READ(4,20) EXPLAN
      READ(4,20) EXPLAN
  180 FORMAT(/' THESE EIGENVALUES WERE COMPUTED USING A T-MATRIX OF
     1ORDER ',I5/' AND SEED FOR RANDOM NUMBER GENERATOR =',I12)
      READ(4,190) NISO
  190 FORMAT(18X,I6)
      READ(4,20) EXPLAN
      READ(4,20) EXPLAN
      READ(4,20) EXPLAN
  200 DO 230 J=1,NGOOD
      ERR(J) = 0.D0
      IF(MP(J).NE.1) GO TO 230
      READ(4,210) EVAL, ERR(J)
  210 FORMAT(10X,E25.16,E14.3)
      IF(DABS(EVAL - GOODEV(J)).LT.1.D-10) GO TO 230
      WRITE(6,220) EVAL,GOODEV(J)
  220 FORMAT(' PROBLEM WITH READ IN OF ERROR ESTIMATES'/' EIGENVALUE REA
     1D IN',E20.12,' DOES NOT MATCH GOODEV(J) ='/E20.12)
      GO TO 1630
C
  230 CONTINUE
C
      WRITE(6,240) (J,GOODEV(J),ERR(J), J=1,NGOOD)
  240 FORMAT(' ERROR ESTIMATES ='/4X,' J',5X,'EIGENVALUE',10X,'ESTIMATE'
     1 /(I6,E20.12,E14.3))
C
      IF(IREAD.EQ.0) GO TO 340
C
C     READ IN THE SIZE OF THE T-MATRIX PROVIDED ON FILE 2.  READ IN
C     THE ORDER OF THE USER-SPECIFIED MATRIX , THE SEED FOR THE
C     RANDOM NUMBER GENERATOR, AND THE MATRIX/TEST IDENTIFICATION
C     NUMBER THAT WERE USED IN THE LANCZOS EIGENVALUE COMPUTATIONS.
C     IF FLAG IREAD = 0, REGENERATE HISTORY.  HISTORY MUST BE
C     STORED IN HEXADECIMAL FORMAT TO AVOID ERRORS INCURRED IN
C     INPUT/OUTPUT CONVERSIONS.
C
      READ(2,250) KMAX,NOLD,SVSOLD,MATOLD
  250 FORMAT(2I6,I12,I8)
C
      WRITE(6,260) KMAX,NOLD,SVSOLD,MATOLD
  260 FORMAT(/' READ IN THE T-MATRICES STORED ON FILE 2'/' FILE 2 HEADER
```

```
     1 IS'/2X,'KMAX',2X,'NOLD',6X,'SVSOLD',2X,'MATOLD'/2I6,I12,I8/)
C
C     CHECK THAT THE ORDER, THE MATRIX/TEST IDENTIFICATION NUMBER
C     AND THE SEED FOR THE RANDOM NUMBER GENERATOR USED IN THE
C     LANCZOS COMPUTATIONS THAT GENERATED THE HISTORY FILE
C     BEING USED AGREE WITH WHAT THE USER HAS SPECIFIED.
      IF (NOLD.NE.N.OR.MATOLD.NE.MATNO.OR.SVSOLD.NE.SVSEED) GO TO 1430
C
      KMAX1 = KMAX + 1
C
C     READ IN THE T-MATRICES FROM FILE 2.  THESE ARE USED TO GENERATE
C     THE T-EIGENVECTORS THAT WILL BE USED IN THE RITZ VECTOR
C     COMPUTATIONS.  ALPHA/BETA HISTORY MUST BE STORED IN
C     MACHINE FORMAT, ((4Z20) FOR IBM/3081).
C
      READ(2,270) (ALPHA(J), J=1,KMAX)
      READ(2,270) (BETA(J), J=1,KMAX1)
  270 FORMAT(4Z20)
C
      READ(2,270) (V1(J), J=1,N)
      READ(2,270) (V2(J), J=1,N)
C
C     ENLARGE KMAX IF THE SIZE AT WHICH THE EIGENVALUE
C     COMPUTATIONS WERE PERFORMED IS ESSENTIALLY KMAX AND
C     THERE IS AT LEAST ONE EIGENVALUE THAT IS T-SIMPLE AND
C     T-ISOLATED, IN THE SENSE THAT IF ITS NEAREST T-NEIGHBOR IS TOO
C     CLOSE THAT NEIGHBOR IS A 'GOOD' T-EIGENVALUE.
      DO 280 J = 1,NGOOD
      IF(MP(J).EQ.1) GO TO 300
  280 CONTINUE
      WRITE(6,290)
  290 FORMAT(/' ALL EIGENVALUES USED ARE T-MULTIPLE OR CLOSE TO SPURIOUS
     1 T-EIGENVALUES'/' SO DO NOT CHANGE KMAX')
      IF(KMAX.LT.MEV) GO TO 1450
      GO TO 320
C
  300 KMAXN= 11*MEV/8 + 12
      IF(MBETA.LE.KMAXN) GO TO 1610
      IF(KMAX.GE.KMAXN )  GO TO 320
      WRITE(6,310) KMAX, KMAXN
  310 FORMAT(' ENLARGE KMAX FROM ',I6,' TO ',I6)
      MOLD1 = KMAX + 1
      KMAX = KMAXN
      GO TO 390
C
  320 WRITE(6,330) KMAX
  330 FORMAT(/' T-MATRICES HAVE BEEN READ IN FROM FILE 2'/' THE LARGEST
     1SIZE T-MATRIX ALLOWED IS',I6/)
C
      IF(IREAD.EQ.1) GO TO 410
C
C     REGENERATE THE ALPHA AND BETA
C
  340 MOLD1 = 1
C
      DO 350 J = 1,NGOOD
      IF(MP(J).EQ.1) GO TO 370
  350 CONTINUE
      KMAX = MEV + 12
      WRITE(6,360) KMAX
  360 FORMAT(/' ALL EIGENVALUES FOR WHICH EIGENVECTORS ARE TO BE COMPUTE
     1D ARE EITHER T-MULTIPLE OR CLOSE TO'/' A SPURIOUS EIGENVALUE. THER
     1EFORE SET KMAX = MEV + 12 = ',I7)
      GO TO 390
C
  370 KMAXN = 11*MEV/8 + 12
      IF(MBETA.LE.KMAXN) GO TO 1610
```

```
      WRITE(6,380) KMAXN
  380 FORMAT(' SET KMAX EQUAL TO ',I6)
      KMAX = KMAXN
C
  390 WRITE(6,400) MOLD1,KMAX
  400 FORMAT(/' LANCZS SUBROUTINE GENERATES ALPHA(J), BETA(J+1), J =',
     1 I6,' TO ', I6/)
C
C-----------------------------------------------------------------------
C
      CALL LANCZS(CMATV,V1,V2,VS,ALPHA,BETA,GR,GC,G,KMAX,MOLD1,N,SVSEED)
C
C-----------------------------------------------------------------------
C
  410 CONTINUE
C
C     THE SUBROUTINE STURMI DETERMINES THE SMALLEST SIZE T-MATRIX FOR
C     WHICH THE EIGENVALUE IN QUESTION IS AN EIGENVALUE (TO WITHIN A
C     GIVEN TOLERANCE) AND IF POSSIBLE THE SMALLEST SIZE T-MATRIX
C     FOR WHICH IT IS A DOUBLE EIGENVALUE (TO WITHIN THE SAME
C     TOLERANCE).  THE SIZE T-MATRIX USED IN THE EIGENVECTOR
C     COMPUTATIONS IS THEN DETERMINED BY LOOPING ON THE SIZES OF THE
C     T-EIGENVECTORS, USING THE INFORMATION FROM STURMI TO OBTAIN
C     STARTING GUESSES AT THE T-SIZES.
C
C
      STUTOL = SCALE0*MULTOL
      IF(IWRITE.EQ.1) WRITE(6,420)
  420 FORMAT(' FROM STURMI')
      DO 460 J = 1,NGOOD
      EVAL = GOODEV(J)
C     COMPUTE THE TOLERANCES USED BY STURMI TO DETERMINE AN INTERVAL
C     CONTAINING THE EIGENVALUE EVAL.
      TEMP = DABS(EVAL)*RELTOL
      TOLN = DMAX1(TEMP,STUTOL)
C
C-----------------------------------------------------------------------
C
      CALL STURMI(ALPHA,BETA,EVAL,TOLN,EPSM,KMAX,MK1,MK2,IC,IWRITE)
C
C-----------------------------------------------------------------------
C
C     STORE THE COMPUTED ORDERS OF T-MATRICES FOR LATER PRINTOUT
      M1(J) = MK1
      M2(J) = MK2
      ML(J) = (MK1 + 3*MK2)/4
      IF(MK2.EQ.KMAX)  ML(J) = KMAX
C
      IF(IC.GT.0) GO TO 440
C     IC = 0 MEANS THERE WAS NO T-EIGENVALUE IN THE DESIGNATED INTERVAL
C     BY T-SIZE KMAX.  THIS MEANS THAT THE T-EIGENVALUE PROVIDED HAS
C     NOT YET CONVERGED AS AN EIGENVALUE OF THE TRIDIAGONAL MATRICES
C     SO PROGRAM SHOULD NOT COMPUTE ITS EIGENVECTOR.
      WRITE(6,430) J,GOODEV(J),MK1,MK2
  430 FORMAT(I6,'TH EIGENVALUE',E20.12,' HAS NOT CONVERGED '/
     1' SO DO NOT COMPUTE ANY T-EIGENVECTOR OR RITZ VECTOR FOR IT'
     1/' MK1 AND MK2 FOR THIS EIGENVALUE WERE',2I6)
      MP(J) = MPMIN
      MA(J) = -2*KMAX
      GO TO 460
C     COMPUTE AN APPROPRIATE SIZE T-MATRIX FOR THE GIVEN EIGENVALUE.
  440 IF(M2(J).EQ.KMAX) GO TO 450
C     M1 AND M2 WERE BOTH DETERMINED
      MA(J) = (3*M1(J) + M2(J))/4  + 1
      GO TO 460
C     M2 NOT DETERMINED
```

```
  450 MA(J) = 5*M1(J)/4  + 1
C
  460 CONTINUE
C
      IF (IWRITE.EQ.1) WRITE(6,470) (MA(JJ), JJ=1,NGOOD)
  470 FORMAT(/' 1ST GUESS AT APPROPRIATE SIZE T-MATRICES'/
     1 ' ACTUAL VALUES WILL PROBABLY BE 1/4 AGAIN AS MUCH'/(13I6))
C
C     PRINT OUT TO FILE 10 1ST GUESSES AT SIZES OF T-MATRICES TO
C     BE USED IN THE EIGENVECTOR COMPUTATIONS.
C     ACTUAL SIZES MAY BE 1/4 OR MORE LARGER THAN THESE SIZES.
      WRITE(10,480) N,KMAX
  480 FORMAT(2I8,' = ORDER OF USER MATRIX AND MAX ORDER OF T(1,MEV)')
C
      WRITE(10,490)
  490 FORMAT(/' 1ST GUESS AT APPROPRIATE SIZE T-MATRICES'/
     1 ' ACTUAL VALUES WILL PROBABLY BE 1/4 AGAIN AS MUCH'/)
C
      WRITE(10,500)
  500 FORMAT(4X,'J',7X,'GOODEV(J)',4X,'M1(J)',1X,'M2(J)',1X,'MA(J)')
C
      WRITE(10,510) (J,GOODEV(J),M1(J),M2(J), MA(J), J=1,NGOOD)
  510 FORMAT(I5,E19.12,3I6)
C
      IF(MBOUND.EQ.1) WRITE(10,520)
  520 FORMAT(/' GOODEV(J) IS A GOOD EIGENVALUE OF T(1,MEV)'/
     1 ' M1 = SMALLEST VALUE OF M SUCH THAT T(1,M) HAS AT LEAST'/
     1 '      ONE EIGENVALUE IN THE INTERVAL (EV-TOLN,EV+TOLN)'/
     1 '      TOLN(J) = DMAX1(GOODEV(J)*RELTOL, SCALE0*MULTOL)'/
     1 ' M2 = SMALLEST M (IF ANY) SUCH THAT IN THE ABOVE INTERVAL'/
     1 '      T(1,M) HAS AT LEAST TWO EIGENVALUES '/
     1 ' INITIAL VALUE OF MA(J) IS CHOSEN HEURISTICALLY'/
     1 ' PROGRAM LOOPS ON SIZE OF T-MATRIX TO GET BETTER SIZE'/
     1 ' END OF SIZES OF T-MATRICES FILE 10'///)
C
C
C     TERMINATE AFTER COMPUTING 1ST GUESSES AT  SIZES OF THE
C     T-MATRICES REQUIRED FOR THE GIVEN EIGENVALUES?
      IF(MBOUND.EQ.1) GO TO 1470
C
C
C     IS THERE ROOM FOR ALL OF THE REQUESTED T-EIGENVECTORS?
      MTOL = 0
      DO 530 J = 1,NGOOD
      IF(MP(J).EQ.MPMIN) GO TO 530
      MTOL = MTOL + IABS(MA(J))
  530 CONTINUE
      MTOL = (5*MTOL)/4
      IF(MTOL.GT.MDIMTV.AND.NTVCON.EQ.0) GO TO 1490
C
C-----------------------------------------------------------------------
C     GENERATE A RANDOM VECTOR TO BE USED REPEATEDLY BY
C     SUBROUTINE INVERM
C
      IIL = RHSEED
      CALL GENRAN(IIL,G,KMAX)
C
C-----------------------------------------------------------------------
C
C     LOOP ON GIVEN EIGENVALUES TO COMPUTE THE CORRESPONDING
C     T-EIGENVECTOR.
C
      MTOL = 0
      NTVEC = 0
      ILBIS = 0
      DO 720 J = 1,NGOOD
      ICOUNT = 0
```

```
      ERRMIN = 10.D0
      MABEST = MPMIN
      IF(MP(J).EQ.MPMIN) GO TO 720
      TFLAG = 0
      EVAL = GOODEV(J)
      TEMP = RELTOL* DABS(EVAL)
      UB = EVAL + DMAX1(STUTOL,TEMP)
      LB = EVAL - DMAX1(STUTOL,TEMP)
  540 KMAXU = IABS(MA(J))
C
C     SELECT A SUITABLE INCREMENT FOR THE ORDERS OF THE T-MATRICES
C     TO BE CONSIDERED IN DETERMINING APPROPRIATE SIZES FOR THE RITZ
C     VECTOR COMPUTATIONS.
      IF(ICOUNT.GT.0) GO TO 560
C     SELECT IDELTA(J) BASED UPON THE T-MULTIPLICITY OBTAINED
      IF(M2(J).EQ.KMAX) GO TO 550
C     M2 DETERMINED
      IDELTA(J) = ((3*M1(J) + 5*M2(J))/8  +  1 - IABS(MA(J)))/10 + 1
      GO TO 560
C     M2 NOT DETERMINED
  550 MAMAX = MIN0((11*MEV)/8 + 12, (13*M1(J))/8 + 1)
      IDELTA(J) = (MAMAX - IABS(MA(J)))/10  +  1
  560 ICOUNT = ICOUNT + 1
C
C-----------------------------------------------------------------------
C     TO MIMIMIZE THE EFFECT OF THE ONE-SIDED ACCEPTANCE TEST FOR
C     T-EIGENVALUES IN THE BISEC SUBROUTINE, RECOMPUTE THE GIVEN
C     EIGENVALUE AT THE SPECIFIED KMAXU
C
      CALL LBISEC(ALPHA,BETA,EPSM,EVAL,EVALN,LB,UB,TTOL,KMAXU,NEVT)
C
C-----------------------------------------------------------------------
C
C     CHECK WHETHER OR NOT GIVEN T-MATRIX HAS AN EIGENVALUE IN THE
C     SPECIFIED INTERVAL AND IF SO WHAT ITS T-MULTIPLICITY IS.
C
      IF(NEVT.EQ.1) GO TO 600
      IF(NEVT.NE.0) GO TO 580
      ILBIS = 1
      WRITE(6,570) EVAL,KMAXU
  570 FORMAT(/' PROBLEM ENCOUNTERED IN RECOMPUTATION OF USER-SUPPLIED EI
     1GENVALUE',E20.12/' THE SIZE T-MATRIX SPECIFIED',I6,' DOES NOT
     1HAVE AN EIGENVALUE IN THE INTERVAL SPECIFIED'/' THEREFORE NO EIGEN
     1VECTOR WILL BE COMPUTED FOR THIS PARTICULAR EIGENVALUE'/)
      GO TO 620
C
  580 IF(NEVT.GT.1)  WRITE(6,590) EVAL,KMAXU
  590 FORMAT(/' PROBLEM ENCOUNTERED IN RECOMPUTATION OF USER-SUPPLIED
     1EIGENVALUE',E20.12/' FOR THE SIZE T-MATRIX SPECIFIED =',I6,' THE
     1GIVEN EIGENVALUE IS T-MULTIPLE IN THE INTERVAL SPECIFIED'/' SOMETH
     1ING IS WRONG, THEREFORE NO EIGENVECTOR WILL BE COMPUTED FOR THIS E
     1IGENVALUE'/)
C
      MP(J) = MPMIN
      MA(J) = -2*KMAX
      GO TO 720
C
  600 CONTINUE
      ILBIS = 0
C
      EVNEW(J) = EVALN
      EVAL = EVALN
      MTOL = MTOL+KMAXU
C
C     IS THERE ROOM IN TVEC ARRAY FOR THE NEXT T-EIGENVECTOR?
C     IF NOT, SKIP TO RITZ VECTOR COMPUTATIONS.
```

```
      IF (MTOL.GT.MDIMTV) GO TO 730
C
      IT = 3
      KINT = MTOL - KMAXU +1
C
C     RECORD THE BEGINNING AND END OF THE T-EIGENVECTOR BEING COMPUTED
      MINT(J) = KINT
      MFIN(J) = MTOL
C
C-----------------------------------------------------------------------
C     SUBROUTINE INVERM DOES INVERSE ITERATION, I.E. SOLVES
C     (T(1,KMAXU) - EVAL)*U = RHS  FOR EACH EIGENVALUE TO OBTAIN THE
C     DESIRED T-EIGENVECTOR.
C
      IF(IWRITE.EQ.1)  WRITE(6,610) J
  610 FORMAT(/I6,'TH EIGENVALUE')
C
      CALL INVERM(ALPHA,BETA,V1,TVEC(KINT),EVAL,ERROR,TERROR,EPSM,
     1 G,KMAXU,IT,IWRITE)
C
C-----------------------------------------------------------------------
C
      TERR(J) = TERROR
      TLAST(J) = ERROR
      KMAXU1 = KMAXU + 1
      TBETA(J) = BETA(KMAXU1)*ERROR
C
C     AFTER COMPUTING EACH OF THE T-EIGENVECTORS,
C     CHECK THE SIZE OF THE ERROR ESTIMATE, ERROR.
C     IF THIS ESTIMATE IS NOT AS SMALL AS DESIRED AND
C     |MA(J)| < ML(J), ATTEMPT TO INCREASE THE SIZE OF |MA(J)|
C     AND REPEAT THE T-EIGENVECTOR COMPUTATIONS.
C
      IF(ERROR.LT.ERTOL.OR.TFLAG.EQ.1)  GO TO 710
C
      IF(ERROR.GE.ERRMIN) GO TO 620
C     LAST COMPONENT IS LESS THAN MINIMAL TO DATE
      ERRMIN = ERROR
      MABEST = MA(J)
  620 CONTINUE
C
      IF(MA(J).GT.0)  ITEST = MA(J) + IDELTA(J)
      IF(MA(J).LT.0)  ITEST = -(IABS(MA(J)) + IDELTA(J))
      IF(IABS(ITEST).LE.ML(J).AND.ICOUNT.LE.10) GO TO 640
C     NEW MA(J) IS GREATER THAN MAXIMUM ALLOWED.
      IF(ERCONT.EQ.0.OR.MABEST.EQ.MPMIN) GO TO 660
      TFLAG = 1
      MA(J) = MABEST
      IF(ILBIS.EQ.0)  MTOL = MTOL - KMAXU
      WRITE(6,630) MA(J)
  630 FORMAT(' 10 ORDERS WERE CONSIDERED.  NONE SATISFIED THE ERROR TEST
     1'/' THEREFORE USE THE BEST ORDER OBTAINED FOR THE EIGENVECTORS'
     1,I6)
      GO TO 540
C
  640 MA(J) = ITEST
C
      MT = IABS(MA(J))
      IF(IWRITE.EQ.1) WRITE(6,650)  MT
  650 FORMAT(/' CHANGE SIZE OF T-MATRIX TO ',I6,' RECOMPUTE T-EIGENVECTO
     1R')
C
      IF(ILBIS.EQ.0)  MTOL = MTOL - KMAXU
C
      GO TO 540
C
C       APPROPRIATE SIZE T-MATRIX WAS NOT OBTAINED
```

```
  660 CONTINUE
      WRITE(10,670) J,EVAL,MP(J)
  670 FORMAT(/' ON 10 INCREMENTS NOT ABLE TO IDENTIFY APPROPRIATE SIZE
     1T-MATRIX FOR'/
     1' EIGENVALUE(',I4,') = ',E20.12,' T-MULTIPLICITY =',I4/)
      IF(M2(J).EQ.KMAX) WRITE(10,680)
      IF(M2(J).LT.KMAX) WRITE(10,690)
  680 FORMAT(/' ORDERS TESTED RANGED FROM 5*M1(J)/4 TO APPROXIMATELY'/
     1'    MIN(11*MEV/8, 13*M1(J)/8)'/)
  690 FORMAT(/' ORDERS TESTED RANGED FROM (3*M1(J)+M2(J)/4 TO APPROXIMAT
     1ELY'/'  (3*M1(J) + 5*M2(J))/8'/)
      WRITE(10,700)
  700 FORMAT(' ALLOWING LARGER ORDERS FOR THE T-MATRICES MAY RESULT IN
     1 SUCCESS'/' BUT PROBABLY WILL NOT.  PROBLEM IS PROBABLY DUE TO'
     1 /' LACK OF CONVERGENCE OF GIVEN EIGENVALUE, CHECK THE ERROR ESTIM
     1ATE')
      MP(J) = MPMIN
      IF(ILBIS.EQ.0) MTOL = MTOL - KMAXU
      GO TO 720
  710 NTVEC = NTVEC + 1
C
  720 CONTINUE
      NGOODC = NGOOD
      GO TO 750
C
C     COME HERE IF THERE IS NOT ENOUGH ROOM FOR ALL OF T-EIGENVECTORS
  730 NGOODC = J-1
      WRITE(6,740) J,MTOL,MDIMTV
  740 FORMAT(/' NOT ENOUGH ROOM IN TVEC ARRAY FOR ',I4,'TH T-EIGENVECTOR
     1'/'  TVEC DIMENSION REQUESTED = ',I6,' BUT TVEC HAS DIMENSION ',I6
     1/)
      IF(NGOODC.EQ.0)  GO TO 1510
      MTOL = MTOL-KMAXU
C
  750 CONTINUE
C
C     THE LOOP ON T-EIGENVECTOR COMPUTATIONS IS COMPLETE.
C     WRITE OUT THE SIZE T-MATRICES THAT WILL BE USED FOR
C     THE RITZ VECTOR COMPUTATIONS.
C
      WRITE(10,760)
  760 FORMAT(/' SIZES OF T-MATRICES THAT WILL BE USED IN THE RITZ COMPUT
     1ATIONS'/5X,'J',16X,'GOODEV(J)',1X,'MA(J)')
C
      WRITE(10,770)   (J,GOODEV(J),MA(J), J=1,NGOOD)
  770 FORMAT(I6,E25.14,I6)
      WRITE(10,520)
C
      WRITE(6,780) MTOL
  780 FORMAT(/' THE CUMULATIVE LENGTH OF THE T-EIGENVECTORS IS',I18)
C
      WRITE(6,790) NTVEC,NGOOD
  790 FORMAT(/I6,' T-EIGENVECTORS OUT OF',I6,' REQUESTED WERE COMPUTED')
C
C     SAVE THE T-EIGENVECTORS ON FILE 11?
      IF(TVSTOP.NE.1.AND.SVTVEC.EQ.0) GO TO 850
C
      WRITE(11,800) NTVEC,MTOL,MATNO,SVSEED
  800 FORMAT(I6,3I12,' = NTVEC,MTOL,MATNO,SVSEED')
C
      DO 830 J=1,NGOODC
C     IF MP(J) = MPMIN THEN NO SUITABLE T-EIGENVECTOR IS AVAILABLE
C     FOR THAT EIGENVALUE.
      IF(MP(J).EQ.MPMIN) WRITE(11,810) J,MA(J),GOODEV(J),MP(J)
  810 FORMAT(2I6,E20.12,I6/' TH EIGVAL,T-SIZE,EVALUE,FLAG,NO EIGVEC')
      IF(MP(J).NE.MPMIN) WRITE(11,820) J,MA(J),GOODEV(J),MP(J)
  820 FORMAT(I6,I6,E20.12,I6/' T-EIGVEC,SIZE T,EVALUE OF A,MP(J)')
```

```
      IF(MP(J).EQ.MPMIN) GO TO 830
      KI = MINT(J)
      KF = MFIN(J)
C
      WRITE(11,270) (TVEC(K), K=KI,KF)
C
  830 CONTINUE
C
      IF(TVSTOP.NE.1) GO TO 850
C
      WRITE(6,840) TVSTOP, NTVEC,NGOOD
  840 FORMAT(/' USER SET TVSTOP = ',I1/
     1' THEREFORE PROGRAM TERMINATES AFTER T-EIGENVECTOR COMPUTATIONS'/
     1' T-EIGENVECTORS THAT WERE COMPUTED ARE SAVED ON FILE 11'/
     1I8,' T-EIGENVECTORS WERE COMPUTED OUT OF',I7,' REQUESTED'/)
C
      GO TO 1630
C
  850 CONTINUE
C     IF NOT ABLE TO COMPUTE ALL THE REQUESTED T-EIGENVECTORS
C     CONTINUE WITH THE LANCZOS VECTOR COMPUTATIONS ANYWAY?
      IF(NTVEC.NE.NGOOD.AND.LVCONT.EQ.0) GO TO 1530
C
C     COMPUTE THE MAXIMUM SIZE OF THE T-MATRIX USED FOR THOSE
C     EIGENVALUES WITH GOOD ERROR ESTIMATES.
C
      KMAXU = 0
      DO 860 J = 1,NGOODC
      MT = IABS(MA(J))
      IF(MT.LT.KMAXU.OR.MP(J).EQ.MPMIN) GO TO 860
      KMAXU = MT
  860 CONTINUE
C
      IF(KMAXU.EQ.0) GO TO 1570
C
      WRITE(6,870) KMAXU
  870 FORMAT(/I6,' = LARGEST SIZE T-MATRIX TO BE USED IN THE RITZ VECTOR
     1 COMPUTATIONS')
C
C     COUNT THE NUMBER OF RITZ VECTORS NOT BEING COMPUTED
      MREJEC = 0
      DO 880 J=1,NGOODC
  880 IF(MP(J).EQ.MPMIN)  MREJEC = MREJEC + 1
      MREJET = MREJEC + (NGOOD-NGOODC)
      IF(MREJET.NE.0) WRITE(6,890) MREJET
  890 FORMAT(/' RITZ VECTORS ARE NOT COMPUTED FOR',I6,' OF THE EIGENVALU
     1ES'/)
      NACT = NGOODC - MREJEC
      WRITE(6,900) NGOOD,NTVEC,NACT
  900 FORMAT(/I6,' RITZ VECTORS WERE REQUESTED'/I6,' T-EIGENVECTORS WERE
     1 COMPUTED'/I6,' RITZ VECTORS WILL BE COMPUTED'/)
C     CHECK IF THERE ARE ANY RITZ VECTORS TO COMPUTE
      IF(MREJEC.EQ.NGOODC) GO TO 1550
C
C     CONTINUE WITH THE LANCZOS VECTOR COMPUTATIONS?
      IF(LVCONT.EQ.0.AND.MREJEC.NE.0) GO TO 1530
C
C     NOW COMPUTE THE RITZ VECTORS.  REGENERATE THE
C     LANCZOS VECTORS.
C
      DO 910 I = 1,NMAX
  910 RITVEC(I) = ZEROC
C
C     REGENERATE THE STARTING VECTOR. THIS MUST BE GENERATED AND
C     NORMALIZED PRECISELY THE WAY IT WAS DONE IN THE EIGENVALUE
C     COMPUTATIONS, OTHERWISE THERE WILL BE A MISMATCH BETWEEN
C     THE T-EIGENVECTORS THAT HAVE BEEN COMPUTED FROM THE T-MATRICES
```

```
C     READ IN FROM FILE 2 AND THE LANCZOS VECTORS THAT ARE
C     BEING REGENERATED.
C
C-----------------------------------------------------------------------
C
      IIL = SVSEED
      CALL GENRAN(IIL,G,N)
C
C-----------------------------------------------------------------------
C
      DO 920 I = 1,N
  920 GR(I) = G(I)
C
C-----------------------------------------------------------------------
C
      CALL GENRAN(IIL,G,N)
C
C-----------------------------------------------------------------------
C
      DO 930 I = 1,N
  930 GC(I) = G(I)
C
      DO 940 I = 1,N
  940 V2(I) = DCMPLX(GR(I),GC(I))
C
C-----------------------------------------------------------------------
      CALL CINPRD(V2,V2,SUM,N)
C-----------------------------------------------------------------------
C
      SUM = ONE/DSQRT(SUM)
      DO 950 I = 1,N
      V1(I) = ZEROC
  950 V2(I) = V2(I)*SUM
C
C     LOOP FOR GENERATING REQUIRED RITZ VECTORS (IVEC = 1,KMAXU)
C     USES GRAM-SCHMIDT ORTHOGONALIZATION WITHOUT MODIFICATION
C
      IVEC = 1
      BATA = ZERO
C
      GO TO 1010
C
  960 CONTINUE
C
C-----------------------------------------------------------------------
C     CMATV(V2,VS,SUM) CALCULATES  VS = A*V2 - SUM*VS
      SUM = ZERO
      CALL CMATV(V2,VS,SUM)
      CALL CINPRD(V2,VS,ALFA,N)
C
C-----------------------------------------------------------------------
C
      DO 970 J=1,N
  970 V1(J) = (VS(J) - BATA*V1(J)) - ALFA*V2(J)
C
C-----------------------------------------------------------------------
      CALL CINPRD(V1,V1,BATA,N)
C-----------------------------------------------------------------------
C
      BATA = DSQRT(BATA)
      SUM = ONE/BATA
C
      TEMP = BETA(IVEC)
      TEMP = DABS(BATA - TEMP)/TEMP
      IF (TEMP.LT.1.0D-10)GO TO 990
C
C     THE BETA BEING REGENERATED DO NOT MATCH THE HISTORY FILE
```

```
C     SOMETHING IS WRONG IN THE LANCZOS VECTOR GENERATION
C     PROGRAM TERMINATES FOR USER TO CORRECT THE PROBLEM
C     WHICH MUST BE IN THE STARTING VECTOR GENERATION OR IN
C     THE MATRIX-VECTOR MULTIPLY SUBROUTINE CMATV SUPPLIED.
C     THIS SUBROUTINE MUST BE THE SAME ONE USED IN THE
C     EIGENVALUE COMPUTATIONS OR A MISMATCH WILL ENSUE.
C
      WRITE(6,980) IVEC,BATA,BETA(IVEC),TEMP
  980 FORMAT(/2X,'IVEC',16X,'BATA',10X,'BETA(IVEC)',14X,'RELDIF'/I6,
     13E20.12/' IN LANCZOS VECTOR REGENERATION THE ENTRIES OF THE TRIDIA
     1GONAL MATRICES BEING'/' GENERATED ARE NOT THE SAME AS THOSE IN THE
     1 MATRIX SUPPLIED ON FILE 2.'/' THEREFORE SOMETHING IS BEING INITIA
     1LIZED OR COMPUTED DIFFERENTLY FROM THE WAY'/' IT WAS COMPUTED IN TH
     1HE EIGENVALUE COMPUTATIONS'/' THE PROGRAM TERMINATES FOR THE USER
     1TO DETERMINE WHAT THE PROBLEM IS'/)
      GO TO 1630
C
C
  990 CONTINUE
      DO 1000 J = 1,N
      TEMPC = SUM*V1(J)
      V1(J) = V2(J)
 1000 V2(J) = TEMPC
C
 1010 CONTINUE
C
      LFIN = 0
      DO 1030 J = 1,NGOODC
      LL = LFIN
      LFIN = LFIN + N
C
      IF(IABS(MA(J)).LT.IVEC.OR.MP(J).EQ.MPMIN) GO TO 1030
      II = IVEC + MINT(J) - 1
      TEMP = TVEC(II)
C     II IS THE (IVEC)TH COMPONENT OF THE T-EIGENVECTOR CONTAINED
C     IN TVEC(MINT(J)).
C
      DO 1020 K = 1,N
      LL = LL + 1
 1020 RITVEC(LL) = TEMP*V2(K) + RITVEC(LL)
C
 1030 CONTINUE
C
      IVEC = IVEC + 1
      IF (IVEC.LE.KMAXU) GO TO 960
C
C
C     RITZVECTOR GENERATION IS COMPLETE. NORMALIZE EACH RITZVECTOR.
C     NOTE THAT IF CERTAIN RITZ VECTORS WERE NOT COMPUTED THEN THAT
C     PORTION OF THE RITVEC ARRAY WAS NOT UTILIZED.
C
      LFIN = 0
      DO 1130 J = 1,NGOODC
C
      KK = LFIN
      LFIN = LFIN + N
      IF(MP(J).EQ.MPMIN) GO TO 1130
C
      DO 1040 K = 1,N
      KK = KK + 1
 1040 V2(K) = RITVEC(KK)
C
C-----------------------------------------------------------------------
      CALL CINPRD(V2,V2,SUM,N)
C-----------------------------------------------------------------------
C
      SUM = DSQRT(SUM)
```

```
      RNORM(J) = SUM                                                    HLE09320
      TEMP = DABS(ONE-SUM)                                              HLE09330
      SUM = ONE/SUM                                                     HLE09340
C                                                                       HLE09350
      KK = LFIN - N                                                     HLE09360
      DO 1050 K = 1,N                                                   HLE09370
      KK = KK + 1                                                       HLE09380
      V2(K) = SUM*V2(K)                                                 HLE09390
 1050 RITVEC(KK) = V2(K)                                                HLE09400
C                                                                       HLE09410
C     ONLY ENTER NEXT PORTION IF GIVEN MATRIX IS REAL.                  HLE09420
       IF(ISREAL.NE.1) GO TO 1100                                       HLE09430
C                                                                       HLE09440
C     AT THIS POINT RITZ VECTOR IS IN V2.                               HLE09450
C     THIS PROGRAM CAN BE USED ON REAL MATRICES TO DETERMINE            HLE09460
C     WHICH IF ANY EIGENVALUES ARE A-MULTIPLE AND IF SO TO COMPUTE      HLE09470
C     TWO EIGENVECTORS FOR THOSE EIGENVALUES THAT ARE MULTIPLE AND ONE  HLE09480
C     FOR THOSE THAT ARE NOT MULTIPLE. HERE ONLY IDENTIFIES WHETHER     HLE09490
C     EIGENVALUE IS AT LEAST DOUBLE.  THIS IS DONE BY CHECKING THE      HLE09500
C     RATIOS OF SUCCEEDING REAL AND IMAGINARY PARTS OF THE COMPUTED     HLE09510
C     RITZ VECTORS.                                                     HLE09520
C                                                                       HLE09530
      SUM = DIMAG(V2(1))/DREAL(V2(1))                                   HLE09540
      DO 1060 K=2,N                                                     HLE09550
      TEMP = DREAL(V2(K))                                               HLE09560
      IF(DABS(TEMP).LT.1.D-9) GO TO 1060                                HLE09570
      TEMP = DIMAG(V2(K))/DREAL(V2(K))                                  HLE09580
      IF(DABS(TEMP - SUM).LE.1.D-6) GO TO 1060                          HLE09590
      MULEVA(J) = 2                                                     HLE09600
      GO TO 1070                                                        HLE09610
 1060 CONTINUE                                                          HLE09620
      MULEVA(J) = 1                                                     HLE09630
 1070 IF(MULEVA(J).EQ.2) WRITE(6,1090) J,GOODEV(J)                      HLE09640
      IF(MULEVA(J).EQ.1) WRITE(6,1080) J,GOODEV(J)                      HLE09650
 1080 FORMAT(I6,'TH EIGENVALUE CONSIDERED =',E20.12,' IS SIMPLE')       HLE09660
 1090 FORMAT(I6,'TH EIGENVALUE CONSIDERED =',E20.12,' IS MULTIPLE')     HLE09670
C                                                                       HLE09680
 1100 CONTINUE                                                          HLE09690
C                                                                       HLE09700
      IF (IWRITE.NE.0) WRITE(6,1110) J,GOODEV(J)                        HLE09710
 1110 FORMAT(/I5,' TH EIGENVALUE CONSIDERED = ',E20.12/)                HLE09720
C                                                                       HLE09730
      IF (IWRITE.NE.0) WRITE(6,1120) TERR(J),TBETA(J),TEMP              HLE09740
 1120 FORMAT(' NORM OF ERROR IN T-EIGENVECTOR = ',E14.3/                HLE09750
     1 ' BETA(MA(J)+1)*U(MA(J))  = ',E14.3/                             HLE09760
     1 ' ABS(NORM(RITVEC) - 1.0)  = ',E14.3/)                           HLE09770
C                                                                       HLE09780
      LINT = LFIN - N + 1                                               HLE09790
      EVAL = EVNEW(J)                                                   HLE09800
C                                                                       HLE09810
C-----------------------------------------------------------------------HLE09820
C                                                                       HLE09830
      CALL CMATV(RITVEC(LINT),V2,EVAL)                                  HLE09840
C                                                                       HLE09850
C-----------------------------------------------------------------------HLE09860
C                                                                       HLE09870
C     COMPUTE ERROR IN RITZ VECTOR CONSIDERED AS A EIGENVECTOR OF A.    HLE09880
C     V2 = A*RITVEC - EVAL*RITVEC                                       HLE09890
C                                                                       HLE09900
C-----------------------------------------------------------------------HLE09910
      CALL CINPRD(V2,V2,SUM,N)                                          HLE09920
C-----------------------------------------------------------------------HLE09930
C                                                                       HLE09940
      SUM = DSQRT(SUM)                                                  HLE09950
      ERR(J) = SUM                                                      HLE09960
      GAP = ABS(AMINGP(J))                                              HLE09970
```

```
      ERRDGP(J) = SUM/GAP
C
 1130 CONTINUE
C
C
C     RITZVECTORS ARE NORMALIZED AND ERROR ESTIMATES ARE IN ERR ARRAY
C     AND IN ERRDGP ARRAY. STORE EVERYTHING
C
C
      WRITE(9,1140)
 1140 FORMAT(6X,'GOODEV(J)',1X,'MA(J)',4X,'A MINGAP',6X,'AERROR',2X,
     1 'AERROR/GAP',6X,'TERROR')
C
      WRITE(13,1150)
 1150 FORMAT(16X,'GOODEV(J)',5X,'RITZNORM',6X,'AMINGAP',5X,
     1 'TBETA(J)',5X,'TLAST(J)')
C
      DO 1180 J=1,NGOODC
C
      IF(MP(J).EQ.MPMIN) GO TO 1180
C
      WRITE(9,1160)EVNEW(J),MA(J),AMINGP(J),ERR(J),ERRDGP(J),TERR(J)
 1160 FORMAT(E15.8,I6,4E12.4)
C
      WRITE(13,1170) EVNEW(J),RNORM(J),AMINGP(J),TBETA(J),TLAST(J)
 1170 FORMAT(E25.14,4E13.5)
C
 1180 CONTINUE
C
      IF(MREJEC.EQ.0) GO TO 1260
      WRITE(9,1190)
 1190 FORMAT(/' RITZ VECTORS WERE NOT COMPUTED FOR THE FOLLOWING EIGENVA
     1LUES'/' EITHER BECAUSE THEY HAD NOT CONVERGED OR BECAUSE THE ERROR
     1 ESTIMATE'/'  WAS NOT AS SMALL AS DESIRED'/)
C
      DO 1250 J = 1,NGOODC
      IF(MP(J).NE.MPMIN) GO TO 1250
C     WRITE OUT MESSAGE FOR EACH EIGENVALUE FOR WHICH NO EIGENVECTOR
C     WAS COMPUTED.
C
      WRITE(9,1200)
 1200 FORMAT(6X,'GOODEV(J)',3X,'MA(J)',5X,'AMINGP(J)',6X,'TLAST(J)',3X,
     1'MP(J)')
      WRITE(9,1210) GOODEV(J),MA(J),AMINGP(J),TBETA(J),MP(J)
 1210 FORMAT(E15.8,I8,2E14.4,I8)
C
      WRITE(13,1220)
 1220 FORMAT(/' RITZ VECTORS WERE NOT COMPUTED FOR THE FOLLOWING EIGENVA
     1LUES'/' EITHER BECAUSE THEY HAD NOT CONVERGED OR BECAUSE'/' THE ER
     1ROR ESTIMATE WAS NOT AS SMALL AS DESIRED'/)
C
      WRITE(13,1230)
 1230 FORMAT(6X,'GOODEV(J)',3X,'MA(J)',3X,'M1(J)',3X,'M2(J)',3X,'MP(J)'
     1/)
      WRITE(13,1240) GOODEV(J),MA(J),M1(J),M2(J),MP(J)
 1240 FORMAT(E15.8,4I8)
C
 1250 CONTINUE
 1260 CONTINUE
C
      WRITE(9,1270)
 1270 FORMAT(/' ABOVE ARE ERROR ESTIMATES FOR THE A AND T EIGENVECTORS'/
     1 ' ASSOCIATED WITH THE GOODEV LISTED IN COLUMN 1'/
     1 ' AERROR = NORM(A*X - EV*X)  TERROR = NORM(T*Y - EV*Y) '/
     1 ' WHERE T = T(1,MA(J))    X = RITZ VECTOR = V*Y  V = SUCCESSIVE'/
     1 ' LANCZOS VECTORS. A MINGAP = GAP TO NEAREST A-EIGENVALUE'//)
C
```

```
      WRITE(13,1280)                                                    HLE10650
 1280 FORMAT(/' ABOVE ARE ERROR ESTIMATES ASSOCIATED WITH THE GOODEV'/  HLE10660
     1 ' RITZNORM = NORM(RITZ VECTOR)'/                                 HLE10670
     1 ' TBETA(J) = CDABS(BETA(MA(J)+1)*Y(MA(J))),  T*Y = GOODEV*Y'/    HLE10680
     1 ' TLAST(J) = CDABS(Y(MA(J))'/                                    HLE10690
     1 ' AMINGAP = DISTANCE TO CLOSEST COMPUTED GOOD T-EIGENVALUE'/)     HLE10700
C                                                                       HLE10710
C     NUMBER OF RITZ VECTORS COMPUTED                                   HLE10720
      NCOMPU = NGOODC - MREJEC                                          HLE10730
      WRITE(12,1290) N,NCOMPU,NGOODC,MATNO                              HLE10740
 1290 FORMAT(3I6,I12,' SIZE A, NO.RITZVECS, NO.EVALUES,MATNO')          HLE10750
C                                                                       HLE10760
      LFIN = 0                                                          HLE10770
      DO 1350 J = 1,NGOODC                                              HLE10780
      LINT = LFIN + 1                                                   HLE10790
      LFIN = LFIN + N                                                   HLE10800
C                                                                       HLE10810
      IF(MP(J).EQ.MPMIN) GO TO 1330                                     HLE10820
C     RITZ VECTOR WAS COMPUTED                                          HLE10830
      WRITE(12,1300) J, GOODEV(J), MP(J)                                HLE10840
 1300 FORMAT(I6,4X,E20.12,I6,' J, EIGENVAL, MP(J)')                     HLE10850
C                                                                       HLE10860
      WRITE(12,1310) ERR(J),ERRDGP(J)                                   HLE10870
 1310 FORMAT(2E15.5,' = NORM(A*Z-EVAL*Z) AND  NORM(A*Z-EVAL*Z)/MINGAP') HLE10880
C                                                                       HLE10890
      WRITE(12,1320) (RITVEC(LL), LL=LINT,LFIN)                         HLE10900
 1320 FORMAT(4Z20)                                                      HLE10910
      GO TO 1350                                                        HLE10920
C     NO RITZ VECTOR WAS COMPUTED FOR THIS EIGENVALUE                   HLE10930
 1330 WRITE(12,1340) J,GOODEV(J),MP(J)                                  HLE10940
 1340 FORMAT(I6,4X,E20.12,I6,' J,EIGVALUE,NO RITZ VECTOR COMPUTED')     HLE10950
C                                                                       HLE10960
 1350 CONTINUE                                                          HLE10970
C                                                                       HLE10980
C     DID ANY T-MATRICES INCLUDE OFF-DIAGONAL ENTRIES SMALLER THAN      HLE10990
C     DESIRED, AS SPECIFIED BY BTOL?                                    HLE11000
C                                                                       HLE11010
      IF(IB.GT.0) GO TO 1380                                            HLE11020
      WRITE(6,1360) KMAXU                                               HLE11030
 1360 FORMAT(/' FOR LARGEST T-MATRIX CONSIDERED',I7,' CHECK THE SIZE OF HLE11040
     1BETAS')                                                           HLE11050
C                                                                       HLE11060
C-----------------------------------------------------------------------HLE11070
C                                                                       HLE11080
      CALL TNORM(ALPHA,BETA,BKMIN,TEMP,KMAXU,IBMT)                      HLE11090
C                                                                       HLE11100
C-----------------------------------------------------------------------HLE11110
C                                                                       HLE11120
      IF(IBMT.LT.0) WRITE (6,1370)                                      HLE11130
 1370 FORMAT(/' WARNING THE T-MATRICES FOR ONE OR MORE OF THE EIGENVALUEHLE11140
     1S CONSIDERED'/' HAD AN OFF-DIAGONAL ENTRY THAT WAS SMALLER THAN THHLE11150
     1E BETA TOLERANCE THAT WAS SPECIFIED'/)                            HLE11160
 1380 CONTINUE                                                          HLE11170
C                                                                       HLE11180
       GO TO 1630                                                       HLE11190
C                                                                       HLE11200
 1390 WRITE(6,1400) NGOOD,NMAX,MDIMRV                                   HLE11210
 1400 FORMAT(/I4,' RITZ VECTORS WERE REQUESTED BUT THE REQUIRED DIMENSIOHLE11220
     1N',I6/' IS LARGER THAN THE USER-SPECIFIED DIMENSION OF RITVEC',I6 HLE11230
     1/' THEREFORE, THE EIGENVECTOR PROCEDURE TERMINATES FOR THE USER TOHLE11240
     1 INTERVENE')                                                      HLE11250
C                                                                       HLE11260
      GO TO 1630                                                        HLE11270
C                                                                       HLE11280
 1410 WRITE(6,1420) NOLD,N,MATOLD,MATNO                                 HLE11290
 1420 FORMAT(/' PARAMETERS READ FROM FILE 3 DO NOT AGREE WITH THOSE SPECHLE11300
     1 IFIED'/' BY THE USER.   NOLD,N,MATOLD,MATNO = '/2I6,2I12/        HLE11310
```

```
     1' THEREFORE, PROGRAM TERMINATES FOR USER TO RESOLVE THE DIFFERENCE
     1S'/)
C
      GO TO 1630
C
 1430 WRITE(6,1440)
 1440 FORMAT(/' PARAMETERS IN ALPHA,BETA FILE READ IN DO NOT AGREE WITH
     1 THOSE'/' SPECIFIED BY THE USER.  THEREFORE, THE PROCEDURE TERMINA
     1TES'/' FOR THE USER TO RESOLVE THE DIFFERENCES.'/)
C
      GO TO 1630
C
 1450 WRITE(6,1460) KMAX,MEV
 1460 FORMAT(/' ON ALPHA,BETA HEADER KMAX = ',I6/
     1' BUT EIGENVALUES WERE COMPUTED AT MEV = ',I6,' PROGRAM STOPS'/)
C
      GO TO 1630
C
 1470 WRITE(6,1480)
 1480 FORMAT(/' PROGRAM COMPUTED 1ST GUESSES ON T-MATRIX SIZES, READ THE
     1M TO FILE 10'/' THEN TERMINATED AS REQUESTED.')
      GO TO 1630
C
 1490 WRITE(6,1500) MTOL, MDIMTV
 1500 FORMAT(/' PROGRAM TERMINATES BECAUSE THE TVEC DIMENSION ANTICIPATE
     1D',I7/'  IS LARGER THAN THE TVEC DIMENSION',I7,' SPECIFIED BY THE
     1USER.'/'  USER MAY RESET THE TVEC DIMENSION AND RESTART THE PROGRA
     1M')
      GO TO 1630
C
 1510 WRITE(6,1520)
 1520 FORMAT(/' PROGRAM TERMINATES BECAUSE NO SUITABLE T-EIGENVECTORS WE
     1RE IDENTIFIED'/'  FOR ANY OF THE EIGENVALUES SUPPLIED.  PROBLEM CO
     1ULD BE CAUSED'/'  BY TOO SMALL A TVEC DIMENSION OR SIMPLY BE THAT
     1IT WAS NOT POSSIBLE'/'  TO IDENTIFY T-VECTORS.  USER SHOULD CHECK
     1OUTPUT'/)
      GO TO 1630
C
 1530 WRITE(6,1540) LVCONT,NTVEC,NGOOD
 1540 FORMAT(/' LVCONT FLAG =',I2,' AND NUMBER ',I5,' OF T-EIGENVECTORS
     1 COMPUTED N.E.'/' NUMBER',I5,' REQUESTED SO PROGRAM TERMINATES'/)
      GO TO 1630
 1550 WRITE(6,1560)
 1560 FORMAT(/' PROGRAM TERMINATES WITHOUT COMPUTING ANY RITZ VECTORS'/
     1 ' BECAUSE ALL T-EIGENVECTORS WERE REJECTED AS NOT SUITABLE'/
     1 ' PROBABLE CAUSE IS LACK OF CONVERGENCE OF THE EIGENVALUES'/)
       GO TO 1630
C
 1570 WRITE(6,1580)
 1580 FORMAT(/' PROGRAM INDICATES THAT IT IS NOT POSSIBLE TO COMPUTE ANY
     1 OF THE'/' REQUESTED EIGENVECTORS. THEREFORE PROGRAM TERMINATES')
      DO 1590 J=1,NGOODC
 1590 WRITE(6,1600)  J,GOODEV(J),MP(J)
 1600 FORMAT(/4X,' J',11X,'GOODEV(J)',4X,'MP(J)'/I6,E20.12,I9)
      GO TO 1630
C
 1610 WRITE(6,1620) MBETA,KMAXN
 1620 FORMAT(/' PROGRAM TERMINATES BECAUSE THE STORAGE ALLOTTED FOR THE
     1BETA ARRAY',I8/' IS NOT SUFFICIENT FOR THE ENLARGED KMAX =',I8,' T
     1HAT THE PROGRAM WANTS'/' USER CAN ENLARGE THE ALPHA AND BETA ARRAY
     1S AND RERUN THE PROGRAM.'/)
C
 1630 CONTINUE
C
      STOP
C-----END OF MAIN PROGRAM FOR LANCZOS HERMITIAN EIGENVECTOR COMPUTATIONS
      END
```

SECTION 3.4 LANCZS AND SAMPLE MATRIX-VECTOR MULTIPLY SUBROUTINES

```
C-----HLEMULT----HERMITIAN MATRICES------------------------------------
C
C     CONTAINS SUBROUTINE LANCZS AND SAMPLE USPEC, CMATV
C     USED BY THE HERMITIAN VERSION OF THE LANCZOS ALGORITHMS
C
C     PORTABILITY:
C     THESE PROGRAMS ARE NOT PORTABLE DUE TO THE USE OF COMPLEX*16
C     VARIABLES.  MOREOVER, THE PFORT VERIFIER IDENTIFIED THE
C     FOLLOWING ADDITIONAL NONPORTABLE CONSTRUCTIONS:
C     1.  THE ENTRY MECHANISM USED TO PASS THE STORAGE
C         LOCATIONS OF THE USER-SPECIFIED MATRIX FROM THE
C         SUBROUTINE USPEC TO THE MATRIX-VECTOR SUBROUTINE CMATV.
C     2.  IN THE PROGRAMS PROVIDED FOR 'HERMITIAN POISSON' TEST MATRICES
C         USPEC CONTAINS FREE FORMAT (8,*), AND FORMAT (20A4); AND
C         EXACT ERROR SUBROUTINE CONTAINS DATA/MACHEP DEFINITION.
C
C
C-----LANCZS-COMPUTE THE LANCZOS TRIDIAGONAL MATRICES-------------------
C
C     GRAM-SCHMIDT ORTHOGONALIZATION WITHOUT MODIFICATION
C     REQUIRES EXTRA VECTOR VS IN LANCZS.  MODIFICATION IS NOT
C     PERMISSIBLE IN THE HERMITIAN CASE BECAUSE COMPLEX PORTION
C     OF THE MODIFICATION COULD NOT BE INCORPORATED.
C
      SUBROUTINE LANCZS(MATVEC,V1,V2,VS,ALPHA,BETA,GR,GC,G,KMAX,MOLD1,N,
     1 IIX)
C
C-----------------------------------------------------------------------
      COMPLEX*16 V1(1), V2(1), VS(1), ZEROC, TEMP
      DOUBLE PRECISION  ALPHA(1), BETA(1), BATA, SUM, ONE, ZERO
      DOUBLE PRECISION GR(1),GC(1)
      REAL  G(1)
      EXTERNAL MATVEC
      DOUBLE PRECISION  DSQRT
C-----------------------------------------------------------------------
C
      ZERO = 0.D0
      ONE = 1.D0
      ZEROC = DCMPLX(ZERO,ZERO)
C
      IF(MOLD1.GT.1)GO TO 50
C
C     ALPHA/BETA GENERATION STARTS AT I = 1
C     MOLD1 = 1 SET V1 = 0. AND V2 = RANDOM UNIT VECTOR
      IIL=IIX
C
C-----------------------------------------------------------------------
      CALL GENRAN(IIL,G,N)
C-----------------------------------------------------------------------
C
      DO 10 I = 1,N
   10 GR(I) = G(I)
C
C-----------------------------------------------------------------------
      CALL GENRAN(IIL,G,N)
C-----------------------------------------------------------------------
C
      DO 20 I = 1,N
   20 GC(I) = G(I)
C
      DO 30 I = 1,N
   30 V2(I) = DCMPLX(GR(I),GC(I))
C
C-----------------------------------------------------------------------
```

```
      CALL CINPRD(V2,V2,SUM,N)
C-----------------------------------------------------------------------
C
      SUM = ONE/DSQRT(SUM)
      DO 40 I = 1,N
      V1(I) = ZEROC
   40 V2(I) = V2(I)*SUM
      BETA(1) = ZERO
C
C     ALPHA BETA GENERATION LOOP
   50 CONTINUE
C
      DO 80 I=MOLD1,KMAX
      SUM = ZERO
C
C-----------------------------------------------------------------------
C     MATVEC(V2,VS,SUM) CALCULATES  VS = A*V2 - SUM*VS
      CALL MATVEC(V2,VS,SUM)
      CALL CINPRD(V2,VS,SUM,N)
C-----------------------------------------------------------------------
C
      ALPHA(I) = SUM
      BATA = BETA(I)
      DO 60 J=1,N
   60 V1(J) = (VS(J)-BATA*V1(J)) - SUM*V2(J)
C
C-----------------------------------------------------------------------
      CALL CINPRD(V1,V1,SUM,N)
C-----------------------------------------------------------------------
C
      IN = I+1
      BETA(IN) = DSQRT(SUM)
      SUM = ONE/BETA(IN)
      DO 70 J=1,N
      TEMP = SUM*V1(J)
      V1(J) = V2(J)
   70 V2(J) = TEMP
   80 CONTINUE
C     END ALPHA, BETA GENERATION LOOP
C
C-----END OF LANCZS-----------------------------------------------------
C
      RETURN
      END
C
C-----USPEC-GENERAL SPARSE, HERMITIAN MATRIX----------------------------
C
C     SUBROUTINE USPEC(N,MATNO)
      SUBROUTINE GUSPEC(N,MATNO)
C
C-----------------------------------------------------------------------
      COMPLEX*16 A(3000)
      DOUBLE PRECISION  AD(1000)
      INTEGER  IROW(3000),ICOL(1000)
C-----------------------------------------------------------------------
C    DIMENSION ARRAYS NEEDED TO DEFINE MATRIX, READ IN VALUES FOR
C    ARRAYS AND THEN PASS THE STORAGE LOCATIONS OF THESE ARRAYS TO
C    THE MATRIX-VECTOR MULTIPLY SUBROUTINE CMATV.
C
C    USER-SUPPLIED MATRIX IS STORED IN FOLLOWING SPARSE FORMAT:
C    N = ORDER OF A-MATRIX
C    NZS = NUMBER OF NONZERO SUBDIAGONAL ENTRIES IN A
C    NZL = INDEX OF LAST COLUMN CONTAINING NONZERO SUBDIAGONAL ENTRIES
C    ICOL(J), J=1,NZL IS THE NUMBER OF NONZERO SUBDIAGONAL ELEMENTS
C            IN COLUMN J.
C    IROW(K), K = 1,NZS, IS THE ROW INDEX FOR CORRESPONDING A(K).
C    AD(I), I=1,N ARE DIAGONAL ENTRIES (INCLUDING ANY 0 DIAGONAL
```

```
C           ENTRIES)                                                    HLE01320
C     A(K), K=1,NZS ARE NONZERO SUBDIAGONAL ENTRIES, LISTED BY COLUMN.  HLE01330
C     FOR J > NZL THERE ARE NO NONZERO SUBDIAGONAL ELEMENTS IN COLUMN J. HLE01340
C     ICOL(J) = 0 IS ALLOWED                                            HLE01350
C                                                                       HLE01360
C-----------------------------------------------------------------------HLE01370
C      IN THIS SAMPLE SUBROUTINE THE ARRAYS ARE READ IN FROM FILE 8     HLE01380
C                                                                       HLE01390
      READ(8,10) NZS,NOLD,NZL,MATOLD                                    HLE01400
   10 FORMAT(I10,2I6,I8)                                                HLE01410
C                                                                       HLE01420
      WRITE(6,20) NZS,NOLD,NZL,MATOLD                                   HLE01430
   20 FORMAT(I10,2I6,I8,' = NZS,NOLD,NZL,MATOLD'/)                      HLE01440
C                                                                       HLE01450
C     TEST OF PARAMETER CORRECTNESS                                     HLE01460
      ITEMP = (NOLD-N)**2 + (MATNO-MATOLD)**2                           HLE01470
C                                                                       HLE01480
      IF(ITEMP.EQ.0) GO TO 40                                           HLE01490
C                                                                       HLE01500
      WRITE(6,30)                                                       HLE01510
   30 FORMAT(/' PROGRAM TERMINATES BECAUSE EITHER ORDERS OF OR LABELS FOHLE01520
     1R MATRIX DISAGREE'/)                                              HLE01530
      GO TO 80                                                          HLE01540
C                                                                       HLE01550
   40 CONTINUE                                                          HLE01560
C                                                                       HLE01570
C     NUMBER OF NONZERO SUBDIAGONAL ENTRIES IN EACH COLUMN IS READ      HLE01580
C     THEN THE CORRESPONDING ROW INDEX FOR EACH SUCH ENTRY IS READ      HLE01590
      READ(8,50) (ICOL(K), K=1,NZL)                                     HLE01600
      READ(8,50) (IROW(K), K=1,NZS)                                     HLE01610
   50 FORMAT(13I6)                                                      HLE01620
C     DIAGONAL IS READ FIRST, THEN NONZERO BELOW DIAGONAL ENTRIES       HLE01630
      READ(8,60) (AD(K), K=1,N)                                         HLE01640
   60 FORMAT(4E20.12)                                                   HLE01650
      READ(8,70) (A(K), K=1,NZS)                                        HLE01660
C  50 FORMAT(4Z20)                                                      HLE01670
   70 FORMAT(4E20.12)                                                   HLE01680
C                                                                       HLE01690
C-----------------------------------------------------------------------HLE01700
C     PASS STORAGE LOCATIONS OF ARRAYS THAT DEFINE THE MATRIX TO        HLE01710
C     THE MATRIX-VECTOR MULTIPLY SUBROUTINE CMATV                       HLE01720
      CALL CMATVE(A,AD,ICOL,IROW,N,NZL)                                 HLE01730
C-----------------------------------------------------------------------HLE01740
C                                                                       HLE01750
      RETURN                                                            HLE01760
   80 STOP                                                              HLE01770
C                                                                       HLE01780
C-----END OF USPEC FOR GENERAL, SPARSE HERMITIAN MATRICES---------------HLE01790
      END                                                               HLE01800
C                                                                       HLE01810
C-----START OF MATRIX-VECTOR MULTIPLY-GENERAL SPARSE HERMITIAN--------- HLE01820
C                                                                       HLE01830
C     SUBROUTINE CMATV(W,U,SUM)                                         HLE01840
      SUBROUTINE GCMATV(W,U,SUM)                                        HLE01850
C                                                                       HLE01860
C-----------------------------------------------------------------------HLE01870
      COMPLEX*16  U(1),W(1),A(1)                                        HLE01880
      DOUBLE PRECISION  AD(1),SUM                                       HLE01890
      INTEGER  IROW(1),ICOL(1)                                          HLE01900
C-----------------------------------------------------------------------HLE01910
C    SPARSE MATRIX-VECTOR MULTIPLY FOR LANCZS  U = A*W - SUM*U          HLE01920
C    SEE USPEC SUBROUTINE FOR DESCRIPTION OF THE ARRAYS THAT DEFINE     HLE01930
C    THE MATRIX                                                         HLE01940
C                                                                       HLE01950
C     COMPUTE THE DIAGONAL TERMS                                        HLE01960
      DO 10 I = 1,N                                                     HLE01970
```

```
   10 U(I) = AD(I)*W(I)-SUM*U(I)
C
C     COMPUTE BY COLUMN
      LLAST = 0
      DO 30 J = 1,NZL
C
      IF (ICOL(J).EQ.0) GO TO 30
      LFIRST = LLAST + 1
      LLAST = LLAST + ICOL(J)
C
      DO 20 L = LFIRST,LLAST
      I = IROW(L)
C
      U(I) = U(I) + A(L)*W(J)
      U(J) = U(J) + DCONJG(A(L))*W(I)
C
   20 CONTINUE
C
   30 CONTINUE
C
      RETURN
C
C-----------------------------------------------------------------------
C     STORAGE LOCATIONS OF ARRAYS ARE PASSED TO CMATV FROM USPEC
      ENTRY CMATVE(A,AD,ICOL,IROW,N,NZL)
C-----------------------------------------------------------------------
C
C-----END OF CMATV-GENERAL, SPARSE, HERMITIAN MATRICES ------------------
      RETURN
      END
C
C-----USPEC, CMATV, EXEVG, AND HEXVEC FOR HERMITIAN 'POISSON' MATRICES--
C
C-----USPEC (HERMITIAN POISSON MATRICES)---------------------------------
C
C     SUBROUTINE HUSPEC(N,MATNO)
      SUBROUTINE USPEC(N,MATNO)
C
C-----------------------------------------------------------------------
      DOUBLE PRECISION  C0,C1,C2,HALF,ONE,SCR,SCI,ANGLE,TEMP
      COMPLEX*16 SC,TC,CL0,CL1,CL3,CL4
      REAL EXPLAN(20)
      DOUBLE PRECISION EIGVAL(1000)
      REAL GAPS(1000)
      INTEGER MULTS(1000)
C-----------------------------------------------------------------------
      HALF = 0.5D0
      ONE  = 1.0D0
C
C     READ IN PARAMETERS TO DEFINE MATRIX
C     MATRIX IS COMPLEX DIAGONAL SIMILITARY TRANSFORM OF REAL SYMMETRIC
C     POISSON MATRIX WHICH HAS SYMMETRIC TOEPLITZ BLOCKS ALONG
C     THE DIAGONAL, EACH ONE OF WHICH HAS THE PARAMETER C2 ALONG THE
C     DIAGONAL AND -C0 ABOVE AND BELOW THE DIAGONAL, AND OFF-DIAGONAL
C     BLOCKS THAT ARE DIAGONAL WITH DIAGONAL ENTRIES -C1.  EACH BLOCK
C     IS KX*KX AND THERE ARE KY BLOCKS.  THE HERMITIAN VERSION IS
C     OBTAINED BY APPLYING A DIAGONAL SIMILARITY TRANSFORM TO THE
C     REAL MATRIX WHERE THIS TRANSFORMATION IS SUCH THAT ITS
C     DIAGONAL ENTRIES ARE (SC)**(K-1), K = 1,...,N, WHERE SC
C     HAS MODULUS 1.
C
      READ(8,10) EXPLAN
      WRITE(6,10) EXPLAN
      READ(8,10) EXPLAN
   10 FORMAT(20A4)
C     IF MTYPE = 0 WE HAVE ZERO BOUNDARY CONDITIONS
C     IF MTYPE = 1 WE HAVE NORMAL DERIVATIVE BOUNDARY CONDITIONS
```

```
C     NOTE THAT SUBROUTINES EXEVG AND HEXVEC ARE VALID ONLY FOR
C     MTYPE = 0.
      READ(8,*) NOLD,MATOLD,IVEC,MTYPE
      WRITE(6,20) NOLD,MATOLD
   20 FORMAT(' ORDER OF MATRIX READ FROM FILE =',I6/' MATRIX NUMBER =',
     1I8/)
      IF(MTYPE.EQ.0) WRITE(6,30)
   30 FORMAT(/' HERMITIAN POISSON CORRESPONDING TO ZERO BOUNDARY CONDITI
     1ONS'/)
      IF(MTYPE.EQ.1) WRITE(6,40)
   40 FORMAT(/' HERMITIAN POISSON CORRESPONDING TO NORMAL DERIVATIVE BOU
     1NDARY CONDITIONS'/)
      IF(IVEC.NE.0.AND.MTYPE.EQ.0) WRITE(6,50)
   50 FORMAT(' COMPUTE THE TRUE EIGENVALUES AND PUT IN FN TRUEEVAL'/)
C
C     TEST OF PARAMETER CORRECTNESS
      ITEMP = (NOLD-N)**2 + (MATNO-MATOLD)**2
C
      IF(ITEMP.EQ.0) GO TO 70
C
      WRITE(6,60)
   60 FORMAT(' PROGRAM TERMINATES BECAUSE EITHER ORDERS OF OR LABELS FOR
     1 MATRIX DISAGREE')
      GO TO 150
C
   70 CONTINUE
C
      READ(8,10) EXPLAN
      READ(8,*)  CO,KX,KY
      IF (KX.GT.4.AND.KY.GT.4) GO TO 90
      WRITE(6,80) KX,KY
   80 FORMAT(2I6,' = KX KY  ONE OR BOTH OF KX KY TOO SMALL SO STOP'/)
      GO TO 150
   90 CONTINUE
      READ(8,10) EXPLAN
C     BELOW SC = COS(ANGLE) + I SIN(ANGLE)
C     READ IN DESIRED COSINE, COMPUTE ANGLE, THEN SINE
      READ(8,*) SCR
      ANGLE = DARCOS(SCR)
      SCI = DSIN(ANGLE)
      SC = DCMPLX(SCR,SCI)
      WRITE(6,100) SC
C     IF (IVEC.NE.0.AND.MTYPE.EQ.0) WRITE(9,7) SC
  100 FORMAT(' GENERATOR OF DIAGONAL TRANSFORMATION ='/2E20.12)
C
      TC = SC
      DO 110 J=2,KX
  110 TC = SC*TC
      WRITE(6,120) TC
  120 FORMAT(' TC = ',2E20.12)
C
      N = KX*KY
      C2 = ONE
      C1 = HALF-CO
      TEMP = DSQRT(2.0D0)
      IF (MTYPE.EQ.0) TEMP = ONE
      CL0 = -SC*CO
      CL1 = -TC*C1
      CL3 = -SC*CO*TEMP
      CL4 = -TC*C1*TEMP
C
      WRITE(6,130) N,MTYPE,KX,KY,C2,CO,C1
  130 FORMAT(/5X,'N',1X,'MTYPE',4X,'KX',4X,'KY',7X,'DIAGONAL',
     1 3X,'X-CODIAGONAL',3X,'Y-CODIAGONAL'/4I6,3E15.8/)
C
C-----------------------------------------------------------------------
```

```
      CALL HMATVE(C2,CL0,CL1,CL3,CL4,KX,KY)
C----------------------------------------------------------------------
C
      IF(IVEC.EQ.0.OR.MTYPE.NE.0) GO TO 140
C
C     COMPUTE THE EXACT EIGENVALUES
C
C----------------------------------------------------------------------
      CALL EXEVG(EIGVAL,C0,C1,C2,GAPS,MULTS,KX,KY)
C----------------------------------------------------------------------
C
      IF(IVEC.LT.0) GO TO 150
C
  140 CONTINUE
      RETURN
C
C-----END OF USPEC-----------------------------------------------------
  150 STOP
      END
C
C-----START OF CMATV FOR HERMITIAN POISSON MATRICES--------------------
C
C     SUBROUTINE HMATV(W,U,SUM)
      SUBROUTINE CMATV(W,U,SUM)
C
C----------------------------------------------------------------------
      DOUBLE PRECISION  C2,SUM
      COMPLEX*16 U(1),W(1)
      COMPLEX*16 CL0,CL1,CL3,CL4,CR0,CR1,CR3,CR4
C----------------------------------------------------------------------
C     CALCULATES U = A*W - SUM*U
C
      N = KK*LL
      CR0 = DCONJG(CL0)
      CR1 = DCONJG(CL1)
      CR3 = DCONJG(CL3)
      CR4 = DCONJG(CL4)
C
C----------------------------------------------------------------------
C    FIRST AND LAST BLOCKS
      J = 1
      U(J)=(C2*W(J)+CR3*W(J+1)+CR1*W(J+KK)) - SUM*U(J)
      J = 2
      U(J)=(C2*W(J)+CL3*W(J-1)+CR0*W(J+1)+CR1*W(J+KK))-SUM*U(J)
      J = KK
      U(J)=(C2*W(J)+CL3*W(J-1)+CR1*W(J+KK))-SUM*U(J)
      J = KK - 1
      U(J)=(C2*W(J)+CR3*W(J+1)+CL0*W(J-1)+CR1*W(J+KK))-SUM*U(J)
      J = N - KK + 1
      U(J)=(C2*W(J)+CR3*W(J+1)+CL4*W(J-KK))-SUM*U(J)
      J = N - KK + 2
      U(J)=(C2*W(J)+CL3*W(J-1)+CR0*W(J+1)+CL4*W(J-KK))-SUM*U(J)
      J = N
      U(J)=(C2*W(J)+CL3*W(J-1)+CL4*W(J-KK))-SUM*U(J)
      J = N - 1
      U(J)=(C2*W(J)+CL0*W(J-1)+CR3*W(J+1)+CL4*W(J-KK))-SUM*U(J)
C
      KK2 = KK - 2
      DO 10 JJ = 3,KK2
      J = JJ
      U(J)=(C2*W(J)+CL0*W(J-1)+CR0*W(J+1)+CR1*W(J+KK))-SUM*U(J)
      J = N - KK + JJ
   10 U(J)=(C2*W(J)+CL0*W(J-1)+CR0*W(J+1)+CL4*W(J-KK))-SUM*U(J)
C
C    START BLOCKS 2 AND LL-1
      J = KK + 1
      U(J)=(C2*W(J)+CR3*W(J+1)+CL1*W(J-KK)+CR1*W(J+KK))-SUM*U(J)
```

```
      J = KK + 2
      U(J)=(C2*W(J)+CL3*W(J-1)+CR0*W(J+1)+CL1*W(J-KK)+CR1*W(J+KK))
     1 -SUM*U(J)
      J = KK + KK
      U(J)=(C2*W(J)+CL3*W(J-1)+CL1*W(J-KK)+CR1*W(J+KK))-SUM*U(J)
      J = KK + KK - 1
      U(J)=(C2*W(J)+CR3*W(J+1)+CL0*W(J-1)+CL1*W(J-KK)+CR1*W(J+KK))
     1  -SUM*U(J)
      J = N - 2*KK + 1
      U(J)=(C2*W(J)+CR3*W(J+1)+CR4*W(J+KK)+CL1*W(J-KK))
     1 -SUM*U(J)
      J = N - 2*KK + 2
      U(J)=(C2*W(J)+CL3*W(J-1)+CR0*W(J+1)+CR4*W(J+KK)+CL1*W(J-KK))
     1 -SUM*U(J)
      J = N - KK
      U(J)=(C2*W(J)+CL3*W(J-1)+CR4*W(J+KK)+CL1*W(J-KK))-SUM*U(J)
      J = N - KK - 1
      U(J)=(C2*W(J)+CR3*W(J+1)+CL0*W(J-1)+CR4*W(J+KK)+CL1*W(J-KK))
     1 -SUM*U(J)
C
      DO 20 JJ = 3,KK2
      J = KK + JJ
      U(J)=(C2*W(J)+CL0*W(J-1)+CR0*W(J+1)+CL1*W(J-KK)+CR1*W(J+KK))
     1 -SUM*U(J)
      J = N - 2*KK + JJ
      U(J)=(C2*W(J)+CL0*W(J-1)+CR0*W(J+1)+CR4*W(J+KK)+CL1*W(J-KK))
     1 -SUM*U(J)
   20 CONTINUE
C
C     MIDDLE BLOCKS
      LL2 = LL - 2
      JP = KK
      DO 40 JJ = 3,LL2
      JP = JP + KK
C     JP = (JJ-1)*KK
      J = JP + 1
      U(J)=(C2*W(J)+CR3*W(J+1)+CL1*W(J-KK)+CR1*W(J+KK))-SUM*U(J)
      J = J + 1
      U(J)=(C2*W(J)+CL3*W(J-1)+CR0*W(J+1)+CL1*W(J-KK)+
     1 CR1*W(J+KK))-SUM*U(J)
      J = J + KK - 2
      U(J) = (C2*W(J)+CL3*W(J-1)+CL1*W(J-KK)+CR1*W(J+KK))-SUM*U(J)
      J = J - 1
      U(J)=(C2*W(J)+CR3*W(J+1)+CL0*W(J-1)+CL1*W(J-KK)+
     1 CR1*W(J+KK))-SUM*U(J)
C
      DO 30 II = 3,KK2
      J = JP + II
      U(J)=(C2*W(J)+CL0*W(J-1)+CR0*W(J+1)+CL1*W(J-KK)+CR1*W(J+KK))
     1 -SUM*U(J)
   30 CONTINUE
C
   40 CONTINUE
C
      RETURN
C
      ENTRY HMATVE(C2,CL0,CL1,CL3,CL4,KK,LL)
C
C-----END OF HMATV--------------------------------------------------
      RETURN
      END
C
C-----START OF EXEVG------------------------------------------------
C
C     FOR MTYPE = 0, ZERO BOUNDARY CONDITIONS:
C     COMPUTES EXACT EIGENVALUES OF HERMITIAN POISSON MATRIX,
C     THEIR MULTIPLICITIES, AND THE GAPS BETWEEN THE EIGENVALUES AND
```

```
C     PUTS THEM RESPECTIVELY INTO VECTORS U, MP, AND G.  THESE
C     QUANTITIES ARE ALL WRITTEN TO FILE 9.
C
      SUBROUTINE EXEVG(U,C0,C1,C2,G,MP,KX,KY)
C
C-----------------------------------------------------------------------
      DOUBLE PRECISION  U(1),MACHEP
      DOUBLE PRECISION  EPSM,C0,C1,C2,T0,T1,PIK,PIL,ONE,TWO,ATOLN,EE
      REAL G(1)
      INTEGER MP(1)
C-----------------------------------------------------------------------
      DATA MACHEP/Z3410000000000000/
      EPSM = 2.0D0*MACHEP
C-----------------------------------------------------------------------
      N = KX*KY
      ONE  = 1.0D0
      TWO  = 2.0D0
      T0 = DARCOS(-ONE)
      T1 = DFLOAT(KX+1)
      PIK = T0/T1
      T1 = DFLOAT(KY+1)
      PIL = T0/T1
C     GENERATE EXACT EIGENVALUES
      KP = 0
      DO 20 J = 1,KY
      T1 = PIL*DFLOAT(J)
      T0 = C2 - TWO*C1*DCOS(T1)
      DO 10 I = 1,KX
      KP = KP+1
      T1 = PIK*DFLOAT(I)
   10 U(KP) = T0 - TWO*C0*DCOS(T1)
   20 CONTINUE
C
C     ORDER U VECTOR BY INCREASING ALGEBRAIC SIZE
      DO 40 K = 2,N
      KM1 = K-1
      DO 30 L = 1,KM1
      JJ = K-L
      IF (U(JJ+1).GE.U(JJ)) GO TO 40
      T0 = U(JJ)
      U(JJ) = U(JJ+1)
   30 U(JJ+1) = T0
   40 CONTINUE
      ATOLN = DMAX1(DABS(U(1)),DABS(U(N)))*EPSM
C
      WRITE(9,50)
   50 FORMAT(' TRUE EIGENVALUES FOR HERMITIAN POISSON')
C
      WRITE(9,60)N,KX,KY,C2,C0,C1,ATOLN
      WRITE(6,60) N,KX,KY,C2,C0,C1,ATOLN
   60 FORMAT(1X,'A-SIZE',2X,'X-DIM',2X,'Y-DIM'/3I7/
     1 5X,'A-DIAGONAL',3X,'X-CODIAGONAL',3X,'Y-CODIAGONAL',10X,'ATOLN'/
     2 4E15.8)
C
C     DETERMINE TRUE MULTIPLICITIES FOR EXACT EIGENVALUES
      I = 1
      IDEX = 1
      J = 1
      NEXACT = 0
   70 J = J+1
      IF (J.GT.N) GO TO 80
      EE = DABS(U(J)-U(I))
      IF (EE.GT.ATOLN) GO TO 80
      IDEX = IDEX+1
      GO TO 70
   80 NEXACT = NEXACT+1
      U(NEXACT) = U(I)
```

```
      MP(NEXACT) = IDEX
C     MP(K) = MULTIPLICITY OF KTH EIGENVALUE CLUSTER FOR A
      IDEX = 1
      I = J
      IF (I.GT.N) GO TO 90
      GO TO 70
   90 CONTINUE
C
C     MULTIPLICITIES HAVE BEEN DETERMINED
C     NEXACT = NUMBER OF DISTINCT A-EIGENVALUES
C
      WRITE(9,100)NEXACT
      WRITE(6,100)NEXACT
  100 FORMAT(I6,' = NUMBER OF TRUE A-EIGENVALUES WHICH ARE DISTINCT'/)
C
C     MINGAP CALCULATION FOR DISTINCT A-EIGENVALUES
      NM1 = NEXACT - 1
      G(NEXACT) = U(NM1)-U(NEXACT)
      G(1) = U(2)-U(1)
C
      DO 110 J = 2,NM1
      T0 = U(J)-U(J-1)
      T1 = U(J+1)-U(J)
      G(J) = T1
      IF (T0.LT.T1) G(J) = -T0
  110 CONTINUE
C
C     NEXACT DISTINCT A-EIGENVALUES ARE IN U IN ASCENDING ORDER
C     MP = MULTIPLICITIES OF THE DISTINCT EIGENVALUES OF A
C     G = TRUE MINIMUM GAP IN A FOR EACH OF THESE EIGENVALUES
C     G < 0 INDICATES THE LEFT-HAND GAP WAS MINIMAL.
C     OUTPUT MULTIPLICITIES, DISTINCT EVS, AND MINGAPS TO FILE 11
C
      WRITE(9,120)
  120 FORMAT(5X,'I',1X,'AMULT',5X,'TRUE A-EIGENVALUE(I)',
     1 3X,'A-MINGAP(I)')
C
      WRITE(9,130)(J,MP(J),U(J),G(J), J=1,NEXACT)
  130 FORMAT(2I6,E25.16,E14.3)
C
      WRITE(9,140)
  140 FORMAT(' NEXACT DISTINCT A-EIGENVALUES ARE IN ASCENDING ORDER'/
     1 ' AMULT = MULTIPLICITIES OF THE DISTINCT EIGENVALUES OF A.'/
     2 ' A-MINGAP(I) = TRUE MINIMUM GAP IN A FOR EACH EIGENVALUE.'/
     3 ' A-MINGAP(I) LT 0 INDICATES THE LEFT-HAND GAP WAS MINIMAL.'//)
C
C     WE ORDER U VECTOR BY INCREASING SIZE OF THE GAPS
C
      DO 150 K = 1,N
  150 MP(K) = K
C
      DO 170 K = 2,N
      KM1 = K-1
C
      DO 160 L = 1,KM1
      JJ = K - L
      IF (ABS(G(JJ+1)).GE.ABS(G(JJ))) GO TO 170
      EE = U(JJ)
      U(JJ) = U(JJ+1)
      U(JJ+1) = EE
      GG = G(JJ)
      G(JJ) = G(JJ+1)
      G(JJ+1) = GG
      IEE = MP(JJ)
      MP(JJ) = MP(JJ+1)
  160 MP(JJ+1) = IEE
C
```

```
  170 CONTINUE
C
      WRITE(9,180)
  180 FORMAT(5X,'K',6X,'A-MINGAP',5X,'TRUE A-EIGENVALUE(I)',2X,'A-EVNO')
C
      WRITE(9,190)(J,G(J),U(J),MP(J), J=1,NEXACT)
  190 FORMAT(I6,E14.3,E25.16,I8)
C
      WRITE(9,200)
  200 FORMAT(' NEXACT DISTINCT A-EIGENVALUES. GAPS IN ASCENDING ORDER'/
     2 ' A-MINGAP(I) = TRUE MINIMUM GAP IN A FOR EACH EIGENVALUE.'/
     3 ' A-MINGAP(I) LT 0 INDICATES THE LEFT-HAND GAP WAS MINIMAL.'/
     3 ' A-MATRIX IS BLOCK TRIDIAGONAL AND EACH DIAGONAL BLOCK IS OF ORD
     3ER NX.'/
     4 ' NX = NUMBER OF POINTS ON EACH X-LINE. THERE ARE NY DIAGONAL BLO
     4CKS.'/
     5 ' NY = NUMBER OF POINTS ON EACH Y-LINE.'/
     5 ' A-DIAGONAL    = A(K,K)'/
     6 ' X-CODIAGONAL = A(I,I+1)'/
     7 ' Y-CODIAGONAL = A(I,I+NX)'/
     8 ' ----- END OF FILE 9 EXACTEV---------------------------'//)
C
C-----END OF EXEVG-----------------------------------------------------
C
      RETURN
      END
C
C-----START OF HEXVEC--------------------------------------------------
C
C     FOR THE HERMITIAN POISSON TEST CASES WITH MTYPE = 0 ONLY:
C     FOR A GIVEN RITZ VECTOR V AND EIGENVALUE X1, COMPUTES
C     THE CLOSEST TRUE EIGENVALUE Y1 AND CORRESPONDING TRUE
C     EIGENVECTOR Z, CALCULATES THE NORM OF V-Z AND THE MAXIMAL
C     DIFFERENCE OF THE COMPONENTS.  USER WOULD HAVE TO
C     INCORPORATE ENTRY AND CALL TO THIS SUBROUTINE INTO
C     HLEVEC PROGRAM IF THESE QUANTITIES ARE DESIRED.
C     U CONTAINS THE COMPUTED TRUE EIGENVALUES.
C     W CONTAINS THE TRUE EIGENVECTOR FOR THE REAL POISSON MATRIX
C
      SUBROUTINE HEXVEC(Z,V,U,W,X1,Y1,MP,JNUM)
C
C----------------------------------------------------------------------
      DOUBLE PRECISION  U(1),W(1)
      DOUBLE PRECISION WI(110),WJ(110),WII(110)
      DOUBLE PRECISION X1,Y1,EV,EE,WS,PIK,PIL,SUM,TEMP
      DOUBLE PRECISION ATOLN,EPSM,ZERO,HALF,ONE,TWO,MACHEP
      DOUBLE PRECISION C0,C1,C2,T0,T1,T2
      COMPLEX*16 CONE,S,SB,STEMP,V(1),Z(1)
      INTEGER MP(1)
C----------------------------------------------------------------------
      DATA MACHEP/Z3410000000000000/
      EPSM = 2.0D0*MACHEP
C----------------------------------------------------------------------
C     THIS PROGRAM CALCULATES THE EXACT EIGENVALUES AND EIGENVECTORS
C     OF THE HERMITIAN POISSON MATRIX A OF ORDER  N = KX BY KY
C     A CONSISTS OF KY TRIDIAGONAL BLOCKS OF ORDER KX
C     KX = X-DIMENSION     KY = Y-DIMENSION.
C
C     C2 = DIAGONAL OF KX BY KX MATRIX
C    -C0 = CO-DIAGONAL OF THE KX BY KX MATRIX.
C    -C1 = Y-CODIAGONAL.
C
C     NOTE THAT THE VECTORS WI,WJ,WII ARE DIMENSIONED INTERNALLY
C     THEY ARE USED JUST TO KEEP FROM REGENERATING INFORMATION.
C     WI,WII = REAL*8 ARRAYS OF DIMENSION AT LEAST KX
C     WJ     = REAL*8 ARRAY  OF DIMENSION AT LEAST KY.
C
```

```
C     NOTATION USED IN PROGRAM
C
C     PIK = ARCOS(-1)/(KX+1)     PIL = ARCOS(-1)/(KY+1)
C     WI(I) = PIK*I          WJ(J) = PIL*J
C
C     TO = C2 - 2*C1*COS(PIL*J)     EV(I,J) = TO - 2*CO*COS(PIK*I)
C     I = 1,KX     J = 1,KY     KP = (J-1)*KX + I
C
C     W(KV) = SIN(PIK*I*IK)*SIN(PIL*J*JK)
C     IK = 1,KX     JK = 1,KY     KV = (JK-1)*KX + IK
C     W IS UNSCALED EIGENVECTOR FOR EV(I,J)
C     WS = 1/||W||: ||W|| = .5*DSQRT(T2*T3)  T2 = KX+1  T3 = KY+1
C     U(K) IS A-EV ORDERED BY INCREASING SIZE, K = 1,N
C
C     GIVEN X1 FIND Y1 AND KVEC SUCH THAT
C           Y1 = EV(KVEC) AND |X1-Y1| = MIN
C           ALSO GIVEN UNIT RITZ VECTOR ASSOCIATED WITH X1
C           CALCULATE UNIT EIGENVECTOR W, A*W = Y1*W
C           T2 = ||V-W||   T1 = MAX(|V(K)-W(K)|, K= 1,N)
C           MAX OCCURS FIRST AT K = KK
C
C-----------------------------------------------------------------------
C     C2 = A(K,K)
C     CO = A(K,K+1) = A(K+1,K)
C     C1 = A(K,K+KX) = A(K+KX,K)
C     CO + C1 = HALF
C
C     SPECIFY PARAMETERS
      N = KX*KY
      ZERO = 0.0D0
      HALF = 0.5D0
      ONE  = 1.0D0
      TWO  = 2.0D0
      TO = DARCOS(-ONE)
      T1 = DFLOAT(KX+1)
      PIK = TO/T1
      T2 = DFLOAT(KY+1)
      PIL = TO/T2
      WS = TWO/DSQRT(T1*T2)
C
C     GENERATE WI WJ VECTORS
      KP = 0
      DO 20 J = 1,KY
      T1 = PIL*DFLOAT(J)
      WJ(J) = T1
      TO = C2 - TWO*C1*DCOS(T1)
      DO 10 I = 1,KX
      KP = KP+1
      T1 = PIK*DFLOAT(I)
      WI(I) = T1
   10 U(KP) = TO - TWO*CO*DCOS(T1)
   20 CONTINUE
C     U(KP) = EV(I,J) = C2 - 2*C1*COS(PIL*J) - 2*CO*COS(PIK*I)
C
C     INITIALIZE MP VECTOR
      DO 30 K = 1,N
   30 MP(K) = K
C
C     WE ORDER U VECTOR BY INCREASING SIZE OF THE EVS
      DO 50 K = 2,N
      KM1 = K-1
C
      DO 40 L = 1,KM1
      JJ = K - L
      IF (U(JJ+1).GE.U(JJ)) GO TO 50
      EE = U(JJ)
      U(JJ) = U(JJ+1)
```

```
      U(JJ+1) = EE
      IEE = MP(JJ)
      MP(JJ) = MP(JJ+1)
   40 MP(JJ+1) = IEE
C
   50 CONTINUE
C
      ATOLN = DMAX1(DABS(U(1)),DABS(U(N)))*EPSM
C
      WRITE(6,60) N,KX,KY,C2,C0,C1,ATOLN
   60 FORMAT(/' EXACT ERRORS FOR CONVERGED GOODEV'/
     1 4I6,' = N KX KY'//
     1 4E12.5,' = C2 C0 C1 ATOLN'//)
C
C     KP = MP(K) MEANS EIGENVALUE U(K) CORRESPONDS TO EIGENVECTOR W(KP)
C     COMPUTE TOLERANCE USED IN COMPUTING TRUE MULTIPLICITIES
C
C     X1 IS AN INPUT PARAMETER. WE CALCULATE EXACT
C     A-EIGENVALUE WHICH IS CLOSEST TO X1, LABEL IT Y1 AND CALCULATE
C     UNIT EIGENVECTOR OF A ASSOCIATED WITH Y1. A*W = Y1*W, ||W|| = 1.
C     Y1 = U(KEV). EIGENVALUES OF A ARE ORDERED BY INCREASING SIZE.
C     V = COMPLEX RITZ VECTOR ASSOCIATED WITH GOODEV X1
C     WE SHOULD HAVE V = D*W WHERE D = DIAG(D(1),D(2),..,D(N))
C     D(1) = ONE, D(K+1)/D(K) = SB, |SB| = ONE
C
      KX1 = 0
      IF (X1.LE.U(1)) KX1 = 1
      IF (X1.GE.U(N)) KX1 = N
      NM1 = N-1
      IF (KX1.NE.0) GO TO 80
C
      DO 70 KVEC = 2,N
      IF (X1.GE.U(KVEC)) GO TO 70
C     U(KVEC-1).LE.X1.LT.U(KVEC)
      T1 = X1 - U(KVEC-1)
      T2 = U(KVEC) - X1
      KX1 = KVEC - 1
      IF (T1.GT.T2) KX1 = KVEC
      GO TO 80
   70 CONTINUE
C
   80 Y1 = U(KX1)
C
      IF (KX1.EQ.1) EE = U(2) - U(1)
      IF (KX1.EQ.N) EE = U(N) - U(NM1)
      IF (KX1.EQ.1.OR.KX1.EQ.N) GO TO 90
      EE = DMIN1(U(KX1+1)-U(KX1),U(KX1)-U(KX1-1))
   90 CONTINUE
C
      T0 = DABS(ONE - X1/Y1)
C
      WRITE(6,100) N,KX1,JNUM,Y1,X1,T0,EE
  100 FORMAT(3I8,' = N, A-EV NUMBER,GOODEV NO'//
     1 18X,'EXACTEV',19X,'GOODEV',4X,'RELERROR',4X,'A-MINGAP'/
     1 2E25.16,2E12.3/)
C
      IF (EE.GT.ATOLN) GO TO 120
C
      WRITE(6,110)
  110 FORMAT(' Y1 IS A MULTIPLE EIGENVALUE OF A SO WE EXIT'/)
C
      GO TO 200
C
C     Y1 IS TOEPLITZ EIGENVALUE CLOSEST TO X1.
C     CALCULATION OF EIGENVECTOR ASSOCIATED WITH EIGENVALUE Y1
C     A*W = Y1*W
```

```
C
C     DETERMINE I J FROM K: MP(K) = KP = (J-1)*KX+I
  120 CONTINUE
      K = KX1
      KP = MP(K)
      I = MOD(KP,KX)
      IF (I.EQ.0) I = KX
      T1 = WI(I)
      J = 1 + (KP-1)/KX
      T2 = WJ(J)
C
      DO 130 II = 1,KX
      TO = T1*DFLOAT(II)
  130 WII(II) = WS*DSIN(TO)
C
      KV = 0
      DO 150 JJ = 1,KY
      TO = T2*DFLOAT(JJ)
      TO = DSIN(TO)
C
      DO 140 II = 1,KX
      KV = KV + 1
  140 W(KV) = TO*WII(II)
C
  150 CONTINUE
C
C     W IS UNIT EXACT EIGENVECTOR OF A ASSOCIATED WITH Y1
C     V IS UNIT COMPLEX RITZVECTOR OF B ASSOCIATED WITH X1
C
      CONE = DCMPLX(ONE,ZERO)
      STEMP = CONE
      DO 160 K = 1,N
      Z(K) = STEMP*W(K)
  160 STEMP = STEMP*SB
C
      T1 = ZERO
      S = ONE
      KK = 0
      DO 170 K = 1,N
      IF (CDABS(Z(K)).LE.T1) GO TO 170
      T1 = CDABS(Z(K))
      KK = K
  170 CONTINUE
C
      S = V(KK)/Z(KK)
C
      KK = 0
      T1 = ZERO
      T2 = ZERO
      DO 180 K = 1,N
      TEMP = CDABS(S*Z(K) - V(K))
      T2 = T2 + TEMP**2
      IF (TEMP.LE.T1) GO TO 180
      KK = K
      T1 = TEMP
  180 CONTINUE
C
      T2 = DSQRT(T2)
      WRITE(6,190) KK,T1,T2
  190 FORMAT(' EIGENVECTOR ERROR. MAX ERROR AT COMPONENT = ',I6/
     1 ' MAX CDABS(EXACTVEC(K)-RITZVEC(K)) = ',E12.5/
     1 ' NORM(EXACTVEC-RITZVEC) = ',E12.5/)
C
  200 CONTINUE
C
```

```
      RETURN                                                            HLE08640
C                                                                       HLE08650
C-----------------------------------------------------------------------HLE08660
      ENTRY EXVECP(SB,C0,C1,C2,KX,KY)                                   HLE08670
C-----------------------------------------------------------------------HLE08680
C                                                                       HLE08690
C-----END OF HEXVEC-----------------------------------------------------HLE08700
      RETURN                                                            HLE08710
      END                                                               HLE08720
C                                                                       HLE08730
C-----USPEC (TRIDIAGONAL HERMITIAN MATRICES)----------------------------HLE08740
C                                                                       HLE08750
C     SUBROUTINE USPEC(N,MATNO)                                         HLE08760
      SUBROUTINE TSPEC(N,MATNO)                                         HLE08770
C                                                                       HLE08780
C-----------------------------------------------------------------------HLE08790
      DOUBLE PRECISION  D(100), DAR(100),DAI(100), PI, EIGVAL(100)      HLE08800
      DOUBLE PRECISION SPACE                                            HLE08810
      COMPLEX*16   DA(100),DB(100)                                      HLE08820
      REAL EXPLAN(20)                                                   HLE08830
C-----------------------------------------------------------------------HLE08840
C    DIMENSION ARRAYS NEEDED TO DEFINE MATRIX.  THEN                    HLE08850
C    PASS THE STORAGE LOCATIONS OF THESE ARRAYS TO THE MATRIX-VECTOR    HLE08860
C    MULTIPLY SUBROUTINE CMATV.                                         HLE08870
C                                                                       HLE08880
C    DIAGONAL ENTRY = D, ABOVE DIAGONAL ENTRY = DA, BELOW DIAGONAL = DB.HLE08890
C                                                                       HLE08900
      READ(8,10) EXPLAN                                                 HLE08910
   10 FORMAT(20A4)                                                      HLE08920
      READ(8,*) NOLD,MATOLD                                             HLE08930
C                                                                       HLE08940
      WRITE(6,20) N,MATOLD                                              HLE08950
   20 FORMAT(I10,2I6,I8,' = N,MATOLD'/)                                 HLE08960
C                                                                       HLE08970
C     TEST OF PARAMETER CORRECTNESS                                     HLE08980
      ITEMP = (NOLD-N)**2 + (MATNO-MATOLD)**2                           HLE08990
C                                                                       HLE09000
      IF(ITEMP.EQ.0) GO TO 40                                           HLE09010
C                                                                       HLE09020
      WRITE(6,30)                                                       HLE09030
   30 FORMAT(' PROGRAM TERMINATES BECAUSE EITHER ORDERS OF OR LABELS FORHLE09040
     1 MATRIX DISAGREE')                                                HLE09050
      GO TO 250                                                         HLE09060
C                                                                       HLE09070
   40 CONTINUE                                                          HLE09080
C                                                                       HLE09090
C     IF ITOEP = 1 THEN MATRIX IS TOEPLITZ AND WE PRINT OUT TRUE        HLE09100
C     EIGENVALUES                                                       HLE09110
      READ(8,10) EXPLAN                                                 HLE09120
      READ(8,*) ITOEP                                                   HLE09130
      READ(8,10) EXPLAN                                                 HLE09140
C                                                                       HLE09150
      IF(ITOEP.EQ.1) WRITE(6,50)                                        HLE09160
   50 FORMAT(/' TEST MATRIX IS HERMITIAN TOEPLITZ'/)                    HLE09170
      IF(ITOEP.NE.1) GO TO 110                                          HLE09180
C                                                                       HLE09190
      READ(8,*) DAR(1),DAI(1),D(1)                                      HLE09200
      DA(1) = DCMPLX(DAR(1),DAI(1))                                     HLE09210
      DB(1) = DCONJG(DA(1))                                             HLE09220
      DO 60 J=2,N                                                       HLE09230
      D(J) = D(1)                                                       HLE09240
      DA(J) = DA(1)                                                     HLE09250
   60 DB(J) = DB(1)                                                     HLE09260
      WRITE(6,70) DB(1),D(1),DA(1)                                      HLE09270
      WRITE(9,70) DB(1),D(1),DA(1)                                      HLE09280
   70 FORMAT(' HERMITIAN TOEPLITZ MATRIX IS USED.'/' BELOW DIAGONAL ENTRHLE09290
     1Y = ',2E12.3/' DIAGONAL ENTRY = ',E12.3/' ABOVE DIAGONAL ENTRY =' HLE09300
```

```
     1,2E12.3)
C
C     COMPUTE THE TRUE EIGENVALUES.  FORMULA IS CORRECT ONLY FOR THOSE
C     MATRICES WHOSE DIAGONAL = 2., ABOVE DIAGONAL = A, BELOW DIAGONAL
C     = A-CONJUGATE, AND A HAS NORM 1.
C
      PI = DARCOS(-1.D0)
      DO 80 J=1,N
   80 EIGVAL(J) = 2.D0 * (1.D0 -DCOS(PI*DFLOAT(J)/DFLOAT(N+1)))
      WRITE(9,90) N
   90 FORMAT(I6, ' = ORDER OF MATRIX'/' TRUE EIGENVALUES ARE'/)
      WRITE(9,100) (J, EIGVAL(J), J=1,N)
  100 FORMAT(I5,4X,E25.16,6X,I5,4X,E25.16)
      GO TO 240
C
C     NONTOEPLITZ HERMITIAN.  DIAGONAL ENTRIES ARE EQUALLY-SPACED.
C     ABOVE DIAGONAL ENTRIES ARE GENERATED BY GENERATING EQUALLY-SPACED
C     REAL PARTS, AND EQUALLY-SPACED IMAGINARY PARTS.  THE BELOW
C     DIAGONAL ENTRIES ARE THEN OBTAINED BY TAKING THE COMPLEX CONJUGATE
C     OF THE ABOVE DIAGONAL ENTRIES
C
  110 READ(8,*) D(1), SPACE
      WRITE(6,120) D(1),SPACE
  120 FORMAT(' 1ST DIAGONAL ENTRY =',E20.12,' SPACING =',E20.12)
      DO 130 J=2,N
  130 D(J) = D(J-1) + SPACE
      WRITE(6,140) (D(J), J=1,3)
  140 FORMAT(' 1ST THREE DIAGONAL ENTRIES ='/(2E20.12))
      READ(8,10) EXPLAN
      READ(8,*) DAR(1), SPACE
      WRITE(6,150) DAR(1),SPACE
  150 FORMAT(' REAL PART OF 1ST ABOVE DIAGONAL ENTRY =',E20.12,/
     1' SPACING = ',E20.12)
      DO 160 J=2,N
  160 DAR(J) = DAR(J-1) + SPACE
      WRITE(6,170) (DAR(J), J=1,3)
  170 FORMAT(' REAL PARTS OF 1ST THREE ABOVE DIAGONAL ENTRIES ='/
     1(2E20.12))
      READ(8,10) EXPLAN
      READ(8,*) DAI(1), SPACE
      WRITE(6,180) DAI(1),SPACE
  180 FORMAT(' IMAGINARY PART OF 1ST ABOVE =',E20.12,/' SPACING =',
     1 E20.12)
      DO 190 J=2,N
  190 DAI(J) = DAI(J-1) + SPACE
      WRITE(6,200) (DAI(J), J = 1,3)
  200 FORMAT(' IMAGINARY PARTS OF 1ST THREE ABOVE DIAGONAL ENTRIES ='/
     1 (2E20.12))
      DO 210 J=1,N
      DA(J) = DCMPLX(DAR(J),DAI(J))
  210 DB(J) = DCONJG(DA(J))
C
      WRITE(9,220) (D(J), J=1,N)
  220 FORMAT(' DIAGONAL ENTRIES ='/(4E20.12))
      WRITE(9,230) (DA(J), J=1,N)
  230 FORMAT(' ABOVE DIAGONAL ENTRIES'/(4E20.12))
C
C     PASS STORAGE LOCATIONS OF ARRAYS THAT DEFINE THE MATRIX TO
C     THE MATRIX-VECTOR MULTIPLY SUBROUTINE CMATV
C
  240 CONTINUE
C
C-----------------------------------------------------------------------
      CALL TMATVE(DA,DB,D,N)
C-----------------------------------------------------------------------
C
      RETURN
```

```
  250 STOP
C
C-----END OF USPEC-----------------------------------------------------
      END
C
C-----START OF MATRIX-VECTOR MULTIPLY (HERMITIAN TRIDIAGONAL)-----------
C
C     SUBROUTINE CMATV(W,U,SUM)
      SUBROUTINE TMATV(W,U,SUM)
C
C----------------------------------------------------------------------
      COMPLEX*16   U(1),W(1),DA(1),DB(1)
      DOUBLE PRECISION  D(1),SUM
C----------------------------------------------------------------------
C    HERMITIAN MATRIX-VECTOR MULTIPLY FOR LANCZS  U = A*W - SUM*U
C    MATRIX IS TRIDIAGONAL HERMITIAN TOEPLITZ
C----------------------------------------------------------------------
C
C     COMPUTE A*W - SUM*U
      U(1) = D(1)*W(1) + DA(1)*W(2) - SUM*U(1)
      N1 = N-1
      DO 10 I = 2,N1
   10 U(I) = DB(I-1)*W(I-1)+D(I)*W(I) + DA(I)*W(I+1) -SUM*U(I)
      U(N) = DB(N-1)*W(N-1) + D(N)*W(N) - SUM*U(N)
C
      RETURN
C
C    STORAGE LOCATIONS ARE PASSED TO CMATV FROM USPEC
C
C----------------------------------------------------------------------
      ENTRY TMATVE(DA,DB,D,N)
C----------------------------------------------------------------------
C
C-----END OF CMATV-----------------------------------------------------
      RETURN
      END
```

SECTION 3.5 FILE DEFINITIONS AND SAMPLE INPUT FILE

Below is a listing of the input/output files which are accessed by the Hermitian Lanczos eigenvalue program, HLEVAL. Included also is a sample of the input file which HLEVAL requires on file 5. The parameters supplied in this file are in free format. HLEVAL computes eigenvalues of Hermitian matrices on user-supplied intervals.

SAMPLE DEFINITIONS OF THE INPUT/OUTPUT FILES FOR HLEVAL

```
-----------------------------------------------------------------
*HLEVAL EXEC HERMITIAN EIGENVALUE CALCULATION
FI 06 TERM
FILEDEF  1 DISK &1        NHISTORY  A (RECFM F LRECL 80 BLOCK 80
FILEDEF  2 DISK &1        HISTORY   A (RECFM F LRECL 80 BLOCK 80
FILEDEF  3 DISK &1        GOODEV    A (RECFM F LRECL 80 BLOCK 80
FILEDEF  4 DISK &1        ERRINV    A (RECFM F LRECL 80 BLOCK 80
FILEDEF  5 DISK HLEVAL    INPUT     A (RECFM F LRECL 80 BLOCK 80
FILEDEF  8 DISK &1        INPUT     A (RECFM F LRECL 80 BLOCK 80
FILEDEF 11 DISK &1        DISTINCT  A (RECFM F LRECL 80 BLOCK 80
LOAD  HLEVAL   LESUB   HLEMULT
-----------------------------------------------------------------
```

SAMPLE INPUT FILE FOR HLEVAL

```
-----------------------------------------------------------------
 HLEVAL INPUT EIGENVALUE COMPUTATION, NO REORTHOGONALIZATION
 HERMITIAN TEST MATRIX
LINE 1      N     KMAX     NMEVS      MATNO
          528     2640         3     721830
LINE 2   SVSEED     RHSEED        MXINIT     MXSTUR
       49302312   5731029             5     100000
LINE 3    ISTART      ISTOP
              0          1
LINE 4      IHIS      IDIST  IWRITE
              1          0        1
LINE 5   RELTOL (RELATIVE TOLERANCE IN 'COMBINING' GOODEV)
    .0000000001
LINE 6   MB(1)   MB(2)   MB(3)   MB(4)    (ORDERS OF T(1,MEV) )
          528    1056    1584
LINE 7    NINT     (NUMBER OF SUB-INTERVALS FOR BISEC)
             1
LINE 8    LB(1)    LB(2)   LB(3)   LB(4)  (INTERVAL LOWER BOUNDS)
            1.3
LINE 9    UB(1)    UB(2)   UB(3)   UB(4)  (INTERVAL UPPER BOUNDS)
           1.34
-----------------------------------------------------------------
```

Below is a listing of the input/output files which are accessed by the Hermitian Lanczos eigenvector program, HLEVEC. Included also is a sample of the input file which HLEVEC requires on file 5. The parameters in this file are supplied in free format. HLEVEC computes eigenvectors for each of a user-selected subset of the eigenvalues computed by the eigenvalue program HLEVAL.

```
SAMPLE DEFINITIONS OF THE INPUT/OUTPUT FILES FOR HLEVEC
-----------------------------------------------------------------
*HLEVEC EXEC TO RUN LANCZOS EIGENVECTOR PROGRAM, HERMITIAN MATRICES
FI 06 TERM
FILEDEF  2 DISK &1        HISTORY   A (RECFM F LRECL 80 BLOCK 80
FILEDEF  3 DISK &1        GOODEV    A (RECFM F LRECL 80 BLOCK 80
FILEDEF  4 DISK &1        ERRINV    A (RECFM F LRECL 80 BLOCK 80
FILEDEF  5 DISK HLEVEC    INPUT     A (RECFM F LRECL 80 BLOCK 80
FILEDEF  8 DISK &1        INPUT     A (RECFM F LRECL 80 BLOCK 80
FILEDEF  9 DISK &1        ERREST    A (RECFM F LRECL 80 BLOCK 80
FILEDEF 10 DISK &1        BOUNDS    A (RECFM F LRECL 80 BLOCK 80
FILEDEF 11 DISK &1        TEIGVECS  A (RECFM F LRECL 80 BLOCK 80
FILEDEF 12 DISK &1        RITZVECS  A (RECFM F LRECL 80 BLOCK 80
FILEDEF 13 DISK &1        PAIGE     A (RECFM F LRECL 80 BLOCK 80
LOAD  HLEVEC  LESUB  HLEMULT
-----------------------------------------------------------------
```

```
SAMPLE INPUT FILE FOR HLEVEC
-----------------------------------------------------------------
 HLEVEC EIGENVECTORS OF HERMITIAN MATRIX, NO REORTHOGONALIZATION
LINE 1  MDIMTV    MDIMRV  MBETA(MAX.DIMENSIONS, TVEC, RITVEC AND BETA
          10000     10000   2000
LINE 2        RELTOL
        .0000000001
LINE 3  MBOUND   NTVCON SVTVEC IREAD (FLAGS
            0         1      0     1
LINE 4  TVSTOP   LVCONT ERCONT  IWRITE (FLAGS
            0         1      1     1
LINE 5    RHSEED  (RANDOM GENERATOR SEED FOR STARTING VECTOR IN INVERM)
        45329517
LINE 6  MATNO        N
          100     100
--------------------------------------------------------------------
```

CHAPTER 4

FACTORED INVERSES OF REAL SYMMETRIC MATRICES

SECTION 4.1 INTRODUCTION

The FORTRAN codes in this chapter address the question of computing distinct eigenvalues and corresponding eigenvectors of a real symmetric matrix by applying a single-vector Lanczos procedure to the inverse of an associated matrix $B \equiv PCP^T$, where $C = S0*A + SHIFT*I$. The scalars S0 and SHIFT are specified by the user, selected in such a way that the resulting matrix C (or B) has a reasonable numerical condition. The permutation matrix P is chosen so that for a sparse matrix A, the resulting factorization of B is also sparse.

For a given real symmetric matrix A, these codes compute real scalars λ and corresponding real-valued vectors $x \neq 0$ such that

$$B^{-1}x = \lambda x , \tag{4.1.1}$$

where B is as defined above. Note that the eigenvectors of B^{-1} are simple permutations of the eigenvectors of A. The eigenvalues of A are obtained from those of B by a simple scalar modification, which is incorporated in the codes. These codes do not require the matrix A. The Lanczos computations use only the user-supplied factorization of the associated matrix B, the scalars S0 and SHIFT, and the permutation P (if any).

Real symmetric matrices and factorizations of such matrices are discussed in Stewart [1973]. See also Bunch and Kaufman [1977] and George and Liu [1981]. Chapter 2, Section 2.1 contains a brief summary of the properties of real symmetric matrices which we use in these codes.

Given a real symmetric matrix A, the user may decide to use the codes in this chapter rather than those in Chapter 2 if the eigenvalues to be computed are 'small' with 'small' gaps between them and the required factorization can be obtained with a reasonable amount of computation and storage. The user should note however that this type of transformation of the given matrix may not yield an eigenvalue distribution which is better for these Lanczos codes. Such a transformation will accelerate the Lanczos computations only if the desired eigenvalues either become larger in size relative to the other eigenvalues and/or the gaps between the desired eigenvalues become larger relative to the gaps between the other eigenvalues. This type of transformation can be very effective in compressing the big end of the spectrum of a given matrix and enhancing the small end of the spectrum. The Lanczos procedure, however, does not require large gaps between the desired eigenvalues, all it really requires is a reasonable overall gap ratio. That is, the ratio of the largest gap between two neighboring eigenvalues to the smallest such gap must be a reasonable size.

The single-vector Lanczos codes in this chapter can be used to compute either a very few or very many of the distinct eigenvalues of the given real symmetric matrix. The documentation for these codes is contained in Chapter 2, Section 2.2. As in the direct real symmetric case (Chapter 2, Section 2.1), the A-multiplicity of a given computed eigenvalue can be obtained only with additional computation, and the modifications required to do this additional computation are not included in these versions of the codes. This implementation uses the basic Lanczos recursion contained in Eqns(1.2.1) - (1.2.2) to generate a family of real symmetric tridiagonal matrices (T-matrices) for the matrix B^{-1}, whose sizes are specified by the user. Specifically, for i = 1,2,...,m and a randomly-generated starting vector v_1 with $\|v_1\| = 1$, generate Lanczos vectors v_i using the following recursion and Eqn(1.2.2) applied to the matrix B^{-1}.

$$\beta_{i+1}v_{i+1} = B^{-1}v_i - \alpha_i v_i - \beta_i v_{i-1}. \tag{4.1.2}$$

B is the matrix defined above in terms of the scalars S0 and SHIFT and the permutation P, and each $B^{-1}v_i$ is evaluated by solving the system of equations $Bz = v_i$.

LIVAL, the main program for the factored inverse computations, calls the subroutine BISEC to compute eigenvalues of the specified Lanczos tridiagonal matrices on the user-specified intervals. BISEC simultaneously computes these T-eigenvalues with their T-multiplicities and sorts the computed T-eigenvalues into two classes, the 'good' T-eigenvalues and the 'spurious' T-eigenvalues. The 'good' T-eigenvalues are accepted as approximations to eigenvalues of the B^{-1} matrix associated with the user-specified matrix A, scalars S0 and SHIFT, and the permutation matrix P (if any). The accuracy of these 'good' T-eigenvalues as eigenvalues of B^{-1} is then estimated using error estimates computed by subroutine INVERR. Error estimates are computed only for isolated 'good' T-eigenvalues. All other 'good' T-eigenvalues are assumed to have converged. Convergence is then checked. If convergence has not yet occurred and a larger T-matrix has been specified by the user, the program will continue on to the larger T-matrix, repeating the above procedure on this larger matrix. After each T-matrix eigenvalue computation, the corresponding approximations to the eigenvalues of the user-specified matrix A are computed and included in the output.

Once the eigenvalues of B^{-1} have been computed accurately enough, the user can select a subset of the 'converged' eigenvalues for which eigenvectors are to be computed. The main program LIVEC, for computing eigenvectors of the inverse of a real symmetric matrix given a factorization, is used to compute the desired eigenvectors. If the matrix B is a permutation of the matrix C, then LIVEC unwinds the permutation to obtain the corresponding eigenvectors of the user-supplied A-matrix.

All of the computations are done in double precision real arithmetic. Once the Lanczos T-matrices have been computed, the remaining computations use the same subroutines that are used in the real symmetric case discussed in Chapter 2. In addition to the programs and subroutines provided here, the user must supply a subroutine USPEC which defines and initializes the factorization of the scaled, shifted, and permuted version B of the original matrix A, and a subroutine BSOLV which computes matrix-vector multiplies $B^{-1}x$ for any given vector x. These subroutines must be constructed in such a way as to take advantage of the sparsity (and/or structure) of the user-supplied A-matrix and such that these computations are done accurately.

The sample subroutines USPEC and BSOLV provided assume that the associated matrix B is positive definite and that its Cholesky factorization

$$B = LL^T, \quad L \text{ a lower triangular matrix,} \tag{4.1.3}$$

is used to compute $B^{-1}y$, for any given y. Thus, the sample USPEC subroutine provided for this chapter defines and initializes arrays which define the Cholesky factor L of the associated matrix B. The sample BSOLV subroutine provided computes the required matrix-vector multiplies $u = B^{-1}y$ by solving sequentially the two equations $Lz = y$ and $L^Tu = z$. These two equations are very easy to solve since L is a triangular matrix. The main portions of these Lanczos codes do not however require that the B-matrix be positive definite, only that a factorization be available. Therefore, the user could replace the sample USPEC and BSOLV subroutines by subroutines which use a more general factorization of B, for example $B = LDL^T$, where D is a diagonal matrix. All that is necessary is that the BSOLV subroutine provide the matrix-vector products $B^{-1}x$, rapidly and accurately. The information supplied to the Lanczos procedures about the matrix being processed must be consistent.

Several optional preprocessing programs are provided, PERMUT, LORDER, LFACT, and LTEST. PERMUT calls the SPARSPAK Library [1979] to attempt to identify a reordering or permutation P of the given matrix A for which sparseness will be preserved under factorization of the permuted matrix. LORDER takes a given matrix C and permutation P and computes the sparse matrix format for the permuted matrix, $B \equiv PCP^T$. LFACT computes the Cholesky factors of a given positive definite matrix. LTEST performs a very crude check on the numerical condition of the matrix supplied to it, by solving a system of equations with and without iterative refinement LINPACK [1979]. More details about the single-vector Lanczos procedures are given in Chapter 4 of Volume 1 of this book.

SECTION 4.2 MAIN PROGRAM, EIGENVALUE COMPUTATIONS

```
C-----LIVAL---(EIGENVALUES OF INVERSES OF REAL SYMMETRIC MATRICES)------
C
C     CONTAINS MAIN PROGRAM FOR COMPUTING DISTINCT EIGENVALUES OF
C     INVERSES OF REAL SYMMETRIC MATRICES USING REORDERING
C     AND SPARSE FACTORIZATION.  THE LANCZOS RECURSION IS APPLIED
C     TO A SCALED, SHIFTED, AND REORDERED VERSION B OF THE
C     ORIGINAL A-MATRIX.  THE PROCEDURE USES LANCZOS
C     TRIDIAGONALIZATION WITHOUT REORTHOGONALIZATION
C
C     PFORT VERIFIER IDENITIFIED THE FOLLOWING NONPORTABLE
C     CONSTRUCTIONS
C
C     1.  DATA/MACHEP/ STATEMENT
C     2.  ALL READ(5,*) STATEMENTS (FREE FORMAT)
C     3.  FORMAT(20A4) USED WITH EXPLANATORY HEADER EXPLAN.
C     4.  HEXADECIMAL FORMAT (4Z20) USED IN ALPHA/BETA FILES 1 AND 2.
C
C-----------------------------------------------------------------------
C
      DOUBLE PRECISION  ALPHA(3000),BETA(3001)
      DOUBLE PRECISION  V1(3001),V2(3000),VS(3000)
      DOUBLE PRECISION  LB(20),UB(20)
      DOUBLE PRECISION  BTOL,GAPTOL,TTOL,MACHEP,EPSM,SHIFT,SHIFTO,RELTOL
      DOUBLE PRECISION  SCALE1,SCALE2,SCALE3,SCALE4,BISTOL,CONTOL,MULTOL
      DOUBLE PRECISION  ONE,ZERO,TEMP,TKMAX,BETAM,BKMIN,TO,T1,SO
      REAL  G(3000),GG(3000),EXPLAN(20)
      INTEGER  MP(3000),NMEV(20)
      INTEGER  SVSEED,RHSEED,SVSOLD
      INTEGER IABS
      REAL  ABS
      DOUBLE PRECISION  DABS, DSQRT, DFLOAT
      EXTERNAL BSOLV
C
C-----------------------------------------------------------------------
      DATA MACHEP/Z3410000000000000/
      EPSM = 2.0D0*MACHEP
C-----------------------------------------------------------------------
C
C     ARRAYS MUST BE DIMENSIONED AS FOLLOWS:
C     1.  ALPHA: >= KMAX,    BETA: >= (KMAX+1) WHERE KMAX MAY
C         IS THE LARGEST SIZE T-MATRIX TO BE CONSIDERED.
C     2.  V1:   >= MAX(N,KMAX+1)
C     3.  V2,VS:    >= MAX(N,KMAX)
C     4.  GG:  >=  KMAX
C     5.  G:  >= MAX(N,2*KMAX)
C     6.  MP:  >= KMAX
C     7.  LB,UB:  >= NUMBER OF SUBINTERVALS SUPPLIED TO BISEC.
C     8.  NMEV:  >= NUMBER OF T-MATRICES ALLOWED.
C     9.  EXPLAN:  DIMENSION IS 20.
C
C
C     IMPORTANT TOLERANCES OR SCALES THAT ARE USED REPEATEDLY
C     THROUGHOUT THE PROGRAM ARE THE FOLLOWING:
C     SCALED MACHINE EPSILON:  TTOL = TKMAX*EPSM WHERE
C     EPSM = 2*MACHINE EPSILON AND
C     TKMAX = MAX(|ALPHA(J)|,BETA(J), J = 1,MEV)
C     BISEC CONVERGENCE TOLERANCE:  BISTOL  = DSQRT(1000+MEV)*TTOL
C     BISEC MULTIPLICITY TOLERANCE:  MULTOL = (1000+MEV)*TTOL
C     LANCZOS CONVERGENCE TOLERANCE: CONTOL = BETA(MEV+1)*1.D-10
C
C-----------------------------------------------------------------------
C     OUTPUT HEADER
      WRITE(6,10)
   10 FORMAT(/' LANCZOS PROCEDURE FOR FACTORED INVERSES OF REAL SYMMETRI
```

```
     1C MATRICES')
C
C     SET PROGRAM PARAMETERS
C     SCALEK ARE USED IN TOLERANCES NEEDED IN SUBROUTINES LUMP,
C     ISOEV AND PRTEST.  USER MUST NOT MODIFY THESE SCALES.
      SCALE1 = 5.0D2
      SCALE2 = 5.0D0
      SCALE3 = 5.0D0
      SCALE4 = 1.0D4
      ONE  = 1.0D0
      ZERO = 0.0D0
C     BTOL = 1.0D-8
      BTOL = EPSM
      GAPTOL = 1.0D-8
      ICONV = 0
      MOLD = 0
      MOLD1 = 1
      ICT = 0
      MMB = 0
      IPROJ = 0
C-----------------------------------------------------------------------
C     READ USER-SPECIFIED PARAMETERS FROM INPUT FILE 5 (FREE FORMAT)
C
C     READ USER-PROVIDED HEADER FOR RUN
      READ(5,20) EXPLAN
      WRITE(6,20) EXPLAN
      READ(5,20) EXPLAN
      WRITE(6,20) EXPLAN
   20 FORMAT(20A4)
C
C     READ ORDER OF MATRICES (N) , MAXIMUM ORDER OF T-MATRIX (KMAX),
C     NUMBER OF T-MATRICES ALLOWED (NMEVS), AND MATRIX IDENTIFICATION
C     NUMBERS (MATNO), SHIFT APPLIED TO MATRIX (SHIFT) AND
C     SCALE (SO).
      READ(5,20) EXPLAN
      READ(5,*) N,KMAX,NMEVS,MATNO,SO,SHIFT
C
C     READ SEEDS FOR LANCZS AND INVERR SUBROUTINES (SVSEED AND RHSEED)
C     READ MAXIMUM NUMBER OF ITERATIONS ALLOWED FOR EACH INVERSE
C     ITERATION (MXINIT) AND MAXIMUM NUMBER OF STURM SEQUENCES
C     ALLOWED (MXSTUR)
      READ(5,20) EXPLAN
      READ(5,*) SVSEED,RHSEED,MXINIT,MXSTUR
C
C     ISTART = (0,1):  ISTART = 0 MEANS ALPHA/BETA FILE IS NOT
C     AVAILABLE.  ISTART = 1 MEANS ALPHA/BETA FILE IS AVAILABLE ON
C     FILE 2.
C     ISTOP = (0,1):  ISTOP = 0 MEANS PROCEDURE GENERATES ALPHA/BETA
C     FILE AND THEN TERMINATES.  ISTOP = 1 MEANS PROCEDURE GENERATES
C     ALPHAS/BETAS IF NEEDED AND THEN COMPUTES EIGENVALUES AND ERROR
C     ESTIMATES AND THEN TERMINATES.
      READ(5,20) EXPLAN
      READ(5,*) ISTART,ISTOP
C
C     IHIS = (0,1):  IHIS = 0 MEANS ALPHA/BETA FILE IS NOT WRITTEN
C     TO FILE 1.  IHIS = 1 MEANS ALPHA/BETA FILE IS WRITTEN TO FILE 1.
C     IDIST = (0,1):  IDIST = 0 MEANS DISTINCT T-EIGENVALUES
C     ARE NOT WRITTEN TO FILE 11.  IDIST = 1 MEANS DISTINCT
C     T-EIGENVALUES ARE WRITTEN TO FILE 11.
C     IWRITE = (0,1):  IWRITE = 0 MEANS NO INTERMEDIATE OUTPUT
C     FROM THE COMPUTATIONS IS WRITTEN TO FILE 6.  IWRITE = 1 MEANS
C     T-EIGENVALUES AND ERROR ESTIMATES ARE WRITTEN TO FILE 6
C     AS THEY ARE COMPUTED.
      READ(5,20) EXPLAN
      READ(5,*) IHIS,IDIST,IWRITE
C
```

```
C     READ IN THE RELATIVE TOLERANCE (RELTOL) FOR USE IN THE
C     SPURIOUS, T-MULTIPLICITY, AND PRTESTS.
      READ(5,20) EXPLAN
      READ(5,*) RELTOL
C
C     READ IN THE SIZES OF THE T-MATRICES TO BE CONSIDERED.
      READ(5,20) EXPLAN
      READ(5,*) (NMEV(J), J=1,NMEVS)
C
C     READ IN THE NUMBER OF SUBINTERVALS TO BE CONSIDERED.
      READ(5,20) EXPLAN
      READ(5,*) NINT
C
C     READ IN THE LEFT-END POINTS OF THE SUBINTERVALS TO BE CONSIDERED.
C     THESE MUST BE IN ALGEBRAICALLY-INCREASING ORDER
      READ(5,20) EXPLAN
      READ(5,*) (LB(J), J=1,NINT)
C
C     READ IN THE RIGHT-END POINTS OF THE SUBINTERVALS TO BE CONSIDERED.
C     THESE MUST BE IN ALGEBRAICALLY-INCREASING ORDER
      READ(5,20) EXPLAN
      READ(5,*) (UB(J), J=1,NINT)
C
C-----------------------------------------------------------------------
C     INITIALIZE THE ARRAYS FOR THE FACTORIZATION OF THE ASSOCIATED
C     SCALED, SHIFTED AND PERMUTED VERSION OF THE A-MATRIX.
C     THE STORAGE LOCATIONS OF THESE ARRAYS ARE PASSED TO THE BSOLV
C     SUBROUTINE WHICH WILL BE CALLED FROM LANCZS FOR THE T-MATRIX
C     GENERATION.
C
      CALL USPEC(N,MATNO)
C
C-----------------------------------------------------------------------
C
C     MASKS UNDERFLOW AND OVERFLOW, USER MUST SUPPLY OR COMMENT OUT.
      CALL MASK
C
C-----------------------------------------------------------------------
C
C     WRITE TO FILE 6, A SUMMARY OF THE PARAMETERS FOR THIS RUN
C
      WRITE(6,30) MATNO,N,KMAX,SHIFT,SO
   30 FORMAT(/3X,'MATRIX ID',4X,'ORDER OF A',4X,'MAX ORDER OF T'//
     1 I12,I14,I18//8X,' SHIFT',8X,'SCALE'/2E15.6//
     1 ' C = SCALE*A + SHIFT*I '/
     1 ' B = P*C*P-TRANSPOSE WHERE P IS A REORDERING OF C'/
     1 ' LANCZOS PROCEDURE USES THE FACTORIZATION OF B'/)
C
      WRITE(6,40) ISTART,ISTOP
   40 FORMAT(/2X,'ISTART',3X,'ISTOP'/2I8/)
C
      WRITE(6,50) IHIS,IDIST,IWRITE
   50 FORMAT(/4X,'IHIS',3X,'IDIST',2X,'IWRITE'/3I8/)
C
      WRITE(6,60) SVSEED,RHSEED
   60 FORMAT(/' SEEDS FOR RANDOM NUMBER GENERATOR'//
     1 4X,'LANCZS SEED',4X,'INVERR SEED'/2I15/)
C
      WRITE(6,70) (NMEV(J), J=1,NMEVS)
   70 FORMAT(/' SIZES OF T-MATRICES TO BE CONSIDERED'/(6I12))
C
      WRITE(6,80) RELTOL,GAPTOL,BTOL
   80 FORMAT(/' RELATIVE TOLERANCE USED TO COMBINE COMPUTED T-EIGENVALUE
     1S'/E15.3/' RELATIVE GAP TOLERANCES USED IN INVERSE ITERATION'/
     1E15.3/' RELATIVE TOLERANCE FOR CHECK ON SIZE OF BETAS'/E15.3/)
C
      WRITE(6,90) (J,LB(J),UB(J), J=1,NINT)
```

```
   90 FORMAT(/' BISEC WILL BE USED ON THE FOLLOWING INTERVALS'/
     1 (I6,2E20.6))
C
      IF (ISTART.EQ.0) GO TO 140
C
C     READ IN ALPHA BETA HISTORY
C
      READ(2,100)MOLD,NOLD,SVSOLD,MATOLD,SHIFTO
  100 FORMAT(2I6,I12,I8,E13.4)
C
      IF (KMAX.LT.MOLD) KMAX = MOLD
      KMAX1 = KMAX + 1
C
C     CHECK THAT ORDER N, MATRIX ID MATNO, AND RANDOM SEED SVSEED
C     AGREE WITH THOSE IN THE HISTORY FILE.  IF NOT PROCEDURE STOPS.
C
      ITEMP = (NOLD-N)**2+(MATNO-MATOLD)**2+(SVSEED-SVSOLD)**2
C
      IF (ITEMP.EQ.0.AND.SHIFT.EQ.SHIFTO) GO TO 120
C
      WRITE(6,110)
  110 FORMAT(' PROGRAM TERMINATES'/    ' READ FROM FILE 2 CORRESPONDS TO
     1 DIFFERENT MATRIX THAN MATRIX SPECIFIED'/)
      GO TO 700
C
  120 CONTINUE
      MOLD1 = MOLD+1
C
      READ(2,130)(ALPHA(J), J=1,MOLD)
      READ(2,130)(BETA(J), J=1,MOLD1)
  130 FORMAT(4Z20)
C
      IF (KMAX.EQ.MOLD) GO TO 170
C
      READ(2,130)(V1(J), J=1,N)
      READ(2,130)(V2(J), J=1,N)
C
  140 CONTINUE
      IIX = SVSEED
C
      WRITE(6,150)
  150 FORMAT(' ENTERING LANCZS'/)
C
C-----------------------------------------------------------------------
C
      CALL LANCZS(BSOLV,ALPHA,BETA,V1,V2,VS,G,KMAX,MOLD1,N,IIX)
C
C-----------------------------------------------------------------------
C
C     ALPHA BETA WRITE
      KMAX1 = KMAX + 1
C
      IF(IHIS.EQ.0.AND.ISTOP.GT.0) GO TO 170
C
      WRITE(1,160) KMAX,N,SVSEED,MATNO,SHIFT
  160 FORMAT(2I6,I12,I8,E13.4,' = KMAX,N,SVSEED,MATNO,SHIFT')
C
      WRITE(1,130)(ALPHA(I), I=1,KMAX)
      WRITE(1,130)(BETA(I), I=1,KMAX1)
C
      WRITE(1,130)(V1(I), I=1,N)
      WRITE(1,130)(V2(I), I=1,N)
C
      IF (ISTOP.EQ.0) GO TO 600
C
  170 CONTINUE
      KMAX1 = KMAX + 1
```

```
      BKMIN = BTOL
C
      WRITE(6,180)
  180 FORMAT(/' T-MATRICES (ALPHA AND BETA) ARE NOW AVAILABLE'/)
C
C-----------------------------------------------------------------------
C     SUBROUTINE TNORM CHECKS MIN(BETA)/(ESTIMATED NORM(A)) > BTOL .
C     IF THIS IS VIOLATED IB IS SET EQUAL TO THE NEGATIVE OF THE INDEX
C     OF THE MINIMAL BETA.  IF(IB < 0) THEN SUBROUTINE TNORM IS
C     CALLED FOR EACH VALUE OF MEV TO DETERMINE WHETHER OR NOT THERE
C     IS A BETA IN THE T-MATRIX SPECIFIED THAT VIOLATES THIS TEST.
C     IF THERE IS SUCH A BETA THE PROGRAM TERMINATES FOR THE USER
C     TO DECIDE WHAT TO DO.  THIS TEST CAN BE OVER-RIDDEN BY
C     SIMPLY MAKING BTOL SMALLER, BUT THEN THERE IS THE POSSIBILITY
C     THAT LOSSES IN THE LOCAL ORTHOGONALITY MAY HURT THE COMPUTATIONS.
C     BTOL = 1.D-8 IS HOWEVER A CONSERVATIVE CHOICE FOR BTOL.
C
C     TNORM ALSO COMPUTES TKMAX = MAX(|ALPHA(K)|,BETA(K), K=1,KMAX).
C     TKMAX IS USED TO SCALE THE TOLERANCES USED IN THE
C     T-MULTIPLICITY AND SPURIOUS TESTS IN BISEC. TKMAX IS ALSO USED IN
C     THE PROJECTION TEST FOR HIDDEN EIGENVALUES THAT HAD 'TOO SMALL'
C     A PROJECTION ON THE STARTING VECTOR.
C
      CALL TNORM(ALPHA,BETA,BKMIN,TKMAX,KMAX,IB)
C
C-----------------------------------------------------------------------
      TTOL = EPSM*TKMAX
C
C     LOOP ON THE SIZE OF THE T-MATRIX
  190 CONTINUE
      MMB = MMB + 1
      MEV = NMEV(MMB)
C     IS MEV TOO LARGE ?
      IF(MEV.LE.KMAX) GO TO 210
C
      WRITE(6,200) MMB, MEV, KMAX
  200 FORMAT(/' TERMINATE PRIOR TO CONSIDERING THE',I6,'TH T-MATRIX'/
     1' BECAUSE THE SIZE REQUESTED',I6,' IS GREATER THAN THE MAXIMUM SIZ
     1E ALLOWED',I6/)
      GO TO 600
C
  210 MP1 = MEV + 1
      BETAM = BETA(MP1)
      WRITE(6,220) MEV,MEV,BETA(MEV),MEV,BETAM
  220 FORMAT(/' AT T-SIZE = ',I6,' BETA(',I4,') = ',E13.4/' BETA(',I4,'+
     11) =',E13.4)
      IF (IB.GE.0) GO TO 230
      T0 = BTOL
C-----------------------------------------------------------------------
C
      CALL TNORM(ALPHA,BETA,T0,T1,MEV,IBMEV)
C
C-----------------------------------------------------------------------
      TEMP = T0/TKMAX
      IBMEV = IABS(IBMEV)
      IF (TEMP.GE.BTOL) GO TO 230
      IBMEV = -IBMEV
      GO TO 660
  230 CONTINUE
      IC = MXSTUR-ICT
C
C-----------------------------------------------------------------------
C     BISEC LOOP. THE SUBROUTINE BISEC INCORPORATES DIRECTLY THE
C     T-MULTIPLICITY AND SPURIOUS TESTS. T-EIGENVALUES WILL BE
C     CALCULATED BY BISEC SEQUENTIALLY ON INTERVALS
C     (LB(J),UB(J)), J = 1,NINT).
C
```

```
C     ON RETURN FROM BISEC
C     NDIS = NUMBER OF DISTINCT EIGENVALUES OF T(1,MEV) ON UNION
C            OF THE (LB,UB) INTERVALS
C     VS = DISTINCT T-EIGENVALUES IN ALGEBRAICALLY INCREASING ORDER
C     MP = T-MULTIPLICITIES OF THE T-EIGENVALUES IN VS
C     MP(I) = (0,1,MI), MI>1, I=1,NDIS  MEANS:
C        (0)  VS(I) IS SPURIOUS
C        (1)  VS(I) IS T-SIMPLE AND GOOD
C        (MI) VS(I) IS MULTIPLE AND IS THEREFORE NOT ONLY GOOD BUT
C             ALSO A CONVERGED GOOD T-EIGENVALUE.
C
      CALL BISEC(ALPHA,BETA,V1,V2,VS,LB,UB,EPSM,TTOL,MP,NINT,
     1 MEV,NDIS,IC,IWRITE)
C
C-----------------------------------------------------------------------
      IF (NDIS.EQ.0) GO TO 680
C
C     COMPUTE THE TOTAL NUMBER OF STURM SEQUENCES USED TO DATE
C     COMPUTE THE BISEC CONVERGENCE AND T-MULTIPLICITY TOLERANCES USED.
C     COMPUTE THE CONVERGENCE TOLERANCE FOR EIGENVALUES OF A.
      ICT = ICT + IC
      TEMP = DFLOAT(MEV+1000)
      MULTOL = TEMP*TTOL
      TEMP = DSQRT(TEMP)
      BISTOL = TTOL*TEMP
      CONTOL = BETAM*1.D-10
C
C-----------------------------------------------------------------------
C    SUBROUTINE LUMP 'COMBINES' T-EIGENVALUES THAT ARE 'TOO CLOSE'.
C    NOTE HOWEVER THAT CLOSE SPURIOUS T-EIGENVALUES ARE NOT AVERAGED
C    WITH GOOD ONES. HOWEVER, THEY MAY BE USED TO INCREASE THE
C    T-MULTIPLICITY OF A GOOD T-EIGENVALUE.
C
      LOOP = NDIS
      CALL LUMP(VS,RELTOL,MULTOL,SCALE2,MP,LOOP)
C
C-----------------------------------------------------------------------
      IF(NDIS.EQ.LOOP) GO TO 250
C
      WRITE(6,240) NDIS, MEV, LOOP
  240 FORMAT(/I6,' DISTINCT T-EIGENVALUES WERE COMPUTED IN BISEC AT MEV
     1',I6/ 2X,' LUMP SUBROUTINE REDUCES NUMBER OF DISTINCT T-EIGENVALUE
     1S TO',I6)
C
  250 CONTINUE
      NDIS = LOOP
      BETA(MP1) = BETAM
C-----------------------------------------------------------------------
C     THE SUBROUTINE ISOEV LABELS THOSE SIMPLE EIGENVALUES OF T(1,MEV)
C     WITH VERY SMALL GAPS BETWEEN NEIGHBORING EIGENVALUES OF T(1,MEV)
C     TO AVOID COMPUTING ERROR ESTIMATES FOR ANY SIMPLE GOOD
C     T-EIGENVALUE THAT IS TOO CLOSE TO A SPURIOUS EIGENVALUE.
C     ON RETURN FROM ISOEV, G CONTAINS CODED MINIMAL GAPS
C     BETWEEN THE DISTINCT EIGENVALUES OF T(1,MEV). (G IS REAL).
C     G(I) < 0 MEANS MINGAP IS DUE TO LEFT GAP G(I) > 0 MEANS DUE TO
C     RIGHT GAP. MP(I) = -1 MEANS THAT THE GOOD T-EIGENVALUE IS SIMPLE
C     AND HAS A VERY SMALL MINGAP IN T(1,MEV) DUE TO A SPURIOUS
C     T-EIGENVALUE.  NG = NUMBER OF GOOD T-EIGENVALUES.
C     NISO = NUMBER OF ISOLATED GOOD T-EIGENVALUES.
C
      CALL ISOEV(VS,GAPTOL,MULTOL,SCALE1,G,MP,NDIS,NG,NISO)
C
C-----------------------------------------------------------------------
C
      WRITE(6,260)NG,NISO,NDIS
  260 FORMAT(/I6,' GOOD T-EIGENVALUES HAVE BEEN COMPUTED'/
     1 I6,' OF THESE ARE T-ISOLATED'/
```

```
     2 I6,' = NUMBER OF DISTINCT T-EIGENVALUES COMPUTED'/)
C
C     DO WE WRITE DISTINCT EIGENVALUES OF T-MATRIX TO FILE 11?
      IF (IDIST.EQ.0) GO TO 310
C
      WRITE(11,270) NDIS,NISO,MEV,N,SVSEED,MATNO
  270 FORMAT(/4I6,I12,I8,' = NDIS,NISO,MEV,N,SVSEED,MATNO'/)
C
      WRITE(11,280)
  280 FORMAT(/1X,'MP',21X,'EVBI',5X,'TMINGAP',1X,'MP',21X,'EVBI',5X,
     1'TMINGAP'/)
C
      WRITE(11,290) (MP(I),VS(I),G(I), I=1,NDIS)
  290 FORMAT(2(I3,E25.16,E12.3))
C
      WRITE(11,300) NDIS, (MP(I), I=1,NDIS)
  300 FORMAT(/I6,' = NDIS, T-MULTIPLICITIES (0 MEANS  SPURIOUS)'/(20I4))
C
  310 CONTINUE
      IF (NISO.NE.0) GO TO 340
C
      WRITE(4,320) MEV
  320 FORMAT(/' AT MEV = ',I6,' THERE ARE NO ISOLATED T-EIGENVALUES'/
     1' SO NO ERROR ESTIMATES WERE COMPUTED/')
C
      WRITE(6,330)
  330 FORMAT(/' ALL COMPUTED GOOD T-EIGENVALUES ARE MULTIPLE'/
     1 ' THEREFORE ALL SUCH EIGENVALUES ARE ASSUMED TO HAVE CONVERGED')
C
      ICONV = 1
      GO TO 380
  340 CONTINUE
C-----------------------------------------------------------------------
C     SUBROUTINE INVERR COMPUTES ERROR ESTIMATES FOR ISOLATED GOOD
C     T-EIGENVALUES USING INVERSE ITERATION ON T(1,MEV). ON RETURN
C     G(J) = MINIMUM GAP IN T(1,MEV) FOR EACH VS(J), J=1,NDIS
C     G(MEV+I) = BETAM*|U(MEV)| = ERROR ESTIMATE FOR ISOLATED GOOD
C              T-EIGENVALUES, WHERE I = 1, NISO AND  BETAM = BETA(MEV+1)
C              U(MEV) IS MEVTH COMPONENT OF THE UNIT EIGENVECTOR OF T
C              CORRESPONDING TO THE ITH ISOLATED GOOD T-EIGENVALUE.
C     A NEGATIVE ERROR ESTIMATE MEANS THAT FOR THAT PARTICULAR
C     EIGENVALUE THE INVERSE ITERATION DID NOT CONVERGE IN <= MXINIT
C     STEPS AND THAT THE CORRESPONDING ERROR ESTIMATE IS QUESTIONABLE.
C
C     V2 CONTAINS THE ISOLATED GOOD T-EIGENVALUES
C     V1 CONTAINS THE MINGAPS TO THE NEAREST DISTINCT  EIGENVALUE
C        OF T(1,MEV) FOR EACH ISOLATED GOOD T-EIGENVALUE IN V2.
C     VS CONTAINS THE NDIS DISTINCT EIGENVALUES OF T(1,MEV)
C     MP CONTAINS THE CORRESPONDING CODED T-MULTIPLICITIES
C
      IT = MXINIT
      CALL INVERR(ALPHA,BETA,V1,V2,VS,EPSM,G,MP,MEV,MMB,NDIS,NISO,N,
     1 RHSEED,IT,IWRITE)
C
C-----------------------------------------------------------------------
C     SIMPLE CHECK FOR CONVERGENCE. CHECKS TO SEE IF ALL OF THE
C     LAST COMPONENTS OF EIGENVECTORS ARE L.T. CONTOL.
C     IF THIS TEST IS SATISFIED, THEN CONVERGENCE FLAG, ICONV IS SET
C     TO 1.  TYPICALLY ERROR ESTIMATES ARE VERY CONSERVATIVE.
C
      WRITE(6,350) CONTOL
  350 FORMAT(/' CONVERGENCE IS TESTED USING THE CONVERGENCE TOLERANCE',
     1E13.4/)
C
      II = MEV +1
      IF = MEV+NISO
      DO 360 I = II,IF
```

```
      IF (ABS(G(I)).GT.CONTOL) GO TO 380
  360 CONTINUE
      ICONV = 1
      MMB = NMEVS
C
      WRITE(6,370) CONTOL
  370 FORMAT(' ALL COMPUTED ERROR ESTIMATES WERE LESS THAN',E15.4/
     1 ' THEREFORE PROCEDURE TERMINATES'/)
C
  380 CONTINUE
C
      IF (ICONV.EQ.0) GO TO 510
C
C-----------------------------------------------------------------------
C     IF CONVERGENCE IS INDICATED, THAT IS ICONV = 1 ,THEN
C     THE SUBROUTINE PRTEST IS CALLED TO CHECK FOR ANY CONVERGED
C     T-EIGENVALUES THAT HAVE BEEN MISLABELLED AS SPURIOUS BECAUSE
C     THE PROJECTION OF THEIR EIGENVECTOR(S) ON THE STARTING
C     VECTOR WAS(WERE) TOO SMALL.
C     NUMERICAL TESTS INDICATE THAT SUCH EIGENVALUES ARE RARE.
C     IF FOR SOME REASON MANY OF THESE HIDDEN EIGENVALUES APPEAR
C     ON SOME RUN, YOU CAN BE CERTAIN THAT SOMETHING IS FOULED UP.
C
      CALL PRTEST(ALPHA,BETA,VS,TKMAX,EPSM,RELTOL,SCALE3,SCALE4,
     1 MP,NDIS,MEV,IPROJ)
C
C-----------------------------------------------------------------------
C
      IF(IPROJ.EQ.0) GO TO 500
C
      IF(IDIST.EQ.1)  WRITE(11,390) IPROJ
  390 FORMAT(' SUBROUTINE PRTEST WANTS TO RELABEL',I6,' SPURIOUS T-EIGEN
     1VALUES'/' WE ACCEPT RELABELLING ONLY IF LAST COMPONENT OF T-EIGENV
     1ECTOR IS L.T. 1.D-10'/)
C
      IIX = RHSEED
C
C-----------------------------------------------------------------------
C
      CALL GENRAN(IIX,G,MEV)
C
C-----------------------------------------------------------------------
C
      ITEN = -10
      NISOM = NISO + MEV
      IWRITO = IWRITE
      IWRITE = 0
C
      DO 420 J = 1,NDIS
      IF(MP(J).NE.ITEN) GO TO 420
      TO = VS(J)
C
C-----------------------------------------------------------------------
C
      IT = MXINIT
      CALL INVERM(ALPHA,BETA,V1,V2,TO,TEMP,T1,EPSM,G,MEV,IT,IWRITE)
C
C-----------------------------------------------------------------------
C
      IF(TEMP.LE.1.D-10) GO TO 410
C     ERROR ESTIMATE WAS NOT SMALL REJECT RELABELLING OF THIS
C     T-EIGENVALUE
      IF(IDIST.EQ.1)  WRITE(11,400) J,TO,TEMP
  400 FORMAT(/' LAST COMPONENT FOR',I6,'TH T-EIGENVALUE',E20.12/' IS TOO
     1 LARGE = ',E15.6,' SO DO NOT ACCEPT PRTEST RELABELLING'/)
      MP(J) = 0
      IPROJ = IPROJ - 1
```

```
      GO TO 420
C     RELABELLING ACCEPTED
  410 NISOM = NISOM + 1
      G(NISOM) = BETAM*TEMP
  420 CONTINUE
      IWRITE = IWRITO
C
      IF(IPROJ.EQ.0) GO TO 460
      WRITE(6,430) IPROJ
  430 FORMAT(/I6,' T-EIGENVALUES WERE RECLASSIFIED AS GOOD.'/
     1' THESE ARE IDENTIFIED IN FILE 3 BY A T-MULTIPLICITY OF -10'/' USE
     2R SHOULD INSPECT EACH TO MAKE SURE NEIGHBORS HAVE CONVERGED'/)
C
      IF(IDIST.EQ.1)  WRITE(11,440) IPROJ
  440 FORMAT(/I6,' T-EIGENVALUES WERE RELABELLED AS GOOD'/
     1' BELOW IS CORRECTED T-MULTIPLICITY PATTERN'/)
C
      WRITE(6,450) NDIS, (MP(I), I=1,NDIS)
      IF(IDIST.EQ.1)  WRITE(11,450) NDIS, (MP(I), I=1,NDIS)
  450 FORMAT(/I6,' = NDIS, T-MULTIPLICITIES (0 MEANS  SPURIOUS)'/
     1 6X, ' (-10) MEANS SPURIOUS T-EIGENVALUE RELABELLED AS GOOD'/(20I4
     1))
C
C     RECALCULATE MINGAPS FOR DISTINCT T(1,MEV) EIGENVALUES.
  460 NM1 = NDIS - 1
      G(NDIS) = VS(NM1)-VS(NDIS)
      G(1) = VS(2)-VS(1)
C
      DO 470 J = 2,NM1
      T0 = VS(J)-VS(J-1)
      T1 = VS(J+1)-VS(J)
      G(J) = T1
      IF (T0.LT.T1) G(J) = -T0
  470 CONTINUE
      IF(IPROJ.EQ.0) GO TO 500
C     WRITE TO FILE 4 ERROR ESTIMATES FOR THOSE T-EIGENVALUES RELABELLED
      NGOOD = 0
      DO 480 J = 1,NDIS
      IF(MP(J).EQ.0) GO TO 480
      NGOOD = NGOOD + 1
      IF(MP(J).NE.ITEN) GO TO 480
      T0 = VS(J)
      NISO = NISO + 1
      NISOM = MEV + NISO
      WRITE(4,490) NGOOD,T0,G(NISOM),G(J)
  480 CONTINUE
  490 FORMAT(I10,E25.16,2E14.3)
C
  500 CONTINUE
C
C     WRITE THE GOOD T-EIGENVALUES TO FILE 3.  FIRST TRANSFER THEM
C     TO V2 AND THEIR T-MULTIPLICITIES TO THE CORRESPONDING POSITIONS
C     IN MP AND COMPUTE THE A-MINGAPS, THE MINIMAL GAPS BETWEEN THE
C     GOOD T-EIGENVALUES.  THESE GAPS WILL BE PUT IN THE ARRAY G.
C     SINCE G CURRENTLY CONTAINS THE MINIMAL GAPS BETWEEN THE DISTINCT
C     EIGENVALUES OF THE T-MATRIX, THESE GAPS WILL FIRST BE
C     TRANSFERRED TO V1.  NOTE THAT V1<0 MEANS THAT THAT MINIMAL GAP
C     IN THE T-MATRIX IS DUE TO A SPURIOUS T-EIGENVALUE.
C     ALL THIS INFORMATION IS PRINTED TO FILE 3
C
  510 CONTINUE
      NG = 0
      DO 520 I = 1,NDIS
      IF (MP(I).EQ.0) GO TO 520
      NG = NG+1
      MP(NG) = MP(I)
      V2(NG) = VS(I)
```

```
      TEMP = G(I)
      TEMP = DABS(TEMP)
      J = I+1
      IF (G(I).LT.ZERO) J = I-1
      IF (MP(J).EQ.0) TEMP = -TEMP
      V1(NG) = TEMP
  520 CONTINUE
C
      WRITE(6,530)MEV
  530 FORMAT(//' T-EIGENVALUE CALCULATION AT MEV = ',I6,'  IS COMPLETE'/
     1)
C
C     NG = NUMBER OF COMPUTED DISTINCT GOOD T-EIGENVALUES.  NEXT
C     GENERATE GAPS BETWEEN GOOD T-EIGENVALUES (BIMINGAPS) AND PUT THEM
C     G.  G(J) < 0 MEANS THE MINIMAL GAP IS DUE TO THE LEFT-HAND GAP.
C
C     GG(J) = BIMINGAP FOR EIGENVALUES OF B-INVERSE MATRIX.
      NGM1 = NG - 1
      GG(NG) = V2(NGM1)-V2(NG)
      GG(1) = V2(2)-V2(1)
C
      DO 540 J = 2,NGM1
      T0 = V2(J)-V2(J-1)
      T1 = V2(J+1)-V2(J)
      GG(J) = T1
      IF (T0.LT.T1) GG(J) = -T0
  540 CONTINUE
C
C     WRITE GOOD BI EIGENVALUES TO FILE 3.
      WRITE(3,550)NG,NDIS,MEV,N,SVSEED,MATNO,MULTOL,IB,BTOL,SHIFT
  550 FORMAT(4I6,I12,I8, ' = NG,NDIS,MEV,N,SVSEED,MATNO'/
     1 E20.12,I6,2E10.3,' = MULTOL,I(MINBETA),BTOL,SHIFT')
C
C     CALCULATE EIGENVALUES OF ORIGINAL INPUT MATRIX CORRESPONDING
C     TO COMPUTED GOOD T-EIGENVALUES.
      TEMP = -ONE/S0
      DO 560 K = 1,NG
      VS(K) = (SHIFT - (ONE/V2(K)))*TEMP
  560 CONTINUE
C
      NGM1 = NG - 1
      G(NG) = DABS(VS(NGM1)-VS(NG))
      G(1) = DABS(VS(2)-VS(1))
C
      DO 570 J = 2,NGM1
      T0 = DABS(VS(J)-VS(J-1))
      T1 = DABS(VS(J+1)-VS(J))
      G(J) = T1
      IF (T0.LT.T1) G(J)=-T0
  570 CONTINUE
C
      WRITE(3,580)
  580 FORMAT(' EVNO',1X,'TMULT',20X,'EVBI',5X,'BIGAP',6X,'AGAP',6X,
     1'TGAP',12X,'EVA')
C
      WRITE(3,590)(I,MP(I),V2(I),GG(I),G(I),V1(I),VS(I),  I=1,NG)
  590 FORMAT(2I5,E25.16,3E10.3,E15.8)
C
C     IF CONVERGENCE FLAG ICONV.NE.1 AND NUMBER OF T-MATRICES
C     CONSIDERED TO DATE IS LESS THAN NUMBER ALLOWED, INCREMENT MEV.
C     AND LOOP BACK TO 210 TO REPEAT COMPUTATIONS.  RESTORE BETA(MEV+1).
C
      BETA(MP1) = BETAM
      IF (MMB.LT.NMEVS.AND.ICONV.NE.1) GO TO 190
C     END OF LOOP ON DIFFERENT SIZE T-MATRICES ALLOWED.
  600 CONTINUE
C
```

```
      IF(ISTOP.EQ.0)  WRITE(6,610)
  610 FORMAT(/' T-MATRICES (ALPHA AND BETA) ARE NOW AVAILABLE, TERMINATE
     1')
      IF (IHIS.EQ.1.AND.KMAX.NE.MOLD) WRITE(1,620)
  620 FORMAT(/' ABOVE ARE THE FOLLOWING VECTORS '/
     1 '  ALPHA(I), I = 1,KMAX'/
     2 '  BETA(I), I = 1,KMAX+1'/
     3 ' FINAL TWO LANCZOS VECTORS OF ORDER N FOR I = KMAX,KMAX+1'/
     4 ' ALL VECTORS IN THIS FILE HAVE HEX FORMAT 4Z20'/
     5 ' ----- END OF FILE 1 NEW ALPHA, BETA HISTORY---------------'///)
C
      IF (ISTOP.EQ.0) GO TO 700
C
      WRITE(3,630)
  630 FORMAT(/' ABOVE ARE COMPUTED GOOD T-EIGENVALUES'/
     1 ' NG = NUMBER OF GOOD T-EIGENVALUES COMPUTED'/
     2 ' NDIS = NUMBER OF COMPUTED DISTINCT EIGENVALUES OF T(1,MEV)'/
     3 ' N = ORDER OF A,  MATNO = MATRIX IDENT'/
     3 ' THERE ARE TWO SETS OF EIGENVALUES, THOSE FOR A AND THOSE FOR'/
     3 ' B-INVERSE WHERE C=SO*A + SHIFT*I, B = P*C*P-TRANS = L*L-TRANS'/
     3 ' THE LANCZOS RECURSIONS ARE APPLIED TO B-INVERSE, USING L'/
     3 ' IF EVBI IS A GOOD EIGENVALUE OF B-INVERSE, THEN EVA IS A'/
     3 ' GOOD EIGENVALUE OF A WHERE EVA = (SHIFT-ONE/EVBI)(-ONE/SO)'/
     4 ' MULTOL = T-MULTIPLICITY TOLERANCE FOR T-EIGENVALUES IN BISEC'/
     4 ' TMULT IS THE T-MULTIPLICITY OF GOOD T-EIGENVALUE'/
     5 ' TMULT = -1 MEANS SPURIOUS T-EIGENVALUE TOO CLOSE'/
     6 ' DO NOT COMPUTE ERROR ESTIMATES FOR SUCH EIGENVALUES'/
     7 ' AMINGAP = MINIMAL GAP BETWEEN THE COMPUTED A-EIGENVALUES'/
     8 ' AMINGAP .LT. 0. MEANS MINIMAL GAP IS DUE TO LEFT-HAND GAP'/
     9 ' TMINGAP= MINIMAL GAP W.R.T.  DISTINCT EIGENVALUES IN T(1,MEV)'/
     1 ' TMINGAP .LT. 0. MEANS MINGAP IS DUE TO SPURIOUS T-EIGENVALUE'/
     2 ' ----- END OF FILE 3 GOODEIGENVALUES-----------------------'///)
C
      IF (IDIST.EQ.1) WRITE(11,640)
  640 FORMAT(/' ABOVE ARE THE DISTINCT EIGENVALUES OF T(1,MEV).'/
     2 ' THE FORMAT IS      T-MULTIPLICITY    T-EIGENVALUE   TMINGAP'/
     3 '          THIS FORMAT IS REPEATED TWICE ON EACH LINE.'/
     4 ' T-MULTIPLICITY = -1 MEANS THAT THE SUBROUTINE ISOEV HAS TAGGED'
     5/'   THIS SIMPLE T-EIGENVALUE AS HAVING A VERY CLOSE SPURIOUS'/
     6 '   T-EIGENVALUE SO THAT NO ERROR ESTIMATE WILL BE COMPUTED'/
     7 '   FOR THAT EIGENVALUE IN SUBROUTINE INVERR.'/
     8 ' TMINGAP .LT. 0, TMINGAP IS DUE TO LEFT GAP .GT. 0, RIGHT GAP.'/
     9 ' EACH OF THE DISTINCT T-EIGENVALUE TABLES IS FOLLOWED'/
     9 ' BY THE T-MULTIPLICITY PATTERN.'/
     1 ' NDIS = NUMBER OF COMPUTED DISTINCT EIGENVALUES OF T(1,MEV).'/
     2 ' NG = NUMBER OF GOOD T-EIGENVALUES. '/
     3 ' NISO = NUMBER OF ISOLATED GOOD T-EIGENVALUES. '/
     4 ' NISO ALSO IS THE COUNT OF +1 ENTRIES IN T-MULTIPLICITY PATTERN.
     5 '/' ----- END OF FILE 11 DISTINCT T-EIGENVALUES--------------'///
     6)
C
      IF(NIOS.NE.0)  WRITE(4,650)
  650 FORMAT(/' ABOVE ARE THE ERROR ESTIMATES OBTAINED FOR THE ISOLATED
     1GOOD T-EIGENVALUES'/
     1' OBTAINED VIA INVERSE ITERATION IN THE SUBROUTINE INVERR.'/
     1' ALL OTHER GOOD T-EIGENVALUES HAVE CONVERGED.'/
     2' ERROR ESTIMATE = BETAM*ABS(UM)'/
     2' WHERE BETAM = BETA(MEV+1) AND UM = U(MEV).'/
     3' U = UNIT EIGENVECTOR OF T WHERE T*U = EV*U AND EV = ISOLATED GOO
     3D T-EIGENVALUE.'/
     4' TMINGAP = GAP TO NEAREST DISTINCT EIGENVALUE OF T(1,MEV).'/
     5' TMINGAP .LT. 0. MEANS MINGAP IS DUE TO LEFT NEIGHBOR'/
     6' ERROR ESTIMATE L.T. 0 MEANS INVERSE ITERATION DID NOT CONVERGE'/
     7' ------ END OF FILE 4 ERRINV -------------------------------'//)
      GO TO 700
C
```

```
  660 CONTINUE
C
      IBB = IABS(IBMEV)
      IF (IBMEV.LT.0) WRITE(6,670) MEV,IBB,BETA(IBB)
  670 FORMAT(/' PROGRAM TERMINATES BECAUSE MEV REQUESTED = ',I6,' IS .GT
     1',I6/' AT WHICH AN ABNORMALLY SMALL BETA = ' , E13.4,' OCCURRED'/)
      GO TO 700
C
  680 IF (NDIS.EQ.0.AND.ISTOP.GT.0) WRITE(6,690)
  690 FORMAT(/' INTERVALS SPECIFIED FOR BISECT DID NOT CONTAIN ANY T-EIG
     1ENVALUES'/' PROGRAM TERMINATES')
C
  700 CONTINUE
C
      STOP
C-----END OF LIVAL (INVERSES OF REAL SYMMETRIC MATRICES)----------------
      END
```

SECTION 4.3 MAIN PROGRAM, EIGENVECTOR COMPUTATIONS

```
C-----LIVEC (EIGENVECTORS OF INVERSES OF REAL SYMMETRIC MATRICES)-------
C
C     CONTAINS MAIN PROGRAM FOR COMPUTING AN EIGENVECTOR CORRESPONDING
C     TO EACH OF A SET OF EIGENVALUES WHICH HAVE BEEN COMPUTED
C     ACCURATELY BY THE CORRESPONDING LANCZOS EIGENVALUE PROGRAM
C     (LIVAL) FOR FACTORED INVERSES OF REAL, SYMMETRIC MATRICES.
C     THIS PROGRAM COULD BE MODIFIED TO COMPUTE ADDITIONAL EIGENVECTORS
C     FOR ANY EIGENVALUES WHICH ARE MULTIPLE EIGENVALUES OF THE
C     A-MATRIX.  THE AMOUNT OF ADDITIONAL COMPUTATION REQUIRED BY
C     SUCH A MODIFICATION DEPENDS UPON THE GIVEN A-MATRIX AND UPON
C     WHICH PORTION OF THE SPECTRUM IS INVOLVED.
C
C     THESE LANCZOS EIGENVECTOR COMPUTATIONS ASSUME THAT EACH
C     EIGENVALUE THAT IS BEING CONSIDERED HAS CONVERGED AS AN
C     EIGENVALUE OF THE LANCZOS TRIDIAGONAL MATRICES.
C
C     PFORT VERIFIER IDENTIFIED THE FOLLOWING NONPORTABLE
C     CONSTRUCTIONS
C
C     1.  DATA/MACHEP/ STATEMENT
C     2.  ALL READ(5,*) STATEMENTS (FREE FORMAT)
C     3.  FORMAT(20A4) USED WITH THE EXPLANATORY HEADER, EXPLAN
C     4.  HEXADECIMAL FORMAT (4Z20) USED FOR ALPHA/BETA FILES 1 AND 2.
C
C     IMPORTANT NOTE:  PROGRAM ALLOWS ENLARGEMENT OF THE ALPHA, BETA
C     ARRAYS.  IN PARTICULAR, IF ANY ONE OF THE EIGENVALUES SUPPLIED
C     IS T-SIMPLE AND NOT CLOSE TO A SPURIOUS EIGENVALUE, THE PROGRAM
C     REQUIRES THAT KMAX BE AT LEAST 11*MEV/8 + 12.  IF KMAX IS NOT
C     THIS LARGE, THEN THE PROGRAM RESETS KMAX TO THIS SIZE
C     AND EXTENDS THE ALPHA, BETA HISTORY IF REQUIRED.
C     THUS, THE DIMENSIONS OF THE ALPHA AND BETA ARRAYS MUST BE
C     LARGE ENOUGH TO ALLOW FOR THIS POSSIBILITY.
C     REMEMBER THAT THE BETA ARRAY, BETA(J), IS SUCH THAT
C     J = 1,..., KMAX+1,  SO IF THE KMAX USED BY THE PROGRAM
C     IS TO BE 3000, THEN BETA MUST BE OF LENGTH AT LEAST 3001.
C
C-----------------------------------------------------------------------
      DOUBLE PRECISION  ALPHA(1000),BETA(1001)
      DOUBLE PRECISION  V1(2200),V2(2200),VS(2200)
      DOUBLE PRECISION  RITVEC(40000),TVEC(5000)
      DOUBLE PRECISION  GOODA(50),GOODBI(50),EVNEW(50),TLAST(50)
      DOUBLE PRECISION  EVAL,EVALN,TOLN,TTOL,ERTOL,ALFA,BATA
      DOUBLE PRECISION  MULTOL,SCALEO,STUTOL,BTOL,LB,UB,SO,RNORME
      DOUBLE PRECISION  ONE,ZERO,MACHEP,EPSM,TEMP,SUM,SHIFT,SHIFTO
      DOUBLE PRECISION  RELTOL,ERROR,TERROR,BKMIN,ERRMIN
      REAL G(5000),AMINGP(50),TMINGP(50),BIERR(50),BIEVER(50),BIERRG(50)
      REAL TERR(50),RNORM(50),TBETA(50),BIMING(50)
      REAL EXPLAN(20)
      INTEGER  MP(50),IDELTA(50)
      INTEGER  M1(50),M2(50),MA(50),ML(50),MINT(50),MFIN(50)
      INTEGER  SVSEED,SVSOLD,RHSEED
      INTEGER  MBOUND,NTVCON,SVTVEC,TVSTOP,LVCONT,ERCONT,TFLAG
      DOUBLE PRECISION  FINPRO
      DOUBLE PRECISION  DABS, DMAX1, DSQRT, DFLOAT
      REAL ABS
      INTEGER  IABS
      EXTERNAL BSOLV
C-----------------------------------------------------------------------
      DATA MACHEP/Z3410000000000000/
      EPSM = 2.D0*MACHEP
C-----------------------------------------------------------------------
C     ARRAYS MUST BE DIMENSIONED AS FOLLOWS:
C     1.  ALPHA:  >= KMAXN,  BETA: >= (KMAXN+1) WHERE KMAXN, THE
C                 LARGEST SIZE T-MATRIX CONSIDERED BY THE PROGRAM,
```

```
C               IS THE LARGER OF THE SIZE OF THE ALPHA, BETA HISTORY
C               PROVIDED ON FILE 2 (IF ANY ) AND THE SIZE WHICH THE
C               PROGRAM SPECIFIES INTERNALLY, THIS LATTER IS ALWAYS
C               < = 11*MEV / 8  +  12, WHERE MEV IS THE SIZE
C               T-MATRIX THAT WAS USED IN THE CORRESPONDING EIGENVALUE
C               COMPUTATIONS.
C     2.  V1:  >= MAX(N,KMAX)
C     3.  V2, VS:   >= N
C     4.  G:  >= MAX(N,KMAX)
C     5.  RITVEC:  >= N*NGOOD, WHERE NGOOD IS NUMBER OF EIGENVALUES
C                 SUPPLIED TO THIS PROGRAM.
C     6.  TVEC:  >= CUMULATIVE LENGTH OF ALL THE T-EIGENVECTORS
C         NEEDED TO GENERATE THE DESIRED RITZ VECTORS.  AN EDUCATED
C         GUESS AT AN APPROPRIATE LENGTH CAN BE OBTAINED BY RUNNING THE
C         PROGRAM WITH THE FLAG MBOUND = 1 AND MULTIPLYING THE
C         RESULTING SIZE BY 5/4.
C     7.  GOODA, GOODBI, EVNEW, AMINGP, TMINGP, TERR, RNORM,
C         TBETA, TLAST, BIERR, BIERRG, MP, MA, M1, M2, MINT,
C         MFIN AND IDELTA MUST BE OF DIMENSION AT LEAST NGOOD.
C
C-----------------------------------------------------------------------
C     OUTPUT HEADER
      WRITE(6,10)
   10 FORMAT(/' LANCZOS PROCEDURE FOR FACTORED INVERSES OF REAL SYMMETRI
     1C MATRICES'/'      COMPUTE EIGENVECTORS'/)
C
C     SET PROGRAM PARAMETERS
C     USER MUST NOT MODIFY SCALEO
      SCALEO = 5.0D0
      ZERO = 0.0D0
      ONE = 1.0D0
      MPMIN = -1000
C     CONVERGENCE TOLERANCE FOR T-EIGENVECTORS FOR RITZ COMPUTATIONS
      ERTOL = 1.D-10
C
C     READ USER-SPECIFIED PARAMETER FROM INPUT FILE 5 (FREE FORMAT)
C
C     READ USER-PROVIDED HEADER FOR RUN
      READ(5,20) EXPLAN
      WRITE(6,20) EXPLAN
   20 FORMAT(20A4)
C
C     READ IN MATNO = MATRIX/RUN IDENTIFICATION NUMBER AND
C     N = ORDER OF A-MATRIX
C     READ IN SCALE (SO) AND SHIFT (SHIFT) APPLIED TO GIVEN
C     MATRIX AND FLAG JPERM.  JPERM = (0,1):
C     JPERM = 1 MEANS THAT A-MATRIX HAS BEEN PERMUTED.
C
      READ(5,20) EXPLAN
      READ(5,*) MATNO,N,SO,SHIFT,JPERM
C
C     READ IN THE MAXIMUM PERMISSIBLE DIMENSIONS FOR THE TVEC ARRAY
C     (MDIMTV), FOR THE RITVEC ARRAY (MDIMRV), AND FOR THE BETA
C     ARRAY (MBETA).
C
      READ(5,20) EXPLAN
      READ(5,*) MDIMTV, MDIMRV, MBETA
C
C     READ IN RELATIVE TOLERANCE (RELTOL) USED IN DETERMINING
C     APPROPRIATE SIZES FOR THE T-MATRICES USED IN THE EIGENVECTOR
C     COMPUTATIONS
C
      READ(5,20) EXPLAN
      READ(5,*) RELTOL
C
C     SET FLAGS TO 0 OR 1:
C     MBOUND = 1:  PROGRAM TERMINATES AFTER COMPUTING 1ST GUESSES
```

```
C                   ON APPROPRIATE T-SIZES FOR USE IN THE RITZ VECTOR
C                   COMPUTATIONS
C     NTVCON = 0:   PROGRAM TERMINATES IF THE TVEC ARRAY IS NOT
C                   LARGE ENOUGH TO HOLD ALL THE T-EIGENVECTORS REQUIRED.
C     SVTVEC = 0:   THE T-EIGENVECTORS ARE NOT WRITTEN TO FILE 11
C                   UNLESS TVSTOP = 1
C     SVTVEC = 1:   WRITE THE T-EIGENVECTORS TO FILE 11.
C     TVSTOP = 1:   PROGRAM TERMINATES AFTER COMPUTING THE
C                   T-EIGENVECTORS
C     LVCONT = 0:   PROGRAM TERMINATES IF THE NUMBER OF T-EIGENVECTORS
C                   COMPUTED IS NOT EQUAL TO THE NUMBER OF RITZ
C                   VECTORS REQUESTED.
C     ERCONT = 0:   MEANS FOR ANY GIVEN EIGENVALUE, A RITZ VECTOR
C                   WILL NOT BE COMPUTED FOR THAT EIGENVALUE UNLESS
C                   A T-EIGENVECTOR HAS BEEN IDENTIFIED WITH A LAST
C                   COMPONENT WHICH SATISFIES THE SPECIFIED
C                   CONVERGENCE CRITERION.
C     ERCONT = 1:   MEANS FOR ANY GIVEN EIGENVALUE, A RITZ VECTOR
C                   WILL BE COMPUTED.  IF A T-EIGENVECTOR CANNOT
C                   BE IDENTIFIED WHICH SATISFIES THE LAST
C                   COMPONENT CRITERION, THEN THE PROGRAM WILL
C                   USE THE T-VECTOR THAT CAME CLOSEST TO
C                   SATISFYING THE CRITERION
C     IWRITE = 1:   EXTENDED OUTPUT OF INTERMEDIATE COMPUTATIONS
C                   IS WRITTEN TO FILE 6
C     IREAD = 0:    ALPHA/BETA FILE IS REGENERATED.
C     IREAD = 1:    ALPHA/BETA FILE USED IN EIGENVALUE COMPUTATIONS
C                   IS READ IN AND EXTENDED IF NECESSARY.  IN BOTH
C                   CASES IREAD = 0 OR 1, THE LANCZOS VECTORS ARE
C                   ALWAYS REGENERATED FOR THE RITZ VECTOR
C                   COMPUTATIONS
C
      READ(5,20) EXPLAN
      READ(5,*) MBOUND,NTVCON,SVTVEC,IREAD
C
      READ(5,20) EXPLAN
      READ(5,*) TVSTOP,LVCONT,ERCONT,IWRITE
      IF (TVSTOP.EQ.1) SVTVEC = 1
C
C     READ IN SEED (RHSEED) FOR GENERATING RANDOM STARTING VECTOR
C     FOR THE INVERSE ITERATION ON THE T-MATRICES.
C
      READ(5,20) EXPLAN
      READ(5,*) RHSEED
C
C-----------------------------------------------------------------------
C     INITIALIZE THE ARRAYS THAT DEFINE THE FACTORIZATION OF
C     THE B-MATRIX AND PASS THE STORAGE LOCATIONS OF THESE ARRAYS
C     TO THE SUBROUTINE BSOLV.
C
      CALL USPEC(N,MATNO)
C-----------------------------------------------------------------------
C     MASK UNDERFLOW AND OVERFLOW
C
      CALL MASK
C-----------------------------------------------------------------------
C     WRITE RUN PARAMETERS OUT TO FILE 6
C
      WRITE(6,30) MATNO,N,JPERM
   30 FORMAT(/4X,'MATRIX IDENTIFICATION NO.',4X,'SIZE OF A-MATRIX',4X,
     1'JPERM'/I29,I21,I9)
C
      WRITE(6,40) SO,SHIFT
   40 FORMAT(/4X,'SCALE APPLIED TO MATRIX',4X,'SHIFT APPLIED TO MATRIX'/
     1E27.4,E27.4)
C
      WRITE(6,50) MBOUND,NTVCON,SVTVEC,IREAD
```

```
   50 FORMAT(/3X,'MBOUND',3X,'NTVCON',3X,'SVTVEC',3X,'IREAD'/3I9,I8/)
C
      WRITE(6,60) TVSTOP,LVCONT,ERCONT,IWRITE
   60 FORMAT(3X,'TVSTOP',3X,'LVCONT',3X,'ERCONT',3X,'IWRITE'/4I9)
C
      WRITE(6,70) MDIMTV,MDIMRV,MBETA
   70 FORMAT(/3X,'MDIMTV',3X,'MDIMRV',3X,'MBETA'/2I9,I8)
C
      WRITE(6,80) RELTOL,RHSEED
   80 FORMAT(/7X,'RELTOL',3X,'RHSEED'/E13.4,I9)
C
C     FROM FILE 3 READ IN THE NUMBER OF EIGENVALUES (NGOOD) FOR WHICH
C     EIGENVECTORS ARE REQUESTED, THE ORDER (MEV) OF THE LANCZOS
C     TRIDIAGONAL MATRIX USED IN COMPUTING THESE EIGENVALUES, THE
C     ORDER (NOLD) OF THE USER-SPECIFIED MATRIX USED IN THE EIGENVALUE
C     COMPUTATIONS, THE SEED (SVSEED) USED FOR GENERATING THE STARTING
C     VECTOR THAT WAS USED IN THOSE LANCZOS EIGENVALUE COMPUTATIONS,
C     AND THE MATRIX/RUN IDENTIFICATION NUMBER (MATOLD) USED IN THOSE
C     COMPUTATIONS.  ALSO READ IN THE NUMBER (NDIS) OF DISTINCT
C     EIGENVALUES OF T(1,MEV) THAT WERE COMPUTED BUT THIS VALUE IS
C     NOT USED IN THE EIGENVECTOR COMPUTATIONS.
C
      READ(3,90) NGOOD,NDIS,MEV,NOLD,SVSEED,MATOLD
   90 FORMAT(4I6,I12,I8)
C
C     READ IN THE MULTIPLICITY TOLERANCE USED IN THE BISEC SUBROUTINE
C     DURING THE COMPUTATION OF THE GIVEN EIGENVALUES.
C     ALSO READ IN THE FLAG IB.  IF IB < 0, THEN SOME BETA(I) IN THE
C     T-MATRIX FILE PROVIDED ON FILE 2 FAILED THE ORTHOGONALITY
C     TEST IN THE TNORM SUBROUTINE.  USER SHOULD NOTE THAT THIS VECTOR
C     PROGRAM PROCEEDS INDEPENDENTLY OF THE SIZE OF THE BETA USED.
C
      READ(3,100) MULTOL,IB,BTOL,SHIFTO
  100 FORMAT(E20.12,I6,2E10.3)
C
      TEMP = DFLOAT(MEV+1000)
      TTOL = MULTOL/TEMP
C
      WRITE(6,110) MULTOL,TTOL
  110 FORMAT(/' T-MULTIPLICITY TOLERANCE USED IN THE EIGENVALUE COMPUTAT
     1IONS WAS',E13.4/' SCALED MACHINE EPSILON TTOL IS',E13.4)
C
C     CONTINUE WRITE TO FILE 6 OF THE PARAMETERS FOR THIS RUN
C
      NG = NGOOD
      WRITE(6,120)NG,NDIS,MEV,NOLD,MATOLD,SVSEED,IB,MULTOL,BTOL,SHIFTO
  120 FORMAT(/' EIGENVALUES ARE READ IN FROM FILE 3.  THE HEADER IS'/
     1 4X,'NG',2X,'NDIS',3X,'MEV',2X,'NOLD',2X,'MATOLD',6X,'SVSEED'/
     1 4I6,I8,I12/
     1 6X,'IB',6X,'MULTOL',8X,'BTOL',6X,'SHIFTO'/
     1 I8,E12.3,E12.3,E12.3/)
C
C     IS THE ARRAY RITVEC LONG ENOUGH TO HOLD ALL OF THE DESIRED
C     RITZ VECTORS (APPROXIMATE EIGENVECTORS)?
C
      NMAX = NGOOD*N
      IF(MBOUND.EQ.1) GO TO 130
      IF(TVSTOP.NE.1.AND.NMAX.GT.MDIMRV) GO TO 1430
C
C     CHECK THAT THE ORDER N AND THE MATRIX IDENTIFICATION NUMBER
C     MATNO SPECIFIED BY THE USER AGREE WITH THOSE READ IN FROM FILE 3.
C
  130 ITEMP = (NOLD-N)**2+(MATOLD-MATNO)**2
      IF (ITEMP.NE.0.OR.SHIFTO.NE.SHIFT) GO TO 1450
C
C     READ IN FROM FILE 3, THE T-MULTIPLICITIES OF THE EIGENVALUES
C     WHOSE EIGENVECTORS ARE TO BE COMPUTED, THE VALUES OF THESE
```

```
C     EIGENVALUES AND THEIR MINIMAL GAPS AS EIGENVALUES OF THE
C     USER-SPECIFIED MATRIX AND AS EIGENVALUES OF THE T-MATRIX.
C
      READ(3,20) EXPLAN
      READ(3,140) (MP(J),GOODBI(J),BIMING(J),AMINGP(J),TMINGP(J),
     1 J = 1,NGOOD)
  140 FORMAT(5X,I5,E25.16,3E10.3)
C
C
      DO 150 J=1,NGOOD
  150 GOODA(J) = (ONE/GOODBI(J) - SHIFT)/SO
C
      WRITE(6,160) (J,GOODA(J),MP(J),GOODBI(J), J=1,NGOOD)
  160 FORMAT(/' EIGENVALUES READ IN, T-MULTIPLICITIES'/
     1 4X,' J ',5X,' A-EIGENVALUE',6X,'TMULT',3X,'B-INVERSE EIGENVALUE'/
     1(I6,E25.16,I4,E25.16))
      WRITE(6,170) (J,GOODBI(J),TMINGP(J),BIMING(J), J=1,NGOOD)
  170 FORMAT(/' B(INVERSE) EIGENVALUES READ IN, T-GAPS AND B(INVERSE)-GA
     1PS'/4X,' J ',3X,'B-INVERSE EIGENVALUE',6X,'  TMINGAP ',6X,
     1' BIMINGAP '/(I6,E25.16,2E15.4))
      WRITE(6,180) (J,GOODA(J),AMINGP(J), J=1,NGOOD)
  180 FORMAT(/' A-EIGENVALUES READ IN AND A-GAPS'/
     1 4X,' J ',5X,'A-EIGENVALUE',10X,'  AMINGAP '
     1/(I6,E25.16,E15.4))
C
C     READ IN ERROR ESTIMATES
      WRITE(6,210) MEV,SVSEED
C     CHECK WHETHER OR NOT THERE ARE ANY T-ISOLATED EIGENVALUES IN
C     THE EIGENVALUES PROVIDED
      DO 190 J=1,NGOOD
      IF(MP(J).EQ.1) GO TO 200
  190 CONTINUE
      GO TO 230
  200 READ(4,20) EXPLAN
      READ(4,20) EXPLAN
      READ(4,20) EXPLAN
  210 FORMAT(/' THESE EIGENVALUES WERE COMPUTED USING A T-MATRIX OF
     1ORDER ',I5/' AND SEED FOR RANDOM NUMBER GENERATOR =',I12)
      READ(4,220) NISO
  220 FORMAT(18X,I6)
      READ(4,20) EXPLAN
      READ(4,20) EXPLAN
      READ(4,20) EXPLAN
  230 DO 260 J=1,NGOOD
      BIERR(J) = 0.D0
      IF(MP(J).NE.1) GO TO 260
      READ(4,240) EVAL, BIERR(J)
  240 FORMAT(10X,E25.16,E14.3)
      IF(DABS(EVAL - GOODBI(J)).LT.1.D-10) GO TO 260
      WRITE(6,250) EVAL,GOODBI(J)
  250 FORMAT(' PROBLEM WITH READ IN OF ERROR ESTIMATES'/' EIGENVALUE REA
     1D IN',E20.12,' DOES NOT MATCH GOODBI(J) ='/E20.12)
      GO TO 1670
C
  260 CONTINUE
C
      WRITE(6,270) (J,GOODBI(J),BIERR(J), J=1,NGOOD)
  270 FORMAT(' B(INVERSE) ERROR ESTIMATES '/4X,' J',5X,'EIGENVALUE',10X
     1,'ESTIMATE'/(I6,E20.12,E14.3))
C
      IF(IREAD.EQ.0)  GO TO 370
C
C     READ IN THE SIZE OF THE T-MATRIX PROVIDED ON FILE 2.  READ IN
C     THE ORDER OF THE USER-SPECIFIED MATRIX , THE SEED FOR THE
C     RANDOM NUMBER GENERATOR, AND THE MATRIX/TEST IDENTIFICATION
C     NUMBER THAT WERE USED IN THE LANCZOS EIGENVALUE COMPUTATIONS.
```

```
C      IF FLAG IREAD = 0, REGENERATE ALPHA,BETA ARRAYS
C
       READ(2,280) KMAX,NOLD,SVSOLD,MATOLD,SHIFTO
  280 FORMAT(2I6,I12,I8,E13.4)
C
       WRITE(6,290) KMAX,NOLD,SVSOLD,MATOLD,SHIFTO
  290 FORMAT(/' READ IN THE T-MATRICES STORED ON FILE 2'/' FILE 2 HEADER
     1 IS'/2X,'KMAX',2X,'NOLD',6X,'SVSOLD',2X,'MATOLD',4X,'SHIFTO'/
     1 2I6,I12,I8,E10.3/)
C
C      CHECK THAT THE ORDER, THE MATRIX/TEST IDENTIFICATION NUMBER
C      AND THE SEED FOR THE RANDOM NUMBER GENERATOR USED IN THE
C      LANCZOS COMPUTATIONS THAT GENERATED THE ALPHA,BETA FILE
C      BEING USED AGREE WITH WHAT THE USER HAS SPECIFIED.
       IF (NOLD.NE.N.OR.MATOLD.NE.MATNO.OR.SVSOLD.NE.SVSEED) GO TO 1470
C
       KMAX1 = KMAX + 1
C
C      READ IN THE T-MATRICES FROM FILE 2.  THESE ARE USED TO GENERATE
C      THE T-EIGENVECTORS THAT WILL BE USED IN THE RITZ VECTOR
C      COMPUTATIONS.  ALPHA,BETA MUST BE STORED IN MACHINE FORMAT
C      ((4Z20) ON IBM/3081)
C
       READ(2,300) (ALPHA(J), J=1,KMAX)
       READ(2,300) (BETA(J), J=1,KMAX1)
  300 FORMAT(4Z20)
C
       READ(2,300) (V1(J), J=1,N)
       READ(2,300) (V2(J), J=1,N)
C
C      ENLARGE KMAX IF THE SIZE AT WHICH THE EIGENVALUE
C      COMPUTATIONS WERE PERFORMED IS ESSENTIALLY KMAX AND
C      THERE IS AT LEAST ONE EIGENVALUE THAT IS T-SIMPLE AND
C      T-ISOLATED IN THE SENSE THAT IF ITS NEAREST NEIGHBOR IS
C      TOO CLOSE THEN THAT NEIGHBOR IS A GOOD T-EIGENVALUE.
       DO 310 J = 1,NGOOD
       IF(MP(J).EQ.1) GO TO 330
  310 CONTINUE
       WRITE(6,320)
  320 FORMAT(/' ALL EIGENVALUES USED ARE T-MULTIPLE OR CLOSE TO SPURIOUS
     1 T-EIGENVALUES'/' SO DO NOT CHANGE KMAX')
       IF(KMAX.LT.MEV) GO TO 1490
       GO TO 350
C
  330 KMAXN= 11*MEV/8 + 12
       IF(MBETA.LE.KMAXN) GO TO 1650
       IF(KMAX.GE.KMAXN )  GO TO 350
       WRITE(6,340) KMAX, KMAXN
  340 FORMAT(' ENLARGE KMAX FROM ',I6,' TO ',I6)
       MOLD1 = KMAX + 1
       KMAX = KMAXN
       GO TO 420
C
  350 WRITE(6,360) KMAX
  360 FORMAT(/' T-MATRICES HAVE BEEN READ IN FROM FILE 2'/' THE LARGEST
     1SIZE T-MATRIX ALLOWED IS',I6/)
C
       IF(IREAD.EQ.1) GO TO 440
C
C      REGENERATE THE ALPHA AND BETA
C
  370 MOLD1 = 1
C
       DO 380 J = 1,NGOOD
       IF(MP(J).EQ.1) GO TO 400
  380 CONTINUE
       KMAX = MEV + 12
```

```
      WRITE(6,390) KMAX
  390 FORMAT(/' ALL EIGENVALUES FOR WHICH EIGENVECTORS ARE TO BE COMPUTE
     1D ARE EITHER T-MULTIPLE OR CLOSE TO'/' A SPURIOUS T-EIGENVALUE. TH
     1EREFORE SET KMAX = MEV + 12 = ',I7)
      GO TO 420
C
  400 KMAXN = 11*MEV/8 + 12
      IF(MBETA.LE.KMAXN) GO TO 1650
      WRITE(6,410) KMAXN
  410 FORMAT(' SET KMAX EQUAL TO ',I6)
      KMAX = KMAXN
C
  420 WRITE(6,430) MOLD1,KMAX
  430 FORMAT(/' LANCZS SUBROUTINE GENERATES ALPHA(J), BETA(J+1), J =',
     1 I6,' TO ', I6/)
C
C-----------------------------------------------------------------------
C
      CALL LANCZS(BSOLV,ALPHA,BETA,V1,V2,VS,G,KMAX,MOLD1,N,SVSEED)
C
C-----------------------------------------------------------------------
C
  440 CONTINUE
C
C     THE SUBROUTINE STURMI DETERMINES THE SMALLEST SIZE T-MATRIX FOR
C     WHICH THE EIGENVALUE IN QUESTION IS AN EIGENVALUE (TO WITHIN A
C     GIVEN TOLERANCE) AND IF POSSIBLE THE SMALLEST SIZE T-MATRIX
C     FOR WHICH IT IS A DOUBLE EIGENVALUE (TO WITHIN THE SAME
C     TOLERANCE).  THE SIZE T-MATRIX USED IN THE EIGENVECTOR
C     COMPUTATIONS IS THEN DETERMINED BY LOOPING ON SIZE OF THE
C     T-EIGENVECTORS, USING THE VALUES FROM STURMI TO DETERMINE
C     FIRST GUESSES AT THE APPROPRIATE T-SIZES.
C
C
      STUTOL = SCALE0*MULTOL
      IF(IWRITE.EQ.1) WRITE(6,450)
  450 FORMAT(' FROM STURMI')
      DO 490 J = 1,NGOOD
      EVAL = GOODBI(J)
C     COMPUTE THE TOLERANCES USED BY STURMI TO DETERMINE AN INTERVAL
C     CONTAINING THE EIGENVALUE EVAL.
      TEMP = DABS(EVAL)*RELTOL
      TOLN = DMAX1(TEMP,STUTOL)
C
C-----------------------------------------------------------------------
C
      CALL STURMI(ALPHA,BETA,EVAL,TOLN,EPSM,KMAX,MK1,MK2,IC,IWRITE)
C
C-----------------------------------------------------------------------
C
C     STORE THE COMPUTED ORDERS OF T-MATRICES FOR LATER PRINTOUT
      M1(J) = MK1
      M2(J) = MK2
      ML(J) = (MK1 + 3*MK2)/4
      IF(MK2.EQ.KMAX)  ML(J) = KMAX
C
      IF(IC.GT.0) GO TO 470
C     IC = 0 MEANS THERE WAS NO T-EIGENVALUE IN THE DESIGNATED INTERVAL
C     BY T-SIZE KMAX.  THIS MEANS THAT THE EIGENVALUE PROVIDED HAS
C     NOT YET CONVERGED SO ITS EIGENVECTOR SHOULD NOT BE COMPUTED.
      WRITE(6,460) J,GOODBI(J),MK1,MK2
  460 FORMAT(I6,'TH EIGENVALUE',E20.12,' HAS NOT CONVERGED '/
     1' SO DO NOT COMPUTE ANY T-EIGENVECTOR OR RITZ VECTOR FOR IT'
     1/' MK1 AND MK2 FOR THIS EIGENVALUE WERE',2I6)
      MP(J) = MPMIN
      MA(J) = -2*KMAX
```

```
      GO TO 490
C     COMPUTE AN APPROPRIATE SIZE T-MATRIX FOR THE GIVEN EIGENVALUE.
  470 IF(M2(J).EQ.KMAX) GO TO 480
C     M1 AND M2 WERE BOTH DETERMINED
      MA(J) = (3*M1(J) + M2(J))/4 + 1
      GO TO 490
C     M2 NOT DETERMINED
  480 MA(J) = (5*M1(J))/4 + 1
C
  490 CONTINUE
C
      IF (IWRITE.EQ.1) WRITE(6,500) (MA(JJ), JJ=1,NGOOD)
  500 FORMAT(/' 1ST GUESS AT APPROPRIATE SIZE T-MATRICES'/
     1 ' ACTUAL VALUES WILL PROBABLY BE 1/4 AGAIN AS MUCH'/(13I6))
C
C     PRINT OUT TO FILE 10 1ST GUESSES AT SIZES OF THE T-MATRICES TO
C     BE USED IN THE EIGENVECTOR COMPUTATIONS.
C     ACTUAL VALUES USED MAY BE 1/4 OR MORE LARGER THAN THESE VALUES.
      WRITE(10,510) N,KMAX
  510 FORMAT(2I8,' = ORDER OF USER MATRIX AND MAX ORDER OF T(1,MEV)')
C
      WRITE(10,520)
  520 FORMAT(/' 1ST GUESS AT APPROPRIATE SIZE T-MATRICES'/
     1 ' ACTUAL VALUES WILL PROBABLY BE 1/4 AGAIN AS MUCH'/)
C
      WRITE(10,530)
  530 FORMAT(4X,'J',7X,'GOODBI(J)',4X,'M1(J)',1X,'M2(J)',1X,'MA(J)')
C
      WRITE(10,540) (J,GOODBI(J),M1(J),M2(J), MA(J), J=1,NGOOD)
  540 FORMAT(I5,E19.12,3I6)
C
      IF(MBOUND.EQ.1) WRITE(10,550)
  550 FORMAT(/' EV = GOODBI(J) IS A GOOD EIGENVALUE OF T(1,MEV)'/
     1 ' M1 = SMALLEST VALUE OF M SUCH THAT T(1,M) HAS AT LEAST'/
     1 '      ONE EIGENVALUE IN THE INTERVAL (EV-TOLN,EV+TOLN)'/
     1 ' TOLN(J) = DMAX1(GOODBI(J)*RELTOL, SCALEO*MULTOL)'/
     1 ' M2 = SMALLEST M (IF ANY) SUCH THAT IN THE ABOVE INTERVAL'/
     1 '      T(1,M) HAS AT LEAST TWO EIGENVALUES '/
     1 ' INITIAL VALUE OF MA(J) IS CHOSEN HEURISTICALLY'/
     1 ' PROGRAM LOOPS ON SIZE OF T-MATRIX TO GET APPROPRIATE SIZE'/
     1 ' END OF SIZES OF T-MATRICES FILE 10'///)
C
C
C     TERMINATE AFTER COMPUTING 1ST GUESSES AT  SIZES OF THE
C     T-MATRICES REQUIRED FOR THE GIVEN EIGENVALUES?
      IF(MBOUND.EQ.1) GO TO 1510
C
C
C     WILL THERE BE ROOM FOR ALL OF THE REQUESTED T-EIGENVECTORS?
      MTOL = 0
      DO 560 J = 1,NGOOD
      IF(MP(J).EQ.MPMIN) GO TO 560
      MTOL = MTOL + IABS(MA(J))
  560 CONTINUE
      MTOL = (5*MTOL)/4
      IF(MTOL.GT.MDIMTV.AND.NTVCON.EQ.0) GO TO 1530
C
C-----------------------------------------------------------------------
C     GENERATE A RANDOM VECTOR TO BE USED REPEATEDLY BY
C     SUBROUTINE INVERM
C
      IIL = RHSEED
      CALL GENRAN(IIL,G,KMAX)
C
C-----------------------------------------------------------------------
C
C     FOR EACH EIGENVALUE LOOP ON T-EIGENVECTOR COMPUTATIONS TO
```

```
C     COMPUTE AN APPROPRIATE T-EIGENVECTOR TO USE IN THE RITZ
C     VECTOR COMPUTATIONS.
C
      MTOL = 0
      NTVEC = 0
      ILBIS = 0
      DO 750 J = 1,NGOOD
      ICOUNT = 0
      ERRMIN = 10.D0
      MABEST = MPMIN
      IF(MP(J).EQ.MPMIN) GO TO 750
      TFLAG = 0
      EVAL = GOODBI(J)
      TEMP = RELTOL*DABS(EVAL)
      UB = EVAL + DMAX1(STUTOL,TEMP)
      LB = EVAL - DMAX1(STUTOL,TEMP)
  570 KMAXU = IABS(MA(J))
C
C     SELECT A SUITABLE INCREMENT FOR THE ORDERS OF THE T-MATRICES
C     TO BE CONSIDERED IN DETERMINING APPROPRIATE SIZES FOR THE RITZ
C     VECTOR COMPUTATIONS.
      IF(ICOUNT.GT.0) GO TO 590
C     SELECT IDELTA(J) BASED UPON THE T-MULTIPLICITY OBTAINED
      IF(M2(J).EQ.KMAX) GO TO 580
C     M2 DETERMINED
      IDELTA(J) = ((3*M1(J) + 5*M2(J))/8  +  1 - IABS(MA(J)))/10 +  1
      GO TO 590
C     M2 NOT DETERMINED
  580 MAMAX = MIN0((11*MEV)/8 + 12, (13*M1(J))/8 + 1)
      IDELTA(J) = (MAMAX - IABS(MA(J)))/10 +  1
  590 ICOUNT = ICOUNT + 1
C
C-----------------------------------------------------------------------
C     TO MIMIMIZE THE EFFECT OF THE ONE-SIDED ACCEPTANCE TEST FOR
C     EIGENVALUES IN THE BISEC SUBROUTINE, RECOMPUTE THE GIVEN
C     EIGENVALUE AT THE SPECIFIED KMAXU
C
      CALL LBISEC(ALPHA,BETA,EPSM,EVAL,EVALN,LB,UB,TTOL,KMAXU,NEVT)
C
C-----------------------------------------------------------------------
C
C     CHECK WHETHER OR NOT GIVEN T-MATRIX HAS AN EIGENVALUE IN THE
C     SPECIFIED INTERVAL AND IF SO WHAT ITS T-MULTIPLICITY IS.
C
      IF(NEVT.EQ.1) GO TO 630
      IF(NEVT.NE.0) GO TO 610
      ILBIS = 1
      WRITE(6,600) EVAL,KMAXU
  600 FORMAT(/' PROBLEM ENCOUNTERED IN RECOMPUTATION OF USER-SUPPLIED EI
     1GENVALUE',E20.12/' THE SIZE T-MATRIX SPECIFIED',I6,' DOES NOT
     1HAVE AN EIGENVALUE IN THE INTERVAL SPECIFIED'/' THEREFORE NO EIGEN
     1VECTOR WILL BE COMPUTED FOR THIS PARTICULAR EIGENVALUE'/)
      GO TO 650
C
  610 IF(NEVT.GT.1)  WRITE(6,620) EVAL,KMAXU
  620 FORMAT(/' PROBLEM ENCOUNTERED IN RECOMPUTATION OF USER-SUPPLIED
     1EIGENVALUE',E20.12/' FOR THE SIZE T-MATRIX SPECIFIED =',I6,' THE
     1GIVEN EIGENVALUE IS MULTIPLE IN THE INTERVAL SPECIFIED'/' SOMETHIN
     1G IS WRONG, THEREFORE NO EIGENVECTOR WILL BE COMPUTED FOR THIS EIG
     1NVALUE'/)
C
      MP(J) = MPMIN
      MA(J) = -2*KMAX
      GO TO 750
C
  630 CONTINUE
```

```
      ILBIS = 0
C
      EVNEW(J) = EVALN
      EVAL = EVALN
      MTOL = MTOL+KMAXU
C
C     IS THERE ROOM IN TVEC ARRAY FOR THE NEXT T-EIGENVECTOR?
C     IF NOT, SKIP TO RITZ VECTOR COMPUTATIONS.
      IF (MTOL.GT.MDIMTV) GO TO 760
C
      IT = 3
      KINT = MTOL - KMAXU +1
C
C     RECORD THE BEGINNING AND END OF THE T-EIGENVECTOR BEING COMPUTED
      MINT(J) = KINT
      MFIN(J) = MTOL
C
C-----------------------------------------------------------------------
C     SUBROUTINE INVERM DOES INVERSE ITERATION, I.E. SOLVES
C     (T(1,KMAXU) - EVAL)*U = RHS  FOR EACH EIGENVALUE TO OBTAIN
C     THE DESIRED T-EIGENVECTOR.
C
      IF(IWRITE.EQ.1)   WRITE(6,640) J
  640 FORMAT(/I6,'TH EIGENVALUE')
C
      CALL INVERM(ALPHA,BETA,V1,TVEC(KINT),EVAL,ERROR,TERROR,EPSM,
     1 G,KMAXU,IT,IWRITE)
C
C-----------------------------------------------------------------------
C
      TERR(J) = TERROR
      TLAST(J) = ERROR
      KMAXU1 = KMAXU + 1
      TBETA(J) = BETA(KMAXU1)*ERROR
C
C     AFTER COMPUTING EACH OF THE T-EIGENVECTORS,
C     CHECK THE SIZE OF THE ERROR ESTIMATE, ERROR.
C     IF THIS ESTIMATE IS NOT AS SMALL AS DESIRED AND
C     |MA(J)| < ML(J), ATTEMPT TO INCREASE THE SIZE OF |MA(J)|
C     AND REPEAT THE T-EIGENVECTOR COMPUTATIONS.
C
      IF(ERROR.LT.ERTOL.OR.TFLAG.EQ.1)  GO TO 740
C
      IF(ERROR.GE.ERRMIN) GO TO 650
C     LAST COMPONENT IS LESS THAN MINIMAL TO DATE
      ERRMIN = ERROR
      MABEST = MA(J)
  650 CONTINUE
C
      IF(MA(J).GT.0)  ITEST = MA(J) + IDELTA(J)
      IF(MA(J).LT.0)  ITEST = -(IABS(MA(J)) + IDELTA(J))
      IF(IABS(ITEST).LE.ML(J).AND.ICOUNT.LE.10) GO TO 670
C     NEW MA(J) IS GREATER THAN MAXIMUM ALLOWED.
      IF(ERCONT.EQ.0.OR.MABEST.EQ.MPMIN) GO TO 690
      TFLAG = 1
      MA(J) = MABEST
      IF(ILBIS.EQ.0) MTOL = MTOL - KMAXU
      WRITE(6,660) MA(J)
  660 FORMAT(' 10 ORDERS WERE CONSIDERED.  NONE SATISFIED THE ERROR TEST
     1'/' THEREFORE USE THE BEST ORDER OBTAINED FOR THE EIGENVECTORS'
     1,I6)
      GO TO 570
C
  670 MA(J) = ITEST
C
      MT = IABS(MA(J))
      IF(IWRITE.EQ.1) WRITE(6,680)  MT
```

```
  680 FORMAT(/' CHANGE SIZE OF T-MATRIX TO ',I6,' RECOMPUTE T-EIGENVECTO
     1R')
C
      IF(ILBIS.EQ.0)  MTOL = MTOL - KMAXU
C
      GO TO 570
C
C     APPROPRIATE SIZE T-MATRIX WAS NOT OBTAINED
  690 CONTINUE
      WRITE(10,700) J,EVAL,MP(J)
  700 FORMAT(/' ON 10 INCREMENTS NOT ABLE TO IDENTIFY APPROPRIATE SIZE
     1T-MATRIX FOR'/
     1' EIGENVALUE(',I4,') = ',E20.12,' T-MULTIPLICITY =',I4/)
      IF(M2(J).EQ.KMAX) WRITE(10,710)
      IF(M2(J).LT.KMAX) WRITE(10,720)
  710 FORMAT(' ORDERS TESTED RANGED FROM 5*M1(J)/4 TO APPROXIMATELY
     1 MIN(11*MEV/8,13*M1(J)/8)')
  720 FORMAT(' ORDERS TESTED RANGED FROM APPROX. (3*M1(J)+M2(J))/4 TO (3
     1*M1(J)+5*M2(J))/8')
      WRITE(10,730)
  730 FORMAT(' ALLOWING LARGER ORDERS FOR THE T-MATRICES MAY RESULT IN
     1 SUCCESS'/' BUT PROBABLY WILL NOT.  PROBLEM IS PROBABLY DUE TO'
     1 /' LACK OF CONVERGENCE OF GIVEN EIGENVALUE, CHECK THE ERROR ESTIM
     1ATE')
      MP(J) = MPMIN
      IF(ILBIS.EQ.0)  MTOL = MTOL - KMAXU
      GO TO 750
  740 NTVEC = NTVEC + 1
C
  750 CONTINUE
      NGOODC = NGOOD
      GO TO 780
C
C     COME HERE IF THERE IS NOT ENOUGH ROOM FOR ALL OF T-EIGENVECTORS
  760 NGOODC = J-1
      WRITE(6,770)  J,MTOL,MDIMTV
  770 FORMAT(/' NOT ENOUGH ROOM IN TVEC ARRAY FOR ',I4,'TH T-EIGENVECTOR
     1'/'  TVEC DIMENSION REQUESTED = ',I6,' BUT TVEC HAS DIMENSION ',I6
     1/)
      IF(NGOODC.EQ.0)  GO TO 1550
      MTOL = MTOL-KMAXU
C
  780 CONTINUE
C
C     THE LOOP ON T-EIGENVECTOR COMPUTATIONS IS COMPLETE.
C     WRITE OUT THE SIZE T-MATRICES THAT WILL BE USED FOR
C     THE RITZ VECTOR COMPUTATIONS.
C
      WRITE(10,790)
  790 FORMAT(/' SIZES OF T-MATRICES THAT WILL BE USED IN THE RITZ COMPUT
     1ATIONS'/5X,'J',8X,'  GOODBI(J) ',13X,'  GOODA(J) ',7X,'MA(J)')
C
      WRITE(10,800)   (J,GOODBI(J),GOODA(J),MA(J), J=1,NGOOD)
  800 FORMAT(I6,2E25.14,I6)
      WRITE(10,550)
C
      WRITE(6,810) MTOL
  810 FORMAT(/' THE CUMULATIVE LENGTH OF THE T-EIGENVECTORS IS',I18)
C
      WRITE(6,820) NTVEC,NGOOD
  820 FORMAT(/I6,' T-EIGENVECTORS OUT OF',I6,' REQUESTED WERE COMPUTED')
C
C     SAVE THE T-EIGENVECTORS ON FILE 11?
      IF(TVSTOP.NE.1.AND.SVTVEC.EQ.0) GO TO 880
C
      WRITE(11,830) NTVEC,MTOL,MATNO,SVSEED
```

```
  830 FORMAT(I6,3I12,' = NTVEC,MTOL,MATNO,SVSEED')
C
      DO 860 J=1,NGOODC
C     IF MP(J) = MPMIN THEN NO SUITABLE T-EIGENVECTOR IS AVAILABLE
C     FOR THAT EIGENVALUE.
      IF(MP(J).EQ.MPMIN) WRITE(11,840) J,MA(J),GOODBI(J),MP(J)
  840 FORMAT(2I6,E20.12,I6/' TH EIGVAL,T-SIZE,EVALUE,FLAG,NO EIGVEC')
      IF(MP(J).NE.MPMIN) WRITE(11,850) J,MA(J),GOODBI(J),MP(J)
  850 FORMAT(I6,I6,E20.12,I6/'  T-EIGENVECTOR, T-SIZE , BI-EIGENVALUE, T
     1-MULTIPLICITY')
      IF(MP(J).EQ.MPMIN) GO TO 860
      KI = MINT(J)
      KF = MFIN(J)
C
      WRITE(11,300) (TVEC(K), K=KI,KF)
C
  860 CONTINUE
C
      IF(TVSTOP.NE.1) GO TO 880
C
      WRITE(6,870) TVSTOP, NTVEC,NGOOD
  870 FORMAT(/' USER SET TVSTOP = ',I1/
     1' THEREFORE PROGRAM TERMINATES AFTER T-EIGENVECTOR COMPUTATIONS'/
     1' T-EIGENVECTORS THAT WERE COMPUTED ARE SAVED ON FILE 11'/
     1I8,' T-EIGENVECTORS WERE COMPUTED OUT OF',I7,' REQUESTED'/)
C
      GO TO 1670
C
  880 CONTINUE
C     IF NOT ABLE TO COMPUTE ALL THE REQUESTED T-EIGENVECTORS
C     CONTINUE WITH THE LANCZOS VECTOR COMPUTATIONS ANYWAY?
      IF(NTVEC.NE.NGOOD.AND.LVCONT.EQ.0) GO TO 1570
C
C     COMPUTE THE MAXIMUM SIZE OF THE T-MATRIX USED FOR THOSE
C     EIGENVALUES WITH GOOD ERROR ESTIMATES.
C
      KMAXU = 0
      DO 890 J = 1,NGOODC
      MT = IABS(MA(J))
      IF(MT.LT.KMAXU.OR.MP(J).EQ.MPMIN) GO TO 890
      KMAXU = MT
  890 CONTINUE
C
      IF(KMAXU.EQ.0) GO TO 1610
C
      WRITE(6,900) KMAXU
  900 FORMAT(/I6,' = LARGEST SIZE T-MATRIX TO BE USED IN THE RITZ VECTOR
     1 COMPUTATIONS')
C
C     COUNT THE NUMBER OF RITZ VECTORS NOT BEING COMPUTED
      MREJEC = 0
      DO 910 J=1,NGOODC
  910 IF(MP(J).EQ.MPMIN)  MREJEC = MREJEC + 1
      MREJET = MREJEC + (NGOOD-NGOODC)
      IF(MREJET.NE.0) WRITE(6,920) MREJET
  920 FORMAT(/' RITZ VECTORS ARE NOT COMPUTED FOR',I6,' OF THE EIGNEVALU
     1ES'/)
      NACT = NGOODC - MREJEC
      WRITE(6,930) NGOOD,NTVEC,NACT
  930 FORMAT(/I6,' RITZ VECTORS WERE REQUESTED'/I6,' T-EIGENVECTORS WERE
     1 COMPUTED'/I6,' RITZ VECTORS WILL BE COMPUTED'/)
C     CHECK IF THERE ARE ANY RITZ VECTORS TO COMPUTE
      IF(MREJEC.EQ.NGOODC) GO TO 1590
C
C     CONTINUE WITH THE LANCZOS VECTOR COMPUTATIONS?
      IF(LVCONT.EQ.0.AND.MREJEC.NE.0) GO TO 1570
C
```

```
C     NOW COMPUTE THE RITZ VECTORS.  REGENERATE THE
C     LANCZOS VECTORS.
C
      DO 940 I = 1,NMAX
  940 RITVEC(I) = ZERO
C
C-----------------------------------------------------------------------
C     REGENERATE THE STARTING VECTOR. THIS MUST BE GENERATED AND
C     NORMALIZED PRECISELY THE WAY IT WAS DONE IN THE EIGENVALUE
C     COMPUTATIONS, OTHERWISE THERE WILL BE A MISMATCH BETWEEN
C     THE T-EIGENVECTORS THAT HAVE BEEN COMPUTED FROM THE T-MATRICES
C     READ IN FROM FILE 2 AND THE LANCZOS VECTORS THAT ARE
C     BEING REGENERATED.
C
      CALL GENRAN(SVSEED,G,N)
C
C-----------------------------------------------------------------------
C
      DO 950 J = 1,N
  950 V2(J) = G(J)
C
C-----------------------------------------------------------------------
      SUM = FINPRO(N,V2(1),1,V2(1),1)
C-----------------------------------------------------------------------
C
      SUM = ONE/DSQRT(SUM)
C
      DO 960 I = 1,N
      V1(I) = ZERO
  960 V2(I) = V2(I)*SUM
C
      WRITE(6,970)
  970 FORMAT(' STARTING LANCZOS VECTOR HAS BEEN CALCULATED'/)
C
C     LOOP FOR GENERATING RITZ VECTORS  (IVEC = 1,KMAXU)
      IVEC = 1
      BATA = ZERO
C
      GO TO 1050
C
  980 CONTINUE
C
C     SOLVE B*VS = V2 FOR VS
      DO 990 K = 1,N
  990 VS(K) = V2(K)
C
C-----------------------------------------------------------------------
      JBSOLV = 2
      CALL BSOLV(VS,VS,JBSOLV)
C-----------------------------------------------------------------------
C
C     VS = BI*V2   BI = B(INVERSE)
C     COMPUTE V1 = BI*V2 - BATA*V1
      DO 1000 K = 1,N
 1000 V1(K) = VS(K) - BATA*V1(K)
C
C-----------------------------------------------------------------------
      ALFA = FINPRO(N,V1(1),1,V2(1),1)
C-----------------------------------------------------------------------
C
      DO 1010 J = 1,N
 1010 V1(J) = V1(J)-ALFA*V2(J)
C
C-----------------------------------------------------------------------
      BATA = FINPRO(N,V1(1),1,V1(1),1)
C-----------------------------------------------------------------------
C
```

```
      BATA = DSQRT(BATA)
      SUM = ONE/BATA
C
      TEMP = BETA(IVEC)
      TEMP = DABS(BATA - TEMP)/TEMP
      IF (TEMP.LT.1.0D-10)GO TO 1030
C
C     THE BETA BEING REGENERATED DO NOT MATCH THE HISTORY FILE
C     SOMETHING IS WRONG IN THE LANCZOS VECTOR GENERATION
C     PROGRAM TERMINATES FOR USER TO CORRECT THE PROBLEM
C     WHICH MUST BE IN THE STARTING VECTOR GENERATION OR IN
C     THE MATRIX-VECTOR MULTIPLY SUBROUTINE CMATV SUPPLIED.
C     THIS SUBROUTINE MUST BE THE SAME ONE USED IN THE
C     EIGENVALUE COMPUTATIONS OR AGAIN A MISMATCH WILL ENSUE.
C
      WRITE(6,1020) IVEC,BATA,BETA(IVEC),TEMP
 1020 FORMAT(/2X,'IVEC',16X,'BATA',10X,'BETA(IVEC)',14X,'RELDIF'/I6,
     13E20.12/' IN LANCZOS VECTOR REGENERATION THE ENTRIES OF THE TRIDIAL
     1GONAL MATRICES BEING'/' GENERATED ARE NOT THE SAME AS THOSE IN THEL
     1 MATRIX SUPPLIED ON FILE 2.'/' THEREFORE SOMETHING IS BEING INITIAL
     1LIZED OR COMPUTED DIFFERENTLY FROM THE WAY'/' IT WAS COMPUTED IN TL
     1HE EIGENVALUE COMPUTATIONS'/' THE PROGRAM TERMINATES FOR THE USER 
     1TO DETERMINE WHAT THE PROBLEM IS'/)
      GO TO 1670
C
 1030 CONTINUE
      DO 1040 J = 1,N
      TEMP = SUM*V1(J)
      V1(J) = V2(J)
 1040 V2(J) = TEMP
C
 1050 CONTINUE
C
      LFIN = 0
      DO 1070 J = 1,NGOODC
      LL = LFIN
      LFIN = LFIN + N
C
      IF(IABS(MA(J)).LT.IVEC.OR.MP(J).EQ.MPMIN) GO TO 1070
      II = IVEC + MINT(J) - 1
      TEMP = TVEC(II)
C     II IS THE (IVEC)TH COMPONENT OF THE T-EIGENVECTOR CONTAINED
C     IN TVEC(MINT(J)).
C
      DO 1060 K = 1,N
      LL = LL + 1
 1060 RITVEC(LL) = TEMP*V2(K) + RITVEC(LL)
C
 1070 CONTINUE
C
      IVEC = IVEC + 1
      IF (IVEC.LE.KMAXU) GO TO 980
C
C     RITZVECTOR GENERATION IS COMPLETE. NORMALIZE EACH RITZVECTOR.
C     NOTE THAT IF CERTAIN RITZ VECTORS WERE NOT COMPUTED THEN THAT
C     PORTION OF THE RITVEC ARRAY WAS NOT UTILIZED.
C
      LFIN = 0
      DO 1140 J = 1,NGOODC
C
      KK = LFIN
      LFIN = LFIN + N
      IF(MP(J).EQ.MPMIN) GO TO 1140
C
      DO 1080 K = 1,N
      KK = KK + 1
      V1(K) = RITVEC(KK)
```

```
 1080 VS(K) = V1(K)
C
      IF(JPERM.EQ.0)  GO TO 1090
C
C-----------------------------------------------------------------------
C     V2 = V1 = (L-TRANSPOSE)*V1
      IPERM = 2
      CALL LPERM(V1,V2,IPERM)
C-----------------------------------------------------------------------
C
C     V2 CONTAINS RITZ VECTOR FOR A, VS CONTAINS THE RITZ VECTOR FOR B
C
 1090 CONTINUE
C
C-----------------------------------------------------------------------
      SUM = FINPRO(N,V1(1),1,V1(1),1)
C-----------------------------------------------------------------------
C
      SUM = DSQRT(SUM)
      RNORM(J) = SUM
      RNORME = DABS(ONE-SUM)
      SUM = ONE/SUM
C
      KK = LFIN - N
      DO 1100 K = 1,N
      KK = KK + 1
      VS(K) = SUM*VS(K)
 1100 RITVEC(KK) = SUM*V1(K)
C
C     VS IS RITZ VECTOR FOR BI:  RITVEC IS RITZ VECTOR FOR A-MATRIX
C     B = SO*P*A*P' + SHIFT*I
C     BIERR = ||BI*VS - GOODBI(J)*VS||
C     BIEVER = |(VS-TRANS)*BI*VS - GOODBI(J)|
C
C-----------------------------------------------------------------------
C     V1 = (B-INVERSE)*VS
      JBSOLV = 2
      CALL BSOLV(VS,V1,JBSOLV)
C-----------------------------------------------------------------------
C
      EVALN = EVNEW(J)
C
C-----------------------------------------------------------------------
      TEMP = FINPRO(N,V1(1),1,VS(1),1)
C-----------------------------------------------------------------------
C
      TEMP = DABS(TEMP - EVALN)
      BIEVER(J) = TEMP
      DO 1110 K = 1,N
 1110 V1(K) = V1(K) - EVALN*VS(K)
C
C-----------------------------------------------------------------------
      SUM = FINPRO(N,V1(1),1,V1(1),1)
C-----------------------------------------------------------------------
C
      SUM = DSQRT(SUM)
      BIERR(J) = SUM
      BIERRG(J) = SUM/ABS(BIMING(J))
C
      LINT = LFIN - N + 1
      EVAL = (ONE/EVALN - SHIFT)/SO
      GOODA(J) = EVAL
      TEMP = BIEVER(J)
C
      IF(IWRITE.EQ.0)  GO TO 1140
      WRITE(6,1120) J,GOODBI(J)
```

```
 1120 FORMAT(/I5,' TH B-INVERSE EIGENVALUE COMPUTED = ',E20.12/)
C
      WRITE(6,1130)  TERR(J),TBETA(J),RNORME
 1130 FORMAT(' NORM OF ERROR IN T-EIGENVECTOR = ',E14.3/
     1' BETA(MA(J)+1)*U(MA(J)) = ',E14.3/
     1' ABS(NORM(RITVEC) - 1.0) = ',E14.3/)
C
 1140 CONTINUE
C
C     RITZVECTORS ARE NORMALIZED AND ERROR ESTIMATES ARE IN BIERR
C     AND BIERRG ARRAYS. STORE EVERYTHING
C
      WRITE(13,1150)
 1150 FORMAT(6X,'BIEIGENVALUE',6X,'RITZNORM',7X,'TBETA',7X,'TLAST',5X,
     1 'BIERROR',6X,'BIEVER')
C
      WRITE(9,1160)
 1160 FORMAT(5X,'BIEIGENVALUE',4X,'MA(J)',4X,'BIMINGAP',5X,'BIERROR',3X
     1 ,'BIERR/GAP',6X,'TERROR')
C
      DO 1190 J=1,NGOODC
C
      IF(MP(J).EQ.MPMIN) GO TO 1190
C
      WRITE(9,1170) GOODBI(J),MA(J),BIMING(J),BIERR(J),BIERRG(J),TERR(J)
 1170 FORMAT(E20.12,I6,4E12.4)
C
      WRITE(13,1180) EVNEW(J),RNORM(J),TBETA(J),TLAST(J),BIERR(J),
     1 BIEVER(J)
 1180 FORMAT(E20.12,5E12.4)
C
 1190 CONTINUE
C
      WRITE(9,1200)
 1200 FORMAT(/5X, 'J',7X,'AEIGENVALUE',3X,'MA(J)',5X,'AMINGAP')
C
      DO 1210 J = 1,NGOOD
      IF(MP(J).EQ.MPMIN) GO TO 1210
      WRITE(9,1220) J,GOODA(J),MA(J),AMINGP(J)
 1210 CONTINUE
 1220 FORMAT(I6,E20.12,I6,E12.4)
C
      IF (MREJEC.EQ.0) GO TO 1300
C
      WRITE(9,1230)
 1230 FORMAT(/' RITZ VECTORS WERE NOT COMPUTED FOR THE FOLLOWING EIGENVAL
     1LUES'/' EITHER BECAUSE THEY HAD NOT CONVERGED OR BECAUSE THE ERROR
     1 ESTIMATE'/'  WAS NOT AS SMALL AS DESIRED'/)
C
      WRITE(9,1240)
 1240 FORMAT(6X,'GOODBI(J)',3X,'MA(J)',5X,'BIMING(J)',6X,'TBETA(J)',3X,
     1'MP(J)')
C
      WRITE(13,1250)
 1250 FORMAT(/' RITZ VECTORS WERE NOT COMPUTED FOR THE FOLLOWING EIGENVAL
     1LUES'/' EITHER BECAUSE THEY HAD NOT CONVERGED OR BECAUSE'/' THE ER
     1ROR ESTIMATE WAS NOT AS SMALL AS DESIRED'/)
C
      WRITE(13,1260)
 1260 FORMAT(3X,'BIEIGENVALUE',3X,'MA(J)',3X,'M1(J)',3X,'M2(J)',3X,'MP(J
     1)')
C
      DO 1290 J = 1,NGOODC
C
      IF(MP(J).NE.MPMIN) GO TO 1290
C
C     WRITE OUT MESSAGE FOR EACH EIGENVALUE FOR WHICH NO EIGENVECTOR
```

```
C     WAS COMPUTED.
C
      WRITE(9,1270) GOODBI(J),MA(J),BIMING(J),TBETA(J),MP(J)
 1270 FORMAT(E15.8,I8,2E14.4,I8)
C
      WRITE(13,1280) GOODBI(J),MA(J),M1(J),M2(J),MP(J)
 1280 FORMAT(E15.8,4I8)
C
 1290 CONTINUE
C
 1300 CONTINUE
C
      WRITE(9,1310)
 1310 FORMAT(/' ABOVE ARE ERROR ESTIMATES FOR THE BI AND T EIGENVECTORS'
     1/ ' ASSOCIATED WITH THE GOODBI LISTED, DENOTED BY EV '/
     1 ' BIERROR = NORM(BI*X-EV*X),  TERROR = NORM(T*Y - EV*Y)'/
     1 ' WHERE T = T(1,MA(J)),  P*X = RITZVEC = V*Y, T*Y = GOODBI*Y'/
     1 ' BIMINGAP = GAP TO NEAREST BI-EIGENVALUE'/)
C
      WRITE(13,1320)
 1320 FORMAT(/' ABOVE ARE ERROR ESTIMATES FOR THE EIGENVECTORS'/
     1 ' ASSOCIATED WITH THE BI-EIGENVALUES'/
     1 ' RITZNORM = NORM(COMPUTED RITZ VECTOR FOR B-INVERSE)'/
     1 ' TBETA(J) = BETA(MA(J)+1)*Y(MA(J)),  T*Y = BIEVAL*Y'/
     1 ' TLAST(J) = DABS(Y(MA(J)))'/
     1 ' BIERROR = NORM(BI*X - BIEVAL*X) WHERE X = V*Y'/
     1 ' BIEVER = DABS(BIEIGENVALUE - (X-TRANSPOSE*BINVERSE*X))'/)
C
C     NUMBER OF RITZ VECTORS COMPUTED
      NCOMPU = NGOODC - MREJEC
      WRITE(12,1330) N,NCOMPU,NGOODC,MATNO
 1330 FORMAT(3I6,I8,' = SIZE A, NO.RITZVECS, NO.GOODEVALUES,MATNO')
C
      LFIN = 0
      DO 1390 J = 1,NGOODC
      LINT = LFIN + 1
      LFIN = LFIN + N
C
      IF(MP(J).EQ.MPMIN) GO TO 1370
C     RITZ VECTOR WAS COMPUTED
      WRITE(12,1340) J, EVNEW(J), GOODA(J),MP(J)
 1340 FORMAT(I6,4X,2E20.12,I6,' J,GOODBI,GOODA,MP(J)')
C
      WRITE(12,1350) BIERR(J), BIERRG(J), BIMING(J),AMINGP(J)
 1350 FORMAT(4X,' BIRESIDUAL  ',2X,'BIRESIDUAL/GAP',
     12X,'BIMINGAP',3X,' AMINGAP'/
     1 E15.5,E16.5,2E11.3)
C
      WRITE(12,1360) (RITVEC(LL), LL=LINT,LFIN)
 1360 FORMAT(4E20.12)
      GO TO 1390
C     NO RITZ VECTOR WAS COMPUTED FOR THIS EIGENVALUE
 1370 CONTINUE
      WRITE(12,1380) J,GOODBI(J),GOODA(J),MP(J)
 1380 FORMAT(/I5,E20.12,E20.12,I6,' = J,GOODBI,GOODA,MP'/' NO RITZ VECTO
     1R WAS COMPUTED FOR THIS EIGENVALUE'/)
C
 1390 CONTINUE
C
C     DID ANY T-MATRICES INCLUDE OFF-DIAGONAL ENTRIES SMALLER THAN
C     DESIRED, AS SPECIFIED BY BTOL?
C
      IF(IB.GT.0) GO TO 1420
      WRITE(6,1400) KMAXU
 1400 FORMAT(/' FOR LARGEST T-MATRIX CONSIDERED',I7,' CHECK THE SIZE OF
     1BETAS')
C
```

```
C-----------------------------------------------------------------------
C
      CALL TNORM(ALPHA,BETA,BKMIN,TEMP,KMAXU,IBMT)
C
C-----------------------------------------------------------------------
C
      IF(IBMT.LT.0) WRITE(6,1410)
 1410 FORMAT(/' WARNING THE T-MATRICES FOR ONE OR MORE OF THE EIGENVALUE
     1S CONSIDERED'/' HAD AN OFF DIAGONAL ENTRY THAT WAS SMALLER THAN TH
     1E BETA TOLERANCE THAT WAS SPECIFIED'/)
 1420 CONTINUE
C
       GO TO 1670
C
 1430 WRITE(6,1440) NGOOD,NMAX,MDIMRV
 1440 FORMAT(/I4,' RITZ VECTORS WERE REQUESTED BUT THE REQUIRED DIMENSIO
     1N',I6/' IS LARGER THAN USER-SPECIFIED DIMENSION OF RITVEC',I6/
     1' THEREFORE, THE EIGENVECTOR PROCEDURE TERMINATES FOR THE USER TO
     1 INTERVENE'/)
C
      GO TO 1670
C
 1450 WRITE(6,1460) NOLD,N,MATOLD,MATNO,SHIFTO,SHIFT
 1460 FORMAT(/' PARAMETERS READ FROM FILE 3 DO NOT AGREE WITH WHAT USER
     1SPECIFIED'/ ' NOLD,N,MATOLD,MATNO,SHIFTO,SHIFT = '/2I6,2I8,2E10.3
     1/' THEREFORE PROGRAM TERMINATES FOR USER TO RESOLVE THE DIFFERENCE
     1S'/)
C
      GO TO 1670
C
 1470 WRITE(6,1480)
 1480 FORMAT(/' PARAMETERS READ FROM ALPHA,BETA FILE DO NOT AGREE WITH W
     1HAT USER SPECIFIED'/' PROGRAM TERMINATES FOR USER TO RESOLVE THE D
     1IFFERENCES'/)
C
      GO TO 1670
C
 1490 WRITE(6,1500) KMAX,MEV
 1500 FORMAT(/' IN ALPHA, BETA FILE KMAX = ',I6/
     1' BUT EIGENVALUES WERE COMPUTED AT MEV =',I6,' PROGRAM STOPS'/)
C
      GO TO 1670
C
 1510 WRITE(6,1520)
 1520 FORMAT(/' PROGRAM COMPUTED 1ST GUESSES ON T-MATRIX SIZES AND READ
     1THEM TO FILE 10'/' THEN TERMINATED AS REQUESTED.'/)
      GO TO 1670
C
 1530 WRITE(6,1540) MTOL, MDIMTV
 1540 FORMAT(/' PROGRAM TERMINATES BECAUSE THE TVEC DIMENSION ANTICIPATE
     1D',I7/'  IS LARGER THAN THE TVEC DIMENSION',I7,' SPECIFIED BY THE
     1USER.'/'  USER MAY RESET THE TVEC DIMENSION AND RESTART THE PROGRA
     1M')
      GO TO 1670
C
 1550 WRITE(6,1560)
 1560 FORMAT(/' PROGRAM TERMINATES BECAUSE NO SUITABLE T-EIGENVECTORS WE
     1RE IDENTIFIED'/'  FOR ANY OF THE EIGENVALUES SUPPLIED.  PROBLEM CO
     1ULD BE CAUSED'/'  BY TOO SMALL A TVEC DIMENSION OR SIMPLY THAT SUI
     1TABLE T-VECTORS COULD'/' NOT BE IDENTIFIED.  USER SHOULD EXAMINE O
     1UTPUT'/)
      GO TO 1670
C
 1570 WRITE(6,1580) LVCONT,NTVEC,NGOOD
 1580 FORMAT(/' LVCONT FLAG =',I2,' AND NUMBER ',I5,' OF T-EIGENVECTORS
     1 COMPUTED N.E.'/' NUMBER',I5,' REQUESTED SO PROGRAM TERMINATES'/)
```

```
      GO TO 1670                                                        LIV11980
C                                                                       LIV11990
 1590 WRITE(6,1600)                                                     LIV12000
 1600 FORMAT(/' PROGRAM TERMINATES WITHOUT COMPUTING RITZ VECTORS'/     LIV12010
     1' BECAUSE ALL T-EIGENVECTORS WERE REJECTED AS NOT SUITABLE FOR THELIV12020
     1RITZ VECTOR'/' COMPUTATIONS.  PROBABLE CAUSE IS LACK OF CONVERGENCLIV12030
     1E OF EIGENVALUES SUPPLIED'/)                                      LIV12040
       GO TO 1670                                                       LIV12050
C                                                                       LIV12060
 1610 WRITE(6,1620)                                                     LIV12070
 1620 FORMAT(/' PROGRAM INDICATES THAT IT IS NOT POSSIBLE TO COMPUTE ANYLIV12080
     1 OF THE REQUESTED EIGENVECTORS.'/' THEREFORE PROGRAM TERMINATES') LIV12090
      DO 1630 J=1,NGOODC                                                LIV12100
 1630 WRITE(6,1640)  J,GOODBI(J),MP(J)                                  LIV12110
 1640 FORMAT(/4X,' J',11X,'GOODBI(J)',4X,'MP(J)'/I6,E20.12,I9/)         LIV12120
      GO TO 1670                                                        LIV12130
C                                                                       LIV12140
 1650 WRITE(6,1660) MBETA,KMAXN                                         LIV12150
 1660 FORMAT(/' PROGRAM TERMINATES BECAUSE THE STORAGE ALLOTTED FOR THE LIV12160
     1BETA ARRAY',I8,/' IS NOT SUFFICIENT FOR THE ENLARGED KMAX =',I8,' LIV12170
     1THAT THE PROGRAM WANTS.'/' USER CAN ENLARGE THE ALPHA,BETA ARRAYS LIV12180
     1 AND RERUN THE PROGRAM'/)                                         LIV12190
C                                                                       LIV12200
 1670 CONTINUE                                                          LIV12210
C                                                                       LIV12220
      STOP                                                              LIV12230
C-----END EIGENVECTOR COMPUTATIONS FOR INVERSES OF REAL SYMMETRIC-------LIV12240
      END                                                               LIV12250
```

SECTION 4.4 LANCZS AND SAMPLE MATRIX-VECTOR SOLVE SUBROUTINES

```
C---LIMULT-(INVERSES OF REAL SYMMETRIC MATRICES)------------------------
C
C     CONTAINS SUBROUTINE LANCZS AND SAMPLE USPEC AND BSOLV
C     USED BY THE VERSION OF THE LANCZOS ALGORITHMS FOR
C     FACTORED INVERSES OF REAL SYMMETRIC MATRICES, LIVAL AND LIVEC.
C
C     NONPORTABLE CONSTRUCTIONS:
C     1.  THE ENTRY MECHANISM USED TO PASS THE STORAGE LOCATIONS
C         OF THE FACTORIZATION OF THE MATRIX TO BE USED BY
C         LANCZS TO THE SOLVE SUBROUTINE BSOLV.
C     2.  IN THE SAMPLE USPEC SUBROUTINES PROVIDED:
C         THE FREE FORMAT (7,*) AND FORMATS (20A4) AND  (4Z20)
C         USED IN DEFINING THE MATRICES.
C
C-----LANCZS-COMPUTE LANCZOS TRIDIAGONAL MATRICES-----------------------
C
      SUBROUTINE LANCZS(MATVEC,ALPHA,BETA,V1,V2,VS,G,KMAX,MOLD1,N,IIX)
C
C-----------------------------------------------------------------------
      DOUBLE PRECISION  ALPHA(1), BETA(1), V1(1), V2(1), VS(1)
      DOUBLE PRECISION SUM, ONE, ZERO, TEMP
      REAL  G(1)
      EXTERNAL MATVEC
      DOUBLE PRECISION  FINPRO, DSQRT
C-----------------------------------------------------------------------
C     ALPHA, BETA, LANCZOS VECTOR GENERATION
C     ALPHA BETA GENERATION STARTS WITH IVEC = 1, BETA(1) = ZERO
C     V2 =  RANDOM UNIT VECTOR AND V1 = ZERO, OR EXTENDS
C     AN EXISTING ALPHA/BETA FILE.
C
      ZERO = 0.0D0
      ONE  = 1.0D0
      IF (MOLD1.GT.1) GO TO 30
      BETA(1) = ZERO
      IIL = IIX
C
C-----------------------------------------------------------------------
      CALL GENRAN(IIL,G,N)
C-----------------------------------------------------------------------
C
      DO 10 K = 1,N
   10 V2(K) = G(K)
C
C-----------------------------------------------------------------------
      SUM = FINPRO(N,V2(1),1,V2(1),1)
C-----------------------------------------------------------------------
C
      SUM = ONE/DSQRT(SUM)
C
      DO 20 K = 1,N
      V1(K) = ZERO
   20 V2(K) = SUM*V2(K)
C
   30 CONTINUE
C
      DO 80 IVEC = MOLD1,KMAX
C
      DO 40 K = 1,N
   40 VS(K) = V2(K)
C
C-----------------------------------------------------------------------
      JBSOLV = 2
      CALL MATVEC(VS,VS,JBSOLV)
C-----------------------------------------------------------------------
```

```
C
C     VS = B(INVERSE)*V2
C
      SUM = BETA(IVEC)
C
      DO 50 K = 1,N
   50 V1(K) = VS(K)-SUM*V1(K)
C
C-----------------------------------------------------------------------
      SUM = FINPRO(N,V1(1),1,V2(1),1)
C-----------------------------------------------------------------------
C
      ALPHA(IVEC) = SUM
C
      DO 60 K = 1,N
   60 V1(K) = V1(K)-SUM*V2(K)
C
C-----------------------------------------------------------------------
      SUM = FINPRO(N,V1(1),1,V1(1),1)
C-----------------------------------------------------------------------
C
      IN = IVEC+1
C
      BETA(IN) = DSQRT(SUM)
      SUM = ONE/BETA(IN)
C
      DO 70 K = 1,N
      TEMP = SUM*V1(K)
      V1(K) = V2(K)
   70 V2(K) = TEMP
C
   80 CONTINUE
C
      RETURN
C-----END LANCZS--------------------------------------------------------
      END
C
C-----USPEC FOR FACTORED INVERSES OF REAL SYMMETRIC MATRICES------------
C
      SUBROUTINE CUSPEC(N,MATNO)
C     SUBROUTINE USPEC(N,MATNO)
C
C-----------------------------------------------------------------------
      DOUBLE PRECISION  BD(2200),BSD(10000)
      INTEGER  KCOL(2200),KROW(10000),IPR(2200),IPT(2200)
C-----------------------------------------------------------------------
C     NOTE THAT THIS SUBROUTINE ASSUMES THAT B IS POSITIVE DEFINITE.
C     USER COULD REPLACE THIS SUBROUTINE AND CORRESPONDING SAMPLE
C     USPEC SUBROUTINE BY ONE THAT WORKS WITH GENERAL FACTORIZATION.
C
C     DIMENSIONS ARRAYS NEEDED TO DEFINE CHOLESKY FACTOR OF B-MATRIX,
C     READS CHOLESKY FACTOR FROM FILE 7, AND THEN PASSES STORAGE
C     LOCATIONS OF THESE ARRAYS TO THE B-MATRIX SOLVE SUBROUTINE BSOLV.
C
C     HERE WE HAVE B = P*C*P' = L*L'  WHERE  C = SO*A + SHIFT*I.
C     P IS A PERMUTATION MATRIX DEFINED BY THE VECTOR MAPS IPR AND IPT.
C     THE ITH ROW OF B CORRESPONDS TO THE JTH ROW OF C (A) WHERE
C     J = IPR(I) AND I = IPT(J).   A IS THE ORIGINAL MATRIX.
C
C     THE B-CHOLESKY FACTOR IS STORED IN THE FOLLOWING SPARSE FORMAT:
C     N = ORDER OF THE B-MATRIX.
C     NZT = NUMBER OF NONZERO SUBDIAGONAL ENTRIES IN THE CHOLESKY
C           FACTOR, L.
C     KCOL(J), J=1,N IS THE NUMBER OF NONZERO SUBDIAGONAL ELEMENTS IN
C              COLUMN J OF L.
C     KROW(K), K=1,NZT IS THE ROW INDEX FOR CORRESPONDING ENTRY BSD(K).
C     BD(J), J = 1,N CONTAINS THE DIAGONAL ENTRIES OF L.
```

```
C     BSD(K), K =1,NZT CONTAINS THE NONZERO SUBDIAGONAL ENTRIES OF L
C           BY COLUMN.
C     JPERM = (0,1):  1 MEANS CHOLEKSY FACTOR CORRESPONDS TO
C                     PERMUTED C.  0 MEANS NO PERMUTATION WAS USED.
C-----------------------------------------------------------------------
C     READ CHOLESKY FACTOR FROM FILE 7.  MUST BE STORED
C     IN SPARSE MATRIX FORMAT.
      READ(7,10) NZT,NOLD,NZL,MATOLD,JPERM
   10 FORMAT(I10,2I6,I8,I6)
C
      WRITE(6,20) NZT,NZL,N,NOLD,MATOLD,JPERM
   20 FORMAT(' HEADER, CHOLESKY FACTOR FILE'/
     1 3X,'NZT',3X,'NZL',5X,'N',2X,'NOLD',2X,'MATOLD',1X,'JPERM'/
     1 4I6,I8,I6/)
C
      IF (N.NE.NOLD.OR.MATNO.NE.MATOLD) GO TO 70
C
      READ(7,30) (KCOL(K), K = 1,NZL)
      READ(7,30) (KROW(K), K = 1,NZT)
   30 FORMAT(13I6)
      READ(7,40) (BD(K), K = 1,N)
      READ(7,40) (BSD(K), K = 1,NZT)
   40 FORMAT(4Z20)
C  20 FORMAT(3E25.16)
C
C     DOES CHOLESKY FACTOR CORRESPOND TO PERMUTED B?
      IF(JPERM.EQ.0)  GO TO 60
      READ(7,30) (IPR(K), K = 1,N)
C
      DO 50 K = 1,N
      J = IPR(K)
   50 IPT(J) = K
C
C-----------------------------------------------------------------------
      CALL LPERME(IPR,IPT,N)
C-----------------------------------------------------------------------
C
   60 CONTINUE
C
C-----------------------------------------------------------------------
C     PASS STORAGE LOCATIONS OF FACTORS TO INVERSION SUBROUTINE BSOLV
      CALL BSOLVE(BSD,BD,KCOL,KROW,N,NZT,NZL)
C-----------------------------------------------------------------------
C
      GO TO 90
C
   70 CONTINUE
C     DEFAULT EXIT
      WRITE(6,80)
   80 FORMAT(' TERMINATE.  PARAMETERS IN CHOLESKY FACTOR FILE'/
     1' DO NOT AGREE WITH THOSE SPECIFIED BY THE USER'/)
      STOP
C
   90 CONTINUE
C-----END OF USPEC------------------------------------------------------
      RETURN
      END
C
C-----BSOLV-(FACTORED INVERSE OR L*L-TRANS MULTIPLY)--------------------
C           (FOR POSITIVE DEFINITE SYMMETRIC SPARSE MATRICES)
C
C     SUBROUTINE BSOLV(V,U,JBSOLV)
      SUBROUTINE CBSOLV(V,U,JBSOLV)
C
C-----------------------------------------------------------------------
      DOUBLE PRECISION  BD(1),BSD(1),U(1),V(1),TEMP,ZERO,ONE
```

```
      INTEGER  KCOL(1),KROW(1)
C-----------------------------------------------------------------------
C     JBSOLV = 2 MEANS SOLVE B*U = V
C     JBSOLV = 1 MEANS COMPUTE U = B*V:  NOTE THAT IN THIS CASE V IS
C     DESTROYED.  LANCZOS PROGRAMS AS WRITTEN DO NOT USE JBSOLV = 1
C     PATH.
      ZERO = 0.0D0
      ONE  = 1.0D0
      IF (JBSOLV .EQ.2) GO TO 60
C     U = B*V WHERE  B = L*L'
      KL = 0
      DO 20 J = 1,N
      TEMP = V(J)*BD(J)
      IF (KCOL(J).EQ.0.OR.J.EQ.N) GO TO 20
      KF = KL + 1
      KL  = KL + KCOL(J)
      DO 10 K = KF,KL
      IK = KROW(K)
   10 TEMP = BSD(K)*V(IK) + TEMP
   20 V(J) = TEMP
C     V = L'*V
      DO 30 K = 1,N
   30 U(K) = V(K)*BD(K)
      KL = 0
      DO 50 K = 1,N
      TEMP = V(K)
      IF (KCOL(K).EQ.0.OR.K.EQ.N) GO TO 50
      KF = KL + 1
      KL = KL + KCOL(K)
      DO 40 KK = KF,KL
      KR = KROW(KK)
   40 U(KR) = U(KR) + TEMP*BSD(KK)
   50 CONTINUE
      GO TO 120
C     U = B*V
C--------------------------------------------------------------------
   60 CONTINUE
C     SOLVE B*U = V FOR U WHERE  B = L*L'
C     SET U = V. FIRST SOLVE L*U = U FOR U, THEN SOLVE L'*U = U FOR U
      KL = 0
      DO 70 K = 1,N
   70 U(K) = V(K)
      DO 90 K = 1,N
      TEMP = U(K)/BD(K)
      U(K) = TEMP
      IF (KCOL(K).EQ.0.OR.K.EQ.N) GO TO 90
      KF = KL + 1
      KL = KL + KCOL(K)
      DO 80 KK = KF,KL
      KR = KROW(KK)
   80 U(KR) = U(KR) - TEMP*BSD(KK)
   90 CONTINUE
      NP1 = N+1
      KF = NZT + 1
      DO 110 K = 1,N
      L = NP1 - K
      TEMP = U(L)
      IF (KCOL(L).EQ.0.OR.L.EQ.N) GO TO 110
      KL = KF - 1
      KF = KF - KCOL(L)
      DO 100 LL = KF,KL
      LR = KROW(LL)
  100 TEMP = TEMP - BSD(LL)*U(LR)
  110 U(L) = TEMP/BD(L)
  120 CONTINUE
C
```

```
      RETURN
C
C-----------------------------------------------------------------------
      ENTRY BSOLVE(BSD,BD,KCOL,KROW,N,NZT,NZL)
C-----------------------------------------------------------------------
C
      RETURN
C-----END OF BSOLV------------------------------------------------------
      END
C
C-----SUBROUTINES FOR DIAGONAL TEST MATRICES----------------------------
C
C     BSOLV AND USPEC SUBROUTINES FOR DIAGONAL TEST MATRICES
C
C-----BSOLV DIAGONAL TEST MATRIX----------------------------------------
C
C     SUBROUTINE DBSOLV(V,U,JBSOLV)
      SUBROUTINE BSOLV(V,U,JBSOLV)
C
C-----------------------------------------------------------------------
      DOUBLE PRECISION  V(1),U(1),D(1)
C-----------------------------------------------------------------------
C     JBSOLV = 1, COMPUTE U = D*V. (NOTE THIS IS NOT USED)
C     JBSOLV = 2, COMPUTE U = (D-INVERSE)*V
      IF(JBSOLV.EQ.2) GO TO 20
      DO 10 I=1,N
   10 U(I) = D(I)*V(I)
      GO TO 40
C
   20 DO 30 I=1,N
   30 U(I)= V(I)/D(I)
C
   40 CONTINUE
      RETURN
C
C-----------------------------------------------------------------------
C     BELOW ENTRY IS FOR A DIAGONAL TEST MATRIX
      ENTRY DSOLVE(D,N)
C-----------------------------------------------------------------------
C
      RETURN
C-----END OF BSOLV FOR DIAGONAL TEST MATRIX ----------------------------
      END
C
C-----START OF USPEC FOR DIAGONAL TEST MATRIX---------------------------
C
      SUBROUTINE USPEC(N,MATNO)
C     SUBROUTINE DUSPEC(N,MATNO)
C
C-----------------------------------------------------------------------
      DOUBLE PRECISION  D(1000), DI(1000), SHIFT, SPACE
      DOUBLE PRECISION  DABS, DFLOAT
      REAL  EXPLAN(20)
C-----------------------------------------------------------------------
C
      READ(7,10) EXPLAN
   10 FORMAT(20A4)
      READ(7,*) NOLD,NUNIF,SPACE,D(1),SHIFT
      NNUNIF = NOLD - NUNIF
      WRITE(6,20) NOLD,SPACE,NNUNIF,D(1),SHIFT
   20 FORMAT(/' DIAGONAL TEST MATRIX, SIZE = ',I4/' IS THE INVERSE OF MA
     1TRIX WITH MOST ENTRIES',E10.3/' UNITS APART AND WITH ',I3,' ENTRIE
     1S IRREGULARLY SPACED'/' FIRST ENTRY WAS ',E13.4,' SHIFT = ',E10.3
     1/)
C
      IF(N.NE.NOLD) GO TO 100
C     COMPUTE THE UNIFORM PORTION OF THE SPECTRUM
```

```
      DO 30 J=2,NUNIF
   30 D(J) = D(1) - DFLOAT(J-1)*SPACE
      NUNIF1=NUNIF + 1
      READ(7,10)  EXPLAN
      DO 40 J=NUNIF1,N
   40 READ(7,*) D(J)
      NB = NUNIF - 2
C
      IF(SHIFT.EQ.0.) GO TO 60
      DO 50 J=1,N
   50 D(J) = D(J) + SHIFT
C
C     COMPUTE EIGENVALUES OF INVERSE FOR PRINTOUT ONLY
   60 DO 70 J = 1,N
   70 DI(J) = 1.D0/D(J)
      WRITE(6,80) (DI(I), I=1,10 )
      WRITE(6,90) (DI(I), I = NB,N)
   80 FORMAT(/' INVERSE LANCZOS TEST, LANCZS USES INVERSE OF GIVEN MATRI
     1X'/' 1ST 10 ENTRIES OF INVERSE OF DIAGONAL TEST MATRIX = '/(3E22.1
     14))
   90 FORMAT(/' MIDDLE (ORIGINALLY UNIFORM) PORTION OF MATRIX IS NOT PRI
     1NTED OUT'/' END OF (UNIFORM) PLUS NONUNIFORM SECTION = '/(3E25.16)
     1)
C
C     DIAGONAL GENERATION COMPLETE
C
C-----------------------------------------------------------------------
C    PASS STORAGE LOCATIONS OF D AND N TO DSOLV SUBROUTINE
        CALL DSOLVE(D,N)
C-----------------------------------------------------------------------
C
      RETURN
  100 WRITE(6,110) NOLD,N
  110 FORMAT(' PROGRAM TERMINATES BECAUSE NOLD = ',I5,'DOES NOT EQUAL N
     1 =',I5)
C-----END OF USPEC SUBROUTINE FOR DIAGONAL TEST MATRICES----------------
      STOP
      END
```

SECTION 4.5 OPTIONAL PREPROCESSING PROGRAMS FOR CHAPTERS 4,5,9

```
C-----PERMUT-(USES SPARSPAK PACKAGE)------------------------------------
C
C     OPTIONAL PREPROCESSING PROGRAM FOR USE WITH LANCZOS CODES.
C     GIVEN A REAL SYMMETRIC A-MATRIX IN SPARSE MATRIX FORMAT, PERMUT
C     CALLS THE SPARSPAK PACKAGE (A. GEORGE, J. LIU, E. NG, U. WATERLOO)
C     TO DETERMINE A REORDERING OF A, THAT IS A PERMUTATION MATRIX
C     P, SUCH THAT SPARSITY IS PRESERVED IN THE FACTORIZATION OF
C     THE PERMUTED MATRIX.  PERMUT ALSO MODIFIES THE GIVEN A-MATRIX
C     TO FORM THE MATRIX C = SO*A + SHIFT*I, WHERE SO AND SHIFT
C     ARE SCALARS PROVIDED BY THE USER, AND THEN WRITES THIS
C     C-MATRIX OUT TO FILE 9 ALONG WITH THE PERMUTATION P WHICH
C     IS DEFINED BY THE VECTOR IPR.  IPR IS ALSO WRITTEN SEPARATELY
C     TO FILE 14.
C
C     NONPORTABLE CONSTRUCTIONS:
C     1.  INTEGER*2 VARIABLE NPERM.  NOTE THAT THIS VARIABLE CANNOT
C         BE CHANGED TO INTEGER*4.
C     2.  FREE FORMAT (5,*) AND THE FORMAT (20A4).
C     3.  TO AVOID COMPOUNDING FORMAT CONVERSION ERRORS, THE MATRIX
C         ENTRIES SHOULD BE STORED IN MACHINE FORMAT, ((4Z20) FOR
C         IBM/3081)
C
C-----------------------------------------------------------------------
C     SYMMETRIC A-MATRIX IS READ FROM FILE 8.  MATRIX IS STORED
C     IN FOLLOWING SPARSE FORMAT:
C
C     NZL = INDEX OF LAST COLUMN CONTAINING NONZEROS BELOW THE DIAGONAL.
C     NZS = NUMBER OF NONZERO SUBDIAGONAL ENTRIES
C     ICOL(K), K=1,NZL CONTAINS THE NUMBER OF NONZERO SUBDIAGONAL
C              ENTRIES IN COLUMN K.
C     IROW(K), K=1,NZS CONTAINS ROW INDEX OF KTH NONZERO SUBDIAGONAL
C              ENTRY, ENTRIES STORED COLUMN BY COLUMN.
C     AD(K), K=1,N CONTAINS THE DIAGONAL ENTRIES OF A, INCLUDING ANY
C            ZERO ENTRIES.
C     ASD(K), K=1,NZS CONTAINS THE NONZERO SUBDIAGONAL ENTRIES OF A,
C             COLUMN BY COLUMN.
C
C-----INPUT/OUTPUT FILES ------------------------------------------------
C
C    INPUT FILES:
C    FILE 5      CONTAINS THE PROGRAM PARAMETERS SET BY USER
C    FILE 8      CONTAINS THE SPARSE A-MATRIX
C
C    OUTPUT FILES:
C    FILE 6      INTERACTIVE TERMINAL FILE
C    FILE 9      CONTAINS THE SPARSE DATA FOR C = SO*A + SHIFT*I.
C    FILE 14     CONTAINS PERMUTATION IPR DEFINING THE REORDERING.
C                IN PARTICULAR J = IPR(I) MEANS ROW(COL) I OF
C                B = P*C*(P-TRANSPOSE) CORRESPONDS TO ROW(COL) J
C                OF THE A-MATRIX.
C
C-----SPARSPAK-----------------------------------------------------------
C     ARRAYS AND PARAMETERS THAT ARE REQUIRED BY SPARSPAK.
C     NOTE THAT THE CALL FOR SPARSPAK IS SPRSPK.  SUBROUTINES
C     IJBEGN, INIJ, IJEND, ORDRB5, AND PSTATS ARE SPARSPAK
C     SUBROUTINES.
C
C     S = VECTOR WHOSE ACTUAL DIMENSION IS DETERMINED BY SPARSPAK
C         WHEN THE REORDERING IS OBTAINED.  USER SPECIFIES MAXIMUM
C         DIMENSION MAXS ALLOWED; SPARSPAK DEFAULTS IF THIS MAXIMUM
C         IS EXCEEDED.  SPARSPAK IS DESIGNED FOR SOLVING SYSTEMS
C         OF EQUATIONS, THUS THE VECTOR S IS DESIGNED TO CONTAIN
C         THE SOLUTION VECTOR IF THERE IS ONE, FOLLOWED BY THE
C         PERMUTATION VECTOR IPR, FOLLOWED BY OTHER INFORMATION
```

```
C           GENERATED BY SPARSPAK.  A CORRECT SIZE FOR MAXS CAN BE
C           DETERMINED ONLY AFTER THE FACT.  AS A FIRST GUESS ONE
C           CAN SET MAXS = K*N WHERE K >= 10.
C
C       MSGLVL = CONTROL FOR WRITES TO FILE 6
C       NEQNS = ORDER OF A, THIS IS COMPUTED BY SPARSPAK
C       IERR = CONTROLS WRITING OF ERROR MESSAGES BY SPARSPAK.
C       MAXS = USER-SPECIFIED MAXIMUM ALLOWED DIMENSION OF S-ARRAY.
C
C
C-----------------------------------------------------------------------
      DOUBLE PRECISION AD(3000),ASD(10000),SO,SHIFT
      DOUBLE PRECISION S(30000),STEMP
      REAL EXPLAN(20)
      INTEGER ICOL(3000),IROW(10000),IPR(3000)
      INTEGER*2 NPERM(4)
      COMMON /SPKUSR/ MSGLVL,IERR,MAXS,NEQNS
      EQUIVALENCE (STEMP,NPERM(1))
C----------------------------------------------------------------------
C
C       ARRAYS MUST BE DIMENSIONED AS FOLLOWS:
C       1.  AD:  >= N, THE ORDER OF A-MATRIX.
C       2.  ASD: >= NZS, THE NUMBER OF NONZERO SUBDIAGONAL ENTRIES IN A.
C       4.  ICOL:    >= N
C       5.  IROW:    >= NZS
C       6.  IPR:     >= N + 4
C       7.  S:   >= MAXS
C
C----------------------------------------------------------------------
C
      WRITE(6,10)
   10 FORMAT(/' CALL SPARSPAK TO FIND REORDERING OF THE GIVEN MATRIX'/
     1' THAT PRESERVES SPARSITY IN THE FACTORIZATION'/)
C
C     READ IN USER-SPECIFIED PARAMETERS
C     SPECIFY THE MAXIMUM DIMENSION (MAXS) ALLOWED FOR ARRAY S,
C     AND WHETHER OR NOT A-MATRIX IS BEING SCALED BY SO OR SHIFTED
C     ISCALE = 0, THEN NO SCALING OR SHIFTING
      READ (5,20) EXPLAN
      WRITE (6,20) EXPLAN
      READ (5,20) EXPLAN
   20 FORMAT(20A4)
      READ (5,*) MAXS,ISCALE,SO,SHIFT
C
C     READ IN INDICES FOR A-MATRIX STORED IN SPARSE FORMAT ON FILE 8.
C     ONLY THE NONZERO STRUCTURE IS NEEDED TO OBTAIN THE PERMUTATION.
C
      READ(8,30) NZS,N,NZL,MATNO
   30 FORMAT(I10,2I6,I8)
C
      WRITE(6,40) NZS,N,NZL,MATNO,MAXS,ISCALE,SHIFT,SO
   40 FORMAT(/I10,2I6,I10,I10,' = NZS,N,NZL,MATNO,MAXS'/
     1 I6,2E12.5,' = ISCALE,SHIFT,SO'/)
C
      READ(8,50) (ICOL(K), K=1,NZL)
      READ(8,50) (IROW(K), K=1,NZS)
   50 FORMAT(13I6)
C
C     DIAGONAL IS READ (INCLUDING ANY ZERO ENTRIES), THEN NONZERO
C     BELOW DIAGONAL ENTRIES ARE READ IN
      READ(8,60) (AD(K), K=1,N)
      READ(8,60) (ASD(K), K=1,NZS)
   60 FORMAT(4E19.10)
C
      IF (ISCALE.EQ.0) GO TO 90
C
C     CALCULATE C = SO*A + SHIFT*I AND PUT IN A-ARRAY
```

```
      DO 70 K = 1,N
   70 AD(K) = SO*AD(K) + SHIFT
      DO 80 K = 1,NZS
   80 ASD(K) = SO*ASD(K)
   90 CONTINUE
C
C-----------------------------------------------------------------------
C     INPUT THE SPARSENESS STRUCTURE OF GIVEN A-MATRIX TO SPARSPAK
      CALL SPRSPK
C-----------------------------------------------------------------------
C
      MSGLVL = 4
C
C-----------------------------------------------------------------------
      CALL IJBEGN
C-----------------------------------------------------------------------
C
      LLAST = 0
      DO 110 J = 1,NZL
      IF (ICOL(J).EQ.0) GO TO 110
      JJ = J
      LFIRST = LLAST + 1
      LLAST = LLAST + ICOL(J)
      DO 100 L = LFIRST,LLAST
      II = IROW(L)
C
C-----------------------------------------------------------------------
      CALL INIJ(II,JJ,S)
C-----------------------------------------------------------------------
C
  100 CONTINUE
C
  110 CONTINUE
C
C     SPARSENESS STRUCTURE HAS BEEN INPUTED TO SPARSPAK.
C
C-----------------------------------------------------------------------
      CALL IJEND(S)
C-----------------------------------------------------------------------
C
      WRITE(6,120) N,NEQNS
  120 FORMAT(/2I6,' = N,NEQNS'/)
      IF (N.NE.NEQNS) GO TO 230
C
C-----------------------------------------------------------------------
C     USE SPARSPAK TO GENERATE REORDERING OF A THAT PRESERVES
C     SPARSITY.  CORRESPONDING FACTORIZATION CAN BE COMPUTED BY
C     PREPROCESSING PROGRAM LFACT WHEN C = SO*A + SHIFT*I IS POSITIVE
C     DEFINITE. BELOW CALLS THE MINIMUM DEGREE ALGORITHM PROVIDED
C     IN SPARSPAK.
      CALL ORDRB5(S)
      CALL PSTATS
C-----------------------------------------------------------------------
C
C     EXTRACT THE REORDERING FROM SPARSPAK S VECTOR AND STORE IN FILE 14
      L = 1
      KNUM = N
      DO 130 K = 1,N
      KNUM = KNUM + 1
      STEMP = S(KNUM)
      IPR(L) = NPERM(1)
      IPR(L+1) = NPERM(2)
      IPR(L+2) = NPERM(3)
      IPR(L+3) = NPERM(4)
      L = L+4
      IF (L.GT.N) GO TO 140
  130 CONTINUE
```

```
  140 CONTINUE
C
      WRITE(14,150) N,MATNO
  150 FORMAT(I6,I8,' = N MATNO  K IPR(K)   A-MATRIX PERMUTATION')
      WRITE(14,160) (K,IPR(K), K = 1,N)
  160 FORMAT(6(1X,2I6))
C
C
C     WRITE C = SO*A + SHIFT*I WITH THE PERMUTATION IPR TO FILE 9.
C
      JPERM = 1
      WRITE(9,170) NZS,N,NZL,MATNO,JPERM
  170 FORMAT(I10,2I6,I8,I6,' = NZS,N,NZL,MATNO,JPERM. ACOMPAC')
C
C     NUMBER OF NONZERO SUBDIAGONAL ENTRIES IN EACH COLUMN IS WRITTEN
C     THEN THE CORRESPONDING ROW INDEX FOR EACH SUCH ENTRY IS WRITTEN
      WRITE(9,180) (ICOL(K), K=1,NZL)
      WRITE(9,180) (IROW(K), K=1,NZS)
  180 FORMAT(13I6)
C     DIAGONAL IS WRITTEN FIRST, THEN NONZERO BELOW DIAGONAL ENTRIES
      WRITE(9,190) (AD(K), K=1,N)
      WRITE(9,190) (ASD(K), K=1,NZS)
  190 FORMAT(4E19.10)
      WRITE(9,180) (IPR(K), K=1,N)
C
      IF(ISCALE.NE.0) GO TO 200
C     ISCALE = 0, SET DEFAULT VALUES OF SO AND SHIFT
      SO = 1.D0
      SHIFT = 0.D0
  200  WRITE(9,210) SO,SHIFT
  210 FORMAT(2E12.5,' = SO SHIFT'/
     1 ' ABOVE IS SPARSE DATA FOLLOWED BY PERMUTATION IPR'/
     1 ' FOR THE MATRIX C = SO*A+SHIFT*I '/
     1 ' B = P*C*PTRANS CAN BE GENERATED IN SUBROUTINE LORDER'/
     1 ' ROW(COL) I OF B CORRESPONDS TO ROW(COL) J OF C, J = IPR(I)'/
     1 ' NZS = TOTAL NUMBER OF SUBDIAGONAL NONZEROS IN C'/
     1 ' KCOL(K) = NUMBER OF SUBDIAGONAL NONZEROS IN COL K OF C'/
     1 ' KROW(K) = ROW INDEX OF SUBDIAGONAL NONZERO'/
     1 ' SUBDIAGONAL NONZEROS IN C ARE STORED COLUMN BY COLUMN'/
     1 ' AD(K) = THE KTH DIAGONAL ELEMENT OF C'/
     1 ' ASD(K) =  KTH SUBDIAGONAL NONZERO IN C'/)
C
      WRITE(6,220)
  220 FORMAT(/' PERMUT IS FINISHED MATRIX IS ON FILE 9'/)
C
  230 CONTINUE
      STOP
C-----END PERMUT------------------------------------------------------
      END
```

```
C-----LORDER-(STAND ALONE PROGRAM)---------------------------------------
C
C     ACCORDING TO THE PFORT VERIFIER THIS PROGRAM IS PORTABLE.
C     HOWEVER TO AVOID COMPOUNDING FORMAT CONVERSION ERRORS,
C     MATRIX ENTRIES SHOULD BE STORED IN MACHINE FORMAT, ((4Z20)
C     FOR IBM/3081).
C
C     LORDER TAKES A SPARSE MATRIX C AND A PERMUTATION P GIVEN BY
C     THE VECTOR IPR AND COMPUTES THE PERMUTED MATRIX B = P*C*P' ,
C     AND THEN WRITES B TO FILE 9 ALONG WITH IPR AND ANY SCALE SO
C     AND SHIFT THAT WERE USED TO OBTAIN THE INPUT MATRIX C. (HERE
C     ROW(COL) I OF B CORRESPONDS TO ROW(COL) J OF A WHERE J = IPR(I),
C     AND INPUT MATRIX C = SO*A + SHIFT*I.
C
C-----------------------------------------------------------------------
      DOUBLE PRECISION ASD(10000),AD(3000),BSD(10000),BD(3000)
      DOUBLE PRECISION SHIFT,SO
      INTEGER  IPR(3000),IPT(3000)
      INTEGER  IROW(10000),INUM(10000),ICOL(3000)
      INTEGER  KROW(10000),KNUM(10000),KCOL(3000)
C-----------------------------------------------------------------------
C
C     ARRAYS MUST BE DIMENSIONED AS FOLLOWS:
C     1.  AD, BD:  >= N, THE ORDER OF C-MATRIX.
C     2.  ASD: >= NZS, THE NUMBER OF NONZERO SUBDIAGONAL ENTRIES IN C.
C     3.  BSD:   >= NZS, THE NUMBER OF NONZERO SUBDIAGONAL ENTRIES IN
C                B = P*C*P-TRANSPOSE
C     4.  IPR, IPT:    >= N
C     5.  ICOL, KCOL:    >= N
C     6.  IROW, KROW, INUM, KNUM:   >=  NZ = 2*NZS + N
C
C-----------------------------------------------------------------------
C     OUTPUT HEADER
      WRITE(6,10)
   10 FORMAT(/' LORDER PROGRAM, COMPUTE B = P*C*(P-TRANSPOSE), STORE ON
     1FILE 9'/)
C
C     READ NUMBER OF NONZERO SUBDIAGONAL ENTRIES (NZS), ORDER OF MATRIX
C     (N),INDEX OF LAST COLUMN CONTAINING NONZERO ENTRIES BELOW THE
C     DIAGONAL (NZL), MATRIX IDENTIFICATION NUMBER (MATNO), PERMUTATION
C     FLAG (JPERM).
      READ(8,20) NZS,N,NZL,MATNO,JPERM
   20 FORMAT(I10,2I6,I8,I6)
C
      WRITE(6,30) NZS,N,NZL,MATNO,JPERM
   30 FORMAT(/I10,2I6,I8,I3,' = NZS,N,NZL,MATNO,JPERM'/)
C
C     NUMBER OF NONZERO SUBDIAGONAL ENTRIES IN EACH COLUMN IS READ
C     THEN THE CORRESPONDING ROW INDEX FOR EACH SUCH ENTRY IS READ
      READ(8,40) (ICOL(K), K=1,NZL)
      READ(8,40) (IROW(K), K=1,NZS)
   40 FORMAT(13I6)
C
      NZL1 = NZL + 1
      DO 50 K = NZL1,N
   50 ICOL(K) = 0
C
C     DIAGONAL OF C-MATRIX IS READ (INCLUDING ANY ZERO ENTRIES), THEN
C     NONZERO SUBDIAGONAL ENTRIES ARE READ IN
      READ(8,60) (AD(K), K=1,N)
      READ(8,60) (ASD(K), K=1,NZS)
   60 FORMAT(4E19.10)
C
      IF(JPERM.EQ.0) GO TO 390
C     READ PERMUTATION
```

```
      READ(8,40) (IPR(K), K = 1,N)
C
      DO 70 K = 1,N
      J = IPR(K)
   70 IPT(J) = K
C
      READ(8,80) SO,SHIFT
   80 FORMAT(2E12.5)
C
      WRITE(6,90)
   90 FORMAT(/' MATRIX HAS BEEN READ IN FROM FILE 8'/
     1 ' PERMUTATION IPR HAS BEEN READ IN'/)
C
C     EXPAND IROW AND ICOL TO INCLUDE DIAGONAL AND SUPER DIAGONAL
      KCOL(1) = 1 + ICOL(1)
      KNUM(1) = -1
      KROW(1) = 1
      IF (ICOL(1).EQ.0) GO TO 110
      KL = ICOL(1)
      DO 100 K = 1,KL
      KP1 = K+1
      KROW(KP1) = IROW(K)
  100 KNUM(KP1) = K
  110 KCOUNT = KCOL(1)
C
      DO 160 K = 2,N
      K1 = MIN(K-1,NZL)
      JL = 0
      JCOUNT = 0
      DO 140 J = 1,K1
      IF (ICOL(J).EQ.0) GO TO 140
      JF = JL + 1
      JL = JL + ICOL(J)
      DO 130 JJ = JF,JL
      IF (IROW(JJ)-K) 130,120,140
  120 KCOUNT = KCOUNT + 1
      JCOUNT = JCOUNT + 1
      KROW(KCOUNT) = J
      KNUM(KCOUNT) = JJ
      GO TO 140
  130 CONTINUE
  140 CONTINUE
      KCOUNT = KCOUNT + 1
      KROW(KCOUNT) = K
      KNUM(KCOUNT) = -K
      ITEMP = 0
      IF (K.LE.NZL) ITEMP = ICOL(K)
      KCOL(K) = JCOUNT + 1 + ITEMP
      IF (K.GT.NZL.OR.ICOL(K).EQ.0) GO TO 160
      KF = 1 + KL
      KL = KL + ICOL(K)
      DO 150 J = KF,KL
      KCOUNT = KCOUNT + 1
      KROW(KCOUNT) = IROW(J)
  150 KNUM(KCOUNT) = J
  160 CONTINUE
C     NTOTAL = N + 2*NZS
C     A-MATRIX INDEX LISTS HAVE BEEN EXPANDED
C
      WRITE(6,170)
  170 FORMAT(/' EXPANSION OF INDEX LISTS FOR C-MATRIX IS COMPLETED'/)
C
C     DETERMINE STRUCTURE OF B = P*C*P-TRANSPOSE
      IL = 0
      KCOUNT = 0
      DO 180 K = 1,N
  180 ICOL(K) = 0
```

```
      DO 270 K = 1,N
      J = IPR(K)
      JL = 0
      IF (J.EQ.1) GO TO 200
      JM1 = J - 1
      DO 190 JJ = 1,JM1
  190 JL = JL + KCOL(JJ)
  200 CONTINUE
      JF = JL + 1
      JL = JL + KCOL(J)
      ICOL(K) = KCOL(J)
      IF = IL + 1
      IL = IL + ICOL(K)
C
      DO 210 JJ = JF,JL
      KCOUNT = KCOUNT + 1
      JR = KROW(JJ)
      JK = IPT(JR)
      INUM(KCOUNT) = KNUM(JJ)
  210 IROW(KCOUNT) = JK
C
C     ORDER IROW VECTOR BY INCREASING SIZE
      IF (IF.EQ.IL) GO TO 240
      IF1 = IF + 1
      DO 230 I = IF1,IL
      IM1 = I-1
      IMF = IM1 + IF
      DO 220 L = IF,IM1
      II = IMF - L
      IF (IROW(II+1).GE.IROW(II)) GO TO 230
      IO = IROW(II)
      IROW(II) = IROW(II+1)
      IROW(II+1) = IO
      IO = INUM(II)
      INUM(II) = INUM(II+1)
      INUM(II+1) = IO
  220 CONTINUE
  230 CONTINUE
  240 CONTINUE
C
      DO 250 I = IF,IL
      IF (INUM(I).LT.0) GO TO 260
  250 CONTINUE
  260 INUM(I) = -J
  270 CONTINUE
C
C     GENERATE SPARSE MATRIX REPRESENTATION OF B-MATRIX
      KCOUNT = 0
      DO 280 K = 1,N
  280 KCOL(K) = 0
      DO 320 K = 1,N
      KL = 0
      DO 290 KK = 1,K
  290 KL = KL + ICOL(KK)
      KK = KL+ 1
  300 KK = KK - 1
      IF (INUM(KK).GE.0) GO TO 300
      KCOL(K) = KL - KK
      J = IPR(K)
      BD(K) = AD(J)
      KF - KK + 1
      IF (KCOL(K).EQ.0) GO TO 320
      DO 310 JJ = KF,KL
      KCOUNT = KCOUNT + 1
      KROW(KCOUNT) = IROW(JJ)
      KK = INUM(JJ)
  310 BSD(KCOUNT) = ASD(KK)
```

```
  320 CONTINUE
      NZL = 0
      DO 330 K = 1,N
      IF (KCOL(K).NE.0) NZL = K
  330 CONTINUE
C     WE NOW HAVE B = P*A*P-TRANSPOSE IN SPARSE MATRIX FORMAT, WRITE TO
C     FILE 9
C
      JPERM = 1
      WRITE(9,340) NZS,N,NZL,MATNO,JPERM
  340 FORMAT(I10,2I6,I8,I6,' = NZS,N,NZL,MATNO,JPERM. BCOMPAC')
C
C     NUMBER OF NONZERO SUBDIAGONAL ENTRIES IN EACH COLUMN IS WRITTEN
C     THEN THE CORRESPONDING ROW INDEX FOR EACH SUCH ENTRY IS WRITTEN
      WRITE(9,350) (KCOL(K), K=1,NZL)
      WRITE(9,350) (KROW(K), K=1,NZS)
  350 FORMAT(13I6)
C     DIAGONAL IS WRITTEN FIRST, THEN NONZERO BELOW DIAGONAL ENTRIES
      WRITE(9,360) (BD(K), K=1,N)
      WRITE(9,360) (BSD(K), K=1,NZS)
  360 FORMAT(4E19.10)
C
C     WRITE PERMUTATION
      WRITE(9,350) (IPR(K), K=1,N)
C
      WRITE(9,370) SO,SHIFT
  370 FORMAT(2E12.5,' = SO SHIFT'/
     1 ' ABOVE IS REORDERED MATRIX, B'/
     1 ' INPUT MATRIX SUPPLIED WAS C = SO*A + SHIFT*I'/
     1 ' B = P*C*(P-TRANSPOSE),  B IS STORED IN SPARSE MATRIX FORMAT'/
     1 ' ROW(COL) I OF B CORRESPONDS TO ROW(COL) J OF C, J = IPR(I)'/
     1 ' NZS = TOTAL NUMBER OF SUBDIAGONAL NONZEROS IN B-MATRIX'/
     1 ' KCOL(K) = NUMBER OF SUBDIAGONAL NONZEROS IN COL K OF B'/
     1 ' KROW(K) = ROW INDEX OF SUBDIAGONAL NONZERO'/
     1 ' SUBDIAGONAL NONZEROS IN B ARE STORED COLUMN BY COLUMN'/
     1 ' BD(K) = THE KTH DIAGONAL ELEMENT OF B'/
     1 ' BSD(K) = NUMERICAL VALUE OF KTH SUBDIAGONAL NONZERO IN B'/
     1 ' IPR(K) = J MEANS THAT ROW J OF C CORRESPONDS TO ROW K OF B'/)
C
      WRITE(6,380)
  380 FORMAT(' SPARSE FORMAT FOR B-MATRIX HAS BEEN WRITTEN TO FILE 9'/)
      GO TO 410
C
  390 WRITE(6,400)
  400 FORMAT(/' LORDER PROGRAM TERMINATES BECAUSE MATRIX FILE SUPPLIED D
     1ID NOT'/' CONTAIN A PERMUTATION'/)
C
  410 CONTINUE
C
      STOP
C-----END OF LORDER-----------------------------------------------------
      END
```

```
C-----LFACT-----------------------------------------------------------
C
C     NONPORTABLE CONSTRUCTIONS:
C     1.  FORMAT (4Z20).  TO AVOID COMPOUNDING FORMAT CONVERSION
C         ERRORS, THE MATRIX ENTRIES SHOULD BE IN MACHINE FORMAT,
C         (4Z20) FOR IBM/3081.
C
C     LFACT COMPUTES THE CHOLESKY FACTOR L FOR THE MATRIX B AND STORES
C     THIS FACTOR ON FILE 7.  B MUST BE A POSITIVE DEFINITE MATRIX.
C     THE PERMUTATION P (IN IPR), THE SCALE SO AND THE SHIFT (IF ANY)
C     USED TO OBTAIN B FROM THE ORIGINAL MATRIX A ARE STORED AT THE END
C     OF FILE 7.   THAT IS,  B = SO*P*A*P' + SHIFT*I.  THE PROGRAM
C     ASSUMES THAT THE DATA READ FROM FILE 9 IS FOR THE B-MATRIX.
C
C-----------------------------------------------------------------------
C
C     ARRAYS MUST BE DIMENSIONED AS FOLLOWS:
C     1.  AD:  >= N, THE ORDER OF A-MATRIX.
C     3.  ASD:   >= NZT, THE NUMBER OF NONZERO SUBDIAGONAL ENTRIES
C                IN THE CHOLESKY FACTOR OF B.
C     4.  ICOL,IPR:   >= N
C     5.  IROW:   >=  NZT
C
C-----------------------------------------------------------------------
      DOUBLE PRECISION ASD(10000),AD(3000)
      DOUBLE PRECISION ZERO,ONE,TEMP,SO,SHIFT
      INTEGER IROW(10000),ICOL(3000),IPR(3000)
      DOUBLE PRECISION DSQRT
C-----------------------------------------------------------------------
C     OUTPUT HEADER
      WRITE(6,5)
    5 FORMAT(/' LFACT PROGRAM, COMPUTE CHOLESKY FACTOR FOR POSITIVED DEF
     1INITE B-MATRIX'/' AND STORE THE FACTOR ON FILE 7'/)
C
C     SET PROGRAM PARAMETERS
      ONE = 1.0D0
      ZERO = 0.0D0
C
C     READ NUMBER OF NONZERO BELOW DIAGONAL ENTRIES, ORDER OF MATRIX,
C     INDEX OF LAST COLUMN CONTAINING NONZERO ENTRIES BELOW THE
C     DIAGONAL, MATRIX IDENTIFICATION NUMBER
      READ(9,15) NZS,N,NZL,MATNO,JPERM
   15 FORMAT(I10,2I6,I8,I6)
C
      WRITE(6,20) NZS,N,NZL,JPERM,MATNO
   20 FORMAT(I10,3I6,I8,' = NZS,N,NZL,JPERM,MATNO'/)
C
C     NUMBER OF NONZERO SUBDIAGONAL ENTRIES IN EACH COLUMN IS READ
C     THEN THE CORRESPONDING ROW INDEX FOR EACH SUCH ENTRY IS READ
      READ(9,30) (ICOL(K), K=1,NZL)
      READ(9,30) (IROW(K), K=1,NZS)
   30 FORMAT(13I6)
C
C
      NZL1 = NZL + 1
      DO 40 K = NZL1,N
   40 ICOL(K) = 0
C
C     DIAGONAL IS READ (INCLUDING ANY ZERO ENTRIES), THEN NONZERO
C     BELOW DIAGONAL ENTRIES ARE READ IN
      READ(9,50) (AD(K), K=1,N)
      READ(9,50) (ASD(K), K=1,NZS)
   50 FORMAT(4E19.10)
C  50 FORMAT(4Z20)
C
```

```
      IF (JPERM.NE.0) READ(9,30) (IPR(K), K = 1,N)
C
      READ(9,55) SO,SHIFT
   55 FORMAT(2E12.5)
C
      WRITE(6,60)
   60 FORMAT(/' B-MATRIX HAS BEEN READ IN FROM FILE 9'/)
C
      IF (JPERM.NE.0) WRITE(6,65)
   65 FORMAT( ' PERMUTATION IPR HAS BEEN READ IN'/)
C
C     CALCULATE CHOLESKY FACTOR,  B = BL*(BL-TRANSPOSE)
      NZT = NZS
      NZL = N-1
      KL = 0
      DO 70 K = 1,N
C     CALCULATE KTH PIVOT FOR BL
      TEMP = AD(K)
C
      IF (AD(K).GT.ZERO) GO TO 80
C
      WRITE(6,90) K,AD(K)
   90 FORMAT(/I6,E15.8,' = K,AD(K)'/
     1' PIVOT IS NEGATIVE SO B-MATRIX IS NOT POSITIVE DEFINITE'/
     1' THEREFORE COMPUTATION OF CHOLESKY FACTOR TERMINATES'/)
      GO TO 240
C
   80 CONTINUE
      TEMP = DSQRT(TEMP)
      AD(K) = TEMP
      TEMP = ONE/TEMP
      IF(K.EQ.N.OR.ICOL(K).EQ.0) GO TO 70
      KF = KL + 1
      KL = KL + ICOL(K)
      DO 100 KK = KF,KL
      KR = IROW(KK)
      ASD(KK) = TEMP*ASD(KK)
  100 AD(KR) = AD(KR) - ASD(KK)**2
      IF (KF.EQ.KL) GO TO 70
      K1 = K+1
      DO 110 KK = KF,KL
      KR = IROW(KK)
      IF (KK.EQ.KL) GO TO 110
      KE = KL
      DO 120 KC = K1,KR
  120 KE= KE + ICOL(KC)
      KB = KE - ICOL(KR) + 1
      KK1 = KK + 1
      L = KB
      DO 130 LL = KK1,KL
      LR = IROW(LL)
      IF (ICOL(KR).EQ.0.OR.L.GT.KE) GO TO 140
  150 LC = IROW(L)
      IF (LC - LR) 160,170,140
  160 L = L + 1
      IF (L.LE.KE) GO TO 150
C     NEW NONZERO IN CHOLESKY FACTOR L
  140 NZT = NZT + 1
      L1 = L + 1
      NT = NZT + L1
      DO 180 KM = L1,NZT
      MK = NT - KM
      ASD(MK) = ASD(MK-1)
  180 IROW(MK) = IROW(MK-1)
      ICOL(KR) = ICOL(KR) + 1
      KE = KE + 1
      ASD(L) = -ASD(KK)*ASD(LL)
```

```
      IROW(L) = LR
      GO TO 130
C     UPDATE EXISTING ELEMENT
  170 ASD(L) = ASD(L) - ASD(KK)*ASD(LL)
  130 L = L + 1
  110 CONTINUE
   70 CONTINUE
C
C
C     FACTOR L HAS BEEN COMPUTED, STORE IN SPARSE FORMAT ON FILE 7
C
      WRITE(7,190) NZT,N,NZL,MATNO,JPERM
  190 FORMAT(I10,2I6,I8,I6,' = NZT,N,NZL,MATNO,JPERM. LCOMPAC')
C
C     NUMBER OF NONZERO SUBDIAGONAL ENTRIES IN EACH COLUMN IS WRITTEN
C     THEN THE CORRESPONDING ROW INDEX FOR EACH SUCH ENTRY IS WRITTEN
      WRITE(7,200) (ICOL(K), K=1,NZL)
      WRITE(7,200) (IROW(K), K=1,NZT)
  200 FORMAT(13I6)
C     DIAGONAL IS WRITTEN FIRST, THEN NONZERO BELOW DIAGONAL ENTRIES
      WRITE(7,210) (AD(K), K=1,N)
      WRITE(7,210) (ASD(K), K=1,NZT)
  210 FORMAT(4Z20)
C 210 FORMAT(3E25.16)
      IF (JPERM.NE.0) WRITE(7,200) (IPR(K), K=1,N)
C
      WRITE(7,220) SO,SHIFT
  220 FORMAT(2E12.5,' = SO SHIFT'/
     1 ' ABOVE IS CHOLESKY FACTOR FOR B-MATRIX'/
     1 ' IF JPERM = 0, THEN P = I.  C = SO*A * SHIFT*I'/
     1 ' B = P*C*P-TRANS = L*L-TRANS,  L IS STORED IN SPARSE FORMAT'/
     1 ' ROW(COL) I OF B CORRESPONDS TO ROW(COL) J OF C, J = IPR(I)'/
     1 ' NZT = TOTAL NUMBER OF SUBDIAGONAL NONZEROS IN L'/
     1 ' ICOL(K) = NUMBER OF SUBDIAGONAL NONZEROS IN COL K OF L'/
     1 ' IROW(K) = ROW INDEX OF SUBDIAGONAL NONZERO'/
     1 ' SUBDIAGONAL NONZEROS IN L ARE STORED COLUMN BY COLUMN'/
     1 ' AD(K) = KTH DIAGONAL ELEMENT OF L'/
     1 ' ASD(K) = KTH SUBDIAGONAL NONZERO IN L'/)
C
      WRITE(6,230)
  230 FORMAT(' CHOLESKY FACTOR HAS BEEN WRITTEN TO FILE 7 '/)
C
  240 CONTINUE
      STOP
C-----END OF IFACT--------------------------------------------------
      END
```

```
C-----LTEST-----------------------------------------------------------
C
C     CONTAINS MAIN PROGRAM LTEST AND SAMPLE CMATS, CMATV, BSOLV
C     LTEST ALSO REQUIRES A RANDOM NUMBER GENERATOR.
C
C     LTEST GIVES A ROUGH CHECK ON THE CONDITION OF A MATRIX B BY
C     SOLVING B*X = B*V1 FOR X WHERE V1 IS A KNOWN, RANDOMLY-GENERATED
C     VECTOR.  SOLVING IS DONE, WITH AND WITHOUT ITERATIVE REFINEMENT.
C     IN BOTH CASES, X IS COMPARED WITH V1 AND THE ERRORS ARE
C     WRITTEN TO FILE 6.
C
C     VECTORS V0, V1, V2, VS, AND G ARE USED IN THE COMPUTATIONS.
C     NOTE THAT THE SUBROUTINE CMATS USED TO COMPUTE THE RESIDUAL
C     IN EXTENDED PRECISION FOR THE ITERATIVE REFINEMENT CALCULATION
C     REQUIRES AN EXTRA LONG V0 VECTOR OF LENGTH TWICE THE SIZE OF B.
C
C     NONPORTABLE CONSTRUCTIONS:
C     1.  THE ENTRY MECHANISM WHICH PASSES THE STORAGE LOCATIONS OF
C         ARRAYS AND PARAMETERS THAT DEFINE THE B-MATRIX TO THE
C         SUBROUTINES CMATV, CMATS, AND BSOLV.
C     2.  FORMATS (20A4) AND (4Z20).  TO AVOID COMPOUNDING FORMAT
C         CONVERSION ERRORS, MATRIX ENTRIES SHOULD BE STORED IN
C         MACHINE FORMAT, ((4Z20) FOR IBM/3081). ALSO FREE FORMAT
C         (5,*).
C     3.  REAL*16 VARIABLES IN CMATS SUBROUTINE.
C
C
C-----------------------------------------------------------------------
      DOUBLE PRECISION ASD(10000),AD(3000),BSD(20000),BD(3000)
      DOUBLE PRECISION V0(6000),V1(3000),V2(3000),VS(3000)
      DOUBLE PRECISION ZERO,ONE,TEMP,SUM
      DOUBLE PRECISION ERROR0,ERROR1,ENORM0,ENORM1
      REAL EXPLAN(20),G(3000)
      INTEGER  IROW(20000),ICOL(3000),KROW(30000),KCOL(3000),SVSEED
      DOUBLE PRECISION FINPRO
      DOUBLE PRECISION DABS, DMAX1, DSQRT
C-----------------------------------------------------------------------
C
C     ARRAYS MUST BE DIMENSIONED AS FOLLOWS:
C     1.  AD, BD:  >= N, THE ORDER OF A-MATRIX.
C     2.  ASD:    >= NZS, THE NUMBER OF NONZERO SUBDIAGONAL ENTRIES IN B.
C     3.  BSD:    >= NZT, THE NUMBER OF NONZERO SUBDIAGONAL ENTRIES
C                 IN THE CHOLESKY FACTOR OF B.
C     5.  ICOL, KCOL:   >= N
C     6.  KROW:   >=  NZS
C     7.  IROW:   >=  NZT
C     8.  V1,V2,VS:   >=   N
C     9.  V0:   >=   2*N
C
C-----------------------------------------------------------------------
C     OUTPUT HEADER
      WRITE(6,10)
   10 FORMAT(/' LTEST PROGRAM, ROUGH CHECK ON NUMERICAL CONDITION OF GIV
     1EN MATRIX'/)
C
C     SET PROGRAM PARAMETERS
      ONE = 1.0D0
      ZERO = 0.0D0
C
C     READ INPUT HEADER
      READ(5,20) EXPLAN
      WRITE(6,20) EXPLAN
   20 FORMAT(20A4)
C
C     READ IN IN FREE FORMAT USER-SPECIFIED PARAMETERS FROM FILE 5
      READ(5,20) EXPLAN
```

```
      READ(5,*) SVSEED
C
C     READ NUMBER OF NONZERO BELOW DIAGONAL ENTRIES, ORDER OF MATRIX,
C     INDEX OF LAST COLUMN CONTAINING NONZERO ENTRIES BELOW THE
C     DIAGONAL, MATRIX IDENTIFICATION NUMBER
      READ(9,30) NZS,N,NZL,MATNO,JPERM
   30 FORMAT(I10,2I6,I8,I6)
C
      WRITE(6,40) NZS,N,NZL,JPERM,MATNO,SVSEED
   40 FORMAT(I10,3I6,' = NZS,N,NZL,JPERM'/
     1 I8,I12,' = MATNO,SVSEED'/)
C
C     NUMBER OF NONZERO SUBDIAGONAL ENTRIES IN EACH COLUMN IS READ
C     THEN THE CORRESPONDING ROW INDEX FOR EACH SUCH ENTRY IS READ
      READ(9,50) (KCOL(K), K=1,NZL)
      READ(9,50) (KROW(K), K=1,NZS)
   50 FORMAT(13I6)
C
C
      NZL1 = NZL + 1
      DO 60 K = NZL1,N
   60 KCOL(K) = 0
C
C     DIAGONAL IS READ (INCLUDING ANY ZERO ENTRIES), THEN NONZERO
C     BELOW DIAGONAL ENTRIES ARE READ IN
      READ(9,70) (AD(K), K=1,N)
      READ(9,70) (ASD(K), K=1,NZS)
   70 FORMAT(4E19.10)
C
      WRITE(6,80)
   80 FORMAT(/' B-MATRIX HAS BEEN READ IN FROM FILE 9'/)
C
C-----------------------------------------------------------------------
C     ENTRIES TO CMATS AND CMATV SUBROUTINES
      CALL CMATSE(ASD,AD,KCOL,KROW,N,NZL)
      CALL CMATVE(ASD,AD,KCOL,KROW,N,NZL)
C-----------------------------------------------------------------------
C
C     READ CHOLESKY FACTOR FROM FILE 7
C
      READ(7,90) NZT,N,NZL,MATNO,JPERM
   90 FORMAT(I10,2I6,I8,I6)
C
C     NUMBER OF NONZERO SUBDIAGONAL ENTRIES IN EACH COLUMN IS READ
C     THEN THE CORRESPONDING ROW INDEX FOR EACH SUCH ENTRY IS READ
      READ(7,100) (ICOL(K), K=1,NZL)
      READ(7,100) (IROW(K), K=1,NZT)
  100 FORMAT(13I6)
C     DIAGONAL IS READ FIRST, THEN NONZERO BELOW DIAGONAL ENTRIES
      READ(7,110) (BD(K), K=1,N)
      READ(7,110) (BSD(K), K=1,NZT)
  110 FORMAT(4Z20)
C  90 FORMAT(3E25.16)
C
C-----------------------------------------------------------------------
C     ENTRY TO BSOLV SUBROUTINE, PASS FACTOR OF B
      CALL BSOLVE(BSD,BD,ICOL,IROW,N,NZT,NZL)
C-----------------------------------------------------------------------
C
C     SOLVE B*X = B*V1 WITH AND WITHOUT ITERATIVE REFINEMENT, COMPARE
C     ERRORS IN SOLVING AS A ROUGH CHECK ON THE CONDITION OF THE
C     MATRIX B.
C
      IIX = SVSEED
C
C-----------------------------------------------------------------------
C     COMPUTES RANDOM VECTOR FOR USE IN RIGHT-HAND SIDE
```

```
      CALL GENRAN(IIX,G,N)
C-----------------------------------------------------------------------
C
      DO 120 K = 1,N
  120 V1(K) = G(K)
C
C-----------------------------------------------------------------------
      SUM = FINPRO(N,V1(1),1,V1(1),1)
C-----------------------------------------------------------------------
      SUM = ONE/DSQRT(SUM)
C
      DO 130 K = 1,N
  130 V1(K) = V1(K)*SUM
C
      SUM = ZERO
C
C----------------------------------------------------------------------
C     COMPUTE V2 = RHS = B*V1  C = SO*A + SHIFT*I  B = P*C*P'
C     VS = B(INVERSE)*V2
      CALL CMATV(V1,V2,SUM)
      CALL BSOLV(VS,V2)
C----------------------------------------------------------------------
C
      SUM = ZERO
      ERRORO = ZERO
      DO 140 K = 1,N
      TEMP = DABS(V1(K) - VS(K))
      SUM = SUM + TEMP*TEMP
  140 ERRORO = DMAX1(ERRORO,TEMP)
      ENORMO = DSQRT(SUM)
C
      WRITE(6,150) ENORMO,ERRORO
  150 FORMAT(6X,'ENORMO',6X,'ERRORO'/2E12.4/
     1 ' ENORMO  = NORM (V1 - VS),     VS = BI*(B*V1)'/
     1 ' ERRORO = MAX DABS(V1(K) - VS(K)),  K = 1,N'/)
C
      SUM = ONE
C
C----------------------------------------------------------------------
C     CALCULATE RESIDUAL IN EXTENDED PRECISION V2 = B*VS - V2
C     THEN DO ITERATIVE REFINEMENT
      CALL CMATS(VS,V2,VO,SUM)
      CALL BSOLV(V2,V2)
C----------------------------------------------------------------------
C
      DO 160 K = 1,N
  160 VS(K) = VS(K) - V2(K)
C
      SUM = ZERO
      ERROR1 = ZERO
      DO 170 K = 1,N
      TEMP = DABS(V1(K) - VS(K))
      SUM = SUM + TEMP*TEMP
  170 ERROR1 = DMAX1(ERROR1,TEMP)
      ENORM1 = DSQRT(SUM)
C
      WRITE(6,180) ENORM1,ERROR1
  180 FORMAT(6X,'ENORM1',6X,'ERROR1'/2E12.4/
     1 ' ERROR AFTER ITERATIVE REFINEMENT'/
     1 ' ENORM1  = NORM (V1 - VS),     VS = BI*(B*V1)'/
     1 ' ERROR1 = MAX DABS(V1(K) - VS(K)),  K = 1,N'/)
C
      STOP
C-----END OF LTEST------------------------------------------------------
      END
C
C----CMATS--------------------------------------------------------------
```

```
C
C     REAL, SYMMETRIC, SPARSE MATRIX-VECTOR MULTIPLY USING EXTENDED
C     PRECISION.  CALCULATES U = B*W - SUM*U FOR USE IN ITERATIVE
C     REFINEMENT.  MATRIX B STORED IN SPARSE FORMAT.
C
      SUBROUTINE CMATS(W,U,Z,SUM)
C
C-----------------------------------------------------------------------
      DOUBLE PRECISION  U(1),W(1),BSD(1),BD(1),SUM
      REAL*16 Z(1),T0,T1,T2,S0
      INTEGER  IROW(1),ICOL(1)
C-----------------------------------------------------------------------
      S0 = SUM
C
      DO 10 I = 1,N
      T0 = BD(I)
      T1 = W(I)
      T2 = U(I)
   10 Z(I) = T0*T1-S0*T2
C
      LLAST = 0
C
      DO 30 J = 1,NZL
C
      IF (ICOL(J).EQ.0) GO TO 30
C
      LFIRST = LLAST + 1
      LLAST = LLAST + ICOL(J)
C
      DO 20 L = LFIRST,LLAST
      I = IROW(L)
      T0 = BSD(L)
      T1 = W(J)
      T2 = W(I)
C
      Z(I) = Z(I) + T0*T1
      Z(J) = Z(J) + T0*T2
C
   20 CONTINUE
C
   30 CONTINUE
C
      DO 40 I =1,N
   40 U(I) = Z(I)
C
      RETURN
C
C-----------------------------------------------------------------------
      ENTRY CMATSE(BSD,BD,ICOL,IROW,N,NZL)
C-----------------------------------------------------------------------
C
      RETURN
C-----END OF CMATS------------------------------------------------------
      END
C
C-----CMATV-------------------------------------------------------------
C
C     SYMMETRIC, SPARSE MATRIX-VECTOR MULTIPLY, B MATRIX STORED
C     IN SPARSE FORMAT.  CMATV CALCULATES U = B*W - SUM*U
C
      SUBROUTINE CMATV(W,U,SUM)
C
C-----------------------------------------------------------------------
      DOUBLE PRECISION  U(1),W(1),BSD(1),BD(1),SUM
      INTEGER  KROW(1),KCOL(1)
C-----------------------------------------------------------------------
C
```

```
      DO 10 I = 1,N
   10 U(I) = BD(I)*W(I) - SUM*U(I)
C
      LLAST = 0
C
      DO 30 J = 1,NZL
C
      IF (KCOL(J).EQ.0) GO TO 30
C
      LFIRST = LLAST + 1
      LLAST = LLAST + KCOL(J)
C
      DO 20 L = LFIRST,LLAST
      I = KROW(L)
C
      U(I) = U(I) + BSD(L)*W(J)
      U(J) = U(J) + BSD(L)*W(I)
C
   20 CONTINUE
C
   30 CONTINUE
C
      RETURN
C
C-----------------------------------------------------------------------
      ENTRY CMATVE(BSD,BD,KCOL,KROW,N,NZL)
C-----------------------------------------------------------------------
C
      RETURN
C-----END OF CMATV------------------------------------------------------
      END
C
C-----BSOLV-----------------------------------------------------------
C
C     SOLVES B*U = V WHERE  B = L*L'.
C     FIRST SOLVES L*U = V FOR U, THEN SOLVES L'*U = U FOR U
C
      SUBROUTINE BSOLV(U,V)
C
C-----------------------------------------------------------------------
      DOUBLE PRECISION   AD(1),ASD(1),U(1),V(1),TEMP
      INTEGER  ICOL(1),IROW(1)
C-----------------------------------------------------------------------
      KL = 0
      DO 10 K = 1,N
   10 U(K) = V(K)
      DO 30 K = 1,N
      TEMP = U(K)/AD(K)
      U(K) = TEMP
      IF (ICOL(K).EQ.0.OR.K.EQ.N) GO TO 30
      KF = KL + 1
      KL = KL + ICOL(K)
      DO 20 KK = KF,KL
      KR = IROW(KK)
   20 U(KR) = U(KR) - TEMP*ASD(KK)
   30 CONTINUE
C
      NP1 = N+1
      KF = NZT + 1
      DO 50 K = 1,N
      L = NP1 - K
      TEMP = U(L)
      IF (ICOL(L).EQ.0.OR.L.EQ.N) GO TO 50
      KL = KF - 1
      KF = KF - ICOL(L)
      DO 40 LL = KF,KL
      LR = IROW(LL)
```

```
   40 TEMP = TEMP - ASD(LL)*U(LR)
   50 U(L) = TEMP/AD(L)
C
      RETURN
C
C-----------------------------------------------------------------------
      ENTRY BSOLVE(ASD,AD,ICOL,IROW,N,NZT,NZL)
C-----------------------------------------------------------------------
C
      RETURN
C-----END OF BSOLV------------------------------------------------------
      END
```

SECTION 4.6 FILE DEFINITIONS AND SAMPLE INPUT FILE

Below is a listing of the input/output files which are accessed by the real symmetric Lanczos eigenvalue program, LIVAL. Included also is a sample of the input file which LIVAL requires on file 5. The parameters in this file are supplied in free format. LIVAL computes eigenvalues of real symmetric matrices BI on user-specified intervals where BI is the inverse of the matrix B = P*C*P' with C = SO*A + SHIFT*I and SO is a scale and SHIFT is a shift. The sample codes assume that C is positive definite and has a reasonable condition number. The permutation matrix P Is used to preserve the sparseness of the given matrix in the Cholesky factorization, B = L*L'. The user could replace the BSOLVE subroutine provided here by another more general factorization subroutine.

```
SAMPLE DEFINITIONS OF THE INPUT/OUTPUT FILES FOR LIVAL
-----------------------------------------------------------------
*LIVAL EXEC LANCZOS EIGENVALUE CALCULATION USING FACTORIZATION
FI 06 TERM
FILEDEF  1 DISK &1        NHISTORY  A (RECFM F LRECL 80 BLOCK 80
FILEDEF  2 DISK &1        HISTORY   A (RECFM F LRECL 80 BLOCK 80
FILEDEF  3 DISK &1        GOODEV    A (RECFM F LRECL 80 BLOCK 80
FILEDEF  4 DISK &1        ERRINV    A (RECFM F LRECL 80 BLOCK 80
FILEDEF  5 DISK LIVAL     INPUT     A (RECFM F LRECL 80 BLOCK 80
FILEDEF  7 DISK &1        LCOMPAC   A (RECFM F LRECL 80 BLOCK 80
FILEDEF 11 DISK &1        DISTINCT  A (RECFM F LRECL 80 BLOCK 80
LOAD   LIVAL   LESUB   LIMULT
------------------------------------------------------------------
```

```
SAMPLE INPUT FILE FOR LIVAL
------------------------------------------------------------------
 LIVAL EIGENVALUE COMPUTATION, NO REORTHOGONALIZATION
 USING INVERSE OF REAL SYMMETRIC MATRIX VIA FACTORIZATION
LINE 1      N     KMAX     NMEVS      MATNO      SO     SHIFT
          528     2640         2     721830      1.0        0.
LINE 2   SVSEED     RHSEED        MXINIT     MXSTUR
       49302312   5731029             5     100000
LINE 3    ISTART       ISTOP
               0           1
LINE 4      IHIS       IDIST  IWRITE
               1           0       1
LINE 5   RELTOL (RELATIVE TOLERANCE IN 'COMBINING' GOODEV)
    .0000000001
LINE 6   MB(1)   MB(2)   MB(3)   MB(4)     (ORDERS OF T(1,MEV) )
          100     125
LINE 7    NINT      (NUMBER OF SUB-INTERVALS FOR BISEC)
            1
LINE 8    LB(1)     LB(2)    LB(3)    LB(4)  (INTERVAL LOWER BOUNDS)
           1.0
LINE 9     UB(1)     UB(2)    UB(3)    UB(4)  (INTERVAL UPPER BOUNDS)
          100.0
--------------------------------------------------------------------
```

Below is a listing of the input/output files which are accessed by the real symmetric Lanczos eigenvector program, LIVEC. Included also is a sample of the input file which LIVEC requires on file 5. The parameters in this file are supplied in free format. LIVEC computes eigenvectors for each of a user-specified subset of the eigenvalues computed by the companion program LIVAL. The matrix used in the eigenvector computation is a scaled, shifted and inverted version of a given matrix. Inversion is accomplished via matrix factorization.

SAMPLE DEFINITIONS OF THE INPUT/OUTPUT FILES FOR LIVEC

```
-----------------------------------------------------------------------
*LIVEC EXEC, CALCULATE EIGENVECTORS FOR INVERSE OF REAL SYMMETRIC MATRIX
FI 06 TERM
FILEDEF  2 DISK &1       HISTORY   A (RECFM F LRECL 80 BLOCK 80
FILEDEF  3 DISK &1       GOODEV    A (RECFM F LRECL 80 BLOCK 80
FILEDEF  4 DISK &1       ERRINV    A (RECFM F LRECL 80 BLOCK 80
FILEDEF  5 DISK LIVEC    INPUT     A (RECFM F LRECL 80 BLOCK 80
FILEDEF  7 DISK &1       LCOMPAC   A (RECFM F LRECL 80 BLOCK 80
FILEDEF  9 DISK &1       ERREST    A (RECFM F LRECL 80 BLOCK 80
FILEDEF 10 DISK &1       BOUNDS    A (RECFM F LRECL 80 BLOCK 80
FILEDEF 11 DISK &1       TEIGVECS  A (RECFM F LRECL 80 BLOCK 80
FILEDEF 12 DISK &1       RITZVECS  A (RECFM F LRECL 80 BLOCK 80
FILEDEF 13 DISK &1       PAIGE     A (RECFM F LRECL 80 BLOCK 80
LOAD  LIVEC  LESUB  LIMULT
-----------------------------------------------------------------------
```

SAMPLE INPUT FILE FOR LIVEC

```
-----------------------------------------------------------------------
 LIVEC INPUT LANCZOS EIGENVECTOR COMPUTATIONS, NO REORTHOGONALIZATION
LINE 1  MATNO     N    SO   SHIFT  JPERM (ID,SIZE,SCALE,SHIFT,PERMUT?
            20  2161   -1.0     0.01      0
LINE 2  MDIMTV    MDIMRV  MBETA (MAX.DIMENSIONS,TVEC,RITVEC AND BETA
         10000     10000   2000
LINE 3       RELTOL
        .0000000001
LINE 4  MBOUND   NTVCON SVTVEC IREAD (FLAGS
             0        1      0     1
LINE 5  TVSTOP   LVCONT ERCONT  IWRITE (FLAGS
             0        1      1      1
LINE 6    RHSEED  (RANDOM GENERATOR SEED FOR STARTING VECTOR IN INVERM)
        45329517
-----------------------------------------------------------------------
```

CHAPTER 5

REAL SYMMETRIC GENERALIZED PROBLEMS

SECTION 5.1 INTRODUCTION

The FORTRAN codes in this Chapter address the question of computing distinct eigenvalues and corresponding eigenvectors of a real symmetric generalized eigenvalue problem. Given two real symmetric matrices A and B, where B is positive definite and its Cholesky factors are available, these codes compute real scalars λ and corresponding real-valued vectors $x \neq 0$ such that

$$Ax = \lambda Bx . \tag{5.1.1}$$

Given a real symmetric positive definite matrix B, the Cholesky decomposition of B has the form

$$B = LL^T \text{ where L is a lower triangular matrix.} \tag{5.1.2}$$

Real symmetric matrices and Cholesky factorizations are discussed in detail in Stewart [1973]. See Section 2.1 for a brief summary of the properties of real symmetric matrices which we use.

Theoretically, this type of real symmetric generalized problem is equivalent to the following real symmetric problem:

$$L^{-1}AL^{-T}y = \lambda y, \text{ where } y = L^T x . \tag{5.1.3}$$

Therefore, we could solve this type of generalized problem by applying the real symmetric Lanczos procedure given in Chapter 2 directly to the composite matrix $C \equiv L^{-1}AL^{-T}$ given in Eqn(5.1.3). However, we prefer to work directly with the generalized problem. In this setting the role of the B-matrix in the single-vector Lanczos computations is clearly displayed.

The single-vector Lanczos codes in this chapter can be used to compute either a very few or very many of the distinct eigenvalues of the given real symmetric generalized problem. The documentation for these codes is contained in Section 2.2. As in the real symmetric case, the AB-multiplicity of a given computed 'good' Lanczos eigenvalue can be obtained only with additional computation, and the modifications required to do this additional computation are not included in the enclosed versions of these codes.

We use the following 'generalized' Lanczos recursion. For $i = 1,2,...,m$ and a randomly-generated starting vector v_1 with $\|v_1\|_B = 1$, generate Lanczos vectors v_i using the following recursion.

$$\beta_{i+1}Bv_{i+1} = Av_i - \alpha_i Bv_i - \beta_i Bv_{i-1} \tag{5.1.4}$$

where

$$\alpha_i \equiv v_i^T(Av_i - \beta_i Bv_{i-1}) \text{ and } \beta_{i+1} \equiv \| L^{-1}(Av_i - \alpha_i Bv_i - \beta_i Bv_{i-1}) \| \qquad (5.1.5)$$

By construction, the B-norm of each Lanczos vector is one. That is, for all i, $\| v_i \|_B \equiv [v_i^T Bv_i]^{1/2} = 1$.

The B-norm is used because it is the 'natural' norm for real symmetric generalized problems when the B-matrix is positive definite. Given any two distinct eigenvalues λ and μ of Eqn(5.1.1), and corresponding eigenvectors x and y, we have that $x^T By = 0$. That is, the eigenvectors are orthogonal w.r.t. the B-norm, and the eigenvectors form a complete set of vectors. The positive definiteness of B is essential. The closer B is to being singular or indefinite, the less stable these computations will be. The generalized Lanczos recursion in Eqns(5.1.4) and (5.1.5) generates a family of real symmetric tridiagonal matrices (T-matrices) whose sizes are specified by the user.

LGVAL, the main program for the real symmetric generalized computations, calls the subroutine BISEC to compute eigenvalues of the specified tridiagonal T-matrices on the user-specified intervals. BISEC simultaneously computes these T-eigenvalues with their T-multiplicities and sorts the computed T-eigenvalues into two classes, the 'good' T-eigenvalues and the 'spurious' T-eigenvalues. The 'good' T-eigenvalues are accepted as approximations to eigenvalues of the generalized problem. The accuracy of these 'good' T-eigenvalues as eigenvalues of the generalized problem is then estimated using error estimates computed by the subroutine INVERR. Error estimates are computed only for isolated 'good' T-eigenvalues. All other 'good' T-eigenvalues are assumed to have converged. Convergence is then checked. If convergence has not yet occurred and a larger T-matrix has been specified by the user, the program will continue on to the larger T-matrix, repeating the above procedure on this larger matrix. After each T-matrix eigenvalue computation the corresponding approximations to the eigenvalues of the user-specified matrix A are computed and included in the output.

Once the eigenvalues have been computed accurately enough, the user can select a subset of the 'converged' eigenvalues for which eigenvectors are to be computed. The main program LGVEC, for computing eigenvectors of the real symmetric generalized problem using a factorization of B, is used to compute the desired eigenvectors.

All of the computations are done in double precision arithmetic. Once the Lanczos matrices have been computed, the remaining computations use the same subroutines which are used in the real symmetric case discussed in Chapter 2. In addition to the programs and subroutines provided here, the user must supply a subroutine USPECA which defines and initializes the A-matrix and a subroutine USPECB which defines and initializes the factors of the B-matrix. A subroutine AMATV which computes matrix-vector multiplies Ax for the A-matrix, and a subroutine BSOLV which solves the system of equations Bz=v must also be supplied. These subroutines must be constructed in such a way as to take advantage of the sparsity (and/or structure) of the two user-supplied matrices A and B and such that they are accurate.

The optional preprocessing programs PERMUT, LORDER, LFACT, and LTEST listed in Chapter 4 can also be used with the codes in this chapter. PERMUT calls the SPARSPAK Library [1979] to attempt to identify a reordering or permutation P of the given matrix B for which the sparseness of B is preserved under the factorization of the permuted matrix. LORDER takes a given matrix C and permutation P and computes the sparse format for the permuted matrix, PCP^T. LFACT computes the Cholesky factors of a given positive definite matrix. LTEST performs a very crude check on the numerical condition of the matrix supplied to it, by solving a system of equations with and without iterative refinement, LINPACK [1979]. Obviously, if the B-matrix is permuted then the A-matrix must be subjected to the same permutation. These codes assume that the Cholesky factor supplied in the subroutine USPECB corresponds to the permuted B-matrix and that the AMATV subroutine supplied corresponds to the corresponding permuted A-matrix. Thus, the Lanczos codes compute the eigenvalues and eigenvectors of the permuted problem. The permutation (if any) is then unwrapped in the eigenvector program LGVEC.

SECTION 5.2 MAIN PROGRAM, EIGENVALUE CALCULATIONS

```
C-----LGVAL  (EIGENVALUES, GENERALIZED SYMMETRIC PROBLEM)---------------
C
C     CONTAINS MAIN PROGRAM FOR COMPUTING DISTINCT EIGENVALUES OF
C     A*X = EVAL*B*X WHERE A AND B ARE REAL SYMMETRIC MATRICES,
C     B IS POSITIVE DEFINITE, AND THE CHOLESKY FACTORS OF B
C     ARE AVAILABLE FOR USE IN THE PROCEDURE.  PROCEDURE USES
C     GENERALIZATION OF LANCZOS TRIDIAGONALIZATION WITHOUT ANY
C     REORTHGONALIZATION.
C
C     PFORT VERIFIER IDENTIFIED THE FOLLOWING NONPORTABLE
C     CONSTRUCTIONS
C
C     1.  DATA/MACHEP/ STATEMENT
C     2.  ALL READ(5,*) STATEMENTS (FREE FORMAT)
C     3.  FORMAT(20A4) USED WITH EXPLANATORY HEADER EXPLAN.
C     4.  HEXADECIMAL FORMAT (4Z20) USED IN ALPHA/BETA FILES 1 AND 2.
C
C-----------------------------------------------------------------------
C
      DOUBLE PRECISION  ALPHA(5000),BETA(5001)
      DOUBLE PRECISION  V1(5000),V2(5000),VS(5000)
      DOUBLE PRECISION  LB(20),UB(20)
      DOUBLE PRECISION  BTOL,GAPTOL,TTOL,MACHEP,EPSM,RELTOL
      DOUBLE PRECISION  SCALE1,SCALE2,SCALE3,SCALE4,BISTOL,CONTOL,MULTOL
      DOUBLE PRECISION  ONE,ZERO,TEMP,TKMAX,BETAM,BKMIN,TO,T1
      REAL  G(5000),EXPLAN(20)
      INTEGER  MP(5000),NMEV(20)
      INTEGER  SVSEED,RHSEED,SVSOLD
      INTEGER IABS
      REAL  ABS
      DOUBLE PRECISION  DABS, DSQRT, DFLOAT
      EXTERNAL LSOLV, AMATV
C
C-----------------------------------------------------------------------
      DATA MACHEP/Z3410000000000000/
      EPSM = 2.0D0*MACHEP
C-----------------------------------------------------------------------
C
C     ARRAYS MUST BE DIMENSIONED AS FOLLOWS:
C     DIMENSION OF V2 ASSUMES THAT NO MORE THAN KMAX/2 EIGENVALUES
C     OF THE LANCZOS T-MATRICES ARE BEING COMPUTED IN ANY ONE OF THE
C     SUB-INTERVALS BEING CONSIDERED.  V2 CONTAINS THE UPPER AND LOWER
C     BOUNDS FOR EACH T-EIGENVALUE BEING COMPUTED BY BISEC IN ANY ONE
C     GIVEN INTERVAL.
C
C     1.  ALPHA: >= KMAX,    BETA: >= (KMAX+1)
C     2.  V1:  >= MAX(N,KMAX+1)
C     3.  V2,VS:    >= MAX(N,KMAX)
C     4.  G:  >= MAX(N,2*KMAX)
C     5.  MP:     >= KMAX
C     6.  LB,UB:  >= NUMBER OF SUBINTERVALS SUPPLIED TO BISEC.
C     7.  NMEV:  >= NUMBER OF T-MATRICES ALLOWED.
C     8.  EXPLAN:  DIMENSION IS 20.
C
C
C     IMPORTANT TOLERANCES OR SCALES THAT ARE USED REPEATEDLY
C     THROUGHOUT THE PROGRAM ARE THE FOLLOWING:
C     SCALED MACHINE EPSILON:  TTOL = TKMAX*EPSM WHERE
C     EPSM = 2*MACHINE EPSILON AND
C     TKMAX = MAX(|ALPHA(J)|,BETA(J), J = 1,MEV)
C     BISEC CONVERGENCE TOLERANCE:  BISTOL = DSQRT(1000+MEV)*TTOL
C     BISEC T-MULTIPLICITY TOLERANCE:  MULTOL = (1000+MEV)*TTOL
C     LANCZOS CONVERGENCE TOLERANCE:   CONTOL = BETA(MEV+1)*1.D-10
```

```
C-----------------------------------------------------------------------
C     OUTPUT HEADER
      WRITE(6,10)
   10 FORMAT(/' LANCZOS EIGENVALUE PROCEDURE FOR REAL SYMMETRIC GENERALI
     1ZED PROBLEMS,'/' A*X = EVAL*B*X, B POSITIVE DEFINITE WITH CHOLESKY
     1 FACTORS AVAILABLE'/)
C
C     SET PROGRAM PARAMETERS
C     SCALEK ARE USED IN TOLERANCES NEEDED IN SUBROUTINES LUMP,
C     ISOEV AND PRTEST.  USER MUST NOT MODIFY THEM.
      SCALE1 = 5.0D2
      SCALE2 = 5.0D0
      SCALE3 = 5.0D0
      SCALE4 = 1.0D4
      ONE  = 1.0D0
      ZERO = 0.0D0
      BTOL = 1.0D-8
C     BTOL = EPSM
      GAPTOL = 1.0D-8
      ICONV = 0
      MOLD = 0
      MOLD1 = 1
      ICT = 0
      MMB = 0
      IPROJ = 0
C-----------------------------------------------------------------------
C     READ USER-SPECIFIED PARAMETERS FROM INPUT FILE 5 (FREE FORMAT)
C
C     READ USER-PROVIDED HEADER FOR RUN
      READ(5,20) EXPLAN
      WRITE(6,20) EXPLAN
      READ(5,20) EXPLAN
      WRITE(6,20) EXPLAN
   20 FORMAT(20A4)
C
C     READ ORDER OF MATRICES (N) , MAXIMUM ORDER OF T-MATRIX (KMAX),
C     NUMBER OF T-MATRICES ALLOWED (NMEVS), AND MATRIX IDENTIFICATION
C     NUMBERS (MATNOA AND MATNOB)
      READ(5,20) EXPLAN
      READ(5,*) N,KMAX,NMEVS,MATNOA,MATNOB
C
C     READ SEEDS FOR LANCZS AND INVERR SUBROUTINES (SVSEED AND RHSEED)
C     READ MAXIMUM NUMBER OF ITERATIONS ALLOWED FOR EACH INVERSE
C     ITERATION (MXINIT) AND MAXIMUM NUMBER OF STURM SEQUENCES
C     ALLOWED (MXSTUR)
      READ(5,20) EXPLAN
      READ(5,*) SVSEED,RHSEED,MXINIT,MXSTUR
C
C     ISTART = (0,1):  ISTART = 0 MEANS ALPHA/BETA FILE IS NOT
C     AVAILABLE.  ISTART = 1 MEANS ALPHA/BETA FILE IS AVAILABLE ON
C     FILE 2.
C     ISTOP = (0,1):  ISTOP = 0 MEANS PROCEDURE GENERATES ALPHA/BETA
C     FILE AND THEN TERMINATES.  ISTOP = 1 MEANS PROCEDURE GENERATES
C     ALPHAS/BETAS IF NEEDED AND THEN COMPUTES EIGENVALUES AND ERROR
C     ESTIMATES AND THEN TERMINATES.
      READ(5,20) EXPLAN
      READ(5,*) ISTART,ISTOP
C
C     IHIS = (0,1):  IHIS = 0 MEANS ALPHA/BETA FILE IS NOT WRITTEN
C     TO FILE 1.  IHIS = 1 MEANS ALPHA/BETA FILE IS WRITTEN TO FILE 1.
C     IDIST = (0,1):  IDIST = 0 MEANS DISTINCT EIGENVALUES OF
C     ARE NOT WRITTEN TO FILE 11.  IDIST = 1 MEANS DISTINCT
C     EIGENVALUES ARE WRITTEN TO FILE 11.
C     IWRITE = (0,1):  IWRITE = 0 MEANS NO INTERMEDIATE OUTPUT
C     FROM THE COMPUTATIONS IS WRITTEN TO FILE 6.  IWRITE = 1 MEANS
C     EIGENVALUES AND ERROR ESTIMATES ARE WRITTEN TO FILE 6
C     AS THEY ARE COMPUTED.
```

```
      READ(5,20) EXPLAN                                                 LGV01310
      READ(5,*) IHIS,IDIST,IWRITE                                       LGV01320
C                                                                       LGV01330
C     READ IN THE RELATIVE TOLERANCE (RELTOL) FOR USE IN THE            LGV01340
C     SPURIOUS, T-MULTIPLICITY, AND PRTESTS.                            LGV01350
      READ(5,20) EXPLAN                                                 LGV01360
      READ(5,*) RELTOL                                                  LGV01370
C                                                                       LGV01380
C     READ IN THE SIZES OF THE T-MATRICES TO BE CONSIDERED.             LGV01390
      READ(5,20) EXPLAN                                                 LGV01400
      READ(5,*) (NMEV(J), J=1,NMEVS)                                    LGV01410
C                                                                       LGV01420
C     READ IN THE NUMBER OF SUBINTERVALS TO BE CONSIDERED.              LGV01430
      READ(5,20) EXPLAN                                                 LGV01440
      READ(5,*) NINT                                                    LGV01450
C                                                                       LGV01460
C     READ IN THE LEFT-END POINTS OF THE SUBINTERVALS TO BE CONSIDERED. LGV01470
C     THESE MUST BE IN ALGEBRAICALLY-INCREASING ORDER                   LGV01480
      READ(5,20) EXPLAN                                                 LGV01490
      READ(5,*) (LB(J), J=1,NINT)                                       LGV01500
C                                                                       LGV01510
C     READ IN THE RIGHT-END POINTS OF THE SUBINTERVALS TO BE CONSIDERED.LGV01520
C     THESE MUST BE IN ALGEBRAICALLY-INCREASING ORDER                   LGV01530
      READ(5,20) EXPLAN                                                 LGV01540
      READ(5,*) (UB(J), J=1,NINT)                                       LGV01550
C                                                                       LGV01560
C-----------------------------------------------------------------------LGV01570
C     INITIALIZE THE ARRAYS FOR THE USER-SPECIFIED MATRICES             LGV01580
C     AND PASS THE STORAGE LOCATIONS OF THESE ARRAYS TO THE             LGV01590
C     MATRIX-VECTOR MULTIPLY SUBROUTINE AMATV AND THE SOLVE             LGV01600
C     SUBROUTINE LSOLV.                                                 LGV01610
C                                                                       LGV01620
      CALL USPECA(N,MATNOA)                                             LGV01630
      CALL USPECB(N,MATNOB)                                             LGV01640
C                                                                       LGV01650
C-----------------------------------------------------------------------LGV01660
C                                                                       LGV01670
C     MASK UNDERFLOW AND OVERFLOW                                       LGV01680
      CALL MASK                                                         LGV01690
C                                                                       LGV01700
C-----------------------------------------------------------------------LGV01710
C                                                                       LGV01720
C     WRITE TO FILE 6, A SUMMARY OF THE PARAMETERS FOR THIS RUN         LGV01730
C                                                                       LGV01740
      WRITE(6,30) MATNOA,MATNOB,N,KMAX                                  LGV01750
   30 FORMAT(/3X,'A-MATRIX ID',3X,'B-MATRIX ID',4X,'ORDER OF A',4X,     LGV01760
     1'MAX ORDER OF T'/I14,I14,I14,I18/)                                LGV01770
C                                                                       LGV01780
      WRITE(6,40) ISTART,ISTOP                                          LGV01790
   40 FORMAT(/2X,'ISTART',3X,'ISTOP'/2I8/)                              LGV01800
C                                                                       LGV01810
      WRITE(6,50) IHIS,IDIST,IWRITE                                     LGV01820
   50 FORMAT(/4X,'IHIS',3X,'IDIST',2X,'IWRITE'/3I8/)                    LGV01830
C                                                                       LGV01840
      WRITE(6,60) SVSEED,RHSEED                                         LGV01850
   60 FORMAT(/' SEEDS FOR RANDOM NUMBER GENERATOR'//                    LGV01860
     1 4X,'LANCZS SEED',4X,'INVERR SEED'/2I15/)                         LGV01870
C                                                                       LGV01880
      WRITE(6,70) (NMEV(J), J=1,NMEVS)                                  LGV01890
   70 FORMAT(/' SIZES OF T-MATRICES TO BE CONSIDERED'/(6I12))           LGV01900
C                                                                       LGV01910
      WRITE(6,80) RELTOL,GAPTOL,BTOL                                    LGV01920
   80 FORMAT(/' RELATIVE TOLERANCE USED TO COMBINE COMPUTED T-EIGENVALUELGV01930
     1S'/E15.3/' RELATIVE GAP TOLERANCES USED IN INVERSE ITERATION'/    LGV01940
     1E15.3/' RELATIVE TOLERANCE FOR CHECK ON SIZE OF BETAS'/E15.3/)    LGV01950
C                                                                       LGV01960
      WRITE(6,90) (J,LB(J),UB(J), J=1,NINT)                             LGV01970
```

```
   90 FORMAT(/' BISEC WILL BE USED ON THE FOLLOWING INTERVALS'/
     1 (I6,2E20.6))
C
      IF (ISTART.EQ.0) GO TO 140
C
C     READ IN ALPHA BETA HISTORY
C
      READ(2,100)MOLD,NOLD,SVSOLD,MATAO,MATBO
  100 FORMAT(2I6,I12,2I8)
C
      IF (KMAX.LT.MOLD) KMAX = MOLD
      KMAX1 = KMAX + 1
C
C     CHECK THAT ORDER N, MATRIX IDS (MATNOA AND MATNOB), AND RANDOM
C     SEED (SVSEED) AGREE WITH THOSE IN THE HISTORY FILE.  IF NOT
C     PROCEDURE STOPS.
C
      ITEMP = (NOLD-N)**2 + (MATNOA-MATAO)**2 + (SVSEED-SVSOLD)**2
     1 + (MATNOB-MATBO)**2
C
      IF (ITEMP.EQ.0) GO TO 120
C
      WRITE(6,110)
  110 FORMAT(' PROGRAM TERMINATES'/    ' READ FROM FILE 2 CORRESPONDS TO
     1 DIFFERENT MATRIX THAN MATRIX SPECIFIED'/)
      GO TO 640
C
  120 CONTINUE
      MOLD1 = MOLD+1
C
      READ(2,130)(ALPHA(J), J=1,MOLD)
      READ(2,130)(BETA(J), J=1,MOLD1)
  130 FORMAT(4Z20)
C
      IF (KMAX.EQ.MOLD) GO TO 160
C
C     SAVE V1 = B*V(KMAX), VS = B*V(KMAX+1), V2 = V(KMAX+1)
      READ(2,130) (V1(J), J=1,N)
      READ(2,130) (VS(J), J=1,N)
      READ(2,130) (V2(J), J=1,N)
C
  140 CONTINUE
      IIX = SVSEED
C
C-----------------------------------------------------------------------
C
      CALL LANCZS(LSOLV,AMATV,ALPHA,BETA,V1,V2,VS,G,KMAX,MOLD1,N,IIX)
C
C-----------------------------------------------------------------------
C
      KMAX1 = KMAX + 1
C
      IF (IHIS.EQ.0.AND.ISTOP.GT.0) GO TO 160
C
      WRITE(1,150) KMAX,N,SVSEED,MATNOA,MATNOB
  150 FORMAT(2I6,I12,2I8,' = KMAX,N,SVSEED,MATNOA,MATNOB')
C
      WRITE(1,130)(ALPHA(I), I=1,KMAX)
      WRITE(1,130)(BETA(I), I=1,KMAX1)
C
C     SAVE V1 = B*V(KMAX), VS = B*V(KMAX+1), V2 = V(KMAX+1)
      WRITE(1,130) (V1(I), I=1,N)
      WRITE(1,130) (VS(I), I=1,N)
      WRITE(1,130) (V2(I), I=1,N)
C
      IF (ISTOP.EQ.0) GO TO 540
C
```

```
  160 CONTINUE
      BKMIN = BTOL
      WRITE(6,170)
  170 FORMAT(/' T-MATRICES (ALPHA AND BETA) ARE NOW AVAILABLE'/)
C
C-----------------------------------------------------------------------
C     SUBROUTINE TNORM CHECKS MIN(BETA)/(ESTIMATED NORM(A)) > BTOL .
C     IF THIS IS VIOLATED IB IS SET EQUAL TO THE NEGATIVE OF THE INDEX
C     OF THE MINIMAL BETA.  IF(IB < 0) THEN SUBROUTINE TNORM IS
C     CALLED FOR EACH VALUE OF MEV TO DETERMINE WHETHER OR NOT THERE
C     IS A BETA IN THE T-MATRIX SPECIFIED THAT VIOLATES THIS TEST.
C     IF THERE IS SUCH A BETA THE PROGRAM TERMINATES FOR THE USER
C     TO DECIDE WHAT TO DO.  THIS TEST CAN BE OVER-RIDDEN BY
C     SIMPLY MAKING BTOL SMALLER, BUT THEN THERE IS THE POSSIBILITY
C     THAT LOSSES IN THE LOCAL ORTHOGONALITY MAY HURT THE COMPUTATIONS.
C     BTOL = 1.D-8 IS HOWEVER A CONSERVATIVE CHOICE FOR BTOL.
C
C     TNORM ALSO COMPUTES TKMAX = MAX(|ALPHA(K)|,BETA(K), K=1,KMAX).
C     TKMAX IS USED TO SCALE THE TOLERANCES USED IN THE
C     T-MULTIPLICITY AND SPURIOUS TESTS IN BISEC. TKMAX IS ALSO USED IN
C     THE PROJECTION TEST FOR HIDDEN EIGENVALUES THAT HAD 'TOO SMALL'
C     A PROJECTION ON THE STARTING VECTOR.
C
      CALL TNORM(ALPHA,BETA,BKMIN,TKMAX,KMAX,IB)
C
C-----------------------------------------------------------------------
C
      TTOL = EPSM*TKMAX
C
C     LOOP ON THE SIZE OF THE T-MATRIX
C
  180 CONTINUE
      MMB = MMB + 1
      MEV = NMEV(MMB)
C     IS MEV TOO LARGE ?
      IF(MEV.LE.KMAX) GO TO 200
      WRITE(6,190) MMB, MEV, KMAX
  190 FORMAT(/' TERMINATE PRIOR TO CONSIDERING THE',I6,'TH T-MATRIX'/
     1' BECAUSE THE SIZE REQUESTED',I6,' IS GREATER THAN THE MAXIMUM SIZ
     1E ALLOWED',I6/)
      GO TO 540
C
  200 MP1 = MEV + 1
      BETAM = BETA(MP1)
C
      IF (IB.GE.0) GO TO 210
C
      T0 = BTOL
C
C-----------------------------------------------------------------------
C
      CALL TNORM(ALPHA,BETA,T0,T1,MEV,IBMEV)
C
C-----------------------------------------------------------------------
C
      TEMP = T0/TKMAX
      IBMEV = IABS(IBMEV)
      IF (TEMP.GE.BTOL) GO TO 210
      IBMEV = -IBMEV
      GO TO 600
C
  210 CONTINUE
      IC = MXSTUR-ICT
C
C-----------------------------------------------------------------------
C     BISEC LOOP. THE SUBROUTINE BISEC INCORPORATES DIRECTLY THE
C     T-MULTIPLICITY AND SPURIOUS TESTS. T-EIGENVALUES WILL BE
```

```
C     CALCULATED BY BISEC SEQUENTIALLY ON INTERVALS
C     (LB(J),UB(J)), J = 1,NINT).
C
C     ON RETURN FROM BISEC
C     NDIS = NUMBER OF DISTINCT EIGENVALUES OF T(1,MEV) ON UNION
C            OF THE (LB,UB) INTERVALS
C     VS = DISTINCT T-EIGENVALUES IN ALGEBRAICALLY INCREASING ORDER
C     MP = MULTIPLICITIES OF THE T-EIGENVALUES IN VS
C     MP(I) = (0,1,MI), MI>1, I=1,NDIS  MEANS:
C        (0)  VS(I) IS SPURIOUS
C        (1)  VS(I) IS T-SIMPLE AND GOOD
C        (MI) VS(I) IS MULTIPLE AND IS THEREFORE NOT ONLY GOOD BUT
C             ALSO A CONVERGED GOOD T-EIGENVALUE.
C
C
      CALL BISEC(ALPHA,BETA,V1,V2,VS,LB,UB,EPSM,TTOL,MP,NINT,
     1 MEV,NDIS,IC,IWRITE)
C
C-----------------------------------------------------------------------
C
      IF (NDIS.EQ.0) GO TO 620
C
C     COMPUTE THE TOTAL NUMBER OF STURM SEQUENCES USED TO DATE
C     COMPUTE THE BISEC CONVERGENCE AND T-MULTIPLICITY TOLERANCES USED.
C     COMPUTE THE CONVERGENCE TOLERANCE FOR EIGENVALUES OF A.
      ICT = ICT + IC
      TEMP = DFLOAT(MEV+1000)
      MULTOL = TEMP*TTOL
      TEMP = DSQRT(TEMP)
      BISTOL = TTOL*TEMP
      CONTOL = BETAM*1.D-10
C
C-----------------------------------------------------------------------
C    SUBROUTINE LUMP 'COMBINES' T-EIGENVALUES THAT ARE 'TOO CLOSE'.
C    NOTE HOWEVER THAT CLOSE SPURIOUS T-EIGENVALUES ARE NOT AVERAGED
C    WITH GOOD ONES. HOWEVER, THEY MAY BE USED TO INCREASE THE
C    MULTIPLICITY OF A GOOD T-EIGENVALUE.
C
      LOOP = NDIS
      CALL LUMP(VS,RELTOL,MULTOL,SCALE2,MP,LOOP)
C
C-----------------------------------------------------------------------
C
      IF(NDIS.EQ.LOOP) GO TO 230
C
      WRITE(6,220) NDIS, MEV, LOOP
  220 FORMAT(/I6,' DISTINCT T-EIGENVALUES WERE COMPUTED IN BISEC AT MEV
     1',I6/ 2X,' LUMP SUBROUTINE REDUCES NUMBER OF DISTINCT EIGENVALUES
     1O',I6)
C
  230 CONTINUE
      NDIS = LOOP
      BETA(MP1) = BETAM
C
C-----------------------------------------------------------------------
C     THE SUBROUTINE ISOEV LABELS THOSE SIMPLE EIGENVALUES OF T(1,MEV)
C     WITH VERY SMALL GAPS BETWEEN NEIGHBORING EIGENVALUES OF T(1,MEV)
C     TO AVOID COMPUTING ERROR ESTIMATES FOR ANY SIMPLE GOOD
C     T-EIGENVALUE THAT IS TOO CLOSE TO A SPURIOUS EIGENVALUE.
C     ON RETURN FROM ISOEV, G CONTAINS CODED MINIMAL GAPS
C     BETWEEN THE DISTINCT EIGENVALUES OF T(1,MEV). (G IS REAL).
C     G(I) < 0 MEANS MINGAP IS DUE TO LEFT GAP G(I) > 0 MEANS DUE TO
C     RIGHT GAP. MP(I) = -1 MEANS THAT THE GOOD T-EIGENVALUE IS SIMPLE
C     AND HAS A VERY SMALL MINGAP IN T(1,MEV) DUE TO A SPURIOUS
C     EIGENVALUE.  NG = NUMBER OF GOOD T-EIGENVALUES.
C     NISO = NUMBER OF ISOLATED GOOD T-EIGENVALUES.
C
```

```
      CALL ISOEV(VS,GAPTOL,MULTOL,SCALE1,G,MP,NDIS,NG,NISO)
C
C-----------------------------------------------------------------------
C
      WRITE(6,240)NG,NISO,NDIS
  240 FORMAT(/I6,' GOOD T-EIGENVALUES HAVE BEEN COMPUTED'/
     1 I6,' OF THESE ARE T-ISOLATED'/
     2 I6,' = NUMBER OF DISTINCT T-EIGENVALUES COMPUTED'/)
C
C     DO WE WRITE DISTINCT EIGENVALUES OF T-MATRIX TO FILE 11?
      IF (IDIST.EQ.0) GO TO 280
C
      WRITE(11,250) NDIS,NISO,MEV,N,SVSEED,MATNOA,MATNOB
  250 FORMAT(/4I6,I12,I8,'=ND,NIS,MEV,N,SEED,MNA,MNB'/)
C
      WRITE(11,260) (MP(I),VS(I),G(I), I=1,NDIS)
  260 FORMAT(2(I3,E25.16,E12.3))
C
      WRITE(11,270) NDIS, (MP(I), I=1,NDIS)
  270 FORMAT(/I6,' = NDIS, T-MULTIPLICITIES (0 MEANS  SPURIOUS)'/(20I4))
C
  280 CONTINUE
C
      IF (NISO.NE.0) GO TO 310
C
      WRITE(4,290) MEV
  290 FORMAT(/' AT MEV = ',I6,' THERE ARE NO ISOLATED T-EIGENVALUES'/
     1' SO NO ERROR ESTIMATES WERE COMPUTED/')
C
      WRITE(6,300)
  300 FORMAT(/' ALL COMPUTED GOOD T-EIGENVALUES ARE MULTIPLE'/
     1 ' THEREFORE ALL SUCH EIGENVALUES ARE ASSUMED TO HAVE CONVERGED')
C
      ICONV = 1
      GO TO 350
C
  310 CONTINUE
C
C-----------------------------------------------------------------------
C     SUBROUTINE INVERR COMPUTES ERROR ESTIMATES FOR ISOLATED GOOD
C     T-EIGENVALUES USING INVERSE ITERATION ON T(1,MEV). ON RETURN
C     G(J) = MINIMUM GAP IN T(1,MEV) FOR EACH VS(J), J=1,NDIS
C     G(MEV+I) = BETAM*|U(MEV)| = ERROR ESTIMATE FOR ISOLATED GOOD
C              T-EIGENVALUES, WHERE I = 1, NISO AND  BETAM = BETA(MEV+1)
C              U(MEV) IS MEVTH COMPONENT OF THE UNIT EIGENVECTOR OF T
C              CORRESPONDING TO THE ITH ISOLATED GOOD T-EIGENVALUE.
C     A NEGATIVE ERROR ESTIMATE MEANS THAT FOR THAT PARTICULAR
C     EIGENVALUE THE INVERSE ITERATION DID NOT CONVERGE IN <= MXINIT
C     STEPS AND THAT THE CORRESPONDING ERROR ESTIMATE IS QUESTIONABLE.
C
C     V2 CONTAINS THE ISOLATED GOOD T-EIGENVALUES
C     V1 CONTAINS THE MINGAPS TO THE NEAREST DISTINCT  EIGENVALUE
C        OF T(1,MEV) FOR EACH ISOLATED GOOD T-EIGENVALUE IN V2.
C     VS CONTAINS THE NDIS DISTINCT EIGENVALUES OF T(1,MEV)
C     MP CONTAINS THE CORRESPONDING CODED T-MULTIPLICITIES
C
      IT = MXINIT
      CALL INVERR(ALPHA,BETA,V1,V2,VS,EPSM,G,MP,MEV,MMB,NDIS,NISO,N,
     1 RHSEED,IT,IWRITE)
C
C-----------------------------------------------------------------------
C
C     SIMPLE CHECK FOR CONVERGENCE. CHECKS TO SEE IF ALL OF THE ERROR
C     ESTIMATES ARE SMALLER THAN CONTOL = BETAM*1.D-10.
C     IF THIS TEST IS SATISFIED, THEN CONVERGENCE FLAG, ICONV IS SET
C     TO 1.  TYPICALLY ERROR ESTIMATES ARE VERY CONSERVATIVE.
C
```

```
      WRITE(6,320) CONTOL
  320 FORMAT(/' CONVERGENCE IS TESTED USING THE CONVERGENCE TOLERANCE',
     1E13.4/)
C
      II = MEV +1
      IF = MEV+NISO
      DO 330 I = II,IF
      IF (ABS(G(I)).GT.CONTOL) GO TO 350
  330 CONTINUE
      ICONV = 1
      MMB = NMEVS
C
      WRITE(6,340) CONTOL
  340 FORMAT(' ALL COMPUTED ERROR ESTIMATES WERE LESS THAN',E15.4/
     1 ' THEREFORE PROCEDURE TERMINATES'/)
C
  350 CONTINUE
C
C     IF CONVERGENCE IS INDICATED, THAT IS ICONV = 1 ,THEN
C     THE SUBROUTINE PRTEST IS CALLED TO CHECK FOR ANY CONVERGED
C     T-EIGENVALUES THAT HAVE BEEN MISLABELLED AS SPURIOUS BECAUSE
C     THE PROJECTION OF THEIR EIGENVECTOR(S) ON THE STARTING
C     VECTOR WERE TOO SMALL.
C     NUMERICAL TESTS INDICATE THAT SUCH EIGENVALUES ARE RARE.
C     IF FOR SOME REASON MANY OF THESE HIDDEN EIGENVALUES APPEAR
C     ON SOME RUN, YOU CAN BE CERTAIN THAT SOMETHING IS FOULED UP.
C
      IF (ICONV.EQ.0) GO TO 480
C
C-----------------------------------------------------------------------
C
      CALL PRTEST (ALPHA,BETA,VS,TKMAX,EPSM,RELTOL,SCALE3,SCALE4,
     1 MP,NDIS,MEV,IPROJ)
C
C-----------------------------------------------------------------------
C
      IF(IPROJ.EQ.0) GO TO 470
C
      IF(IDIST.EQ.1)  WRITE(11,360) IPROJ
  360 FORMAT(' SUBROUTINE PRTEST WANTS TO RELABEL',I6,' SPURIOUS T-EIGEN
     1VALUES'/' WE ACCEPT RELABELLING ONLY IF LAST COMPONENT OF T-EIGENV
     1ECTOR IS L.T. 1.D-10'/)
C
      IIX = RHSEED
C
C-----------------------------------------------------------------------
C
      CALL GENRAN(IIX,G,MEV)
C
C-----------------------------------------------------------------------
C
      ITEN = -10
      NISOM = NISO + MEV
      IWRITO = IWRITE
      IWRITE = 0
C
      DO 390 J = 1,NDIS
      IF(MP(J).NE.ITEN) GO TO 390
      TO = VS(J)
C
C-----------------------------------------------------------------------
C
      IT = MXINIT
      CALL INVERM(ALPHA,BETA,V1,V2,TO,TEMP,T1,EPSM,G,MEV,IT,IWRITE)
C
C-----------------------------------------------------------------------
C
```

```
      IF(TEMP.LE.1.D-10) GO TO 380                                      LGV05330
C     ERROR ESTIMATE WAS NOT SMALL REJECT RELABELLING OF THIS EIGENVALUELGV05340
      IF(IDIST.EQ.1)  WRITE(11,370) J,TO,TEMP                           LGV05350
  370 FORMAT(/' LAST COMPONENT FOR',I6,'TH EIGENVALUE',E20.12/' IS TOO LLGV05360
     1ARGE = ',E15.6,' SO DO NOT ACCEPT PRTEST RELABELLING'/)           LGV05370
      MP(J) = 0                                                         LGV05380
      IPROJ = IPROJ - 1                                                 LGV05390
      GO TO 390                                                         LGV05400
C     RELABELLING ACCEPTED                                              LGV05410
  380 NISOM = NISOM + 1                                                 LGV05420
      G(NISOM) = BETAM*TEMP                                             LGV05430
  390 CONTINUE                                                          LGV05440
      IWRITE = IWRITO                                                   LGV05450
C                                                                       LGV05460
      IF(IPROJ.EQ.0) GO TO 430                                          LGV05470
      WRITE(6,400) IPROJ                                                LGV05480
  400 FORMAT(/I6,' T-EIGENVALUES WERE RECLASSIFIED AS GOOD.'/           LGV05490
     1' THESE ARE IDENTIFIED IN FILE 3 BY A T-MULTIPLICITY OF -10'/' USELGV05500
     2R SHOULD INSPECT EACH TO MAKE SURE NEIGHBORS HAVE CONVERGED'/)    LGV05510
C                                                                       LGV05520
      IF(IDIST.EQ.1)  WRITE(11,410) IPROJ                               LGV05530
  410 FORMAT(/I6,' T-EIGENVALUES WERE RELABELLED AS GOOD'/              LGV05540
     1' BELOW IS CORRECTED T-MULTIPLICITY PATTERN'/)                    LGV05550
C                                                                       LGV05560
      WRITE(6,420) NDIS, (MP(I), I=1,NDIS)                              LGV05570
      IF(IDIST.EQ.1)  WRITE(11,420) NDIS, (MP(I), I=1,NDIS)             LGV05580
  420 FORMAT(/I6,' = NDIS, T-MULTIPLICITIES (0 MEANS  SPURIOUS)'/       LGV05590
     1 6X, ' (-10) MEANS SPURIOUS T-EIGENVALUE RELABELLED AS GOOD'/(20I4LGV05600
     1))                                                                LGV05610
C                                                                       LGV05620
C     RECALCULATE MINGAPS FOR DISTINCT T(1,MEV) EIGENVALUES.            LGV05630
  430 NM1 = NDIS - 1                                                    LGV05640
      G(NDIS) = VS(NM1)-VS(NDIS)                                        LGV05650
      G(1) = VS(2)-VS(1)                                                LGV05660
C                                                                       LGV05670
      DO 440 J = 2,NM1                                                  LGV05680
      TO = VS(J)-VS(J-1)                                                LGV05690
      T1 = VS(J+1)-VS(J)                                                LGV05700
      G(J) = T1                                                         LGV05710
      IF (TO.LT.T1) G(J) = -TO                                          LGV05720
  440 CONTINUE                                                          LGV05730
      IF(IPROJ.EQ.0) GO TO 470                                          LGV05740
C     WRITE TO FILE 4 ERROR ESTIMATES FOR THOSE T-EIGENVALUES RELABELLEDLGV05750
      NGOOD = 0                                                         LGV05760
      DO 450 J = 1,NDIS                                                 LGV05770
      IF(MP(J).EQ.0) GO TO 450                                          LGV05780
      NGOOD = NGOOD + 1                                                 LGV05790
      IF(MP(J).NE.ITEN) GO TO 450                                       LGV05800
      TO = VS(J)                                                        LGV05810
      NISO = NISO + 1                                                   LGV05820
      NISOM = MEV + NISO                                                LGV05830
      WRITE(4,460) NGOOD,TO,G(NISOM),G(J)                               LGV05840
  450 CONTINUE                                                          LGV05850
  460 FORMAT(I10,E25.16,2E14.3)                                         LGV05860
C                                                                       LGV05870
  470 CONTINUE                                                          LGV05880
C                                                                       LGV05890
C     WRITE THE GOOD T-EIGENVALUES TO FILE 3.  FIRST TRANSFER THEM      LGV05900
C     TO V2 AND THEIR T-MULTIPLICITIES TO THE CORRESPONDING POSITIONS   LGV05910
C     IN MP AND COMPUTE THE AB-MINGAPS, THE MINIMAL GAPS BETWEEN THE    LGV05920
C     GOOD T-EIGENVALUES.  THESE GAPS WILL BE PUT IN THE ARRAY G.       LGV05930
C     SINCE G CURRENTLY CONTAINS THE MINIMAL GAPS BETWEEN THE DISTINCT  LGV05940
C     EIGENVALUES OF THE T-MATRIX, THESE GAPS WILL FIRST BE             LGV05950
C     TRANSFERRED TO V1.  NOTE THAT V1<0 MEANS THAT THAT MINIMAL GAP    LGV05960
C     IN THE T-MATRIX IS DUE TO A SPURIOUS T-EIGENVALUE.                LGV05970
C     ALL THIS INFORMATION IS PRINTED TO FILE 3                         LGV05980
C                                                                       LGV05990
```

```
  480 CONTINUE
C
      NG = 0
      DO 490 I = 1,NDIS
      IF (MP(I).EQ.0) GO TO 490
      NG = NG+1
      MP(NG) = MP(I)
      V2(NG) = VS(I)
      TEMP = G(I)
      TEMP = DABS(TEMP)
      J = I+1
      IF (G(I).LT.ZERO) J = I-1
      IF (MP(J).EQ.0) TEMP = -TEMP
      V1(NG) = TEMP
  490 CONTINUE
C
      WRITE(6,500)MEV
  500 FORMAT(//' T-EIGENVALUE CALCULATION AT MEV = ',I6,'    IS COMPLETE
     1')
C
C     NG = NUMBER OF COMPUTED DISTINCT GOOD T-EIGENVALUES.  NEXT
C     GENERATE GAPS BETWEEN GOOD T-EIGENVALUES (ABMINGAPS) AND PUT THEM
C     IN G.  G(J) < 0 MEANS THE ABMINGAP IS DUE TO THE LEFT-HAND GAP.
C
      NGM1 = NG - 1
      G(NG) = V2(NGM1)-V2(NG)
      G(1) = V2(2)-V2(1)
C
      DO 510 J = 2,NGM1
      T0 = V2(J)-V2(J-1)
      T1 = V2(J+1)-V2(J)
      G(J) = T1
      IF (T0.LT.T1) G(J) = -T0
  510 CONTINUE
C
C     WRITE GOOD T-EIGENVALUES OUT TO FILE 3.
C
      WRITE(3,520)NG,NDIS,MEV,N,SVSEED,MATNOA,MATNOB,MULTOL,IB,BTOL
  520 FORMAT(4I6,I12,2I8,'=NG,ND,MEV,N,SEED,MNA,MNB'/
     1 E20.12,I6,E13.4,' = MUTOL,INDEX MINIMAL BETA,BTOL'/
     1' EV NO',1X,'TMULT',10X,'GOOD EIGENVALUE',7X,'TMINGAP',6X,'ABMINGA
     1P')
C
      WRITE(3,530)(I,MP(I),V2(I),V1(I),G(I), I=1,NG)
  530 FORMAT(2I6,E25.16,2E14.3)
C
C     IF CONVERGENCE FLAG ICONV.NE.1 AND NUMBER OF T-MATRICES
C     CONSIDERED TO DATE IS LESS THAN NUMBER ALLOWED, INCREMENT MEV.
C     AND LOOP BACK TO 210 TO REPEAT COMPUTATIONS.  RESTORE BETA(MEV+1).
C
      BETA(MP1) = BETAM
C
      IF (MMB.LT.NMEVS.AND.ICONV.NE.1) GO TO 180
C
C     END OF LOOP ON DIFFERENT SIZE T-MATRICES ALLOWED.
C
  540 CONTINUE
C
      IF(ISTOP.EQ.0)  WRITE(6,550)
  550 FORMAT(/' T-MATRICES (ALPHA AND BETA) ARE NOW AVAILABLE, TERMINATE
     1')
      IF (IHIS.EQ.1.AND.KMAX.NE.MOLD) WRITE(1,560)
  560 FORMAT(/' ABOVE ARE THE FOLLOWING VECTORS '/
     1 '   ALPHA(I), I = 1,KMAX'/
     2 '   BETA(I), I = 1,KMAX+1'/
     3 ' FINAL THREE VECTORS USED IN LANCZS SUBROUTINE'/
     3 ' V1 = B*V(KMAX), VS = B*V(KMAX+1), V2 = V(KMAX+1)'/
```

```
     4 ' ALL VECTORS IN THIS FILE HAVE HEX FORMAT 4Z20'/                LGV06670
     5 ' ----- END OF FILE 1 NEW ALPHA, BETA HISTORY---------------'///)LGV06680
C                                                                       LGV06690
      IF (ISTOP.EQ.0) GO TO 640                                         LGV06700
C                                                                       LGV06710
      WRITE(3,570)                                                      LGV06720
  570 FORMAT(/' ABOVE ARE COMPUTED GOOD T-EIGENVALUES'/                 LGV06730
     1 ' NG = NUMBER OF GOOD T-EIGENVALUES COMPUTED'/                   LGV06740
     2 ' ND = NUMBER OF COMPUTED DISTINCT EIGENVALUES OF T(1,MEV)'/     LGV06750
     3 ' N = ORDER OF A AND B-MATRIX,  MNA, MNB = MATRIX IDENTS'/       LGV06760
     4 ' MULTOL = T-MULTIPLICITY TOLERANCE FOR T-EIGENVALUES IN BISEC'/ LGV06770
     4 ' TMULT IS THE T-MULTIPLICITY OF GOOD T-EIGENVALUE'/             LGV06780
     5 ' TMULT = -1 MEANS SPURIOUS T-EIGENVALUE TOO CLOSE'/             LGV06790
     6 ' DO NOT COMPUTE ERROR ESTIMATES FOR SUCH EIGENVALUES'/          LGV06800
     7 ' ABMINGAP = MINIMAL GAP BETWEEN THE COMPUTED EIGENVALUES'/      LGV06810
     8 ' ABMINGAP .LT. 0. MEANS MINIMAL GAP IS DUE TO LEFT-HAND GAP'/   LGV06820
     9 ' TMINGAP= MINIMAL GAP W.R.T.  DISTINCT EIGENVALUES IN T(1,MEV)'/LGV06830
     1 ' TMINGAP .LT. 0. MEANS MINGAP IS DUE TO SPURIOUS T-EIGENVALUE'/ LGV06840
     2 ' ----- END OF FILE 3 GOODEIGENVALUES-----------------------'///)LGV06850
C                                                                       LGV06860
      IF (IDIST.EQ.1) WRITE(11,580)                                     LGV06870
  580 FORMAT(/' ABOVE ARE THE DISTINCT EIGENVALUES OF T(1,MEV).'/       LGV06880
     2 ' THE FORMAT IS      T-MULTIPLICITY    T-EIGENVALUE   TMINGAP'/  LGV06890
     3 '          THIS FORMAT IS REPEATED TWICE ON EACH LINE.'/         LGV06900
     4 ' T-MULTIPLICITY = -1 MEANS THAT THE SUBROUTINE ISOEV HAS TAGGED'LGV06910
     5/'    THIS SIMPLE T-EIGENVALUE AS HAVING A VERY CLOSE SPURIOUS'/  LGV06920
     6 '    T-EIGENVALUE SO THAT NO ERROR ESTIMATE WILL BE COMPUTED'/   LGV06930
     7 '    FOR THAT EIGENVALUE IN SUBROUTINE INVERR.'/                 LGV06940
     8 ' TMINGAP .LT. 0, TMINGAP IS DUE TO LEFT GAP .GT. 0, RIGHT GAP.'/LGV06950
     9 ' EACH OF THE DISTINCT T-EIGENVALUE TABLES IS FOLLOWED'/         LGV06960
     9 ' BY THE T-MULTIPLICITY PATTERN.'/                               LGV06970
     1 ' NDIS = NUMBER OF COMPUTED DISTINCT EIGENVALUES OF T(1,MEV).'/  LGV06980
     2 ' NG = NUMBER OF GOOD T-EIGENVALUES. '/                          LGV06990
     3 ' NISO = NUMBER OF ISOLATED GOOD T-EIGENVALUES. '/               LGV07000
     4 ' NISO ALSO IS THE COUNT OF +1 ENTRIES IN T-MULTIPLICITY PATTERN.LGV07010
     5 '/' ----- END OF FILE 11 DISTINCT T-EIGENVALUES--------------'///LGV07020
     6 )                                                                LGV07030
C                                                                       LGV07040
      IF(NISO.NE.0)  WRITE(4,590)                                       LGV07050
  590 FORMAT(/' ABOVE ARE THE ERROR ESTIMATES OBTAINED FOR THE ISOLATED LGV07060
     1GOOD T-EIGENVALUES'/                                              LGV07070
     1' OBTAINED VIA INVERSE ITERATION IN THE SUBROUTINE INVERR.'/      LGV07080
     1' ALL OTHER GOOD T-EIGENVALUES HAVE CONVERGED.'/                  LGV07090
     2' ERROR ESTIMATE = BETAM*ABS(UM)'/                                LGV07100
     2' WHERE BETAM = BETA(MEV+1) AND UM = U(MEV).'/                    LGV07110
     3' U = UNIT EIGENVECTOR OF T WHERE T*U = EV*U AND EV = ISOLATED GOOLGV07120
     3D T-EIGENVALUE.'/                                                 LGV07130
     4' TMINGAP = GAP TO NEAREST DISTINCT EIGENVALUE OF T(1,MEV).'/     LGV07140
     5' TMINGAP .LT. 0. MEANS MINGAP IS DUE TO A LEFT NEIGHBOR.'/       LGV07150
     6' ERROR ESTIMATE L.T. 0 MEANS INVERSE ITERATION DID NOT CONVERGE'/LGV07160
     7' ------ END OF FILE 4 ERRINV -------------------------------'//) LGV07170
      GO TO 640                                                         LGV07180
C                                                                       LGV07190
  600 CONTINUE                                                          LGV07200
C                                                                       LGV07210
      IBB = IABS(IBMEV)                                                 LGV07220
      IF (IBMEV.LT.0) WRITE(6,610) MEV,IBB,BETA(IBB)                    LGV07230
  610 FORMAT(/' PROGRAM TERMINATES BECAUSE MEV REQUESTED = ',I6,' IS .GTLGV07240
     1',I6/' AT WHICH AN ABNORMALLY SMALL BETA = ' , E13.4,' OCCURRED'/)LGV07250
      GO TO 640                                                         LGV07260
C                                                                       LGV07270
  620 IF (NDIS.EQ.0.AND.ISTOP.GT.0) WRITE(6,630)                        LGV07280
  630 FORMAT(/' INTERVALS SPECIFIED FOR BISECT DID NOT CONTAIN ANY T-EIGLGV07290
     1ENVALUES'/' PROGRAM TERMINATES')                                  LGV07300
C                                                                       LGV07310
  640 CONTINUE                                                          LGV07320
C                                                                       LGV07330
```

```
      STOP                                                              LGV07340
C-----END OF MAIN PROGRAM FOR LANCZOS EIGENVALUE COMPUTATIONS-----------LGV07350
      END                                                               LGV07360
```

SECTION 5.3 MAIN PROGRAM, EIGENVECTOR CALCULATIONS

```
C-----LGVEC (EIGENVECTORS OF A*X = EVAL*B*X)---------------------------
C
C     CONTAINS MAIN PROGRAM FOR COMPUTING AN EIGENVECTOR CORRESPONDING
C     TO EACH OF A SET OF EIGENVALUES WHICH HAVE BEEN COMPUTED
C     ACCURATELY BY THE CORRESPONDING LANCZOS EIGENVALUE PROGRAM
C     (LGVAL) FOR THE SYMMETRIC, GENERALIZED PROBLEM A*X = EVAL*B*X.
C     LGVAL AND LGVEC ASSUME THAT B IS POSITIVE DEFINITE AND THAT THE
C     CHOLESKY FACTORS OF B (OR OF A PERMUTATION OF B) ARE AVAILABLE
C     FOR USE IN THE LANCZOS PROCEDURES.  IF B HAS BEEN PERMUTED,
C     THEN THESE PROCEDURES ASSUME THAT THE DATA PRESENTED FOR THE
C     A-MATRIX HAS BEEN SUBJECTED TO THE SAME PERMUTATION.  THAT
C     PERMUTATION WILL THEN BE USED AFTER THE RITZ VECTORS FOR THE
C     PERMUTED VERSION OF THE ORIGINAL PROBLEM HAVE BEEN COMPUTED,
C     TO OBTAIN THE ASSOCIATED RITZ VECTORS FOR THE ORIGINAL PROBLEM.
C     NOTE THAT THIS PROGRAM COULD BE MODIFIED TO COMPUTE ADDITIONAL
C     EIGENVECTORS FOR ANY COMPUTED EIGENVALUE WHICH IS A MULTIPLE
C     EIGENVALUE OF THE GIVEN A-MATRIX.  THE AMOUNT OF ADDITIONAL
C     COMPUTATION REQUIRED WOULD DEPEND UPON THE PARTICULAR
C     A-MATRIX AND B-MATRIX USED AND UPON WHAT PART OF THE
C     SPECTRUM OF EIGENVALUES IS BEING CONSIDERED.
C
C     THE LANCZOS EIGENVECTOR COMPUTATIONS ASSUME THAT EACH
C     EIGENVALUE THAT IS BEING CONSIDERED HAS CONVERGED AS AN
C     EIGENVALUE OF THE ASSOCIATED LANCZOS TRIDIAGONAL MATRICES.
C
C     PFORT VERIFIER IDENTIFIED THE FOLLOWING NONPORTABLE
C     CONSTRUCTIONS
C
C     1.  DATA/MACHEP/ STATEMENT
C     2.  ALL READ(5,*) STATEMENTS (FREE FORMAT)
C     3.  FORMAT(20A4) USED WITH THE EXPLANATORY HEADER, EXPLAN
C     4.  HEXADECIMAL FORMAT (4Z20) USED IN ALPHA/BETA FILES 1 AND 2.
C
C     IMPORTANT NOTE:  PROGRAM ALLOWS ENLARGEMENT OF THE ALPHA, BETA
C     ARRAYS.  IN PARTICULAR, IF ANY ONE OF THE EIGENVALUES SUPPLIED
C     IS T-SIMPLE AND NOT CLOSE TO A SPURIOUS EIGENVALUE, THE PROGRAM
C     REQUIRES THAT KMAX BE AT LEAST 11*MEV/8 + 12.  IF KMAX IS NOT
C     THIS LARGE, THEN THE PROGRAM RESETS KMAX TO THIS SIZE
C     AND EXTENDS THE ALPHA, BETA HISTORY IF REQUIRED.
C     THUS, THE DIMENSIONS OF THE ALPHA AND BETA ARRAYS MUST BE
C     LARGE ENOUGH TO ALLOW FOR THIS POSSIBILITY.
C     REMEMBER THAT THE BETA ARRAY, BETA(J), IS SUCH THAT
C     J = 1,..., KMAX+1.  SO IF THE KMAX USED BY THE PROGRAM
C     IS TO BE 3000, THEN BETA MUST BE OF LENGTH AT LEAST 3001.
C
C-----------------------------------------------------------------------
      DOUBLE PRECISION  ALPHA(5000),BETA(5001)
      DOUBLE PRECISION  V1(5000),V2(5000),VS(5000)
      DOUBLE PRECISION  RITVEC(30000),TVEC(30000),GOODEV(50),EVNEW(50)
      DOUBLE PRECISION  EVAL,EVALN,TOLN,TTOL,ERTOL,ALFA,BATA
      DOUBLE PRECISION  MULTOL,SCALEO,STUTOL,BTOL,LB,UB
      DOUBLE PRECISION  ONE,ZERO,MACHEP,EPSM,TEMP,SUM,ERRMIN,BKMIN
      DOUBLE PRECISION  RELTOL,ERROR,TERROR,TLAST(50)
      REAL  G(10000),AMINGP(50),TMINGP(50),EXPLAN(20)
      REAL  TERR(50),ERR(50),ERRDGP(50),RNORM(50),TBETA(50)
      INTEGER  MP(50),M1(50),M2(50),MA(50),ML(50),MINT(50),MFIN(50)
      INTEGER  SVSEED,SVSOLD,RHSEED,IDELTA(50)
      INTEGER  MBOUND,NTVCON,SVTVEC,TVSTOP,LVCONT,ERCONT,TFLAG
      DOUBLE PRECISION  FINPRO
      DOUBLE PRECISION  DABS, DMAX1, DSQRT, DFLOAT
      REAL ABS
      INTEGER  IABS
      EXTERNAL LSOLV, AMATV
C-----------------------------------------------------------------------
```

```
      DATA MACHEP/Z341000000000000/
      EPSM = 2.D0*MACHEP
C-----------------------------------------------------------------------
C
C     ARRAYS MUST BE DIMENSIONED AS FOLLOWS:
C     1.  ALPHA:  >= KMAXN,  BETA: >= (KMAXN+1) WHERE KMAXN, THE
C               LARGEST SIZE T-MATRIX CONSIDERED BY THE PROGRAM,
C               IS THE LARGER OF THE SIZE OF THE ALPHA, BETA HISTORY
C               PROVIDED ON FILE 2 (IF ANY ) AND THE SIZE WHICH THE
C               PROGRAM SPECIFIES INTERNALLY, THIS LATTER IS ALWAYS
C               < = 11*MEV / 8  +  12, WHERE MEV IS THE SIZE
C               T-MATRIX THAT WAS USED IN THE CORRESPONDING EIGENVALUE
C               COMPUTATIONS.
C     2.  V1:  >= MAX(N,KMAX)
C     3.  V2,VS:   >= N
C     4.  G:  >= MAX(N,KMAX)
C     5.  RITVEC:  >= N*NGOOD, WHERE NGOOD IS NUMBER OF EIGENVALUES
C                 SUPPLIED TO THIS PROGRAM.
C     6.  TVEC:  >= CUMULATIVE LENGTH OF ALL THE T-EIGENVECTORS
C         NEEDED TO GENERATE THE DESIRED RITZ VECTORS.  AN EDUCATED
C         GUESS AT AN APPROPRIATE LENGTH CAN BE OBTAINED BY RUNNING THE
C         PROGRAM WITH THE FLAG MBOUND = 1 AND MULTIPLYING THE
C         RESULTING SIZE BY 5/4.
C     7.  GOODEV, AMINGP, TMINGP, TERR, ERR, ERRGDP, RNORM, TBETA,
C         TLAST, EVNEW, MP, MA, M1, M2, MINT, MFIN AND IDELTA ALL MUST
C         BE >= NGOOD.
C
C-----------------------------------------------------------------------
C     OUTPUT HEADER
      WRITE(6,10)
   10 FORMAT(/' LANCZOS EIGENVECTOR PROCEDURE FOR REAL SYMMETRIC MATRICE
     1S'/)
C
C     SET PROGRAM PARAMETERS
C     USER MUST NOT MODIFY SCALEO
      SCALEO = 5.0D0
      ZERO = 0.0D0
      ONE = 1.0D0
      MPMIN = -1000
C     SET CONVERGENCE CRITERION FOR T-EIGENVECTORS.
      ERTOL = 1.D-10
C
C     READ USER-SPECIFIED PARAMETER FROM INPUT FILE 5 (FREE FORMAT)
C
C     READ USER-PROVIDED HEADER FOR RUN
      READ(5,20) EXPLAN
      WRITE(6,20) EXPLAN
   20 FORMAT(20A4)
C
C     READ IN THE MAXIMUM PERMISSIBLE DIMENSIONS FOR THE TVEC ARRAY
C     (MDIMTV), FOR THE RITVEC ARRAY (MDIMRV), AND FOR THE BETA
C     ARRAY (MBETA).
      READ(5,20) EXPLAN
      READ(5,*) MDIMTV, MDIMRV, MBETA
C
C     READ IN RELATIVE TOLERANCE (RELTOL) USED IN DETERMINING
C     APPROPRIATE SIZES FOR THE T-MATRICES TO BE USED IN THE RITZ
C     VECTOR COMPUTATIONS.
      READ(5,20) EXPLAN
      READ(5,*) RELTOL
C
C     SET FLAGS TO 0 OR 1:
C     MBOUND = 1:  PROGRAM TERMINATES AFTER COMPUTING 1ST GUESSES
C                  ON APPROPRIATE T-SIZES FOR USE IN THE RITZ VECTOR
C                  COMPUTATIONS
C     NTVCON = 0:  PROGRAM TERMINATES IF THE TVEC ARRAY IS NOT
C                  LARGE ENOUGH TO HOLD ALL THE T-EIGENVECTORS REQUIRED.
```

```
C      SVTVEC = 0:  THE T-EIGENVECTORS ARE NOT WRITTEN TO FILE 11
C                   UNLESS TVSTOP = 1
C      SVTVEC = 1:  WRITE THE T-EIGENVECTORS TO FILE 11.
C      TVSTOP = 1:  PROGRAM TERMINATES AFTER COMPUTING THE
C                   T-EIGENVECTORS
C      LVCONT = 0:  PROGRAM TERMINATES IF THE NUMBER OF T-EIGENVECTORS
C                   COMPUTED IS NOT EQUAL TO THE NUMBER OF RITZ
C                   VECTORS REQUESTED.
C      ERCONT = 0:  MEANS FOR ANY GIVEN EIGENVALUE, A RITZ VECTOR
C                   WILL NOT BE COMPUTED FOR THAT EIGENVALUE UNLESS
C                   A T-EIGENVECTOR HAS BEEN IDENTIFIED WITH A LAST
C                   COMPONENT WHICH SATISFIES THE SPECIFIED
C                   CONVERGENCE CRITERION.
C      ERCONT = 1:  MEANS FOR ANY GIVEN EIGENVALUE, A RITZ VECTOR
C                   WILL BE COMPUTED.  IF A T-EIGENVECTOR CANNOT
C                   BE IDENTIFIED WHICH SATISFIES THE LAST
C                   COMPONENT CRITERION, THEN THE PROGRAM WILL
C                   USE THE T-VECTOR THAT CAME CLOSEST TO
C                   SATISFYING THE CRITERION
C      IWRITE = 1:  EXTENDED OUTPUT OF INTERMEDIATE COMPUTATIONS
C                   IS WRITTEN TO FILE 6
C      IREAD = 0:   ALPHA/BETA FILE IS REGENERATED.
C      IREAD = 1:   ALPHA/BETA FILE USED IN EIGENVALUE COMPUTATIONS
C                   IS READ IN AND EXTENDED IF NECESSARY.  IN BOTH
C                   CASES IREAD = 0 OR 1, THE LANCZOS VECTORS ARE
C                   ALWAYS REGENERATED FOR THE RITZ VECTOR
C                   COMPUTATIONS
C
      READ(5,20) EXPLAN
      READ(5,*) MBOUND,NTVCON,SVTVEC,IREAD
C
      READ(5,20) EXPLAN
      READ(5,*) TVSTOP,LVCONT,ERCONT,IWRITE
      IF (TVSTOP.EQ.1) SVTVEC = 1
C
C      READ IN SEED (RHSEED) FOR GENERATING RANDOM STARTING VECTOR
C      FOR INVERSE ITERATION ON THE T-MATRICES.
      READ(5,20) EXPLAN
      READ(5,*) RHSEED
C
C      READ IN MATNOA, MATNOB = MATRIX/RUN IDENTIFICATION NUMBERS,
C      N = ORDER OF A-MATRIX AND B-MATRIX AND FLAG, JPERM.
C      JPERM = (0,1):  1 MEANS PERMUTED A AND B ARE BEING USED, 0
C      MEANS A AND B HAVE NOT BEEN PERMUTED.
      READ(5,20) EXPLAN
      READ(5,*) MATNOA,MATNOB,N,JPERM
C
C-----------------------------------------------------------------------
C      INITIALIZE THE ARRAYS FOR THE USER-SPECIFIED MATRICES
C      AND PASS THE STORAGE LOCATIONS OF THESE ARRAYS TO THE
C      MATRIX-VECTOR MULTIPLY SUBROUTINE AMATV AND THE SOLVE
C      SUBROUTINE LSOLV.
      CALL USPECA(N,MATNOA)
      CALL USPECB(N,MATNOB)
C
C-----------------------------------------------------------------------
C
C      MASK UNDERFLOW AND OVERFLOW
      CALL MASK
C
C-----------------------------------------------------------------------
C      WRITE RUN PARAMETERS OUT TO FILE 6
C
      WRITE(6,30) MATNOA,MATNOB,N,JPERM
   30 FORMAT(/4X,'A-MATRIX ID',4X,'B-MATRIX ID',4X,'SIZES OF MATRICES',
     14X,'JPERM'/I15,I15,I21,I9)
C
```

```
      WRITE(6,40) MBOUND,NTVCON,SVTVEC,IREAD                            LGV01990
   40 FORMAT(/3X,'MBOUND',3X,'NTVCON',3X,'SVTVEC',3X,'IREAD'/3I9,I8)    LGV02000
C                                                                       LGV02010
      WRITE(6,50) TVSTOP,LVCONT,ERCONT,IWRITE                           LGV02020
   50 FORMAT(/3X,'TVSTOP',3X,'LVCONT',3X,'ERCONT',3X,'IWRITE'/4I9)      LGV02030
C                                                                       LGV02040
      WRITE(6,60) MDIMTV,MDIMRV,MBETA                                   LGV02050
   60 FORMAT(/3X,'MDIMTV',3X,'MDIMRV',3X,'MBETA'/2I9,I8)                LGV02060
C                                                                       LGV02070
      WRITE(6,70) RELTOL,RHSEED                                         LGV02080
   70 FORMAT(/7X,'RELTOL',3X,'RHSEED'/E13.4,I9)                         LGV02090
C                                                                       LGV02100
C                                                                       LGV02110
C     FROM FILE 3 READ IN THE NUMBER OF EIGENVALUES (NGOOD) FOR WHICH   LGV02120
C     EIGENVECTORS ARE REQUESTED, THE ORDER (MEV) OF THE LANCZOS        LGV02130
C     TRIDIAGONAL MATRIX USED IN COMPUTING THESE EIGENVALUES, THE       LGV02140
C     ORDER (NOLD) OF THE USER-SPECIFIED MATRIX USED IN THE EIGENVALUE  LGV02150
C     COMPUTATIONS, THE SEED (SVSEED) USED FOR GENERATING THE STARTING  LGV02160
C     VECTOR THAT WAS USED IN THOSE LANCZOS EIGENVALUE COMPUTATIONS,    LGV02170
C     AND THE MATRIX/RUN IDENTIFICATION NUMBERS (MATA, MATB) USED IN    LGV02180
C     THOSE COMPUTATIONS.  ALSO READ IN THE NUMBER (NDIS) OF DISTINCT   LGV02190
C     EIGENVALUES OF T(1,MEV) THAT WERE COMPUTED BUT THIS VALUE IS      LGV02200
C     NOT USED IN THE EIGENVECTOR COMPUTATIONS.                         LGV02210
C                                                                       LGV02220
      READ(3,80) NGOOD,NDIS,MEV,NOLD,SVSEED,MATA,MATB                   LGV02230
   80 FORMAT(4I6,I12,2I8)                                               LGV02240
C                                                                       LGV02250
C     READ IN THE T-MULTIPLICITY TOLERANCE USED IN THE BISEC SUBROUTINE LGV02260
C     DURING THE COMPUTATION OF THE GIVEN EIGENVALUES.                  LGV02270
C     ALSO READ IN THE FLAG IB.  IF IB < 0, THEN SOME BETA(I) IN THE    LGV02280
C     T-MATRIX FILE PROVIDED ON FILE 2 FAILED THE ORTHOGONALITY         LGV02290
C     TEST IN THE TNORM SUBROUTINE.  USER SHOULD NOTE THAT THIS VECTOR  LGV02300
C     PROGRAM PROCEEDS INDEPENDENTLY OF THE SIZE OF THE BETA USED.      LGV02310
C                                                                       LGV02320
      READ(3,90) MULTOL,IB,BTOL                                         LGV02330
   90 FORMAT(E20.12,I6,E13.4)                                           LGV02340
C                                                                       LGV02350
      TEMP = DFLOAT(MEV+1000)                                           LGV02360
      TTOL = MULTOL/TEMP                                                LGV02370
      WRITE(6,100) MULTOL,TTOL                                          LGV02380
  100 FORMAT(/' T-MULTIPLICITY TOLERANCE USED IN THE EIGENVALUE COMPUTATLGV02390
     1IONS WAS',E13.4/' SCALED MACHINE EPSILON IS',E13.4)               LGV02400
C                                                                       LGV02410
C     CONTINUE WRITE TO FILE 6 OF THE PARAMETERS FOR THIS RUN           LGV02420
C                                                                       LGV02430
      WRITE(6,110)NGOOD,NDIS,MEV,NOLD,MATA,MATB,SVSEED,MULTOL,IB,BTOL   LGV02440
  110 FORMAT(/' EIGENVALUES SUPPLIED ARE READ IN FROM FILE 3'/' FILE 3  LGV02450
     1HEADER IS'/4X,'NG',2X,'NDIS',3X,'MEV',2X,'NOLD',2X,'MATNOA',2X,   LGV02460
     1'MATNOB'/(4I6,2I8)/4X,'SVSEED',6X,'MULTOL',6X,'IB',9X,'BTOL'/     LGV02470
     1I12,E12.3,I8,E13.4/)                                              LGV02480
C                                                                       LGV02490
C     IS THE ARRAY RITVEC LONG ENOUGH TO HOLD ALL OF THE DESIRED        LGV02500
C     RITZ VECTORS (APPROXIMATE EIGENVECTORS)?                          LGV02510
      NMAX = NGOOD*N                                                    LGV02520
      IF(MBOUND.NE.0) GO TO 120                                         LGV02530
      IF(TVSTOP.NE.1.AND.NMAX.GT.MDIMRV) GO TO 1350                     LGV02540
C                                                                       LGV02550
C     CHECK THAT THE ORDER N AND THE MATRIX IDENTIFICATION NUMBERS      LGV02560
C     MATNOA AND MATNOB SPECIFIED BY THE USER AGREE WITH THOSE READ     LGV02570
C     IN FROM FILE 3.                                                   LGV02580
  120 ITEMP = (NOLD-N)**2 + (MATA-MATNOA)**2 + (MATB-MATNOB)**2         LGV02590
      IF (ITEMP.NE.0) GO TO 1370                                        LGV02600
C                                                                       LGV02610
C     READ IN FROM FILE 3, THE T-MULTIPLICITIES OF THE EIGENVALUES      LGV02620
C     WHOSE EIGENVECTORS ARE TO BE COMPUTED, THE VALUES OF THESE        LGV02630
C     EIGENVALUES AND THEIR MINIMAL GAPS AS EIGENVALUES OF THE          LGV02640
```

```
C     USER-SPECIFIED MATRIX AND AS EIGENVALUES OF THE T-MATRIX.
C
      READ(3,20) EXPLAN
      READ(3,130) (MP(J),GOODEV(J),TMINGP(J),AMINGP(J), J=1,NGOOD)
  130 FORMAT(6X,I6,E25.16,2E14.3)
C
      WRITE(6,140) (J,GOODEV(J),MP(J),TMINGP(J),AMINGP(J), J=1,NGOOD)
  140 FORMAT(/' EIGENVALUES READ IN, T-MULTIPLICITIES, T-GAPS AND A-GAPS
     1 '/4X,' J ',5X,'GOOD EIGENVALUE',5X,'MULT',4X,'  TMINGAP ',4X,
     1'  ABMINGAP '/(I6,E25.16,I4,2E15.4))
C
C     READ IN ERROR ESTIMATES
      WRITE(6,150) MEV,SVSEED
  150 FORMAT(/' THESE EIGENVALUES WERE COMPUTED USING A T-MATRIX OF
     1ORDER ',I5/' AND SEED FOR RANDOM NUMBER GENERATOR =',I12)
C     CHECK WHETHER OR NOT THERE ARE ANY T-ISOLATED EIGENVALUES IN
C     THE EIGENVALUES PROVIDED
      DO 160 J=1,NGOOD
      IF(MP(J).EQ.1) GO TO 170
  160 CONTINUE
      GO TO 190
  170 READ(4,20) EXPLAN
      READ(4,20) EXPLAN
      READ(4,20) EXPLAN
      READ(4,180) NISO
  180 FORMAT(18X,I6)
      READ(4,20) EXPLAN
      READ(4,20) EXPLAN
      READ(4,20) EXPLAN
  190 DO 220 J=1,NGOOD
      ERR(J) = 0.D0
      IF(MP(J).NE.1) GO TO 220
      READ(4,200) EVAL, ERR(J)
  200 FORMAT(10X,E25.16,E14.3)
      IF(DABS(EVAL - GOODEV(J)).LT.1.D-10) GO TO 220
      WRITE(6,210) EVAL,GOODEV(J)
  210 FORMAT(' PROBLEM WITH READ IN OF ERROR ESTIMATES'/' EIGENVALUE REA
     1D IN',E20.12,' DOES NOT MATCH GOODEV(J) ='/E20.12)
      GO TO 1590
C
  220 CONTINUE
C
      WRITE(6,230) (J,GOODEV(J),ERR(J), J=1,NGOOD)
  230 FORMAT(' ERROR ESTMATES ='/4X,' J',5X,'EIGENVALUE',10X,' ESTIMATE
     1'/(I6,E20.12,E14.3))
C
      IF(IREAD.EQ.0)  GO TO 330
C
C     READ IN THE SIZE OF THE T-MATRIX PROVIDED ON FILE 2.  READ IN
C     THE ORDER OF THE USER-SPECIFIED MATRIX , THE SEED FOR THE
C     RANDOM NUMBER GENERATOR, AND THE MATRIX/TEST IDENTIFICATION
C     NUMBERS THAT WERE USED IN THE LANCZOS EIGENVALUE COMPUTATIONS.
C     THESE ARE USED IN A CONSISTENCY CHECK
C     IF FLAG IREAD = 0 REGENERATE ALPHA, BETA
C
      READ(2,240) KMAX,NOLD,SVSOLD,MATA,MATB
  240 FORMAT(2I6,I12,2I8)
C
      WRITE(6,250) KMAX,NOLD,SVSOLD,MATA,MATB
  250 FORMAT(/' READ IN THE T-MATRICES STORED ON FILE 2'/' FILE 2 HEADER
     1 IS'/2X,'KMAX',2X,'NOLD',6X,'SVSOLD',2X,'MATNOA',2X,'MATNOB'/
     1 2I6,I12,2I8/)
C
C     CHECK THAT THE ORDER, THE MATRIX/TEST IDENTIFICATION NUMBERS
C     AND THE SEED FOR THE RANDOM NUMBER GENERATOR USED IN THE
C     LANCZOS COMPUTATIONS THAT GENERATED THE HISTORY FILE
C     BEING USED AGREE WITH WHAT THE USER HAS SPECIFIED.
```

```
      IF (NOLD.NE.N.OR.MATA.NE.MATNOA.OR.MATNOB.NE.MATB.OR.SVSOLD.NE.
     1 SVSEED) GO TO 1390
C
      KMAX1 = KMAX + 1
C
C     READ IN THE T-MATRICES FROM FILE 2.  THESE ARE USED TO GENERATE
C     THE T-EIGENVECTORS THAT WILL BE USED IN THE RITZ VECTOR
C     COMPUTATIONS.  HISTORY MUST BE STORED IN MACHINE FORMAT
C     ((4Z20) FOR IBM/3081).
C
      READ(2,260) (ALPHA(J), J=1,KMAX)
      READ(2,260) (BETA(J), J=1,KMAX1)
  260 FORMAT(4Z20)
C
      READ(2,260) (V1(J), J=1,N)
      READ(2,260) (VS(J), J=1,N)
      READ(2,260) (V2(J), J=1,N)
C
C     KMAX MAY BE ENLARGED IF THE SIZE AT WHICH THE EIGENVALUE
C     COMPUTATIONS WERE PERFORMED IS ESSENTIALLY KMAX AND
C     THERE IS AT LEAST ONE EIGENVALUE THAT IS T-SIMPLE AND
C     T-ISOLATED, IN THE SENSE THAT IF ITS NEAREST NEIGHBOR IS TOO
C     CLOSE THAT NEIGHBOR IS A 'GOOD' T-EIGENVALUE.
      DO 270 J = 1,NGOOD
      IF(MP(J).EQ.1) GO TO 290
  270 CONTINUE
      WRITE(6,280)
  280 FORMAT(/' ALL EIGENVALUES USED ARE T-MULTIPLE OR CLOSE TO SPURIOUS
     1 T-EIGENVALUES'/' SO DO NOT CHANGE KMAX')
      IF(KMAX.LT.MEV) GO TO 1410
      GO TO 310
C
  290 KMAXN= 11*MEV/8 + 12
      IF(MBETA.LE.KMAXN) GO TO 1570
      IF(KMAX.GE.KMAXN )  GO TO 310
      WRITE(6,300) KMAX, KMAXN
  300 FORMAT(' ENLARGE KMAX FROM ',I6,' TO ',I6)
      MOLD1 = KMAX + 1
      KMAX = KMAXN
      GO TO 380
C
  310 WRITE(6,320) KMAX
  320 FORMAT(/' T-MATRICES HAVE BEEN READ IN FROM FILE 2'/' THE LARGEST
     1SIZE T-MATRIX ALLOWED IS',I6/)
C
      IF(IREAD.EQ.1) GO TO 400
C
C     REGENERATE THE ALPHA AND BETA
C
  330 MOLD1 = 1
C
      DO 340 J = 1,NGOOD
      IF(MP(J).EQ.1) GO TO 360
  340 CONTINUE
      KMAX = MEV + 12
      WRITE(6,350) KMAX
  350 FORMAT(/' ALL EIGENVALUES FOR WHICH EIGENVECTORS ARE TO BE COMPUTE
     1D ARE EITHER T-MULTIPLE OR CLOSE TO'/' A SPURIOUS T-EIGENVALUE. TH
     1EREFORE SET KMAX = MEV + 12 = ',I7)
      GO TO 380
C
  360 KMAXN = 11*MEV/8 + 12
      IF(MBETA.LE.KMAXN) GO TO 1570
      WRITE(6,370) KMAXN
  370 FORMAT(' SET KMAX EQUAL TO ',I6)
      KMAX = KMAXN
C
```

```
  380 WRITE(6,390) MOLD1,KMAX
  390 FORMAT(/' LANCZS SUBROUTINE GENERATES ALPHA(J), BETA(J+1), J =',
     1 I6,' TO ', I6/)
C
C-----------------------------------------------------------------------
C
      IIX = SVSEED
      CALL LANCZS(LSOLV,AMATV,ALPHA,BETA,V1,V2,VS,G,KMAX,MOLD1,N,IIX)
C
C-----------------------------------------------------------------------
C
  400 CONTINUE
C
C     THE SUBROUTINE STURMI DETERMINES THE SMALLEST SIZE T-MATRIX FOR
C     WHICH THE EIGENVALUE IN QUESTION IS A T-EIGENVALUE (TO WITHIN A
C     GIVEN TOLERANCE) AND IF POSSIBLE THE SMALLEST SIZE T-MATRIX
C     FOR WHICH IT IS A DOUBLE T-EIGENVALUE (TO WITHIN THE SAME
C     TOLERANCE).  THE SIZE T-MATRIX USED IN THE RITZ VECTOR
C     COMPUTATIONS IS THEN DETERMINED BY LOOPING ON SIZE OF THE
C     T-EIGENVECTORS, STARTING WITH A T-SIZE DETERMINED BY STURMI.
C
C
      STUTOL = SCALE0*MULTOL
      IF(IWRITE.EQ.1) WRITE(6,410)
  410 FORMAT(' FROM STURMI')
      DO 450 J = 1,NGOOD
      EVAL = GOODEV(J)
C     COMPUTE THE TOLERANCES USED BY STURMI TO DETERMINE AN INTERVAL
C     CONTAINING THE EIGENVALUE EVAL.
      TEMP = DABS(EVAL)*RELTOL
      TOLN = DMAX1(TEMP,STUTOL)
C
C-----------------------------------------------------------------------
C
      CALL STURMI(ALPHA,BETA,EVAL,TOLN,EPSM,KMAX,MK1,MK2,IC,IWRITE)
C
C-----------------------------------------------------------------------
C
C     STORE THE COMPUTED ORDERS OF T-MATRICES FOR LATER PRINTOUT
      M1(J) = MK1
      M2(J) = MK2
      ML(J) = (MK1 + 3*MK2)/4
      IF(MK2.EQ.KMAX)  ML(J) = KMAX
C
      IF(IC.GT.0) GO TO 430
C     IC = 0 MEANS THERE WAS NO EIGENVALUE IN THE DESIGNATED INTERVAL
C     BY T-SIZE KMAX.  THIS MEANS THAT THE EIGENVALUE PROVIDED HAS
C     NOT YET CONVERGED SO ITS EIGENVECTOR IS NOT COMPUTED.
      WRITE(6,420) J,GOODEV(J),MK1,MK2
  420 FORMAT(I6,'TH EIGENVALUE',E20.12,' HAS NOT CONVERGED '/
     1' SO DO NOT COMPUTE ANY T-EIGENVECTOR OR RITZ VECTOR FOR IT'
     1/' MK1 AND MK2 FOR THIS EIGENVALUE WERE',2I6)
      MP(J) = MPMIN
      MA(J) = -2*KMAX
      GO TO 450
C     COMPUTE AN APPROPRIATE SIZE T-MATRIX FOR THE GIVEN EIGENVALUE.
  430 IF(M2(J).EQ.KMAX) GO TO 440
C     M1 AND M2 WERE BOTH DETERMINED
      MA(J) = (3*M1(J) + M2(J))/4  + 1
      GO TO 450
C     M2 NOT DETERMINED
  440 MA(J) = (5*M1(J))/4  + 1
C
  450 CONTINUE
C
      IF (IWRITE.EQ.1) WRITE(6,460) (MA(JJ), JJ=1,NGOOD)
  460 FORMAT(/' 1ST GUESS AT APPROPRIATE SIZE T-MATRICES'/
```

```
     1 ' ACTUAL VALUES WILL PROBABLY BE 1/4 AGAIN AS MUCH'/(1316))
C
C     PRINT OUT TO FILE 10 1ST GUESSES AT SIZES OF THE T-MATRICES TO
C     BE USED IN THE EIGENVECTOR COMPUTATIONS.
C     PROGRAM LOOPS ON T-SIZE TO DETERMINE APPROPRIATE SIZE T-MATRIX.
      WRITE(10,470) N,KMAX
  470 FORMAT(2I8,' = ORDER OF USER MATRIX AND MAX ORDER OF T(1,MEV)')
C
      WRITE(10,480)
  480 FORMAT(/' 1ST GUESS AT APPROPRIATE SIZE T-MATRICES'/
     1 ' ACTUAL VALUES WILL PROBABLY BE 1/4 AGAIN AS MUCH'/)
C
      WRITE(10,490)
  490 FORMAT(4X,'J',3X,'AB-EIGENVALUE',4X,'M1(J)',1X,'M2(J)',1X,'MA(J)')
C
      WRITE(10,500) (J,GOODEV(J),M1(J),M2(J), MA(J), J=1,NGOOD)
  500 FORMAT(I5,E19.12,3I6)
C
      IF(MBOUND.EQ.1) WRITE(10,510)
  510 FORMAT(/'  EV = AB-EIGENVALUE IS A GOOD EIGENVALUE OF T(1,MEV)'/
     1' M1 = SMALLEST VALUE OF M SUCH THAT T(1,M) HAS AT LEAST'/
     1 '      ONE EIGENVALUE IN THE INTERVAL (EV-TOLN,EV+TOLN)'/
     1 '      TOLN(J) = DMAX1(EV(J)*RELTOL, SCALEO*MULTOL)'/
     1 ' M2 = SMALLEST M (IF ANY) SUCH THAT IN THE ABOVE INTERVAL'/
     1 '       T(1,M) HAS AT LEAST TWO EIGENVALUES '/
     1 ' IABS(MA(J)) = APPROPRIATE SIZE T-MATRIX FOR EV(J)'/
     1 ' INITIAL VALUE OF MA(J) IS CHOSEN HEURISTICALLY'/
     1 ' PROGRAM LOOPS ON SIZE OF T-MATRIX TO GET BETTER SIZE'/
     1 ' END OF SIZES OF T-MATRICES FILE 10'///)
C
C
C     TERMINATE AFTER COMPUTING 1ST GUESSES AT  SIZES OF THE
C     T-MATRICES REQUIRED FOR THE GIVEN EIGENVALUES?
      IF(MBOUND.EQ.1) GO TO 1430
C
C
C     IS THERE ROOM FOR ALL OF THE REQUESTED T-EIGENVECTORS?
      MTOL = 0
      DO 520 J = 1,NGOOD
      IF(MP(J).EQ.MPMIN) GO TO 520
      MTOL = MTOL + IABS(MA(J))
  520 CONTINUE
      MTOL = (5*MTOL)/4
      IF(MTOL.GT.MDIMTV.AND.NTVCON.EQ.0) GO TO 1450
C
C-----------------------------------------------------------------------
C     GENERATE A RANDOM VECTOR TO BE USED REPEATEDLY BY
C     SUBROUTINE INVERM
C
      IIL = RHSEED
      CALL GENRAN(IIL,G,KMAX)
C
C----------------------------------------------------------------------
C
C     LOOP ON GIVEN EIGENVALUES TO COMPUTE THE CORRESPONDING
C     T-EIGENVECTOR.
C
      MTOL = 0
      NTVEC = 0
      ILBIS = 0
      DO 710 J = 1,NGOOD
      ICOUNT = 0
      ERRMIN = 10.D0
      MABEST = MPMIN
      IF(MP(J).EQ.MPMIN) GO TO 710
      TFLAG = 0
      EVAL = GOODEV(J)
```

```
      TEMP = DABS(EVAL)*RELTOL
      UB = EVAL + DMAX1(STUTOL,TEMP)
      LB = EVAL - DMAX1(STUTOL,TEMP)
  530 KMAXU = IABS(MA(J))
C
C     SELECT A SUITABLE INCREMENT FOR THE ORDERS OF THE T-MATRICES
C     TO BE CONSIDERED IN DETERMINING APPROPRIATE SIZES FOR THE RITZ
C     VECTOR COMPUTATIONS.
      IF(ICOUNT.GT.0) GO TO 550
C     SELECT IDELTA(J) BASED UPON THE T-MULTIPLICITY OBTAINED
      IF(M2(J).EQ.KMAX) GO TO 540
C     M2 DETERMINED
      IDELTA(J) = ((3*M1(J) + 5*M2(J))/8  +  1 - IABS(MA(J)))/10  +  1
      GO TO 550
C     M2 NOT DETERMINED
  540 MAMAX = MINO((11*MEV)/8 + 12, (13*M1(J))/8 + 1)
      IDELTA(J) = (MAMAX - IABS(MA(J)))/10  +  1
  550 ICOUNT = ICOUNT + 1
C
C-----------------------------------------------------------------------
C     TO MIMIMIZE THE EFFECT OF THE ONE-SIDED ACCEPTANCE TEST FOR
C     EIGENVALUES IN THE BISEC SUBROUTINE, RECOMPUTE THE GIVEN
C     EIGENVALUE AT THE SPECIFIED KMAXU
C
      CALL LBISEC(ALPHA,BETA,EPSM,EVAL,EVALN,LB,UB,TTOL,KMAXU,NEVT)
C
C-----------------------------------------------------------------------
C
C     CHECK WHETHER OR NOT GIVEN T-MATRIX HAS AN EIGENVALUE IN THE
C     SPECIFIED INTERVAL AND IF SO WHAT ITS T-MULTIPLICITY IS.
C
      IF(NEVT.EQ.1) GO TO 590
      IF(NEVT.NE.0) GO TO 570
      ILBIS = 1
      WRITE(6,560) EVAL,KMAXU
  560 FORMAT(/' PROBLEM ENCOUNTERED IN RECOMPUTATION OF USER-SUPPLIED EI
     1GENVALUE',E20.12/' THE SIZE T-MATRIX SPECIFIED',I6,' DOES NOT
     1HAVE AN EIGENVALUE IN THE INTERVAL SPECIFIED'/' THEREFORE NO EIGEN
     1VECTOR WILL BE COMPUTED FOR THIS PARTICULAR EIGENVALUE'/)
      GO TO 610
C
  570 IF(NEVT.GT.1)  WRITE(6,580) EVAL,KMAXU
  580 FORMAT(/' PROBLEM ENCOUNTERED IN RECOMPUTATION OF USER-SUPPLIED
     1EIGENVALUE',E20.12/' FOR THE SIZE T-MATRIX SPECIFIED =',I6,' THE
     1GIVEN EIGENVALUE IS T-MULTIPLE IN THE INTERVAL SPECIFIED'/' SOMETH
     1ING IS WRONG, THEREFORE NO EIGENVECTOR WILL BE COMPUTED FOR THIS E
     1IGENVALUE'/)
C
      MP(J) = MPMIN
      MA(J) = -2*KMAX
      GO TO 710
C
  590 CONTINUE
      ILBIS = 0
C
      EVNEW(J) = EVALN
      EVAL = EVALN
      MTOL = MTOL+KMAXU
C
C     IS THERE ROOM IN TVEC ARRAY FOR THE NEXT T-EIGENVECTOR?
C     IF NOT, SKIP TO RITZ VECTOR COMPUTATIONS.
      IF (MTOL.GT.MDIMTV) GO TO 720
C
      IT = 3
      KINT = MTOL - KMAXU +1
C
C     RECORD THE BEGINNING AND END OF THE T-EIGENVECTOR BEING COMPUTED
```

```
      MINT(J) = KINT
      MFIN(J) = MTOL
C
C-----------------------------------------------------------------------
C     SUBROUTINE INVERM DOES INVERSE ITERATION, I.E. SOLVES
C     (T(1,KMAXU) - EVAL)*U = RHS  FOR EACH EIGENVALUE TO OBTAIN THE
C     DESIRED T-EIGENVECTOR.
C
      IF(IWRITE.EQ.1)  WRITE(6,600) J
  600 FORMAT(/I6,'TH EIGENVALUE')
C
      CALL INVERM(ALPHA,BETA,V1,TVEC(KINT),EVAL,ERROR,TERROR,EPSM,
     1 G,KMAXU,IT,IWRITE)
C
C-----------------------------------------------------------------------
C
      TERR(J) = TERROR
      TLAST(J) = ERROR
      KMAXU1 = KMAXU + 1
      TBETA(J) = BETA(KMAXU1)*ERROR
C
C     AFTER COMPUTING EACH OF THE T-EIGENVECTORS,
C     CHECK THE SIZE OF THE ERROR ESTIMATE, ERROR.
C     IF THIS ESTIMATE IS NOT AS SMALL AS DESIRED AND
C     |MA(J)| < ML(J), ATTEMPT TO INCREASE THE SIZE OF |MA(J)|
C     AND REPEAT THE T-EIGENVECTOR COMPUTATIONS.
C
      IF(ERROR.LT.ERTOL.OR.TFLAG.EQ.1)  GO TO 700
C
      IF(ERROR.GE.ERRMIN) GO TO 610
C     LAST COMPONENT IS LESS THAN MINIMAL TO DATE
      ERRMIN = ERROR
      MABEST = MA(J)
  610 CONTINUE
C
      IF(MA(J).GT.0)  ITEST = MA(J) + IDELTA(J)
      IF(MA(J).LT.0)  ITEST = -(IABS(MA(J)) + IDELTA(J))
      IF(IABS(ITEST).LE.ML(J).AND.ICOUNT.LE.10) GO TO 630
C     NEW MA(J) IS GREATER THAN MAXIMUM ALLOWED.
      IF(ERCONT.EQ.0.OR.MABEST.EQ.MPMIN) GO TO 650
      TFLAG = 1
      MA(J) = MABEST
      IF(ILBIS.EQ.0)  MTOL = MTOL - KMAXU
      WRITE(6,620) MA(J)
  620 FORMAT(' 10 ORDERS WERE CONSIDERED.  NONE SATISFIED THE ERROR TEST
     1'/' THEREFORE USE THE BEST ORDER OBTAINED FOR THE EIGENVECTORS'
     1,I6)
      GO TO 530
C
  630 MA(J) = ITEST
C
      MT = IABS(MA(J))
      IF(IWRITE.EQ.1) WRITE(6,640)  MT
  640 FORMAT(/' CHANGE SIZE OF T-MATRIX TO ',I6,' RECOMPUTE T-EIGENVECTO
     1R')
C
      IF(ILBIS.EQ.0)  MTOL = MTOL - KMAXU
C
      GO TO 530
C
C     APPROPRIATE SIZE T-MATRIX WAS NOT OBTAINED
  650 CONTINUE
      WRITE(10,660) J,EVAL,MP(J)
  660 FORMAT(/' ON 10 INCREMENTS NOT ABLE TO IDENTIFY APPROPRIATE SIZE
     1T-MATRIX FOR'/
     1' EIGENVALUE(',I4,') = ',E20.12,' T-MULTIPLICITY =',I4/)
      IF(M2(J).EQ.KMAX) WRITE(10,670)
```

```
      IF(M2(J).LT.KMAX) WRITE(10,680)                                   LGV06670
  670 FORMAT(/' ORDERS TESTED RANGED FROM 5*M1(J)/4 TO APPROXIMATELY    LGV06680
     1 '/'  MIN(11*MEV/8,13*M1(J)/8)'/)                                 LGV06690
  680 FORMAT(/' ORDERS TESTED RANGED FROM (3*M1(J)+M2(J))/4 TO APPROXIMALGV06700
     1TELY'/'  (3*M1(J) + 5*M2(J))/8.'/)                                LGV06710
      WRITE(10,690)                                                     LGV06720
  690 FORMAT(' ALLOWING LARGER ORDERS FOR THE T-MATRICES MAY RESULT IN  LGV06730
     1 SUCCESS'/' BUT PROBABLY WILL NOT.  PROBLEM IS PROBABLY DUE TO'   LGV06740
     1 /' LACK OF CONVERGENCE OF GIVEN EIGENVALUE, CHECK THE ERROR ESTIMLGV06750
     1ATE'/)                                                            LGV06760
      MP(J) = MPMIN                                                     LGV06770
      IF(ILBIS.EQ.0)  MTOL = MTOL - KMAXU                               LGV06780
      GO TO 710                                                         LGV06790
  700 NTVEC = NTVEC + 1                                                 LGV06800
C                                                                       LGV06810
  710 CONTINUE                                                          LGV06820
      NGOODC = NGOOD                                                    LGV06830
      GO TO 740                                                         LGV06840
C                                                                       LGV06850
C     COME HERE IF THERE IS NOT ENOUGH ROOM FOR ALL OF T-EIGENVECTORS   LGV06860
  720 NGOODC = J-1                                                      LGV06870
      WRITE(6,730)  J, MTOL, MDIMTV                                     LGV06880
  730 FORMAT(/'  NOT ENOUGH ROOM IN TVEC FOR ',I4,'TH T-VECTOR'/'  T-DIMLGV06890
     1ENSION REQUESTED = ',I6,' BUT TVEC HAS DIMENSION = ',I6/)         LGV06900
      IF(NGOODC.EQ.0)  GO TO 1470                                       LGV06910
      MTOL = MTOL-KMAXU                                                 LGV06920
C                                                                       LGV06930
  740 CONTINUE                                                          LGV06940
C                                                                       LGV06950
C     THE LOOP ON T-EIGENVECTOR COMPUTATIONS IS COMPLETE.                LGV06960
C     WRITE OUT THE SIZE T-MATRICES THAT WILL BE USED FOR               LGV06970
C     THE RITZ VECTOR COMPUTATIONS.                                     LGV06980
C                                                                       LGV06990
      WRITE(10,750)                                                     LGV07000
  750 FORMAT(/' SIZES OF T-MATRICES THAT WILL BE USED IN THE RITZ COMPUTLGV07010
     1ATIONS'/5X,'J',16X,'GOODEV(J)',1X,'MA(J)')                        LGV07020
C                                                                       LGV07030
      WRITE(10,760)   (J,GOODEV(J),MA(J), J=1,NGOOD)                    LGV07040
  760 FORMAT(I6,E25.14,I6)                                              LGV07050
      WRITE(10,510)                                                     LGV07060
C                                                                       LGV07070
      WRITE(6,770) MTOL                                                 LGV07080
  770 FORMAT(/' THE CUMULATIVE LENGTH OF THE T-EIGENVECTORS IS',I18)    LGV07090
C                                                                       LGV07100
      WRITE(6,780) NTVEC,NGOOD                                          LGV07110
  780 FORMAT(/I6,' T-EIGENVECTORS OUT OF',I6,' REQUESTED WERE COMPUTED')LGV07120
C                                                                       LGV07130
C     SAVE THE T-EIGENVECTORS ON FILE 11?                               LGV07140
      IF(TVSTOP.NE.1.AND.SVTVEC.EQ.0) GO TO 840                         LGV07150
C                                                                       LGV07160
      WRITE(11,790) NTVEC,MTOL,MATNOA,MATNOB,SVSEED                     LGV07170
  790 FORMAT(I6,3I8,I12,' = NTVEC,MTOL,MATNOA,MATNOB,SVSEED')           LGV07180
C                                                                       LGV07190
      DO 820 J=1,NGOODC                                                 LGV07200
C     IF MP(J) = MPMIN THEN NO SUITABLE T-EIGENVECTOR IS AVAILABLE      LGV07210
C     FOR THAT EIGENVALUE.                                              LGV07220
      IF(MP(J).EQ.MPMIN) WRITE(11,800) J,MA(J),GOODEV(J),MP(J)          LGV07230
  800 FORMAT(2I6,E20.12,I6/' TH EIGVAL,T-SIZE,EVALUE,FLAG,NO EIGVEC')   LGV07240
      IF(MP(J).NE.MPMIN) WRITE(11,810) J,MA(J),GOODEV(J),MP(J)          LGV07250
  810 FORMAT(I6,I6,E20.12,I6/' T-EIGVEC,SIZE T,EVALUE OF A,MP(J)')      LGV07260
      IF(MP(J).EQ.MPMIN) GO TO 820                                      LGV07270
      KI = MINT(J)                                                      LGV07280
      KF = MFIN(J)                                                      LGV07290
C                                                                       LGV07300
      WRITE(11,260) (TVEC(K), K=KI,KF)                                  LGV07310
C                                                                       LGV07320
```

```
  820 CONTINUE
C
      IF(TVSTOP.NE.1) GO TO 840
C
      WRITE(6,830) TVSTOP, NTVEC,NGOOD
  830 FORMAT(/' USER SET TVSTOP = ',I1/
     1' THEREFORE PROGRAM TERMINATES AFTER T-EIGENVECTOR COMPUTATIONS'/
     1' T-EIGENVECTORS THAT WERE COMPUTED ARE SAVED ON FILE 11'/
     1I8,' T-EIGENVECTORS WERE COMPUTED OUT OF',I7,' REQUESTED'/)
C
      GO TO 1590
C
  840 CONTINUE
C     IF NOT ABLE TO COMPUTE ALL THE REQUESTED T-EIGENVECTORS,
C     CONTINUE WITH THE LANCZOS VECTOR COMPUTATIONS ANYWAY?
      IF(NTVEC.NE.NGOOD.AND.LVCONT.EQ.0) GO TO 1490
C
C     COMPUTE THE MAXIMUM SIZE OF THE T-MATRIX USED FOR THOSE
C     EIGENVALUES WITH GOOD ERROR ESTIMATES.
C
      KMAXU = 0
      DO 850 J = 1,NGOODC
      MT = IABS(MA(J))
      IF(MT.LT.KMAXU.OR.MP(J).EQ.MPMIN) GO TO 850
      KMAXU = MT
  850 CONTINUE
C
      IF(KMAXU.EQ.0) GO TO 1530
C
      WRITE(6,860) KMAXU
  860 FORMAT(/I6,' = LARGEST SIZE T-MATRIX TO BE USED IN THE RITZ VECTOR
     1 COMPUTATIONS')
C
C     COUNT THE NUMBER OF RITZ VECTORS NOT BEING COMPUTED
      MREJEC = 0
      DO 870 J=1,NGOODC
  870 IF(MP(J).EQ.MPMIN)  MREJEC = MREJEC + 1
      MREJET = MREJEC + (NGOOD-NGOODC)
      IF(MREJET.NE.0) WRITE(6,880) MREJET
  880 FORMAT(/' RITZ VECTORS ARE NOT COMPUTED FOR',I6,' OF THE EIGENVALU
     1ES'/)
      NACT = NGOODC - MREJEC
      WRITE(6,890) NGOOD,NTVEC,NACT
  890 FORMAT(/I6,' RITZ VECTORS WERE REQUESTED'/I6,' T-EIGENVECTORS WERE
     1 COMPUTED'/I6,' RITZ VECTORS WILL BE COMPUTED'/)
C     CHECK IF THERE ARE ANY RITZ VECTORS TO COMPUTE
      IF(MREJEC.EQ.NGOODC) GO TO 1510
C
C     CONTINUE WITH THE LANCZOS VECTOR COMPUTATIONS?
      IF(LVCONT.EQ.0.AND.MREJEC.NE.0) GO TO 1490
C
C     NOW COMPUTE THE RITZ VECTORS.  REGENERATE THE
C     LANCZOS VECTORS.
C
      DO 900 I = 1,NMAX
  900 RITVEC(I) = ZERO
C
C-----------------------------------------------------------------------
C     REGENERATE THE STARTING VECTOR. THIS MUST BE GENERATED AND
C     NORMALIZED PRECISELY THE WAY IT WAS DONE IN THE EIGENVALUE
C     COMPUTATIONS, OTHERWISE THERE WILL BE A MISMATCH BETWEEN
C     THE T-EIGENVECTORS THAT HAVE BEEN COMPUTED FROM THE T-MATRICES
C     READ IN FROM FILE 2 AND THE LANCZOS VECTORS THAT ARE
C     BEING REGENERATED.
C
      IIL = SVSEED
```

```
      CALL GENRAN(IIL,G,N)
C
C-----------------------------------------------------------------------
C
      DO 910 K = 1,N
  910 V2(K) = G(K)
C
C-----------------------------------------------------------------------
C     COMPUTE L-TRANSPOSE*V2 AND ITS NORM
      ISOLV = 2
      CALL LSOLV(V2,VS,ISOLV)
      SUM = FINPRO(N,VS(1),1,VS(1),1)
C-----------------------------------------------------------------------
C
C     NORMALIZE STARTING VECTORS:  (V2-TRANSPOSE*B*V2) = 1
      SUM = ONE/DSQRT(SUM)
      DO 920 K = 1,N
      VS(K) = SUM*VS(K)
  920 V2(K) = SUM*V2(K)
C
C-----------------------------------------------------------------------
C     INITIALIZE V1 = B*V2 = L*VS
      ISOLV = 1
      CALL LSOLV(VS,V1,ISOLV)
C-----------------------------------------------------------------------
C
      DO 930 K = 1,N
      VS(K) = V1(K)
  930 V1(K) = ZERO
C
      IVEC = 1
      BATA = ZERO
C
      GO TO 1000
C
C     VS = B*V(I),  V1 = B*V(I-1),  V2 = V(I)
  940 CONTINUE
      SUM = BATA
C
C-----------------------------------------------------------------------
C     COMPUTE V1 = A*V2 - SUM*V1
      CALL MATVEC(V2,V1,SUM)
C     COMPUTE ALFA
      ALFA = FINPRO(N,V1(1),1,V2(1),1)
C-----------------------------------------------------------------------
C
      DO 950 K = 1,N
  950 V1(K) = V1(K)-ALFA*VS(K)
C
C     SET V1 = B*V(IVEC) AND VS = (NEW BATA)*B*V(IVEC+1)
      DO 960 K = 1,N
      TEMP = V1(K)
      V1(K) = VS(K)
  960 VS(K) = TEMP
C
C-----------------------------------------------------------------------
C     COMPUTE V2 = (L-INVERSE)*VS
      ISOLV = 3
      CALL LSOLV(VS,V2,ISOLV)
C     COMPUTE NEXT BATA
      SUM = FINPRO(N,V2(1),1,V2(1),1)
C-----------------------------------------------------------------------
C
      BATA = DSQRT(SUM)
      TEMP = BETA(IVEC)
      TEMP = DABS(BATA - TEMP)/TEMP
```

```
      IF (TEMP.LT.1.0D-10)GO TO 980                                     LGV08650
C                                                                       LGV08660
C     THE BETA BEING REGENERATED DO NOT MATCH THE BETA IN FILE 2.       LGV08670
C     SOMETHING IS WRONG IN THE LANCZOS VECTOR GENERATION.              LGV08680
C     PROGRAM TERMINATES FOR USER TO CORRECT THE PROBLEM                LGV08690
C     WHICH MUST BE IN THE STARTING VECTOR GENERATION OR IN             LGV08700
C     THE SUBROUTINES AMATV AND LSOLV SUPPLIED.                         LGV08710
C     THESE SUBROUTINES MUST BE THE SAME ONES USED IN THE               LGV08720
C     EIGENVALUE COMPUTATIONS OR A MISMATCH WILL ENSUE.                 LGV08730
C                                                                       LGV08740
      WRITE(6,970) IVEC,BATA,BETA(IVEC),TEMP                            LGV08750
  970 FORMAT(/2X,'IVEC',16X,'BATA',10X,'BETA(IVEC)',14X,'RELDIF'/I6,    LGV08760
     13E20.12/' IN LANCZOS VECTOR REGENERATION THE ENTRIES OF THE TRIDIALGV08770
     1GONAL MATRICES BEING'/' GENERATED ARE NOT THE SAME AS THOSE IN THELGV08780
     1 MATRIX SUPPLIED ON FILE 2.'/' THEREFORE SOMETHING IS BEING INITIALGV08790
     1LIZED OR COMPUTED DIFFERENTLY FROM THE WAY'/' IT WAS COMPUTED IN TLGV08800
     1HE EIGENVALUE COMPUTATIONS'/' THE PROGRAM TERMINATES FOR THE USER LGV08810
     1TO DETERMINE WHAT THE PROBLEM IS'/)                               LGV08820
      GO TO 1590                                                        LGV08830
  980 CONTINUE                                                          LGV08840
C                                                                       LGV08850
C-----------------------------------------------------------------------LGV08860
      ISOLV = 4                                                         LGV08870
      CALL LSOLV(V2,V2,ISOLV)                                           LGV08880
C-----------------------------------------------------------------------LGV08890
C                                                                       LGV08900
      SUM = ONE/BATA                                                    LGV08910
      DO 990 K = 1,N                                                    LGV08920
      V2(K) = SUM*V2(K)                                                 LGV08930
  990 VS(K) = SUM*VS(K)                                                 LGV08940
C                                                                       LGV08950
 1000 CONTINUE                                                          LGV08960
C                                                                       LGV08970
      LFIN = 0                                                          LGV08980
      DO 1020 J = 1,NGOODC                                              LGV08990
      LL = LFIN                                                         LGV09000
      LFIN = LFIN + N                                                   LGV09010
C                                                                       LGV09020
      IF(IABS(MA(J)).LT.IVEC.OR.MP(J).EQ.MPMIN) GO TO 1020              LGV09030
      II = IVEC + MINT(J) - 1                                           LGV09040
      TEMP = TVEC(II)                                                   LGV09050
C     II IS THE (IVEC)TH COMPONENT OF THE T-EIGENVECTOR CONTAINED       LGV09060
C     IN TVEC(MINT(J)).                                                 LGV09070
C                                                                       LGV09080
      DO 1010 K = 1,N                                                   LGV09090
      LL = LL + 1                                                       LGV09100
 1010 RITVEC(LL) = TEMP*V2(K) + RITVEC(LL)                              LGV09110
C                                                                       LGV09120
 1020 CONTINUE                                                          LGV09130
C                                                                       LGV09140
      IVEC = IVEC + 1                                                   LGV09150
      IF (IVEC.LE.KMAXU) GO TO 940                                      LGV09160
C                                                                       LGV09170
C     RITZVECTOR GENERATION IS COMPLETE. B-NORMALIZE EACH RITZVECTOR.   LGV09180
C     NOTE THAT IF CERTAIN RITZ VECTORS WERE NOT COMPUTED THEN THAT     LGV09190
C     PORTION OF THE RITVEC ARRAY WAS NOT UTILIZED.                     LGV09200
C                                                                       LGV09210
      LFIN = 0                                                          LGV09220
      DO 1090 J = 1,NGOODC                                              LGV09230
C                                                                       LGV09240
      KK = LFIN                                                         LGV09250
      LFIN = LFIN + N                                                   LGV09260
      IF(MP(J).EQ.MPMIN) GO TO 1090                                     LGV09270
C                                                                       LGV09280
      DO 1030 K = 1,N                                                   LGV09290
      KK = KK + 1                                                       LGV09300
```

```
 1030 V2(K) = RITVEC(KK)
C
C-----------------------------------------------------------------------
      ISOLV = 2
      CALL LSOLV(V2,VS,ISOLV)
      SUM = FINPRO(N,VS(1),1,VS(1),1)
C-----------------------------------------------------------------------
C
      SUM = DSQRT(SUM)
      RNORM(J) = SUM
      TEMP = DABS(ONE-SUM)
      SUM = ONE/SUM
C
      DO 1040 K = 1,N
      VS(K) = SUM*VS(K)
      V2(K) = SUM*V2(K)
 1040 CONTINUE
C
C-----------------------------------------------------------------------
      ISOLV = 1
      CALL LSOLV(VS,V1,ISOLV)
C-----------------------------------------------------------------------
C
C     V1 = B*V2
      EVAL = EVNEW(J)
C
C     COMPUTE ERROR IN RITZ VECTOR CONSIDERED AS A EIGENVECTOR OF A.
C     V1 = A*RITVEC - EVAL*B*RITVEC
C
C-----------------------------------------------------------------------
      CALL AMATV(V2,V1,EVAL)
      SUM = FINPRO(N,V1(1),1,V1(1),1)
C-----------------------------------------------------------------------
C
      SUM = DSQRT(SUM)
      ERR(J) = SUM
      GAP = ABS(AMINGP(J))
      ERRDGP(J) = SUM/GAP
C
C
      IF (JPERM.EQ.0) GO TO 1050
C
C-----------------------------------------------------------------------
C     ON RETURN V2 = P(TRANSPOSE)*V2
      IPERM = 2
      CALL LPERM(V2,V1,IPERM)
C-----------------------------------------------------------------------
C
 1050 CONTINUE
      KK = LFIN - N
      DO 1060 K = 1,N
      KK = KK + 1
 1060 RITVEC(KK) = V2(K)
C
      IF (IWRITE.NE.0) WRITE(6,1070) J,GOODEV(J)
 1070 FORMAT(/I5,' TH EIGENVALUE CONSIDERED = ',E20.12/)
C
      IF (IWRITE.NE.0) WRITE(6,1080) TERR(J),TBETA(J),TEMP
 1080 FORMAT(' NORM OF ERROR IN T-EIGENVECTOR = ',E14.3/
     1 ' BETA(MA(J)+1)*U(MA(J)) = ',E14.3/
     1 ' ABS(NORM(RITVEC) - 1.0)  = ',E14.3/)
C
 1090 CONTINUE
C
C     RITZVECTORS ARE NORMALIZED AND ERROR ESTIMATES ARE IN ERR ARRAY
C     AND IN ERRDGP ARRAY. STORE EVERYTHING
```

```
C
C
      WRITE(9,1100)
 1100 FORMAT(2X,'AB-EIGENVALUE',2X,'MA(J)',2X,'AB-MINGAP',5X,'ABERROR',1
     1X, 'ABERROR/GAP',6X,'TERROR')
C
      WRITE(13,1110)
 1110 FORMAT(12X,'AB-EIGENVALUE',5X,'RITZNORM',5X,'ABMINGAP',5X,
     1 'TBETA(J)',5X,'TLAST(J)')
C
      DO 1140 J=1,NGOODC
C
      IF(MP(J).EQ.MPMIN) GO TO 1140
C
      WRITE(9,1120)EVNEW(J),MA(J),AMINGP(J),ERR(J),ERRDGP(J),TERR(J)
 1120 FORMAT(E15.8,I6,4E12.4)
C
      WRITE(13,1130) EVNEW(J),RNORM(J),AMINGP(J),TBETA(J),TLAST(J)
 1130 FORMAT(E25.14,4E13.5)
C
 1140 CONTINUE
C
      IF(MREJEC.EQ.0) GO TO 1220
      WRITE(9,1150)
 1150 FORMAT(/' RITZ VECTORS WERE NOT COMPUTED FOR THE FOLLOWING EIGENVA
     1LUES'/' EITHER BECAUSE THEY HAD NOT CONVERGED OR BECAUSE THE ERROR
     1 ESTIMATE'/'  WAS NOT AS SMALL AS DESIRED'/)
C
      DO 1210 J = 1,NGOODC
      IF(MP(J).NE.MPMIN) GO TO 1210
C     WRITE OUT MESSAGE FOR EACH EIGENVALUE FOR WHICH NO EIGENVECTOR
C     WAS COMPUTED.
C
      WRITE(9,1160)
 1160 FORMAT(2X,'AB-EIGENVALUE',3X,'MA(J)',5X,'AMINGP(J)',6X,'TLAST(J)',
     13X,'MP(J)')
      WRITE(9,1170) GOODEV(J),MA(J),AMINGP(J),TBETA(J),MP(J)
 1170 FORMAT(E15.8,I8,2E14.4,I8)
C
      WRITE(13,1180)
 1180 FORMAT(/' RITZ VECTORS WERE NOT COMPUTED FOR THE FOLLOWING EIGENVA
     1LUES'/' BECAUSE THEY HAD NOT CONVERGED'/)
C
      WRITE(13,1190)
 1190 FORMAT(2X,'AB-EIGENVALUE',3X,'MA(J)',3X,'M1(J)',3X,'M2(J)',3X,'MP(
     1J)'/)
      WRITE(13,1200) GOODEV(J),MA(J),M1(J),M2(J),MP(J)
 1200 FORMAT(E15.8,4I8)
C
 1210 CONTINUE
 1220 CONTINUE
C
      WRITE(9,1230)
 1230 FORMAT(/' ABOVE ARE ERROR ESTIMATES FOR THE AB AND T EIGENVECTORS'
     1 /' ASSOCIATED WITH THE AB-EIGENVALUES LISTED IN COLUMN 1'/
     1 ' ABERROR = NORM(A*X - EV*B*X)  TERROR = NORM(T*Y - EV*Y)
     1 '/' WHERE T = T(1,MA(J))    X = RITZ VECTOR = V*Y  V = SUCCESSIVE
     1 '/' LANCZOS VECTORS. ABMINGAP = GAP TO NEAREST AB-EIGENVALUE'//)
C
      WRITE(13,1240)
 1240 FORMAT(/' ABOVE ARE ERROR ESTIMATES ASSOCIATED WITH THE AB-EIGVALS
     1 '/' RITZNORM = NORM(COMPUTED RITZ VECTOR)'/
     1 ' TBETA(J) = BETA(MA(J)+1)*Y(MA(J)),  T*Y = EVAL*Y'/
     1 ' TLAST(J) = Y(MA(J))'/
     1 ' ABMINGAP = GAP TO NEAREST AB-EIGENVALUE'/)
C
C     NUMBER OF RITZ VECTORS COMPUTED
```

```
      NCOMPU = NGOODC - MREJEC
      WRITE(12,1250) N,NCOMPU,NGOODC,MATNOA,MATNOB
 1250 FORMAT(3I6,2I8,' SIZE A, NO.RITZVECS, NO.EVALS,MATNOA,MATNOB')
C
      LFIN = 0
      DO 1310 J = 1,NGOODC
      LINT = LFIN + 1
      LFIN = LFIN + N
C
      IF(MP(J).EQ.MPMIN) GO TO 1290
C     RITZ VECTOR WAS COMPUTED
      WRITE(12,1260) J, GOODEV(J), MP(J)
 1260 FORMAT(I6,4X,E20.12,I6,' J, AB-EIGENVAL, MP(J)')
C
      WRITE(12,1270) ERR(J),ERRDGP(J)
 1270 FORMAT(2E15.5,'= NORM(A*Z-EVAL*B*Z), NORM(A*Z-EVAL*B*Z)/ABMINGAP')
C
      WRITE(12,1280) (RITVEC(LL), LL=LINT,LFIN)
 1280 FORMAT(4E20.12)
      GO TO 1310
C     NO RITZ VECTOR WAS COMPUTED FOR THIS EIGENVALUE
 1290 WRITE(12,1300) J,GOODEV(J),MP(J)
 1300 FORMAT(I6,4X,E20.12,I6,' J,AB-EIGVALUE,NO RITZ VECTOR COMPUTED')
C
 1310 CONTINUE
C
C     DID ANY T-MATRICES INCLUDE OFF-DIAGONAL ENTRIES SMALLER THAN
C     DESIRED, AS SPECIFIED BY BTOL?
C
      IF(IB.GT.0) GO TO 1340
C
      WRITE(6,1320) KMAXU
 1320 FORMAT(/' FOR LARGEST T-MATRIX CONSIDERED',I7,' CHECK THE SIZE OF
     1BETAS')
C
C-----------------------------------------------------------------------
C
      CALL TNORM(ALPHA,BETA,BKMIN,TEMP,KMAXU,IBMT)
C
C-----------------------------------------------------------------------
C
      IF(IBMT.LT.0) WRITE (6,1330)
 1330 FORMAT(/' WARNING THE T-MATRICES FOR ONE OR MORE OF THE EIGENVALUE
     1S CONSIDERED'/' HAD AN OFF-DIAGONAL ENTRY THAT WAS SMALLER THAN TH
     1E BETA TOLERANCE THAT WAS SPECIFIED'/)
 1340 CONTINUE
C
       GO TO 1590
C
 1350 WRITE(6,1360) NGOOD,NMAX,MDIMRV
 1360 FORMAT(/I4,' RITZ VECTORS WERE REQUESTED BUT THE REQUIRED DIMENSIO
     1N',I6/' IS LARGER THAN THE USER-SPECIFIED DIMENSION OF RITVEC',I6
     1/' THEREFORE, THE EIGENVECTOR PROCEDURE TERMINATES FOR THE USER TO
     1 INTERVENE')
C
      GO TO 1590
C
 1370 WRITE(6,1380) NOLD,N,MATA,MATNOA,MATB,MATNOB
 1380 FORMAT(/' PARAMETERS READ FROM FILE 3 DO NOT AGREE WITH USER-SPECI
     1FIED VALUES'/' NOLD,N,MATA,MATNOA,MATB,MATNOB = '/2I6,4I12/
     1' THEREFORE PROGRAM TERMINATES FOR USER TO RESOLVE DIFFERENCES'/)
C
      GO TO 1590
C
 1390 WRITE(6,1400)
 1400 FORMAT(/' PARAMETERS IN ALPHA,BETA FILE READ IN DO NOT AGREE WITH
     1THOSE'/' SPECIFIED BY THE USER.  THEREFORE PROGRAM TERMINATES FOR'
```

```
     1/' USER TO RESOLVE DIFFERENCES'/)
C
      GO TO 1590
C
 1410 WRITE(6,1420) KMAX,MEV
 1420 FORMAT(/' ALPHA,BETA HEADER HAS KMAX = ',I6/
     1' BUT EIGENVALUES WERE COMPUTED AT MEV = ',I6,' PROGRAM STOPS'/)
C
      GO TO 1590
C
 1430 WRITE(6,1440)
 1440 FORMAT(/' PROGRAM COMPUTED 1ST GUESSES AT T-MATRIX SIZES AND READ
     1THEM TO FILE 10'/' THEN TERMINATED AS REQUESTED.')
      GO TO 1590
C
 1450 WRITE(6,1460) MTOL, MDIMTV
 1460 FORMAT(/' PROGRAM TERMINATES BECAUSE THE TVEC DIMENSION ANTICIPATE
     1D',I7/'  IS LARGER THAN THE TVEC DIMENSION',I7,' SPECIFIED BY THE
     1USER.'/'  USER MAY RESET THE TVEC DIMENSION AND RESTART THE PROGRA
     1M')
      GO TO 1590
C
 1470 WRITE(6,1480)
 1480 FORMAT(/' PROGRAM TERMINATES BECAUSE NO SUITABLE T-EIGENVECTORS WE
     1RE IDENTIFIED'/'  FOR ANY OF THE EIGENVALUES SUPPLIED.  PROBLEM CO
     1ULD BE CAUSED'/'  BY TOO SMALL A TVEC DIMENSION OR SIMPLY THAT SUI
     1TABLE T-VECTORS COULD'/'  NOT BE IDENTIFIED.  USER SHOULD CHECK OU
     1TPUT'/)
      GO TO 1590
C
 1490 WRITE(6,1500) LVCONT,NTVEC,NGOOD
 1500 FORMAT(/' LVCONT FLAG =',I2,' AND NUMBER ',I5,' OF T-EIGENVECTORS
     1 COMPUTED N.E.'/' NUMBER',I5,' REQUESTED SO PROGRAM TERMINATES'/)
      GO TO 1590
C
 1510 WRITE(6,1520)
 1520 FORMAT(/' PROGRAM TERMINATES WITHOUT COMPUTING RITZ VECTORS'/
     1' BECAUSE ALL T-EIGENVECTORS WERE REJECTED AS NOT SUITABLE FOR THE
     1 RITZ VECTOR'/' COMPUTATIONS.  PROBABLE CAUSE IS LACK OF CONVERGEN
     1CE OF THE EIGENVALUES SUPPLIED'/)
       GO TO 1590
C
 1530 WRITE(6,1540)
 1540 FORMAT(/' PROGRAM INDICATES THAT IT IS NOT POSSIBLE TO COMPUTE ANY
     1 OF THE'/' REQUESTED EIGENVECTORS. THEREFORE PROGRAM TERMINATES')
      DO 1550 J=1,NGOODC
 1550 WRITE(6,1560)  J,GOODEV(J),MP(J)
 1560 FORMAT(/4X,' J',9X,'AB-EIGENVALUE',4X,'MP(J)'/I6,E20.12,I9)
      GO TO 1590
C
 1570 WRITE(6,1580) MBETA,KMAXN
 1580 FORMAT(/' PROGRAM TERMINATES BECAUSE THE STORAGE ALLOTTED FOR THE
     1BETA ARRAY',I8/' IS NOT SUFFICIENT FOR THE ENLARGED KMAX =',I8,' T
     1HAT THE PROGRAM WANTS'/' USER CAN ENLARGE THE ALPHA AND BETA ARRAY
     1S AND RERUN THE PROGRAM'/)
C
 1590 CONTINUE
C
      STOP
C-----END OF MAIN PROGRAM FOR LANCZOS EIGENVECTOR COMPUTATIONS----------
      END
```

SECTION 5.4 LANCZS AND SAMPLE MATRIX-VECTOR MULTIPLY AND SOLVE SUBROUTINES

```
C-----LGMULT------------------------------------------------------------
C
C     CONTAINS SUBROUTINES  LANCZS, USPECA, USPECB, AMATV, AND LSOLV.
C     TO BE USED WITH THE LANCZOS CODES FOR THE GENERALIZED EIGENVALUE
C     PROBLEM, A*X = EVAL*B*X, WHERE A AND B ARE REAL SYMMETRIC, AND
C     B IS POSITIVE DEFINITE WITH ITS CHOLESKY FACTORS AVAILABLE.
C
C     NONPORTABLE CONSTRUCTIONS:
C     1.  THE ENTRY MECHANISM USED TO PASS THE STORAGE
C         LOCATIONS OF THE USER-SPECIFIED MATRICES FROM THE
C         SUBROUTINES USPECA AND USPECB TO THE MATRIX-VECTOR
C         SUBROUTINE, AMATV AND TO THE SOLVE SUBROUTINE, LSOLV.
C     2.  IN SAMPLE USPECA AND USPECB:  FREE FORMAT (8,*); FORMAT
C         (20A4), AND FORMAT (4Z20).
C
C-----LANCZS-COMPUTE LANCZOS TRIDIAGONAL MATRICES-----------------------
C
      SUBROUTINE LANCZS(LSOLV,MATVEC,ALPHA,BETA,V1,V2,VS,G,KMAX,MOLD1,N,
     1 IIX)
C
C-----------------------------------------------------------------------
      DOUBLE PRECISION  ALPHA(1), BETA(1), V1(1), V2(1), VS(1)
      DOUBLE PRECISION SUM, ONE, ZERO, TEMP
      REAL  G(1)
      DOUBLE PRECISION  FINPRO,DSQRT
      EXTERNAL MATVEC, LSOLV
C-----------------------------------------------------------------------
C     ALPHA, BETA, AND LANCZOS VECTOR GENERATION
C     ALPHA BETA GENERATION STARTS WITH IVEC = 1, BETA(1) = ZERO,
C     V2 =  RANDOM VECTOR WITH UNIT B-NORM, VS = B*V2, AND V1 = 0.;
C     OR STARTS WITH AN EXISTING ALPHA/BETA FILE AND THE MOST
C     RECENTLY GENERATED V2, VS, AND V1.
C
      ZERO = 0.0D0
      ONE  = 1.0D0
      IF (MOLD1.GT.1) GO TO 40
      BETA(1) = ZERO
      IIL = IIX
C
C-----------------------------------------------------------------------
      CALL GENRAN(IIL,G,N)
C-----------------------------------------------------------------------
C
      DO 10 K = 1,N
   10 V2(K) = G(K)
C
C-----------------------------------------------------------------------
C     COMPUTE L-TRANSPOSE*V2 AND ITS NORM
      ISOLV = 2
      CALL LSOLV(V2,VS,ISOLV)
      SUM = FINPRO(N,VS(1),1,VS(1),1)
C-----------------------------------------------------------------------
C
C     NORMALIZE STARTING VECTORS:  (V2-TRANSPOSE*B*V2) = 1
      SUM = ONE/DSQRT(SUM)
      DO 20 K = 1,N
      VS(K) = SUM*VS(K)
   20 V2(K) = SUM*V2(K)
C
C-----------------------------------------------------------------------
C     INITIALIZE V1 = B*V2 = L*VS
      ISOLV = 1
```

```
      CALL LSOLV(VS,V1,ISOLV)
C-----------------------------------------------------------------------
C
      DO 30 K = 1,N
      VS(K) = V1(K)
   30 V1(K) = 0.D0
   40 CONTINUE
C
C     INITIALIZATIONS ARE:  VS = B*V(I),  V1 = B*V(I-1),  V2 = V(I)
C
      DO 80 IVEC = MOLD1,KMAX
      SUM = BETA(IVEC)
C
C-----------------------------------------------------------------------
C     COMPUTE V1 = A*V2 - SUM*V1
      CALL MATVEC(V2,V1,SUM)
C     COMPUTE ALPHA(I)
      SUM = FINPRO(N,V1(1),1,V2(1),1)
C-----------------------------------------------------------------------
C
      ALPHA(IVEC) = SUM
      DO 50 K = 1,N
   50 V1(K) = V1(K)-SUM*VS(K)
C
C     SET V1 = B*V(IVEC) AND VS = BETA(IVEC+1)*B*V(IVEC+1)
      DO 60 K = 1,N
      TEMP = V1(K)
      V1(K) = VS(K)
   60 VS(K) = TEMP
C
C-----------------------------------------------------------------------
C     COMPUTE V2 = (L-INVERSE)*VS
      ISOLV = 3
      CALL LSOLV(VS,V2,ISOLV)
C     COMPUTE BETA(IVEC+1)
      SUM = FINPRO(N,V2(1),1,V2(1),1)
C-----------------------------------------------------------------------
C
      IN = IVEC+1
      BETA(IN) = DSQRT(SUM)
C
C-----------------------------------------------------------------------
      ISOLV = 4
      CALL LSOLV(V2,V2,ISOLV)
C-----------------------------------------------------------------------
C
      SUM = ONE/BETA(IN)
      DO 70 K = 1,N
      V2(K) = SUM*V2(K)
   70 VS(K) = SUM*VS(K)
C
   80 CONTINUE
C
      RETURN
C-----END LANCZS--------------------------------------------------------
      END
C
C-----USPEC (GENERAL SYMMETRIC SPARSE MATRICES)-------------------------
C
C     SUBROUTINE USPECA(N,MATNOA)
      SUBROUTINE GUSPEC(N,MATNOA)
C
C-----------------------------------------------------------------------
      DOUBLE PRECISION  ASD(10000),AD(5010)
      INTEGER  IROW(10000),ICOL(5010)
C-----------------------------------------------------------------------
C    USPEC DIMENSIONS AND INITIALIZES THE ARRAYS NEEDED TO DEFINE
```

```
C     THE USER-SPECIFIED A-MATRIX AND THEN PASSES THE STORAGE LOCATIONS
C     OF THESE ARRAYS TO THE MULTIPLY SUBROUTINE AMATV.
C
C     MATRIX IS STORED IN FOLLOWING SPARSE MATRIX FORMAT:
C     N = ORDER OF A-MATRIX,
C     NZS = NUMBER OF NONZERO SUBDIAGONAL ENTRIES,
C     NZL = INDEX OF LAST COLUMN CONTAINING NONZERO SUBDIAGONAL ENTRIES,
C     ICOL(J), J=1,NZL IS THE NUMBER OF NONZERO SUBDIAGONAL ELEMENTS
C              IN COLUMN J.
C     IROW(K), K = 1,NZS IS THE CORRESPONDING ROW INDEX FOR ASD(K).
C     AD(I), I=1,N CONTAINS DIAGONAL ENTRIES (INCLUDING ANY 0
C              DIAGONAL ENTRIES).
C     ASD(K), K=1,NZS CONTAINS NONZERO SUBDIAGONAL ENTRIES, BY COLUMN
C     FOR J > NZL THERE ARE NO NONZERO SUBDIAGONAL ELEMENTS IN COLUMN J.
C     ICOL(J) = 0 IS ALLOWED
C
C-----------------------------------------------------------------------
C      ARRAYS THAT DEFINE THE A-MATRIX ARE READ IN FROM FILE 8.  NOTE
C      THAT IF THE B-MATRIX IS PERMUTED, THEN LANCZOS PROGRAM ASSUMES
C      THAT THE DATA ON FILE 8 CORRESPONDS TO THE CORRESPONDING
C      PERMUTED A-MATRIX.  LANCZOS PROCEDURE WORKS DIRECTLY WITH THE
C      PERMUTED MATRICES.  EIGENVECTOR CODE, LGVEC, THEN PERMUTES THE
C      COMPUTED EIGENVECTORS TO GET THOSE CORRESPONDING TO THE ORIGINAL
C      MATRICES.
C
       READ(8,10) NZS,NOLD,NZL,MATOLD
    10 FORMAT(I10,2I6,I8)
C
       WRITE(6,20) NZS,NOLD,NZL,MATOLD
    20 FORMAT(I10,2I6,I8,' = NZS,NOLD,NZL,MATOLD'/)
C
C      TEST OF PARAMETER CORRECTNESS
       ITEMP = (NOLD-N)**2 + (MATNOA-MATOLD)**2
C
       IF(ITEMP.EQ.0) GO TO 40
C
       WRITE(6,30)
    30 FORMAT(' PROGRAM TERMINATES BECAUSE EITHER ORDERS OF OR LABELS FOR
     1 MATRIX DISAGREE')
       GO TO 70
C
    40 CONTINUE
C
C      NUMBER OF NONZERO SUBDIAGONAL ENTRIES IN EACH COLUMN IS READ
C      THEN THE CORRESPONDING ROW INDEX FOR EACH SUCH ENTRY IS READ
       READ(8,50) (ICOL(K), K=1,NZL)
       READ(8,50) (IROW(K), K=1,NZS)
    50 FORMAT(13I6)
C
C      DIAGONAL IS READ FIRST, THEN NONZERO BELOW DIAGONAL ENTRIES
       READ(8,60) (AD(K), K=1,N)
       READ(8,60) (ASD(K), K=1,NZS)
    60 FORMAT(4E19.10)
C
C-----------------------------------------------------------------------
C      PASS STORAGE LOCATIONS OF ARRAYS THAT DEFINE THE A-MATRIX TO
C      THE MATRIX-VECTOR MULTIPLY SUBROUTINE AMATV
       CALL AMATVE(ASD,AD,ICOL,IROW,N,NZL)
C-----------------------------------------------------------------------
C
       RETURN
    70 STOP
C-----END OF USPECA-----------------------------------------------------
       END
C
C-----USPECB FOR CHOLESKY FACTORS OF GENERAL SPARSE SYMMETRIC MATRIX----
```

```
C
C     SUBROUTINE USPECB(N,MATNOB)
      SUBROUTINE CUSPEC(N,MATNOB)
C
C-----------------------------------------------------------------------
      DOUBLE PRECISION  BD(2200),BSD(10000)
      INTEGER  KCOL(2200),KROW(10000),IPR(2200),IPT(2200)
C-----------------------------------------------------------------------
C     DIMENSIONS ARRAYS NEEDED TO DEFINE CHOLESKY FACTOR OF B-MATRIX,
C     READS CHOLESKY FACTOR FROM FILE 7, AND THEN PASSES STORAGE
C     LOCATIONS OF THESE ARRAYS TO THE MATRIX SOLVE SUBROUTINE LSOLV
C
C     THE LANCZOS PROCEDURE LGVAL WILL USE THE CHOLESKY FACTORS ON
C     FILE 7.  THESE FACTORS MAY CORRESPOND TO A PERMUTED VERSION OF
C     THE GIVEN B-MATRIX IN WHICH CASE THIS PERMUTATION WILL BE STORED
C     IN IPR. THE ITH ROW OF THE PERMUTED B WILL CORRESPOND TO THE
C     JTH ROW OF B WHERE J = IPR(I) AND I = IPT(J).  IF B IS
C     PERMUTED, THE LANCZOS PROCEDURE ASSUMES THAT THE USER-PROVIDED
C     A-MATRIX IS IN FACT, THE CORRESPONDING PERMUTED VERSION OF THE
C     ORIGINAL A-MATRIX.
C
C     THE CHOLESKY FACTOR IS STORED IN THE FOLLOWING SPARSE FORMAT:
C     N = ORDER OF THE B-MATRIX.
C     NZT = NUMBER OF NONZERO SUBDIAGONAL ENTRIES IN THE CHOLESKY
C          FACTOR, L.
C     KCOL(J), J=1,N IS THE NUMBER OF NONZERO SUBDIAGONAL ELEMENTS IN
C             COLUMN J OF L.
C     KROW(K), K=1,NZT IS THE ROW INDEX FOR CORRESPONDING ENTRY BSD(K).
C     BD(J), J = 1,N CONTAINS THE DIAGONAL ENTRIES OF L.
C     BSD(K), K =1,NZT CONTAINS THE NONZERO SUBDIAGONAL ENTRIES OF L
C            BY COLUMN.
C-----------------------------------------------------------------------
C
C     READ CHOLESKY FACTOR FROM FILE 7.  MUST BE STORED
C     IN SPARSE MATRIX FORMAT.
      READ(7,10) NZT,NOLD,NZL,MATOLD,JPERM
   10 FORMAT(I10,2I6,I8,I6)
C
      WRITE(6,20) NZT,NZL,N,NOLD,MATOLD,JPERM
   20 FORMAT(' HEADER, CHOLESKY FACTOR FILE'/
     1 3X,'NZT',3X,'NZL',5X,'N',2X,'NOLD',2X,'MATOLD',1X,'JPERM'/
     1 4I6,I8,I6/)
C
      IF (N.NE.NOLD.OR.MATNOB.NE.MATOLD) GO TO 70
C
      READ(7,30) (KCOL(K), K = 1,NZL)
      READ(7,30) (KROW(K), K = 1,NZT)
   30 FORMAT(13I6)
      READ(7,40) (BD(K), K = 1,N)
      READ(7,40) (BSD(K), K = 1,NZT)
   40 FORMAT(4Z20)
C  20 FORMAT(3E25.16)
C
      IF(JPERM.EQ.0)  GO TO 60
C
      READ(7,30) (IPR(K), K = 1,N)
      DO 50 K = 1,N
      J = IPR(K)
   50 IPT(J) = K
C
C-----------------------------------------------------------------------
      CALL LPERME(IPR,IPT,N)
C-----------------------------------------------------------------------
C
   60 CONTINUE
C
```

```
C-----------------------------------------------------------------------
C     PASS STORAGE LOCATIONS OF FACTORS TO SUBROUTINE LSOLV
      CALL LSOLVE(BSD,BD,KCOL,KROW,N,NZT,NZL)
C-----------------------------------------------------------------------
C
      GO TO 90
C
   70 CONTINUE
C     DEFAULT EXIT
      WRITE(6,80) MATNOB,MATOLD
   80 FORMAT(' TERMINATE.  PARAMETERS IN CHOLESKY FACTOR FILE'/
     1' DO NOT AGREE WITH THOSE SPECIFIED BY THE USER'/
     1' MATNOB = ',I8,' MATOLD = ',I8/)
      STOP
C
   90 CONTINUE
C-----END OF USPECB-----------------------------------------------------
      RETURN
      END
C
C-----MATRIX-VECTOR MULTIPLY FOR REAL SPARSE SYMMETRIC MATRICES---------
C
C     SUBROUTINE AMATV(W,U,SUM)
      SUBROUTINE GCMATV(W,U,SUM)
C
C-----------------------------------------------------------------------
      DOUBLE PRECISION   U(1),W(1),ASD(1),AD(1),SUM
      INTEGER  IROW(1),ICOL(1)
C-----------------------------------------------------------------------
C    SPARSE MATRIX-VECTOR MULTIPLY FOR LANCZS  U = A*W - SUM*U
C    SEE USPECA SUBROUTINE FOR DESCRIPTION OF THE ARRAYS
C    THAT DEFINE THE A-MATRIX
C-----------------------------------------------------------------------
C
C     COMPUTE THE DIAGONAL TERMS
      DO 10 I = 1,N
   10 U(I) = AD(I)*W(I)-SUM*U(I)
C
C     COMPUTE BY COLUMN
      LLAST = 0
      DO 30 J = 1,NZL
C
      IF (ICOL(J).EQ.0) GO TO 30
      LFIRST = LLAST + 1
      LLAST = LLAST + ICOL(J)
C
      DO 20 L = LFIRST,LLAST
      I = IROW(L)
C
      U(I) = U(I) + ASD(L)*W(J)
      U(J) = U(J) + ASD(L)*W(I)
C
   20 CONTINUE
C
   30 CONTINUE
C
      RETURN
C
C-----------------------------------------------------------------------
C     STORAGE LOCATIONS OF ARRAYS ARE PASSED TO AMATV FROM USPECA
      ENTRY AMATVE(ASD,AD,ICOL,IROW,N,NZL)
C-----------------------------------------------------------------------
C
      RETURN
C-----END OF AMATV------------------------------------------------------
      END
C
```

```
C-----LSOLV-GENERAL SPARSE, POSITIVE DEFINITE B-MATRIX------------------
C     (USES THE CHOLESKY FACTORS OF B, B = L*(L-TRANSPOSE))
C
      SUBROUTINE TLSOLV(W,U,ISOLV)
C     SUBROUTINE LSOLV(W,U,ISOLV)
C
C-----------------------------------------------------------------------
      DOUBLE PRECISION  U(1),W(1),BD(1),BSD(1), TEMP
      INTEGER  KCOL(1),KROW(1)
C-----------------------------------------------------------------------
C     SUBROUTINE HAS 4 BRANCHES:  ISOLV = (1,2,3,4)  CALCULATES
C     ISOLV = 1   U = L*W
C     ISOLV = 2   U = L'*W
C     ISOLV = 3   SOLVE FOR U IN  L*U = W
C     ISOLV = 4   SOLVE FOR U IN  L'*U = W
C-----------------------------------------------------------------------
      GO TO (10,50,80,120), ISOLV
C
C     ISOLV = 1,  U=L*W
   10 CONTINUE
      KL = 0
      DO 20 K = 1,N
   20 U(K) = W(K)*BD(K)
      DO 40 K = 1,N
      TEMP = W(K)
      IF (KCOL(K).EQ.0.OR.K.EQ.N) GO TO 40
      KF = KL + 1
      KL = KL + KCOL(K)
      DO 30 KK = KF,KL
      KR = KROW(KK)
   30 U(KR) = U(KR) + TEMP*BSD(KK)
   40 CONTINUE
      GO TO 150
C
C     ISOLV = 2,  U = (L-TRANSPOSE)*W
   50 CONTINUE
      KL = 0
      DO 70 J = 1,N
      TEMP = W(J)*BD(J)
      IF (KCOL(J).EQ.0.OR.J.EQ.N) GO TO 70
      KF = KL + 1
      KL  = KL + KCOL(J)
      DO 60 K = KF,KL
      IK = KROW(K)
   60 TEMP = BSD(K)*W(IK) + TEMP
   70 U(J) = TEMP
      GO TO 150
C
C     ISOLV = 3,  U = (L-INVERSE)*W
   80 CONTINUE
      DO 90 K = 1,N
   90 U(K) = W(K)
      KL = 0
      DO 110 K = 1,N
      TEMP = U(K)/BD(K)
      U(K) = TEMP
      IF (KCOL(K).EQ.0.OR.K.EQ.N) GO TO 110
      KF = KL + 1
      KL = KL + KCOL(K)
      DO 100 KK = KF,KL
      KR = KROW(KK)
  100 U(KR) = U(KR) - TEMP*BSD(KK)
  110 CONTINUE
      GO TO 150
C
C     ISOLV = 4,  U = (L-TRANSPOSE)-INVERSE*W
  120 CONTINUE
```

```
      NP1 = N+1
      KF = NZT + 1
      DO 140 K = 1,N
      L = NP1 - K
      TEMP = W(L)
      IF (KCOL(L).EQ.0.OR.L.EQ.N) GO TO 140
      KL = KF - 1
      KF = KF - KCOL(L)
      DO 130 LL = KF,KL
      LR = KROW(LL)
  130 TEMP = TEMP - BSD(LL)*U(LR)
  140 U(L) = TEMP/BD(L)
      GO TO 150
  150 CONTINUE
C
      RETURN
C
C-----------------------------------------------------------------------
      ENTRY LSOLVE(BSD,BD,KCOL,KROW,N,NZT,NZL)
C-----------------------------------------------------------------------
C
C-----END OF LSOLV-------------------------------------------------
      RETURN
      END
C
C-----START OF USPEC FOR DIAGONAL TEST A-MATRIX-------------------------
C
      SUBROUTINE USPECA(N,MATNO)
C     SUBROUTINE DUSPEC(N,MATNO)
C
C-----------------------------------------------------------------------
      DOUBLE PRECISION  D(1000), SPACE, SHIFT
      DOUBLE PRECISION  DABS, DFLOAT
      REAL  EXPLAN(20)
C-----------------------------------------------------------------------
C
      READ(8,10) EXPLAN
   10 FORMAT(20A4)
      READ(8,*) NOLD,NUNIF,SPACE,D(1),SHIFT
      NNUNIF = NOLD - NUNIF
      WRITE(6,20) NOLD,SPACE,NNUNIF,D(1),SHIFT
   20 FORMAT(/' DIAGONAL TEST A-MATRIX, SIZE = ',I4/' MOST ENTRIES ARE '
     1,E10.3,' UNITS APART.',I3,' ENTRIES'/' ARE IRREGULARLY SPACED. FIR
     1ST ENTRY IS ',E10.3,' SHIFT = ',E10.3/)
C
      IF(N.NE.NOLD) GO TO 90
C     COMPUTE THE UNIFORM PORTION OF THE SPECTRUM
      DO 30 J=2,NUNIF
   30 D(J) = D(1) - DFLOAT(J-1)*SPACE
      NUNIF1=NUNIF + 1
      READ(8,10)  EXPLAN
      DO 40 J=NUNIF1,N
   40 READ(8,*) D(J)
      NB = NUNIF - 2
C
      IF(SHIFT.EQ.0.) GO TO 60
      DO 50 J=1,N
   50 D(J) = D(J) + SHIFT
C
C     PRINT OUT A-MATRIX
   60 WRITE(6,70) (D(I), I=1,10 )
      WRITE(6,80) (D(I), I = NB,N)
   70 FORMAT(/' GENERALIZED LANCZOS TEST, 1ST 10 ENTRIES OF DIAGONAL A-M
     1ATRIX = '/(3E22.14))
   80 FORMAT(/' MIDDLE UNIFORM PORTION OF MATRIX IS NOT PRINTED OUT'/
     1' END OF UNIFORM PLUS NONUNIFORM SECTION = '/(3E25.16))
C
```

```
C     DIAGONAL GENERATION COMPLETE
C
C-----------------------------------------------------------------------
C    CALL ENTRY TO MATRIX-VECTOR MULTIPLY SUBROUTINE  TO PASS
C    STORAGE LOCATION OF D-ARRAY AND ORDER OF A-MATRIX.
        CALL MVDIAE(D,N)
C-----------------------------------------------------------------------
C
      RETURN
   90 WRITE(6,100) NOLD,N
  100 FORMAT(' PROGRAM TERMINATES BECAUSE NOLD = ',I5,'DOES NOT EQUAL N
     1 =',I5)
C-----END OF USPECA SUBROUTINE FOR DIAGONAL TEST MATRICES---------------
      STOP
      END
C
C-----USPECB--DIAGONAL TEST B-MATRIX------------------------------------
C
      SUBROUTINE USPECB(N,MATNO)
C     SUBROUTINE USPECD(N,MATNO)
C
C-----------------------------------------------------------------------
      DOUBLE PRECISION  D(1000), DS(1000), SPACE, SHIFT
      DOUBLE PRECISION  DFLOAT, DSQRT
      REAL  EXPLAN(20)
C-----------------------------------------------------------------------
C
      READ(7,10) EXPLAN
   10 FORMAT(20A4)
      READ(7,*) NOLD,NUNIF,SPACE,D(1),SHIFT
      NNUNIF = NOLD - NUNIF
      WRITE(6,20) NOLD,SPACE,NNUNIF,D(1),SHIFT
   20 FORMAT(/' DIAGONAL TEST B-MATRIX, SIZE = ',I4/' MOST ENTRIES ARE '
     1,E10.3,' UNITS APART.',I3,' ENTRIES'/' ARE IRREGULARLY SPACED. FIR
     1ST ENTRY IS ',E10.3,' SHIFT = ',E10.3/)
C
      IF(N.NE.NOLD) GO TO 100
C     COMPUTE THE UNIFORM PORTION OF THE SPECTRUM
      DO 30 J=2,NUNIF
   30 D(J) = D(1) - DFLOAT(J-1)*SPACE
      NUNIF1=NUNIF + 1
      READ(7,10)  EXPLAN
      DO 40 J=NUNIF1,N
   40 READ(7,*) D(J)
      NB = NUNIF - 2
C
      IF(SHIFT.EQ.0.) GO TO 60
      DO 50 J=1,N
   50 D(J) = D(J) + SHIFT
C
C     PRINT OUT B-MATRIX
   60 WRITE(6,70) (D(I), I=1,10 )
      WRITE(6,80) (D(I), I = NB,N)
   70 FORMAT(/' GENERALIZED LANCZOS TEST, 1ST 10 ENTRIES OF DIAGONAL B-M
     1ATRIX = '/(3E22.14))
   80 FORMAT(/' MIDDLE UNIFORM PORTION OF MATRIX IS NOT PRINTED OUT'/
     1' END OF UNIFORM PLUS NONUNIFORM SECTION = '/(3E25.16))
C
C     DIAGONAL GENERATION COMPLETE
C
      DO 90 K = 1,N
   90 DS(K) = DSQRT(D(K))
C
C-----------------------------------------------------------------------
C    PASS STORAGE LOCATION OF THE L-FACTOR (THE DS-ARRAY) AND ORDER OF
C    B-MATRIX TO LSOLV SUBROUTINE.
```

```
      CALL DSOLVE(DS,N)                                                 LGM05290
C-----------------------------------------------------------------------LGM05300
C                                                                       LGM05310
      RETURN                                                            LGM05320
  100 WRITE(6,110) NOLD,N                                               LGM05330
  110 FORMAT(' PROGRAM TERMINATES BECAUSE NOLD = ',I5,'DOES NOT EQUAL N LGM05340
     1 =',I5)                                                           LGM05350
C-----END OF USPECB SUBROUTINE FOR DIAGONAL TEST MATRICES---------------LGM05360
      STOP                                                              LGM05370
      END                                                               LGM05380
C                                                                       LGM05390
C-----MATRIX-VECTOR MULTIPLY FOR DIAGONAL TEST MATRICES-----------------LGM05400
C                                                                       LGM05410
      SUBROUTINE AMATV(W,U,SUM)                                         LGM05420
C     SUBROUTINE DCMATV(W,U,SUM)                                        LGM05430
C                                                                       LGM05440
C     AMATV COMPUTES  U = (DIAGONAL MATRIX) * W - SUM * U               LGM05450
C-----------------------------------------------------------------------LGM05460
      DOUBLE PRECISION  W(1),U(1),D(1),SUM                              LGM05470
C-----------------------------------------------------------------------LGM05480
C                                                                       LGM05490
      DO 10 I=1,N                                                       LGM05500
   10 U(I)= D(I)*W(I) - SUM*U(I)                                        LGM05510
C                                                                       LGM05520
      RETURN                                                            LGM05530
C                                                                       LGM05540
C-----------------------------------------------------------------------LGM05550
      ENTRY MVDIAE(D,N)                                                 LGM05560
C-----------------------------------------------------------------------LGM05570
C                                                                       LGM05580
      RETURN                                                            LGM05590
C-----END OF DIAGONAL TEST MATRIX  MULTIPLY-----------------------------LGM05600
      END                                                               LGM05610
C                                                                       LGM05620
C-----LSOLV FOR DIAGONAL MATRIX-----------------------------------------LGM05630
C                                                                       LGM05640
      SUBROUTINE LSOLV(W,U,ISOLV)                                       LGM05650
C     SUBROUTINE DSOLV(W,U,ISOLV)                                       LGM05660
C                                                                       LGM05670
C-----------------------------------------------------------------------LGM05680
      DOUBLE PRECISION U(1), W(1),  DS(1)                               LGM05690
C-----------------------------------------------------------------------LGM05700
      GO TO (10,30,50,70), ISOLV                                        LGM05710
C                                                                       LGM05720
C     ISOLV = 1                                                         LGM05730
   10 CONTINUE                                                          LGM05740
      DO 20 K = 1,N                                                     LGM05750
   20 U(K) = DS(K)*W(K)                                                 LGM05760
      GO TO 90                                                          LGM05770
C                                                                       LGM05780
C     ISOLV = 2                                                         LGM05790
   30 CONTINUE                                                          LGM05800
      DO 40 K = 1,N                                                     LGM05810
   40 U(K) = DS(K)*W(K)                                                 LGM05820
      GO TO 90                                                          LGM05830
C                                                                       LGM05840
C     ISOLV = 3                                                         LGM05850
   50 CONTINUE                                                          LGM05860
      DO 60 K = 1,N                                                     LGM05870
   60 U(K) = W(K)/DS(K)                                                 LGM05880
      GO TO 90                                                          LGM05890
C                                                                       LGM05900
C     ISOLV = 4                                                         LGM05910
   70 CONTINUE                                                          LGM05920
      DO 80 K = 1,N                                                     LGM05930
   80 U(K) = W(K)/DS(K)                                                 LGM05940
C                                                                       LGM05950
```

```
   90 CONTINUE
C
      RETURN
C
C-----------------------------------------------------------------------
      ENTRY DSOLVE(DS,N)
C-----------------------------------------------------------------------
C
      RETURN
C-----END OF DSOLV------------------------------------------------------
      END
```

SECTION 5.5 FILE DEFINITIONS AND SAMPLE INPUT FILES

Below is a listing of the input/output files which are accessed by the Lanczos eigenvalue program LGVAL for real symmetric generalized problems where one of the two matrices is positive definite. Included also is a sample of the input file which LGVAL requires on file 5. The parameters in this file are supplied in free format. LGVAL computes eigenvalues of the matrix eigenvalue problem AX = EV*BX on user-specified intervals where A and B are both real symmetric matrices and B is positive definite. The program uses the Cholesky factor L OF B = L*L'.

```
SAMPLE DEFINITIONS OF THE INPUT/OUTPUT FILES FOR LGVAL
-----------------------------------------------------------------
*LGVAL EXEC LANCZOS EIGENVALUE CALCULATION AX = EV*BX CASE
FI 06 TERM
FILEDEF  1 DISK &1        NHISTORY  A (RECFM F LRECL 80 BLOCK 80
FILEDEF  2 DISK &1        HISTORY   A (RECFM F LRECL 80 BLOCK 80
FILEDEF  3 DISK &1        GOODEV    A (RECFM F LRECL 80 BLOCK 80
FILEDEF  4 DISK &1        ERRINV    A (RECFM F LRECL 80 BLOCK 80
FILEDEF  5 DISK LGVAL     INPUT     A (RECFM F LRECL 80 BLOCK 80
FILEDEF  7 DISK &1        LCOMPAC   A (RECFM F LRECL 80 BLOCK 80
FILEDEF  8 DISK &1        ACOMPAC   A (RECFM F LRECL 80 BLOCK 80
FILEDEF 11 DISK &1        DISTINCT  A (RECFM F LRECL 80 BLOCK 80
LOAD  LGVAL  LESUB  LGMULT
-----------------------------------------------------------------
```

```
SAMPLE INPUT FOR LGVAL
-----------------------------------------------------------------
 LGVAL INPUT LANCZOS EIGENVALUE COMPUTATION, NO REORTHOGONALIZATION
 AX = EV*BX GENERALIZED EIGENVALUE PROBLEM
LINE 1      N     KMAX    NMEVS     MATNOA     MATNOB
          100      300        1        100        100
LINE 2   SVSEED    RHSEED       MXINIT    MXSTUR
       49302312   5731029          5     100000
LINE 3   ISTART      ISTOP
              0          1
LINE 4     IHIS      IDIST  IWRITE
              1          0       1
LINE 5   RELTOL (RELATIVE TOLERANCE IN 'COMBINING' GOODEV)
    .0000000001
LINE 6   MB(1)   MB(2)   MB(3)   MB(4)    (ORDERS OF T(1,MEV) )
           300
LINE 7    NINT      (NUMBER OF SUB-INTERVALS FOR BISEC)
             1
LINE 8    LB(1)     LB(2)   LB(3)   LB(4)  (INTERVAL LOWER BOUNDS)
            1.5
LINE 9    UB(1)     UB(2)   UB(3)   UB(4)  (INTERVAL UPPER BOUNDS)
          2100.
-----------------------------------------------------------------
```

Below is a listing of the input/output files which are accessed by the Lanczos eigenvector program for real symmetric generalized problems, LGVEC. Also included below is a sample of the input file which LGVEC requires on file 5. The parameters in this file are supplied in free format. LGVEC computes eigenvectors for each of a user-specified subset of the eigenvalues computed by the companion program LGVAL.

SAMPLE DEFINITIONS OF THE INPUT/OUTPUT FILES FOR LGVEC

```
-----------------------------------------------------------------
*LGVEC EXEC TO RUN LANCZOS EIGENVECTOR PROGRAM, REAL SYMMETRIC MATRICES
FI 06 TERM
FILEDEF  2 DISK &1        HISTORY   A (RECFM F LRECL 80 BLOCK 80
FILEDEF  3 DISK &1        GOODEV    A (RECFM F LRECL 80 BLOCK 80
FILEDEF  4 DISK &1        ERRINV    A (RECFM F LRECL 80 BLOCK 80
FILEDEF  5 DISK LGVEC     INPUT     A (RECFM F LRECL 80 BLOCK 80
FILEDEF  7 DISK &1        LCOMPAC   A (RECFM F LRECL 80 BLOCK 80
FILEDEF  8 DISK &1        ACOMPAC   A (RECFM F LRECL 80 BLOCK 80
FILEDEF  9 DISK &1        ERREST    A (RECFM F LRECL 80 BLOCK 80
FILEDEF 10 DISK &1        BOUNDS    A (RECFM F LRECL 80 BLOCK 80
FILEDEF 11 DISK &1        TEIGVECS  A (RECFM F LRECL 80 BLOCK 80
FILEDEF 12 DISK &1        RITZVECS  A (RECFM F LRECL 80 BLOCK 80
FILEDEF 13 DISK &1        PAIGE     A (RECFM F LRECL 80 BLOCK 80
LOAD  LGVEC  LESUB  LGMULT
------------------------------------------------------------------
```

SAMPLE INPUT FILE FOR LGVEC

```
-----------------------------------------------------------------
 LGVEC EIGENVECTOR COMPUTATIONS AX = EV*BX NO REORTHOGONALIZATION
LINE 1  MDIMTV    MDIMRV  MBETA (MAX.DIMENSIONS, TVEC, RITVEC AND BETA
          10000     10000   2000
LINE 2        RELTOL
        .0000000001
LINE 3  MBOUND   NTVCON SVTVEC IREAD (FLAGS
            0        1      0     1
LINE 4  TVSTOP   LVCONT ERCONT  IWRITE (FLAGS
            0        1      1      1
LINE 5    RHSEED  (RANDOM GENERATOR SEED FOR STARTING VECTOR IN INVERM)
        45329517
LINE 6  MATNOA      MATNOB      N     JPERM
           100         100    100         0
------------------------------------------------------------------------
```

CHAPTER 6

REAL RECTANGULAR MATRICES

SECTION 6.1 INTRODUCTION

The FORTRAN codes in this Chapter address the question of computing distinct singular values and corresponding left and right singular vectors of real rectangular matrices, using a single-vector Lanczos procedure. For a given real rectangular ℓxn matrix A, these codes compute nonnegative scalars σ and corresponding real vectors $x \neq 0$ and $y \neq 0$ such that

$$Ax = \sigma y \text{ and } A^T y = \sigma x. \tag{6.1.1}$$

Every real rectangular ℓxn matrix, where $\ell \geq n$, has a singular value decomposition,

$$A = Y \Sigma X^T \text{ with } X^T X = I, \; Y^T Y = I, \text{ and } \Sigma = \begin{bmatrix} \Sigma_1 \\ 0 \end{bmatrix} \tag{6.1.2}$$

where Σ is $\ell \times n$ and $\Sigma_1 = \mathrm{diag}\{\sigma_1,...,\sigma_n\}$ with σ_i, $1 \leq i \leq n$, the singular values of A. X is a nxn orthogonal matrix, Y is a ℓxℓ orthogonal matrix, and the columns of X and of Y are respectively, right and left singular vectors of A. There are many applications for this type of decomposition. Singular values and vectors are discussed in detail for example in Stewart [1973].

Using Eqn(6.1.1), it is not difficult to demonstrate that the singular values of a given real matrix A are just the nonnegative square roots of the eigenvalues of the associated real symmetric matrix A^TA (or AA^T). Thus from the perturbation theorems for real symmetric matrices, we have that a small perturbation in the given matrix A causes small perturbations in the singular values. The same arguments demonstrate that the right singular vectors of a matrix A are eigenvectors of the matrix A^TA, and the left singular vectors are eigenvectors of the matrix AA^T. Therefore, we also have that the perturbation theorems for eigenvectors of real symmetric matrices apply to the singular vectors.

The Lanczos recursion as presented in Eqns(1.2.1) - (1.2.2) is only applicable to real symmetric matrices. Therefore we ask the question: How do we construct a real symmetric matrix which will give us the desired singular values? Obviously, we could just apply the real symmetric Lanczos recursion to either A^TA or to AA^T. However in general, these matrices are not suitable because of the effects that squaring a matrix can have on the eigenvalues. Small singular values of A which are close together correspond to eigenvalues of A^TA which are smaller and even closer together. Large singular values of A which are far apart correspond to eigenvalues of A^TA which are larger and further apart. When a matrix A has both small and large singular values, dealing numerically with the square of that matrix is difficult. Lanczos [1961] suggested the use of an alternative real symmetric matrix. He proposed that the following larger but real symmetric $[\ell + n]x[\ell + n]$ matrix be used.

$$B \equiv \begin{bmatrix} 0 & A \\ A^T & 0 \end{bmatrix} . \qquad (6.1.3)$$

The relationships between the eigenvalues and the eigenvectors of B and the singular values and singular vectors of A are discussed in detail in Section 5.4 of Chapter 5 in Volume 1.

We could apply the real symmetric version of the Lanczos recursion directly to the matrix B in Eqn(6.1.3). However, because this matrix is considerably larger than the A-matrix, we use a modification of the real symmetric Lanczos recursion which incorporates the following choice of starting vector suggested by Golub and Kahan [1965]. We choose a starting vector either of the form $(0,u^T)^T$ or of the form $(v^T,0)^T$ where u is of length n, the column order of the A-matrix, and v is of length ℓ, the row order of the A-matrix. If we use such a starting vector in the basic Lanczos recursion in Eqns(1.2.1) - (1.2.2), we obtain a version of the Lanczos recursion designed specifically for the B-matrix in Eqn(6.1.3). The Lanczos vectors generated by this recursion alternate in form from either $(0,u^T)^T$ to $(v^T,0)^T$ or vice-versa, as the iterations proceed. Furthermore, on each iteration of this recursion it is only necessary to either compute Au_i or A^Tv_i. Therefore, the amount of work per iteration of this recursion is no more than applying the real symmetric Lanczos recursion to a real symmetric matrix of order $\max(\ell,n)$. For details on the corresponding Lanczos recursion see Section 5.4 of Chapter 5 in Volume 1.

These codes can compute either a very few or very many of the distinct singular values of a given real rectangular matrix. As the documentation in Section 6.2 indicates, the A-multiplicity of a computed singular value can be obtained only with additional computation, and the modifications required to do this additional computation are not included in these versions of the codes.

The Lanczos recursions which we use generate a family of real symmetric, tridiagonal matrices (T-matrices). The diagonal entries of each of these T-matrices are all 0. The eigenvalues of any even-ordered T-matrix occur in $\pm$ pairs. This latter property is inherited from the B-matrix whose eigenvalues are just $\pm\sigma_i$, the $\pm$ pairs of singular values plus $\ell - n$ additional zero eigenvalues if $\ell \neq n$. Only even-ordered T-matrices may be used in the Lanczos computations. There is no reorthogonalization of the Lanczos vectors at any stage in any of the computations.

LSVAL, the main program for the single-vector, Lanczos singular value computations, calls the subroutine BISEC to compute eigenvalues of those Lanczos tridiagonal matrices specified by the user and on those sub-intervals specified by the user. The BISEC subroutine used in this Chapter is a modification of the BISEC subroutine given in Chapter 2 which assumes that that the diagonal entries of the T-matrices supplied to it are all 0. BISEC simultaneously computes the T-eigenvalues and T-multiplicities and then sorts the computed T-eigenvalues into two categories, the 'good' T-eigenvalues and the 'spurious' T-eigenvalues. The 'good' T-eigenvalues are accepted as approximations to singular values of the user-specified matrix A. The accuracy of these 'good' T-eigenvalues as singular values of A is then estimated using error estimates computed by subroutine INVERR. The subroutine INVERR in this chapter is a modification of the INVERR subroutine in Chapter 2 which assumes the diagonal entries of the tridiagonal matrices supplied to it are all 0. Error estimates are computed only for isolated 'good' T-eigenvalues. All other 'good' T-eigenvalues are assumed to have converged. If convergence has not yet occurred and a larger Lanczos matrix has been specified by the user, these programs will continue on repeating the above procedure on a larger Lanczos matrix.

Once the singular values been computed accurately enough, the user can select a subset of the 'converged' singular values for which singular vectors are to be computed. The main program LSVEC, for computing singular vectors of real rectangular matrices, is then used to compute these desired singular vectors. These singular vectors are obtained by computing Ritz vectors for the B-matrix and then splitting each of these $(\ell + n)$ – dimensional Ritz vectors into approximate left and right singular vectors of A. The user should note that if the singular value being considered is very small, then LSVEC is not able to accurately compute both a left and a right singular vector approximation simultaneously. In this situation one of the two singular vectors will be more accurate than the other one is. If the starting vector is of the form $(0,u^T)^T$, then the right singular vector will be more accurate than the corresponding left vector. Similarly, if we use a starting vector of the form $(v^T,0)^T$, then the left vector will be more accurate than the right vector will be. This loss in accuracy in one of the two vectors increases as the size of the singular value is decreased, and in the limit for a zero singular value, one of the two computed singular vectors will have no accuracy at all. See Section 5.4 of Chapter 5 in Volume 1.

All computations are in double precision real arithmetic. The user must supply a subroutine USPEC which defines and initializes the user-specified matrix A, and subroutines SVMAT and STRAN which compute respectively, matrix-vector multiplies Ax and A^Ty for any given vectors x and y. These subroutines must be constructed in such a way as to take advantage of the sparsity (and/or structure) of the user-supplied A-matrix and such that these computations are done accurately. More details about these real rectangular, single-vector Lanczos procedures are given in Section 5.4 of Chapter 5 in Volume 1.

SECTION 6.2 DOCUMENTATION FOR PROGRAMS IN CHAPTER 6

```
C-----DOCUMENTATION FOR THE SINGLE-VECTOR-------------------------------
C     LANCZOS SINGULAR VALUE/VECTOR PROGRAMS
C     FOR REAL, RECTANGULAR MATRICES
C
C
C     AUTHORS:  JANE CULLUM AND RALPH A. WILLOUGHBY, IBM RESEARCH,
C               YORKTOWN HEIGHTS, NY 10598.  PHONE: 914-945-2227
C
C
C     GIVEN A REAL RECTANGULAR MATRIX A OF ORDER M X N THE THREE
C     SETS OF FORTRAN FILES LABELLED LSVAL, LSSUB, AND LSMULT
C     CAN BE USED TO COMPUTE DISTINCT SINGULAR VALUES OF A IN
C     USER-SPECIFIED INTERVALS.
C
C     CORRESPONDING SINGULAR VECTORS FOR SELECTED, COMPUTED
C     SINGULAR VALUES CAN BE COMPUTED USING THE SETS OF FILES
C     LABELLED LSVEC, LSSUB AND LSMULT.
C
C     THESE PROGRAMS USE LANCZOS TRIDIAGONALIZATION WITHOUT
C     REORTHOGONALIZATION ON THE ASSOCIATED REAL SYMMETRIC MATRIX
C
C                            ----               ----
C                           | 0                  A |
C               B  =        |                      |
C                           | A-TRANSPOSE        0 |
C                            ----               ----
C
C     OF ORDER M + N  TO GENERATE REAL SYMMETRIC TRIDIAGONAL
C     MATRICES, T(1,MEV), OF ORDER MEV.  SUBSETS OF THE EIGENVALUES OF
C     THESE T-MATRICES, LABELLED AS THE 'GOOD EIGENVALUES' OF T(1,MEV),
C     ARE APPROXIMATIONS TO THE DESIRED SINGULAR VALUES OF A.
C     CORRESPONDING RITZ VECTORS FOR B ARE APPROXIMATIONS TO
C     EIGENVECTORS OF B WHICH IN TURN CONTAIN APPROXIMATIONS TO
C     THE DESIRED LEFT AND RIGHT SINGULAR VECTORS OF A.  THIS
C     PROCEDURE USES A SPECIAL STARTING VECTOR SUGGESTED BY GOLUB
C     AND KAHAN.  THUS, THE STARTING LANCZOS VECTOR IS EITHER OF
C     THE FORM (V1,0) OR (0,V2) WHERE V1 IS MX1 AND V2 IS NX1 AND
C     ALL SUCCEEDING LANCZOS VECTORS GENERATED ALTERNATE BETWEEN
C     THESE 2 FORMS.  THIS SPECIAL CHOICE OF STARTING VECTOR RESULTS
C     IN SIGNIFICANT GAINS IN STORAGE AND OPERATION COUNTS AND
C     ALSO IN CONVERGENCE RELATIVE TO A 'BRUTE FORCE' APPLICATION
C     OF THE REAL SYMMETRIC LANCZOS PROCEDURE DIRECTLY TO THE
C     MATRIX B ABOVE.  FOR MORE DETAILS SEE REFERENCE 1 BELOW.
C     IN THE DISCUSSIONS T(1,MEV) DENOTES THE LANCZOS T-MATRIX
C     OF SIZE MEV.
C
C     THE IDEAS USED IN THESE PROGRAMS ARE DISCUSSED IN THE FOLLOWING
C     REFERENCES.
C
C     1.  JANE CULLUM, RALPH A. WILLOUGHBY AND MARK LAKE, A LANCZOS
C         ALGORITHM FOR COMPUTING SINGULAR VALUES AND VECTORS OF LARGE
C         MATRICES, SIAM J. SCIENTIFIC AND STATISTICAL COMPUTING,
C         VOL. 4, JUNE 1983, PP. 197-215.
C
C     2.  JANE CULLUM AND RALPH A. WILLOUGHBY, LANCZOS ALGORITHMS
C         FOR LARGE SYMMETRIC MATRICES,  PROGRESS IN
C         SCIENTIFIC COMPUTING, EDITORS, G. GOLUB, H.O. KREISS,
C         S. ARBARBANEL, AND R. GLOWINSKI,  BIRKHAUSER BOSTON INC.,
C         CAMBRIDGE, MASSACHUSETTS, 1984.
C
C     3.  JANE CULLUM AND RALPH A. WILLOUGHBY, COMPUTING EIGENVECTORS
C         (AND EIGENVALUES) OF LARGE, SYMMETRIC MATRICES USING
C         LANCZOS TRIDIAGONALIZATION, LECTURE NOTES IN MATHEMATICS,
C         773, NUMERICAL ANALYSIS PROCEEDINGS, DUNDEE 1979, EDITED BY
```

```
C        G. A. WATSON, SPRINGER-VERLAG, (1980), BERLIN, PP.46-63.
C
C     4. IBID, LANCZOS AND THE COMPUTATION IN SPECIFIED INTERVALS OF
C        THE SPECTRUM OF LARGE SPARSE, REAL SYMMETRIC MATRICES, SPARSE
C        MATRIX PROCEEDINGS 1978, ED. I.S. DUFF AND G. W. STEWART,
C        SIAM, PHILADELPHIA, PP.220-255, 1979.
C
C     5. IBID, COMPUTING EIGENVALUES OF VERY LARGE SYMMETRIC MATRICES-
C        AN IMPLEMENTATION OF A LANCZOS ALGORITHM WITHOUT
C        REORTHOGONALIZATION, J. COMPUT. PHYS. 44(1981), 329-358.
C
C
C-----PORTABILITY------------------------------------------------------
C
C
C     PROGRAMS WERE TESTED FOR PORTABILITY USING THE PFORT VERIFIER.
C     FOR DETAILS OF THE VERIFIER SEE FOR EXAMPLE, B. G. RYDER AND
C     A. D. HALL, "THE PFORT VERIFIER", COMPUTING SCIENCE TECHNICAL
C     REPORT 12, BELL LABORATORIES, MURRAY HILL, NEW JERSEY 07974,
C     (REVISED), JANUARY 1981.
C
C     EXCEPT FOR THE FOLLOWING CONSTRUCTIONS WHICH CAN BE EASILY
C     MODIFIED BY THE USER TO MATCH THE PARTICULAR COMPUTER BEING
C     USED, THE PROGRAM STATEMENTS ARE PORTABLE.
C
C    NONPORTABLE CONSTRUCTIONS.
C
C    IN LSVAL AND IN LSVEC
C        1.  DATA/MACHEP STATEMENT
C        2.  ALL READ(5,*) STATEMENTS (FREE FORMAT)
C        3.  FORMAT(20A4) USED FOR THE EXPLANATORY HEADER ARRAY, EXPLAN
C        4.  FORMAT(4Z20) USED TO READ AND WRITE BETA FILES 1 AND 2.
C    IN LSMULT
C        1.  IN SVMAT, STRAN, AND  USPEC THE ENTRY THAT PASSES THE
C            STORAGE LOCATIONS OF THE ARRAYS DEFINING THE
C            USER-SPECIFIED MATRIX.
C        2.  IN SAMPLE USPEC FOR 'DIAGONAL' MATRICES:  THE FREE
C            FORMAT (8,*) AND THE FORMAT (20A4).
C    IN LSSUB
C        1.  ALL STATEMENTS ARE PORTABLE.
C
C
C     IN THE COMMENTS BELOW:
C     COMPLEX*16 = COMPLEX VARIABLE, 16 BYTES OF STORAGE
C     REAL*8 = REAL VARIABLE, 8 BYTES OF STORAGE
C     REAL*4 = REAL VARIABLE, 4 BYTES OF STORAGE
C     INTEGER*4 = INTEGER VARIABLE, 4 BYTES
C
C
C-----A-MATRIX SPECIFICATION-------------------------------------------
C
C
C     SUBROUTINE USPEC IS USED TO SPECIFY THE USER-SUPPLIED A-MATRIX.
C     SUBROUTINES SVMAT AND STRAN ARE, RESPECTIVELY, CORRESPONDING
C     MATRIX-VECTOR MULTIPLE SUBROUTINES FOR A AND FOR A-TRANSPOSE.
C     THESE SUBROUTINES SHOULD BE DESIGNED TO TAKE ADVANTAGE OF
C     ANY SPECIAL PROPERTIES OF THE USER-SUPPLIED MATRIX.  THE
C     MATRIX-VECTOR MULTIPLIES REQUIRED BY THE LANCZOS PROCEDURES
C     MUST BE COMPUTED RAPIDLY AND ACCURATELY.
C
C     SUBROUTINE USPEC HAS THE CALLING SEQUENCE
C
C            CALL USPEC(M,N,MATNO)
C
C     WHERE M IS THE NUMBER OF ROWS IN THE USER-SPECIFIED
C     A-MATRIX AND N IS THE NUMBER OF COLUMNS. MATNO IS A
C     <= 8 DIGIT INTEGER USED AS A MATRIX AND TEST IDENTIFICATION
```

```
C     NUMBER.  THIS SUBROUTINE DEFINES (DIMENSIONS) THE ARRAYS
C     REQUIRED TO SPECIFY THE A-MATRIX.  THIS SUBROUTINE ALSO
C     INITIALIZES THESE ARRAYS AND ANY OTHER PARAMETERS NEEDED TO
C     DEFINE THE MATRIX.  THE STORAGE LOCATIONS OF THESE PARAMETERS
C     AND ARRAYS ARE THEN PASSED TO THE MATRIX-VECTOR MULTIPLY
C     SUBROUTINES SVMAT AND STRAN VIA ENTRIES.  SAMPLE SUBROUTINES
C     ARE INCLUDED IN THE FORTRAN FILE LSMULT.
C
C     IMPORTANT NOTE:
C     THE SAMPLE MATRIX-VECTOR MULTIPLY SUBROUTINES IN LSMULT
C     ASSUME THAT M >= N.  THEY ALSO ASSUME THAT THE USER-SUPPLIED
C     INFORMATION ABOUT THE GIVEN MATRIX IS STORED ON FILE 8.
C     THE USER SHOULD SEE THE LSMULT PROGRAMS FOR MORE DETAILS.
C
C     SUBROUTINE SVMAT HAS THE CALLING SEQUENCE
C
C            CALL SVMAT(W,U,SUM)
C
C     WHERE U AND W ARE REAL*8 VECTORS AND SUM IS A REAL*8
C     SCALAR.  SVMAT CALCULATES U = A*W - SUM*U FOR THE
C     USER-SPECIFIED A-MATRIX.  SUBROUTINE STRAN HAS THE
C     CALLING SEQUENCE
C
C            CALL STRAN(W,U,SUM)
C
C     STRAN CALCULATES U = (A-TRANSPOSE)*W - SUM*U FOR THE
C     TRANSPOSE OF THE USER-SUPPLIED A-MATRIX.  THE ARRAY AND PARAMETER
C     INFORMATION NEEDED TO PERFORM THE MATRIX-VECTOR MULTIPLIES
C     IS PASSED TO THE SVMAT AND THE STRAN SUBROUTINES FROM THE
C     USPEC SUBROUTINE VIA ENTRIES.  ONE SET OF THE SAMPLE SVMAT
C     AND STRAN SUBROUTINES INCLUDED IN LSMULT COMPUTES
C     MATRIX-VECTOR MULTIPLIES FOR AN ARBITRARY SPARSE,
C     RECTANGULAR MATRIX STORED IN THE SPARSE FORMAT SPECIFIED
C     IN THE CORRESPONDING SAMPLE USPEC SUBROUTINE.  THE LANCZS
C     SUBROUTINE CALLS SVMAT AND STRAN IN THE GENERATION OF
C     THE LANCZOS T-MATRICES FOR THE B MATRIX.
C
C     THE DATA FOR THE A-MATRIX IS ASSUMED TO BE ON FILE 8 AND
C     IN THE FOLLOWING SPARSE FORMAT:
C     NZ = NUMBER OF NONZERO ELEMENTS OF A
C     ICOL(K), K = 1,N, NUMBER OF NONZEROS OF A IN COLUMN K.
C     IROW(K), K = 1,NZ, ROW INDEX OF A(K).
C     A(K), K=1,NZ  CONTAINS THE ELEMENTS OF A BY COLUMN.
C
C
C     SVMATV AND STRAN ARE CALLED FROM THE SUBROUTINE LANCZS
C     WHICH GENERATES THE LANCZOS TRIDIAGONAL MATRICES, THE
C     BETA HISTORY.   SIMILARLY, THSE SUBROUTINES ARE CALLED FROM
C     THE CORRESPONDING SINGULAR VECTOR PROGRAM, LSVEC.
C     SVMAT AND STRAN ARE DECLARED AS EXTERNAL VARIABLES.
C     EACH IS AN ARGUMENT FOR THE LANCZS SUBROUTINE.
C
C     USPEC, SVMAT, AND STRAN SUBROUTINES SUITABLE FOR THE
C     USER-SPECIFIED MATRIX MUST BE SUPPLIED BY THE USER.
C
C     THE MAIN PROGRAMS FOR THE SINGULAR VALUE AND SINGULAR VECTOR
C     CALCULATIONS ASSUME THAT INPUT FILE 5 CONTAINS THE ROW ORDER
C     M AND THE COLUMN ORDER N OF THE GIVEN A-MATRIX AND MATNO,
C     AN IDENTIFICATION NUMBER OF <= 8 DIGITS FOR THE GIVEN MATRIX.
C
C
C-----MACHEP------------------------------------------------------------
C
C
C     MACHEP IS A MACHINE DEPENDENT PARAMETER SPECIFYING THE RELATIVE
C     PRECISION OF THE FLOATING POINT ARITHMETIC USED.
C     MACHEP = 2.2 * 10**-16 FOR DOUBLE PRECISION ARITHMETIC ON
```

```
C      IBM 370-3081.
C
C      THE USER WILL HAVE TO RESET THIS PARAMETER TO
C      THE CORRESPONDING VALUE FOR THE MACHINE BEING USED.  NOTE THAT
C      IF A MACHINE WITH A MACHINE EPSILON THAT IS MUCH LARGER THAN THE
C      VALUE GIVEN HERE IS BEING USED, THEN THERE COULD BE
C      PROBLEMS WITH THE TOLERANCES.
C
C
C-----SUBROUTINES AND FUNCTIONS USER MUST SUPPLY-----------------------
C
C
C      GENRAN,  FINPRO,  MASK,  USPEC,  SVMAT AND  STRAN
C
C      GENRAN = COMPUTES K PSEUDO-RANDOM NUMBERS AND STORES THEM IN
C               THE REAL ARRAY, G.  THIS SUBROUTINE IS USED TO
C               GENERATE A STARTING VECTOR FOR THE LANCZOS PROCEDURE
C               IN THE SUBROUTINE LANCZS AND A STARTING RIGHT-HAND SIDE
C               FOR INVERSE ITERATION IN THE SUBROUTINE INVERR.
C
C               TESTS REPORTED IN THE REFERENCES USED EITHER GGL1 OR
C               GGL2 FROM THE IBM LIBRARY  SLMATH.
C               THE EXISTING CALLING SEQUENCE IS:
C
C                     CALL GENRAN(IIX,G,K).
C
C               WHERE IIX =INTEGER SEED, G = REAL*4 ARRAY WHOSE
C               DIMENSION MUST BE >= K.  K RANDOM NUMBERS ARE GENERATED
C               AND PLACED IN G.
C
C      FINPRO = DOUBLE PRECISION FUNCTION WHICH COMPUTES THE INNER
C               PRODUCT OF 2 DOUBLE PRECISION VECTORS OF DIMENSION  K.
C               TESTS REPORTED IN THE REFERENCES USED THE HARWELL
C               LIBRARY SUBROUTINE FM02AD.
C               EXISTING CALLING SEQUENCE IS
C
C                     CALL FINPRO(N,V,J,W,K).
C
C               COMPUTES THE INNER PRODUCT OF DIMENSION N OF THE VECTORS
C               V AND W.  SUCCESSIVE COMPONENTS OF V AND OF W ARE STORED
C               AT LOCATIONS THAT ARE ,RESPECTIVELY, J AND K UNITS APART.
C
C      MASK = MASKS OVERFLOW AND UNDERFLOW.
C             USER MUST SUPPLY OR COMMENT OUT CALL.
C
C     USPEC = DIMENSIONS AND INITIALIZES ARRAYS NEEDED TO SPECIFY
C             USER-SUPPLIED A-MATRIX.  SEE A-MATRIX SPECIFICATION SECTION
C
C     SVMAT = MATRIX-VECTOR MULTIPLY FOR USER-SUPPLIED A-MATRIX.
C             SEE A-MATRIX SPECIFICATION SECTION.
C
C     STRAN = MATRIX-VECTOR MULTIPLY FOR TRANSPOSE OF USER-SUPPLIED
C             A-MATRIX.  SEE A-MATRIX SPECIFICATION SECTION.
C
C
C---------------------------------------------------------------------
C
C      COMMENTS FOR SINGULAR VALUE COMPUTATIONS
C
C---------------------------------------------------------------------
C
C
C-----PARAMETER CONTROLS FOR SINGULAR VALUE PROGRAMS--------------------
C
C
C      PARAMETER CONTROLS ARE INTRODUCED TO ALLOW SEGMENTATION OF THE
C      SINGULAR VALUE COMPUTATIONS AND TO ALLOW VARIOUS COMBINATIONS
```

```
C     OF READ/WRITES.
C
C     THE FLAG ISTART CONTROLS THE T-MATRIX (BETA HISTORY)
C     GENERATION.
C
C     ISTART = (0,1)  MEANS
C
C              (0) THERE IS NO EXISTING BETA HISTORY AND ONE
C                  MUST BE GENERATED.
C
C              (1) THERE IS AN EXISTING BETA HISTORY AND IT IS
C                  TO BE READ IN FROM FILE 2 AND EXTENDED IF NECESSARY.
C
C     THE FLAG ISTOP CAN BE USED IN CONJUNCTION WITH THE FLAG ISTART TO
C     ALLOW SEGMENTATION OF THE SINGULAR VALUE COMPUTATIONS.
C
C     ISTOP  = (0,1)  MEANS
C
C              (0) PROGRAM COMPUTES ONLY THE REQUESTED BETAS,
C                  STORES THEM AND THE LAST 2 LANCZOS VECTORS GENERATED
C                  IN FILE 1 AND THEN TERMINATES.
C
C              (1) PROGRAM COMPUTES REQUESTED BETAS AND THEN
C                  USES THE BISEC SUBROUTINE TO CALCULATE EIGENVALUES
C                  OF THE TRIDIAGONAL MATRICES GENERATED FOR THE ORDERS
C                  SPECIFIED BY THE USER AND ON THE USER-SPECIFIED
C                  INTERVALS.  PROGRAM THEN USES THE SUBROUTINE INVERR
C                  TO COMPUTE ERROR ESTIMATES FOR THE ISOLATED GOOD
C                  T-EIGENVALUES WHICH ARE USED TO CHECK THE
C                  CONVERGENCE OF THESE T-EIGENVALUES.
C
C     CONTROL PARAMETERS FOR WRITES
C
C     IHIS  = (0,1)  MEANS
C
C             (0) IF ISTOP .GT. 0 THEN BETAS ARE NOT SAVED ON FILE 1.
C
C             (1) PROGRAM WRITES BETAS AND LAST 2 LANCZOS
C                 VECTORS TO FILE 1 SO THAT THE T-MATRIX GENERATION
C                 MAY BE REUSED OR CONTINUED LATER IF NECESSARY.
C                 TYPICALLY ONE WOULD ALWAYS DO THIS ON ANY RUN WHERE
C                 A HISTORY FILE IS BEING GENERATED.  HISTORY MUST BE
C                 SAVED IN MACHINE FORMAT ((4Z20) FOR IBM/3081) SO
C                 THAT NO ERRORS DUE TO FORMAT CONVERSIONS OCCUR.
C
C     IDIST  = (0,1)  MEANS
C
C              (0) DISTINCT EIGENVALUES OF T-MATRICES ARE NOT SAVED.
C
C              (1) PROGRAM WRITES COMPUTED DISTINCT EIGENVALUES OF
C                  T-MATRICES ALONG WITH THEIR T-MULTIPLICITIES
C                  TO FILE 11.
C
C     IWRITE = (0,1)   MEANS
C
C              (0) NO EXTENDED OUTPUT FROM SUBROUTINES BISEC AND INVERR
C                  IS SENT TO FILE 6.
C
C              (1) INDIVIDUAL COMPUTED T-EIGENVALUES AND CORRESPONDING
C                  ERROR ESTIMATES FROM THE SUBROUTINES BISEC AND INVERR
C                  ARE PRINTED OUT TO FILE 6 AS THEY ARE COMPUTED.
C
C     THE PROGRAM ALWAYS MAKES A SEPARATE LIST OF THE COMPUTED GOOD
C     EIGENVALUES OF THE LANCZOS MATRICES T(1,MEV) CONSIDERED,
C     THESE ARE THE APPROXIMATIONS TO THE DESIRED SINGULAR VALUES,
C     ALONG WITH THEIR MINIMAL GAPS AS SINGULAR VALUES OF A AND
C     WRITES THEM TO FILE 3.  CORRESPONDING ERROR ESTIMATES FOR ANY
```

```
C     ISOLATED COMPUTED GOOD T-EIGENVALUES (SINGULAR VALUES OF A)
C     ARE ALWAYS WRITTEN TO FILE 4.
C
C
C-----INPUT/OUTPUT FILES FOR SINGULAR VALUE PROGRAMS------------------
C
C     ANY INPUT DATA OTHER THAN THE BETA HISTORY SHOULD BE STORED
C     ON FILE 5. SEE SAMPLE INPUT/OUTPUT FROM TYPICAL RUN.
C     THE READ STATEMENTS IN THE GIVEN FORTRAN PROGRAM ASSUME THAT
C     THE DATA STORED ON FILE 5 IS IN FREE FORMAT.  USER SHOULD NOTE
C     THAT 'FREE FORMAT' IS NOT CLASSIFIED AS PORTABLE BY PFORT SO THAT
C     THE USER MAY HAVE TO MODIFY THE READ STATEMENTS FROM FILE 5 TO
C     CONFORM TO WHAT IS PERMISSIBLE ON THE MACHINE BEING USED.
C
C     FILE 6 WAS USED AS THE INTERACTIVE TERMINAL OUTPUT FILE.
C     THIS FILE PROVIDES A RUNNING ACCOUNT OF THE PROGRESS OF THE
C     COMPUTATIONS.  THE AMOUNT OF INFORMATION PRINTED OUT IS
C     CONTROLLED BY THE PARAMETER IWRITE.
C
C DESCRIPTION OF OTHER I/O FILES
C
C FILE (K)      CONTAINS:
C
C      (1)      OUTPUT FILE:
C               HISTORY FILE OF NEWLY-GENERATED T-MATRIX
C               (BETA VECTOR) AND LAST 2 LANCZOS VECTORS USED
C               IN THE T-MATRIX GENERATION.
C               IF IHIS = 0 AND ISTOP = 1, FILE 1 IS NOT WRITTEN.
C
C      (2)      INPUT FILE:
C               SAME AS FILE 1 EXCEPT THAT IT CONTAINS A
C               PREVIOUSLY-GENERATED T-MATRIX (IF ANY). IF ISTART = 1,
C               PROGRAM ASSUMES THAT THERE IS A HISTORY FILE OF
C               BETAS ON FILE 2.  THESE BETAS AND THE LAST TWO LANCZOS
C               VECTORS USED IN THE T-MATRIX GENERATION ARE READ IN.
C
C      (3)      OUTPUT FILE:
C               COMPUTED GOOD EIGENVALUES OF THE T-MATRICES CONSIDERED.
C               ALSO CONTAINS T-MULTIPLICITIES OF THESE T-EIGENVALUES AS
C               EIGENVALUES OF THE T-MATRIX, AND THEIR GAPS AS
C               EIGENVALUES IN THE B MATRIX AND IN THE T-MATRIX.
C               NOTE THAT THESE GOOD T-EIGENVALUES ARE THE COMPUTED
C               SINGULAR VALUES OF THE A-MATRIX AND THAT THE GAPS
C               OF THESE EIGENVALUES AS EIGENVALUES OF THE B-MATRIX
C               ARE EQUAL TO THEIR GAPS AS SINGULAR VALUES OF A.  FILE
C               3 IS ALWAYS WRITTEN.
C
C      (4)      OUTPUT FILE:
C               ERROR ESTIMATES FOR THE ISOLATED COMPUTED SINGULAR
C               SINGULAR VALUES (ISOLATED GOOD EIGENVALUES OF T(1,MEV))
C               THESE ARE OBTAINED USING THE SUBROUTINE INVERR. THESE
C               ESTIMATES USE THE LAST COMPONENTS OF THE ASSOCIATED
C               T-EIGENVECTORS WHICH ARE COMPUTED USING INVERSE
C               ITERATION.  FILE 4 IS ALWAYS WRITTEN.
C
C
C      (8)      INPUT FILE:
C               SAMPLE USPEC SUBROUTINE ASSUMES THAT THE ARRAYS
C               REQUIRED TO SPECIFY THE USER'S MATRIX ARE STORED ON
C               FILE 8.  USERS MUST MAKE WHATEVER DEFINITIONS ARE
C               APPROPRIATE FOR THEIR MATRICES.
C
C      (9)      OUTPUT FILE:  OPTIONAL
C               CAN BE USED TO STORE THE TRUE SINGULAR VALUES OF
C               A GIVEN TEST MATRIX, WHEN THE SINGULAR VALUE PROCEDURE
C               IS BEING EXERCISED ON A TEST MATRIX.
C
```

```
C     (11)      OUTPUT FILE:
C               COMPUTED DISTINCT EIGENVALUES OF T-MATRICES USED.
C               ALSO CONTAINS THEIR T-MULTIPLICITIES AND T-GAPS TO
C               NEAREST DISTINCT T-EIGENVALUES, AND THE T-MULTIPLICITY
C               PATTERN OF THE GOOD AND THE SPURIOUS T-EIGENVALUES.
C               FILE 11 IS WRITTEN ONLY IF IDIST = 1.
C
C
C-----PARAMETERS SET BY THE SINGULAR VALUE PROGRAMS---------------------
C
C
C     THESE PARAMETERS ARE SET INTERNALLY IN THE PROGRAM
C
C     SCALEK      K = 1,2,3,4
C
C                 THE SCALING FACTORS SCALEK HAVE BEEN INTRODUCED IN AN
C                 ATTEMPT TO MAKE THE TOLERANCES USED IN THE
C                 T-MULTIPLICITY, SPURIOUS, ISOLATION AND PRTESTS ADJUST
C                 TO THE SCALE OF THE GIVEN MATRIX.  THESE FACTORS MUST
C                 NOT BE MODIFIED.
C
C     NOTE:    THE USER SHOULD NOTE THAT IF THE MATRIX BEING
C     PROCESSED IS VERY STIFF, THAT IS THE RATIO OF THE LARGEST
C     SINGULAR VALUE TO THE SMALLEST SINGULAR VALUE IS VERY
C     LARGE, THEN THE TOLERANCES BEING USED IN BISEC, LUMP, ISOEV
C     AND PRTEST MAY NOT TREAT THE SMALLEST SINGULAR VALUES
C     VERY WELL.  IN SOME SUCH CASES A USER-INTRODUCED REDUCTION
C     IN THE SIZE OF TKMAX AND THE SUBSEQUENT RECOMPUTATION OF
C     THE T-MATRIX EIGENVALUES CORRESPONDING TO THE SMALLEST
C     SINGULAR VALUES USING THIS TKMAX MAY RESULT IN IMPROVED
C     COMPUTATIONS AT THE LOW END.
C
C     THE LUMP, ISOEV, AND PRTEST TOLERANCES THAT WERE USED
C     MOST IN  THE TESTING OF THIS ALGORITHM WERE NOT
C     SCALE INVARIANT BUT SEEMED TO WORK WELL ON MATRICES THAT
C     HAD SINGULAR VALUES BOTH GREATER THAN AND LESS
C     THAN 1.  THESE TOLERANCES ARE ALSO INCLUDED IN THESE THREE
C     SUBROUTINES BUT AS COMMENTED OUT STATEMENTS.  THEY CAN BE
C     REVIVED BY COMMENTING OUT THE CORRESPONDING TOLERANCES
C     SPECIFIED IN THE STATEMENT ABOVE EACH OF THESE.
C
C     IMPORTANT TOLERANCES OR SCALES THAT ARE USED REPEATEDLY
C     THROUGHOUT THIS PROGRAM ARE THE FOLLOWING:
C     SCALED MACHINE EPSILON:  TTOL = TKMAX*EPSM WHERE
C     EPSM = 2*MACHINE EPSILON AND
C     TKMAX = MAX(BETA(J), J = 1,MEV)
C     BISEC CONVERGENCE TOLERANCE:  BISTOL = DSQRT(1000+MEV)*TTOL
C     BISEC T-MULTIPLICITY TOLERANCE:  MULTOL = (1000+MEV)*TTOL
C     LANCZOS CONVERGENCE TOLERANCE:   CONTOL = BETA(MEV+1)*1.D-10
C
C
C     BTOL = RELATIVE TOLERANCE USED TO ESTIMATE ANY LOSS OF LOCAL
C            ORTHOGONALITY OF THE LANCZOS VECTORS AFTER THE T-MATRIX
C            HAS BEEN GENERATED.  THE LANCZOS PROCEDURE WORKS WELL
C            ONLY IF LOCAL ORTHOGONALITY BETWEEN SUCCESSIVE LANCZOS
C            VECTORS IS MAINTAINED.  THE TNORM SUBROUTINE TESTS
C            WHETHER OR NOT
C
C                 MINIMUM  |BETA(I)|/||A|| > BTOL.
C                 I=2,KMAX
C
C            IF THIS TEST IS VIOLATED BY SOME BETA AND A T-MATRIX THAT
C            WOULD INCLUDE SUCH A BETA IS REQUESTED, THEN THE LANCZOS
C            PROCEDURE WILL TERMINATE FOR THE USER TO DECIDE WHAT TO
C            DO.  THE USER CAN OVER-RIDE THIS TEST BY SIMPLY DECREASING
C            THE SIZE OF BTOL, BUT THEN CONVERGENCE IS NOT AS CERTAIN.
C            THE PROGRAM SETS BTOL = 1.D-8 WHICH IS A VERY CONSERVATIVE
```

```
C           CHOICE. THE || A || IS ESTIMATED BY USING AN ESTIMATE       LSV04670
C           OF THE NORM OF THE T-MATRIX, T(1,KMAX).                     LSV04680
C                                                                       LSV04690
C     GAPTOL = RELATIVE TOLERANCE USED IN THE SUBROUTINE ISOEV          LSV04700
C             TO DETERMINE FOR WHICH OF THE GOOD T-EIGENVALUES,         LSV04710
C             THE COMPUTED SINGULAR VALUES, ERROR ESTIMATES SHOULD      LSV04720
C             BE COMPUTED.  THE PROGRAM SETS GAPTOL = 1.D-8.            LSV04730
C             IF FOR A GIVEN 'GOOD' T-EIGENVALUE OF THE GIVEN           LSV04740
C             T-MATRIX  THE COMPUTED GAP IN THE T-MATRIX IS TOO         LSV04750
C             SMALL AND IS DUE TO A 'SPURIOUS' EIGENVALUE OF            LSV04760
C             THE T-MATRIX, THEN THE 'GOOD' T-EIGENVALUE IS ASSUMED     LSV04770
C             TO HAVE CONVERGED AND AN ERROR ESTIMATE IS NOT            LSV04780
C             COMPUTED.                                                 LSV04790
C                                                                       LSV04800
C                                                                       LSV04810
C-----USER-SPECIFIED PARAMETERS FOR SINGULAR VALUE PROGRAMS-------------LSV04820
C                                                                       LSV04830
C                                                                       LSV04840
C     RELTOL = RELATIVE TOLERANCE USED IN 'COMBINING' COMPUTED          LSV04850
C             EIGENVALUES OF T(1,MEV) PRIOR TO COMPUTING ERROR          LSV04860
C             ESTIMATES.                                                LSV04870
C                                                                       LSV04880
C     THE LUMPING OF T-EIGENVALUES OCCURS IN SUBROUTINE LUMP.           LSV04890
C     LUMPING IS NECESSARY BECAUSE IT IS IMPOSSIBLE TO ACCURATELY       LSV04900
C     PREDICT THE ACCURACY OF THE BISEC SUBROUTINE.  LUMP 'COMBINES'    LSV04910
C     T-EIGENVALUES THAT HAVE SLIPPED BY THE TOLERANCE THAT WAS USED    LSV04920
C     IN THE T-MULTIPLICITY TESTS.  IN PARTICULAR IF FOR SOME J,        LSV04930
C                                                                       LSV04940
C   |EVALUE(J)-EVALUE(J-1)| < DMAX1(RELTOL*|EVALUE(J)|,SCALE2*MULTOL)   LSV04950
C                                                                       LSV04960
C     THEN THESE T-EIGENVALUES ARE 'COMBINED'.  MULTOL IS THE TOLERANCE LSV04970
C     THAT WAS USED IN THE T-MULTIPLICITY TEST IN BISEC.  SEE THE HEADERLSV04980
C     ON THE LUMP SUBROUTINE FOR MORE DETAILS.                          LSV04990
C                                                                       LSV05000
C     RELTOL IS SET TO 1.D-10.                                          LSV05010
C                                                                       LSV05020
C     MXINIT = MAXIMUM NUMBER OF INVERSE ITERATIONS ALLOWED IN          LSV05030
C             SUBROUTINE INVERR FOR EACH ISOLATED GOOD T-EIGENVALUE     LSV05040
C             CONSIDERED.  TYPICALLY ONLY ONE IS REQUIRED.              LSV05050
C                                                                       LSV05060
C     SEEDS FOR RANDOM NUMBER GENERATORS = INTEGER*4 SCALARS.           LSV05070
C                                                                       LSV05080
C               (1) SVSEED = SEED FOR STARTING VECTOR USED IN           LSV05090
C                   T-MATRIX GENERATION IN LANCZS SUBROUTINE            LSV05100
C                                                                       LSV05110
C               (2) RHSEED = SEED FOR RIGHT-HAND SIDE USED IN           LSV05120
C                   INVERSE ITERATION COMPUTATIONS IN INVERR.           LSV05130
C                                                                       LSV05140
C     BISEC DATA                                                        LSV05150
C                                                                       LSV05160
C     (1) NINT  =  NUMBER OF SUBINTERVALS ON WHICH SINGULAR VALUES      LSV05170
C                  ARE TO BE COMPUTED.                                  LSV05180
C                                                                       LSV05190
C     (2) LB(J) = (J = 1,NINT)  =  LEFT END POINTS OF THESE INTERVALS.  LSV05200
C                MUST BE PROVIDED IN INCREASING ORDER.  THAT IS,        LSV05210
C                LB(J) < LB(J+1) FOR J = 1,NINT.                        LSV05220
C                                                                       LSV05230
C     (3) UB(J) = (J = 1,NINT) =  RIGHT END POINTS OF THESE INTERVALS.  LSV05240
C                MUST BE PROVIDED IN INCREASING ORDER.  THAT IS,        LSV05250
C                UB(J) < UB(J+1) FOR J = 1,NINT.                        LSV05260
C                                                                       LSV05270
C     (4) MXSTUR  =  MAXIMUM NUMBER OF STURM ITERATIONS ALLOWED FOR     LSV05280
C                    ENTIRE SET OF SINGULAR VALUE CALCULATIONS OVER     LSV05290
C                    ALL SPECIFIED SIZE T-MATRICES.  PROGRAM WILL       LSV05300
C                    TERMINATE IF THIS LIMIT IS EXCEEDED.               LSV05310
C                                                                       LSV05320
C     T-MATRICES                                                        LSV05330
```

```
C
C     SIZES OF T-MATRICES
C
C             (1) KMAX= MAXIMUM ORDER FOR T-MATRIX THAT USER IS WILLING
C                       TO CONSIDER.
C
C             (2) NMEVS = MAXIMUM NUMBER OF T-MATRICES THAT WILL BE
C                         CONSIDERED.
C
C             (3) NMEV(J)  (J=1,NMEVS)  = SIZES OF T-MATRIX TO BE
C                                         CONSIDERED SEQUENTIALLY.
C
C     T-MATRIX-GENERATION
C
C     IPAR   = (1,2)   MEANS
C
C              (1) STARTING VECTOR IS OF FORM (0,V2) WHERE V2 IS
C                  NX1.  USE WHEN M > N .
C
C              (2) STARTING VECTOR IF OF FORM (V1,0) WHERE V1 IS
C                  MX1.  USE WHEN M < N .
C
C     USER SHOULD NOTE THAT THIS PROGRAM FIRST COMPUTES A T-MATRIX
C     OF ORDER KMAX AND THEN CYCLES THROUGH THE T-MATRICES SPECIFIED
C     A PRIORI BY THE USER, USING THE SUBROUTINE BISEC TO COMPUTE
C     EIGENVALUES OF THE T-MATRICES ON THE INTERVALS SPECIFIED BY
C     THE USER.  SUBSETS OF THESE T-EIGENVALUES ARE THEN SELECTED
C     AS APPROXIMATIONS TO THE DESIRED SINGULAR VALUES.
C
C     IDEALLY, ONE WOULD COMPUTE THE SINGULAR VALUE APPROXIMATIONS
C     AT A REASONABLE SIZE T-MATRIX, LOOK AT THE ACCURACY OF THE
C     COMPUTED RESULTS AND USE THAT TO DETERMINE AN APPROPRIATE
C     INCREMENT FOR THE SIZE OF THE T-MATRIX BASED UPON WHAT
C     HAS ALREADY CONVERGED AND UPON THE SIZES OF THE ERROR ESTIMATES
C     ON THOSE SINGULAR VALUES THAT ARE DESIRED BUT THAT HAVE NOT
C     YET CONVERGED.  HOWEVER, IN THE INTERESTS OF GENERALITY AND
C     SIMPLICITY WE CHOSE NOT TO DO THAT HERE.
C
C
C-----CONVERGENCE TESTS FOR THE SINGULAR VALUE PROGRAMS-----------------
C
C
C     THE CONVERGENCE TEST INCORPORATED IN THIS PROGRAM IS
C     BASED UPON THE ASSUMPTION THAT THOSE T-EIGENVALUES AND
C     THEIR ASSOCIATED T-EIGENVECTORS THAT CORRESPOND TO
C     THE SINGULAR VALUES AND VECTORS WHICH WE WISH TO COMPUTE
C     CONVERGE AS THE T-SIZE IS INCREASED.
C
C     AS CURRENTLY PROGRAMMED, CONVERGENCE IS CHECKED BY EXAMINING
C     THE SIZES OF ALL OF THE COMPUTED ERROR ESTIMATES ON ALL OF THE
C     INTERVALS SPECIFIED BY THE USER.  IDEALLY CONVERGENCE SHOULD
C     BE CHECKED ONLY ON THOSE SINGULAR VALUES OF INTEREST AND
C     ONCE THE SINGULAR VALUES ON SUB-INTERVALS OF THESE INTERVALS
C     HAVE CONVERGED, ANY SUBSEQUENT SINGULAR VALUE COMPUTATIONS
C     SHOULD BE MADE ONLY ON THE UNCONVERGED PORTIONS.  OBVIOUSLY,
C     IT WOULD BE DIFFICULT TO INCORPORATE CODE TO DO THE ABOVE
C     WITHOUT KNOWING A PRIORI PRECISELY WHAT THE USER IS TRYING
C     TO COMPUTE.  THEREFORE, WE DID NOT ATTEMPT TO DO THIS.  IF
C     ONE WISHES TO MAKE SUCH A MODIFICATION THEN ONE MUST ALSO
C     MODIFY THE PROGRAM SO THAT IT CREATES AN OVERALL LIST OF THE
C     CONVERGED SINGULAR VALUES AS THEY ARE COMPUTED, SINCE
C     CONVERGED SINGULAR VALUES OBTAINED AT A PARTICULAR VALUE OF
C     MEV WOULD NO LONGER BE RECOMPUTED AT LARGER VALUES OF MEV.
C
C     IF ONLY A FEW SINGULAR VALUES ARE TO BE COMPUTED THEN SUCH
C     CHANGES WOULD NOT MAKE MUCH DIFFERENCE IN THE RUNNING TIME.
C
```

```
C
C-----ARRAYS REQUIRED BY THE SINGULAR VALUE PROGRAMS--------------------
C
C
C     BETA(J) = REAL*8 ARRAY.  ITS DIMENSION MUST BE AT LEAST KMAX+1.
C             THE LENGTH OF THE LARGEST T-MATRIX ALLOWED.  THIS
C             ARRAY CONTAINS THE SUBDIAGONAL ENTRIES OF THE
C             T-MATRICES.  THE DIAGONAL ENTRIES ARE ALL ZERO.
C
C             THE BETA VECTOR IS NOT ALTERED DURING THE
C             CALCULATIONS.  IMPORTANT NOTE:  ONLY EVEN ORDER
C             T-MATRICES ARE PERMISSIBLE.
C
C     V1(J),V2(J),VS(J) = REAL*8 ARRAYS.   VS MUST BE OF
C                        DIMENSION AT LEAST KMAX.  V1 MUST BE
C                        OF DIMENSION AT LEAST MAX(M,KMAX+1).
C                        V2 MUST BE OF DIMENSION AT LEAST
C                        MAX(N,KMAX). M IS THE ROW DIMENSION OF
C                        A, AND N IS THE COLUMN DIMENSION.
C                        HOWEVER, THE DIMENSION
C                        FOR V2 IS VALID ONLY IF NO MORE
C                        THAN KMAX/2 EIGENVALUES OF THE GIVEN
C                        T-MATRICES ARE TO BE COMPUTED IN ANY GIVEN
C                        SUBINTERVAL. V2 IS USED IN THE SUBROUTINE
C                        BISEC TO HOLD THE UPPER AND LOWER
C                        ENDPOINTS OF THE SUBINTERVALS GENERATED
C                        DURING THE BISECTIONS.  THEREFORE, ITS
C                        DIMENSION MUST ALWAYS BE AT LEAST 2*Q
C                        WHERE Q IS THE MAXIMUM NUMBER OF
C                        EIGENVALUES OF THE SPECIFIED T-MATRIX IN ANY
C                        ONE OF THE SPECIFIED INTERVALS.
C
C     LB(J),UB(J) = REAL*8 ARRAYS. EACH MUST BE OF DIMENSION AT LEAST
C                 NINT, THE NUMBER OF SUBINTERVALS TO BE CONSIDERED.
C                 LB CONTAINS THE LEFT-END POINTS OF THE INTERVALS
C                 ON WHICH SINGULAR VALUES ARE TO BE COMPUTED.
C                 UB CONTAINS THE RIGHT-END POINTS.
C
C     EXPLAN(J) = REAL*4 ARRAY.  ITS DIMENSION IS 20.  THIS ARRAY IS
C               USED TO ALLOW EXPLANATORY COMMENTS IN THE INPUT FILES.
C
C     G(J) = REAL*4 ARRAY.  ITS DIMENSION MUST BE >= MAX(2*KMAX,M,N)
C          IT IS USED FOR HOLDING THE RANDOM VECTORS GENERATED,
C          HOLDING THE COMPUTED ERROR ESTIMATES AND THE COMPUTED
C          MINIMAL GAPS FOR THE SINGULAR VALUES.
C
C     MP(J) = INTEGER*4 ARRAY.  ITS DIMENSION MUST BE AT LEAST KMAX,
C           THE MAXIMUM SIZE OF THE T-MATRICES ALLOWED.  IT CONTAINS
C           THE T-MULTIPLICITIES OF THE COMPUTED T-EIGENVALUES OF
C           THE T-MATRICES.  NOTE THAT 'SPURIOUS' EIGENVALUES
C           OF THE T-MATRICES ARE DENOTED BY A T-MULTIPLICITY OF
C           0.  T-EIGENVALUES THAT THE SUBROUTINE PRTEST HAS
C           IDENTIFIED AS 'GOOD' BUT HIDDEN ARE IDENTIFIED BY A
C           T-MULTIPLICITY OF -10 AND SUBSEQUENTLY ADDED TO THE LIST
C           OF COMPUTED SINGULAR VALUES.
C
C     NMEV(J) = INTEGER*4 ARRAY.  ITS DIMENSION MUST BE AT LEAST THE
C             NUMBER OF T-MATRICES ALLOWED.  IT CONTAINS THE ORDERS
C             OF THE T-MATRICES TO BE CONSIDERED.
C
C
C     OTHER ARRAYS
C
C     THE USER MUST SPECIFY IN THE SUBROUTINE USPEC WHATEVER ARRAYS
C     ARE REQUIRED TO DEFINE THE MATRIX BEING USED.
C
C
```

```
C-----SUBROUTINES INCLUDED-----------------------------------------------
C
C
C     LANCZS = COMPUTES THE BETA HISTORY. USES SUBROUTINES
C              FINPRO, GENRAN, SVMAT AND STRAN.
C
C     BISEC = COMPUTES EIGENVALUES OF THE SPECIFIED T-MATRIX USING
C             STURM SEQUENCING, ON SEQUENCE OF INTERVALS SPECIFIED
C             BY THE USER.  EACH SUBINTERVAL IS TREATED AS OPEN
C             ON THE LEFT AND CLOSED ON THE RIGHT.  EIGENVALUES
C             ARE COMPUTED WITH SIMULTANEOUS DETERMINATION OF THE
C             T-MULTIPLICITIES AND OF WHICH T-EIGENVALUES ARE SPURIOUS.
C
C     INVERR = USES INVERSE ITERATION ON T-MATRICES TO COMPUTE ERROR
C              ESTIMATES ON COMPUTED SINGULAR VALUES. (USES GENRAN)
C
C     LUMP = 'COMBINES' EIGENVALUES OF T-MATRIX USING THE RELATIVE
C             TOLERANCE RELTOL.
C
C     ISOEV = CALCULATES GAPS BETWEEN DISTINCT EIGENVALUES OF T-MATRIX
C             AND THEN USES THESE GAPS TO LABEL THOSE 'GOOD'
C             T-EIGENVALUES FOR WHICH ERROR ESTIMATES ARE NOT COMPUTED.
C
C     TNORM = COMPUTES THE SCALE TKMAX USED IN DETERMINING THE
C             TOLERANCES FOR THE SPURIOUS, T-MULTIPLICITY AND PRTESTS.
C             IT ALSO CHECKS FOR LOCAL ORTHOGONALITY OF THE LANCZOS
C             VECTORS BY TESTING THE RELATIVE SIZE OF THE BETAS USING
C             THE RELATIVE TOLERANCE BTOL.
C
C     PRTEST = LOOKS FOR 'GOOD' T-EIGENVALUES THAT HAVE BEEN MISLABELLED
C              BY THE SPURIOUS TEST BECAUSE THEY HAD 'TOO SMALL' A
C              PROJECTION ON THE STARTING LANCZOS VECTOR.
C              (LESS THAN SINGLE PRECISION)
C              TESTS INDICATE THAT SUCH T-EIGENVALUES ARE RARE.
C              PRTEST SHOULD BE CALLED ONLY AFTER CONVERGENCE
C              HAS BEEN ESTABLISHED.
C
C     INVERM = USED TO COMPUTE ERROR ESTIMATES FOR ANY T-EIGENVALUES
C              WHICH PRTEST INDICATES MAY HAVE BEEN MISLABELLED.
C              SUCH T-EIGENVALUES ARE RELABELLED ONLY IF THEIR ERROR
C              ESTIMATES ARE SUFFICIENTLY SMALL.  PRIMARY USE OF
C              INVERM IS IN THE CORRESPONDING SINGULAR VECTOR PROGRAM.
C
C     SAMPLE USPEC, SVMAT AND STRAN SUBROUTINES ARE INCLUDED.
C
C     ALSO INCLUDED IS A STAND-ALONE PROGRAM, LSCOMPAC, THAT
C     TRANSLATES A MATRIX GIVEN IN THE I,J, A(I,J) FORMAT INTO
C     THE PARTICULAR SPARSE MATRIX FORMAT USED IN THE SAMPLE USPEC,
C     SVMAT AND STRAN SUBROUTINES PROVIDED.
C
C
C-----OTHER PROGRAMS PROVIDED--------------------------------------------
C
C
C     LSCOMPAC = STAND-ALONE PROGRAM THAT TRANSLATES A SPARSE
C                RECTANGULAR M X N MATRIX A, GIVEN AS I, J, A(I,J),
C                INTO THE SPARSE MATRIX FORMAT REQUIRED BY THE SAMPLE
C                USPEC, STRAN AND SVMAT SUBROUTINES PROVIDED FOR USE
C                IN THE SINGULAR VALUE/VECTOR PROGRAMS.
C                THIS PROGRAM ASSUMES THAT THE MATRIX ENTRIES ARE
C                GIVEN EITHER COLUMN BY COLUMN OR ROW BY ROW.  IT
C                CANNOT HANDLE ANY OTHER ORDERINGS.  IN FACT IF
C                THE ENTRIES ARE GIVEN ROW BY ROW, THE DATA SET
C                CREATED ON FILE 8 CORRESPONDS TO A-TRANSPOSE AND
C                NOT TO A.  THUS, IN THIS SITUATION, IN ANY
C                SUBSEQUENT USE OF THE LANCZOS SINGULAR VALUE/VECTOR
C                PROGRAMS THE USER WILL HAVE TO INTERCHANGE THE
```

```
C              ROLES OF M AND OF N.
C
C
C-----COMMENTS ON THE STORAGE REQUIRED FOR SINGULAR VALUE PROGRAMS------
C
C
C     THE ARRAYS IN THE REAL SINGULAR VALUE PROGRAM REQUIRE
C     APPROXIMATELY THE EQUIVALENT OF ONE REAL*8 ARRAY OF DIMENSION
C
C        2.5*KMAX + MAX(KMAX,M) + MAX(KMAX,N) + .5* MAX(2*KMAX,M,N)
C
C     PLUS WHATEVER IS NEEDED TO GENERATE A*X FOR THE GIVEN MATRIX A.
C     THE ARRAYS BETA, VS AND MP CONSUME 2.5*KMAX*8 BYTES.
C     THE ARRAY V1 CONSUMES MAXIMUM(KMAX+1,M)*8 BYTES,  THE
C     ARRAY V2 CONSUMES MAXIMUM(KMAX,N)*8 BYTES, WITH THE
C     QUALIFICATION STATED ABOVE WHERE V2 IS DEFINED.  THE G-ARRAY
C     CONSUMES .5*MAX(2*KMAX,M,N)*8 BYTES.
C
C
C-----------------------------------------------------------------------
C
C     COMMENTS FOR SINGULAR VECTOR COMPUTATIONS
C
C-----------------------------------------------------------------------
C
C
C     THE SINGULAR VALUES WHOSE SINGULAR VECTORS ARE TO BE COMPUTED
C     MUST HAVE BEEN COMPUTED USING THE CORRESPONDING LANCZOS
C     SINGULAR VALUE PROGRAMS FOR REAL RECTANGULAR MATRICES BECAUSE
C     THESE SINGULAR VECTOR PROGRAMS USE THE SAME FAMILY OF LANCZOS
C     TRIDIAGONAL MATRICES THAT WAS USED IN THE CORRESPONDING
C     SINGULAR VALUE COMPUTATIONS.
C
C     THESE PROGRAMS ASSUME THAT THE SINGULAR VALUES SUPPLIED TO IT
C     HAVE BEEN COMPUTED ACCURATELY, AS MEASURED BY THE
C     ERROR ESTIMATES COMPUTED IN THE CORRESPONDING LANCZOS
C     SINGULAR VALUE COMPUTATIONS, ALTHOUGH THESE ESTIMATES
C     ARE TYPICALLY CONSERVATIVE.   THE SINGULAR VALUES SUPPLIED
C     ARE STORED IN THE ARRAY GOODSV(J), J=1,NGOOD.
C
C     FOR EACH GOODSV(J), THE SUBROUTINE STURMI COMPUTES THE
C     SMALLEST SIZE LANCZOS TRIDIAGONAL MATRIX, T(1,M1(J)), FOR
C     WHICH  GOODSV(J) IS A T-EIGENVALUE TO WITHIN A SPECIFIED
C     TOLERANCE.  IT ALSO ATTEMPTS TO COMPUTE THE SIZE, M2(J),
C     BY WHICH THE GIVEN SINGULAR VALUE BECOMES A DOUBLE
C     T-EIGENVALUE TO WITHIN THE GIVEN TOLERANCE.  THESE SIZES ARE
C     USED TO DETERMINE 1ST GUESSES AT SIZES FOR THE T-EIGENVECTORS
C     THAT WILL BE USED IN THE SINGULAR VECTOR COMPUTATIONS.
C     SUBROUTINE INVERM SUCCESSIVELY COMPUTES CORRESPONDING
C     T-EIGENVECTORS OF ENLARGED T-MATRICES UNTIL A SUITABLE
C     SIZE T-MATRIX IS DETERMINED FOR EACH J.  UP TO 10 SUCH
C     T-EIGENVECTOR COMPUTATIONS ARE ALLOWED FOR EACH SINGULAR
C     VALUE SUPPLIED.
C
C     AFTER APPROPRIATE T-EIGENVECTORS HAVE BEEN COMPUTED,
C     RITZ VECTORS FOR THE MATRIX B CORRESPONDING TO THESE
C     T-EIGENVECTORS ARE THEN COMPUTED.  SECTIONS OF THESE
C     RITZ VECTORS ARE THEN TAKEN AS APPROXIMATE LEFT AND
C     RIGHT SINGULAR VECTORS CORRESPONDING TO THE GIVEN
C     SINGULAR VALUES GOODSV(J),  J = 1,...,NGOOD.
C
C     THIS IMPLEMENTATION FIRST COMPUTES ALL OF THE RELEVANT
C     T-EIGENVECTORS OF THE SYMMETRIC TRIDIAGONAL MATRICES
C     IN THE VECTOR, TVEC.
C
C     THEN, AS EACH OF THE LANCZOS VECTORS IS REGENERATED, ALL
C     OF THE B-MATRIX RITZ VECTORS CORRESPONDING TO THESE
```

```
C     T-EIGENVECTORS ARE UPDATED USING THE CURRENTLY-GENERATED
C     LANCZOS VECTOR.  LANCZOS VECTORS ARE GENERATED (NOTE
C     THAT THEY ARE NOT BEING KEPT), UNTIL ENOUGH HAVE
C     BEEN GENERATED TO MAP THE LONGEST T-EIGENVECTOR INTO ITS
C     CORRESPONDING B-MATRIX  RITZ VECTOR.  THE ARRAY RITVEC
C     CONTAINS THE SUCCESSIVE RITZ VECTORS WHICH ARE THEN
C     SPLIT INTO APPROXIMATIONS TO THE LEFT AND RIGHT SINGULAR
C     VECTORS OF THE USER-SUPPLIED MATRIX A.
C
C
C-----PARAMETER CONTROLS FOR SINGULAR VECTOR PROGRAMS-------------------
C
C
C     PARAMETER CONTROLS ARE INTRODUCED TO ALLOW SEGMENTATION OF THE
C     SINGULAR VECTOR COMPUTATIONS AND TO ALLOW VARIOUS COMBINATIONS
C     OF READ/WRITES.
C
C     THE FLAG MBOUND ALLOWS THE USER TO DETERMINE A FIRST GUESS ON THE
C     STORAGE THAT WILL BE REQUIRED BY THE T-EIGENVECTORS FOR THE
C     SINGULAR VALUES WHOSE SINGULAR VECTORS ARE TO BE COMPUTED.
C     THIS CAN BE USED TO ESTIMATE THE REQUIRED SIZE OF THE TVEC ARRAY.
C
C     MBOUND = (0,1) MEANS
C
C               (0)  PROGRAM COMPUTES FIRST GUESSES AT THE SIZES
C                    OF THE T-MATRICES REQUIRED BY EACH OF THE
C                    SINGULAR VALUES SUPPLIED AND THEN CONTINUES
C                    WITH THE CORRESPONDING T-EIGENVECTOR
C                    COMPUTATIONS.
C
C               (1)  PROGRAM COMPUTES FIRST GUESSES AT THE SIZES
C                    OF THE T-MATRICES REQUIRED BY EACH OF THE
C                    SINGULAR VALUES SUPPLIED, STORES THESE IN FILE
C                    10 AND THEN TERMINATES.  THE USER CAN USE THESE
C                    SIZES TO ESTIMATE THE SIZE TVEC ARRAY NEEDED
C                    FOR THE DESIRED T-EIGENVECTOR COMPUTATIONS.
C
C     THE FLAGS NTVCON, TVSTOP, LVCONT, AND ERCONT CONTROL THE STOPPING
C     CRITERIA FOR INTERMEDIATE POINTS IN THE LANCZOS PROCEDURE.  THEY
C     TERMINATE THE PROCEDURE IF VARIOUS SPECIFIED QUANTITIES COULD
C     NOT BE COMPUTED AS DESIRED.
C
C     NTVCON = (0,1) MEANS
C
C                (0)  IF THE ESTIMATED STORAGE FOR THE T-EIGENVECTORS
C                     EXCEEDS THE USER-SPECIFIED DIMENSION OF THE
C                     TVEC ARRAY PROGRAM DOES NOT CONTINUE WITH THE
C                     T-EIGENVECTOR COMPUTATIONS. TERMINATION OCCURS.
C
C                (1)  CONTINUE WITH THE T-EIGENVECTOR COMPUTATIONS
C                     EVEN IF THE ESTIMATED STORAGE FOR TVEC EXCEEDS
C                     THE USER-SPECIFIED DIMENSION OF THE TVEC ARRAY.
C                     IN THIS SITUATION THE PROGRAM COMPUTES AS MANY
C                     T-EIGENVECTORS AS IT HAS ROOM FOR, IN THE SAME
C                     ORDER IN WHICH THE SINGULAR VALUES ARE SUPPLIED.
C
C      SVTVEC = (0,1) MEANS
C
C                (0)  DO NOT STORE THE COMPUTED T-EIGENVECTORS ON
C                     FILE 11 UNLESS ALSO HAVE THE FLAG TVSTOP = 1,
C                     IN WHICH CASE THE T-EIGENVECTORS ARE ALWAYS
C                     WRITTEN TO FILE 11.
C
C                (1)  STORE THE COMPUTED T-EIGENVECTORS ON FILE 11.
C
C      TVSTOP = (0,1) MEANS
C
```

```
C              (0)  ATTEMPT TO CONTINUE ON TO THE COMPUTATION
C                   OF THE B-MATRIX RITZVECTORS AFTER COMPLETING THE
C                   COMPUTATION OF THE T-EIGENVECTORS.
C
C              (1)  TERMINATE AFTER COMPUTING THE
C                   T-EIGENVECTORS AND STORING THEM ON FILE 11.
C
C       LVCONT = (0,1) MEANS
C
C              (0)  IF SOME OF THE T-EIGENVECTORS THAT WERE
C                   REQUESTED WERE NOT COMPUTED, EXIT
C                   FROM THE PROGRAM WITHOUT COMPUTING THE
C                   CORRESPONDING RITZ VECTORS.
C
C              (1)  CONTINUE ON TO THE RITZ VECTOR COMPUTATIONS
C                   EVEN IF NOT ALL OF THE T-EIGENVECTORS THAT
C                   WERE REQUESTED WERE COMPUTED.
C
C       ERCONT = (0,1) MEANS
C
C              (0)  PROGRAM WILL NOT COMPUTE THE RITZ
C                   VECTOR FOR ANY SINGULAR VALUE FOR WHICH NO
C                   T-EIGENVECTOR WHICH SATISFIES THE ERROR
C                   ESTIMATE TEST (ERTOL) HAS BEEN IDENTIFIED.
C
C              (1)  A RITZ VECTOR WILL BE COMPUTED FOR EVERY
C                   SINGULAR VALUE FOR WHICH A T-EIGENVECTOR HAS BEEN
C                   COMPUTED REGARDLESS OF WHETHER OR NOT THAT
C                   T-EIGENVECTOR SATISFIES THE ERROR ESTIMATE TEST.
C
C
C-----INPUT/OUTPUT FILES FOR THE SINGULAR VECTOR COMPUTATIONS-----------
C
C
C      ANY INPUT DATA OTHER THAN THE T-MATRIX HISTORY FILE AND THE
C      PREVIOUSLY COMPUTED SINGULAR VALUES AND ERROR ESTIMATES
C      SHOULD BE STORED ON FILE 5 IN FREE FORMAT.  SEE SAMPLE
C      INPUT/OUTPUT FOR TYPICAL INPUT FILE.
C
C      FILE 6 WAS USED AS THE INTERACTIVE TERMINAL OUTPUT FILE.
C      THIS FILE PROVIDES A RUNNING ACCOUNT OF THE PROGRESS OF THE
C      COMPUTATIONS.  ADDITIONAL PRINTOUT IS GENERATED WHEN
C      THE FLAG IWRITE = 1.
C
C
C DESCRIPTION OF OTHER I/O FILES
C
C FILE (K)      CONTAINS:
C
C       (2)     INPUT FILE:
C               PREVIOUSLY-GENERATED T-MATRICES (BETA ARRAY)
C               AND THE FINAL TWO LANCZOS VECTORS USED ON THAT
C               COMPUTATION.  THIS PROGRAM ALLOWS ENLARGEMENT
C               OF ANY T-MATRICES PROVIDED ON FILE 2.
C
C       (3)     INPUT FILE:
C               THE SINGULAR VALUES FOR WHICH CORRESPONDING
C               SINGULAR VECTORS ARE REQUESTED.  FILE 3 ALSO
C               CONTAINS THE T-MULTIPLICITIES OF THESE SINGULAR
C               VALUES (AS T-EIGENVALUES) AND THEIR COMPUTED GAPS
C               BOTH THE T-MATRICES AND IN THE USER-SUPPLIED MATRIX.
C               THIS FILE IS CREATED IN THE LANCZOS SINGULAR
C               VALUE COMPUTATIONS.
C
C       (4)     INPUT FILE:
C               ERROR ESTIMATES FOR THE ISOLATED SINGULAR VALUES
C               OF FILE 3.  THIS FILE IS CREATED DURING THE LANCZOS
```

```
C               SINGULAR VALUE COMPUTATIONS.
C
C       (8)     INPUT FILE:
C               USPEC SUBROUTINE ASSUMES THAT THE USER-
C               SUPPLIED MATRIX IS ON FILE 8.
C
C       (9)     OUTPUT FILE:
C               ERROR ESTIMATES FOR THE COMPUTED RITZ VECTORS CONSIDERED
C               AS EIGENVECTORS OF THE B-MATRIX. THESE ESTIMATES
C               ARE OF THE FORM
C                     BERROR = || B*RITVEC - SVAL*RITVEC ||
C               WHERE B DENOTES THE M+N ORDER SYMMETRIC MATRIX
C               ASSOCIATED WITH THE USER-SUPPLIED MATRIX A, SVAL
C               DENOTES THE SINGULAR VALUE BEING CONSIDERED AND
C               RITVEC DENOTES THE ASSOCIATED COMPUTED RITZ VECTOR.
C
C       (10)    OUTPUT FILE:
C               GUESSES AT APPROPRIATE SIZE T-MATRICES FOR THE
C               T-EIGENVECTORS FOR EACH SUPPLIED SINGULAR VALUE
C               IN THE ARRAY GOODSV(J),  J = 1,...,NGOOD.
C
C       (11)    OUTPUT FILE:
C               COMPUTED T-EIGENVECTORS CORRESPONDING TO SINGULAR
C               VALUES IN THE GOODSV ARRAY.  NOTE THAT IT IS POSSIBLE
C               IN CERTAIN SITUATIONS THAT FOR SOME SINGULAR VALUES
C               SUPPLIED IN THE GOODSV ARRAY A T-EIGENVECTOR WILL
C               NOT BE COMPUTED.
C
C       (12)    OUTPUT FILE:
C               CONTAINS COMPUTED RITZ VECTORS CORRESPONDING TO
C               THE T-EIGENVECTORS ON FILE 11. NOTE THAT IN
C               SOME SITUATIONS THAT FOR SOME SINGULAR VALUES IN
C               THE GOODSV ARRAY FOR WHICH T-EIGENVECTORS HAVE
C               BEEN COMPUTED NO CORRESPONDING RITZ VECTOR WILL
C               HAVE BEEN COMPUTED.
C
C       (13)    OUTPUT FILE:
C               ADDITIONAL INFORMATION ABOUT THE BOUNDS AND ERROR
C               ESTIMATES OBTAINED.
C
C
C-----SEEDS FOR SINGULAR VECTOR PROGRAMS-------------------------------
C
C     SEEDS FOR RANDOM NUMBER GENERATOR GENRAN
C               (1) SVSEED = INTEGER*4 SCALAR USED IN THE SUBROUTINE
C                            GENRAN TO GENERATE THE STARTING VECTOR FOR
C                            THE REGENERATION OF THE LANCZOS VECTORS.
C
C               (2) RHSEED = INTEGER*4 SCALAR USED IN THE SUBROUTINE
C                            GENRAN TO GENERATE A RANDOM VECTOR FOR
C                            USE IN SUBROUTINE INVERM.
C
C     USER SHOULD NOTE THAT SVSEED MUST BE THE SAME SEED THAT
C     WAS USED TO GENERATE THE T-MATRICES THAT WERE USED TO
C     COMPUTE THE SINGULAR VALUES WHOSE SINGULAR VECTORS ARE TO BE
C     COMPUTED.  SVSEED IS READ IN FROM FILE 3.
C
C
C-----USER-SPECIFIED PARAMETERS FOR THE SINGULAR VECTOR PROGRAMS--------
C
C
C     NGOOD   = NUMBER OF SINGULAR VALUES READ INTO THE GOODSV ARRAY
C               READ FROM FILE 3.
C
C     M       = ROW ORDER OF THE USER-SUPPLIED MATRIX.
C
C     N       = COLUMN ORDER OF THE USER-SUPPLIED MATRIX.
```

```
C
C     MEV     = SIZE OF THE T-MATRIX THAT WAS USED TO COMPUTE
C               THE SINGULAR VALUES WHOSE SINGULAR VECTORS ARE
C               REQUESTED.  MEV IS READ IN FROM FILE 3.
C
C     KMAX    =  SIZE OF THE T-MATRIX PROVIDED ON FILE 2.
C
C     MDIMTV  = MAXIMUM CUMULATIVE SIZE OF THE TVEC ARRAY ALLOWED
C               FOR ALL OF THE T-EIGENVECTORS REQUIRED.  MDIMTV
C               MUST NOT EXCEED THE USER-SPECIFIED DIMENSION OF
C               THE TVEC ARRAY.  PROGRAM CAN BE RUN WITH THE FLAG
C               MBOUND = 1 TO DETERMINE AN EDUCATED GUESS ON AN
C               APPROPRIATE DIMENSION FOR THE TVEC ARRAY.
C
C     MDIMRV = MAXIMUM CUMULATIVE SIZE OF THE RITVEC ARRAY ALLOWED
C              FOR ALL OF THE RITZ VECTORS TO BE COMPUTED. MDIMRV
C              MUST NOT EXCEED THE USER-SPECIFIED DIMENSION OF
C              THE RITVEC ARRAY.  MUST BE SELECTED SO THAT
C              THERE IS ENOUGH ROOM FOR A RITZ VECTOR FOR EVERY
C              GOODEV(J) READ INTO PROGRAM.  (>= NGOOD*(M+N))
C
C
C-----ARRAYS REQUIRED BY THE SINGULAR VECTOR PROGRAMS-------------------
C
C
C     BETA(J) = REAL*8 ARRAY.  ITS DIMENSION MUST BE AT LEAST
C                KMAXN+1,  WHERE KMAXN IS THE LARGEST SIZE T-MATRIX
C                CONSIDERED BY THE PROGRAM.  NOTE THAT KMAXN IS THE
C                LARGER OF THE SIZE OF THE BETA HISTORY PROVIDED
C                ON FILE 2 (IF ANY ) AND THE SIZE WHICH THE PROGRAM
C                SPECIFIES INTERNALLY, THIS LATTER IS ALWAYS
C                < = 11*MEV / 8  +  12, WHERE MEV IS THE SIZE
C                T-MATRIX THAT WAS USED IN THE CORRESPONDING
C                SINGULAR VALUE COMPUTATIONS.  BETA CONTAINS THE
C                NONZERO ENTRIES OF THE LANCZOS T-MATRICES.
C                BETA IS NOT DESTROYED IN THE COMPUTATIONS.
C                THE DIAGONAL ENTRIES OF THE T-MATRICES ARE ALL ZERO.
C
C     RITVEC(J) = REAL*8 ARRAY. IT DIMENSION MUST BE > = NGOOD*(M+N)
C                 WHERE THE USER-SUPPLIED MATRIX IS MXN
C                 AND NGOOD IS THE NUMBER OF SINGULAR VALUES WHOSE
C                 SINGULAR VECTORS ARE TO BE COMPUTED.  IT CONTAINS
C                 THE COMPUTED APPROXIMATE SINGULAR VECTORS OF A.
C                 THESE COMPUTED RITZ VECTORS ARE STORED ON FILE 12.
C
C     TVEC(J)    = REAL*8 ARRAY.  ITS DIMENSION MUST BE AT LEAST
C                 MTOL = |MA(1)| + |MA(2)| + ... + |MA(NGOOD)|
C                 WHERE NGOOD IS THE NUMBER OF SINGULAR VALUES BEING
C                 CONSIDERED AND |MA(J)| IS THE SIZE OF THE
C                 T-MATRIX BEING USED FOR THE B-MATRIX RITZ VECTOR
C                 COMPUTATION FOR GOODSV(J).  THESE SIZES
C                 ARE COMPUTED BY THE PROGRAM.  AN ESTIMATE OF
C                 MTOL CAN BE OBTAINED BY SETTING MBOUND = 1,
C                 RUNNING THE PROGRAM, AND THEN MULTIPLYING THE
C                 RESULTING TOTAL T-SIZE SPECIFIED BY 5/4.  THE TVEC
C                 ARRAY CONTAINS THE COMPUTED T-EIGENVECTORS.  IF
C                 THE FLAG SVTVEC = 1 OR THE FLAG TVSTOP = 1, THEN
C                 THESE VECTORS ARE SAVED ON FILE 11.
C
C     V1(J)      = REAL*8 ARRAY.  ITS DIMENSION MUST BE GREATER
C                 THAN THE MAXIMUM OF KMAX AND M, WHERE M IS
C                 THE ROW ORDER OF THE GIVEN MATRIX.  V1 IS USED
C                 IN THE SUBROUTINE INVERM AND IN THE REGENERATION
C                 OF THE LANCZOS VECTORS.
C
C     V2(J)      = REAL*8 ARRAY.  ITS DIMENSION MUST BE GREATER
C                 THAN MAX(KMAX,N), WHERE N IS THE COLUMN ORDER OF
```

```
C                 THE GIVEN MATRIX.  IT IS USED IN THE REGENERATION
C                 OF THE LANCZOS VECTORS AND IN SUBROUTINE INVERM.
C
C       GOODSV(J),  = REAL*8 ARRAYS EACH OF DIMENSION AT LEAST NGOOD.
C       SVNEW(J)       CONTAIN THE SINGULAR VALUES FOR WHICH
C                      SINGULAR VECTORS ARE REQUESTED.  SINGULAR VALUES
C                      IN GOODSV ARE READ IN FROM FILE 3.
C
C       BMINGP(J), = REAL*4 ARRAYS OF DIMENSION AT LEAST NGOOD.
C       TMINGP(J)     CONTAIN, RESPECTIVELY, THE MINIMAL GAPS FOR
C                     CORRESPONDING SINGULAR VALUES IN GOODSV ARRAY IN
C                     B-MATRIX AND IN T-MATRIX.
C
C       TERR(J), ERR(J),       = REAL*4 ARRAYS (EXCEPT TLAST WHICH IS
C       ERRDGP(J), TLAST(J)      REAL*8).  EACH MUST BE OF DIMENSION
C       RNORM(J), TBETA(J)       AT LEAST NGOOD.  USED TO STORE QUANTITIES
C                                GENERATED DURING THE COMPUTATIONS FOR
C                                LATER PRINTOUT.
C
C       G(J)      = REAL*4 ARRAY WHOSE DIMENSION MUST BE AT LEAST
C                   MAX(KMAX,M,N).  USED IN SUBROUTINE GENRAN TO HOLD
C                   RANDOM NUMBERS NEEDED FOR THE LANCZOS VECTOR
C                   REGENERATION AND FOR THE INVERSE ITERATION
C                   COMPUTATIONS IN THE SUBROUTINE INVERM.
C
C       MP(J) = INTEGER*4 ARRAY WHOSE DIMENSION IS AT LEAST NGOOD.
C               INITIALLY CONTAINS THE T-MULTIPLICITY OF THE SINGULAR
C               VALUE GOODSV(J) AS AN EIGENVALUE OF THE T-MATRIX.
C               USED TO FLAG SINGULAR VALUES FOR WHICH NO T-EIGENVECTOR
C               OR NO RITZ VECTOR IS TO BE COMPUTED.
C
C       MA(J)   = INTEGER*4 ARRAYS EACH OF WHOSE DIMENSIONS
C                 IS AT LEAST NGOOD.  USED IN DETERMINING
C                 AN APPROPRIATE T-MATRIX FOR EACH SINGULAR VALUE
C                 IN GOODSV ARRAY.
C
C       MINT(J),MFIN(J) = INTEGER*4 ARRAYS WHOSE DIMENSIONS MUST BE AT
C                         LEAST NGOOD.  USED TO POINT TO THE BEGINNINGS
C                         AND THE ENDS OF THE COMPUTED EIGENVECTOR
C                         OF THE T-MATRIX, T(1,|MA(J)|).
C
C       IDELTA(J)   = INTEGER*4 ARRAY WHOSE DIMENSION MUST BE AT
C                     LEAST NGOOD.  CONTAINS INCREMENTS USED IN LOOPS
C                     ON APPROPRIATE SIZE T-MATRIX FOR THE T-EIGENVECTOR
C                     COMPUTATIONS.
C
C
C-----SUBROUTINES INCLUDED FOR THE SINGULAR VECTOR COMPUTATIONS---------
C
C
C       STURMI = FOR EACH GIVEN SINGULAR VALUE GOODSV(J) DETERMINES
C                THE SMALLEST SIZE T-MATRIX FOR WHICH GOODSV(J) IS
C                A T-EIGENVALUE (TO WITHIN A GIVEN TOLERANCE) AND IF
C                POSSIBLE THE SMALLEST SIZE T-MATRIX FOR WHICH
C                IT IS A DOUBLE T-EIGENVALUE (TO WITHIN THE SAME
C                TOLERANCE).  THE SIZE T-MATRIX USED IN THE
C                T-EIGENVECTOR COMPUTATIONS IS THEN DETERMINED BY
C                STARTING WITH AN INITIAL GUESS BASED ON THE
C                INFORMATION FROM STURMI, AND THEN LOOPING ON THE
C                SIZE OF THE T-EIGENVECTOR COMPUTATIONS.
C
C       LBISEC = RECOMPUTES THE VALUE OF THE GIVEN SINGULAR VALUE
C                AT THE T-SIZE SPECIFIED FOR THE T-EIGENVECTOR
C                COMPUTATION.  LBISEC IS A SIMPLIFICATION OF THE
C                BISEC SUBROUTINE USED IN THE LANCZOS SINGULAR
C                VALUE COMPUTATIONS.
C
```

```
C      INVERM = FOR THE T-SIZES CONSIDERED BY THE PROGRAM COMPUTES
C               THE CORRESPONDING EIGENVECTORS OF THESE T-MATRICES
C               CORRESPONDING TO THE USER-SUPPLIED SINGULAR VALUES
C               IN THE GOODSV ARRAY.
C
C      LANCZS AND TNORM  SUBROUTINES ARE ALSO USED HERE AS WELL AS
C      IN THE CORRESPONDING SINGULAR VALUE COMPUTATIONS.
C
C
C-----------------------------------------------------------------------
```

SECTION 6.3 MAIN PROGRAM, SINGULAR VALUE COMPUTATIONS

```
C-----LSVAL  (SINGULAR VALUES OF REAL, RECTANGULAR MATRICES-------------
C
C     CONTAINS MAIN PROGRAM FOR COMPUTING DISTINCT SINGULAR VALUES OF
C     A REAL M X N MATRIX USING LANCZOS TRIDIAGONALIZATION WITHOUT
C     REORTHOGONALIZATION AND WITH SPECIAL STARTING VECTORS.
C
C     FOR A GIVEN REAL MATRIX A OF ORDER M X N THE LANCZOS RECURSION
C     IS APPLIED TO THE ASSOCIATED REAL SYMMETRIC MATRIX B OF ORDER
C     MN = M + N
C
C                                ----                  ----
C                               | 0                    A  |
C                  B  =         |                         |
C                               |  A-TRANSPOSE         0  |
C                                ----                  ----
C
C     USING SPECIAL STARTING VECTORS.  PLEASE NOTE: ONLY EVEN ORDER
C     LANCZOS TRIDIAGONAL MATRICES AND ONLY NONNEGATIVE SUBINTERVALS
C     ARE PERMISSIBLE.
C
C     PFORT VERIFIER IDENTIFIED THE FOLLOWING NONPORTABLE
C     CONSTRUCTIONS
C
C     1.  DATA/MACHEP/ STATEMENT
C     2.  ALL READ(5,*) STATEMENTS (FREE FORMAT)
C     3.  FORMAT(20A4) USED WITH EXPLANATORY HEADER EXPLAN.
C     4.  HEXADECIMAL FORMAT (4Z20) USED IN BETA FILES.
C
C-----------------------------------------------------------------------
      DOUBLE PRECISION  BETA(5001),V1(5000),V2(5000),VS(5000)
      DOUBLE PRECISION  LB(20),UB(20)
      DOUBLE PRECISION  BTOL,GAPTOL,TTOL,MACHEP,EPSM,RELTOL
      DOUBLE PRECISION  SCALE1,SCALE2,SCALE3,SCALE4,BISTOL,CONTOL,MULTOL
      DOUBLE PRECISION  ONE,ZERO,TEMP,TKMAX,BETAM,BKMIN,T0,T1
      REAL  G(5000),EXPLAN(20)
      INTEGER  MP(5000),NMEV(20)
      INTEGER  SVSEED,RHSEED,SVSOLD
      INTEGER IABS
      REAL  ABS
      DOUBLE PRECISION  DABS, DSQRT, DFLOAT
      EXTERNAL SVMAT,STRAN
C-----------------------------------------------------------------------
      DATA MACHEP/Z3410000000000000/
      EPSM = 2.0D0*MACHEP
C-----------------------------------------------------------------------
C
C     ARRAYS MUST BE DIMENSIONED AS FOLLOWS:
C     1.  BETA: >= (KMAX+1) WHERE KMAX IS READ IN AND IS
C         THE SIZE OF THE LARGEST T-MATRIX THAT CAN BE CONSIDERED.
C     2.  V1:  >= MAX(M,KMAX+1)
C     3.  V2:  >= MAX(N,KMAX)
C     4.  VS:  >= KMAX
C     5.  G:  >= MAX(2*KMAX,M,N)
C     6.  MP:  >= KMAX
C     7.  LB,UB:  >= NUMBER OF SUBINTERVALS SUPPLIED TO BISEC.
C     8.  NMEV:  >= NUMBER OF T-MATRICES ALLOWED.
C     9.  EXPLAN:  DIMENSION IS 20.
C
C
C     IMPORTANT TOLERANCES OR SCALES THAT ARE USED REPEATEDLY
C     THROUGHOUT THIS PROGRAM ARE THE FOLLOWING:
C     SCALED MACHINE EPSILON:  TTOL = TKMAX*EPSM WHERE
C     EPSM = 2*MACHINE EPSILON AND
C     TKMAX = MAX(BETA(J), J = 1,MEV)
```

```
C     BISEC CONVERGENCE TOLERANCE:  BISTOL = DSQRT(1000+MEV)*TTOL       LSV00650
C     BISEC MULTIPLICITY TOLERANCE:  MULTOL = (1000+MEV)*TTOL           LSV00660
C     LANCZOS CONVERGENCE TOLERANCE:   CONTOL = BETA(MEV+1)*1.D-10      LSV00670
C-----------------------------------------------------------------------LSV00680
C     OUTPUT HEADER                                                     LSV00690
      WRITE(6,10)                                                       LSV00700
   10 FORMAT(/' LANCZOS PROCEDURE FOR REAL, RECTANGULAR MATRICES'/)     LSV00710
C                                                                       LSV00720
C     SET PROGRAM PARAMETERS                                            LSV00730
C     SCALEK ARE USED IN TOLERANCES NEEDED IN SUBROUTINES LUMP,         LSV00740
C     ISOEV AND PRTEST.  USER MUST NOT MODIFY THESE SCALES.             LSV00750
      SCALE1 = 5.0D2                                                    LSV00760
      SCALE2 = 5.0D0                                                    LSV00770
      SCALE3 = 5.0D0                                                    LSV00780
      SCALE4 = 1.0D4                                                    LSV00790
      ONE  = 1.0D0                                                      LSV00800
      ZERO = 0.0D0                                                      LSV00810
      BTOL = 1.0D-8                                                     LSV00820
C     BTOL = EPSM                                                       LSV00830
      GAPTOL = 1.0D-8                                                   LSV00840
      ICONV = 0                                                         LSV00850
      MOLD = 0                                                          LSV00860
      MOLD1 = 1                                                         LSV00870
      ICT = 0                                                           LSV00880
      MMB = 0                                                           LSV00890
      IPROJ = 0                                                         LSV00900
C                                                                       LSV00910
C     READ USER-SPECIFIED PARAMETERS FROM INPUT FILE 5 (FREE FORMAT)    LSV00920
C                                                                       LSV00930
C     READ USER-PROVIDED HEADERS FOR RUN                                LSV00940
      READ(5,20) EXPLAN                                                 LSV00950
      WRITE(6,20) EXPLAN                                                LSV00960
      READ(5,20) EXPLAN                                                 LSV00970
      WRITE(6,20) EXPLAN                                                LSV00980
   20 FORMAT(20A4)                                                      LSV00990
C                                                                       LSV01000
C     READ THE ROW ORDER M OF THE MATRIX AND THE COLUMN ORDER N.         LSV01010
C     READ THE MAXIMUM ORDER OF THE T-MATRICES ALLOWED (KMAX),          LSV01020
C     THE NUMBER OF T-MATRICES ALLOWED (NMEVS), AND A                   LSV01030
C     MATRIX IDENTIFICATION NUMBER (MATNO).                             LSV01040
      READ(5,20) EXPLAN                                                 LSV01050
      READ(5,*) M,N,KMAX,NMEVS,MATNO                                    LSV01060
      NM = M + N                                                        LSV01070
C                                                                       LSV01080
C     READ SEEDS FOR LANCZS AND INVERR SUBROUTINES (SVSEED AND RHSEED)  LSV01090
C     READ MAXIMUM NUMBER OF ITERATIONS ALLOWED FOR EACH INVERSE        LSV01100
C     ITERATION (MXINIT) AND MAXIMUM NUMBER OF STURM SEQUENCES          LSV01110
C     ALLOWED (MXSTUR)                                                  LSV01120
      READ(5,20) EXPLAN                                                 LSV01130
      READ(5,*) SVSEED,RHSEED,MXINIT,MXSTUR                             LSV01140
C                                                                       LSV01150
C     ISTART = (0,1):  ISTART = 0 MEANS BETA FILE IS NOT                LSV01160
C     AVAILABLE.  ISTART = 1 MEANS BETA FILE IS AVAILABLE ON            LSV01170
C     FILE 2.                                                           LSV01180
C     ISTOP = (0,1):  ISTOP = 0 MEANS PROCEDURE GENERATES BETA          LSV01190
C     FILE AND THEN TERMINATES.  ISTOP = 1 MEANS PROCEDURE GENERATES    LSV01200
C     BETAS IF NEEDED AND THEN COMPUTES SINGULAR VALUES AND             LSV01210
C     ERROR ESTIMATES AND THEN TERMINATES.                              LSV01220
      READ(5,20) EXPLAN                                                 LSV01230
      READ(5,*) ISTART,ISTOP                                            LSV01240
C                                                                       LSV01250
C     IHIS = (0,1):  IHIS = 0 MEANS BETA FILE IS NOT WRITTEN            LSV01260
C     TO FILE 1.  IHIS = 1 MEANS BETA FILE IS WRITTEN TO FILE 1.        LSV01270
C     IDIST = (0,1):  IDIST = 0 MEANS DISTINCT T-EIGENVALUES            LSV01280
C     ARE NOT WRITTEN TO FILE 11.  IDIST = 1 MEANS DISTINCT             LSV01290
C     T-EIGENVALUES ARE WRITTEN TO FILE 11.                             LSV01300
C     IWRITE = (0,1):  IWRITE = 0 MEANS NO INTERMEDIATE OUTPUT          LSV01310
```

```
C     FROM THE COMPUTATIONS IS WRITTEN TO FILE 6.  IWRITE = 1 MEANS
C     T-EIGENVALUES AND ERROR ESTIMATES ARE WRITTEN TO FILE 6
C     AS THEY ARE COMPUTED.  SPECIFY THE PARITY (IPAR) OF THE
C     LANCZOS STARTING VECTOR.  IF M > N, THEN IPAR = 1,
C     IF M < N, THEN IPAR = 2.
      READ(5,20) EXPLAN
      READ(5,*) IHIS,IDIST,IWRITE,IPAR
      IF(M.GT.N) IPAR = 1
      IF(M.LT.N) IPAR = 2
      IPARO = IPAR
C
C     READ IN THE RELATIVE TOLERANCE (RELTOL) FOR USE IN THE
C     SPURIOUS, T-MULTIPLICITY, AND PRTEST TESTS.
      READ(5,20) EXPLAN
      READ(5,*) RELTOL
C
C     READ IN THE SIZES OF THE T-MATRICES TO BE CONSIDERED.
C     NOTE THAT ONLY EVEN ORDER T-SIZES ARE PERMISSIBLE.
      READ(5,20) EXPLAN
      READ(5,*) (NMEV(J), J=1,NMEVS)
C
C     CHECK TO SEE THAT ALL T-SIZES PROVIDED ARE EVEN ORDERED.
C     TERMINATE IF THAT IS NOT THE CASE.
      DO 30 I = 1,NMEVS
      NMEV2 = NMEV(I)/2
      IF(2*NMEV2.NE.NMEV(I)) GO TO 670
   30 CONTINUE
C
C     READ IN THE NUMBER OF SUBINTERVALS TO BE CONSIDERED.
      READ(5,20) EXPLAN
      READ(5,*) NINT
C
C     READ IN THE LEFT-END POINTS OF THE SUBINTERVALS TO BE CONSIDERED.
C     THESE MUST BE IN ALGEBRAICALLY INCREASING ORDER
      READ(5,20) EXPLAN
      READ(5,*) (LB(J), J=1,NINT)
C
C     READ IN THE RIGHT-END POINTS OF THE SUBINTERVALS TO BE CONSIDERED.
C     THESE MUST BE IN ALGEBRAICALLY INCREASING ORDER
      READ(5,20) EXPLAN
      READ(5,*) (UB(J), J=1,NINT)
C
C-----------------------------------------------------------------------
C     INITIALIZE THE ARRAYS FOR THE USER-SPECIFIED MATRIX
C     AND PASS THE STORAGE LOCATIONS OF THESE ARRAYS TO THE
C     MATRIX-VECTOR MULTIPLY SUBROUTINES SVMAT AND STRAN.
C
      CALL USPEC(M,N,MATNO)
C
C-----------------------------------------------------------------------
C     MASK UNDERFLOW AND OVERFLOW
C
      CALL MASK
C
C-----------------------------------------------------------------------
C
C     WRITE TO FILE 6, A SUMMARY OF THE PARAMETERS FOR THIS RUN
C
      WRITE(6,40) MATNO,M,N,KMAX
   40 FORMAT(/3X,'MATRIX ID',5X,'M',5X,'N',4X,'MAX ORDER OF T'/
     1 I12,2I6,I18/)
C
      WRITE(6,50) ISTART,ISTOP
   50 FORMAT(/2X,'ISTART',3X,'ISTOP'/2I8/)
C
      WRITE(6,60) IHIS,IDIST,IWRITE,IPAR
```

```
   60 FORMAT(/4X,'IHIS',3X,'IDIST',2X,'IWRITE',4X,'IPAR'/4I8/)
C
      WRITE(6,70) SVSEED,RHSEED
   70 FORMAT(/' SEEDS FOR RANDOM NUMBER GENERATOR'//
     1 4X,'LANCZS SEED',4X,'INVERR SEED'/2I15/)
C
      WRITE(6,80) (NMEV(J), J=1,NMEVS)
   80 FORMAT(/' SIZES OF T-MATRICES TO BE CONSIDERED'/(6I12))
C
      WRITE(6,90) RELTOL,GAPTOL,BTOL
   90 FORMAT(/' RELATIVE TOLERANCE USED TO COMBINE COMPUTED T-EIGENVALUE
     1S'/E15.3/' RELATIVE GAP TOLERANCES USED IN INVERSE ITERATION'/
     1E15.3/' RELATIVE TOLERANCE FOR CHECK ON SIZE OF BETAS'/E15.3/)
C
      WRITE(6,100) (J,LB(J),UB(J), J=1,NINT)
  100 FORMAT(/' BISEC WILL BE USED ON THE FOLLOWING INTERVALS'/
     1 (I6,2E20.6)/)
C
      IF (ISTART.EQ.0.AND.IPAR.EQ.1) WRITE(6,110)
      IF (ISTART.EQ.0.AND.IPAR.EQ.2) WRITE(6,120)
  110 FORMAT(/' STARTING VECTOR IS OF FORM (0,V2)'/)
  120 FORMAT(/' STARTING VECTOR IS OF FORM (V1,0)'/)
C
      IF (ISTART.EQ.0) GO TO 170
C
C     READ IN BETA HISTORY FROM FILE 2
C
      READ(2,130)MOLD,MO,NO,IPARO,IPAR,SVSOLD,MATOLD
  130 FORMAT(3I6,2I3,I12,I8)
C
      IF (KMAX.LT.MOLD) KMAX = MOLD
      KMAX1 = KMAX + 1
C
C     CHECK THAT M, N, MATRIX ID MATNO, AND RANDOM SEED SVSEED
C     AGREE WITH THOSE IN THE HISTORY FILE.  IF NOT PROCEDURE STOPS.
C
      ITEMP = (MO-M)**2+(NO-N)**2+(MATNO-MATOLD)**2+(SVSEED-SVSOLD)**2
C
      IF (ITEMP.EQ.0) GO TO 150
C
      WRITE(6,140)
  140 FORMAT(' PROGRAM TERMINATES'/    ' READ FROM FILE 2 CORRESPONDS TO
     1 DIFFERENT MATRIX THAN MATRIX SPECIFIED'/)
      GO TO 690
C
  150 CONTINUE
      MOLD1 = MOLD+1
C
      READ(2,160)(BETA(J), J=1,MOLD1)
  160 FORMAT(4Z20)
C
      IF (KMAX.EQ.MOLD) GO TO 190
C
      READ(2,160)(V1(J), J=1,M)
      READ(2,160)(V2(J), J=1,N)
C
  170 CONTINUE
      IIX = SVSEED
C
C-----------------------------------------------------------------------
C
      CALL LANCZS(SVMAT,STRAN,BETA,V1,V2,G,KMAX,MOLD1,M,N,IPAR,IIX)
C
C-----------------------------------------------------------------------
C
      KMAX1 = KMAX + 1
C
```

```
      IF (IHIS.EQ.0.AND.ISTOP.GT.0) GO TO 190
C
      WRITE(1,180) KMAX,M,N,IPARO,IPAR,SVSEED,MATNO
  180 FORMAT(3I6,2I3,I12,I8,' = KMAX,M,N,IPARO,IPAR,SVSEED,MATNO')
C
      WRITE(1,160)(BETA(I), I=1,KMAX1)
C
      WRITE(1,160)(V1(I), I=1,M)
      WRITE(1,160)(V2(I), I=1,N)
C
      IF (ISTOP.EQ.0) GO TO 570
C
  190 CONTINUE
      BKMIN = BTOL
      WRITE(6,200)
  200 FORMAT(/' T-MATRICES (BETA) ARE NOW AVAILABLE'/)
C
C-----------------------------------------------------------------------
C     SUBROUTINE TNORM CHECKS MIN(BETA)/(ESTIMATED NORM(A)) > BTOL .
C     IF THIS IS VIOLATED IB IS SET EQUAL TO THE NEGATIVE OF THE INDEX
C     OF THE MINIMAL BETA.  IF(IB < 0) THEN SUBROUTINE TNORM IS
C     CALLED FOR EACH VALUE OF MEV TO DETERMINE WHETHER OR NOT THERE
C     IS A BETA IN THE T-MATRIX SPECIFIED THAT VIOLATES THIS TEST.
C     IF THERE IS SUCH A BETA THE PROGRAM TERMINATES FOR THE USER
C     TO DECIDE WHAT TO DO.  THIS TEST CAN BE OVER-RIDDEN BY
C     SIMPLY MAKING BTOL SMALLER, BUT THEN THERE IS THE POSSIBILITY
C     THAT LOSSES IN THE LOCAL ORTHOGONALITY MAY HURT THE COMPUTATIONS.
C     BTOL = 1.D-8 IS HOWEVER A CONSERVATIVE CHOICE FOR BTOL.
C
C     TNORM ALSO COMPUTES TKMAX = MAX(BETA(K), K=1,KMAX).
C     TKMAX IS USED TO SCALE THE TOLERANCES USED IN THE
C     T-MULTIPLICITY AND SPURIOUS TESTS IN BISEC. TKMAX IS ALSO USED IN
C     THE PROJECTION TEST FOR HIDDEN T-EIGENVALUES THAT HAD 'TOO SMALL'
C     A PROJECTION ON THE STARTING VECTOR.
C
      CALL TNORM(BETA,BKMIN,TKMAX,KMAX,IB)
C
C-----------------------------------------------------------------------
C
      TTOL = EPSM*TKMAX
C
C     LOOP ON THE SIZE OF THE T-MATRIX
C
  210 CONTINUE
      MMB = MMB + 1
C     NOTE THAT ONLY EVEN ORDER T-SIZES ARE PERMISSIBLE.
      MEV = NMEV(MMB)
C     IS MEV TOO LARGE ?
      IF(MEV.LE.KMAX) GO TO 230
      WRITE(6,220) MMB, MEV, KMAX
  220 FORMAT(/' TERMINATE PRIOR TO CONSIDERING THE',I6,'TH T-MATRIX'/
     1' BECAUSE THE SIZE REQUESTED',I6,' IS GREATER THAN THE MAXIMUM SIZ
     1E ALLOWED',I6/)
      GO TO 570
C
  230 MP1 = MEV + 1
      BETAM = BETA(MP1)
C
      IF (IB.GE.0) GO TO 240
C
      T0 = BTOL
C
C-----------------------------------------------------------------------
C
      CALL TNORM(BETA,T0,T1,MEV,IBMEV)
C
```

```
C-----------------------------------------------------------------------
C
      TEMP = T0/TKMAX
      IBMEV = IABS(IBMEV)
      IF (TEMP.GE.BTOL) GO TO 240
      IBMEV = -IBMEV
      GO TO 630
C
  240 CONTINUE
      IC = MXSTUR-ICT
C
C-----------------------------------------------------------------------
C     BISEC LOOP. THE SUBROUTINE BISEC INCORPORATES DIRECTLY THE
C     T-MULTIPLICITY AND SPURIOUS TESTS. T-EIGENVALUES WILL BE
C     CALCULATED BY BISEC SEQUENTIALLY ON INTERVALS
C     (LB(J),UB(J)), J = 1,NINT).
C
C     ON RETURN FROM BISEC
C     NDIS = NUMBER OF DISTINCT EIGENVALUES OF T(1,MEV) ON UNION
C            OF THE (LB,UB) INTERVALS
C     VS = DISTINCT T-EIGENVALUES IN ALGEBRAICALLY INCREASING ORDER
C     MP = T-MULTIPLICITIES OF THE T-EIGENVALUES STORED IN VS
C     MP(I) = (0,1,MI), MI>1, I=1,NDIS  MEANS:
C        (0)  VS(I) IS SPURIOUS
C        (1)  VS(I) IS T-SIMPLE AND GOOD
C        (MI) VS(I) IS T-MULTIPLE AND IS THEREFORE NOT ONLY GOOD BUT
C             ALSO A CONVERGED GOOD T-EIGENVALUE.
C
C
      CALL BISEC(BETA,V1,V2,VS,LB,UB,EPSM,TTOL,MP,NINT,
     1 MEV,NDIS,IC,IWRITE)
C
C-----------------------------------------------------------------------
C
      IF (NDIS.EQ.0) GO TO 650
C
C     COMPUTE THE TOTAL NUMBER OF STURM SEQUENCES USED TO DATE
C     COMPUTE THE BISEC CONVERGENCE AND T-MULTIPLICITY TOLERANCES USED.
C     COMPUTE THE CONVERGENCE TOLERANCE FOR T-EIGENVALUES.
      ICT = ICT + IC
      TEMP = DFLOAT(MEV+1000)
      MULTOL = TEMP*TTOL
      TEMP = DSQRT(TEMP)
      BISTOL = TTOL*TEMP
      CONTOL = BETAM*1.D-10
C
C-----------------------------------------------------------------------
C    SUBROUTINE LUMP 'COMBINES' T-EIGENVALUES THAT ARE 'TOO CLOSE'.
C    NOTE HOWEVER THAT CLOSE SPURIOUS T-EIGENVALUES ARE NOT AVERAGED
C    WITH GOOD ONES. HOWEVER, THEY MAY BE USED TO INCREASE THE
C    T-MULTIPLICITY OF A GOOD T-EIGENVALUE.
C
      LOOP = NDIS
      CALL LUMP(VS,RELTOL,MULTOL,SCALE2,MP,LOOP)
C
C-----------------------------------------------------------------------
C
      IF(NDIS.EQ.LOOP) GO TO 260
C
      WRITE(6,250) NDIS, MEV, LOOP
  250 FORMAT(/I6,' DISTINCT T-EIGENVALUES WERE COMPUTED IN BISEC AT MEV
     1=',I6/ 2X,' LUMP SUBROUTINE REDUCES NUMBER OF DISTINCT T-EIGENVALU
     1ES TO',I6)
C
  260 CONTINUE
      NDIS = LOOP
```

```
      BETA(MP1) = BETAM
C
C-----------------------------------------------------------------------
C     THE SUBROUTINE ISOEV LABELS THOSE SIMPLE T-EIGENVALUES OF T(1,MEV)
C     WITH VERY SMALL GAPS BETWEEN NEIGHBORING T-EIGENVALUES OF T(1,MEV)
C     TO AVOID COMPUTING ERROR ESTIMATES FOR ANY SIMPLE GOOD
C     T-EIGENVALUE THAT IS TOO CLOSE TO A SPURIOUS T-EIGENVALUE.
C     ON RETURN FROM ISOEV, G CONTAINS CODED MINIMAL GAPS
C     BETWEEN THE DISTINCT EIGENVALUES OF T(1,MEV). (G IS REAL).
C     G(I) < 0 MEANS MINGAP IS DUE TO LEFT GAP G(I) > 0 MEANS DUE TO
C     RIGHT GAP. MP(I) = -1 MEANS THAT THE GOOD T-EIGENVALUE IS SIMPLE
C     AND HAS A VERY SMALL MINGAP IN T(1,MEV) DUE TO A SPURIOUS
C     T-EIGENVALUE.
C     NG = NUMBER OF GOOD T-EIGENVALUES.
C     NISO = NUMBER OF ISOLATED, GOOD T-EIGENVALUES.
C
      CALL ISOEV(VS,GAPTOL,MULTOL,SCALE1,G,MP,NDIS,NG,NISO)
C
C-----------------------------------------------------------------------
C
      WRITE(6,270)NG,NISO,NDIS
  270 FORMAT(/I6,' SINGULAR VALUES HAVE BEEN COMPUTED'/
     1 I6,' OF THESE ARE ISOLATED'/
     2 I6,' = NUMBER OF DISTINCT T-EIGENVALUES COMPUTED'/)
C
C     DO WE WRITE DISTINCT T-EIGENVALUES TO FILE 11?
      IF (IDIST.EQ.0) GO TO 310
C
      WRITE(11,280) NDIS,NISO,MEV,M,N,SVSEED,MATNO
  280 FORMAT(5I5,I12,I8,' = NDIS,NISO,MEV,M,N,SVSEED,MATNO'/)
C
      WRITE(11,290) (MP(I),VS(I),G(I), I=1,NDIS)
  290 FORMAT(2(I3,E25.16,E12.3))
C
      WRITE(11,300) NDIS, (MP(I), I=1,NDIS)
  300 FORMAT(/I6,' = NDIS, T-MULTIPLICITIES (0 MEANS  SPURIOUS)'/(20I4))
C
  310 CONTINUE
C
      IF (NISO.NE.0) GO TO 340
C
      WRITE(4,320) MEV
  320 FORMAT(/' AT MEV = ',I6,' THERE ARE NO ISOLATED T-EIGENVALUES'/
     1' SO NO ERROR ESTIMATES WERE COMPUTED/')
C
      WRITE(6,330)
  330 FORMAT(/' ALL COMPUTED SINGULAR VALUES ARE T-MULTIPLE'/
     1 ' THEREFORE ALL COMPUTED SINGULAR VALUES ARE ASSUMED TO HAVE CONV
     1ERGED'/)
C
      ICONV = 1
      GO TO 380
C
  340 CONTINUE
C
C-----------------------------------------------------------------------
C     SUBROUTINE INVERR COMPUTES ERROR ESTIMATES FOR ISOLATED GOOD
C     T-EIGENVALUES USING INVERSE ITERATION ON T(1,MEV). ON RETURN
C     G(J) = MINIMUM GAP IN T(1,MEV) FOR EACH VS(J), J=1,NDIS
C     G(MEV+I) = BETAM*|U(MEV)| = ERROR ESTIMATE FOR ISOLATED GOOD
C             T-EIGENVALUES, WHERE I = 1, NISO AND  BETAM = BETA(MEV+1)
C             U(MEV) IS MEVTH COMPONENT OF THE UNIT EIGENVECTOR OF T
C             CORRESPONDING TO THE ITH ISOLATED GOOD T-EIGENVALUE.
C     A NEGATIVE ERROR ESTIMATE MEANS THAT FOR THAT PARTICULAR
C     T-EIGENVALUE THE INVERSE ITERATION DID NOT CONVERGE IN <= MXINIT
C     STEPS AND THAT THE CORRESPONDING ERROR ESTIMATE IS QUESTIONABLE.
C
```

```
C      V2 CONTAINS THE ISOLATED GOOD T-EIGENVALUES
C      V1 CONTAINS THE MINGAPS TO THE NEAREST DISTINCT  EIGENVALUE
C         OF T(1,MEV) FOR EACH ISOLATED GOOD T-EIGENVALUE IN V2.
C      VS CONTAINS THE NDIS DISTINCT EIGENVALUES OF T(1,MEV)
C      MP CONTAINS THE CORRESPONDING CODED T-MULTIPLICITIES
C
       IT = MXINIT
       CALL INVERR(BETA,V1,V2,VS,EPSM,G,MP,MEV,MMB,NDIS,NISO,NM,
     1 RHSEED,IT,IWRITE)
C
C-----------------------------------------------------------------------
C
C      SIMPLE CHECK FOR CONVERGENCE. CHECKS TO SEE IF ALL OF THE ERROR
C      ESTIMATES ARE SMALLER THAN CONTOL.
C      IF THIS TEST IS SATISFIED, THEN CONVERGENCE FLAG, ICONV IS SET
C      TO 1.  TYPICALLY ERROR ESTIMATES ARE VERY CONSERVATIVE.
C
       WRITE(6,350) CONTOL
  350  FORMAT(/' CONVERGENCE IS TESTED USING THE CONVERGENCE TOLERANCE',
     1E13.4/)
C
       II = MEV +1
       IF = MEV+NISO
       DO 360 I = II,IF
       IF (ABS(G(I)).GT.CONTOL) GO TO 380
  360  CONTINUE
       ICONV = 1
       MMB = NMEVS
C
       WRITE(6,370) CONTOL
  370  FORMAT(' ALL COMPUTED ERROR ESTIMATES WERE LESS THAN',E15.4/
     1 ' THEREFORE PROCEDURE TERMINATES'/)
C
  380  CONTINUE
C
C      IF CONVERGENCE IS INDICATED, THAT IS ICONV = 1 ,THEN
C      THE SUBROUTINE PRTEST IS CALLED TO CHECK FOR ANY CONVERGED
C      T-EIGENVALUES THAT HAVE BEEN MISLABELLED AS SPURIOUS BECAUSE
C      THE PROJECTION OF THEIR SINGULAR VECTOR ON THE STARTING
C      VECTOR WAS TOO SMALL.  NUMERICAL TESTS INDICATE THAT
C      SUCH SINGULAR VALUES ARE RARE.  THEREFORE, IF MANY OF
C      THESE HIDDEN SINGULAR VALUES APPEAR ON SOME RUN, THE USER
C      CAN BE CERTAIN THAT SOMETHING IS FOULED UP.
C
       IF (ICONV.EQ.0) GO TO 510
C
C-----------------------------------------------------------------------
C
       CALL PRTEST (BETA,VS,TKMAX,EPSM,RELTOL,SCALE3,SCALE4,
     1 MP,NDIS,MEV,IPROJ)
C
C-----------------------------------------------------------------------
C
       IF(IPROJ.EQ.0) GO TO 500
C
       IF(IDIST.EQ.1)  WRITE(11,390) IPROJ
  390  FORMAT(' SUBROUTINE PRTEST WANTS TO RELABEL',I6,' SPURIOUS T-EIGEN
     1VALUES'/' WE ACCEPT RELABELLING ONLY IF LAST COMPONENT OF T-EIGENV
     1ECTOR IS L.T. 1.D-10'/)
C
       IIX = RHSEED
C
C-----------------------------------------------------------------------
C
       CALL GENRAN(IIX,G,MEV)
C
```

```
C-----------------------------------------------------------------------
C
      ITEN = -10
      NISOM = NISO + MEV
      IWRITO = IWRITE
      IWRITE = 0
C
      DO 420 J = 1,NDIS
      IF(MP(J).NE.ITEN) GO TO 420
      TO = VS(J)
C
C-----------------------------------------------------------------------
C
      IT = MXINIT
      CALL INVERM(BETA,V1,V2,TO,TEMP,T1,EPSM,G,MEV,IT,IWRITE)
C
C-----------------------------------------------------------------------
C
      IF(TEMP.LE.1.D-10) GO TO 410
C     ERROR ESTIMATE WAS NOT SMALL REJECT RELABELLING OF THIS
C     T-EIGENVALUE.
      IF(IDIST.EQ.1) WRITE(11,400) J,TO,TEMP
  400 FORMAT(/' LAST COMPONENT FOR',I6,'TH T-EIGENVALUE',E20.12/' IS TOO
     1 LARGE = ',E15.6,' SO DO NOT ACCEPT PRTEST RELABELLING'/)
      MP(J) = 0
      IPROJ = IPROJ - 1
      GO TO 420
C     RELABELLING ACCEPTED
  410 NISOM = NISOM + 1
      G(NISOM) = BETAM*TEMP
  420 CONTINUE
      IWRITE = IWRITO
C
      IF(IPROJ.EQ.0) GO TO 460
      WRITE(6,430) IPROJ
  430 FORMAT(/I6,' T-EIGENVALUES WERE RECLASSIFIED AS GOOD.'/
     1' THESE ARE IDENTIFIED IN FILE 3 BY A T-MULTIPLICITY OF -10'/' USE
     2R SHOULD INSPECT EACH TO MAKE SURE NEIGHBORS HAVE CONVERGED'/)
C
      IF(IDIST.EQ.1)  WRITE(11,440) IPROJ
  440 FORMAT(/I6,' T-EIGENVALUES WERE RELABELLED AS GOOD'/
     1' BELOW IS CORRECTED T-MULTIPLICITY PATTERN'/)
C
      WRITE(6,450) NDIS, (MP(I), I=1,NDIS)
      IF(IDIST.EQ.1) WRITE(11,450) NDIS, (MP(I), I=1,NDIS)
  450 FORMAT(/I6,' = NDIS, T-MULTIPLICITIES (0 MEANS  SPURIOUS)'/
     1 6X, ' (-10) MEANS SPURIOUS T-EIGENVALUE RELABELLED AS GOOD'/(20I4
     1))
C
C     RECALCULATE MINGAPS FOR DISTINCT T(1,MEV) EIGENVALUES.
  460 NDIS1 = NDIS - 1
      G(NDIS) = VS(NDIS1)-VS(NDIS)
      G(1) = VS(2)-VS(1)
C
      DO 470 J = 2,NDIS1
      TO = VS(J)-VS(J-1)
      T1 = VS(J+1)-VS(J)
      G(J) = T1
      IF (TO.LT.T1) G(J) = -TO
  470 CONTINUE
      IF(IPROJ.EQ.0) GO TO 500
C     WRITE TO FILE 4 ERROR ESTIMATES FOR THOSE T-EIGENVALUES RELABELLED
      NGOOD = 0
      DO 480 J = 1,NDIS
      IF(MP(J).EQ.0) GO TO 480
      NGOOD = NGOOD + 1
      IF(MP(J).NE.ITEN) GO TO 480
```

```
      TO = VS(J)
      NISO = NISO + 1
      NISOM = MEV + NISO
      WRITE(4,490) NGOOD,TO,G(NISOM),G(J)
  480 CONTINUE
  490 FORMAT(I10,E25.16,2E14.3)
C
  500 CONTINUE
C
C     WRITE THE COMPUTED SINGULAR VALUES TO FILE 3.  FIRST TRANSFER THEM
C     TO V2 AND THEIR T-MULTIPLICITIES TO THE CORRESPONDING POSITIONS
C     IN MP AND COMPUTE THE B-MINGAPS, THE MINIMAL GAPS BETWEEN THE
C     SINGULAR VALUES CONSIDERED AS EIGENVALUES OF THE B-MATRIX.
C     THESE GAPS WILL BE PUT IN THE ARRAY G.
C     SINCE G CURRENTLY CONTAINS THE MINIMAL GAPS BETWEEN THE DISTINCT
C     EIGENVALUES OF THE T-MATRIX, THESE GAPS WILL FIRST BE
C     TRANSFERRED TO V1.  NOTE THAT V1<0 MEANS THAT THAT MINIMAL GAP
C     IN THE T-MATRIX IS DUE TO A SPURIOUS T-EIGENVALUE.
C     ALL THIS INFORMATION IS PRINTED TO FILE 3
C
  510 CONTINUE
C
      NG = 0
      DO 520 I = 1,NDIS
      IF (MP(I).EQ.0) GO TO 520
      NG = NG+1
      MP(NG) = MP(I)
      V2(NG) = VS(I)
      TEMP = G(I)
      TEMP = DABS(TEMP)
      J = I+1
      IF (G(I).LT.ZERO) J = I-1
      IF (MP(J).EQ.0) TEMP = -TEMP
      V1(NG) = TEMP
  520 CONTINUE
C
      WRITE(6,530)MEV
  530 FORMAT(//' SINGULAR VALUE CALCULATION AT MEV = ',I6,'    IS COMPLE
     1TE'//)
C
C     NG = NUMBER OF COMPUTED DISTINCT GOOD T-EIGENVALUES.  NEXT
C     GENERATE GAPS BETWEEN GOOD T-EIGENVALUES (BMINGAPS) AND PUT THEM
C     IN G.  G(J) < 0 MEANS THE BMINGAP IS DUE TO THE LEFT-HAND GAP.
C
      NGM1 = NG - 1
      G(NG) = V2(NGM1)-V2(NG)
      G(1) = V2(2)-V2(1)
C
      DO 540 J = 2,NGM1
      TO = V2(J)-V2(J-1)
      T1 = V2(J+1)-V2(J)
      G(J) = T1
      IF (TO.LT.T1) G(J) = -TO
  540 CONTINUE
C
C     WRITE GOOD T-EIGENVALUES (COMPUTED SINGULAR VALUES) OUT TO FILE 3.
C
      WRITE(3,550)NG,NDIS,MEV,M,N,SVSEED,MATNO,IPARO,MULTOL,IB,BTOL
  550 FORMAT(5I6,I12,I8,I2,'=NG,ND,MEV,M,N,SEED,MN,IPARO'/
     1 E20.12,I6,E13.4,' = MUTOL,INDEX MINIMAL BETA,BTOL'/
     1' SV NO',2X,'T-MULT',10X,'SINGULAR VALUE',7X,'BMINGAP',7X,'TMINGAP
     1')
C
      WRITE(3,560)(I,MP(I),V2(I),G(I),V1(I),  I=1,NG)
  560 FORMAT(I6,I8,E25.16,2E14.3)
C
C     IF CONVERGENCE FLAG ICONV.NE.1 AND NUMBER OF T-MATRICES
```

```
C     CONSIDERED TO DATE IS LESS THAN NUMBER ALLOWED, INCREMENT MEV.
C     AND LOOP BACK TO 210 TO REPEAT COMPUTATIONS.  RESTORE BETA(MEV+1).
C
      BETA(MP1) = BETAM
C
      IF (MMB.LT.NMEVS.AND.ICONV.NE.1) GO TO 210
C
C     END OF LOOP ON DIFFERENT SIZE T-MATRICES ALLOWED.
C
  570 CONTINUE
C
      IF(ISTOP.EQ.0)  WRITE(6,580)
  580 FORMAT(/' T-MATRICES (BETA) ARE NOW AVAILABLE, TERMINATE'/)
      IF (IHIS.EQ.1.AND.KMAX.NE.MOLD) WRITE(1,590)
  590 FORMAT(/' ABOVE ARE THE FOLLOWING VECTORS '/
     2 '   BETA(I), I = 1,KMAX+1'/
     3 ' FINAL TWO LANCZOS VECTORS OF ORDERS M,N FOR I = KMAX,KMAX+1'/
     4 ' ALL VECTORS IN THIS FILE HAVE FORMAT 4Z20'/
     5 ' ----- END OF FILE 1 NEW BETA HISTORY---------------'///)
C
      IF (ISTOP.EQ.0) GO TO 690
C
      WRITE(3,600)
  600 FORMAT(/' ABOVE ARE COMPUTED SINGULAR VALUES'/
     1 ' NG = NUMBER OF SINGULAR VALUES COMPUTED'/
     2 ' NDIS = NUMBER OF COMPUTED DISTINCT EIGENVALUES OF T(1,MEV)'/
     3 ' M = ROW ORDER OF A N = COLUMN ORDER,  MATNO = MATRIX IDENT'/
     4 ' MULTOL = T-MULTIPLICITY TOLERANCE FOR T-EIGENVALUES IN BISEC'/
     4 ' T-MULT IS THE T-MULTIPLICITY OF SINGULAR VALUE'/
     5 ' T-MULT = -1 MEANS SPURIOUS T-EIGENVALUE TOO CLOSE'/
     6 ' DO NOT COMPUTE ERROR ESTIMATES FOR SUCH T-EIGENVALUES'/
     7 ' BMINGAP = MINIMAL GAP BETWEEN THE COMPUTED SINGULAR VALUES'/
     8 ' BMINGAP .LT. 0. MEANS MINIMAL GAP IS DUE TO LEFT-HAND GAP'/
     9 ' TMINGAP= MINIMAL GAP W.R.T.  DISTINCT EIGENVALUES IN T(1,MEV)'/
     1 ' TMINGAP .LT. 0. MEANS MINGAP IS DUE TO SPURIOUS T-EIGENVALUE'/
     2 ' ----- END OF FILE 3 SINGULAR VALUES------------------------'//)
C
      IF (IDIST.EQ.1) WRITE(11,610)
  610 FORMAT(/' ABOVE ARE THE DISTINCT EIGENVALUES OF T(1,MEV).'/
     2 ' THE FORMAT IS      T-MULTIPLICITY   T-EIGENVALUE   TMINGAP'/
     3 '          THIS FORMAT IS REPEATED TWICE ON EACH LINE.'/
     4 ' T-MULTIPLICITY = -1 MEANS THAT THE SUBROUTINE ISOEV HAS TAGGED'
     5 /'   THIS COMPUTED SINGULAR VALUE AS HAVING A VERY CLOSE SPURIOUS
     6 '/'   T-EIGENVALUE SO THAT NO ERROR ESTIMATE WILL BE COMPUTED'/
     7 '   FOR THAT SINGULAR VALUE IN SUBROUTINE INVERR.'/
     8 ' TMINGAP .LT. 0, TMINGAP IS DUE TO LEFT GAP .GT. 0, RIGHT GAP.'/
     9 ' EACH OF THE DISTINCT T-EIGENVALUE TABLES IS FOLLOWED'/
     9 ' BY THE T-MULTIPLICITY PATTERN.'/
     1 ' NDIS = NUMBER OF COMPUTED DISTINCT EIGENVALUES OF T(1,MEV).'/
     2 ' NG = NUMBER OF COMPUTED SINGULAR VALUES. '/
     3 ' NISO = NUMBER OF ISOLATED (IN T-MATRIX) SINGULAR VALUES. '/
     4 ' NISO ALSO IS THE COUNT OF +1 ENTRIES IN T-MULTIPLICITY PATTERN.
     5 '/' ----- END OF FILE 11 DISTINCT T-EIGENVALUES--------------'//)
C
      IF(NISO.NE.0)  WRITE(4,620)
  620 FORMAT(/' ABOVE ARE THE ERROR ESTIMATES OBTAINED FOR THE ISOLATED
     1GOOD T-EIGENVALUES'/
     1' OBTAINED VIA INVERSE ITERATION IN THE SUBROUTINE INVERR.'/
     1' ALL OTHER GOOD T-EIGENVALUES HAVE CONVERGED.'/
     2' ERROR ESTIMATE = BETAM*ABS(UM)'/
     2' WHERE BETAM = BETA(MEV+1) AND UM = U(MEV).'/
     3' U = UNIT EIGENVECTOR OF T WHERE T*U = SV*U AND SV = ISOLATED GOO
     3D T-EIGENVALUE.'/
     4' TMINGAP = GAP TO NEAREST DISTINCT EIGENVALUE OF T(1,MEV).'/
     5' TMINGAP .LT. 0. MEANS MINGAP IS DUE TO A SPURIOUS T-EIGENVALUE.'
     6/' ------ END OF FILE 4 ERRINV ------------------------------'//)
```

```
      GO TO 690
C
  630 CONTINUE
C
      IBB = IABS(IBMEV)
      IF (IBMEV.LT.0) WRITE(6,640) MEV,IBB,BETA(IBB)
  640 FORMAT(/' PROGRAM TERMINATES BECAUSE MEV REQUESTED = ',I6,' IS .GT
     1',I6/' AT WHICH AN ABNORMALLY SMALL BETA = ' , E13.4,' OCCURRED'/)
      GO TO 690
C
  650 IF (NDIS.EQ.0.AND.ISTOP.GT.0) WRITE(6,660)
  660 FORMAT(/' INTERVALS SPECIFIED FOR BISECT DID NOT CONTAIN ANY T-EIG
     1ENVALUES'/' PROGRAM TERMINATES')
      GO TO 690
C
  670 WRITE(6,680) I, NMEV(I)
  680 FORMAT(//I6,'TH T-SIZE REQUESTED ',I6,' IS ODD'/
     1' BUT ONLY EVEN T-SIZES ARE PERMISSIBLE.  PROGRAM TERMINATES FOR U
     1SER TO FIX'//)
      GO TO 690
C
  690 CONTINUE
C
      STOP
C-----END OF MAIN PROGRAM FOR LANCZOS SINGULAR VALUE COMPUTATIONS-------
      END
```

SECTION 6.4 MAIN PROGRAM, SINGULAR VECTOR CALCULATIONS

```
C-----LSVEC (SINGULAR VECTORS OF REAL RECTANGULAR MATRICES)-------------
C
C     CONTAINS MAIN PROGRAM FOR COMPUTING A LEFT AND A
C     RIGHT SINGULAR VECTOR CORRESPONDING TO EACH OF A SET
C     OF SINGULAR VALUES WHICH HAVE BEEN COMPUTED ACCURATELY BY THE
C     CORRESPONDING LANCZOS SINGULAR VALUE PROGRAM (LSVAL)
C     FOR REAL RECTANGULAR MATRICES.  THIS PROGRAM COULD BE
C     MODIFIED TO COMPUTE ADDITIONAL SINGULAR VECTORS FOR ANY
C     SINGULAR VALUE THAT IS A MULTIPLE SINGULAR VALUE OF A.
C     THE AMOUNT OF ADDITIONAL COMPUTATION REQUIRED BY SUCH A
C     MODIFICATION DEPENDS UPON THE GIVEN A-MATRIX AND UPON
C     THE PART OF THE SPECTRUM INVOLVED.
C
C     FOR A GIVEN REAL MATRIX A OF ORDER M X N THE LANCZOS RECURSION
C     IS APPLIED TO THE ASSOCIATED REAL SYMMETRIC MATRIX B OF ORDER
C     MN = M+N
C
C                                 ----              ----
C                                | 0                  A |
C                    B  =        |                      |
C                                | A-TRANSPOSE        0 |
C                                 ----              ----
C     USING SPECIAL STARTING VECTORS.
C
C     THESE SINGULAR VECTOR COMPUTATIONS ASSUME THAT EACH
C     SINGULAR VALUE THAT IS BEING CONSIDERED HAS CONVERGED AS
C     AN EIGENVALUE OF THE LANCZOS TRIDIAGONAL MATRICES GENERATED.
C
C     THE EIGENVALUES OF EACH EVEN-ORDERED LANCZOS MATRIX OCCUR
C     IN + AND - PAIRS, AND THE RITZ VECTOR COMPUTATION RESTS ON
C     AN INVERSE ITERATION COMPUTATION FOR A LANCZOS MATRIX.
C     THIS CAUSES AN ANOMALY IN THE SINGULAR VECTOR COMPUTATIONS
C     FOR VERY SMALL SINGULAR VALUES.  IN PRACTICE WE SEE THAT
C     FOR ANY SUCH SINGULAR VALUE THAT ONE MEMBER OF EACH PAIR OF
C     APPROXIMATE SINGULAR VECTORS WILL BE MORE ACCURATE THAN THE
C     OTHER MEMBER OF THAT PAIR IS.  IF IPAR = 1 (STARTING LANCZOS
C     VECTOR IS OF FORM (0,V2) WHERE V2 IS NX1) THEN THE RIGHT
C     SINGULAR VECTOR WILL BE OBTAINED MORE ACCURATELY THAN THE
C     LEFT SINGULAR VECTOR.  IF IPAR = 2 (STARTING LANCZOS VECTOR
C     IS OF FORM (V1,0) WHERE V1 IS MX1) THEN THE LEFT SINGULAR
C     VECTOR WILL BE MORE ACCURATE THAN THE RIGHT SINGULAR VECTOR.
C     PRIOR TO NORMALIZATION THE SIZES OF THESE INACCURATE VECTORS
C     WILL BE THE SAME AS THE SIZE OF THE ASSOCIATED VERY SMALL
C     SINGULAR VALUE.  IN FACT IN THE LIMIT, FOR A ZERO SINGULAR VALUE
C     AND IPAR = 1, THE VECTOR COMPUTED AS THE APPROXIMATION TO THE
C     LEFT SINGULAR VECTOR WILL BE THE 0 VECTOR.  (IF IPAR = 2 THEN
C     THIS WOULD BE THE RIGHT SINGULAR VECTOR).  THE CORRESPONDING
C     ERROR ESTIMATES WILL REFLECT THE INACCURACY OF THE ONE MEMBER
C     OF EACH SUCH PAIR, SINCE THESE ESTIMATES ARE A SUM OF ESTIMATES
C     FOR THE INDIVIDUAL MEMBERS OF THE PAIR.  THEREFORE, FOR ANY VERY
C     SMALL SINGULAR VALUE A CORRESPONDING SINGULAR VECTOR WILL BE
C     COMPUTED ONLY IF THE USER HAS SET THE FLAG ERCONT TO 1.
C
C-----------------------------------------------------------------------
C
C     PFORT VERIFIER IDENTIFIED THE FOLLOWING NONPORTABLE
C     CONSTRUCTIONS
C
C     1.  DATA/MACHEP/ STATEMENT
C     2.  ALL READ(5,*) STATEMENTS (FREE FORMAT)
C     3.  FORMAT(20A4) USED WITH THE EXPLANATORY HEADER, EXPLAN
C     4.  HEXADECIMAL FORMAT (4Z20) USED FOR BETA HISTORY.
C
C     IMPORTANT NOTE:  THIS PROGRAM ALLOWS ENLARGEMENT OF THE
```

```
C     BETA ARRAY.  IN PARTICULAR, IF ANY ONE OF THE SINGULAR VALUES
C     SUPPLIED IS T-SIMPLE AND AS AN EIGENVALUE OF THE ASSOCIATED
C     LANCZOS TRIDIAGONAL MATRIX IS NOT CLOSE TO A SPURIOUS
C     EIGENVALUE OF THAT MATRIX, THIS PROGRAM WILL REQUIRE
C     THAT KMAX BE AT LEAST THE LARGEST EVEN NUMBER LESS
C     THAN OR EQUAL TO  (11*MEV)/8 + 13.  IF KMAX IS NOT THAT
C     LARGE, THEN THIS PROGRAM WILL RESET KMAX TO THIS SIZE
C     AND EXTEND THE BETA HISTORY IF REQUIRED.
C     THUS, THE DIMENSION OF THE BETA ARRAY MUST BE
C     LARGE ENOUGH TO ALLOW FOR THIS POSSIBILITY.
C     REMEMBER THAT THE BETA ARRAY, BETA(J), IS SUCH THAT
C     J = 1,..., KMAX+1.  SO IF THE KMAX USED BY THE PROGRAM
C     IS TO BE 3000, THEN BETA MUST BE OF LENGTH AT LEAST 3001.
C
C-----------------------------------------------------------------------
      DOUBLE PRECISION  BETA(5001),V1(5000),V2(5000),RITVEC(30000)
      DOUBLE PRECISION  TVEC(30000),GOODSV(50),SVNEW(50),TLAST(50)
      DOUBLE PRECISION  SVAL,SVALN,TOLN,TTOL,ERTOL,BATA
      DOUBLE PRECISION  MULTOL,SCALEO,STUTOL,BTOL,LB,UB
      DOUBLE PRECISION  ONE,ZERO,MACHEP,EPSM,TEMP,SUM
      DOUBLE PRECISION  RELTOL,ERROR,TERROR,ERRMIN,BKMIN
      REAL     G(10000),BMINGP(50),TMINGP(50),EXPLAN(20)
      REAL     TERR(50),BERR(50),BERRGP(50),RNORM(50),TBETA(50)
      INTEGER  MP(50),M1(50),M2(50),MA(50),ML(50),MINT(50),MFIN(50)
      INTEGER  SVSEED,SVSOLD,RHSEED,IDELTA(50)
      INTEGER  MBOUND,NTVCON,SVTVEC,TVSTOP,LVCONT,ERCONT,TFLAG
      DOUBLE PRECISION  FINPRO
      DOUBLE PRECISION  DABS, DMAX1, DSQRT, DFLOAT
      REAL ABS
      INTEGER  IABS
C-----------------------------------------------------------------------
      EXTERNAL SVMAT, STRAN
      DATA MACHEP/Z3410000000000000/
      EPSM = 2.D0*MACHEP
C-----------------------------------------------------------------------
C
C     ARRAYS MUST BE DIMENSIONED AS FOLLOWS:
C     1.  BETA: >= (KMAX+1) WHERE KMAX, THE LARGEST SIZE
C                T-MATRIX CONSIDERED BY THE PROGRAM, IS THE
C                LARGER OF THE SIZE OF THE BETA HISTORY PROVIDED
C                ON FILE 2 (IF ANY ) AND THE SIZE WHICH THE PROGRAM
C                SPECIFIES INTERNALLY, THIS LATTER IS ALWAYS
C                < = (11*MEV)/8 + 13, WHERE MEV IS THE SIZE
C                T-MATRIX THAT WAS USED IN THE CORRESPONDING
C                SINGULAR VALUE COMPUTATIONS.  NOTE THAT ALL
C                T-MATRICES CONSIDERED MUST HAVE EVEN ORDER.
C     2.  V1:  >= MAX(M,KMAX)
C     3.  V2:   >= N
C     4.  G:  >= MAX(M,N,KMAX)
C     5.  RITVEC:  >= (N+M)*NGOOD, WHERE NGOOD IS THE NUMBER OF
C                   SINGULAR VALUES SUPPLIED TO THIS PROGRAM.
C     6.  TVEC:  >= CUMULATIVE LENGTH OF ALL THE T-EIGENVECTORS
C                NEEDED TO GENERATE THE DESIRED RITZ VECTORS.  AN
C                EDUCATED GUESS AT AN APPROPRIATE LENGTH CAN BE
C                OBTAINED BY RUNNING THE PROGRAM WITH THE FLAG
C                MBOUND = 1 AND MULTIPLYING THE RESULTING SIZE BY 5/4.
C     7.  GOODSV, TMINGP, BMINGP,TERR, BERR, BERRGP, RNORM,
C         TBETA, TLAST, SVNEW, MP, MA, M1, M2, MINT, MFIN AND
C         IDELTA MUST ALL BE >= NGOOD.
C
C-----------------------------------------------------------------------
C     OUTPUT HEADER
      WRITE(6,10)
   10 FORMAT(/' LANCZOS PROCEDURE FOR REAL, RECTANGULAR MATRICES'/
     1'       COMPUTE SINGULAR VECTORS'/)
C
```

```
C     SET PROGRAM PARAMETERS
C     USER MUST NOT MODIFY SCALEO
      SCALEO = 5.0D0
      ZERO = 0.0D0
      ONE = 1.0D0
      MPMIN = -1000
C     CONVERGENCE TOLERANCE FOR T-EIGENVECTORS FOR RITZ COMPUTATIONS
      ERTOL = 1.D-10
C
C     READ USER-SPECIFIED PARAMETER FROM INPUT FILE 5 (FREE FORMAT)
C
C     READ USER-PROVIDED HEADER FOR RUN
      READ(5,20) EXPLAN
      WRITE(6,20) EXPLAN
   20 FORMAT(20A4)
C
C     READ IN MATNO = MATRIX/RUN IDENTIFICATION NUMBER, 8 DIGITS OR LESS
C     AND THE ORDER OF THE MATRIX M X N .
C
      READ(5,20) EXPLAN
      READ(5,*) MATNO, M, N
      MN = M + N
C
C     READ IN THE MAXIMUM PERMISSIBLE DIMENSIONS FOR THE TVEC ARRAY
C     (MDIMTV), FOR THE RITVEC ARRAY (MDIMRV), AND FOR THE BETA
C     ARRAY (MBETA).
C
      READ(5,20) EXPLAN
      READ(5,*) MDIMTV, MDIMRV, MBETA
C
C     READ IN RELATIVE TOLERANCE USED IN DETERMINING APPROPRIATE
C     SIZES FOR THE T-MATRICES USED IN THE SINGULAR VECTOR COMPUTATIONS.
C
      READ(5,20) EXPLAN
      READ(5,*) RELTOL
C
C     SET FLAGS TO 0 OR 1:
C     MBOUND = 1:  PROGRAM TERMINATES AFTER COMPUTING 1ST GUESSES
C                  ON APPROPRIATE T-SIZES FOR USE IN THE RITZ VECTOR
C                  COMPUTATIONS
C     NTVCON = 0:  PROGRAM TERMINATES IF THE TVEC ARRAY IS NOT
C                  LARGE ENOUGH TO HOLD ALL THE T-EIGENVECTORS REQUIRED.
C     SVTVEC = 0:  THE T-EIGENVECTORS ARE NOT WRITTEN TO FILE 11
C                  UNLESS TVSTOP = 1
C     SVTVEC = 1:  WRITE THE T-EIGENVECTORS TO FILE 11.
C     TVSTOP = 1:  PROGRAM TERMINATES AFTER COMPUTING THE
C                  T-EIGENVECTORS
C     LVCONT = 0:  PROGRAM TERMINATES IF THE NUMBER OF T-EIGENVECTORS
C                  COMPUTED IS NOT EQUAL TO THE NUMBER OF RITZ
C                  VECTORS (SINGULAR VECTORS) REQUESTED.
C     ERCONT = 0:  MEANS FOR ANY GIVEN SINGULAR VALUE, A RITZ VECTOR
C                  WILL NOT BE COMPUTED FOR THAT SINGULAR VALUE UNLESS
C                  A T-EIGENVECTOR HAS BEEN IDENTIFIED WITH A LAST
C                  COMPONENT WHICH SATISFIES THE SPECIFIED
C                  CONVERGENCE CRITERION.
C     ERCONT = 1:  MEANS FOR ANY GIVEN SINGULAR VALUE, A RITZ VECTOR
C                  WILL BE COMPUTED.  IF A T-EIGENVECTOR CANNOT
C                  BE IDENTIFIED WHICH SATISFIES THE LAST
C                  COMPONENT CRITERION, THEN THE PROGRAM WILL
C                  USE THE T-VECTOR THAT CAME CLOSEST TO
C                  SATISFYING THE CRITERION
C     IWRITE = 1:  EXTENDED OUTPUT OF INTERMEDIATE COMPUTATIONS
C                  IS WRITTEN TO FILE 6
C     IREAD = 0:   BETA FILE IS REGENERATED.
C     IREAD = 1:   BETA FILE USED IN SINGULAR VALUE COMPUTATIONS
C                  IS READ IN AND EXTENDED IF NECESSARY.  IN BOTH
C                  CASES IREAD = 0 OR 1, THE LANCZOS VECTORS ARE
```

```
C                 ALWAYS REGENERATED FOR THE RITZ VECTOR
C                 COMPUTATIONS
C
      READ(5,20) EXPLAN
      READ(5,*) MBOUND,NTVCON,SVTVEC,IREAD
C
      READ(5,20) EXPLAN
      READ(5,*) TVSTOP,LVCONT,ERCONT,IWRITE
      IF (TVSTOP.EQ.1) SVTVEC = 1
C
C     READ IN SEED (RHSEED) FOR GENERATING RANDOM STARTING VECTOR
C     FOR THE INVERSE ITERATION ON THE T-MATRICES.
C
      READ(5,20) EXPLAN
      READ(5,*) RHSEED
C
C-----------------------------------------------------------------------
C     INITIALIZE THE ARRAYS FOR THE USER-SPECIFIED MATRIX AND
C     PASS THE STORAGE LOCATIONS OF THESE ARRAYS TO THE MATRIX-VECTOR
C     MULTIPLY SUBROUTINES SVMAT AND STRAN.
C
      CALL USPEC(M,N,MATNO)
C
C-----------------------------------------------------------------------
C     MASK UNDERFLOW AND OVERFLOW
      CALL MASK
C
C-----------------------------------------------------------------------
C     WRITE RUN PARAMETERS OUT TO FILE 6
C
      WRITE(6,30) M,N,MATNO
   30 FORMAT(/' MATRIX ORDER =',I5,' BY ',I5/
     1 ' A-MATRIX AND CASE IDENTIFIER = ',I10/)
C
      WRITE(6,40) MBOUND,NTVCON,SVTVEC,IREAD
   40 FORMAT(/3X,'MBOUND',3X,'NTVCON',3X,'SVTVEC',3X,'IREAD'/3I9,I8/)
C
      WRITE(6,50) TVSTOP,LVCONT,ERCONT,IWRITE
   50 FORMAT(/3X,'TVSTOP',3X,'LVCONT',3X,'ERCONT',3X,'IWRITE'/4I9)
C
      WRITE(6,60) MDIMTV,MDIMRV,MBETA
   60 FORMAT(/3X,'MDIMTV',3X,'MDIMRV',3X,'MBETA'/2I9,I8)
C
      WRITE(6,70) RELTOL,RHSEED
   70 FORMAT(/7X,'RELTOL',3X,'RHSEED'/E13.4,I9)
C
C     FROM FILE 3 READ IN THE NUMBER OF SINGULAR VALUES (NGOOD)
C     FOR WHICH SINGULAR VECTORS ARE REQUESTED, THE ORDER (MEV) OF
C     THE LANCZOS TRIDIAGONAL MATRIX USED IN COMPUTING THESE
C     SINGULAR VALUES, THE ORDER MOLD X NOLD OF THE USER-SPECIFIED
C     MATRIX USED IN THOSE COMPUTATIONS, THE SEED (SVSEED) USED FOR
C     GENERATING THE STARTING VECTOR THAT WAS USED IN THOSE
C     COMPUTATIONS, AND THE MATRIX/RUN IDENTIFICATION NUMBER (MATOLD)
C     USED IN THOSE COMPUTATIONS.  ALSO READ IN THE NUMBER (NDIS) OF
C     DISTINCT EIGENVALUES OF THE MATRIX T(1,MEV) THAT WERE COMPUTED
C     BUT THIS VALUE IS NOT USED IN THE SINGULAR VECTOR
C     COMPUTATIONS.
C
      READ(3,80) NGOOD,NDIS,MEV,MOLD,NOLD,SVSEED,MATOLD,IPARO
   80 FORMAT(5I6,I12,I8,I2)
C
C     READ IN THE T-MULTIPLICITY TOLERANCE USED IN THE BISEC SUBROUTINE
C     DURING THE COMPUTATION OF THE GIVEN SINGULAR VALUES.
C     ALSO READ IN THE FLAG IB.  IF IB < 0, THEN SOME BETA(I) IN THE
C     T-MATRIX FILE PROVIDED ON FILE 2 FAILED THE ORTHOGONALITY
C     TEST IN THE TNORM SUBROUTINE.  USER SHOULD NOTE THAT THIS
```

```
C     PROGRAM PROCEEDS INDEPENDENTLY OF THE SIZE OF THE BETA USED.
C
      READ(3,90) MULTOL,IB,BTOL
   90 FORMAT(E20.12,I6,E13.4)
C
      TEMP = DFLOAT(MEV+1000)
      TTOL = MULTOL/TEMP
      WRITE(6,100) MULTOL,TTOL
  100 FORMAT(/' T-MULTIPLICITY TOLERANCE USED IN THE SINGULAR VALUE COMP
     1UTATIONS WAS',E13.4/' SCALED MACHINE EPSILON IS',E13.4)
C
C     CONTINUE WRITE TO FILE 6 OF THE PARAMETERS FOR THIS RUN
C
      WRITE(6,110)NGOOD,NDIS,MEV,MOLD,NOLD,MATOLD,SVSEED,MULTOL,IB,
     1BTOL,IPARO
  110 FORMAT(/' SINGULAR VALUES SUPPLIED ARE READ IN FROM FILE 3'/
     1 6X,'NG',2X,'NDIS',3X,'MEV',2X,'MOLD',2X,'NOLD',2X,'MATOLD',4X/
     1I8,4I6,I8//6X,'SVSEED',6X,'MULTOL',9X,'IB',8X,'BTOL',4X,'IPARO'/
     1I12,E12.3,I11,E12.4,I9/)
C
C     IS THE ARRAY RITVEC LONG ENOUGH TO HOLD ALL OF THE DESIRED
C     RITZ VECTORS (APPROXIMATE EIGENVECTORS OF B)?
      MNMAX = NGOOD*MN
      IF(MBOUND.EQ.1) GO TO 120
      IF(TVSTOP.NE.1.AND.MNMAX.GT.MDIMRV) GO TO 1600
C
C     CHECK THAT THE ORDERS M,N AND THE MATRIX IDENTIFICATION NUMBER
C     MATNO SPECIFIED BY THE USER AGREE WITH THOSE READ IN FROM
C     FILE 3.
  120 ITEMP = (MOLD-M)**2+(NOLD-N)**2+(MATOLD-MATNO)**2
      IF (ITEMP.NE.0) GO TO 1620
C
C     READ IN FROM FILE 3, THE T-MULTIPLICITIES OF THE SINGULAR VALUES
C     WHOSE SINGULAR VECTORS ARE TO BE COMPUTED, THE VALUES OF THESE
C     SINGULAR VALUES AND THEIR MINIMAL GAPS AS SINGULAR VALUES OF THE
C     USER-SPECIFIED MATRIX AND OF THE RELATED T-MATRIX.
C
      READ(3,20) EXPLAN
      READ(3,130) (MP(J),GOODSV(J),BMINGP(J),TMINGP(J), J=1,NGOOD)
  130 FORMAT(6X,I8,E25.16,2E14.3)
C
      WRITE(6,140) (J,GOODSV(J),MP(J),BMINGP(J),  J=1,NGOOD)
  140 FORMAT(/' SINGULAR VALUES READ IN FROM FILE 3 AND THEIR T-MULTIPLI
     1CITIES'/4X,' J ',4X,' SINGULAR VALUE',5X,'TMULT',4X,'BMINGP'/
     1(I6,E20.12,I6,E13.4))
C
      WRITE(6,150) MEV,SVSEED
  150 FORMAT(/' THESE SINGULAR VALUES WERE COMPUTED USING A T-MATRIX OF
     1ORDER ',I5/' AND SEED FOR RANDOM NUMBER GENERATOR =',I12)
C
C     READ IN THE ERROR ESTIMATES
C
C     CHECK WHETHER OR NOT THERE ARE ANY ISOLATED T-EIGENVALUES IN
C     THE T-EIGENVALUES PROVIDED (HERE THE SINGULAR VALUES ARE
C     CONSIDERED AS EIGENVALUES OF THE ASSOCIATED LANCZOS TRIDIAGONAL
C     MATRICES.)
      DO 160 J=1,NGOOD
      IF(MP(J).EQ.1) GO TO 170
  160 CONTINUE
      GO TO 190
  170 READ(4,20) EXPLAN
      READ(4,20) EXPLAN
      READ(4,20) EXPLAN
      READ(4,180) NISO
  180 FORMAT(18X,I6)
      READ(4,20) EXPLAN
      READ(4,20) EXPLAN
```

```
      READ(4,20) EXPLAN
  190 DO 220 J=1,NGOOD
      BERR(J) = 0.D0
      IF(MP(J).NE.1) GO TO 220
      READ(4,200) SVAL, BERR(J)
  200 FORMAT(10X,E25.16,E14.3)
      IF(DABS(SVAL - GOODSV(J)).LT.1.D-10) GO TO 220
      WRITE(6,210) SVAL,GOODSV(J)
  210 FORMAT(' PROBLEM WITH READ IN OF ERROR ESTIMATES'/' SINGULAR VALUE
     1READ IN',E20.12,' DOES NOT MATCH GOODSV(J) ='/E20.12)
      GO TO 1860
C
  220 CONTINUE
C
      WRITE(6,230) (J,GOODSV(J),BERR(J), J=1,NGOOD)
  230 FORMAT(' ERROR ESTIMATES ='/4X,' J',3X,'SINGULAR VALUE',8X,
     1'ESTIMATE'/(I6,E20.12,E14.3))
C
      IF(IREAD.EQ.0) IPAR = IPARO
      IF(IREAD.EQ.0)  GO TO 350
C
C     READ IN THE SIZE OF THE T-MATRIX PROVIDED ON FILE 2.  READ IN
C     THE ORDER OF THE USER-SPECIFIED MATRIX , THE FLAGS IPARO
C     AND IPAR WHICH INDICATE RESPECTIVELY THE PARITY OF THE
C     STARTING VECTOR USED IN THE GENERATION OF THE EXISTING
C     BETA AND THE PARITY OF THE NEXT LANCZOS VECTOR THAT
C     HAS TO BE GENERATED IF THE BETA HISTORY IS EXTENDED,
C     THE SEED USED BY THE RANDOM NUMBER GENERATOR WHEN
C     GENERATING THE STARTING VECTOR THAT WAS USED, AND THE
C     MATRIX/TEST IDENTIFICATION NUMBER THAT WERE USED IN
C     THE LANCZOS SINGULAR VALUE COMPUTATIONS.  IF THE FLAG
C     IREAD = 0, REGENERATE HISTORY AND DO NOT READ ANYTHING
C     FROM FILE 2.  HISTORY MUST BE STORED IN MACHINE FORMAT,
C     ((4Z20) FOR IBM 3081).
C
      READ(2,240) KMAX,MOLD,NOLD,IPARO,IPAR,SVSOLD,MATOLD
  240 FORMAT(3I6,2I3,I12,I8)
C
      WRITE(6,250) KMAX,MOLD,NOLD,IPARO,IPAR,SVSOLD,MATOLD
  250 FORMAT(/' READ IN HEADER FROM BETA FILE 2'/
     1 2X,'KMAX',2X,'MOLD',2X,'NOLD',2X,'IPARO',2X,'IPAR',6X,'SVSOLD
     1 ',2X,'MATOLD'/3I6,I7,I6,I12,I12)
C
C     CHECK THAT THE PARAMETERS READ IN AGREE WITH WHAT THE USER
C     HAS SPECIFIED
      IF(MOLD.NE.M.OR.NOLD.NE.N.OR.MATOLD.NE.MATNO.OR.SVSOLD.NE.SVSEED)
     1 GO TO 1640
C
      IF(IPARO.EQ.1) WRITE(6,260)
      IF(IPARO.EQ.2) WRITE(6,270)
  260 FORMAT(/' STARTING VECTOR USED IN EXISTING SINGULAR VALUE HISTORY
     1WAS'/' OF THE FORM (0,V2)')
  270 FORMAT(/' STARTING VECTOR USED IN EXISTING SINGULAR VALUE HISTORY
     1WAS'/' OF THE FORM (V1,0)')
C
      KMAX1 = KMAX + 1
C
C     READ IN THE T-MATRICES FROM FILE 2.  THESE ARE USED TO GENERATE
C     THE T-EIGENVECTORS THAT WILL BE USED IN THE RITZ VECTOR
C     COMPUTATIONS.  HISTORY MUST BE STORED IN 4Z20 FORMAT.
C
      READ(2,280) (BETA(J), J=1,KMAX1)
  280 FORMAT(4Z20)
C
      READ(2,280) (V1(J), J=1,M)
      READ(2,280) (V2(J), J=1,N)
C
```

```
C     KMAX MAY BE ENLARGED IF THE SIZE AT WHICH THE SINGULAR VALUE      LSV03980
C     COMPUTATIONS WERE PERFORMED IS ESSENTIALLY KMAX AND               LSV03990
C     THERE IS AT LEAST ONE SINGULAR VALUE THAT IS SIMPLE AS AN         LSV04000
C     EIGENVALUE OF T(1,MEV), AND IF ITS NEAREST NEIGHBOR IN THE        LSV04010
C     T-MATRIX IS TOO CLOSE, THAT NEIGHBOR IS A 'GOOD' T-EIGENVALUE.    LSV04020
      DO 290 J = 1,NGOOD                                                LSV04030
      IF(MP(J).EQ.1) GO TO 310                                          LSV04040
  290 CONTINUE                                                          LSV04050
      WRITE(6,300)                                                      LSV04060
  300 FORMAT(/' ALL SINGULAR VALUES USED ARE T-MULTIPLE OR CLOSE TO SPURLSV04070
     1IOUS EIGENVALUES'/' (AS EIGENVALUES OF T(1,MEV)) SO KMAX IS NOT CHLSV04080
     1ANGED'/)                                                          LSV04090
      IF(KMAX.LT.MEV) GO TO 1660                                        LSV04100
      GO TO 330                                                         LSV04110
C                                                                       LSV04120
  310 KMAXN= (11*MEV)/8 + 12                                            LSV04130
      IF((KMAXN/2)*2.NE.KMAXN) KMAXN = KMAXN + 1                        LSV04140
      IF(MBETA.LE.KMAXN) GO TO 1840                                     LSV04150
      IF(KMAX.GE.KMAXN )  GO TO 330                                     LSV04160
      WRITE(6,320) KMAX, KMAXN                                          LSV04170
  320 FORMAT(' ENLARGE KMAX FROM ',I6,' TO ',I6)                        LSV04180
      MOLD1 = KMAX + 1                                                  LSV04190
      KMAX = KMAXN                                                      LSV04200
      GO TO 420                                                         LSV04210
C                                                                       LSV04220
  330 WRITE(6,340) KMAX                                                 LSV04230
  340 FORMAT(/' T-MATRICES HAVE BEEN READ IN FROM FILE 2'/' THE LARGEST LSV04240
     1SIZE T-MATRIX ALLOWED IS',I6/)                                    LSV04250
C                                                                       LSV04260
      IF(IREAD.EQ.1) GO TO 460                                          LSV04270
C                                                                       LSV04280
C     REGENERATE THE BETA                                               LSV04290
C                                                                       LSV04300
  350 MOLD1 = 1                                                         LSV04310
C                                                                       LSV04320
      IF(IPAR.EQ.1) WRITE(6,360)                                        LSV04330
      IF(IPAR.EQ.2) WRITE(6,370)                                        LSV04340
  360 FORMAT(/' STARTING VECTOR USED IN HISTORY REGENERATION IS OF THE  LSV04350
     1FORM (0,V2)')                                                     LSV04360
  370 FORMAT(/' STARTING VECTOR USED IN HISTORY REGENERATION IS OF THE  LSV04370
     1FORM (V1,0)')                                                     LSV04380
C                                                                       LSV04390
      DO 380 J = 1,NGOOD                                                LSV04400
      IF(MP(J).EQ.1) GO TO 400                                          LSV04410
  380 CONTINUE                                                          LSV04420
      KMAX = MEV + 12                                                   LSV04430
      IF((KMAX/2)*2.NE.KMAX) GO TO 1680                                 LSV04440
      WRITE(6,390) KMAX                                                 LSV04450
  390 FORMAT(/' ALL SINGULAR VALUES FOR WHICH SINGULAR VECTORS ARE TO BELSV04460
     1COMPUTED ARE EITHER T-MULTIPLE OR CLOSE TO'/' A SPURIOUS T-EIGENVALSV04470
     1LUE THEREFORE SET KMAX = MEV + 12 = ',I7)                         LSV04480
      GO TO 420                                                         LSV04490
C                                                                       LSV04500
  400 KMAXN = (11*MEV)/8 + 12                                           LSV04510
      IF((KMAXN/2)*2.NE.KMAXN) KMAXN = KMAXN + 1                        LSV04520
      IF(MBETA.LE.KMAXN) GO TO 1840                                     LSV04530
      WRITE(6,410) KMAXN                                                LSV04540
  410 FORMAT(' SET KMAX EQUAL TO ',I6)                                  LSV04550
      KMAX = KMAXN                                                      LSV04560
C                                                                       LSV04570
  420 KMAX1 = KMAX + 1                                                  LSV04580
      WRITE(6,430) MOLD1,KMAX1                                          LSV04590
  430 FORMAT(/' LANCZS SUBROUTINE GENERATES BETA(J+1), J =',            LSV04600
     1 I6,' TO ', I6/)                                                  LSV04610
      IF(IREAD.EQ.1.AND.IPAR.EQ.1) WRITE(6,440)                         LSV04620
      IF(IREAD.EQ.1.AND.IPAR.EQ.2) WRITE(6,450)                         LSV04630
  440 FORMAT(/' FIRST LANCZOS VECTOR IN HISTORY EXTENSION IF OF THE FORMLSV04640
```

```
     1 (0,V2)')                                                         LSV04650
  450 FORMAT(/' FIRST LANCZOS VECTOR IN HISTORY EXTENSION IF OF THE FORMLSV04660
     1 (V1,0)')                                                         LSV04670
C                                                                       LSV04680
C-----------------------------------------------------------------------LSV04690
C                                                                       LSV04700
      CALL LANCZS(SVMAT,STRAN,BETA,V1,V2,G,KMAX,MOLD1,M,N,IPAR,SVSEED)  LSV04710
C                                                                       LSV04720
C-----------------------------------------------------------------------LSV04730
C                                                                       LSV04740
  460 CONTINUE                                                          LSV04750
C                                                                       LSV04760
C     THE SUBROUTINE STURMI DETERMINES THE SMALLEST SIZE T-MATRIX FOR   LSV04770
C     WHICH THE SINGULAR VALUE IN QUESTION IS AN EIGENVALUE (TO         LSV04780
C     WITHIN A SPECIFIED TOLERANCE) AND IF POSSIBLE THE SMALLEST        LSV04790
C     SIZE T-MATRIX FOR WHICH THE SINGULAR VALUE IS A DOUBLE            LSV04800
C     EIGENVALUE (TO WITHIN THE SAME TOLERANCE).  THE SIZE              LSV04810
C     T-MATRIX THAT WILL BE USED IN EACH OF THE RITZ VECTOR COMPUTATIONSLSV04820
C     IS THEN DETERMINED BY LOOPING ON THE SIZE OF THE T-EIGENVECTOR    LSV04830
C     COMPUTATIONS, STARTING WITH A SIZE DETERMINED FROM THE            LSV04840
C     INFORMATION OBTAINED FROM STURMI.                                 LSV04850
C                                                                       LSV04860
      STUTOL = SCALEO*MULTOL                                            LSV04870
      IF(IWRITE.EQ.1) WRITE(6,470)                                      LSV04880
  470 FORMAT(' FROM STURMI')                                            LSV04890
      DO 510 J = 1,NGOOD                                                LSV04900
      SVAL = GOODSV(J)                                                  LSV04910
C     COMPUTE THE TOLERANCES USED BY STURMI TO DETERMINE AN INTERVAL    LSV04920
C     CONTAINING THE SINGULAR VALUE SVAL.                               LSV04930
      TEMP = DABS(SVAL)*RELTOL                                          LSV04940
      TOLN = DMAX1(TEMP,STUTOL)                                         LSV04950
C                                                                       LSV04960
C-----------------------------------------------------------------------LSV04970
C                                                                       LSV04980
      CALL STURMI(BETA,SVAL,TOLN,EPSM,KMAX,MK1,MK2,IC,IWRITE)           LSV04990
C                                                                       LSV05000
C-----------------------------------------------------------------------LSV05010
C                                                                       LSV05020
C     STORE THE COMPUTED ORDERS OF T-MATRICES FOR LATER PRINTOUT        LSV05030
      IF(MK1.GT.1) GO TO 475                                            LSV05040
C     SVAL IS VERY SMALL SINGULAR VALUE, RESET MK1 TO CORRECT VALUE     LSV05050
      MK1 = MK2                                                         LSV05060
      MK2 = MIN0(2*MK1,KMAX)                                            LSV05070
      M1(J) = MK1                                                       LSV05080
      M2(J) = MK2                                                       LSV05090
      ML(J) = MK2                                                       LSV05100
      GO TO 476                                                         LSV05110
  475 M1(J) = MK1                                                       LSV05120
      M2(J) = MK2                                                       LSV05130
      ML(J) = (MK1 + 3*MK2)/4                                           LSV05140
      IF(MK2.EQ.KMAX)  ML(J) = KMAX                                     LSV05150
C                                                                       LSV05160
  476 IF(IC.GT.0) GO TO 490                                             LSV05170
C     IC = 0 MEANS THERE WAS NO T-EIGENVALUE IN THE DESIGNATED INTERVAL LSV05180
C     EVEN BY T-SIZE KMAX.  THIS MEANS THAT THE SINGULAR VALUE          LSV05190
C     PROVIDED HAS NOT YET CONVERGED SO PROGRAM DOES NOT COMPUTE        LSV05200
C     A SINGULAR VECTOR FOR IT.                                         LSV05210
      WRITE(6,480) J,GOODSV(J),MK1,MK2                                  LSV05220
  480 FORMAT(I6,'TH SINGULAR VALUE',E20.12,' HAS NOT CONVERGED '/       LSV05230
     1' SO DO NOT COMPUTE ANY T-EIGENVECTOR OR RITZ VECTOR FOR IT'      LSV05240
     1/' MK1 AND MK2 FOR THIS SINGULAR VALUE WERE',2I6)                 LSV05250
      MP(J) = MPMIN                                                     LSV05260
      MA(J) = -2*KMAX                                                   LSV05270
      GO TO 510                                                         LSV05280
C     COMPUTE AN APPROPRIATE SIZE T-MATRIX FOR THE GIVEN SINGULAR       LSV05290
C     VALUE.                                                            LSV05300
```

```
  490 IF(M2(J).EQ.KMAX) GO TO 500
C     M1 AND M2 WERE BOTH DETERMINED
      MAJ = (3*M1(J) + M2(J))/4  + 1
      IF((MAJ/2)*2.NE.MAJ)  MAJ = MAJ + 1
      MA(J) = MAJ
      GO TO 510
C     M2 NOT DETERMINED
  500 MAJ = (5*M1(J))/4  + 1
      IF((MAJ/2)*2.NE.MAJ)  MAJ = MAJ + 1
      MA(J) = MAJ
C
  510 CONTINUE
C
      IF (IWRITE.EQ.1) WRITE(6,520) (MA(JJ), JJ=1,NGOOD)
  520 FORMAT(/' 1ST GUESS AT APPROPRIATE SIZE T-MATRICES'/
     1 ' ACTUAL VALUES WILL PROBABLY BE 1/4 AGAIN AS MUCH'/(13I6))
C
C     PRINT OUT TO FILE 10 1ST GUESSES AT SIZES OF THE T-MATRICES TO
C     BE USED IN THE SINGULAR VECTOR COMPUTATIONS.
C     PROGRAM LOOPS ON T-SIZE TO DETERMINE APPROPRIATE SIZE T-MATRIX.
      WRITE(10,530) N,KMAX
  530 FORMAT(2I8,' = ORDER OF USER MATRIX AND MAX ORDER OF T(1,MEV)')
C
      WRITE(10,540)
  540 FORMAT(/' 1ST GUESS AT APPROPRIATE SIZE T-MATRICES'/
     1 ' ACTUAL VALUES WILL PROBABLY BE 1/4 AGAIN AS MUCH'/)
C
      WRITE(10,550)
  550 FORMAT(4X,'J',7X,'GOODSV(J)',4X,'M1(J)',1X,'M2(J)',1X,'MA(J)')
C
      WRITE(10,560) (J,GOODSV(J),M1(J),M2(J), MA(J), J=1,NGOOD)
  560 FORMAT(I5,E19.12,3I6)
C
      IF(MBOUND.EQ.1) WRITE(10,570)
  570 FORMAT(/' GOODSV(J) IS A GOOD EIGENVALUE OF T(1,MEV)'/
     1 ' M1 = SMALLEST VALUE OF M SUCH THAT T(1,M) HAS AT LEAST'/
     1 '      ONE EIGENVALUE IN THE INTERVAL (SV-TOLN,SV+TOLN)'/
     1 '      TOLN(J) = DMAX1(GOODSV(J)*RELTOL, SCALEO*MULTOL)'/
     1 ' M2 = SMALLEST M (IF ANY) SUCH THAT IN THE ABOVE INTERVAL'/
     1 '      T(1,M) HAS AT LEAST TWO EIGENVALUES '/
     1 ' INITIAL VALUE OF MA(J) IS CHOSEN HEURISTICALLY'/
     1 ' PROGRAM LOOPS ON SIZE OF T-MATRIX TO GET APPROPRIATE SIZE'/
     1 ' END OF SIZES OF T-MATRICES FILE 10'///)
C
C
C     TERMINATE AFTER COMPUTING 1ST GUESSES AT  SIZES OF THE
C     T-MATRICES REQUIRED FOR THE GIVEN SINGULAR VALUES?
      IF(MBOUND.EQ.1) GO TO 1700
C
C
C     IS THERE ROOM FOR ALL OF THE REQUESTED T-EIGENVECTORS?
      MTOL = 0
      DO 580 J = 1,NGOOD
      IF(MP(J).EQ.MPMIN) GO TO 580
      MTOL = MTOL + IABS(MA(J))
  580 CONTINUE
      MTOL = (5*MTOL)/4
      IF(MTOL.GT.MDIMTV.AND.NTVCON.EQ.0) GO TO 1720
C
C-----------------------------------------------------------------------
C     GENERATE A RANDOM VECTOR TO BE USED REPEATEDLY BY
C     SUBROUTINE INVERM
C
      IIL = RHSEED
      CALL GENRAN(IIL,G,KMAX)
C
C-----------------------------------------------------------------------
```

```
C                                                                       LSV05980
C     FOR EACH SINGULAR VALUE LOOP ON T-EIGENVECTOR COMPUTATIONS        LSV05990
C     TO COMPUTE AN APPROPRIATE T-EIGENVECTOR TO USE IN THE             LSV06000
C     RITZ VECTOR COMPUTATIONS.                                         LSV06010
C                                                                       LSV06020
      MTOL = 0                                                          LSV06030
      NTVEC = 0                                                         LSV06040
      ILBIS = 0                                                         LSV06050
      DO 770 J = 1,NGOOD                                                LSV06060
      ICOUNT = 0                                                        LSV06070
      ERRMIN = 10.D0                                                    LSV06080
      MABEST = MPMIN                                                    LSV06090
      IF(MP(J).EQ.MPMIN) GO TO 770                                      LSV06100
      TFLAG = 0                                                         LSV06110
      SVAL = GOODSV(J)                                                  LSV06120
      TEMP = RELTOL*DABS(SVAL)                                          LSV06130
      UB = SVAL + DMAX1(STUTOL,TEMP)                                    LSV06140
      LB = SVAL - DMAX1(STUTOL,TEMP)                                    LSV06150
      LB = DMAX1(LB,ZERO)                                               LSV06160
  590 KMAXU = IABS(MA(J))                                               LSV06170
C                                                                       LSV06180
C     SELECT A SUITABLE INCREMENT FOR THE ORDERS OF THE T-MATRICES      LSV06190
C     TO BE CONSIDERED IN DETERMINING APPROPRIATE SIZES FOR THE RITZ    LSV06200
C     VECTOR COMPUTATIONS.  ALL ORDERS CONSIDERED MUST BE EVEN.         LSV06210
      IF(ICOUNT.GT.0) GO TO 610                                         LSV06220
C     SELECT IDELTA(J) BASED UPON THE T-MULTIPLICITY OBTAINED           LSV06230
      IF(M2(J).EQ.KMAX) GO TO 600                                       LSV06240
C     M2 DETERMINED                                                     LSV06250
      IDEL = ((3*M1(J) + 5*M2(J))/8  +  1 - IABS(MA(J)))/10  +  1       LSV06260
      IF((IDEL/2)*2.NE.IDEL) IDEL = IDEL + 1                            LSV06270
      IDELTA(J) = IDEL                                                  LSV06280
      GO TO 610                                                         LSV06290
C     M2 NOT DETERMINED                                                 LSV06300
  600 MAMAX = MIN0((11*MEV)/8 + 12, (13*M1(J))/8 + 1)                   LSV06310
      IDEL = (MAMAX - IABS(MA(J)))/10  +  1                             LSV06320
      IF((IDEL/2)*2.NE.IDEL)  IDEL = IDEL + 1                           LSV06330
      IDELTA(J) = IDEL                                                  LSV06340
  610 ICOUNT = ICOUNT + 1                                               LSV06350
C                                                                       LSV06360
C-----------------------------------------------------------------------LSV06370
C     TO MIMIMIZE THE EFFECT OF THE ONE-SIDED ACCEPTANCE TEST FOR       LSV06380
C     EIGENVALUES IN THE BISEC SUBROUTINE, RECOMPUTE THE GIVEN          LSV06390
C     SINGULAR VALUE AT THE SPECIFIED KMAXU                             LSV06400
C                                                                       LSV06410
      CALL LBISEC(BETA,EPSM,SVAL,SVALN,LB,UB,TTOL,KMAXU,NEVT)           LSV06420
C                                                                       LSV06430
C-----------------------------------------------------------------------LSV06440
C                                                                       LSV06450
C     CHECK WHETHER OR NOT GIVEN T-MATRIX HAS AN EIGENVALUE IN THE      LSV06460
C     SPECIFIED INTERVAL AND IF SO WHAT ITS T-MULTIPLICITY IS.          LSV06470
C                                                                       LSV06480
      IF(NEVT.EQ.1) GO TO 650                                           LSV06490
      IF(NEVT.NE.0)  GO TO 630                                          LSV06500
      ILBIS = 1                                                         LSV06510
      WRITE(6,620) SVAL,KMAXU                                           LSV06520
  620 FORMAT(/' PROBLEM ENCOUNTERED IN RECOMPUTATION OF USER-SUPPLIED SILSV06530
     1NGULAR VALUE',E20.12/' THE SIZE T-MATRIX SPECIFIED',I6,' DOES NOT LSV06540
     1HAVE A SINGULAR VALUE IN THE INTERVAL SPECIFIED'/' INCREASE SIZE ALSV06550
     1ND TRY AGAIN'/)                                                   LSV06560
      GO TO 670                                                         LSV06570
C                                                                       LSV06580
  630 IF(NEVT.GT.1)  WRITE(6,640) SVAL,KMAXU                            LSV06590
  640 FORMAT(/' PROBLEM ENCOUNTERED IN RECOMPUTATION OF USER-SUPPLIED   LSV06600
     1SINGULAR VALUE',E20.12/' FOR THE SIZE T-MATRIX SPECIFIED =',I6,' TLSV06610
     1HE GIVEN SINGULAR VALUE IS T-MULTIPLE IN THE INTERVAL SPECIFIED'/'LSV06620
     1SOMETHING IS WRONG, THEREFORE NO SINGULAR VECTORS WILL BE COMPUTEDLSV06630
```

```
     1 FOR THIS SINGULAR VALUE'/)
C
      MP(J) = MPMIN
      MA(J) = -2*KMAX
      GO TO 770
C
  650 CONTINUE
      ILBIS = 0
C
C
      SVNEW(J) = SVALN
      SVAL = SVALN
      MTOL = MTOL+KMAXU
C
C     IS THERE ROOM IN TVEC ARRAY FOR THE NEXT T-EIGENVECTOR?
C     IF NOT, SKIP TO RITZ VECTOR COMPUTATIONS.
      IF (MTOL.GT.MDIMTV) GO TO 780
C
      IT = 3
      KINT = MTOL - KMAXU +1
C
C     RECORD THE BEGINNING AND END OF THE T-EIGENVECTOR BEING COMPUTED
      MINT(J) = KINT
      MFIN(J) = MTOL
C
C-----------------------------------------------------------------------
C     SUBROUTINE INVERM DOES INVERSE ITERATION, I.E. SOLVES
C     (T(1,KMAXU) - SVAL)*U = RHS  FOR EACH SINGULAR VALUE TO
C     OBTAIN THE DESIRED T-EIGENVECTOR.
C
      IF(IWRITE.EQ.1)  WRITE(6,660) J
  660 FORMAT(/I6,'TH SINGULAR VALUE ')
C
      CALL INVERM(BETA,V1,TVEC(KINT),SVAL,ERROR,TERROR,EPSM,G,KMAXU,
     1 IT,IWRITE)
C
C-----------------------------------------------------------------------
C
      TERR(J) = TERROR
      TLAST(J) = ERROR
      KMAXU1 = KMAXU + 1
      TBETA(J) = BETA(KMAXU1)*ERROR
C
C     AFTER COMPUTING EACH OF THE T-EIGENVECTORS,
C     CHECK THE SIZE OF THE ERROR ESTIMATE, ERROR.
C     IF THIS ESTIMATE IS NOT AS SMALL AS DESIRED AND
C     |MA(J)| < ML(J), ATTEMPT TO INCREASE THE SIZE OF |MA(J)|
C     AND REPEAT THE T-EIGENVECTOR COMPUTATIONS.
C
      IF(ERROR.LT.ERTOL.OR.TFLAG.EQ.1)  GO TO 760
C
      IF(ERROR.GE.ERRMIN) GO TO 670
C     LAST COMPONENT IS LESS THAN MINIMAL TO DATE
      ERRMIN = ERROR
      MABEST = MA(J)
  670 CONTINUE
C
      IF(MA(J).GT.0)  ITEST = MA(J) + IDELTA(J)
      IF(MA(J).LT.0)  ITEST = -(IABS(MA(J)) + IDELTA(J))
      IF(IABS(ITEST).LE.ML(J).AND.ICOUNT.LE.10) GO TO 690
C     NEW MA(J) IS GREATER THAN MAXIMUM ALLOWED.
      IF(ERCONT.EQ.0.OR.MABEST.EQ.MPMIN) GO TO 710
      TFLAG = 1
      MA(J) = MABEST
      IF(ILBIS.EQ.0)   MTOL = MTOL - KMAXU
      WRITE(6,680) MA(J)
  680 FORMAT(' 10 ORDERS WERE CONSIDERED.  NONE SATISFIED THE ERROR TEST
```

```
     1'/' THEREFORE USE THE BEST ORDER OBTAINED FOR THE T-EIGENVECTORS'
     1,I6)
      GO TO 590
C
  690 MA(J) = ITEST
C
      MT = IABS(MA(J))
      IF(IWRITE.EQ.1.AND.ILBIS.EQ.0) WRITE(6,700) MT
  700 FORMAT(/' CHANGE SIZE OF T-MATRIX TO ',I6,' RECOMPUTE T-EIGENVECTO
     1R')
C
      IF(ILBIS.EQ.0)  MTOL = MTOL - KMAXU
C
      GO TO 590
C
C     APPROPRIATE SIZE T-MATRIX WAS NOT OBTAINED
  710 CONTINUE
      WRITE(10,720) J,SVAL,MP(J)
  720 FORMAT(/' ON 10 INCREMENTS NOT ABLE TO IDENTIFY APPROPRIATE SIZE
     1T-MATRIX FOR'/
     1I4,' TH SINGULAR VALUE  = ',E20.12,' T-MULTIPLICITY =',I4/)
      IF(M2(J).EQ.KMAX) WRITE(10,730)
      IF(M2(J).LT.KMAX) WRITE(10,740)
  730 FORMAT(/' ORDERS TESTED RANGED FROM 5*M1(J)/4 TO APPROXIMATELY'/
     1 '  MIN(11*MEV/8, 13*M1(J)/8)'/)
  740 FORMAT(/' ORDERS TESTED RANGED FROM (3*M1(J)+M2(J))/4  TO APPROXIM
     1ATELY'/'  (3*M1(J)+5*M2(J))/8'/)
      WRITE(10,750)
  750 FORMAT(' ALLOWING LARGER ORDERS FOR THE T-MATRICES MAY RESULT IN
     1 SUCCESS'/' BUT PROBABLY WILL NOT.  PROBLEM IS PROBABLY DUE TO'
     1 /' LACK OF CONVERGENCE OF GIVEN SINGULAR VALUE, CHECK THE ERROR E
     1STIMATE')
      MP(J) = MPMIN
      IF(ILBIS.EQ.0)  MTOL = MTOL - KMAXU
      GO TO 770
  760 NTVEC = NTVEC + 1
C
  770 CONTINUE
      NGOODC = NGOOD
      GO TO 800
C
C     COME HERE IF THERE IS NOT ENOUGH ROOM FOR ALL OF T-EIGENVECTORS
  780 NGOODC = J-1
      WRITE(6,790) J,MTOL,MDIMTV
  790 FORMAT(/' NOT ENOUGH ROOM IN TVEC ARRAY FOR ',I4,'TH T-EIGENVECTOR
     1'/'  TVEC DIMENSION REQUESTED = ',I6,' BUT TVEC HAS DIMENSION ',I6
     1/)
      IF(NGOODC.EQ.0) GO TO 1740
      MTOL = MTOL-KMAXU
C
  800 CONTINUE
C
C     THE LOOP ON T-EIGENVECTOR COMPUTATIONS IS COMPLETE.
C     WRITE OUT THE SIZE T-MATRICES THAT WILL BE USED FOR
C     THE RITZ VECTOR COMPUTATIONS.
C
      WRITE(10,810)
  810 FORMAT(/' SIZES OF T-MATRICES THAT WILL BE USED IN THE RITZ COMPUT
     1ATIONS'/5X,'J',8X,' SINGULAR VALUE  ',1X,'MA(J)')
C
      WRITE(10,820)   (J,GOODSV(J),MA(J),  J=1,NGOOD)
  820 FORMAT(I6,E25.14,I6)
      WRITE(10,570)
C
      WRITE(6,830) MTOL
  830 FORMAT(/' THE CUMULATIVE LENGTH OF THE T-EIGENVECTORS IS',I18)
C
```

```
      WRITE(6,840) NTVEC,NGOOD
  840 FORMAT(/I6,' T-EIGENVECTORS OUT OF',I6,' REQUESTED WERE COMPUTED')
C
C     SAVE THE T-EIGENVECTORS ON FILE 11?
      IF(TVSTOP.NE.1.AND.SVTVEC.EQ.0) GO TO 900
C
      WRITE(11,850) NTVEC,MTOL,MATNO,SVSEED
  850 FORMAT(I6,3I12,' = NTVEC,MTOL,MATNO,SVSEED')
C
      DO 880 J=1,NGOODC
C     IF MP(J) = MPMIN THEN NO SUITABLE T-EIGENVECTOR IS AVAILABLE
C     FOR THAT SINGULAR VALUE.
      IF(MP(J).EQ.MPMIN) WRITE(11,860) J,MA(J),GOODSV(J),MP(J)
  860 FORMAT(2I6,E20.12,I6/' TH SINGVAL,T-SIZE,SVALUE,FLAG,NO EIGVEC')
      IF(MP(J).NE.MPMIN) WRITE(11,870) J,MA(J),GOODSV(J),MP(J)
  870 FORMAT(I6,I6,E20.12,I6/' T-EIGVEC,SIZE T,SVALUE OF A,MP(J)')
      IF(MP(J).EQ.MPMIN) GO TO 880
      KI = MINT(J)
      KF = MFIN(J)
C
      WRITE(11,280) (TVEC(K), K=KI,KF)
C
  880 CONTINUE
C
      IF(TVSTOP.NE.1) GO TO 900
C
      WRITE(6,890) TVSTOP, NTVEC,NGOOD
  890 FORMAT(/' USER SET TVSTOP = ',I1/
     1' THEREFORE PROGRAM TERMINATES AFTER T-EIGENVECTOR COMPUTATIONS'/
     1' T-EIGENVECTORS THAT WERE COMPUTED ARE SAVED ON FILE 11'/
     1I8,' T-EIGENVECTORS WERE COMPUTED OUT OF',I7,' REQUESTED'/)
C
      GO TO 1860
C
  900 CONTINUE
C     IF NOT ALL OF THE REQUESTED T-EIGENVECTORS WERE COMPUTED,
C     ARE THE LANCZOS SINGULAR VECTOR COMPUTATIONS CONTINUED?
C
      IF(NTVEC.NE.NGOOD.AND.LVCONT.EQ.0) GO TO 1760
C
C     COMPUTE THE MAXIMUM SIZE OF THE T-MATRIX USED FOR THOSE
C     SINGULAR VALUES WITH GOOD ERROR ESTIMATES.
C
      KMAXU = 0
      DO 910 J = 1,NGOODC
      MT = IABS(MA(J))
      IF(MT.LT.KMAXU.OR.MP(J).EQ.MPMIN) GO TO 910
      KMAXU = MT
  910 CONTINUE
C
      IF(KMAXU.EQ.0) GO TO 1800
C
      WRITE(6,920) KMAXU
  920 FORMAT(/I6,' = LARGEST SIZE T-MATRIX TO BE USED IN THE RITZ VECTOR
     1 COMPUTATIONS')
C
C     COUNT THE NUMBER OF RITZ VECTORS NOT BEING COMPUTED
      MREJEC = 0
      DO 930 J=1,NGOODC
  930 IF(MP(J).EQ.MPMIN)  MREJEC = MREJEC + 1
      MREJET = MREJEC + (NGOOD-NGOODC)
      IF(MREJET.NE.0) WRITE(6,940) MREJET
  940 FORMAT(/' RITZ VECTORS ARE NOT COMPUTED FOR',I6,' OF THE SINGULAR
     1VALUES'/)
      NACT = NGOODC - MREJEC
      WRITE(6,950) NGOOD,NTVEC,NACT
  950 FORMAT(/I6,' RITZ VECTORS WERE REQUESTED'/I6,' T-EIGENVECTORS WERE
```

```
     1 COMPUTED'/I6,' RITZ VECTORS WILL BE COMPUTED'/)
C     CHECK IF THERE ARE ANY RITZ VECTORS TO COMPUTE
      IF(MREJEC.EQ.NGOODC) GO TO 1780
C
C     CONTINUE WITH THE LANCZOS VECTOR COMPUTATIONS?
      IF(LVCONT.EQ.0.AND.MREJEC.NE.0) GO TO 1760
C
C     NOW COMPUTE THE RITZ VECTORS.  REGENERATE THE
C     LANCZOS VECTORS.
C
      DO 960 I = 1,MNMAX
  960 RITVEC(I) = ZERO
C
C     REGENERATE THE STARTING VECTOR. THIS MUST BE GENERATED AND
C     NORMALIZED PRECISELY THE WAY IT WAS DONE IN THE CORRESPONDING
C     SINGULAR VALUE COMPUTATIONS, OTHERWISE THERE WILL BE A
C     MISMATCH BETWEEN THE T-EIGENVECTORS THAT HAVE BEEN COMPUTED
C     FROM THE T-MATRICES READ IN FROM FILE 2 (IF THEY WERE READ IN)
C     AND THE LANCZOS TRIDIAGONAL MATRICES  THAT ARE BEING REGENERATED.
C
C     STARTING VECTORS ARE OF THE FORM (V1,0) OR (0,V2) WHERE V1 IS
C     OF LENGTH M AND V2 IS OF LENGTH N.  SUCCEEDING LANCZOS VECTORS
C     ALTERNATE BETWEEN THESE TWO FORMS AND THE DIAGONAL ENTRIES OF THE
C     T-MATRICES ALL VANISH.  THE PARAMETER IPARO DETERMINES THE SHAPE
C     OF THE STARTING VECTOR.  IF IPARO=1, THEN STARTING VECTOR WAS
C     OF THE FORM (0,V2).   IF IPARO=2, THEN STARTING VECTOR WAS OF
C     THE FORM (V1,0).
C     REGENERATE STARTING VECTOR
      BATA = ZERO
      IPAR = IPARO
      ITNUM = 1
      IF (IPAR.EQ.2) GO TO 1020
C
C-----------------------------------------------------------------------
C     IPAR = 1 SO SET V2 TO RANDOM UNIT VECTOR AND SET V1 = 0.
C
      IIL = SVSEED
      CALL GENRAN(IIL,G,N)
C
C-----------------------------------------------------------------------
C
      DO 970 J = 1,N
  970 V2(J) = G(J)
C-----------------------------------------------------------------------
      SUM = ONE/DSQRT(FINPRO(N,V2,1,V2,1))
C-----------------------------------------------------------------------
C
      DO 980 J = 1,M
  980 V1(J) = ZERO
C
      DO 990 J = 1,N
  990 V2(J) = V2(J)*SUM
C
C     INITIALIZE RITZ VECTORS
      DO 1010 J = 1,NGOODC
      IF (MP(J).EQ.MPMIN) GO TO 1010
      LL = MN*J - N
      II = MINT(J)
      TEMP = TVEC(II)
C
      DO 1000 K = 1,N
      LL = LL + 1
 1000 RITVEC(LL) = TEMP*V2(K)
C
 1010 CONTINUE
C
```

```
      GO TO 1150
C
 1020 CONTINUE
C
C-----------------------------------------------------------------------
C     IPAR = 2 SO SET V1 TO RANDOM UNIT VECTOR AND SET V2 = 0.
C
      CALL GENRAN(SVSEED,G,M)
C
C-----------------------------------------------------------------------
C
      DO 1030 J = 1,M
 1030 V1(J) = G(J)
C-----------------------------------------------------------------------
      SUM = ONE/DSQRT(FINPRO(M,V1,1,V1,1))
C-----------------------------------------------------------------------
C
      DO 1040 J = 1,N
 1040 V2(J) = ZERO
C
      DO 1050 J = 1,M
 1050 V1(J) = V1(J)*SUM
C
C     INITIALIZE RITZ VECTORS
      DO 1070 J = 1,NGOODC
      IF (MP(J).EQ.MPMIN) GO TO 1070
      LL = MN*(J-1)
      II = MINT(J)
      TEMP = TVEC(II)
C
      DO 1060 K = 1,M
      LL = LL + 1
 1060 RITVEC(LL) = TEMP*V1(K)
C
 1070 CONTINUE
C
 1080 CONTINUE
C
C  DO ONE ITERATION OF LANCZOS WHERE NEW LANCZOS VECTOR WILL HAVE THE
C  FORM  (0,V2).
C
C-----------------------------------------------------------------------
C
      CALL STRAN(V1,V2,BATA)
C
C-----------------------------------------------------------------------
C
C-----------------------------------------------------------------------
      BATA = DSQRT(FINPRO(N,V2,1,V2,1))
C-----------------------------------------------------------------------
      SUM = ONE/BATA
      ITNUM = ITNUM + 1
      IPAR = 2
C
      TEMP = BETA(ITNUM)
      TEMP = DABS(BATA - TEMP)/TEMP
      IF (TEMP.LT.1.0D-10) GO TO 1110
C
C     HISTORY MISMATCH ON REGENERATION THUS DEFAULT
 1090 WRITE(6,1100) ITNUM,IPAR,BATA,BETA(ITNUM),TEMP
 1100 FORMAT(1X,'ITNUM',2X,'IPAR',16X,'BATA',16X,'BETA',14X,'RELERR'/
     1 2I6,3E20.12/' BATA AND BETA DO NOT AGREE SO PROGRAM STOPS'/)
      GO TO 1860
C
 1110 CONTINUE
C     NORMALIZE LANCZOS VECTOR
      DO 1120 J = 1,N
```

```
 1120 V2(J) = V2(J)*SUM
C
C  UPDATE RITZ VECTORS
      DO 1140 J = 1,NGOODC
      IF (IABS(MA(J)).LT.ITNUM.OR.MP(J).EQ.MPMIN) GO TO 1140
      LL = MN*J - N
      II = MINT(J) + ITNUM - 1
      TEMP = TVEC(II)
C
      DO 1130 K = 1,N
      LL = LL + 1
 1130 RITVEC(LL) = TEMP*V2(K) + RITVEC(LL)
C
 1140 CONTINUE
C     HAVE ALL REQUIRED LANCZOS VECTORS BEEN REGENERATED ?
C
      IF(ITNUM.EQ.KMAXU) GO TO 1190
C
 1150 CONTINUE
C
C     DO ONE ITERATION OF LANCZOS WHERE NEW LANCZOS VECTOR WILL HAVE
C     THE FORM  (V1,0).
C
C-----------------------------------------------------------------------
C
      CALL SVMAT(V2,V1,BATA)
C
C-----------------------------------------------------------------------
C
C-----------------------------------------------------------------------
      BATA = DSQRT(FINPRO(M,V1,1,V1,1))
C-----------------------------------------------------------------------
      SUM = ONE/BATA
      ITNUM = ITNUM + 1
      IPAR = 1
C
      TEMP = BETA(ITNUM)
      TEMP = DABS(BATA - TEMP)/TEMP
      IF (TEMP.GE.1.0D-10) GO TO 1090
C
C     NORMALIZE LANCZOS VECTOR
      DO 1160 J = 1,M
 1160 V1(J) = V1(J)*SUM
C
C     UPDATE RITZ VECTORS
      DO 1180 J = 1,NGOODC
      IF (IABS(MA(J)).LT.ITNUM.OR.MP(J).EQ.MPMIN) GO TO 1180
      LL = MN*(J-1)
      II = MINT(J) + ITNUM - 1
      TEMP = TVEC(II)
C
      DO 1170 K = 1,M
      LL = LL + 1
 1170 RITVEC(LL) = TEMP*V1(K) + RITVEC(LL)
C
 1180 CONTINUE
C  HAVE ALL REQUIRED LANCZOS VECTORS BEEN COMPUTED ?
      IF (ITNUM.LT.KMAXU) GO TO 1080
C
 1190 CONTINUE
C
C     RITZVECTOR GENERATION IS COMPLETE. NORMALIZE EACH RITZVECTOR
C     AS AN EIGENVECTOR OF THE ASSOCIATED SYMMETRIC MATRIX B.
C     THEN COMPUTE THE ERRORS IN THESE VECTORS AS EIGENVECTORS
C     OF B AND WRITE THESE OUT TO FILE 9.  THEN INDIVIDUALLY
C     NORMALIZE THE FIRST M AND THE LAST N COMPONENTS OF EACH OF
C     THESE RITZ VECTORS AND TAKE THESE NORMALIZED VECTORS AS
```

```
C     RESPECTIVELY APPROXIMATIONS TO THE LEFT AND TO THE RIGHT
C     SINGULAR VECTORS OF THE CORRESPONDING SINGULAR VALUE OF
C     THE ORIGINAL MATRIX.
C
C
C     NORMALIZE THE RITZ VECTORS AS EIGENVECTORS OF B
      DO 1280 J = 1,NGOODC
      IF (MP(J).EQ.MPMIN) GO TO 1280
      LINT = MN*(J-1) + 1
      LFIN = MN*J
      SUM = ZERO
      SVAL = SVNEW(J)
C
      DO 1200 K = LINT,LFIN
 1200 SUM = SUM + RITVEC(K)*RITVEC(K)
C
      SUM = DSQRT(SUM)
      RNORM(J) = SUM
      TEMP = ONE - SUM
      SUM = ONE/SUM
C
      DO 1210 K = LINT,LFIN
 1210 RITVEC(K) = RITVEC(K)*SUM
C
C     COMPUTE ERROR IN RITZ VECTOR CONSIDERED AS AN EIGENVECTOR OF B.
      LINTM = LINT + M
      L = LINT - 1
      DO 1220 K = 1,M
      L = L + 1
 1220 V1(K) = RITVEC(L)
      DO 1230 K = 1,N
      L = L + 1
 1230 V2(K) = RITVEC(L)
C
C-----------------------------------------------------------------------
C
      CALL SVMAT(RITVEC(LINTM),V1,SVAL)
      CALL STRAN(RITVEC(LINT),V2,SVAL)
C
C-----------------------------------------------------------------------
C
      SUM = ZERO
      DO 1240 JJ = 1,M
 1240 SUM = SUM + V1(JJ)*V1(JJ)
C
      DO 1250 JJ = 1,N
 1250 SUM = SUM + V2(JJ)*V2(JJ)
C
      IF(IWRITE.NE.0)  WRITE(6,1260) J,GOODSV(J)
 1260 FORMAT(/I5,'TH SINGULAR VALUE CONSIDERED =',E20.12/)
C
      IF(IWRITE.NE.0) WRITE(6,1270) TERR(J), TBETA(J), RNORM(J)
 1270 FORMAT(' RESIDUAL FOR T-EIGENVECTOR = ',E14.3/
     1' DABS(BETA(MA(J)+1)*U(MA(J)) = ' ,E14.3/
     1' NORM(RITZVEC) = ', E14.3/)
C
      SUM = DSQRT(SUM)
      BERR(J) = SUM
      BERRGP(J) = SUM/ABS(BMINGP(J))
 1280 CONTINUE
C
C     RITZVECTORS ARE NORMALIZED AND B-MATRIX ESTIMATES ARE IN BERR
C     AND IN BERRGP ARRAYS.   STORE THESE ESTIMATES BUT NOT THE
C     VECTORS.
C
      WRITE(9,1290)
 1290 FORMAT(11X,'GOODSV(J)',3X,'MA(J)',2X,'  BMINGAP',6X,' BERROR',2X,
```

```
     1 'BERROR/BGAP',4X,' TERROR')
C
      WRITE(13,1300)
 1300 FORMAT(11X,'GOODSV(J)',5X,'RITZNORM',5X,' BMINGAP',5X,'TBETA(J)',
     1 5X,'TLAST(J)')
C
      DO 1330 J=1,NGOODC
C
      IF(MP(J).EQ.MPMIN) GO TO 1330
C
      WRITE(9,1310)SVNEW(J),MA(J),BMINGP(J),BERR(J),BERRGP(J),TERR(J)
 1310 FORMAT(E20.12,I6,4E13.5)
C
      WRITE(13,1320) SVNEW(J),RNORM(J),BMINGP(J),TBETA(J),TLAST(J)
 1320 FORMAT(E20.12,4E13.5)
C
 1330 CONTINUE
C
      IF (MREJEC.EQ.0) GO TO 1410
C
      WRITE(9,1340)
 1340 FORMAT(/' RITZ VECTORS WERE NOT COMPUTED FOR THE FOLLOWING SINGULA
     1R VALUES'/' EITHER BECAUSE THEY HAD NOT CONVERGED OR BECAUSE THE E
     1RROR ESTIMATE'/'  WAS NOT AS SMALL AS DESIRED'/)
C
      WRITE(13,1350)
 1350 FORMAT(/' RITZ VECTORS WERE NOT COMPUTED FOR THE FOLLOWING SINGULA
     1R VALUES'/' EITHER BECAUSE THEY HAD NOT CONVERGED OR BECAUSE THE E
     1ERROR ESTIMATE'/' WAS NOT AS SMALL AS DESIRED'/)
C
      DO 1400 J = 1,NGOODC
      IF(MP(J).NE.MPMIN) GO TO 1400
C     EACH SINGULAR VALUE FOR WHICH NO SINGULAR VECTOR WAS CALCULATED
C     HAS INFORMATION OUTPUTTED TO FILES 4 AND 13
C
      WRITE(9,1360)
 1360 FORMAT(6X,'GOODSV(J)',3X,'MA(J)',5X,'BMINGP(J)',6X,'TLAST(J)',
     1 6X,'TBETA(J)',3X,'MP(J)')
      WRITE(9,1370) GOODSV(J),MA(J),BMINGP(J),TLAST(J),TBETA(J),MP(J)
 1370 FORMAT(E15.8,I8,3E14.4,I8)
C
      WRITE(13,1380)
 1380 FORMAT(6X,'GOODSV(J)',3X,'MA(J)',3X,'M1(J)',3X,'M2(J)',3X,'MP(J)'
     1/)
      WRITE(13,1390) GOODSV(J),MA(J),M1(J),M2(J),MP(J)
 1390 FORMAT(E15.8,4I8)
C
 1400 CONTINUE
C
 1410 CONTINUE
C
      WRITE(9,1420)
 1420 FORMAT(/' ABOVE ARE ERROR ESTIMATES FOR THE B AND T EIGENVECTORS'/
     1 ' ASSOCIATED WITH THE GOODSV LISTED IN COLUMN 1'/
     1 ' BERROR = NORM(B*X - SV*X)  TERROR = NORM(T*Y - SV*Y) '/
     1 ' WHERE T = T(1,MA(J))    X = RITZ VECTOR = V*Y  V = SUCCESSIVE'/
     1 ' LANCZOS VECTORS. BMINGAP = GAP TO NEAREST B-EIGENVALUE'//)
C
      WRITE(13,1430)
 1430 FORMAT(/' ABOVE ARE ERROR ESTIMATES FOR THE GOODSVS'/
     1 ' RITZNORM = NORM(COMPUTED RITZ VECTOR FOR B-MATRIX'/
     1 ' TBETA(J) = BETA(MA(J)+1)*Y(MA(J)),  T*Y = SV*Y '/
     1 ' TLAST(J) = DABS(Y(MA(J)) '/)
C
C     NUMBER OF RITZ VECTORS COMPUTED
      NCOMPU = NGOODC - MREJEC
      WRITE(12,1440) N,NCOMPU,NGOODC,MATNO
```

```
 1440 FORMAT(3I6,I12,' SIZE A, NO.RITZVECS, NO.SVALUES,MATNO')
C
C     INDIVIDUALLY NORMALIZE THE FIRST M AND THE LAST N COMPONENTS OF
C     EACH RITZ VECTOR.
C
      LFIN = 0
      DO 1560 J = 1,NGOODC
C
      IF(MP(J).EQ.MPMIN) GO TO 1540
C
C     RITZ VECTOR WAS COMPUTED
      LINT = MN*(J-1) + 1
      LFIN = MN*J
      LFIN1 = LINT + M - 1
      LINT1 = LFIN1 + 1
C
      SUM = 0.D0
      TEMP = 0.D0
      DO 1450 I = LINT,LFIN1
 1450 SUM = SUM + RITVEC(I)*RITVEC(I)
      SUM = ONE/DSQRT(SUM)
      DO 1460 I = LINT,LFIN1
 1460 RITVEC(I) = SUM*RITVEC(I)
      DO 1470 I = LINT1,LFIN
 1470 TEMP = TEMP + RITVEC(I)*RITVEC(I)
      TEMP = ONE/DSQRT(TEMP)
      DO 1480 I = LINT1,LFIN
 1480 RITVEC(I) = TEMP*RITVEC(I)
C
      WRITE(12,1490) J, GOODSV(J), MP(J)
 1490 FORMAT(/I6,4X,E20.12,I6,' J, SINGULAR VALUE, MP(J)')
C
      WRITE(12,1500) BERR(J),BERRGP(J)
 1500 FORMAT(2E15.5,' = NORM(B*Z-SVAL*Z) AND  NORM(B*Z-SVAL*Z)/BMINGAP')
C
      WRITE(12,1510) J
 1510 FORMAT(/I6,'TH LEFT SINGULAR VECTOR'/)
C     WRITE(12,170) (RITVEC(LL), LL=LINT,LFIN1)
      WRITE(12,1520) (RITVEC(LL), LL=LINT,LFIN1)
 1520 FORMAT(4E20.12)
C
      WRITE(12,1530) J
 1530 FORMAT(/I6,'TH RIGHT SINGULAR VECTOR'/)
C     WRITE(12,170) (RITVEC(LL), LL=LINT1,LFIN)
      WRITE(12,1520) (RITVEC(LL), LL=LINT1,LFIN)
C
      GO TO 1560
C
C     NO RITZ VECTOR WAS COMPUTED FOR THIS SINGULAR VALUE
 1540 WRITE(12,1550) J,GOODSV(J),MP(J)
 1550 FORMAT(I6,4X,E20.12,I6,' J,SINGVALUE,MP(J),NO RITZ VECTOR COMPUTED
     1')
C
 1560 CONTINUE
C
C     DID ANY T-MATRICES INCLUDE OFF-DIAGONAL ENTRIES SMALLER THAN
C     DESIRED, AS SPECIFIED BY BTOL?
C
      IF(IB.GT.0) GO TO 1590
      WRITE(6,1570) KMAXU
 1570 FORMAT(/' FOR LARGEST T-MATRIX CONSIDERED',I7,' CHECK THE SIZE OF
     1BETAS')
C
C-----------------------------------------------------------------------
C
      CALL TNORM(BETA,BKMIN,TEMP,KMAXU,IBMT)
C
```

```
C-----------------------------------------------------------------------LSV12660
C                                                                       LSV12670
      IF(IBMT.LT.0) WRITE (6,1580)                                      LSV12680
 1580 FORMAT(/' WARNING THE T-MATRICES FOR ONE OR MORE OF THE SINGULAR VLSV12690
     1ALUES CONSIDERED'/' HAD AN OFF-DIAGONAL ENTRY THAT WAS SMALLER THALSV12700
     1N THE BETA TOLERANCE THAT WAS SPECIFIED'/)                        LSV12710
 1590 CONTINUE                                                          LSV12720
C                                                                       LSV12730
       GO TO 1860                                                       LSV12740
C                                                                       LSV12750
 1600 WRITE(6,1610) NGOOD,MNMAX,MDIMRV                                  LSV12760
 1610 FORMAT(/I4,' RITZ VECTORS WERE REQUESTED BUT THE REQUIRED DIMENSIOLSV12770
     1N',I6/' IS LARGER THAN THE USER-SPECIFIED DIMENSION OF RITVEC',I6 LSV12780
     1/' THEREFORE, THE SINGULAR VECTOR PROCEDURE TERMINATES FOR THE USELSV12790
     1R TO INTERVENE')                                                  LSV12800
C                                                                       LSV12810
      GO TO 1860                                                        LSV12820
C                                                                       LSV12830
 1620 WRITE(6,1630)  MOLD,M,NOLD,N,MATOLD,MATNO                         LSV12840
 1630 FORMAT(/' GOODSV PARAMETERS READ FROM FILE 3 DO NOT AGREE WITH THOLSV12850
     1SE'/' SPECIFIED BY THE USER. MOLD,M,NOLD,N,MATOLD,MATNO ='/       LSV12860
     14I6, 2I12/' THEREFORE PROGRAM TERMINATES FOR USER TO RESOLVE DIFFELSV12870
     1RENCES'/)                                                         LSV12880
C                                                                       LSV12890
      GO TO 1860                                                        LSV12900
C                                                                       LSV12910
 1640 WRITE(6,1650)                                                     LSV12920
 1650 FORMAT(/' PARAMETERS IN BETA FILE DO NOT AGREE WITH THOSE SPECIFIELSV12930
     1D BY THE USER.'/' THEREFORE, THE PROGRAM TERMINATES FOR THE USER TLSV12940
     1O RESOLVE THE DIFFERENCES'/)                                      LSV12950
C                                                                       LSV12960
      GO TO 1860                                                        LSV12970
C                                                                       LSV12980
 1660 WRITE(6,1670) KMAX,MEV                                            LSV12990
 1670 FORMAT(/' IN BETA HISTORY HEADER KMAX =',I6/                      LSV13000
     1' BUT SINGULAR VALUES WERE COMPUTED AT MEV = ',I6,' PROGRAM STOPS'LSV13010
     1)                                                                 LSV13020
C                                                                       LSV13030
      GO TO 1860                                                        LSV13040
C                                                                       LSV13050
 1680 WRITE(6,1690) MEV                                                 LSV13060
 1690 FORMAT(/' SOMETHING IS WRONG.'/' HEADER SAYS THAT SIZE T-MATRIX USLSV13070
     1ED IN THE SINGULAR VALUE COMPUTATIONS WAS = ',I6/' BUT THIS IS AN LSV13080
     1ODD ORDER AND THAT IS NOT ALLOWED.  PROGRAM STOPS'/)              LSV13090
C                                                                       LSV13100
      GO TO 1860                                                        LSV13110
C                                                                       LSV13120
 1700 WRITE(6,1710)                                                     LSV13130
 1710 FORMAT(/' PROGRAM COMPUTED 1ST GUESSES AT T-MATRIX SIZES, READ THELSV13140
     1M TO FILE 10'/' THEN TERMINATED AS REQUESTED.')                   LSV13150
      GO TO 1860                                                        LSV13160
C                                                                       LSV13170
 1720 WRITE(6,1730) MTOL, MDIMTV                                        LSV13180
 1730 FORMAT(/' PROGRAM TERMINATES BECAUSE THE TVEC DIMENSION ANTICIPATELSV13190
     1D',I7/'  IS LARGER THAN THE TVEC DIMENSION',I7,' SPECIFIED BY THE LSV13200
     1USER.'/'  USER MAY RESET THE TVEC DIMENSION AND RESTART THE PROGRALSV13210
     1M'/)                                                              LSV13220
      GO TO 1860                                                        LSV13230
C                                                                       LSV13240
 1740 WRITE(6,1750)                                                     LSV13250
 1750 FORMAT(/' PROGRAM TERMINATES BECAUSE NO SUITABLE T-EIGENVECTORS WELSV13260
     1RE IDENTIFIED'/' FOR ANY OF THE SINGULAR VALUES SUPPLIED.  PROBLEMLSV13270
     1 COULD BE CAUSED BY'/' TOO SMALL A TVEC DIMENSION OR SIMPLY BE THALSV13280
     1T NO SUITABLE T-VECTORS'/' WERE IDENTIFIED.  USER SHOULD CHECK OUTLSV13290
     1PUT'/)                                                            LSV13300
      GO TO 1860                                                        LSV13310
C                                                                       LSV13320
```

```
 1760 WRITE(6,1770) LVCONT,NTVEC,NGOOD
 1770 FORMAT(/' LVCONT FLAG =',I2,' AND NUMBER ',I5,' OF T-EIGENVECTORS
     1 COMPUTED N.E.'/' NUMBER',I5,' REQUESTED SO PROGRAM TERMINATES'/)
      GO TO 1860
 1780 WRITE(6,1790)
 1790 FORMAT(/' PROGRAM TERMINATES WITHOUT COMPUTING ANY RITZ VECTORS'/
     1/' BECAUSE ALL OF THE T-EIGENVECTORS WERE REJECTED AS NOT SUITABLE
     1 FOR'/' THE RITZ VECTOR COMPUTATIONS.  PROBABLE CAUSE WAS LACK OF
     1CONVERGENCE'/' OF THE SINGULAR VALUES'/)
       GO TO 1860
C
 1800 WRITE(6,1810)
 1810 FORMAT(/' PROGRAM INDICATES THAT IT IS NOT POSSIBLE TO COMPUTE ANY
     1 OF THE'/' REQUESTED EIGENVECTORS. THEREFORE PROGRAM TERMINATES')
      DO 1820 J=1,NGOODC
 1820 WRITE(6,1830)  J,GOODSV(J),MP(J)
 1830 FORMAT(/4X,' J',11X,'GOODSV(J)',4X,'MP(J)'/I6,E20.12,I9)
      GO TO 1860
C
 1840 WRITE(6,1850) MBETA,KMAXN
 1850 FORMAT(/' PROGRAM TERMINATES BECAUSE THE STORAGE ALLOTTED FOR THE
     1BETA ARRAY',I8/' IS NOT SUFFICIENT FOR THE ENLARGED KMAX =',I8,' T
     1HAT THE PROGRAM WANTS'/' USER CAN ENLARGE THE BETA ARRAY AND RERUN
     1 THE PROGRAM'/)
C
 1860 CONTINUE
C
      STOP
C-----END OF MAIN PROGRAM FOR LANCZOS SINGULAR VECTOR COMPUTATIONS------
      END
```

SECTION 6.5 LANCZS AND SAMPLE MATRIX-VECTOR MULTIPLY SUBROUTINES

```
C-----LSMULT-----------------------------------------------------------
C
C     CONTAINS SUBROUTINES LANCZS, USPECS, STRAN, AND SVMAT
C     FOR USE WITH THE LANCZOS SINGULAR VALUE/VECTOR PROGRAMS
C
C     NONPORTABLE CONSTRUCTIONS:
C     1.  THE ENTRY MECHANISM USED TO PASS THE STORAGE LOCATIONS
C         OF THE USER-SPECIFIED MATRIX FROM THE SUBROUTINE USPEC
C         TO THE MATRIX-VECTOR MULTIPLY SUBROUTINES SVMAT AND
C         STRAN.
C     2.  IN THE SAMPLE USPEC PROVIDED:  THE FREE FORMAT (8,*),
C         AND THE FORMAT (20A4).
C
C-----START OF LANCZS--------------------------------------------------
C
      SUBROUTINE LANCZS(MATVEC,MTRAN,BETA,V1,V2,G,KMAX,MOLD1,
     1                  M,N,IPAR,IIX)
C
C----------------------------------------------------------------------
      DOUBLE PRECISION BETA(1),V1(1),V2(1),SUM,TEMP,ONE,ZERO
      REAL G(1)
      DOUBLE PRECISION FINPRO
      INTEGER IPAR
      EXTERNAL MATVEC,MTRAN
C----------------------------------------------------------------------
C     COMPUTE T(1,MEV) FOR SYMMETRIZED VERSION OF GIVEN A-MATRIX.
C
C                                 ----                 ----
C                                |  0                    A  |
C                  B   =         |                          |
C                                |  A-TRANSPOSE          0  |
C                                 ----                 ----
C
C     WHERE A IS AN M BY N REAL SPARSE MATRIX, USING STARTING
C     VECTORS OF THE FORM (V1,0) WHEN THE FLAG IPAR = 2 AND
C     OF THE FORM (0,V2) WHEN THE FLAG IPAR = 1.  V1 IS OF
C     DIMENSION M, THE ROW DIMENSION OF A, AND V2 IS OF DIMENSION
C     N, THE COLUMN DIMENSION OF A.
C
C     WITH STARTING VECTORS OF THESE FORMS, THE LANCZOS VECTORS
C     GENERATED ALTERNATE BETWEEN THESE 2 FORMS AND ALL OF THE
C     DIAGONAL ENTRIES OF THE LANCZOS TRIDIAGONAL MATRICES T(1,MEV)
C     GENERATED ARE 0.
C
C     LANCZS USES 2 USER-SUPPLIED SUBROUTINES MATVEC AND MTRAN.
C     MAIN PROGRAM CALLS THESE SVMAT AND STRAN, RESPECTIVELY.
C     CALLING SEQUENCES ARE
C
C           CALL MATVEC(V2,V1,SUM)
C           CALL MTRAN(V1,V2,SUM)
C
C     MATVEC  COMPUTES V1 = A*V2 - SUM*V1.
C     MTRAN COMPUTES V2 = (A-TRANSPOSE)*V1 - SUM*V2.
C
C     ON EXIT V1 AND V2 CONTAIN THE NONZERO PARTS OF THE
C     LAST TWO LANCZOS VECTORS.
C
C     IF MOLD1 = 1 THEN T(1,KMAX) IS GENERATED FROM SCRATCH.
C     IF MOLD1 > 1 THEN A PREVIOUSLY-GENERATED T-MATRIX OF SIZE
C     (MOLD1-1) IS EXTENDED TO ONE OF SIZE KMAX. SINGULAR VALUE
C     PRGORAMS CAN ONLY UTILIZE T-MATRICES OF EVEN ORDER.
C     BETA(KMAX+1) IS ALSO COMPUTED FOR USE IN THE ERROR ESTIMATES.
C
C----------------------------------------------------------------------
```

```
      ONE  = 1.0D0
      ZERO = 0.0D0
      ITNUM = MOLD1
C
      IF (ITNUM .GT. 1) GO TO (80,100), IPAR
C
C     NO PREVIOUS BETA HISTORY
      BETA(1) = ZERO
      IIL = IIX
      IF (IPAR .EQ. 2) GO TO 40
C
C-----------------------------------------------------------------------
C     IPAR = 1 SO SET V2 EQUAL TO A UNIT RANDOM VECTOR AND SET V1 = 0.
      CALL GENRAN(IIL,G,N)
C-----------------------------------------------------------------------
C
      DO 10 J = 1,N
   10 V2(J) = G(J)
C
C-----------------------------------------------------------------------
      TEMP = FINPRO(N,V2(1),1,V2(1),1)
C-----------------------------------------------------------------------
C
      SUM = ONE/DSQRT(TEMP)
      DO 20 J = 1,M
   20 V1(J) = ZERO
C
      DO 30 J = 1,N
   30 V2(J) = V2(J)*SUM
      GO TO 100
C
   40 CONTINUE
C
C-----------------------------------------------------------------------
C     IPAR = 2 SO SET V1 EQUAL TO A UNIT RANDOM VECTOR AND SET V2 = 0.
      CALL GENRAN(IIL,G,M)
C-----------------------------------------------------------------------
C
      DO 50 J=1,M
   50 V1(J) = G(J)
C
C-----------------------------------------------------------------------
      TEMP = FINPRO(M,V1(1),1,V1(1),1)
C-----------------------------------------------------------------------
C
      SUM = ONE/DSQRT(TEMP)
      DO 60 J = 1,N
   60 V2(J) = ZERO
      DO 70 J = 1,M
   70 V1(J) = V1(J)*SUM
C
C     BELOW IS START FOR MOLD1 > 1 AND IPAR = 1
C     DO ONE ITERATION OF LANCZOS TO OBTAIN (0,V2)
C
   80 CONTINUE
      SUM = BETA(ITNUM)
C
C-----------------------------------------------------------------------
      CALL MTRAN(V1,V2,SUM)
C-----------------------------------------------------------------------
C
C-----------------------------------------------------------------------
      SUM = FINPRO(N,V2(1),1,V2(1),1)
C-----------------------------------------------------------------------
C
      ITNUM = ITNUM + 1
      BETA(ITNUM) = DSQRT(SUM)
```

```
      SUM = ONE/BETA(ITNUM)
C
      DO 90 J = 1,N
   90 V2(J) = V2(J)*SUM
C
      IPAR = 2
      IF (ITNUM .GT. KMAX) GO TO 120
C
C     BELOW IS START FOR MOLD1 > 1 AND IPAR = 2
C     DO ONE ITERATION OF LANCZOS TO OBTAIN (V1,0)
C
  100 CONTINUE
      SUM = BETA(ITNUM)
C
C-----------------------------------------------------------------------
      CALL MATVEC(V2,V1,SUM)
C-----------------------------------------------------------------------
C
C-----------------------------------------------------------------------
      SUM = FINPRO(M,V1(1),1,V1(1),1)
C-----------------------------------------------------------------------
C
      ITNUM = ITNUM + 1
      BETA(ITNUM) = DSQRT(SUM)
      SUM = ONE/BETA(ITNUM)
C
      DO 110 J = 1,M
  110 V1(J)= V1(J) * SUM
C
      IPAR = 1
      IF (ITNUM .GT. KMAX) GO TO 120
      GO TO 80
C
  120 CONTINUE
C
      RETURN
C-----END OF LANCZS-----------------------------------------------------
      END
C
C-----START OF USPEC (GENERAL SPARSE, RECTANGULAR MATRIX)---------------
C
C     SUBROUTINE USPEC(M,N,MATNO)
      SUBROUTINE SUSPEC(M,N,MATNO)
C
C---------------------------------------------------------------------
      DOUBLE PRECISION A(10000)
      INTEGER IROW(10000),ICOL(3010)
C---------------------------------------------------------------------
C    DIMENSIONS ARRAYS NEEDED TO DEFINE THE USER-SUPPLIED
C    M X N RECTANGULAR A-MATRIX, READS IN VALUES OF THESE
C    ARRAYS AND THEN PASSES THE STORAGE LOCATIONS OF THESE
C    ARRAYS TO THE CORRESPONDING MATRIX-VECTOR MULTIPLY
C    SUBROUTINES SVMAT AND STRAN.
C
C    THE A-MATRIX IS STORED IN THE FOLLOWING SPARSE FORMAT:
C    M = NUMBER OF ROWS IN A.
C    N = NUMBER OF COLUMNS IN A.
C    NZ = NUMBER OF NONZERO ENTRIES IN A-MATRIX.
C    ICOL(J), J=1,N IS NUMBER OF NONZERO ENTRIES IN COLUMN J.
C    IROW(K), K = 1,NZ IS THE ROW INDEX FOR CORRESPONDING A(K).
C    A(K), K=1,NZ IS NONZERO ENTRIES IN A, COLUMN BY COLUMN.
C    IT IS ASSUMED THAT ICOL(J) > 0 FOR ALL J
C
C     NOTE:  ASSOCIATED SUBROUTINES SVMAT AND STRAN ASSUME THAT
C            M >= N.
C
C-----------------------------------------------------------------------
```

```
C     READ IN  MATRIX FROM  FILE 8
C
      READ(8,10) NZ,MOLD,NOLD,MATOLD
   10 FORMAT(I10,2I6,I8)
C
      WRITE(6,20) NZ,MOLD,NOLD,MATOLD
   20 FORMAT(6X,'NZ',4X,'MOLD',4X,'NOLD',4X,'MATOLD'/I10,2I6,I10/)
C
C     TEST OF PARAMETER CORRECTNESS
      ITEMP = (MOLD-M)**2 + (NOLD-N)**2 + (MATOLD-MATNO)**2
C
      IF (ITEMP.EQ.0) GO TO 40
C
      WRITE(6,30)
   30 FORMAT(' PROGRAM TERMINATES BECAUSE EITHER ORDERS OF OR LABELS FOR
     1 MATRIX DISAGREE')
      GO TO 70
C
   40 CONTINUE
C
C     NUMBER OF NONZERO ENTRIES IN EACH COLUMN IS READ IN
C     THEN THE CORRESPONDING ROW INDEX FOR EACH SUCH ENTRY IS READ
      READ(8,50) (ICOL(K), K=1,N)
      READ(8,50) (IROW(K), K=1,NZ)
   50 FORMAT(13I6)
C
C     READ IN THE NONZERO ENTRIES IN THE MATRIX
      READ(8,60) (A(K), K=1,NZ)
   60 FORMAT(3E25.16)
C  50 FORMAT(4E19.10)
C
C-----------------------------------------------------------------------
C     PASS STORAGE LOCATIONS OF ARRAYS THAT DEFINE THE MATRIX TO
C     THE MATRIX-VECTOR MULTIPLY SUBROUTINES SVMAT AND STRAN
      CALL SMATVE(A,ICOL,IROW,M,N)
      CALL STRANE(A,ICOL,IROW,M,N)
C-----------------------------------------------------------------------
C
C-----END OF USPEC------------------------------------------------------
      RETURN
   70 STOP
      END
C
C-----STRAN (GENERAL SPARSE MATRIX)-------------------------------------
C
C     SUBROUTINE STRAN(W,U,SUM)
      SUBROUTINE SSTRAN(W,U,SUM)
C
C-----------------------------------------------------------------------
      DOUBLE PRECISION W(1),U(1),A(1),SUM,TEMP
      INTEGER IROW(1),ICOL(1)
C-----------------------------------------------------------------------
C     SUBROUTINE TO COMPUTE U = (A-TRANSPOSE)*W - SUM*U WHERE A IS
C     A GENERAL, SPARSE M X N MATRIX WITH M >= N.
C
C     ASSUMES MATRIX IS STORED IN SPARSE FORMAT GIVEN IN
C     CORRESPONDING USPEC SUBROUTINE.
C-----------------------------------------------------------------------
      JLAST = 0
      DO 20 J = 1,N
      JFIRST = JLAST + 1
      JLAST  = JLAST + ICOL(J)
      TEMP   = -SUM*U(J)
C
      DO 10 K = JFIRST,JLAST
      IK = IROW(K)
```

```
   10 TEMP = A(K)*W(IK) + TEMP
C
   20 U(J) = TEMP
C
      RETURN
C
C-----------------------------------------------------------------------
      ENTRY STRANE(A,ICOL,IROW,M,N)
C-----------------------------------------------------------------------
C
C-----END OF STRAN FOR GENERAL SPARSE MATRIX----------------------------
      RETURN
      END
C
C-----SVMAT (GENERAL SPARSE MATRIX)-------------------------------------
C
C     SUBROUTINE SVMAT(W,U,SUM)
      SUBROUTINE SSVMAT(W,U,SUM)
C
C-----------------------------------------------------------------------
      DOUBLE PRECISION  W(1),U(1),A(1),SUM,TEMP
      INTEGER IROW(1),ICOL(1)
C-----------------------------------------------------------------------
C     SUBROUTINE TO COMPUTE U = A*W - SUM*U WHERE A IS A
C     GENERAL, SPARSE M X N MATRIX WITH M >= N.
C
C     ASSUMES THAT THE MATRIX IS STORED IN THE SPARSE FORMAT
C     GIVEN IN THE CORRESPONDING USPEC SUBROUTINE.
C-----------------------------------------------------------------------
      DO 10 I = 1,M
   10 U(I) = -SUM*U(I)
C
C  MAIN LOOP.  PROCESSING PROCEEDS COL BY COL.  JFIRST AND JLAST ARE
C  POINTERS TO THE FIRST AND LAST NONZEROS IN COLUMN J.
C
      JLAST = 0
      DO 30 J = 1,N
      JFIRST = JLAST + 1
      JLAST = JLAST + ICOL(J)
      TEMP = W(J)
C
      DO 20 K = JFIRST,JLAST
      IK = IROW(K)
   20 U(IK) = U(IK) + A(K)*TEMP
C
   30 CONTINUE
C
      RETURN
C
C-----------------------------------------------------------------------
      ENTRY SMATVE(A,ICOL,IROW,M,N)
C-----------------------------------------------------------------------
C
C----END OF SVMAT FOR GENERAL SPARSE MATRICES---------------------------
      RETURN
      END
C
C-----ROUTINES FOR 'DIAGONAL' TEST MATRICES-----------------------------
C     DMATV,DMTRAN,DIAGSP SUBROUTINES ARE FOR RECTANGULAR DIAGONAL
C     TEST MATRICES.
C
C-----START OF USPEC FOR 'DIAGONAL' TEST MATRIX-------------------------
C
      SUBROUTINE USPEC(M,N,MATNO)
C     SUBROUTINE DIAGSP(M,N,MATNO)
C
C     DEFINES 'DIAGONAL' MATRIX OF FOLLOWING FORM
```

```
C
C                         -----         -----
C                        | 0       0       D |
C               A =      | 0       0       0 |
C                        |D-TRANS  0       0 |
C                         -----         -----
C
C     WHERE D IS DIAGONAL MATRIX OF ORDER N, AND IN THE
C     MIDDLE THERE ARE (M-N) ROWS OF ZEROES.
C     CALLS ENTRY TO MATRIX-VECTOR MULTIPLY SUBROUTINE TO PASS
C     STORAGE LOCATION OF THE D-ARRAY AND THE ORDERS M AND N.
C
C     NOTE:  ASSOCIATED MATRIX-VECTOR SUBROUTINES ASSUME THAT
C            M >= N.
C-----------------------------------------------------------------------
      DOUBLE PRECISION  D(1000), SPACE
      REAL  EXPLAN(20)
C-----------------------------------------------------------------------
C
      READ(8,10) EXPLAN
   10 FORMAT(20A4)
      READ(8,*) MOLD,NOLD,NUNIF,SPACE,D(1)
C
      IF(N.NE.NOLD.OR.M.NE.MOLD) GO TO 80
C     COMPUTE THE UNIFORM PORTION OF THE SPECTRUM
      DO 20 J=2,NUNIF
   20 D(J) = D(1) - DFLOAT(J-1)*SPACE
      NUNIF1=NUNIF + 1
      READ(8,10)  EXPLAN
      DO 30 J=NUNIF1,N
   30 READ(8,*) D(J)
      NNUNIF = NOLD - NUNIF
      WRITE(6,40) NOLD,SPACE,NNUNIF,D(1)
   40 FORMAT(/' DIAGONAL TEST MATRIX, SIZE = ',I4/' MOST ENTRIES ARE ',
     1E10.3,' UNITS APART.',I3,' ENTRIES'/' ARE IRREGULARLY SPACED. FIRS
     1T ENTRY IS ',E10.3/)
      NB = NUNIF - 2
C
C
C     PRINT OUT DIAGONAL PORTION OF A-MATRIX
      WRITE(6,50) (D(I), I=1,10 )
      WRITE(6,60) (D(I), I = NB,N)
      MNDIF = MOLD - NOLD
      IF(MNDIF.NE.0) WRITE(6,70) MNDIF
   50 FORMAT(/' SINGULAR VALUE LANCZOS TEST, 1ST 10 ENTRIES OF DIAGONAL
     1A-MATRIX = '/(3E22.14))
   60 FORMAT(/' MIDDLE UNIFORM PORTION OF MATRIX IS NOT PRINTED OUT'/
     1' END OF UNIFORM PLUS NONUNIFORM SECTION = '/(3E22.14))
   70 FORMAT(I4,' ZERO ROWS ARE ADDED TO THE DIAGONAL TO MAKE IT RECTANG
     1ULAR'/)
C
C     DIAGONAL GENERATION COMPLETE
C
C-----------------------------------------------------------------------
C     CALL ENTRY TO MATRIX-VECTOR MULTIPLY SUBROUTINES  TO PASS
C     STORAGE LOCATION OF D-ARRAY AND ORDER OF A-MATRIX.
      CALL DMATVE(D,M,N)
      CALL DMTRAE(D,M,N)
C-----------------------------------------------------------------------
C
      RETURN
   80 WRITE(6,90) MOLD,NOLD,M,N
   90 FORMAT(' PROGRAM TERMINATES MOLD=',I5,' N.E. M=',I5,' OR NOLD=',
     1I5,' N.E. N=',I5)
C-----END OF USPEC SUBROUTINE FOR 'DIAGONAL' TEST MATRICES--------------
      STOP
```

```
      END
C
C-----DSVMAT ('DIAGONAL' TEST MATRICES)---------------------------------
C
C     SUBROUTINE DSVMAT(Z,W,SUM)
      SUBROUTINE SVMAT(Z,W,SUM)
C
C-----------------------------------------------------------------------
      DOUBLE PRECISION  A(1),Z(1),W(1),SUM
C-----------------------------------------------------------------------
C
C     COMPUTES W = A*Z - SUM*W .  ASSUMES THAT M >= N.
      DO 10 I = 1,N
   10 W(I) = A(I)*Z(I) - SUM *W(I)
      IF(M.EQ.N) RETURN
      N1 = N+1
      DO 20 I = N1,M
   20 W(I) = -SUM*W(I)
      RETURN
C
C-----------------------------------------------------------------------
C     STORAGE LOCATIONS OF THE A-ARRAY
C     AND THE ORDER OF THE A-MATRIX ARE PASSED TO THE MATVEC SUBROUTINE.
C     ENTRY MATVE(A,M,N)
      ENTRY DMATVE(A,M,N)
C-----------------------------------------------------------------------
C
C-----END OF MATRIX -VECTOR MULTIPLY 'DIAGONAL' TEST PROBLEMS-----------
      RETURN
      END
C
C-----MATRIX-VECTOR MULTIPLY FOR 'DIAGONAL' TEST MATRICES--------------
C
      SUBROUTINE STRAN(Z,W,SUM)
C     SUBROUTINE DSTRAN(Z,W,SUM)
C
C-----------------------------------------------------------------------
      DOUBLE PRECISION  A(1),Z(1),W(1),SUM
C-----------------------------------------------------------------------
C
C     COMPUTES W = A-TRANSPOSE*Z - SUM*W .  ASSUMES M >= N.
      DO 10 I = 1,N
   10 W(I) = A(I)*Z(I)- SUM*W(I)
      RETURN
C
C-----------------------------------------------------------------------
C     STORAGE LOCATIONS OF THE A-ARRAY AND THE ORDER
C     OF THE A-MATRIX ARE OBTAINED FROM USPEC SUBROUTINE.
C     ENTRY MTRANE(A,M,N)
      ENTRY DMTRAE(A,M,N)
C-----------------------------------------------------------------------
C
C-----END OF SPARSE SYMMETRIC MATRIX-VECTOR MULTIPLY--------------------
      RETURN
      END
```

SECTION 6.6 OTHER SUBROUTINES USED BY THE PROGRAMS IN CHAPTER 6

```
C-----LSSUB-------(SINGULAR VALUES AND VECTORS)-------------------------
C     ACCORDING TO PFORT THESE SUBROUTINES ARE PORTABLE
C
C
C     SUBROUTINES    BISEC, INVERR, TNORM, LUMP, ISOEV, PRTEST, AND
C                    INVERM ARE USED WITH LANCZOS SINGULAR VALUE
C                    PROGRAM LSVAL.  STURMI, INVERM, LBISEC, TNORM
C                    ARE USED WITH THE LANCZOS SINGULAR VECTOR
C                    PROGRAM LSVEC.
C
C
C-----COMPUTE T-EIGENVALUES BY BISECTION-------------------------------
C
      SUBROUTINE BISEC(BETA,BETA2,VB,VS,LBD,UBD,EPS,TTOL,MP,
     1 NINT,MEV,NDIS,IC,IWRITE)
C
C-----------------------------------------------------------------------
      DOUBLE PRECISION  BETA(1),BETA2(1),VB(1),VS(1)
      DOUBLE PRECISION  LBD(1),UBD(1),EPS,EPT,EP0,EP1,TEMP,TTOL
      DOUBLE PRECISION  ZERO,ONE,HALF,YU,YV,LB,UB,XL,XU,X1,X0,XS,BETAM
      INTEGER  MP(1),IDEF(10)
      DOUBLE PRECISION  DABS, DSQRT, DMAX1, DMIN1, DFLOAT
C-----------------------------------------------------------------------
C     COMPUTES EIGENVALUES OF T(1,MEV) BY LOOPING INTERNALLY ON THE
C     USER-SPECIFIED INTERVALS,  (LB(J),UB(J)), J = 1,NINT.  INTERVALS
C     ARE TREATED AS OPEN ON THE LEFT AND CLOSED ON THE RIGHT.
C     THE BISEC SUBROUTINE SIMULTANEOUSLY LABELS SPURIOUS T-EIGENVALUES
C     AND DETERMINES THE T-MULTIPLICITIES OF EACH GOOD T-EIGENVALUE.
C     SPURIOUS T-EIGENVALUES ARE LABELLED BY A T-MULTIPLICITY = 0.
C     ANY T-EIGENVALUE WITH A T-MULTIPLICITY >= 1 IS 'GOOD'.
C
C     IF IWRITE = 0 THEN MOST OF THE WRITES TO FILE 6 ARE NOT
C     ACTIVATED.
C
C     NOTE THAT PROGRAM ASSUMES THAT NO MORE THAN MMAX/2 T-EIGENVALUES
C     OF T(1,MEV) ARE TO BE COMPUTED IN ANY ONE OF THE SUBINTERVALS
C     CONSIDERED, WHERE MMAX = DIMENSION OF VB SPECIFIED BY THE USER
C     IN THE MAIN PROGRAM LEVAL.
C
C     ON ENTRY
C     BETA2(J) IS SET = BETA(J)*BETA(J).  THE STORAGE FOR BETA2 COULD
C     BE ELIMINATED BY RECOMPUTING THE BETA(J)**2 FOR EACH STURM
C     SEQUENCE.
C
C     EPS = 2*MACHEP =  4.4 * 10**-16 ON IBM 3081.
C     TTOL = EPS*TKMAX WHERE
C     TKMAX = MAX(BETA(K), K=1,KMAX)
C
C     ON EXIT
C     NDIS = TOTAL NUMBER OF COMPUTED DISTINCT T-EIGENVALUES OF
C            T(1,MEV) ON THE UNION OF THE (LB,UB) INTERVALS.
C     VS = COMPUTED DISTINCT T-EIGENVALUES OF T(1,MEV) IN ALGEBRAICALLY-
C          INCREASING ORDER
C     MP = CORRESPONDING T-MULTIPLICITIES OF THESE T-EIGENVALUES
C     MP(I) = (0,1,MI), MI>1, I=1,NDIS  MEANS:
C        (0)  V(I) IS SPURIOUS
C        (1)  V(I) IS T-ISOLATED AND GOOD
C        (MI) V(I) IS T-MULTIPLE AND HENCE A CONVERGED GOOD T-EIGENVALUE
C     IC = TOTAL NUMBER OF STURMS USED
C
C     DEFAULTS
C     ISKIP = 0 INITIALLY. IF DEFAULT OCCURS ON J-TH SUB-INTERVAL, SET
C             ISKIP=ISKIP+1 AND IDEF(ISKIP) = J
C             DEFAULTS OCCUR IF THERE ARE NO T-EIGENVALUES  IN THE
```

```
C             SUBINTERVAL SPECIFIED OR IF THE NUMBER
C             OF STURMS SEQUENCES REQUIRED EXCEEDS MXSTUR.
C             WHEN A DEFAULT OCCURS THE PROGRAM
C             SKIPS THE INTERVAL INVOLVED AND GOES ON TO THE NEXT
C             INTERVAL.
C
C-----------------------------------------------------------------------
C     SPECIFY PARAMETERS
      ZERO = 0.0D0
      ONE  = 1.0D0
      HALF = 0.5D0
      MXSTUR = IC
      NDIS = 0
      IC = 0
      ISKIP = 0
      MP1 = MEV+1
C     SAVE THEN SET BETA(MEV+1) = 0. GENERATE BETA**2
      BETAM = BETA(MP1)
      BETA(MP1) = ZERO
C
      DO 10 I = 1,MP1
   10 BETA2(I) = BETA(I)*BETA(I)
C
C     EP0 IS USED IN T-MULTIPLICITY AND SPURIOUS TESTS
C     EP1 AND EPS ARE USED IN THE BISEC CONVERGENCE TEST
C
      TEMP = DFLOAT(MEV+1000)
      EP0  = TEMP*TTOL
      EP1  = DSQRT(TEMP)*TTOL
C
      WRITE(6,20)MEV,NINT
   20 FORMAT(/' BISEC CALCULATION'/' ORDER OF T IS',I6/
     1' NUMBER OF INTERVALS IS',I6/)
C
      WRITE(6,30) EP0,EP1
   30 FORMAT(/' MULTOL, TOLERANCE USED IN T-MULTIPLICITY AND SPURIOUS TE
     1STS = ',E10.3/' BISTOL, TOLERANCE USED IN BISEC CONVERGENCE TEST =
     1',E10.3/)
C
C     LOOP ON THE NINT INTERVALS  (LB(J),UB(J)), J=1,NINT
      DO 430 JIND = 1,NINT
      LB = LBD(JIND)
      UB = UBD(JIND)
C
      WRITE(6,40)JIND,LB,UB
   40 FORMAT(//1X,'BISEC INTERVAL NO',2X,'LOWER BOUND',2X,'UPPER BOUND'/
     1I18,2E13.5/)
C
C     INITIALIZATION AND PARAMETER SPECIFICATION
C     ICT IS TOTAL STURM COUNT ON (LB,UB)
C
      NA = 0
      MD = 0
      NG = 0
      ICT = 0
C
C     START OF T-EIGENVALUE CALCULATIONS
      X1 = UB
      ISTURM = 1
      GO TO 330
C     FORWARD STURM CALCULATION TO DETERMINE NA = NO. T-EIGENVALUES > UB
   50 NA = NEV
C
      X1 = LB
      ISTURM = 2
      GO TO 330
C     FORWARD STURM CALC TO DETERMINE MT = NO. T-EIGENVALUES ON (LB,UB)
```

```
   60 CONTINUE
      MT=NEV
      ICT = ICT +2
C
      WRITE(6,70)MT,NA
   70 FORMAT(/2I6,' = NO. TMEV ON (LB,UB) AND NO. .GT. UB'/)
C
C     DEFAULT TEST: IS ESTIMATED NUMBER OF STURMS > MXSTUR?
      IEST = 30*MT
      IF (IEST.LT.MXSTUR) GO TO 90
C
      WRITE(6,80)
   80 FORMAT(//' ESTIMATED NUMBER OF STURMS REQUIRED EXCEEDS USER LIMIT'
     1/' SKIP THIS SUBINTERVAL')
      GO TO 110
C
   90 CONTINUE
C
      IF (MT.GE.1) GO TO 120
C
      WRITE(6,100)
  100 FORMAT(//' THERE ARE NO T-EIGENVALUES ON THIS INTERVAL)'/)
C
  110 ISKIP = ISKIP+1
      IDEF(ISKIP) = JIND
      GO TO 430
C
C     REGULAR CASE.
  120 CONTINUE
C
      IF (IWRITE.NE.0) WRITE(6,130)
  130 FORMAT(/' DISTINCT T-EIGENVALUES COMPUTED USING BISEC'/
     1 13X,'T-EIGENVALUE',2X,'TMULT',3X,'MD',4X,'NG')
C
C     SET UP INITIAL UPPER AND LOWER BOUNDS FOR T-EIGENVALUES
      DO 140 I=1,MT
      VB(I) = LB
      MTI = MT + I
  140 VB(MTI) = UB
C
C     CALCULATE T-EIGENVALUES FROM LB UP TO UB  K = MT,...,1
C     MAIN LOOP FOR FINDING KTH T-EIGENVALUE
C
      K = MT
  150 CONTINUE
      ICO = 0
      XL = VB(K)
      MTK = MT+K
      XU = VB(MTK)
C
      ISTURM = 3
      X1 = XU
      ICO = ICO + 1
      GO TO 330
C     FORWARD STURM CALCULATION AT XU
  160 NU=NEV
C
C     BISECTION LOOP FOR KTH T-EIGENVALUE. TEST X1=MIDPOINT OF (XL,XU)
      ISTURM = 4
  170 CONTINUE
      X1 = (XL+XU)*HALF
      XS = DABS(XL)+DABS(XU)
      X0 = XU-XL
      EPT = EPS*XS+EP1
C
C     EPT IS CONVERGENCE TOLERANCE FOR KTH T-EIGENVALUE
C
```

```
      IF (X0.LE.EPT) GO TO 230
C
C     T-EIGENVALUE HAS NOT YET CONVERGED
C
      ICO = ICO + 1
      GO TO 330
C     FORWARD STURM CALCULATION AT CURRENT T-EIGENVALUE APPROXIMATION.
  180 CONTINUE
C
C     UPDATE T-EIGENVALUE INTERVAL (XL,XU)
C
      IF (NEV.LT.K) GO TO 190
C
C     NUMBER OF T-EIGENVALUES NEV = K
      XL = X1
      GO TO 170
  190 CONTINUE
C     NUMBER OF T-EIGENVALUES NEV<K
      XU = X1
      NU = NEV
C
C     UPDATE OF T-EIGENVALUE BOUNDS
C
      IF (NEV.EQ.0) GO TO 210
C
      DO 200 I = 1,NEV
  200 VB(I) = DMAX1(X1,VB(I))
C
  210 NEV1 = NEV+1
C
      DO 220 II = NEV1,K
      I = MT+II
  220 VB(I) = DMIN1(X1,VB(I))
C
      GO TO 170
C
C     END (XL,XU) BISECTION LOOP FOR KTH T-EIGENVALUE ON (LB,UB)
C     TEST FOR T-MULTIPLICITY AND IF SIMPLE THEN TEST FOR SPURIOUSNESS
C
  230 CONTINUE
      NDIS = NDIS+1
      MD = MD+1
      VS(NDIS) = X1
C
      JSTURM = 1
      X1 = XL-EP0
      GO TO 370
C     BACKWARD STURM CALCULATION
  240 KL = KEV
      JL = JEV
C
      JSTURM = 2
      ICO = ICO + 2
      X1 = XU+EP0
      GO TO 370
C     BACKWARD STURM CALCULATION
  250 JU = JEV
      KU = KEV
C
C     FOR T(1,MEV)
C     NU - KU = NO. T-EIGENVALUES ON (XU, XU + EP0)
C     KL - KU = NO. T-EIGENVALUES ON (XL - EP0, XU + EP0)
C
C     FOR T(2,MEV)
C     JL -JU = NO. T-EIGENVALUES ON (XL - EP0, XU + EP0)
C
```

```
C     IS THIS A SIMPLE T-EIGENVALUE?
C
      IF (KL-KU-1.EQ.0) GO TO 290
C
C     VS(NDIS) = KTH-T-EIGENVALUE OF (LB,UB) IS T-MULTIPLE AND HENCE
C     GOOD
      IF (KU.EQ.NU) GO TO 280
C     CONTINUE TO CHECK FOR T-MULTIPLICITY
  260 CONTINUE
      ISTURM = 5
      X1 = X1+EPO
      ICO = ICO + 1
      GO TO 330
C     FORWARD STURM CALCULATION
  270 KNE = KU-NEV
      KU = NEV
      IF (KNE.NE.0) GO TO 260
C     SPECIFY T-MULTIPLICITY = MP(NDIS)
  280 MPEV = KL-KU
      KNEW = KU
      GO TO 300
C     END T-MULTIPLE CASE
C
C     T-EIGENVALUE IS SIMPLE   CHECK IF IT IS SPURIOUS
  290 CONTINUE
      MPEV = 1
      IF (JU.LT.JL) MPEV=0
      KNEW = K-1
C
C     X1 >= XU+EPO
C     SPURIOUS TEST AND SIMPLE CASE COMPLETED
C     START OF NEXT T-EIGENVALUE COMPUTATION
C
  300 K = KNEW
      MP(NDIS) = MPEV
      IF (MPEV.GE.1) NG = NG + 1
C
      IF (IWRITE.NE.0) WRITE(6,310) VS(NDIS),MPEV,MD,NG
  310 FORMAT(E25.16,3I6)
C
C     UPDATE STURM COUNT. ICO = STURM COUNT FOR KTH T-EIGENVALUE
      ICT = ICT + ICO
C
C     EXIT TEST FOR K DO LOOP
C
      IF (K.LE.0) GO TO 410
C
C     UPDATE LOWER BOUNDS
      DO 320 I=1,KNEW
  320 VB(I) = DMAX1(X1,VB(I))
C
      GO TO 150
C     END OF BISECTION LOOP FOR KTH EIGENVALUE
C
C     FORWARD STURM CALCULATION
  330 NEV = -NA
      YU  = ONE
C
      DO 360 I = 1,MEV
      IF (YU.NE.ZERO) GO TO 340
      YV = BETA(I)/EPS
      GO TO 350
  340 YV = BETA2(I)/YU
  350 YU = X1 - YV
      IF (YU.GE.ZERO) GO TO 360
      NEV = NEV + 1
```

```
  360 CONTINUE
C     NEV = NUMBER OF T-EIGENVALUES ON (X1,UB)
C
      GO TO (50,60,160,180,270), ISTURM
C
C     BACKWARD STURM CALCULATION FOR T(1,MEV) AND T(2,MEV)
  370 KEV = -NA
      YU  = ONE
C
      DO 400 II = 1,MEV
      I = MP1-II
      IF (YU.NE.ZERO) GO TO 380
      YV = BETA(I+1)/EPS
      GO TO 390
  380 YV = BETA2(I+1)/YU
  390 YU = X1-YV
      JEV = 0
      IF (YU.GE.ZERO) GO TO 400
      KEV = KEV+1
      JEV = 1
  400 CONTINUE
      JEV = KEV-JEV
C
      GO TO (240,250), JSTURM
C
C     KEV = -NA + (NUMBER OF T(1,MEV) T-EIGENVALUES) > X1
C     JEV = -NA + (NUMBER OF T(2,MEV) T-EIGENVALUES) > X1
C
C     SET PARAMETERS FOR NEXT INTERVAL
  410 CONTINUE
      IC = ICT+IC
      MXSTUR = MXSTUR-ICT
C
      WRITE(6,420) JIND,NG,MD,ICT
  420 FORMAT(/' T-EIGENVALUE CALCULATION ON INTERVAL',I6,'  IS COMPLETE'
     1 /3X,'NO. GOOD',3X,'NO. DISTINCT',4X,'STURMS'/I10,I13,I10)
C
  430 CONTINUE
C
C     END LOOP ON THE SUBINTERVALS (LB(J),UB(J)), J=1,NINT
C     ISKIP OUTPUT
C
      IF (ISKIP.GT.0) WRITE(6,440)ISKIP
  440 FORMAT(' BISEC DEFAULTED ON',I3,3X,'INTERVALS'/
     1 ' DEFAULTS OCCUR IF AN INTERVAL HAS NO T-EIGENVALUES'/
     2 ' OR THE STURM ESTIMATE EXCEEDS THE USER-SPECIFIED LIMIT'/)
C
      IF (ISKIP.GT.0) WRITE(6,450)(IDEF(I), I=1,ISKIP)
  450 FORMAT(' BISEC DEFAULTED ON INTERVALS'/(10I8))
C
C     RESET BETA AT I = MP1
      BETA(MP1) = BETAM
C-----END OF BISEC-----------------------------------------------------
      RETURN
      END
C
C-----INVERSE ITERATION ON T(1,MEV)------------------------------------
C
      SUBROUTINE INVERR(BETA,V1,V2,VS,EPS,G,MP,MEV,MMB,NDIS,NISO,
     1 NM,IKL,IT,IWRITE)
C
C---------------------------------------------------------------------
      DOUBLE PRECISION BETA(1),V1(1),V2(1),VS(1)
      DOUBLE PRECISION X1,U,Z,EST,TEMP,T0,T1,RATIO,SUM,XU,NORM,TSUM
      DOUBLE PRECISION  BETAM,EPS,EPS3,EPS4,ZERO,ONE
      REAL G(1)
```

```
      INTEGER  MP(1)
C-----------------------------------------------------------------------
      DOUBLE PRECISION FINPRO
      REAL  ABS
      DOUBLE PRECISION  DABS, DMIN1, DSQRT, DFLOAT
C-----------------------------------------------------------------------
C     COMPUTES ERROR ESTIMATES FOR COMPUTED ISOLATED GOOD T-EIGENVALUES
C     IN VS AND WRITES THESE T-EIGENVALUES AND ESTIMATES TO FILE 4.
C     BY DEFINITION A GOOD T-EIGENVALUE IS ISOLATED IF ITS
C     CLOSEST NEIGHBOR IS ALSO GOOD, OR IF ONE OF ITS NEIGHBORS IS
C     SPURIOUS BUT THAT NEIGHBOR IS FAR ENOUGH AWAY.  SO
C     IN PARTICULAR, WE COMPUTE ESTIMATES FOR GOOD T-EIGENVALUES
C     THAT ARE IN CLUSTERS OF GOOD T-EIGENVALUES.
C
C     USES INVERSE ITERATION ON T(1,MEV) SOLVING THE EQUATION
C     (T - X1*I)V2 = RIGHT-HAND SIDE (RANDOMLY-GENERATED)
C     FOR EACH SUCH GOOD T-EIGENVALUE X1.
C
C     PROGRAM REFACTORS T-X1*I ON EACH ITERATION OF INVERSE ITERATION.
C     TYPICALLY ONLY ONE ITERATION IS NEEDED PER T-EIGENVALUE X1.
C
C     POSSIBLE STORAGE COMPRESSION
C     G STORAGE COULD BE ELIMINATED BY REGENERATING THE RANDOM
C     RIGHT-HAND SIDE ON EACH ITERATION AND PRINTING OUT THE
C     ERROR ESTIMATES AS THEY ARE GENERATED.
C
C     ON ENTRY AND EXIT
C     MEV = ORDER OF T
C     BETA CONTAINS THE NONZERO ENTRIES OF THE T-MATRIX
C     VS = COMPUTED DISTINCT EIGENVALUES OF T(1,MEV)
C     MP = T-MULTIPLICITY OF EACH EIGENVALUE IN VS. MP(I) = -1 MEANS
C          VS(I) IS A GOOD T-EIGENVALUE BUT THAT IT IS SITTING CLOSE TO
C          A SPURIOUS T-EIGENVALUE.  MP(I) = 0 MEANS VS(I) IS SPURIOUS.
C          ESTIMATES ARE COMPUTED ONLY FOR THOSE T-EIGENVALUES
C          WITH MP(I) = 1. FLAGGING WAS DONE IN SUBROUTINE ISOEV
C          PRIOR TO ENTERING INVERR.
C     NISO = NUMBER OF ISOLATED GOOD T-EIGENVALUES CONTAINED IN VS
C     NDIS =  NUMBER OF DISTINCT T-EIGENVALUES IN VS
C     IKL = SEED FOR RANDOM NUMBER GENERATOR
C     EPS = 2. * MACHINE EPSILON
C
C     IN PROGRAM:
C     ITER = MAXIMUM NUMBER OF INVERSE ITERATION STEPS ALLOWED FOR EACH
C            X1.  ITER = IT ON ENTRY.
C     G = ARRAY OF DIMENSION AT LEAST MEV + NISO.  USED TO STORE
C         RANDOMLY-GENERATED RIGHT-HAND SIDE.  THIS IS NOT
C         REGENERATED FOR EACH X1. G IS ALSO USED TO STORE ERROR
C         ESTIMATES AS THEY ARE COMPUTED FOR LATER PRINTOUT.
C     V1,V2 = WORK SPACES USED IN THE FACTORIZATION OF T(1,MEV).
C     AT THE END OF THE INVERSE ITERATION COMPUTATION FOR X1, V2
C     CONTAINS THE UNIT EIGENVECTOR OF T(1,MEV) CORRESPONDING TO X1.
C     V1 AND V2 MUST BE OF DIMENSION AT LEAST MEV.
C
C     ON EXIT
C     G(J) = MINIMUM GAP IN T(1,MEV) FOR EACH VS(J), J=1,NDIS
C     G(MEV+I) = BETAM*|V2(MEV)| = ERROR ESTIMATE FOR ISOLATED GOOD
C              T-EIGENVALUES, WHERE I = 1,NISO  AND  BETAM = BETA(MEV+1)
C              V2(MEV) IS LAST COMPONENT OF THE UNIT T-EIGENVECTOR OF
C              T(1,MEV) CORRESPONDING TO ITH ISOLATED GOOD T-EIGENVALUE.
C
C     IF FOR SOME X1 IT.GT.ITER THEN THE ERROR ESTIMATE IN G IS MARKED
C     WITH A - SIGN.
C
C     V2 = ISOLATED GOOD T-EIGENVALUES
C     V1 = MINIMAL T-GAPS FOR THE EIGENVALUES IN V2.
C     THESE ARE CONSTRUCTED FOR WRITE-OUT PURPOSES ONLY AND NOT
C     NEEDED ELSEWHERE IN THE PROGRAM.
```

```
C-----------------------------------------------------------------------
C
C     LABEL OUTPUT FILE  4
      IF (MMB.EQ.1) WRITE(4,10)
   10 FORMAT(' INVERSE ITERATION ERROR ESTIMATES'/)
C
C     FILE 6 (TERMINAL) OUTPUT OF ERROR ESTIMATES
      IF (IWRITE.NE.0.AND.NISO.NE.0) WRITE(6,20)
   20 FORMAT(/' INVERSE ITERATION ERROR ESTIMATES'/'  JISO',' JDIST',8X
     1,'GOOD T-EIGENVALUE',4X,'BETAM*UM',5X,'TMINGAP')
C
C     INITIALIZATION AND PARAMETER SPECIFICATION
      ZERO = 0.0D0
      ONE = 1.0D0
      NG = 0
      NISO = 0
      ITER = IT
      MP1 = MEV+1
      MM1 = MEV-1
      BETAM = BETA(MP1)
      BETA(MP1) = ZERO
C
C     CALCULATE SCALE AND TOLERANCES
      TSUM = ZERO
      DO 30 I = 2,MEV
   30 TSUM = TSUM + BETA(I)
C
      EPS3 = EPS*TSUM
      EPS4 = DFLOAT(MEV)*EPS3
C
C     GENERATE SCALED RANDOM RIGHT-HAND SIDE
      ILL = IKL
C
C-----------------------------------------------------------------------
      CALL GENRAN(ILL,G,MEV)
C-----------------------------------------------------------------------
C
      GSUM = ZERO
      DO 40 I = 1,MEV
   40 GSUM = GSUM+ABS(G(I))
      GSUM = EPS4/GSUM
C
      DO 50 I = 1,MEV
   50 G(I) = GSUM*G(I)
C
C     LOOP ON ISOLATED GOOD T-EIGENVALUES IN VS (MP(I) = 1) TO
C     CALCULATE CORRESPONDING UNIT T-EIGENVECTOR OF T(1,MEV)
C
      DO 180 JEV = 1,NDIS
      IF (MP(JEV).EQ.0) GO TO 180
      NG = NG + 1
      IF (MP(JEV).NE.1) GO TO 180
      IT = 1
      NISO = NISO + 1
      X1 = VS(JEV)
C
C     INITIALIZE RIGHT HAND SIDE FOR INVERSE ITERATION
      DO 60 I = 1,MEV
   60 V2(I) = G(I)
C
C     TRIANGULAR FACTORIZATION WITH NEAREST NEIGHBOR PIVOT
C     STRATEGY. INTERCHANGES ARE LABELLED BY SETTING BETA < 0.
C
   70 CONTINUE
      U = -X1
      Z = BETA(2)
C
```

```
      DO 90 I = 2,MEV
      IF (BETA(I).GT.DABS(U)) GO TO 80
C     NO INTERCHANGE
      V1(I-1) = Z/U
      V2(I-1) = V2(I-1)/U
      V2(I) = V2(I)-BETA(I)*V2(I-1)
      RATIO = BETA(I)/U
      U = -X1-Z*RATIO
      Z = BETA(I+1)
      GO TO 90
   80 CONTINUE
C     INTERCHANGE CASE
      RATIO = U/BETA(I)
      BETA(I) = -BETA(I)
      V1(I-1) = -X1
      U = Z-RATIO*V1(I-1)
      Z = -RATIO*BETA(I+1)
      TEMP = V2(I-1)
      V2(I-1) = V2(I)
      V2(I) = TEMP-RATIO*V2(I)
   90 CONTINUE
      IF (U.EQ.ZERO) U = EPS3
C
C     SMALLNESS TEST AND DEFAULT VALUE FOR LAST COMPONENT
C     PIVOT(I-1) = |BETA(I)| FOR INTERCHANGE CASE
C     (I-1,I+1) ELEMENT IN RIGHT FACTOR = BETA(I+1)
C     END OF FACTORIZATION AND FORWARD SUBSTITUTION
C
C     BACK SUBSTITUTION
      V2(MEV) = V2(MEV)/U
      DO 110 II = 1,MM1
      I = MEV-II
      IF (BETA(I+1).LT.ZERO) GO TO 100
C     NO INTERCHANGE
      V2(I) = V2(I)-V1(I)*V2(I+1)
      GO TO 110
C     INTERCHANGE CASE
  100 BETA(I+1) = -BETA(I+1)
      V2(I) = (V2(I)-V1(I)*V2(I+1)-BETA(I+2)*V2(I+2))/BETA(I+1)
  110 CONTINUE
C
C     TESTS FOR CONVERGENCE OF INVERSE ITERATION
C     IF SUM |V2| COMPS. LE. 1 AND IT. LE. ITER DO ANOTHER INVIT STEP
C
      NORM = DABS(V2(MEV))
      DO 120 II = 1,MM1
      I = MEV-II
  120 NORM = NORM+DABS(V2(I))
C
      IF (NORM.GE.ONE) GO TO 140
      IT = IT+1
      IF (IT.GT.ITER) GO TO 140
      XU = EPS4/NORM
C
      DO 130 I = 1,MEV
  130 V2(I) = V2(I)*XU
C
      GO TO 70
C     ANOTHER INVERSE ITERATION STEP
C
C     INVERSE ITERATION FINISHED
C     NORMALIZE COMPUTED T-EIGENVECTOR : V2 = V2/||V2||
  140 CONTINUE
      SUM = FINPRO(MEV,V2(1),1,V2(1),1)
      SUM = ONE/DSQRT(SUM)
C
      DO 150 II = 1,MEV
```

```
  150 V2(II) = SUM*V2(II)
C
C     SAVE ERROR ESTIMATE FOR LATER OUTPUT
      EST = BETAM*DABS(V2(MEV))
      IF (IT.GT.ITER) EST = -EST
      MEVPNI = MEV + NISO
      G(MEVPNI) = EST
      IF (IWRITE.EQ.0) GO TO 180
C
C     FILE 6 (TERMINAL) OUTPUT OF ERROR ESTIMATES.
      IF (JEV.EQ.1) GAP = VS(2) - VS(1)
      IF (JEV.EQ.MEV) GAP = VS(MEV) - VS(MEV-1)
      IF (JEV.EQ.MEV.OR.JEV.EQ.1) GO TO 160
      TEMP = DMIN1(VS(JEV+1)-VS(JEV),VS(JEV)-VS(JEV-1))
      GAP = TEMP
  160 CONTINUE
C
      WRITE(6,170) NISO,JEV,X1,EST,GAP
  170 FORMAT(2I6,E25.16,2E12.3)
C
  180 CONTINUE
C
C     END ERROR ESTIMATE LOOP ON ISOLATED GOOD T-EIGENVALUES.
C     GENERATE DISTINCT MINGAPS FOR T(1,MEV).  THIS IS USEFUL AS AN
C     INDICATOR OF THE GOODNESS OF THE INVERSE ITERATION ESTIMATES.
C     TRANSFER ISOLATED GOOD T-EIGENVALUES AND CORRESPONDING TMINGAPS
C     TO V2 AND V1 FOR OUTPUT PURPOSES ONLY.
C
      NM1 = NDIS - 1
      G(NDIS) = VS(NM1)-VS(NDIS)
      G(1) = VS(2)-VS(1)
C
      DO 190 J = 2,NM1
      T0 = VS(J)-VS(J-1)
      T1 = VS(J+1)-VS(J)
      G(J) = T1
      IF (T0.LT.T1) G(J)=-T0
  190 CONTINUE
      ISO = 0
      DO 200 J = 1,NDIS
      IF (MP(J).NE.1) GO TO 200
      ISO = ISO+1
      V1(ISO) = G(J)
      V2(ISO) = VS(J)
  200 CONTINUE
C
      IF(NISO.EQ.0) GO TO 250
C
C     ERROR ESTIMATES ARE WRITTEN TO FILE 4
      WRITE(4,210)MEV,NDIS,NG,NISO,NM,IKL,ITER,BETAM
  210 FORMAT(1X,'TSIZE',2X,'NDIS',1X,'NGOOD',2X,'NISO',3X,'M+N'/5I6/
     1 4X,'RHSEED',2X,'MXINIT',5X,'BETAM'/I10,I8,E10.3/
     2 2X,'GOODEVNO',8X,'GOOD T-EIGENVALUE',6X,'BETAM*UM',7X,'TMINGAP')
C
      ISPUR = 0
      I = 0
      DO 240 J = 1,NDIS
      IF(MP(J).NE.0) GO TO 220
      ISPUR = ISPUR + 1
      GO TO 240
  220 IF(MP(J).NE.1) GO TO 240
      I = I + 1
      MEVI = MEV + I
      IGOOD = J - ISPUR
      WRITE(4,230) IGOOD,V2(I),G(MEVI),V1(I)
  230 FORMAT(I10,E25.16,2E14.3)
  240 CONTINUE
```

```
      GO TO 270
C
  250 WRITE(4,260)
  260 FORMAT(/' THERE ARE NO ISOLATED T-EIGENVALUES SO NO ERROR ESTIMATE
     1S WERE COMPUTED')
C     RESTORE BETA(MEV+1) = BETAM
  270 BETA(MP1) = BETAM
C-----END OF INVERR----------------------------------------------------
      RETURN
      END
C
C-----START OF TNORM---------------------------------------------------
C
      SUBROUTINE TNORM(BETA,BMIN,TMAX,MEV,IB)
C
C----------------------------------------------------------------------
      DOUBLE PRECISION  BETA(1)
      DOUBLE PRECISION  TMAX,BMIN,BSIZE,BTOL
      DOUBLE PRECISION  DABS, DMAX1
C----------------------------------------------------------------------
C     COMPUTE SCALING FACTOR USED IN THE T-MULTIPLICITY, SPURIOUS AND
C     PRTESTS.   CHECK RELATIVE SIZE OF THE BETA(K), K=1,MEV
C     AS A TEST ON THE LOCAL ORTHOGONALITY OF THE LANCZOS VECTORS.
C
C             TMAX = MAX (BETA(I),  I=1,MEV)
C             BMIN = MIN (BETA(I) I=2,MEV)
C             BSIZE = BMIN/TMAX
C             |IB| = INDEX OF MINIMAL(BETA)
C             IB < 0 IF BMIN/TMAX < BTOL
C----------------------------------------------------------------------
C     SPECIFY PARAMETERS
      IB = 2
      BTOL = BMIN
      BMIN = BETA(2)
      TMAX = BETA(2)
C
      DO 20 I = 2,MEV
      IF (BETA(I).GE.BMIN) GO TO 10
      IB = I
      BMIN = BETA(I)
   10 TMAX = DMAX1(TMAX,BETA(I))
   20 CONTINUE
C
C     TEST OF LOCAL ORTHOGONALITY USING SCALED BETAS
      BSIZE = BMIN/TMAX
      IF (BSIZE.GE.BTOL) GO TO 40
C
C     DEFAULT.  BSIZE IS SMALLER THAN TOLERANCE BTOL SPECIFIED IN MAIN
C     PROGRAM.  PROGRAM TERMINATES FOR USER TO DECIDE WHAT TO DO
C     BECAUSE LOCAL ORTHOGONALITY OF THE LANCZOS VECTORS COULD BE
C     LOST.
C
      IB = -IB
      WRITE(6,30) MEV
   30 FORMAT(/' BETA TEST INDICATES POSSIBLE LOSS OF LOCAL ORTHOGONALITY
     1OVER 1ST',I6,' LANCZOS VECTORS'/)
C
   40 CONTINUE
C
      WRITE(6,50) IB
   50 FORMAT(/' MINIMUM BETA RATIO OCCURS AT',I6,' TH BETA'/)
C
      WRITE(6,60) MEV,BMIN,TMAX,BSIZE
   60 FORMAT(/1X,'TSIZE',6X,'MIN BETA',5X,'TKMAX',6X,'MIN RATIO'/
     1 I6,E14.3,E10.3,E15.3/)
C
C-----END OF TNORM-----------------------------------------------------
```

```
      RETURN
      END
C
C
C-----START OF LUMP----------------------------------------------------
C
      SUBROUTINE LUMP(V1,RELTOL,MULTOL,SCALE2,LINDEX,LOOP)
C
C---------------------------------------------------------------------
      DOUBLE PRECISION  V1(1),SUM,RELTOL,MULTOL,THOLD,ZERO,SCALE2
      INTEGER  LINDEX(1)
      DOUBLE PRECISION  DABS, DFLOAT, DMAX1
C---------------------------------------------------------------------
C     LINDEX(J) = T-MULTIPLICITY OF JTH DISTINCT T-EIGENVALUE
C     LOOP = NUMBER OF DISTINCT T-EIGENVALUES
C     LUMP 'COMBINES' COMPUTED 'GOOD' T-EIGENVALUES THAT ARE
C     'TOO CLOSE'.
C     VALUE FOR RELTOL IS 1.D-10.
C
C     IF IN A SET OF T-EIGENVALUES TO BE COMBINED THERE IS AN EIGENVALUE
C     WITH LINDEX=1, THEN THE VALUE OF THE COMBINED T-EIGENVALUES IS SET
C     EQUAL TO THE VALUE OF THAT T-EIGENVALUE.  NOTE THAT IF A SPURIOUS
C     T-EIGENVALUE IS TO BE 'COMBINED' WITH A GOOD T-EIGENVALUE, THEN
C     THIS IS DONE ONLY BY INCREASING THE INDEX, LINDEX, FOR THAT
C     T-EIGENVALUE. NUMERICAL VALUES OF SPURIOUS T-EIGENVALUES ARE
C     NEVER COMBINED WITH THOSE OF GOOD T-EIGENVALUES.
C---------------------------------------------------------------------
      ZERO = 0.0D0
      NLOOP = 0
      J = 0
      ICOUNT = 1
      JI = 1
      THOLD = DMAX1(RELTOL*DABS(V1(1)),SCALE2*MULTOL)
C     THOLD = DMAX1(RELTOL*DABS(V1(1)),RELTOL)
C
   10 J = J+1
      IF (J.EQ.LOOP) GO TO 20
      SUM = DABS(V1(J)-V1(J+1))
      IF (SUM.LT.THOLD) GO TO 60
   20 JF = JI + ICOUNT - 1
      INDSUM = 0
      ISPUR = 0
C
      DO 30 KK = JI,JF
      IF (LINDEX(KK).NE.0) GO TO 30
      ISPUR = ISPUR + 1
      INDSUM = INDSUM + 1
   30 INDSUM = INDSUM + LINDEX(KK)
C
C     IF (JF-JI.GE.1) WRITE(6,40) (V1(KKK), KKK=JI,JF)
   40 FORMAT(/' LUMP LUMPS THE T-EIGENVALUES'/(4E20.13))
C
C     COMPUTE THE 'COMBINED' T-EIGENVALUE AND THE RESULTING
C     T-MULTIPLICITY
      K = JI - 1
   50 K = K+1
      IF (K.GT.JF) GO TO 70
      IF (LINDEX(K) .NE.1) GO TO 50
      NLOOP = NLOOP + 1
      V1(NLOOP) = V1(K)
      GO TO 100
   60 ICOUNT = ICOUNT + 1
      GO TO 10
C
C     ALL INDICES WERE 0 OR >1
   70 NLOOP = NLOOP + 1
      IDIF = INDSUM - ISPUR
```

```
      IF (IDIF.EQ.0) GO TO 90
C
      SUM = ZERO
      DO 80 KK = JI,JF
   80 SUM = SUM + V1(KK) * DFLOAT(LINDEX(KK))
C
      V1(NLOOP) = SUM/DFLOAT(IDIF)
      GO TO 100
   90 V1(NLOOP) = V1(JI)
  100 LINDEX(NLOOP) = INDSUM
      IDIF = INDSUM - ISPUR
      IF (IDIF.EQ.0.AND.ISPUR.EQ.1) LINDEX(NLOOP) = 0
      IF (J.EQ.LOOP) GO TO 110
      ICOUNT = 1
      JI= J+1
      THOLD = DMAX1(RELTOL*DABS(V1(JI)),SCALE2*MULTOL)
C     THOLD = DMAX1(RELTOL*DABS(V1(JI)),RELTOL)
      IF (JI.LT.LOOP) GO TO 10
      NLOOP = NLOOP + 1
      V1(NLOOP)= V1(JI)
      LINDEX(NLOOP) = LINDEX(JI)
  110 CONTINUE
C
C     ON RETURN V1 CONTAINS THE DISTINCT T-EIGENVALUES
C     LINDEX CONTAINS THE CORRESPONDING T-MULTIPLICITIES
C
      LOOP = NLOOP
      RETURN
C-----END OF LUMP-------------------------------------------------------
      END
C
C
C-----START OF ISOEV----------------------------------------------------
C
      SUBROUTINE ISOEV(VS,GAPTOL,MULTOL,SCALE1,G,MP,NDIS,NG,NISO)
C
C-----------------------------------------------------------------------
      DOUBLE PRECISION  VS(1),T0,T1,MULTOL,GAPTOL,SCALE1,TEMP
      REAL   G(1),GAP
      INTEGER  MP(1)
      REAL  ABS
      DOUBLE PRECISION  DABS, DMAX1
C-----------------------------------------------------------------------
C     GENERATE DISTINCT TMINGAPS AND USE THEM TO LABEL THE ISOLATED
C     GOOD T-EIGENVALUES THAT ARE VERY CLOSE TO SPURIOUS ONES.
C     ERROR ESTIMATES WILL NOT BE COMPUTED FOR THESE T-EIGENVALUES.
C
C     ON ENTRY AND EXIT
C     VS CONTAINS THE COMPUTED DISTINCT T-EIGENVALUES OF T(1,MEV)
C     MP CONTAINS THE CORRESPONDING T-MULTIPLICITIES
C     NDIS = NUMBER OF DISTINCT T-EIGENVALUES
C     GAPTOL = RELATIVE GAP TOLERANCE SET IN MAIN
C
C     ON EXIT
C     G CONTAINS THE TMINGAPS.
C     G(I) < 0 MEANS MINGAP IS DUE TO LEFT GAP
C     MP(I) IS NOT CHANGED EXCEPT THAT  MP(I)=-1, IF MP(I)=1,
C     TMINGAP WAS TOO SMALL AND DUE TO A SPURIOUS T-EIGENVALUE.
C
C     IF MP(I)=-1 THAT SIMPLE GOOD T-EIGENVALUE WILL BE SKIPPED
C     IN THE SUBSEQUENT ERROR ESTIMATE COMPUTATIONS IN INVERR
C     THAT IS, WE COMPUTE ERROR ESTIMATES ONLY FOR THOSE GOOD
C     T-EIGENVALUES WITH MP(I)=1.
C-----------------------------------------------------------------------
C     CALCULATE MINGAPS FOR DISTINCT T(1,MEV) EIGENVALUES.
      NM1 = NDIS - 1
      G(NDIS) = VS(NM1)-VS(NDIS)
```

```
      G(1) = VS(2)-VS(1)
C
      DO 10 J = 2,NM1
      T0 = VS(J)-VS(J-1)
      T1 = VS(J+1)-VS(J)
      G(J) = T1
      IF (T0.LT.T1) G(J) = -T0
   10 CONTINUE
C
C     SET MP(I)=-1 FOR SIMPLE GOOD T-EIGENVALUES WHOSE MINGAPS  ARE
C     'TOO SMALL' AND DUE TO SPURIOUS T-EIGENVALUES.
C
      NISO = 0
      NG = 0
      DO 20 J = 1,NDIS
      IF (MP(J).EQ.0) GO TO 20
      NG = NG+1
      IF (MP(J).NE.1) GO TO 20
C     VS(J) IS NEXT TO SIMPLE GOOD T-EIGENVALUE
      NISO = NISO + 1
      I = J+1
      IF (G(J).LT.0.0) I = J-1
      IF (MP(I).NE.0) GO TO 20
      GAP = ABS(G(J))
      T0 = DMAX1(SCALE1*MULTOL,GAPTOL*DABS(VS(J)))
C     T0 = DMAX1(GAPTOL,GAPTOL*DABS(VS(J)))
      TEMP = T0
      IF (GAP.GT.TEMP) GO TO 20
      MP(J) = -MP(J)
      NISO = NISO-1
   20 CONTINUE
C
C-----END OF ISOEV--------------------------------------------------------
      RETURN
      END
C
C-----START OF PRTEST-----------------------------------------------------
C
      SUBROUTINE PRTEST(BETA,TEIG,TKMAX,EPSM,RELTOL,SCALE3,SCALE4,
     1 TMULT,NDIST,MEV,IPROJ)
C
C-----------------------------------------------------------------------
      DOUBLE PRECISION  BETA(1),TEIG(1),SIGMA(4)
      DOUBLE PRECISION  EPSM,RELTOL,PRTOL,TKMAX,LRATIO,URATIO
      DOUBLE PRECISION  EPS,EPS1,BETAM,LBD,UBD,SIG,YU,YV,LRATS,URATS
      DOUBLE PRECISION  ZERO,ONE,TEN,BISTOL,SCALE3,SCALE4,AEV,TEMP
      INTEGER  TMULT(1),ISIGMA(4)
      DOUBLE PRECISION  DABS, DMAX1, DSQRT, DFLOAT
C-----------------------------------------------------------------------
C     AFTER CONVERGENCE HAS BEEN ESTABLISHED, SUBROUTINE PRTEST
C     TESTS COMPUTED EIGENVALUES OF T(1,MEV) THAT HAVE BEEN LABELLED
C     SPURIOUS TO DETERMINE IF ANY SINGULAR VALUES OF A HAVE BEEN
C     MISSED BY LANCZOS PROCEDURE.  A SINGULAR VALUE WHOSE
C     SINGULAR VECTOR(S) HAS A VERY SMALL PROJECTION ON THE
C     STARTING VECTOR (< SINGLE PRECISION) CAN BE MISSED BECAUSE
C     IT WILL THEN ALSO BE AN EIGENVALUE OF T(2,MEV) TO WITHIN
C     THE SQUARE OF THIS ORIGINAL PROJECTION.  HOWEVER,
C     OUR EXPERIENCE IS THAT SUCH SMALL PROJECTIONS OCCUR ONLY
C     VERY INFREQUENTLY.
C
C     THIS SUBROUTINE IS CALLED ONLY AFTER CONVERGENCE HAS BEEN
C     ESTABLISHED. ONCE CONVERGENCE HAS BEEN OBSERVED ON THE
C     OTHER SINGULAR VALUES, THEN ONE CAN EXPECT TO ALSO HAVE
C     CONVERGENCE ON ANY SUCH 'HIDDEN' SINGULAR VALUES. (IF THERE
C     ARE ANY).  PROCEDURE CONSIDERS ONLY SPURIOUS T-EIGENVALUES AND
C     ONLY THOSE SPURIOUS T-EIGENVALUES THAT ARE ISOLATED FROM GOOD
C     T-EIGENVALUES.  FOR EACH SUCH T-EIGENVALUE IT DOES 2 STURM
```

```
C     SEQUENCES AND A FEW SCALAR MULTIPLICATIONS.  UPON RETURN TO MAIN
C     PROGRAM ERROR ESTIMATES WILL BE COMPUTED FOR ANY T-EIGENVALUES
C     THAT HAVE BEEN LABELLED AS 'HIDDEN'.  SUCH T-EIGENVALUES
C     WILL BE RELABELLED AS 'GOOD' ONLY IF THESE ERROR ESTIMATES
C     ARE SUFFICIENTLY SMALL.
C-----------------------------------------------------------------------
      ZERO = 0.0D0
      ONE  = 1.0D0
      TEN  = 10.0D0
      PRTOL = 1.D-6
      TEMP = DFLOAT(MEV+1000)
      TEMP = DSQRT(TEMP)
      BISTOL = TKMAX*EPSM*TEMP
      NSIGMA = 4
      SIGMA(1) = TEN*TKMAX
C
      DO 10 J = 2,NSIGMA
   10 SIGMA(J) = TEN*SIGMA(J-1)
C
      IFIN = 0
      MF = 1
      ML = MEV
      BETAM = BETA(MF)
      BETA(MF) = ZERO
      IPROJ = 0
      J = 1
C
      IF (TMULT(1).NE.0) GO TO 110
C
      AEV = DABS(TEIG(1))
      TEMP = PRTOL*AEV
      EPS1 = DMAX1(TEMP,SCALE4*BISTOL)
C     EPS1 = DMAX1(TEMP,PRTOL)
      TEMP = RELTOL*AEV
      EPS  = DMAX1(TEMP,SCALE3*BISTOL)
C     EPS  = DMAX1(TEMP,RELTOL)
C
      IF (TEIG(2)-TEIG(1).LT.EPS1.AND.TMULT(2).NE.0) GO TO 110
C
   20 LBD = TEIG(J) - EPS
      UBD = TEIG(J) + EPS
      MEVL = 0
      IL = 0
      YU = ONE
C
      DO 50 I=MF,ML
      IF (YU.NE.ZERO) GO TO 30
      YV = BETA(I)/EPSM
      GO TO 40
   30 YV = BETA(I)*BETA(I)/YU
   40 YU = -LBD-YV
      IF (YU.GE.ZERO) GO TO 50
C     MEVL INCREMENTED
      MEVL = MEVL + 1
      IL = I
   50 CONTINUE
C
      LRATIO = YU
      MEV1L = MEVL
      IF (IL.EQ.ML) MEV1L=MEVL-1
C
C     MEVL = NUMBER OF EVS OF T(1,MEV) WHICH ARE < LBD
C     MEV1L = NUMBER OF EVS OF T(1,MEV-1) WHICH ARE < LBD
C     LRATIO = DET(T(1,MEV)-LBD)/DET(T(1,MEV-1)-LBD):
C
      MEVU = 0
      IL = 0
```

```
      YU = ONE
C
      DO 80 I=MF,ML
      IF (YU.NE.ZERO) GO TO 60
      YV = BETA(I)/EPSM
      GO TO 70
   60 YV = BETA(I)*BETA(I)/YU
   70 YU = -UBD-YV
      IF (YU.GE.ZERO) GO TO 80
C     MEVU INCREMENTED
      MEVU = MEVU + 1
      IL = I
   80 CONTINUE
C
      URATIO = YU
      MEV1U = MEVU
      IF (IL.EQ.ML) MEV1U=MEVU-1
C
C     MEVU = NUMBER OF EVS OF T(MEV) WHICH ARE < UBD
C     MEV1U = NUMBER OF EVS OF T(MEV-1) WHICH ARE < UBD
C     URATIO = DET(TM-UBD)/DET(T(M-1)-UBD): TM=T(MF,ML)
C
      NEV1 = MEV1U-MEV1L
C
      DO 90 K=1,NSIGMA
      SIG = SIGMA(K)
      LRATS = LRATIO-SIG
      URATS = URATIO-SIG
C     NOTE THE INCREMENT IS ON NUMBER OF EVALUES OF T(M-1)
      MEVLS = MEV1L
      IF (LRATS.LT.0.) MEVLS=MEV1L+1
      MEVUS = MEV1U
      IF (URATS.LT.0.) MEVUS=MEV1U+1
      ISIGMA(K) = MEVUS - MEVLS
   90 CONTINUE
C
      ICOUNT = 0
      DO 100 K=1,NSIGMA
  100 IF (ISIGMA(K).EQ.1) ICOUNT=ICOUNT + 1
C
      IF (ICOUNT.LT.2.OR.NEV1.EQ.0) GO TO 110
      TMULT(J) = -10
      IPROJ=IPROJ+1
C
  110 J=J+1
C
      IF (J.GE.NDIST) GO TO 120
      IF (TMULT(J).NE.0) GO TO 110
C
      AEV = DABS(TEIG(J))
      TEMP = PRTOL*AEV
      EPS1 = DMAX1(TEMP,SCALE4*BISTOL)
C     EPS1 = DMAX1(TEMP,PRTOL)
      TEMP = RELTOL*AEV
      EPS  = DMAX1(TEMP,SCALE3*BISTOL)
C     EPS  = DMAX1(TEMP,RELTOL)
C
      IF (TEIG(J)-TEIG(J-1).LT.EPS1.AND.TMULT(J-1).NE.0) GO TO 110
      IF (TEIG(J+1)-TEIG(J).LT.EPS1.AND.TMULT(J+1).NE.0) GO TO 110
C
      GO TO 20
C
  120 IF (IFIN.EQ.1) GO TO 130
      IF (TMULT(NDIST).NE.0) GO TO 130
C
      AEV = DABS(TEIG(NDIST))
      TEMP = PRTOL*AEV
```

```
      EPS1 = DMAX1(TEMP,SCALE4*BISTOL)
C     EPS1 = DMAX1(TEMP,PRTOL)
      TEMP = RELTOL*AEV
      EPS  = DMAX1(TEMP,SCALE3*BISTOL)
C     EPS  = DMAX1(TEMP,RELTOL)
C
      NDIST1=NDIST -1
      TEMP = TEIG(NDIST)-TEIG(NDIST1)
      IF (TEMP.LT.EPS1.AND.TMULT(NDIST1).NE.0) GO TO 130
      IFIN = 1
C
      GO TO 20
C
  130 BETA(MF) = BETAM
C
C-----END OF PRTEST-----------------------------------------------------
      RETURN
      END
C
C------START OF STURMI--------------------------------------------------
C
      SUBROUTINE STURMI(BETA,X1,TOLN,EPSM,MMAX,MK1,MK2,IC,IWRITE)
C
C-----------------------------------------------------------------------
      DOUBLE PRECISION  BETA(1)
      DOUBLE PRECISION  EPSM,X1,TOLN,EVL,EVU,BETA2
      DOUBLE PRECISION  U1,U2,V1,V2,ZERO,ONE
      INTEGER I,IC,ICD,IC0,IC1,IC2,MK1,MK2,MMAX
C-----------------------------------------------------------------------
C
C     FOR ANY GOOD T-EIGENVALUE THAT HAS CONVERGED AS AN EIGENVALUE
C     OF THE T-MATRICES THIS SUBROUTINE CALCULATES
C     THE SMALLEST SIZE OF THE T-MATRIX, T(1,MK1) DEFINED
C     BY THE BETA ARRAY SUCH THAT MK1.LE.MMAX
C     AND THE INTERVAL (X1-TOLN,X1+TOLN) CONTAINS AT LEAST ONE
C     EIGENVALUE OF T(1,MK1). IT ALSO CALCULATES MK2 <= MMAX
C     AS THE SMALLEST SIZE T-MATRIX (IF ANY) SUCH THAT THIS INTERVAL
C     CONTAINS AT LEAST TWO EIGENVALUES OF T(1,MK2).
C     IF NO T-MATRIX OF ORDER < MMAX SATISFIES THIS REQUIREMENT
C     THEN MK2 IS SET EQUAL TO MMAX.  THE SINGULAR VECTOR PROGRAM
C     USES THESE VALUES TO DETERMINE A 1ST GUESS AT AN APPROPRIATE
C     SIZE T-MATRIX FOR THE SINGULAR VALUE  X1.
C
C     ON EXIT IC = NUMBER OF EIGENVALUES OF T(1,MK2) IN THIS INTERVAL
C
C     STURMI REGENERATES THE QUANTITIES BETA(I)**2 EACH TIME IT IS
C     CALLED, OBVIOUSLY FOR THE PRICE OF ANOTHER VECTOR OF LENGTH
C     MMAX THIS GENERATION COULD BE DONE ONCE IN THE MAIN
C     PROGRAM BEFORE THE LOOP ON THE CALLS TO SUBROUTINE STURMI.
C
C     IF ANY OF THE GOOD T-EIGENVALUES BEING CONSIDERED WERE MULTIPLE
C     AS SINGULAR VALUES OF THE USER-SPECIFIED MATRIX, THEN
C     THIS SUBROUTINE COULD BE MODIFIED TO COMPUTE ADDITIONAL
C     SIZES MKJ, J = 3, ...  WHICH COULD THEN BE USED IN THE
C     MAIN LANCZOS SINGULAR VECTOR PROGRAM TO COMPUTE ADDITIONAL
C     SINGULAR VECTORS CORRESPONDING TO THESE MULTIPLE SINGULAR
C     VALUES. THE MAIN PROGRAM LSVEC PROVIDED DOES NOT INCLUDE
C     THIS OPTION.
C
C-----------------------------------------------------------------------
C     INITIALIZATION OF PARAMETERS
      MK1 = 0
      MK2 = 0
      ZERO = 0.0D0
      ONE  = 1.0D0
      BETA(1) = ZERO
      EVL = X1-TOLN
```

```
      EVU = X1+TOLN
      U1 = ONE
      U2 = ONE
      ICO = 0
      IC1 = 0
      IC2 = 0
C
C     MAIN LOOP FOR CALCULATING THE SIZES MK1,MK2
      DO 60 I = 1,MMAX
      BETA2 = BETA(I)*BETA(I)
      IF (U1.NE.ZERO) GO TO 10
      V1 = BETA(I)/EPSM
      GO TO 20
   10 V1 = BETA2/U1
   20 U1 = EVL - V1
      IF (U1.LT.ZERO) IC1 = IC1+1
      IF (U2.NE.ZERO) GO TO 30
      V2 = BETA(I)/EPSM
      GO TO 40
   30 V2 = BETA2/U2
   40 U2 = EVU - V2
      IF (U2.LT.ZERO) IC2 = IC2+1
C     TEST FOR CHANGE IN NUMBER OF T-EIGENVALUES ON (EVL,EVU)
      ICD = IC1-IC2
      IC = ICD-ICO
      IF (IC.GE.1) GO TO 50
      GO TO 60
   50 CONTINUE
      IF (ICO.EQ.0) MK1 = I
      ICO = ICO+1
      IF (ICO.GT.1) GO TO 70
   60 CONTINUE
C
      I = I-1
      IF (ICO.EQ.0) MK1 = MMAX
   70 MK2 = I
      IC = ICD
C
      IF (IWRITE.EQ.1) WRITE(6,80) X1,MK1,MK2,IC
   80 FORMAT(' EVAL =',E20.12,' MK1 =',I6,' MK2 =',I6,' IC =',I3/)
C
      RETURN
C-----END OF STURMI-------------------------------------------------------
      END
C
C
C-----START OF INVERM-----------------------------------------------------
C
      SUBROUTINE INVERM(BETA,V1,V2,X1,ERROR,ERRORV,EPS,G,MEV,IT,
     1 IWRITE)
C
C-----------------------------------------------------------------------
      DOUBLE PRECISION  BETA(1),V1(1),V2(1)
      DOUBLE PRECISION  X1,U,Z,TEMP,RATIO,SUM,XU,NORM,TSUM,BETAM
      DOUBLE PRECISION  EPS,EPS3,EPS4,ERROR,ERRORV,ZERO,ONE
      REAL  G(1)
      DOUBLE PRECISION  DABS, DSQRT, DFLOAT
      DOUBLE PRECISION FINPRO
      REAL  ABS
C-----------------------------------------------------------------------
C
C     COMPUTES T-EIGENVECTORS FOR ISOLATED GOOD T-EIGENVALUES X1
C     USING INVERSE ITERATION ON T(1,MEV(X1)) SOLVING EQUATION
C     (T - X1*I)V2 = RIGHT-HAND SIDE (RANDOMLY-GENERATED) .
C     PROGRAM REFACTORS T- X1*I ON EACH ITERATION OF INVERSE ITERATION.
C     TYPICALLY ONLY ONE ITERATION IS NEEDED PER T-EIGENVALUE X1.
C
```

```
C     IF IWRITE = 1 THEN THERE ARE EXTENDED WRITES TO FILE 6 (TERMINAL)
C
C     ON ENTRY G CONTAINS A REAL*4 RANDOM VECTOR WHICH WAS GENERATED
C     IN MAIN PROGRAM.
C
C     ON ENTRY AND EXIT
C     MEV = ORDER OF T
C     BETA CONTAINS THE OFFDIAGONAL ENTRIES OF T.
C     EPS = 2. * MACHINE EPSILON
C
C     IN PROGRAM:
C     ITER = MAXIMUM NUMBER STEPS ALLOWED FOR INVERSE ITERATION
C     ITER = IT ON ENTRY.
C     V1,V2 = WORK SPACES USED IN THE FACTORIZATION OF T(1,MEV).
C     V1 AND V2 MUST BE OF DIMENSION AT LEAST MEV.
C
C     ON EXIT
C     V2 = THE UNIT EIGENVECTOR OF T(1,MEV) CORRESPONDING TO X1.
C     ERROR = |V2(MEV)| = ERROR ESTIMATE FOR CORRESPONDING
C             RITZ VECTOR FOR X1.
C
C     ERRORV = || T*V2 - X1*V2 || = ERROR ESTIMATE ON T-EIGENVECTOR.
C     IF IT.GT.ITER THEN ERRORV = -ERRORV
C     IT = NUMBER OF ITERATIONS ACTUALLY REQUIRED
C-----------------------------------------------------------------------
C     INITIALIZATION AND PARAMETER SPECIFICATION
      ONE  = 1.0D0
      ZERO = 0.0D0
      ITER = IT
      MP1 = MEV+1
      MM1 = MEV-1
      BETAM = BETA(MP1)
      BETA(MP1) = ZERO
C
C     CALCULATE SCALE AND TOLERANCES
      TSUM = ZERO
      DO 10 I = 2,MEV
   10 TSUM = TSUM + BETA(I)
C
      EPS3 = EPS*TSUM
      EPS4 = DFLOAT(MEV)*EPS3
C
C     GENERATE SCALED RANDOM RIGHT-HAND SIDE
      GSUM = ZERO
      DO 20 I = 1,MEV
   20 GSUM = GSUM+ABS(G(I))
      GSUM = EPS4/GSUM
C
C     INITIALIZE RIGHT HAND SIDE FOR INVERSE ITERATION
      DO 30 I = 1,MEV
   30 V2(I) = GSUM*G(I)
      IT = 1
C
C     CALCULATE UNIT EIGENVECTOR OF T(1,MEV) FOR ISOLATED GOOD
C     T-EIGENVALUE X1.
C
C     TRIANGULAR FACTORIZATION WITH NEAREST NEIGHBOR PIVOT
C     STRATEGY. INTERCHANGES ARE LABELLED BY SETTING BETA < 0.
C
   40 CONTINUE
      U = -X1
      Z = BETA(2)
C
      DO 60 I=2,MEV
      IF (BETA(I).GT.DABS(U)) GO TO 50
C     NO PIVOT INTERCHANGE
      V1(I-1) = Z/U
```

```
      V2(I-1) = V2(I-1)/U
      V2(I) = V2(I)-BETA(I)*V2(I-1)
      RATIO = BETA(I)/U
      U = -X1-Z*RATIO
      Z = BETA(I+1)
      GO TO 60
C     PIVOT INTERCHANGE
   50 CONTINUE
      RATIO = U/BETA(I)
      BETA(I) = -BETA(I)
      V1(I-1) = -X1
      U = Z-RATIO*V1(I-1)
      Z = -RATIO*BETA(I+1)
      TEMP = V2(I-1)
      V2(I-1) = V2(I)
      V2(I) = TEMP-RATIO*V2(I)
   60 CONTINUE
C
      IF (U.EQ.ZERO) U=EPS3
C
C     SMALLNESS TEST AND DEFAULT VALUE FOR LAST COMPONENT
C     PIVOT(I-1) = |BETA(I)| FOR INTERCHANGE CASE
C     (I-1,I+1) ELEMENT IN RIGHT FACTOR = BETA(I+1)
C     END OF FACTORIZATION AND FORWARD SUBSTITUTION
C
C     BACK SUBSTITUTION
      V2(MEV) = V2(MEV)/U
      DO 80 II = 1,MM1
      I = MEV-II
      IF (BETA(I+1).LT.ZERO) GO TO 70
C     NO PIVOT INTERCHANGE
      V2(I) = V2(I)-V1(I)*V2(I+1)
      GO TO 80
C     PIVOT INTERCHANGE
   70 BETA(I+1) = -BETA(I+1)
      V2(I) = (V2(I)-V1(I)*V2(I+1)-BETA(I+2)*V2(I+2))/BETA(I+1)
   80 CONTINUE
C
C
C     TESTS FOR CONVERGENCE OF INVERSE ITERATION
C     IF SUM |V2| COMPS. LE. 1 AND IT. LE. ITER DO ANOTHER INVIT STEP
C
      NORM = DABS(V2(MEV))
      DO 90 II = 1,MM1
      I = MEV-II
   90 NORM = NORM+DABS(V2(I))
C
C     IS DESIRED GROWTH IN VECTOR ACHIEVED ?
C     IF NOT, DO ANOTHER INVERSE ITERATION STEP UNLESS NUMBER ALLOWED IS
C     EXCEEDED.
      IF (NORM.GE.ONE) GO TO 110
C
      IT=IT+1
      IF (IT.GT.ITER) GO TO 110
C
      XU = EPS4/NORM
      DO 100 I=1,MEV
  100 V2(I) = V2(I)*XU
C
      GO TO 40
C
C     NORMALIZE COMPUTED T-EIGENVECTOR : V2 = V2/||V2||
C
  110 CONTINUE
C
      SUM = FINPRO(MEV,V2(1),1,V2(1),1)
      SUM = ONE/DSQRT(SUM)
```

```
      DO 120 II = 1,MEV
  120 V2(II) = SUM*V2(II)
C
C     SAVE ERROR ESTIMATE FOR LATER OUTPUT
      ERROR = DABS(V2(MEV))
C
C     GENERATE ERRORV = ||T*V2 - X1*V2||.
      V1(MEV) = BETA(MEV)*V2(MEV-1)-X1*V2(MEV)
      DO 130 J = 2,MM1
      JM = MP1 - J
      V1(JM) = BETA(JM)*V2(JM-1) + BETA(JM+1)*V2(JM+1
     1) - X1*V2(JM)
  130 CONTINUE
C
      V1(1) = BETA(2)*V2(2) - X1*V2(1)
      ERRORV = FINPRO(MEV,V1(1),1,V1(1),1)
      ERRORV = DSQRT(ERRORV)
      IF (IT.GT.ITER) ERRORV = -ERRORV
      IF (IWRITE.EQ.0) GO TO 150
C
C     FILE 6 (TERMINAL) OUTPUT OF ERROR ESTIMATES.
      WRITE(6,140) MEV,X1,ERROR,ERRORV
  140 FORMAT(' INVERSE ITERATION OUTPUT'/
     1 2X,'TSIZE',13X,'T-EIGENVALUE',11X,'U(M)',9X,'ERRORV'/
     1 I6,E25.16,2E15.5)
C
C     RESTORE BETA(MEV+1) = BETAM
  150 CONTINUE
      BETA(MP1) = BETAM
C-----END OF INVERM-----------------------------------------------------
      RETURN
      END
C
C-----START OF LBISEC---------------------------------------------------
C
      SUBROUTINE LBISEC(BETA,EPSM,EVAL,EVALN,LB,UB,TTOL,M,NEVT)
C
C-----------------------------------------------------------------------
      DOUBLE PRECISION  BETA(1),X0,X1,XL,XU,YU,YV,LB,UB
      DOUBLE PRECISION  EPSM,EP1,EVAL,EVALN,EVD,EPT
      DOUBLE PRECISION  ZERO,ONE,HALF,TTOL,TEMP
      DOUBLE PRECISION  DABS,DSQRT,DFLOAT
C-----------------------------------------------------------------------
C     SPECIFY PARAMETERS
      ZERO = 0.0D0
      HALF = 0.5D0
      ONE  = 1.0D0
      XL = LB
      XU = UB
C
C     EP1 = DSQRT(1000+M)*TTOL      TTOL = EPSM*TKMAX
C     TKMAX = MAX(BETA(K), K= 1,KMAX)
C
      TEMP = DFLOAT(1000+M)
      EP1 = DSQRT(TEMP)*TTOL
C
      NA = 0
      X1 = XU
      JSTURM = 1
      GO TO 60
C     FORWARD STURM CALCULATION
   10 NA = NEV
      X1 = XL
      JSTURM = 2
      GO TO 60
C     FORWARD STURM CALCULATION
```

```
   20 NEVT = NEV
C
C     WRITE(6,30) M,EVAL,NEVT,EP1
   30 FORMAT(/3X,'TSIZE',23X,'EV',9X/I8,E25.16/
     1 I6,' = NUMBER OF T(1,M) EIGENVALUES ON TEST INTERVAL'/
     1 E12.3,' = CONVERGENCE TOLERANCE'/)
C
      IF (NEVT.NE.1) GO TO 120
C
C     BISECTION LOOP
      JSTURM = 3
   40 X1 = HALF*(XL+XU)
      X0 = XU-XL
      EPT = EPSM*(DABS(XL) + DABS(XU)) + EP1
C     CONVERGENCE TEST
      IF (X0.LE.EPT) GO TO 100
      GO TO 60
C     FORWARD STURM CALCULATION
   50 CONTINUE
      IF(NEV.EQ.0) XU = X1
      IF(NEV.EQ.1) XL = X1
      GO TO 40
C     NEV = NUMBER OF EIGENVALUES OF T(1,M) ON (X1,XU)
C     THERE IS EXACTLY ONE EIGENVALUE OF T(1,M) ON (XL,XU)
C
C     FORWARD STURM CALCULATION
   60 NEV = -NA
      YU = ONE
      DO 90 I = 1,M
      IF (YU.NE.ZERO) GO TO 70
      YV = BETA(I)/EPSM
      GO TO 80
   70 YV = BETA(I)*BETA(I)/YU
   80 YU = X1 - YV
      IF (YU.GE.ZERO) GO TO 90
      NEV = NEV+1
   90 CONTINUE
      GO TO (10,20,50), JSTURM
C
  100 CONTINUE
C
      EVALN = X1
      EVD = DABS(EVALN-EVAL)
C     WRITE(6,110) EVALN,EVAL,EVD
  110 FORMAT(/20X,'EVALN',21X,'EVAL',6X,'CHANGE'/2E25.16,E12.3/)
C
  120 CONTINUE
      RETURN
C-----END OF LBISEC--------------------------------------------------
      END
```

SECTION 6.7 OPTIONAL PREPROCESSING PROGRAM, FILE DEFINITIONS, AND SAMPLE INPUT FILES

```
C-----LSCOMPAC-(STAND ALONE PROGRAM)-----------------------------------
C
C     THIS PROGRAM TRANSLATES A SPARSE RECTANGULAR M X N MATRIX A,
C     GIVEN AS I, J, A(I,J), INTO THE SPARSE MATRIX FORMAT
C     REQUIRED BY THE SAMPLE USPEC AND CMATV PROGRAMS PROVIDED
C     FOR USE WITH THE LANCZOS SINGULAR VALUE PROCEDURES.
C     THIS PROGRAM ASSUMES THAT THE MATRIX ENTRIES ARE GIVEN EITHER
C     COLUMN BY COLUMN OR ROW BY ROW.  MOREOVER, IF THE ENTRIES
C     ARE GIVEN ROW BY ROW, THEN THE FILE 8 CREATED WILL CORRESPOND
C     TO A-TRANSPOSE NOT A.  THUS, IN ANY SUBSEQUENT USE OF THIS FILE
C     IN THE LANCZOS SINGULAR VALUE/VECTOR PROGRAMS, THE USER WILL HAVE
C     TO INTERCHANGE THE ROLES OF M AND N IN THE INPUT FILES.
C
C     NONPORTABLE STATEMENTS:  PFORT VERIFIER INDICATES THIS PROGRAM
C                              IS PORTABLE
C
C----------------------------------------------------------------------
      DOUBLE PRECISION A(15000)
      INTEGER IROW(15000),ICOL(15000)
C----------------------------------------------------------------------
C     INPUT FILE 7 CONTAINS THE SPARSE REAL MXN MATRIX STORED AS:
C
C                NZ,M,N,MATNO
C                I(K)  J(K)  A(K)  K = 1,NZ
C
C     WHERE NZ IS THE TOTAL NUMBER OF NONZEROS IN THE MATRIX A,
C     M IS THE ROW DIMENSION OF A, N IS THE COLUMN DIMENSION,
C     AND A(K) ARE THE NONZERO ENTRIES STORED ROW BY ROW OR
C     COLUMN BY COLUMN.  PROGRAM READS THIS IN AS IROW(K) = I(K),
C     ICOL(K) = J(K), AND A(K) = A(K).
C
C     OUTPUT FILE = 8 CONTAINS THE A-MATRIX IN SPARSE FORMAT
C
C              NZ,M,N,MATNO
C              ICOL(K)    K = 1,N
C              IROW(K)    K = 1,NZ
C                 A(K)    K = 1,NZ
C
C     ON OUTPUT A(K), K=1,NZ CONTAINS THE KTH NONZERO ELEMENT
C     OF THE GIVEN MATRIX, ORDERED COLUMN BY COLUMN.  IROW(K),
C     K=1,NZ, EQUALS THE ROW INDEX OF THE NONZERO ENTRY A(K).
C     ICOL(K), K=1,N, EQUALS THE NUMBER OF NONZEROS OF A IN
C     COLUMN K.
C
C----------------------------------------------------------------------
C
      READ(7,10) NZ,M,N,MATNO,IIROW
   10 FORMAT(I10,2I6,I8,I6)
C
      WRITE(6,20) NZ,M,N,MATNO,IIROW
   20 FORMAT(I10,2I6,I8,' = NO. NONZERO AIJ,M,N,MATNO'/
     1 I6,' = IIROW IF IIROW=0 ORDERING IS BY COLS IIROW=1 BY ROWS'/)
C
      IF (IIROW.EQ.0) READ(7,30) (IROW(K),ICOL(K),A(K), K=1,NZ)
      IF (IIROW.EQ.0) GO TO 40
C
      READ(7,30) (ICOL(K),IROW(K),A(K), K=1,NZ)
   30 FORMAT(2I5,E14.7)
      ITEMP = M
      M = N
      N = ITEMP
```

```
   40 CONTINUE
C
      LCOUNT = 0
      K = 1
C
C     START OF A NEW COLUMN
   50 CONTINUE
      J = ICOL(K)
      ICOL(J) = 0
   60 CONTINUE
      LCOUNT = LCOUNT + 1
      A(LCOUNT) = A(K)
      IROW(LCOUNT) = IROW(K)
      ICOL(J) = ICOL(J) + 1
C
   70 CONTINUE
      K = K+1
C
      IF(K.GT.NZ) GO TO 80
C
      IF(ICOL(K).GT.J) GO TO 50
C
      GO TO 60
C
   80 CONTINUE
C     NZ = LCOUNT
C
      WRITE(8,90) NZ,M,N,MATNO
      WRITE(6,90) NZ,M,N,MATNO
   90 FORMAT(I10,2I6,I8,' = NZ M N MATNO')
C
      WRITE(8,100) (ICOL(I), I=1,N)
      WRITE(8,100) (IROW(K), K=1,NZ)
  100 FORMAT(13I6)
C
      WRITE(8,110) (A(K), K=1,NZ)
  110 FORMAT(4E19.10)
C
C-----END LSCOMPAC-----------------------------------------------------
      STOP
      END
```

-----INPUT/OUTPUT FILE DEFINITIONS FOR LSVAL AND LSVEC-----

Below is a listing of the input/output files which are accessed by the Lanczos program LSVAL for computing singular values of real rectangular matrices on user-specified intervals. Included also is a sample of the input file which LSVAL requires on file 5. The parameters in this file are supplied in free format.

SAMPLE DEFINITIONS OF THE INPUT/OUTPUT FILES FOR LSVAL

```
-----------------------------------------------------------------
*LSVAL EXEC FOR LANCZOS SINGULAR VALUE CALCULATIONS
FI 06 TERM
FILEDEF  1 DISK &1         NSHISTORY A (RECFM F LRECL 80 BLOCK 80
FILEDEF  2 DISK &1         SVHISTORY A (RECFM F LRECL 80 BLOCK 80
FILEDEF  3 DISK &1         GOODEV    A (RECFM F LRECL 80 BLOCK 80
FILEDEF  4 DISK &1         ERRINV    A (RECFM F LRECL 80 BLOCK 80
FILEDEF  5 DISK LSVAL      INPUT     A (RECFM F LRECL 80 BLOCK 80
FILEDEF  8 DISK &1         ACOMPAC   A (RECFM F LRECL 80 BLOCK 80
FILEDEF 11 DISK &1         DISTINCT  A (RECFM F LRECL 80 BLOCK 80
LOAD   LSVAL   LSSUB    LSMULT
-----------------------------------------------------------------
```

SAMPLE INPUT FILE FOR LSVAL

```
---------------------------------------------------------------------------
 LANCZOS SINGULAR VALUE PROCEDURE,
 WITHOUT REORTHOGONALIZATION BUT WITH BIDIAGONALIZATION.
LINE 1      M     N     KMAX    NMEVS      MATNO
          100   100     300        1       2220
LINE 2    SVSEED     RHSEED     MXINIT      MXSTUR
       49302312    7549309          5      100000
LINE 3       ISTART      ISTOP
                  0          1
LINE 4       IHIS      IDIST      IWRITE      IPAR
                1          0           1          2
LINE 5  RELTOL(RELATIVE TOLERANCE USED IN 'COMBINING' COMPUTED GOOD EIGENVALUES)
   .0000000001
LINE 6  MB(1)    MB(2)    MB(3)    MB(4)   (ORDERS OF T(1,MEV) MUST BE EVEN
         280
LINE 7    NINT     (NUMBER OF BISEC INTERVALS)
            1
LINE 8     LB(1)     LB(2)    LB(3)    LB(4)   (LOWER BOUNDS OF THESE INTERVALS)
            0.0
LINE 9     UB(1)     UB(2)    UB(3)    UB(4)   (UPPER BOUNDS OF THESE INTERVALS)
            1.0
---------------------------------------------------------------------------
```

Below is a listing of the input/output files which are accessed by the Lanczos program for computing singular vectors, LSVEC. Included also is a sample of the input file which LSVEC requires on file 5. The parameters in this file are supplied in free format. LSVEC computes singular vectors for each of a user-specified subset of the singular values computed by the companion program LSVAL.

SAMPLE DEFINITIONS OF THE INPUT/OUTPUT FILES FOR LSVEC

```
-----------------------------------------------------------------
*LSVEC EXEC TO RUN LANCZOS SINGULAR VECTOR PROGRAM
FI 06 TERM
FILEDEF  2 DISK &1       SVHISTORY A (RECFM F LRECL 80 BLOCK 80
FILEDEF  3 DISK &1       GOODSV    A (RECFM F LRECL 80 BLOCK 80
FILEDEF  4 DISK &1       ERRINV    A (RECFM F LRECL 80 BLOCK 80
FILEDEF  5 DISK LSVEC    INPUT     A (RECFM F LRECL 80 BLOCK 80
FILEDEF  8 DISK &1       ACOMPAC   A (RECFM F LRECL 80 BLOCK 80
FILEDEF  9 DISK &1       ERREST    A (RECFM F LRECL 80 BLOCK 80
FILEDEF 10 DISK &1       BOUNDS    A (RECFM F LRECL 80 BLOCK 80
FILEDEF 11 DISK &1       TEIGVECS  A (RECFM F LRECL 80 BLOCK 80
FILEDEF 12 DISK &1       RITZVECS  A (RECFM F LRECL 80 BLOCK 80
FILEDEF 13 DISK &1       PAIGE     A (RECFM F LRECL 80 BLOCK 80
LOAD   LSVEC   LSSUB   LSMULT
-------------------------------------------------------------------
```

SAMPLE INPUT FILE FOR LSVEC

```
-------------------------------------------------------------------
 LSVEC SINGULAR VECTORS, NO REORTHOGONALIZATION BUT BIDIAGONALIZATION
LINE 1  MATNO        M      N
          100      100     80
LINE 2  MDIMTV     MDIMRV  MBETA (MAX.DIMENSIONS,TVEC,RITVEC AND BETA
          10000      10000   2000
LINE 3       RELTOL
        .0000000001
LINE 4  MBOUND   NTVCON SVTVEC IREAD (FLAGS
             0        1      0     1
LINE 5  TVSTOP   LVCONT ERCONT  IWRITE (FLAGS
             0        1      1       1
LINE 6    RHSEED  (RANDOM GENERATOR SEED FOR STARTING VECTOR IN INVERM)
        45329517
---------------------------------------------------------------------
```

CHAPTER 7

NONDEFECTIVE COMPLEX SYMMETRIC MATRICES

SECTION 7.1 INTRODUCTION

The FORTRAN codes in this chapter address the question of computing distinct eigenvalues and eigenvectors of a nondefective, complex symmetric matrix, using a single-vector Lanczos procedure. For a given nondefective, complex symmetric matrix A, these codes compute complex scalars λ and corresponding complex vectors $x \neq 0$ such that

$$Ax = \lambda x. \tag{7.1.1}$$

DEFINITION 7.1.1 A complex nxn matrix $A \equiv (a_{ij})$, $1 \leq i,j \leq n$, is complex symmetric if and only if for every i and j, $a_{ij} = a_{ji}$. A complex symmetric matrix is nondefective if and only if it has a complete set of eigenvectors.

It is straight-forward to show from Definition 7.1.1 that if $A = B + iC$, where A and B are real matrices and $i = \sqrt{-1}$, is a complex symmetric matrix then B and C are real symmetric matrices. It is also easy to prove that if λ and μ are two distinct eigenvalues of A and x and y are corresponding eigenvectors of A, then the Euclidean inner product applied to the complex vectors x and y satisfies

$$x^T y = 0. \tag{7.1.2}$$

In Eqn(7.1.2) the superscript T denotes transpose. Thus, although the eigenvectors of a complex symmetric matrix are not orthogonal with respect to the complex norm, $\| x \|^2_{\mathscr{C}} = \sum_{i=1}^{n} \overline{x(i)} x(i)$, they are real orthogonal in the sense specified in Eqn(7.1.2). Therefore, when we consider generalizing the Lanczos recursion to the complex symmetric case we are led to consider an 'inner product' which is a mixture of real and complex quantities. In fact the Euclidean inner product, which of course is not an inner product for complex vectors, is the natural 'inner product' to use in the complex symmetric case.

Complex symmetric matrices are not 'easy' like real symmetric matrices. They bear little resemblance to real symmetric matrices. Complex symmetric matrices need not have complete sets of eigenvectors. Even if a complete set of eigenvectors exists, eigenvectors corresponding to different eigenvalues are only real orthogonal in the sense of Eqn(7.1.2). If a small perturbation is applied to a complex symmetric matrix, then large perturbations in the eigenvalues may result. See Wilkinson [1965] for a discussion of the properties of complex symmetric matrices.

The Lanczos recursion as presented in Eqns(1.2.1) - (1.2.2) is only applicable to real symmetric matrices so we ask the question: How do we construct a complex symmetric version of the basic Lanczos recursion which will give us the desired eigenvalues? We have used what has been suggested elsewhere, Moro and Freed [1981]. In particular, we use the recursion in Eqn(1.2.1) with the formulas for the scalars α_i and β_{i+1} given in Eqn(1.2.2), except that the quantities involved are now complex-valued, but the real Euclidean inner product is used. See Section 6.3 of Chapter 6 in Volume 1.

There are some fundamental differences between the amount of computation required by the complex symmetric codes versus that required by the real symmetric codes. First, all of the complex symmetric computations are done in double precision complex arithmetic. All the vectors used are complex vectors. Each of the Lanczos matrices generated is a complex symmetric tridiagonal matrix. Unfortunately, there is no simple analog of the bisection procedure used in the real symmetric case which would allow us to compute the eigenvalues of a given complex symmetric tridiagonal matrix on only some small portion of the spectrum. We are therefore forced to do a complete eigenvalue computation on each complex symmetric tridiagonal matrix which we consider. Actually in the complex symmetric case we are forced to do two complete eigenvalue computations for each Lanczos tridiagonal matrix which we consider. Two are required because the identification test for categorizing the eigenvalues of the Lanczos T-matrices into 'good' and 'spurious' ones uses the eigenvalues of the corresponding tridiagonal matrix obtained from the Lanczos T-matrix by crossing out the first row and column of that matrix. This is the same identification test as that used in the procedures for real symmetric problems. However, in the real symmetric cases this test is directly incorporated into the BISEC subroutine which is used to compute the eigenvalues of the Lanczos matrices, and the resulting cost of this test is negligible for those types of problems.

These codes can be used to compute either a very few or very many of the distinct eigenvalues of a nondefective, complex symmetric matrix. As the documentation in the next section indicates, the A-multiplicity of a given computed eigenvalue can be obtained only with additional computation, and the modifications required to do this additional computation are not included in these versions of the codes.

The Lanczos recursions used generate a family of complex symmetric, tridiagonal matrices. A real orthogonal analog of the EISPACK [1976,1977] subroutine IMTQL1 which we call CMTQL1 was developed to compute the eigenvalues of the complex symmetric, tridiagonal Lanczos matrices generated. There is no reorthogonalization of the Lanczos vectors at any stage in any of the computations.

CSLEVAL, the main program for the complex symmetric eigenvalue computations, calls the subroutines COMPEV and CMTQL1 to compute the eigenvalues of the Lanczos T-matrices specified by the user. The eigenvalues of the related complex symmetric tridiagonal matrices obtained by deleting the first row and first column from the given Lanczos T-matrix are also computed. COMPEV then determines the T-multiplicities of the T-eigenvalues and sorts the computed T-eigenvalues into two classes, the 'good' T-eigenvalues and the 'spurious' T-eigenvalues. The 'good' T-eigenvalues are accepted as approximations to eigenvalues of the user-specified matrix A. The accuracy of these 'good' T-eigenvalues as eigenvalues of A is then estimated using error estimates computed by a complex version of the subroutine INVERR. Error estimates are computed only for isolated 'good' T-eigenvalues. All other 'good' T-eigenvalues are assumed to have converged. Convergence is then checked. If convergence has not yet occurred and a larger Lanczos matrix has been specified by the user, the program will continue on to the larger T-matrix, repeating the above procedure on this larger matrix.

Once the eigenvalues been computed accurately enough, the user can select a subset of the 'converged' eigenvalues for which eigenvectors are to be computed. The main program CSLEVEC, for computing eigenvectors of complex symmetric matrices, is then used to compute these desired eigenvectors.

As stated earlier, all computations are in double precision complex arithmetic. The user must supply a subroutine USPEC which defines and initializes the user-specified matrix A and a subroutine CMATV which computes matrix-vector multiplies Ax for any given vector x. These subroutines must be constructed in such a way as to take advantage of the sparsity (and/or structure) of the user-supplied A-matrix and such that these computations are done accurately.

The user should note that the complex symmetric computations are considerably more expensive than the corresponding real symmetric ones. Two complete T-matrix eigenvalue computations must be done for each T-size. Moreover, the accuracy of these computations is noticeably less than that achievable in the real symmetric case. This is to be expected from the perturbation analysis for the complex symmetric case. Therefore we reduced the anticipated accuracy of the computed eigenvalues and used larger tolerances in our multiplicity and spuriousness tests. These larger tolerances decrease the resolution capabilities of these codes. However, these tolerances are realistic. Moreover, these complex symmetric codes cannot be expected to handle stiff problems effectively. More details about these complex symmetric, single-vector Lanczos procedures are included in Chapter 6 of Volume 1.

SECTION 7.2 DOCUMENTATION FOR PROGRAMS IN CHAPTER 7

```
C-----DOCUMENTATION FOR SINGLE-VECTOR-----------------------------------CSL00010
C     LANCZOS EIGENVALUE/EIGENVECTOR PROGRAMS FOR                       CSL00020
C     NONDEFECTIVE COMPLEX SYMMETRIC MATRICES                           CSL00030
C                                                                       CSL00040
C     AUTHORS:  JANE CULLUM AND RALPH A. WILLOUGHBY, IBM RESEARCH,      CSL00050
C               YORKTOWN HEIGHTS, NY 10598.  PHONE: 914-945-2227        CSL00060
C                                                                       CSL00070
C     GIVEN A NONDEFECTIVE COMPLEX SYMMETRIC MATRIX A OF ORDER N        CSL00080
C     THE THREE SETS OF FORTRAN FILES LABELLED CSLEVAL, CSLESUB,        CSL00090
C     AND CSLEMULT CAN BE USED TO COMPUTE DISTINCT EIGENVALUES OF       CSL00100
C     A.  NOTE THAT THESE PROGRAMS DIFFER FROM THE REAL SYMMETRIC       CSL00110
C     AND HERMITIAN PROGRAMS IN THAT IT IS NOT POSSIBLE TO              CSL00120
C     COMPUTE THE EIGENVALUES OF THE LANCZOS TRIDIAGONAL MATRICES       CSL00130
C     ONLY IN SPECIFIED INTERVALS.  THUS, ON ANY GIVEN                  CSL00140
C     ITERATION ALL OF THE EIGENVALUES OF THESE TRIDIAGONAL MATRICES    CSL00150
C     MUST BE COMPUTED.  IN FACT TWO COMPLETE TRIDIAGONAL EIGENVALUE    CSL00160
C     COMPUTATIONS ARE USED.                                            CSL00170
C                                                                       CSL00180
C     CORRESPONDING EIGENVECTORS FOR SELECTED, COMPUTED EIGENVALUES CAN CSL00190
C     BE COMPUTED USING THE CORRESPONDING SETS OF FILES LABELLED        CSL00200
C     CSLEVEC, CSLESUB AND CSLEMULT.                                    CSL00210
C                                                                       CSL00220
C     THESE PROGRAMS ALL USE A GENERALIZATION OF LANCZOS                CSL00230
C     TRIDIAGONALIZATION TO COMPLEX SYMMETRIC MATRICES TO               CSL00240
C     GENERATE COMPLEX SYMMETRIC TRIDIAGONAL MATRICES, T(1,MEV)         CSL00250
C     OF ORDER MEV.  NO REORTHOGONALIZATION IS USED.  SUBSETS OF        CSL00260
C     THE EIGENVALUES OF THESE T-MATRICES, LABELLED AS THE              CSL00270
C     'GOOD EIGENVALUES', YIELD APPROXIMATIONS TO THE DESIRED           CSL00280
C     EIGENVALUES OF A.  CORRESPONDING RITZ VECTORS ARE APPROXIMATIONS  CSL00290
C     TO THE DESIRED EIGENVECTORS OF A.  NOTE THAT IN THE DISCUSSION    CSL00300
C     T(1,MEV) DENOTES THE LANCZOS MATRIX OF ORDER MEV AND T(2,MEV)     CSL00310
C     DENOTES THE MATRIX OF SIZE MEV-1 OBTAINED FROM T(1,MEV) BY        CSL00320
C     DELETING THE FIRST ROW AND COLUMN OF T(1,MEV).                    CSL00330
C                                                                       CSL00340
C     THE IDEAS USED IN THESE PROGRAMS ARE DISCUSSED IN THE FOLLOWING   CSL00350
C     REFERENCES.                                                       CSL00360
C                                                                       CSL00370
C     1.  JANE CULLUM AND RALPH A. WILLOUGHBY, LANCZOS ALGORITHMS       CSL00380
C         FOR LARGE SYMMETRIC MATRICES,  PROGRESS IN                    CSL00390
C         SCIENTIFIC COMPUTING, EDITORS, G. GOLUB, H.O. KREISS,         CSL00400
C         S. ARBARBANEL, AND R. GLOWINSKI,  BIRKHAUSER BOSTON INC.,     CSL00410
C         CAMBRIDGE, MASSACHUSETTS, 1984.                               CSL00420
C                                                                       CSL00430
C     2.  JANE CULLUM AND RALPH A. WILLOUGHBY, COMPUTING EIGENVECTORS   CSL00440
C         (AND EIGENVALUES) OF LARGE, SYMMETRIC MATRICES USING          CSL00450
C         LANCZOS TRIDIAGONALIZATION, LECTURE NOTES IN MATHEMATICS,     CSL00460
C         773, NUMERICAL ANALYSIS PROCEEDINGS, DUNDEE 1979, EDITED BY   CSL00470
C         G. A. WATSON, SPRINGER-VERLAG, (1980), BERLIN, PP.46-63.      CSL00480
C                                                                       CSL00490
C     3. IBID, LANCZOS AND THE COMPUTATION IN SPECIFIED INTERVALS OF    CSL00500
C        THE SPECTRUM OF LARGE SPARSE, REAL SYMMETRIC MATRICES, SPARSE  CSL00510
C        MATRIX PROCEEDINGS 1978, ED. I.S. DUFF AND G. W. STEWART,      CSL00520
C        SIAM, PHILADELPHIA, PP.220-255, 1979.                          CSL00530
C                                                                       CSL00540
C     4. IBID, COMPUTING EIGENVALUES OF VERY LARGE SYMMETRIC MATRICES-  CSL00550
C        AN IMPLEMENTATION OF A LANCZOS ALGORITHM WITHOUT               CSL00560
C        REORTHOGONALIZATION, J. COMPUT. PHYS. 44(1981), 329-358.       CSL00570
C                                                                       CSL00580
C     5. IBID, A LANCZOS ALGORITHM FOR NONDEFECTIVE COMPLEX SYMMETRIC   CSL00590
C        MATRICES, IBM RESEARCH REPORT, 1984.                           CSL00600
C                                                                       CSL00610
C-----PORTABILITY-------------------------------------------------------CSL00620
C                                                                       CSL00630
C     PROGRAMS WERE TESTED FOR PORTABILITY USING THE PFORT VERIFIER.    CSL00640
```

```
C     FOR DETAILS OF THE VERIFIER SEE FOR EXAMPLE, B. G. RYDER AND
C     A. D. HALL, 'THE PFORT VERIFIER', COMPUTING SCIENCE TECHNICAL
C     REPORT 12, BELL LABORATORIES, MURRAY HILL, NEW JERSEY 07974,
C     (REVISED), JANUARY 1981.
C
C     PORTABILITY:
C     THESE PROGRAMS ARE NOT PORTABLE DUE TO THE USE OF COMPLEX*16
C     VARIABLES AND CORRESPONDING COMPLEX FUNCTIONS.  IN ADDITION, THE
C     PFORT VERIFIER IDENTIFIED THE FOLLOWING NONPORTABLE
C     CONSTRUCTIONS.
C     IN CSLEVAL AND IN CSLEVEC
C         1.  DATA/MACHEP STATEMENT
C         2.  ALL READ(5,*) STATEMENTS (FREE FORMAT)
C         3.  FORMAT(20A4) USED FOR THE EXPLANATORY HEADER ARRAY, EXPLAN
C         4.  HEXADECIMAL FORMAT (4Z20) FOR ALPHA/BETA FILES 1 AND 2.
C     IN CSLEMULT
C         1.  IN CMATV AND  USPEC THE ENTRY THAT PASSES THE STORAGE
C             LOCATIONS OF THE ARRAYS DEFINING THE USER-SPECIFIED
C             MATRIX.
C         2.  IN SAMPLE USPEC PROVIDED : FREE FORMAT (8,*), THE
C             FORMAT (20A4), AND THE DATA/MACHEP STATEMENT.
C
C     IN THE COMMENTS BELOW :
C     REAL*16 = COMPLEX VARIABLE, 16 BYTES OF STORAGE
C     REAL*8 = REAL VARIABLE, 8 BYTES OF STORAGE
C     REAL*4 = REAL VARIABLE, 4 BYTES OF STORAGE
C     INTEGER*4 = INTEGER VARIABLE, 4 BYTES OF STORAGE
C
C-----A-MATRIX SPECIFICATION-------------------------------------------
C
C     SUBROUTINE USPEC IS USED TO SPECIFY THE USER-SUPPLIED MATRIX.
C     SUBROUTINE CMATV IS A CORRESPONDING MATRIX-VECTOR MULTIPLY
C     SUBROUTINE WHICH SHOULD BE DESIGNED TO TAKE ADVANTAGE OF
C     ANY SPECIAL PROPERTIES OF THE USER-SUPPLIED MATRIX.  THE
C     MATRIX-VECTOR MULTIPLIES REQUIRED BY THE LANCZOS PROCEDURES
C     MUST BE COMPUTED RAPIDLY AND ACCURATELY.
C
C     SUBROUTINE USPEC HAS THE CALLING SEQUENCE
C
C            CALL USPEC(N,MATNO)
C
C     WHERE N IS THE ORDER OF THE USER-SUPPLIED MATRIX A AND
C     MATNO IS A <= 8 DIGIT INTEGER USED AS A MATRIX AND
C     TEST IDENTIFICATION NUMBER.  THIS SUBROUTINE DEFINES (DIMENSIONS)
C     THE ARRAYS REQUIRED TO SPECIFY THE USER-SUPPLIED MATRIX AND
C     INITIALIZES THESE ARRAYS AND ANY OTHER PARAMETERS NEEDED TO
C     DEFINE THE MATRIX.  THE STORAGE LOCATIONS OF THESE PARAMETERS
C     AND ARRAYS ARE THEN PASSED TO THE MATRIX-VECTOR MULTIPLY
C     SUBROUTINE CMATV VIA AN ENTRY.  A SAMPLE USPEC SUBROUTINE
C     IS INCLUDED.  THIS SAMPLE SUBROUTINE ASSUMES THAT THE MATRIX
C     IS STORED ON FILE 8 IN A TYPICAL SPARSE MATRIX FORMAT.
C     SEE THE HEADER ON THE SUBROUTINE USPEC FOR DETAILS ON THIS
C     PARTICULAR STORAGE FORMAT.
C
C     SUBROUTINE CMATV HAS THE CALLING SEQUENCE
C
C            CALL CMATV(W,U,SUM)
C
C     IN THE COMPLEX SYMMETRIC CASE, U AND W ARE
C     COMPLEX*16 VECTORS AND SUM IS A COMPLEX*16
C     SCALAR.   CMATV CALCULATES U = A*W - SUM*U FOR THE
C     USER-SPECIFIED MATRIX A.  THE ARRAY AND PARAMETER INFORMATION
C     NEEDED TO PERFORM THE MATRIX-VECTOR MULTIPLIES IS PASSED TO
C     THE CMATV SUBROUTINE FROM THE USPEC SUBROUTINE VIA THE CMATVE
C     ENTRY IN CMATV.  A SAMPLE CMATV SUBROUTINE IS INCLUDED WHICH
C     COMPUTES MATRIX-VECTOR MULTIPLIES FOR AN ARBITRARY SPARSE,
C     COMPLEX SYMMETRIC MATRIX STORED IN THE SPARSE FORMAT
```

```
C     SPECIFIED IN THE SAMPLE USPEC SUBROUTINE.
C
C     CMATV IS CALLED FROM THE SUBROUTINE LANCZS WHICH GENERATES
C     THE T-MATRICES IN THE ALPHA, BETA ARRAYS.  IT IS ALSO CALLED
C     FROM THE MAIN PROGRAM CSLEVEC FOR THE EIGENVECTOR COMPUTATIONS.
C     CMATV IS DECLARED AS AN EXTERNAL VARIABLE AND IS AN ARGUMENT
C     FOR THE SUBROUTINE LANCZS.
C
C     THE USPEC AND CMATV SUBROUTINES MUST BE MODIFIED BY THE USER
C     TO ACCOMODATE THE USER'S SPECIFIED MATRIX.
C
C     THE MAIN PROGRAMS FOR THE EIGENVALUE AND EIGENVECTOR
C     CALCULATIONS ASSUME THAT INPUT FILE 5 CONTAINS N = ORDER OF
C     THE MATRIX AND MATNO = AN IDENTIFICATION NUMBER OF <= 8 DIGITS
C     FOR THE MATRIX AND THE RUN.
C
C
C-----MACHEP------------------------------------------------------------
C
C
C     MACHEP IS A MACHINE DEPENDENT PARAMETER SPECIFYING THE RELATIVE
C     PRECISION OF THE FLOATING POINT ARITHMETIC USED.
C     MACHEP = 2.2 * 10**-16 FOR DOUBLE PRECISION ARITHMETIC ON
C     IBM 370-3081.
C
C     THE USER WILL HAVE TO RESET THIS PARAMETER TO
C     THE CORRESPONDING VALUE FOR THE MACHINE BEING USED.  NOTE THAT
C     IF A MACHINE WITH A MACHINE EPSILON THAT IS MUCH LARGER THAN THE
C     VALUE GIVEN HERE IS BEING USED, THEN THERE COULD BE
C     PROBLEMS WITH THE TOLERANCES.
C
C
C-----SUBROUTINES AND FUNCTIONS USER MUST SUPPLY------------------------
C
C     GENRAN,  MASK,  USPEC,  AND  CMATV
C
C     GENRAN = COMPUTES K PSEUDO-RANDOM NUMBERS AND STORES THEM IN
C              THE REAL ARRAY, G.  THIS SUBROUTINE IS USED TO
C              GENERATE A STARTING VECTOR FOR THE LANCZOS PROCEDURE
C              IN THE SUBROUTINE LANCZS AND A STARTING RIGHT-HAND SIDE
C              FOR INVERSE ITERATION IN THE SUBROUTINE INVERR.
C
C              TESTS REPORTED IN THE REFERENCES USED EITHER GGL1 OR
C              GGL2 FROM THE IBM LIBRARY  SLMATH.
C              THE EXISTING CALLING SEQUENCE IS:
C
C                      CALL GENRAN(IIX,G,K).
C
C              WHERE IIX =INTEGER SEED, G = REAL*4 ARRAY WHOSE
C              DIMENSION MUST BE >= K.  K RANDOM NUMBERS ARE GENERATED
C              AND PLACED IN G.
C
C     MASK = MASKS OVERFLOW AND UNDERFLOW.
C            USER MUST SUPPLY OR COMMENT OUT CALL.
C
C     USPEC = DIMENSIONS AND INITIALIZES ARRAYS NEEDED TO SPECIFY
C             USER-SUPPLIED MATRIX.  SEE A-MATRIX SPECIFICATION SECTION.
C
C     CMATV = MATRIX-VECTOR MULTIPLY FOR USER-SUPPLIED MATRIX.
C             SEE A-MATRIX SPECIFICATION SECTION.
C
C
C-----------------------------------------------------------------------
C
C     COMMENTS FOR EIGENVALUE COMPUTATIONS
C
C-----------------------------------------------------------------------
```

```
C
C
C-----PARAMETER CONTROLS FOR EIGENVALUE PROGRAMS-----------------------
C
C     PARAMETER CONTROLS ARE INTRODUCED TO ALLOW SEGMENTATION OF THE
C     EIGENVALUE COMPUTATIONS AND TO ALLOW VARIOUS COMBINATIONS OF
C     READ/WRITES.
C
C     THE FLAG ISTART CONTROLS THE T-MATRIX (ALPHA/BETA HISTORY)
C     GENERATION.
C
C     ISTART = (0,1)  MEANS
C
C              (0) THERE IS NO EXISTING ALPHA/BETA HISTORY AND ONE
C                  MUST BE GENERATED.
C
C              (1) THERE IS AN EXISTING ALPHA/BETA HISTORY AND IT IS
C                  TO BE READ IN FROM FILE 2 AND EXTENDED IF NECESSARY.
C
C     THE FLAG ISTOP CAN BE USED IN CONJUNCTION WITH THE FLAG ISTART TO
C     ALLOW SEGMENTATION OF THE EIGENVALUE COMPUTATIONS.
C
C     ISTOP  = (0,1)  MEANS
C
C              (0) PROGRAM COMPUTES ONLY THE REQUESTED ALPHAS/BETAS,
C                  STORES THEM AND THE LAST 2 LANCZOS VECTORS GENERATED
C                  IN FILE 1 AND THEN TERMINATES.
C
C              (1) PROGRAM COMPUTES REQUESTED ALPHAS/BETAS AND THEN
C                  USES THE CMTQL1 SUBROUTINE TO CALCULATE EIGENVALUES
C                  OF THE TRIDIAGONAL MATRICES GENERATED FOR THE ORDERS
C                  SPECIFIED BY THE USER.  PROGRAM THEN USES THE
C                  SUBROUTINE INVERR TO COMPUTE ERROR ESTIMATES FOR
C                  THE ISOLATED GOOD T-EIGENVALUES WHICH ARE USED TO
C                  CHECK THE CONVERGENCE OF THESE GOOD T-EIGENVALUES.
C
C     CONTROL PARAMETERS FOR WRITES
C
C     IHIS  = (0,1)  MEANS
C
C             (0) IF ISTOP .GT. 0 THEN ALPHAS/BETAS ARE NOT SAVED
C                 ON FILE 1.
C
C             (1) PROGRAM WRITES ALPHAS/BETAS AND LAST 2 LANCZOS
C                 VECTORS TO FILE 1 SO THAT THE T-MATRIX GENERATION
C                 MAY BE REUSED OR CONTINUED LATER IF NECESSARY.
C                 TYPICALLY ONE WOULD ALWAYS DO THIS ON ANY RUN WHERE
C                 A HISTORY FILE IS BEING GENERATED.  HISTORY MUST
C                 BE SAVED IN MACHINE FORMAT ((4Z20) FOR IBM/3081)
C                 SO THAT NO ERRORS ARE INTRODUCED DUE TO FORMAT
C                 CONVERSIONS.
C
C     IDIST  = (0,1)  MEANS
C
C              (0) DISTINCT EIGENVALUES OF T-MATRICES ARE NOT SAVED.
C
C              (1) PROGRAM WRITES COMPUTED DISTINCT EIGENVALUES OF
C                  T-MATRICES ALONG WITH THEIR T-MULTIPLICITIES
C                  TO FILE 11.
C
C     IWRITE = (0,1)   MEANS
C
C              (0) NO EXTENDED OUTPUT FROM SUBROUTINES COMPEV AND INVERR
C                  IS SENT TO FILE 6.
C
C              (1) INDIVIDUAL COMPUTED EIGENVALUES AND CORRESPONDING
C                  ERROR ESTIMATES FROM THE SUBROUTINES COMPEV AND
```

```
C                   INVERR ARE PRINTED OUT TO FILE 6 AS THEY ARE COMPUTED
C
C     SAVTEV = (-1,0,1)   MEANS
C
C              (-1) NO T-EIGENVALUE COMPUTATIONS.  PREVIOUSLY-COMPUTED
C                    EIGENVALUES OF T(1,MEV) AND T(2,MEV) ARE TO
C                    BE READ IN FROM FILE 10.
C
C              (0) COMPUTED EIGENVALUES OF T(1,MEV) AND OF T(2,MEV)
C                   ARE NOT TO BE SAVED ON FILE 10.  THIS IS NOT
C                   RECOMMENDED IF THE T-MATRICES BEING USED ARE VERY
C                   LARGE BECAUSE IN THAT CASE THE TRIDIAGONAL
C                   EIGENVALUE COMPUTATIONS ARE VERY EXPENSIVE.
C
C              (1) COMPUTED EIGENVALUES OF T(1,MEV) AND OF T(2,MEV)
C                   WILL BE SAVED ON FILE 10.  THIS IS RECOMMENDED
C                   BECAUSE ONCE THESE T-EIGENVALUES ARE COMPUTED THE
C                   LATTER PORTION OF THE EIGENVALUE PROGRAM IS EASILY
C                   RESTARTED FROM THE POINT OF THESE EIGENVALUE
C                   COMPUTATIONS.
C
C     THE PROGRAM ALWAYS MAKES A SEPARATE LIST OF THE COMPUTED GOOD
C     T-EIGENVALUES ALONG WITH THEIR MINIMAL GAPS AND WRITES THEM OUT
C     TO FILE 3.  CORRESPONDING ERROR ESTIMATES FOR ANY ISOLATED
C     GOOD T-EIGENVALUES ARE ALWAYS WRITTEN TO FILE 4.
C
C
C-----INPUT/OUTPUT FILES FOR EIGENVALUE PROGRAMS-----------------------
C
C     ANY INPUT DATA OTHER THAN THE ALPHA/BETA HISTORY OR PREVIOUSLY-
C     COMPUTED EIGENVALUES OF T(1,MEV) AND T(2,MEV) SHOULD BE STORED
C     ON FILE 5. SEE SAMPLE INPUT/OUTPUT FROM TYPICAL RUN.
C     THE READ STATEMENTS IN THE GIVEN FORTRAN PROGRAM ASSUME THAT
C     THE DATA STORED ON FILE 5 IS IN FREE FORMAT.  USER SHOULD NOTE
C     THAT 'FREE FORMAT' IS NOT CLASSIFIED AS PORTABLE BY PFORT SO THAT
C     THE USER MAY HAVE TO MODIFY THE READ STATEMENTS FROM FILE 5 TO
C     CONFORM TO WHAT IS PERMISSIBLE ON THE MACHINE BEING USED.
C
C     FILE 6 WAS USED AS THE INTERACTIVE TERMINAL OUTPUT FILE.
C     THIS FILE PROVIDES A RUNNING ACCOUNT OF THE PROGRESS OF THE
C     COMPUTATIONS.  THE AMOUNT OF INFORMATION PRINTED OUT IS
C     CONTROLLED BY THE PARAMETER IWRITE.
C
C DESCRIPTION OF OTHER I/O FILES
C
C FILE (K)      CONTAINS:
C
C      (1)      OUTPUT FILE:
C               HISTORY FILE OF NEWLY-GENERATED T-MATRIX (ALPHA AND
C               BETA VECTORS) AND LAST 2 LANCZOS VECTORS USED
C               IN THE T-MATRIX GENERATION.
C               IF IHIS = 0 AND ISTOP = 1, FILE 1 IS NOT WRITTEN.
C
C      (2)      INPUT FILE:
C               SAME AS FILE 1 EXCEPT THAT IT CONTAINS A
C               PREVIOUSLY-GENERATED T-MATRIX (IF ANY).  IF ISTART = 1,
C               PROGRAM ASSUMES THAT THERE IS A HISTORY FILE OF ALPHAS
C               AND BETAS ON FILE 2.  THESE ALPHAS AND BETAS ARE
C               READ IN ALONG WITH THE LAST TWO LANCZOS VECTORS
C               USED IN THE T-MATRIX GENERATION.
C
C      (3)      OUTPUT FILE:
C               COMPUTED GOOD EIGENVALUES OF THE T-MATRICES USED. ALSO
C               CONTAINS T-MULTIPLICITIES OF THESE EIGENVALUES AS
C               EIGENVALUES OF THE T-MATRIX, AND THEIR GAPS AS
C               EIGENVALUES IN THE A MATRIX AND IN THE T-MATRIX.
C               FILE 3 IS ALWAYS WRITTEN.
```

```
C
C      (4)       OUTPUT FILE:
C                ERROR ESTIMATES FOR THE ISOLATED GOOD T-EIGENVALUES WHICH
C                ARE OBTAINED USING THE SUBROUTINE INVERR. THESE
C                ESITMATES USE THE LAST COMPONENTS OF THE ASSOCIATED
C                T-EIGENVECTORS WHICH ARE COMPUTED USING INVERSE
C                ITERATION.  FILE 4 IS ALWAYS WRITTEN.
C
C      (8)       INPUT FILE:
C                SAMPLE USPEC SUBROUTINE ASSUMES THAT THE ARRAYS
C                REQUIRED TO SPECIFY THE USER'S-MATRIX ARE STORED ON
C                FILE 8.  USERS MUST MAKE WHATEVER DEFINITIONS ARE
C                APPROPRIATE FOR THEIR MATRICES.
C
C     (10)       OUTPUT OR INPUT FILE DEPENDING UPON VALUE OF SAVTEV:
C                COMPUTED EIGENVALUES OF EACH T(1,MEV) FOLLOWED
C                BY THE COMPUTED EIGENVALUES OF THE CORRESPONDING
C                T(2,MEV).
C
C     (11)       OUTPUT FILE:
C                COMPUTED DISTINCT EIGENVALUES OF T-MATRICES USED.
C                ALSO CONTAINS THEIR T-MULTIPLICITIES AND T-GAPS TO
C                NEAREST DISTINCT EIGENVALUES, AND THE T-MULTIPLICITY
C                PATTERN OF THE GOOD AND THE SPURIOUS T-EIGENVALUES.
C                FILE 11 IS WRITTEN ONLY IF IDIST = 1.
C
C
C-----PARAMETERS SET BY THE EIGENVALUE PROGRAM-------------------------
C
C     THESE PARAMETERS ARE SET INTERNALLY IN THE PROGRAM
C
C     SCALEK      K = 1,2,3,4
C
C                  THE SCALING FACTORS SCALEK HAVE BEEN INTRODUCED IN AN
C                  ATTEMPT TO MAKE THE TOLERANCES USED IN THE
C                  T-MULTIPLICITY, SPURIOUS, AND ISOLATION TESTS ADJUST
C                  TO THE SCALE OF THE GIVEN MATRIX.  THESE FACTORS MUST
C                  NOT BE MODIFIED.
C
C     BTOL = RELATIVE TOLERANCE USED TO ESTIMATE ANY LOSS OF LOCAL
C            ORTHOGONALITY OF THE LANCZOS VECTORS AFTER THE T-MATRIX
C            HAS BEEN GENERATED.  THE LANCZOS PROCEDURE WORKS WELL
C            ONLY IF LOCAL ORTHOGONALITY BETWEEN SUCCESSIVE LANCZOS
C            VECTORS IS MAINTAINED.  THE TNORM SUBROUTINE TESTS
C            WHETHER OR NOT
C
C                 MINIMUM  |BETA(I)|/||A|| > BTOL.
C                 I=2,KMAX
C
C            IF THIS TEST IS VIOLATED BY SOME BETA AND A T-MATRIX THAT
C            WOULD INCLUDE SUCH A BETA IS REQUESTED, THEN THE LANCZOS
C            PROCEDURE WILL TERMINATE FOR THE USER TO DECIDE WHAT TO
C            DO.  THE USER CAN OVER-RIDE THIS TEST BY SIMPLY DECREASING
C            THE SIZE OF BTOL, BUT THEN CONVERGENCE IS NOT AS CERTAIN.
C            THE PROGRAM SETS BTOL = 1.D-8 WHICH IS A VERY CONSERVATIVE
C            CHOICE. THE || A || IS ESTIMATED BY USING
C            AN ESTIMATE OF THE NORM OF THE T-MATRIX, T(1,KMAX).
C
C     GAPTOL = RELATIVE TOLERANCE USED IN THE SUBROUTINE ISOEV
C              TO DETERMINE WHICH OF THE GOOD T-EIGENVALUES NEED
C              ERROR ESTIMATES.  THE PROGRAM SETS GAPTOL = 1.D-7.
C              IF FOR A GIVEN 'GOOD' T-EIGENVALUE THE COMPUTED GAP
C              IS TOO SMALL AND IS DUE TO A 'SPURIOUS' T-EIGENVALUE
C              THEN THE 'GOOD' T-EIGENVALUE IS ASSUMED TO HAVE CONVERGED
C              AND NO ERROR ESTIMATES ARE COMPUTED.
C
C-----USER-SPECIFIED PARAMETERS FOR EIGENVALUE PROGRAMS-----------------
```

```
C
C     RELTOL = RELATIVE TOLERANCE USED IN 'COMBINING' COMPUTED
C              EIGENVALUES OF T(1,MEV) PRIOR TO COMPUTING ERROR
C              ESTIMATES.
C
C     THE LUMPING OF T-EIGENVALUES OCCURS IN SUBROUTINE LUMP.
C     LUMPING IS NECESSARY BECAUSE IT IS IMPOSSIBLE TO ACCURATELY
C     PREDICT THE ACCURACY OF THE CMTQL1 SUBROUTINE.  LUMP 'COMBINES'
C     T-EIGENVALUES THAT HAVE SLIPPED BY THE TOLERANCE THAT WAS USED
C     IN THE T-MULTIPLICITY TESTS.  IN PARTICULAR IF FOR SOME J,
C
C   |EVALUE(J)-EVALUE(J-1)| < DMAX1(RELTOL*|EVALUE(J)|,SCALE2*MULTOL)
C
C     THEN THESE T-EIGENVALUES ARE 'COMBINED'.  MULTOL IS THE TOLERANCE
C     THAT WAS USED IN THE T-MULTIPLICITY TEST IN COMPEV.  SEE THE
C     HEADER ON THE LUMP SUBROUTINE FOR MORE DETAILS.
C
C     THE RECOMMENDED VALUE OF RELTOL (ONLY IN THE COMPLEX SYMMETRIC
C     CASE) IS 1.D-8 BECAUSE THE OBSERVED ACCURACY OF THE
C     COMPUTED EIGENVALUES OF THE T-MATRICES IS SEVERAL DIGITS
C     LESS THAN THAT OBSERVED IN THE REAL SYMMETRIC CASE.
C     THUS, THE OBSERVED RESOLUTION OF THE COMPLEX SYMMETRIC
C     VERSION IS LESS THAN THAT OBTAINABLE IN THE REAL SYMMETRIC CASE.
C
C     MXINIT = MAXIMUM NUMBER OF INVERSE ITERATIONS ALLOWED IN
C              SUBROUTINE INVERR FOR EACH ISOLATED GOOD T-EIGENVALUE.
C              TYPICALLY ONLY ONE ITERATION IS REQUIRED.
C
C     SEEDS FOR RANDOM NUMBER GENERATORS = INTEGER*4 SCALARS.
C
C               (1) SVSEED = SEED FOR STARTING VECTOR USED IN
C                   T-MATRIX GENERATION IN LANCZS SUBROUTINE
C
C               (2) RHSEED = SEED FOR RIGHT-HAND SIDE USED IN
C                   INVERSE ITERATION COMPUTATIONS IN INVERR.
C
C
C     T-MATRICES
C
C     SIZES OF T-MATRICES
C
C             (1) KMAX= MAXIMUM ORDER FOR T-MATRIX THAT USER IS WILLING
C                       TO CONSIDER.
C
C             (2) NMEVS = MAXIMUM NUMBER OF T-MATRICES THAT WILL BE
C                         CONSIDERED.
C
C             (3) NMEV(J)  (J=1,NMEVS)  = SIZES OF T-MATRIX TO BE
C                                         CONSIDERED SEQUENTIALLY.
C
C     T-MATRIX-GENERATION
C
C     USER SHOULD NOTE THAT THIS PROGRAM FIRST COMPUTES A T-MATRIX
C     OF ORDER KMAX AND THEN CYCLES THROUGH THE T-MATRICES SPECIFIED
C     A PRIORI BY THE USER, USING THE SUBROUTINE CMTQL1 TO COMPUTE THE
C     EIGENVALUES OF THE T-MATRICES.  THE EIGENVALUE COMPUTATION
C     FOR THE COMPLEX SYMMETRIC CASE WILL BE
C     CONSIDERABLY MORE EXPENSIVE THAN FOR THE REAL SYMMETRIC OR
C     HERMITIAN CASES BECAUSE WE DO NOT HAVE AN ANALOG OF
C     THE BISECTION SUBROUTINE FOR THE COMPLEX SYMMETRIC CASE.
C     THUS, ANY RECYCLING AND SUBSEQUENT ENLARGEMENT OF THE T-MATRIX
C     REQUIRES THE RECOMPUTATION OF ALL OF THE EIGENVALUES OF
C     THE RESULTING T-MATRIX.  WE CANNOT GO IN AND COMPUTE ONLY THOSE
C     T-EIGENVALUES ON SOME SUBINTERVAL OF THE SPECTRUM OF THE
C     T-MATRIX AS WE DID IN THE REAL SYMMETRIC AND HERMITIAN CASES.
C     OF COURSE, IF THE T-MATRICES BEING CONSIDERED ARE NOT
C     VERY LARGE, THEN THIS IS NOT REALLY A PROBLEM.  HOWEVER, IF THEY
```

```
C     ARE VERY LARGE, THEN THE USER SHOULD PROBABLY DO ONE EIGENVALUE   CSL04670
C     COMPUTATION OF A LARGE T-MATRIX RATHER THAN START WITH            CSL04680
C     A SMALLER T-MATRIX AND WORK UP TO A BIG ONE.                      CSL04690
C                                                                       CSL04700
C-----CONVERGENCE TESTS FOR THE EIGENVALUE PROGRAMS---------------------CSL04710
C                                                                       CSL04720
C     THE CONVERGENCE TEST INCORPORATED IN THIS PROGRAM IS              CSL04730
C     BASED UPON THE ASSUMPTION THAT THOSE T-EIGENVALUES AND            CSL04740
C     THEIR ASSOCIATED T-EIGENVECTORS WHICH CORRESPOND TO THE           CSL04750
C     EIGENVALUES AND RITZVECTORS WHICH ARE TO BE COMPUTED              CSL04760
C     CONVERGE AS THE T-SIZE IS INCREASED.                              CSL04770
C                                                                       CSL04780
C-----ARRAYS REQUIRED BY THE EIGENVALUE PROGRAM-------------------------CSL04790
C                                                                       CSL04800
C     ALPHA(J) = COMPLEX*16  ARRAY. ITS DIMENSION MUST BE AT LEAST      CSL04810
C                KMAX, THE LENGTH OF THE LARGEST T-MATRIX ALLOWED.      CSL04820
C                THIS ARRAY CONTAINS THE DIAGONAL ENTRIES OF THE        CSL04830
C                T-MATRICES GENERATED.                                  CSL04840
C                                                                       CSL04850
C     BETA(J) = COMPLEX*16 ARRAY.  ITS DIMENSION MUST BE AT LEAST       CSL04860
C               KMAX+1. THIS ARRAY CONTAINS THE SUBDIAGONAL ENTRIES OF  CSL04870
C               THE T-MATRICES.                                         CSL04880
C                                                                       CSL04890
C               THE ALPHA AND BETA VECTORS ARE NOT ALTERED              CSL04900
C               DURING THE CALCULATIONS.                                CSL04910
C                                                                       CSL04920
C     V1(J),V2(J),VS(J) = COMPLEX*16 ARRAYS.  V1 AND V2                 CSL04930
C                         MUST BE OF DIMENSION AT LEAST MAX(KMAX,N).    CSL04940
C                         VS MUST BE OF DIMENSION AT LEAST KMAX.        CSL04950
C                                                                       CSL04960
C     GR(J),GC(J) = REAL*8 ARRAYS. USED FOR RANDOM VECTOR GENERATION.   CSL04970
C                   EACH MUST BE OF DIMENSION AT LEAST MAX(KMAX,N).     CSL04980
C                                                                       CSL04990
C     EXPLAN(J) = REAL*4 ARRAY.  ITS DIMENSION IS 20.  THIS ARRAY IS    CSL05000
C                 USED TO ALLOW EXPLANATORY COMMENTS IN THE INPUT FILES.CSL05010
C                                                                       CSL05020
C     G(J),GG(J) = REAL*4 ARRAYS.  G MUST BE OF DIMENSION AT LEAST      CSL05030
C                  MAX(N,KMAX).  GG MUST BE OF DIMENSION AT LEAST       CSL05040
C                  KMAX.  G AND GG ARE USED IN RANDOM VECTOR GENERATIONSCSL05050
C                  AND TO STORE GAPS IN T-MATRIX, GAPS IN A-MATRIX,     CSL05060
C                  AND ERROR ESTIMATES.                                 CSL05070
C                                                                       CSL05080
C     MP(J),MP2(J) = INTEGER*4 ARRAYS.  EACH MUST HAVE DIMENSION        CSL05090
C                    AT LEAST KMAX, THE MAXIMUM SIZE OF THE T-MATRICES. CSL05100
C                    MP CONTAINS THE T-MULTIPLICITIES OF THE COMPUTED   CSL05110
C                    T-EIGENVALUES. 'SPURIOUS' T-EIGENVALUES ARE DENOTEDCSL05120
C                    BY A T-MULTIPLICITY OF 0.  NOTE THAT WE DO NOT HAVECSL05130
C                    AN ANALOG OF THE SUBROUTINE PRTEST FOR THE         CSL05140
C                    COMPLEX SYMMETRIC CASE, SO NO RELABELLING OF       CSL05150
C                    MP OCCURS.  MP2 IS USED TO KEEP TRACK OF WHICH     CSL05160
C                    EIGENVALUES OF T(1,MEV) HAVE BEEN USED IN THE      CSL05170
C                    T-MULTIPLICITY TEST AND WHICH EIGENVALUES OF       CSL05180
C                    T(2,MEV) HAVE BEEN USED IN THE SPURIOUS TEST.      CSL05190
C                                                                       CSL05200
C     NMEV(J) = INTEGER*4 ARRAY.  ITS DIMENSION MUST BE AT LEAST THE    CSL05210
C               NUMBER OF T-MATRICES ALLOWED.  IT CONTAINS THE ORDERS   CSL05220
C               OF THE T-MATRICES TO BE CONSIDERED.                     CSL05230
C                                                                       CSL05240
C     OTHER ARRAYS                                                      CSL05250
C                                                                       CSL05260
C     THE USER MUST SPECIFY IN THE SUBROUTINE USPEC WHATEVER ARRAYS     CSL05270
C     ARE REQUIRED TO DEFINE THE MATRIX BEING USED.                     CSL05280
C                                                                       CSL05290
C                                                                       CSL05300
C-----SUBROUTINES INCLUDED FOR EIGENVALUE COMPUTATIONS------------------CSL05310
C                                                                       CSL05320
C     LANCZS = COMPUTES THE ALPHA/BETA HISTORY. USES SUBROUTINES        CSL05330
```

```
C               CINPRD, INPRDC, GENRAN, AND CMATV.
C
C      COMPEV = CALLS CMTQL1 TO COMPUTE THE EIGENVALUES OF T(1,MEV)
C               AND OF T(2,MEV), THEN DETERMINES T-MULTIPLE AND
C               SPURIOUS T-EIGENVALUES.
C
C      COMGAP = COMPUTES MINIMAL GAPS BETWEEN T-EIGENVALUES
C               SUPPLIED.
C
C      CMTQL1 = COMPUTES EIGENVALUES OF THE SPECIFIED T-MATRIX USING
C               A REAL ORTHOGONAL ANALOG OF THE QL ALGORITHM IMTQL1
C               IN EISPACK.
C
C      INVERR = USES INVERSE ITERATION ON T-MATRICES TO COMPUTE ERROR
C               ESTIMATES ON COMPUTED T-EIGENVALUES. (USES GENRAN)
C
C      LUMP = 'COMBINES' EIGENVALUES OF T-MATRIX USING THE RELATIVE
C             TOLERANCE RELTOL.
C
C      ISOEV = CALCULATES GAPS BETWEEN DISTINCT EIGENVALUES OF T-MATRIX
C              AND THEN USES THESE GAPS TO LABEL THOSE 'GOOD'
C              T-EIGENVALUES FOR WHICH ERROR ESTIMATES ARE NOT COMPUTED.
C
C      TNORM = COMPUTES THE SCALE TKMAX USED IN CHECKING
C              FOR LOCAL ORTHOGONALITY OF THE LANCZOS VECTORS
C              BY TESTING THE RELATIVE SIZE OF THE BETAS USING
C              THE RELATIVE TOLERANCE BTOL.
C
C     CINPRD = COMPUTES THE HERMITIAN INNER PRODUCT OF TWO
C              COMPLEX*16 VECTORS, USED IN SUBROUTINE INVERR
C              AND IN THE MAIN PROGRAM.
C
C     INPRDC = COMPUTES THE EUCLIDEAN INNER PRODUCT OF TWO
C              COMPLEX*16 VECTORS. USED IN SUBROUTINE LANCZS.
C
C
C-----OTHER PROGRAMS SUPPLIED-------------------------------------------
C
C
C      LCCOMPAC = PROGRAM TO TRANSLATE A SPARSE, COMPLEX SYMMETRIC
C                 MATRIX GIVEN AS I, J, A(I,J), INTO THE SPARSE MATRIX
C                 FORMAT USED IN THE SAMPLE USPEC AND CMATV SUBROUTINES
C                 PROVIDED.  PROGRAM ASSUMES THAT THE MATRIX ENTRIES
C                 ARE GIVEN EITHER COLUMN BY COLUMN OR ROW BY ROW.
C
C
C-----COMMENTS ON THE STORAGE REQUIRED FOR EIGENVALUE COMPUTATIONS------
C
C      THE ARRAYS USED IN THIS EIGENVALUE PROGRAM USE THE EQUIVALENT OF
C      ONE REAL*8 ARRAY OF DIMENSION
C
C         8*KMAX + 4*MAX(KMAX,N)
C
C      PLUS WHATEVER IS NEEDED TO GENERATE A*X FOR THE GIVEN MATRIX A.
C      THE ARRAYS ALPHA, BETA, VS, G, GG, MP, AND MP2 CONSUME
C      8*KMAX*8 BYTES.  THE ARRAYS V1 AND V2 CONSUME
C      4*MAXIMUM(KMAX,N)*8 BYTES.
C
C
C-----------------------------------------------------------------------
C
C      COMMENTS FOR EIGENVECTOR COMPUTATIONS
C
C-----------------------------------------------------------------------
C
C
C      THE EIGENVALUES WHOSE EIGENVECTORS ARE TO BE COMPUTED MUST
```

```
C     HAVE BEEN COMPUTED USING THE CORRESPONDING LANCZOS EIGENVALUE     CSL06010
C     FILES:  CSLEVAL + CSLESUB + CSLEMULT, FOR COMPLEX SYMMETRIC       CSL06020
C     MATRICES BECAUSE THE EIGENVECTOR PROGRAMS WILL USE THE SAME       CSL06030
C     FAMILY OF LANCZOS TRIDIAGONAL MATRICES AND LANCZOS VECTORS        CSL06040
C     THAT WAS USED IN THE EIGENVALUE COMPUTATIONS.                     CSL06050
C                                                                       CSL06060
C     THESE PROGRAMS ASSUME THAT THE EIGENVALUES SUPPLIED TO IT         CSL06070
C     HAVE BEEN COMPUTED ACCURATELY, AS MEASURED BY THE                 CSL06080
C     ERROR ESTIMATES COMPUTED IN THE CORRESPONDING LANCZOS             CSL06090
C     EIGENVALUE COMPUTATIONS, ALTHOUGH THESE ESTIMATES ARE             CSL06100
C     TYPICALLY CONSERVATIVE.  THE EIGENVALUES OF INTEREST              CSL06110
C     ARE IN THE ARRAY GOODEV(J), J=1,NGOOD.                            CSL06120
C                                                                       CSL06130
C     FOR EACH GOODEV(J), AN INITIAL ESTIMATE IS MADE OF AN             CSL06140
C     APPROPRIATE ORDER, MA(J), J=1,NGOOD, FOR A LANCZOS TRIDIAGONAL    CSL06150
C     FOR THE JTH EIGENVECTOR COMPUTATION.  THEN FOR EACH J,            CSL06160
C     SUBROUTINE INVERM SUCCESSIVELY COMPUTES CORRESPONDING             CSL06170
C     EIGENVECTORS OF ENLARGED T-MATRICES UNTIL A SUITABLE              CSL06180
C     SIZE T-MATRIX IS DETERMINED FOR EACH J.  UP TO 10 SUCH            CSL06190
C     EIGENVECTOR COMPUTATIONS ARE ALLOWED FOR EACH EIGENVALUE.         CSL06200
C                                                                       CSL06210
C     ONCE SUITABLE T-EIGENVECTORS HAVE BEEN OBTAINED THEN THE          CSL06220
C     RITZ VECTOR CORRESPONDING TO THESE T-EIGENVECTORS ARE             CSL06230
C     COMPUTED AND TAKEN AS APPROXIMATE EIGENVECTORS OF A FOR THE       CSL06240
C     GIVEN EIGENVALUES, GOODEV(J), J = 1, ..., NGOOD.                  CSL06250
C                                                                       CSL06260
C     THIS IMPLEMENTATION FIRST COMPUTES ALL OF THE RELEVANT            CSL06270
C     EIGENVECTORS OF THE COMPLEX SYMMETRIC TRIDIAGONAL MATRICES        CSL06280
C     IN THE VECTOR, TVEC.                                              CSL06290
C                                                                       CSL06300
C     THEN, AS EACH OF THE LANCZOS VECTORS IS REGENERATED, ALL          CSL06310
C     OF THE RITZ VECTORS CORRESPONDING TO THESE                        CSL06320
C     T-EIGENVECTORS ARE UPDATED USING THE CURRENTLY-GENERATED          CSL06330
C     LANCZOS VECTOR.  LANCZOS VECTORS ARE GENERATED (NOTE              CSL06340
C     THAT THEY ARE NOT BEING KEPT), UNTIL ENOUGH HAVE                  CSL06350
C     BEEN GENERATED TO MAP THE LONGEST T-EIGENVECTOR INTO ITS          CSL06360
C     CORRESPONDING RITZ VECTOR.  THE ARRAY RITVEC CONTAINS THE         CSL06370
C     SUCCESSIVE RITZ VECTORS WHICH ARE THE APPROXIMATE                 CSL06380
C     EIGENVECTORS OF A.                                                CSL06390
C                                                                       CSL06400
C                                                                       CSL06410
C-----PARAMETER CONTROLS FOR EIGENVECTOR PROGRAMS-----------------------CSL06420
C                                                                       CSL06430
C                                                                       CSL06440
C     PARAMETER CONTROLS ARE INTRODUCED TO ALLOW SEGMENTATION OF THE    CSL06450
C     EIGENVECTOR COMPUTATIONS AND TO ALLOW VARIOUS COMBINATIONS OF     CSL06460
C     READ/WRITES.                                                      CSL06470
C                                                                       CSL06480
C     THE FLAG MBOUND ALLOWS THE USER TO DETERMINE A FIRST GUESS ON THE CSL06490
C     STORAGE THAT WILL BE REQUIRED BY THE T-EIGENVECTORS FOR THE       CSL06500
C     EIGENVALUES WHOSE EIGENVECTORS ARE TO BE COMPUTED.                CSL06510
C     THIS CAN BE USED TO ESTIMATE THE REQUIRED SIZE OF THE TVEC ARRAY. CSL06520
C                                                                       CSL06530
C     MBOUND = (0,1) MEANS                                              CSL06540
C                                                                       CSL06550
C              (0)  PROGRAM COMPUTES FIRST GUESSES AT THE SIZES         CSL06560
C                   OF THE T-MATRICES REQUIRED BY EACH OF THE           CSL06570
C                   EIGENVALUES SUPPLIED AND THEN CONTINUES WITH        CSL06580
C                   THE CORRESPONDING T-EIGENVECTOR COMPUTATIONS.       CSL06590
C                                                                       CSL06600
C              (1)  PROGRAM COMPUTES FIRST GUESSES AT THE SIZES         CSL06610
C                   OF THE T-MATRICES REQUIRED BY EACH OF THE           CSL06620
C                   EIGENVALUES SUPPLIED, STORES THESE IN FILE 10       CSL06630
C                   AND THEN TERMINATES.  THE USER CAN USE THESE        CSL06640
C                   SIZES TO ESTIMATE THE SIZE TVEC ARRAY NEEDED        CSL06650
C                   FOR THE DESIRED T-EIGENVECTOR COMPUTATIONS.         CSL06660
C                                                                       CSL06670
```

```
C     THE FLAGS NTVCON, TVSTOP, LVCONT, AND ERCONT CONTROL THE STOPPING
C     CRITIERIA FOR INTERMEDIATE POINTS IN THE LANCZOS PROCEDURE.
C     THEY CAUSE TERMINATION OF THE LANCZOS PROCEDURE IF VARIOUS
C     QUANTITIES CANNOT BE COMPUTED AS DESIRED.
C
C     NTVCON = (0,1) MEANS
C
C             (0)  IF THE ESTIMATED STORAGE FOR THE T-EIGENVECTORS
C                  EXCEEDS THE USER-SPECIFIED DIMENSION OF THE
C                  TVEC ARRAY PROGRAM DOES NOT CONTINUE WITH THE
C                  T-EIGENVECTOR COMPUTATIONS. TERMINATION OCCURS.
C
C             (1)  CONTINUE WITH THE T-EIGENVECTOR COMPUTATIONS
C                  EVEN IF THE ESTIMATED STORAGE FOR TVEC EXCEEDS
C                  THE USER-SPECIFIED DIMENSION OF THE TVEC ARRAY.
C                  IN THIS SITUATION THE PROGRAM COMPUTES AS MANY
C                  T-EIGENVECTORS AS IT HAS ROOM FOR, IN THE SAME
C                  ORDER IN WHICH THE EIGENVALUES ARE PROVIDED.
C
C      SVTVEC = (0,1) MEANS
C
C              (0)  DO NOT STORE THE COMPUTED T-EIGENVECTORS ON
C                   FILE 11 UNLESS ALSO HAVE THE FLAG TVSTOP = 1,
C                   IN WHICH CASE THE T-EIGENVECTORS ARE ALWAYS
C                   WRITTEN TO FILE 11.
C
C              (1)  STORE THE COMPUTED T-EIGENVECTORS ON FILE 11.
C
C      TVSTOP = (0,1) MEANS
C
C              (0)  ATTEMPT TO CONTINUE ON TO THE COMPUTATION
C                   OF THE RITZVECTORS AFTER COMPLETING THE
C                   COMPUTATION OF THE T-EIGENVECTORS.
C
C              (1)  TERMINATE AFTER COMPUTING THE
C                   T-EIGENVECTORS AND STORING THEM ON FILE 11.
C
C      LVCONT = (0,1) MEANS
C
C              (0)  IF SOME OF THE T-EIGENVECTORS THAT WERE
C                   REQUIRED WERE NOT COMPUTED, EXIT
C                   FROM THE PROGRAM WITHOUT COMPUTING THE
C                   CORRESPONDING RITZ VECTORS.
C
C              (1)  CONTINUE ON TO THE RITZ VECTOR COMPUTATIONS
C                   EVEN IF NOT ALL OF THE T-EIGENVECTORS THAT
C                   WERE REQUESTED WERE COMPUTED.
C
C      ERCONT = (0,1) MEANS
C
C              (0)  PROGRAM WILL NOT COMPUTE THE RITZ
C                   VECTOR FOR ANY EIGENVALUE FOR WHICH NO
C                   T-EIGENVECTOR WHICH SATISFIES THE ERROR ESTIMATE
C                   TEST (ERTOL) HAS BEEN IDENTIFIED.
C
C              (1)  A RITZ VECTOR WILL BE COMPUTED FOR EVERY
C                   EIGENVALUE FOR WHICH A T-EIGENVECTOR HAS BEEN
C                   COMPUTED REGARDLESS OF WHETHER OR NOT THAT
C                   T-EIGENVECTOR SATISFIED THE ERROR ESTIMATE TEST.
C
C
C-----INPUT/OUTPUT FILES FOR THE EIGENVECTOR COMPUTATIONS---------------
C
C
C     INPUT DATA OTHER THAN THE T-MATRIX HISTORY FILE AND THE
C     EIGENVALUES AND ERROR ESTIMATES SUPPLIED SHOULD BE STORED ON
C     FILE 5 IN FREE FORMAT.  SEE SAMPLE INPUT/OUTPUT FOR TYPICAL
```

```
C      INPUT/OUTPUT FILE.                                               CSL07350
C                                                                       CSL07360
C      FILE 6 WAS USED AS THE INTERACTIVE TERMINAL OUTPUT FILE.         CSL07370
C      THIS FILE PROVIDES A RUNNING ACCOUNT OF THE PROGRESS OF THE      CSL07380
C      COMPUTATIONS.  ADDITIONAL PRINTOUT IS GENERATED WHEN             CSL07390
C      THE FLAG IWRITE = 1.                                             CSL07400
C                                                                       CSL07410
C                                                                       CSL07420
C DESCRIPTION OF OTHER I/O FILES                                        CSL07430
C                                                                       CSL07440
C FILE (K)      CONTAINS:                                               CSL07450
C                                                                       CSL07460
C       (2)     INPUT FILE:                                             CSL07470
C               PREVIOUSLY-GENERATED T-MATRICES (ALPHA/BETA ARRAYS)     CSL07480
C               AND THE FINAL TWO LANCZOS VECTORS USED ON THAT          CSL07490
C               COMPUTATION.  THIS PROGRAM ALLOWS ENLARGEMENT           CSL07500
C               OF ANY T-MATRICES PROVIDED ON FILE 2.                   CSL07510
C                                                                       CSL07520
C       (3)     INPUT FILE:                                             CSL07530
C               THE GOOD EIGENVALUES OF THE T-MATRIX  T(1,MEV)          CSL07540
C               FOR WHICH EIGENVECTORS ARE REQUESTED.                   CSL07550
C               FILE 3 ALSO CONTAINS THE T-MULTIPLICITIES OF THESE      CSL07560
C               EIGENVALUES AND THEIR COMPUTED GAPS IN THE              CSL07570
C               T-MATRICES AND IN THE USER-SUPPLIED MATRIX. THIS        CSL07580
C               FILE IS CREATED IN THE LANCZOS EIGENVALUE COMPUTATIONS. CSL07590
C                                                                       CSL07600
C       (4)     INPUT FILE:                                             CSL07610
C               ERROR ESTIMATES FOR THE ISOLATED GOOD T-EIGENVALUES     CSL07620
C               IN FILE 3.  THIS FILE IS CREATED DURING THE LANCZOS     CSL07630
C               EIGENVALUE COMPUTATIONS.                                CSL07640
C                                                                       CSL07650
C       (8)     INPUT FILE:                                             CSL07660
C               SAMPLE USPEC SUBROUTINE ASSUMES THAT THE ARRAYS         CSL07670
C               REQUIRED TO SPECIFY THE USER'S-MATRIX ARE STORED ON     CSL07680
C               FILE 8.  USERS MUST MAKE WHATEVER DEFINITIONS ARE       CSL07690
C               APPROPRIATE FOR THEIR MATRICES.                         CSL07700
C                                                                       CSL07710
C       (9)     OUTPUT FILE:                                            CSL07720
C               ERROR ESTIMATES FOR THE COMPUTED RITZ VECTORS CONSIDERED CSL07730
C               AS EIGENVECTORS OF THE ORIGINAL MATRIX. THESE ESTIMATES CSL07740
C               ARE OF THE FORM                                         CSL07750
C                     AERROR = || A*RITVEC - EVAL*RITVEC ||             CSL07760
C               WHERE A DENOTES THE USER-SUPPLIED MATRIX, EVAL DENOTES  CSL07770
C               THE EIGENVALUE BEING CONSIDERED AND RITVEC DENOTES      CSL07780
C               THE COMPUTED RITZ VECTOR.                               CSL07790
C                                                                       CSL07800
C       (10)    OUTPUT FILE:                                            CSL07810
C               GUESSES AT APPROPRIATE SIZE T-MATRICES FOR THE          CSL07820
C               T-EIGENVECTORS FOR EACH SUPPLIED EIGENVALUE GOODEV(J).  CSL07830
C                                                                       CSL07840
C       (11)    OUTPUT FILE:                                            CSL07850
C               COMPUTED T-EIGENVECTORS CORRESPONDING TO EIGENVALUES    CSL07860
C               IN THE GOODEV ARRAY.  NOTE THAT IT IS POSSIBLE IN       CSL07870
C               CERTAIN SITUATIONS THAT FOR SOME EIGENVALUES IN THE     CSL07880
C               GOODEV ARRAY A T-EIGENVECTOR WILL NOT BE COMPUTED.      CSL07890
C               (WRITTEN ONLY IF FLAG SVTVEC = 1).                      CSL07900
C                                                                       CSL07910
C       (12)    OUTPUT FILE:                                            CSL07920
C               CONTAINS COMPUTED RITZ VECTORS CORRESPONDING TO         CSL07930
C               THE T-EIGENVECTORS ON FILE 11. NOTE THAT IN             CSL07940
C               SOME SITUATIONS THAT FOR SOME EIGENVALUES IN            CSL07950
C               THE GOODEV ARRAY FOR WHICH T-EIGENVECTORS HAVE          CSL07960
C               BEEN COMPUTED NO RITZ VECTOR WILL HAVE BEEN             CSL07970
C               COMPUTED.                                               CSL07980
C                                                                       CSL07990
C       (13)    OUTPUT FILE:                                            CSL08000
C               ADDITIONAL INFORMATION ABOUT THE BOUNDS AND ERROR       CSL08010
```

```
C             ESTIMATES OBTAINED.
C
C
C-----SEEDS FOR EIGENVECTOR PROGRAMS-----------------------------------
C
C     SEEDS FOR RANDOM NUMBER GENERATOR GENRAN
C             (1) SVSEED = INTEGER*4 SCALAR USED IN THE SUBROUTINE
C                          GENRAN TO GENERATE THE STARTING VECTOR FOR
C                          THE REGENERATION OF THE LANCZOS VECTORS.
C
C             (2) RHSEED = INTEGER*4 SCALAR USED IN THE SUBROUTINE
C                          GENRAN TO GENERATE A RANDOM VECTOR FOR
C                          USE IN SUBROUTINE INVERM.
C
C     USER SHOULD NOTE THAT SVSEED MUST BE THE SAME SEED THAT
C     WAS USED TO GENERATE THE T-MATRICES THAT WERE USED TO
C     COMPUTE THE EIGENVALUES WHOSE EIGENVECTORS ARE TO BE COMPUTED.
C     SVSEED IS READ IN FROM FILE 3.
C
C
C-----USER-SPECIFIED PARAMETERS FOR THE EIGENVECTOR PROGRAMS-----------
C
C     NGOOD   = NUMBER OF EIGENVALUES READ INTO THE GOODEV ARRAY
C               READ FROM FILE 3.
C
C     N       = SIZE OF THE USER-SUPPLIED MATRIX.
C
C     MEV     = SIZE OF THE T-MATRIX THAT WAS USED TO COMPUTE
C               THE EIGENVALUES WHOSE EIGENVECTORS ARE REQUESTED.
C               MEV IS READ IN FROM FILE 3.
C
C     KMAX    =  SIZE OF THE T-MATRIX PROVIDED ON FILE 2.
C
C     MDIMTV  = MAXIMUM CUMULATIVE SIZE OF THE TVEC ARRAY ALLOWED
C               FOR ALL OF THE T-EIGENVECTORS REQUIRED.  MDIMTV
C               MUST NOT EXCEED THE USER-SPECIFIED DIMENSION OF
C               THE TVEC ARRAY.  PROGRAM CAN BE RUN WITH THE FLAG
C               MBOUND = 1 TO DETERMINE AN EDUCATED GUESS ON AN
C               APPROPRIATE DIMENSION FOR THE TVEC ARRAY.
C
C     MDIMRV = MAXIMUM CUMULATIVE SIZE OF THE RITVEC ARRAY ALLOWED
C              FOR ALL OF THE RITZ VECTORS TO BE COMPUTED. MDIMRV
C              MUST NOT EXCEED THE USER-SPECIFIED DIMENSION OF
C              THE RITVEC ARRAY.  MUST BE SELECTED SO THAT
C              THERE IS ENOUGH ROOM FOR A RITZ VECTOR FOR EVERY
C              GOODEV(J) READ INTO PROGRAM.  (>= NGOOD*N)
C
C
C-----ARRAYS REQUIRED BY THE EIGENVECTOR PROGRAMS----------------------
C
C
C     ALPHA(J) = COMPLEX*16 ARRAY WHOSE DIMENSION MUST BE AT LEAST
C                KMAXN, THE LARGEST SIZE T-MATRIX CONSIDERED BY
C                THE PROGRAM.  NOTE THAT KMAXN IS THE LARGER OF
C                THE SIZE OF THE ALPHA, BETA HISTORY PROVIDED
C                ON FILE 2 (IF ANY ) AND THE SIZE WHICH THE PROGRAM
C                SPECIFIES INTERNALLY, THIS LATTER IS ALWAYS
C                < = 11*MEV / 8  +  12, WHERE MEV IS THE SIZE
C                T-MATRIX THAT WAS USED IN THE CORRESPONDING EIGENVALUE
C                COMPUTATIONS.  ALPHA CONTAINS THE DIAGONAL ENTRIES
C                OF THE LANCZOS T-MATRICES.  ALPHA IS NOT DESTROYED
C                IN THE COMPUTATIONS.
C
C     BETA(J) = COMPLEX*16 ARRAY WHOSE DIMENSION MUST BE AT LEAST 1
C               MORE THAN THAT OF ALPHA.  DIMENSION COMMENTS ABOVE
C               ABOUT ALPHA APPLY ALSO TO THE BETA ARRAY.  BETA
C               CONTAINS THE SUBDIAGONAL ENTRIES OF THE T-MATRICES.
```

```
C              BETA IS NOT DESTROYED IN THE COMPUTATIONS.                 CSL08690
C                                                                         CSL08700
C      RITVEC(J) = COMPLEX*16 ARRAY WHOSE DIMENSION MUST BE AT LEAST      CSL08710
C                  NGOOD*N WHERE N IS THE ORDER OF THE USER-SUPPLIED      CSL08720
C                  MATRIX AND NGOOD IS THE NUMBER OF EIGENVALUES          CSL08730
C                  WHOSE EIGENVECTORS ARE TO BE COMPUTED.  IT CONTAINS    CSL08740
C                  THE COMPUTED RITZ VECTORS  (THE APPROXIMATE            CSL08750
C                  EIGENVECTORS OF A).  THESE VECTORS ARE STORED          CSL08760
C                  ON FILE 12.                                            CSL08770
C                                                                         CSL08780
C      TVEC(J)   = COMPLEX*16 ARRAY WHOSE DIMENSION MUST BE AT LEAST      CSL08790
C                  MTOL = |MA(1)| + |MA(2)| + ... + |MA(NGOOD)|           CSL08800
C                  WHERE NGOOD IS THE NUMBER OF EIGENVALUES BEING         CSL08810
C                  CONSIDERED AND |MA(J)| IS THE SIZE OF THE              CSL08820
C                  T-MATRIX BEING USED IN THE RITZ VECTOR COMPUTATIONS    CSL08830
C                  FOR GOODEV(J).  THESE SIZES ARE DETERMINED BY THE      CSL08840
C                  PROGRAM.  AN ESTIMATE OF MTOL CAN BE OBTAINED BY       CSL08850
C                  SETTING MBOUND = 1, RUNNING THE PROGRAM, AND           CSL08860
C                  MULTIPLYING THE RESULTING TOTAL T-SIZES BY 5/4.        CSL08870
C                  THE ARRAY TVEC IS USED TO HOLD THE COMPUTED            CSL08880
C                  T-EIGENVECTORS.  IF THE FLAG SVTVEC = 1 OR THE         CSL08890
C                  FLAG TVSTOP = 1, THESE VECTORS ARE SAVED ON FILE 11.   CSL08900
C                                                                         CSL08910
C      V1(J)     = COMPLEX*16 ARRAY WHOSE DIMENSION MUST BE AT LEAST      CSL08920
C                  MAX(KMAX,N)  WHERE KMAX IS THE                         CSL08930
C                  LARGEST SIZE T-MATRIX THAT CAN BE CONSIDERED           CSL08940
C                  IN THE T-EIGENVECTOR COMPUTATIONS.  V1 IS USED         CSL08950
C                  IN THE SUBROUTINE INVERM AND IN THE REGENERATION       CSL08960
C                  OF THE LANCZOS VECTORS.                                CSL08970
C                                                                         CSL08980
C      V2(J)     = COMPLEX*16 ARRAY WHOSE DIMENSION MUST BE AT LEAST      CSL08990
C                  MAX(KMAX,N). IT IS USED IN THE REGENERATION OF         CSL09000
C                  THE LANCZOS VECTORS AND IN THE SUBROUTINE INVERM.      CSL09010
C                                                                         CSL09020
C      GOODEV(J) = COMPLEX*16 ARRAY OF DIMENSION AT LEAST NGOOD.          CSL09030
C                  CONTAINS THE EIGENVALUES FOR WHICH EIGENVECTORS        CSL09040
C                  ARE REQUESTED. THESE EIGENVALUES ARE READ IN           CSL09050
C                  FROM FILE 3.                                           CSL09060
C                                                                         CSL09070
C      GR(J),GC(J)      = REAL*8 ARRAYS WHOSE DIMENSION MUST BE AT        CSL09080
C                         LEAST MAX(N,KMAX).  USED TO HOLD RANDOMLY-      CSL09090
C                         GENERATED STARTING VECTORS FOR LANCZS           CSL09100
C                         COMPUTATIONS AND FOR THE INVERM SUBROUTINE.     CSL09110
C                                                                         CSL09120
C      AMINGP(J), = REAL*4 ARRAYS OF DIMENSION AT LEAST NGOOD.            CSL09130
C      TMINGP(J)    CONTAIN, RESPECTIVELY, THE MINIMAL GAPS FOR           CSL09140
C                   CORRESPONDING EIGENVALUES IN GOODEV ARRAY IN          CSL09150
C                   A-MATRIX AND IN T-MATRIX.                             CSL09160
C                                                                         CSL09170
C      TERR(J), ERR(J),      = REAL*4 ARRAYS (EXCEPT TLAST WHICH IS       CSL09180
C      ERRDGP(J), TLAST(J)     REAL*8) EACH OF WHOSE DIMENSIONS MUST BE   CSL09190
C      RNORM(J), TBETA(J)      AT LEAST NGOOD.  USED TO STORE QUANTITIES  CSL09200
C                              GENERATED DURING THE COMPUTATIONS FOR      CSL09210
C                              LATER PRINTOUT.                            CSL09220
C                                                                         CSL09230
C      G(J)     = REAL*4 ARRAY WHOSE DIMENSION MUST BE AT LEAST           CSL09240
C                 MAX(KMAX,N).  USED IN SUBROUTINE GENRAN TO HOLD         CSL09250
C                 RANDOM NUMBERS NEEDED FOR THE LANCZOS VECTORS           CSL09260
C                 REGENERATION AND FOR THE INVERSE ITERATION              CSL09270
C                 COMPUTATIONS IN THE SUBROUTINE INVERM.                  CSL09280
C                                                                         CSL09290
C      MP(J) = INTEGER*4 ARRAY WHOSE DIMENSION IS AT LEAST NGOOD.         CSL09300
C              INITIALLY CONTAINS THE T-MULTIPLICITY OF THE EIGENVALUE    CSL09310
C              GOODEV(J) AS AN EIGENVALUE OF THE T-MATRIX T(1,MEV).       CSL09320
C              USED TO FLAG EIGENVALUES FOR WHICH NO T-EIGENVECTOR        CSL09330
C              OR NO RITZ VECTOR IS TO BE COMPUTED.                       CSL09340
C                                                                         CSL09350
```

```
C      MA(J)    = INTEGER*4 ARRAYS EACH OF WHOSE DIMENSIONS             CSL09360
C                 IS AT LEAST NGOOD.  USED IN DETERMINING               CSL09370
C                 AN APPROPRIATE T-MATRIX FOR EACH EIGENVALUE           CSL09380
C                 IN GOODEV ARRAY.                                      CSL09390
C                                                                       CSL09400
C      MINT(J),MFIN(J) = INTEGER*4 ARRAYS WHOSE DIMENSIONS MUST BE AT   CSL09410
C                        LEAST NGOOD.  USED TO POINT TO THE BEGINNINGS  CSL09420
C                        AND THE ENDS OF THE COMPUTED EIGENVECTOR       CSL09430
C                        OF THE T-MATRIX, T(1,|MA(J)|).                 CSL09440
C                                                                       CSL09450
C      IDELTA(J)  = INTEGER*4 ARRAY WHOSE DIMENSION MUST BE AT          CSL09460
C                   LEAST NGOOD.  CONTAINS INCREMENTS USED IN LOOPS     CSL09470
C                   ON APPROPRIATE SIZE T-MATRIX FOR THE T-EIGENVECTOR  CSL09480
C                   COMPUTATIONS.                                       CSL09490
C                                                                       CSL09500
C                                                                       CSL09510
C      INTERC(J)  = INTEGER*4 ARRAY WHOSE DIMENSION MUST BE AT          CSL09520
C                   LEAST KMAX.  WORK SPACE USED IN INVERM.             CSL09530
C                                                                       CSL09540
C-----SUBROUTINES INCLUDED FOR THE EIGENVECTOR COMPUTATIONS------------CSL09550
C                                                                       CSL09560
C                                                                       CSL09570
C      INVERM = FOR THE T-SIZES CONSIDERED BY THE PROGRAM COMPUTES      CSL09580
C               THE CORRESPONDING EIGENVECTORS OF THESE T-MATRICES      CSL09590
C               CORRESPONDING TO THE USER-SUPPLIED EIGENVALUES IN       CSL09600
C               THE GOODEV ARRAY.                                       CSL09610
C                                                                       CSL09620
C      LANCZS, TNORM , CINPRD, INPRDC, CMATV AND GENRAN ARE USED        CSL09630
C      HERE AS WELL AS IN THE EIGENVALUE COMPUTATIONS.                  CSL09640
C                                                                       CSL09650
C                                                                       CSL09660
C-----------------------------------------------------------------------CSL09670
```

SECTION 7.3 MAIN PROGRAM, EIGENVALUE CALCULATIONS

```
C-----CSLEVAL (EIGENVALUES OF COMPLEX SYMMETRIC MATRICES)---------------
C
C     CONTAINS MAIN PROGRAM FOR COMPUTING DISTINCT EIGENVALUES OF
C     A NONDEFECTIVE COMPLEX SYMMETRIC MATRIX USING LANCZOS
C     TRIDIAGONALIZATION WITHOUT REORTHOGONALIZATION
C
C     PORTABILITY:
C     THESE PROGRAMS ARE NOT PORTABLE DUE TO THE USE OF COMPLEX*16
C     VARIABLES AND CORRESPONDING COMPLEX FUNCTIONS SUCH AS DCMPLX
C     AND CDABS.  FURTHERMORE, OTHER NONPORTABLE CONSTRUCTIONS
C     IDENTIFIED BY THE PFORT VERIFIER ARE THE FOLLOWING:
C
C     1.  DATA/MACHEP/ STATEMENT THAT DEFINES MACHINE EPSILON
C     2.  ALL READ(5,*) INPUT STATEMENTS IN FREE FORMAT
C     3.  FORMAT(20A4) USED WITH EXPLANATORY HEADER EXPLAN.
C     4.  HEXADECIMAL FORMAT (4Z20) USED WITH ALPHA/BETA FILES 1 AND 2.
C
C-----------------------------------------------------------------------
C
      COMPLEX*16  ALPHA(3000),BETA(3000),VS(3000)
      COMPLEX*16  V1(3000),V2(3000),ZEROC,BETAM,Z
      DOUBLE PRECISION  GR(3000),GC(3000)
      DOUBLE PRECISION  BTOL,GAPTOL,TTOL,MACHEP,EPSM,RELTOL
      DOUBLE PRECISION  SCALE1,SCALE2,SPUTOL,CONTOL,MULTOL,EVMAX
      DOUBLE PRECISION  ONE,ZERO,TEMP,TKMAX,EVMAX,BKMIN,TO,T1
      REAL  G(3000),GG(3000),EXPLAN(20),GTEMP
      INTEGER  MP(3000),MP2(3000),NMEV(20)
      INTEGER  SVSEED,RHSEED,SVSOLD,SAVTEV
      INTEGER IABS
      REAL  ABS
      DOUBLE PRECISION  DABS, DFLOAT
      EXTERNAL CMATV
C
C-----------------------------------------------------------------------
      DATA MACHEP/Z3410000000000000/
      EPSM = 2.0D0*MACHEP
C-----------------------------------------------------------------------
C
C     ARRAYS MUST BE DIMENSIONED AS FOLLOWS:
C          1. ALPHA AND VS:  >= KMAX.    BETA:  >= (KMAX+1)
C          2. V1, V2, GR, GC:  >= MAX(N,KMAX)
C          3. G:  >= MAX(N,KMAX).    GG:  >= KMAX.
C          4. MP, MP2:   >= KMAX
C          5. NMEV:  >= NUMBER OF T-MATRICES ALLOWED
C          6. EXPLAN:  DIMENSION IS 20.
C
C     NOTE:   THE OBSERVED ACHIEVABLE ACCURACY FOR THE COMPLEX
C     SYMMETRIC MATRICES TESTED WAS SIGNIFICANTLY LESS THAN THAT
C     OBTAINED WITH THE REAL SYMMETRIC AND HERMITIAN VERSIONS
C     OF THESE LANCZOS CODES AND IT IS DOUBTFUL THAT THIS CODE
C     CAN HANDLE VERY STIFF COMPLEX SYMMETRIC MATRICES.
C
C     IMPORTANT TOLERANCES OR SCALES THAT ARE USED REPEATEDLY
C     THROUGHOUT THE PROGRAM ARE THE FOLLOWING:
C     SCALED MACHINE EPSILON:  TTOL = EVMAX*EPSM WHERE
C     EPSM = 2*MACHINE EPSILON AND
C     EVMAX = MAX(|LAMBDA(J)|), J =1,MEV OF EIGENVALUES OF T(1,MEV).
C     TOLERANCE: T-MULTIPLICITY TESTS:  MULTOL = 500*(1000+MEV)*TTOL
C     TOLERANCE: SPURIOUS TESTS   SPUTOL = MULTOL
C     NOTE THAT IN THE MAIN PROGRAM THESE TOLERANCES ARE INITIALIZED
C     TO QUANTITIES THAT ARE NOT A FUNCTION OF THE SIZE OF THE
C     T-EIGENVALUES AND THEN THE SIZES OF THE T-EIGENVALUES ARE
C     INTRODUCED IN THE SUBROUTINE COMPEV.
C
```

```
C     LANCZOS CONVERGENCE TOLERANCE: CONTOL = CDABS(BETA(MEV+1)*1.D-10  CSL00650
C-----------------------------------------------------------------------CSL00660
C     OUTPUT HEADER                                                     CSL00670
      WRITE(6,10)                                                       CSL00680
   10 FORMAT(/' LANCZOS EIGENVALUE PROCEDURE FOR COMPLEX SYMMETRIC MATRICSL00690
     1CES'/)                                                            CSL00700
C                                                                       CSL00710
C     SET PROGRAM PARAMETERS                                            CSL00720
C     SCALEK ARE USED IN TOLERANCES NEEDED IN SUBROUTINES LUMP          CSL00730
C     AND ISOEV.  USER MUST NOT MODIFY THESE SCALES.                    CSL00740
      SCALE1 = 5.0D2                                                    CSL00750
      SCALE2 = 5.0D0                                                    CSL00760
      ONE  = 1.0D0                                                      CSL00770
      ZERO = 0.0D0                                                      CSL00780
      ZEROC = DCMPLX(ZERO,ZERO)                                         CSL00790
      BTOL = 1.0D-8                                                     CSL00800
C     BTOL = MACHEP                                                     CSL00810
      GAPTOL = 1.0D-7                                                   CSL00820
      ICONV = 0                                                         CSL00830
      MOLD = 0                                                          CSL00840
      MOLD1 = 1                                                         CSL00850
      MMB = 0                                                           CSL00860
C                                                                       CSL00870
C     READ USER-SPECIFIED PARAMETERS FROM INPUT FILE 5 (FREE FORMAT)    CSL00880
C                                                                       CSL00890
C     READ USER-PROVIDED HEADER FOR RUN                                 CSL00900
      READ(5,20) EXPLAN                                                 CSL00910
      WRITE(6,20) EXPLAN                                                CSL00920
      READ(5,20) EXPLAN                                                 CSL00930
      WRITE(6,20) EXPLAN                                                CSL00940
   20 FORMAT(20A4)                                                      CSL00950
C                                                                       CSL00960
C     READ ORDER OF MATRICES (N) , MAXIMUM ORDER OF T-MATRIX (KMAX),    CSL00970
C     NUMBER OF T-MATRICES ALLOWED (NMEVS), AND MATRIX IDENTIFICATION   CSL00980
C     NUMBERS (MATNO)                                                   CSL00990
      READ(5,20) EXPLAN                                                 CSL01000
      READ(5,*) N,KMAX,NMEVS,MATNO                                      CSL01010
C                                                                       CSL01020
C     READ SEEDS FOR LANCZS AND INVERR SUBROUTINES (SVSEED AND RHSEED)  CSL01030
C     READ MAXIMUM NUMBER OF ITERATIONS ALLOWED FOR EACH INVERSE        CSL01040
C     ITERATION (MXINIT).                                               CSL01050
      READ(5,20) EXPLAN                                                 CSL01060
      READ(5,*) SVSEED,RHSEED,MXINIT                                    CSL01070
C                                                                       CSL01080
C     ISTART = (0,1):  ISTART = 0 MEANS ALPHA/BETA FILE IS NOT          CSL01090
C     AVAILABLE.  ISTART = 1 MEANS ALPHA/BETA FILE IS AVAILABLE ON      CSL01100
C     FILE 2.  COMPLEX SYMMETRIC HISTORIES MUST BE STORED               CSL01110
C     IN HEX FORMAT (4Z20).                                             CSL01120
C     ISTOP = (0,1):  ISTOP = 0 MEANS PROCEDURE GENERATES ALPHA/BETA    CSL01130
C     FILE AND THEN TERMINATES.  ISTOP = 1 MEANS PROCEDURE GENERATES    CSL01140
C     ALPHAS/BETAS IF NEEDED AND THEN COMPUTES EIGENVALUES AND ERROR    CSL01150
C     ESTIMATES AND THEN TERMINATES.                                    CSL01160
      READ(5,20) EXPLAN                                                 CSL01170
      READ(5,*) ISTART,ISTOP                                            CSL01180
C                                                                       CSL01190
C     IHIS = (0,1):  IHIS = 0 MEANS ALPHA/BETA FILE IS NOT WRITTEN      CSL01200
C     TO FILE 1.  IHIS = 1 MEANS ALPHA/BETA FILE IS WRITTEN TO FILE 1.  CSL01210
C     IDIST = (0,1):  IDIST = 0 MEANS DISTINCT T(1,MEV)-EIGENVALUES     CSL01220
C     ARE NOT WRITTEN TO FILE 11.  IDIST = 1 MEANS DISTINCT             CSL01230
C     T(1,MEV)-EIGENVALUES ARE WRITTEN TO FILE 11.                      CSL01240
C     SAVTEV = (-1,0,1):  SAVTEV = - 1 MEANS T(1,MEV) AND T(2,MEV)      CSL01250
C     EIGENVALUES ARE AVAILABLE ON FILE 10 FROM AN EARLIER RUN.         CSL01260
C     IN THIS CASE, ALPHA/BETA FILE FROM THAT RUN MUST BE               CSL01270
C     AVAILABLE ON FILE 2.                                              CSL01280
C     SAVTEV = 0 MEANS WE WILL NOT SAVE THE T(1,MEV) AND T(2,MEV)       CSL01290
C     EIGENVALUES.  SAVTEV = 1 MEANS WE WRITE THE T(1,MEV) AND          CSL01300
C     T(2,MEV) EIGENVALUES TO FILE 10.                                  CSL01310
```

```
C     IWRITE = (0,1):  IWRITE = 0 MEANS NO INTERMEDIATE OUTPUT
C     FROM THE COMPUTATIONS IS WRITTEN TO FILE 6.  IWRITE = 1 MEANS
C     EIGENVALUES AND ERROR ESTIMATES ARE WRITTEN TO FILE 6
C     AS THEY ARE COMPUTED.
      READ(5,20) EXPLAN
      READ(5,*) IHIS,IDIST,SAVTEV,IWRITE
C
      IF(SAVTEV.GE.0) GO TO 30
      NMEVS = 1
      IF(ISTART.EQ.0) GO TO 610
C
   30 CONTINUE
C     READ IN THE RELATIVE TOLERANCE (RELTOL) FOR USE IN THE LUMP
C     SUBROUTINE
      READ(5,20) EXPLAN
      READ(5,*) RELTOL
C
C     READ IN THE SIZES OF THE T(1,MEV) MATRICES TO BE CONSIDERED.
      READ(5,20) EXPLAN
      READ(5,*) (NMEV(J), J=1,NMEVS)
C
C-----------------------------------------------------------------------
C     INITIALIZE THE ARRAYS FOR THE USER-SPECIFIED MATRIX
C     AND PASS THE STORAGE LOCATIONS OF THESE ARRAYS TO THE
C     MATRIX-VECTOR MULTIPLY SUBROUTINE CMATV.
C
      CALL USPEC(N,MATNO)
C
C-----------------------------------------------------------------------
C     MASK UNDERFLOW AND OVERFLOW
C
      CALL MASK
C
C-----------------------------------------------------------------------
C
C     WRITE TO FILE 6, A SUMMARY OF THE PARAMETERS FOR THIS RUN
C
      WRITE(6,40) MATNO,N,KMAX
   40 FORMAT(/3X,'MATRIX ID',4X,'ORDER OF A',4X,'MAX ORDER OF T'/
     1 I12,I14,I18/)
C
      WRITE(6,50) ISTART,ISTOP
   50 FORMAT(/2X,'ISTART',3X,'ISTOP'/2I8/)
C
      WRITE(6,60) IHIS,IDIST,SAVTEV,IWRITE
   60 FORMAT(/4X,'IHIS',3X,'IDIST',3X,'SAVTEV',2X,'IWRITE'/2I8,I9,I8/)
C
      WRITE(6,70) SVSEED,RHSEED
   70 FORMAT(/' SEEDS FOR RANDOM NUMBER GENERATOR'//
     1 4X,'LANCZS SEED',4X,'INVERR SEED'/2I15/)
C
      WRITE(6,80) (NMEV(J), J=1,NMEVS)
   80 FORMAT(/' SIZES OF T-MATRICES TO BE CONSIDERED'/(6I12))
C
      WRITE(6,90) RELTOL,GAPTOL,BTOL
   90 FORMAT(/' RELATIVE TOLERANCE USED TO COMBINE COMPUTED T-EIGENVALUE
     1S'/E15.3/' RELATIVE GAP TOLERANCES USED IN INVERSE ITERATION'/
     1E15.3/' RELATIVE TOLERANCE FOR CHECK ON SIZE OF BETAS'/E15.3/)
C
      IF (ISTART.EQ.0) GO TO 140
C
C     READ IN ALPHA BETA HISTORY
C     HISTORY MUST BE STORED IN MACHINE FORMAT TO PREVENT
C     ERRORS CAUSED BY INPUT/OUTPUT CONVERSIONS.
C
      READ(2,100)MOLD,NOLD,SVSOLD,MATOLD
```

```
  100 FORMAT(2I6,I12,I8)
C
      IF (KMAX.LT.MOLD) KMAX = MOLD
      KMAX1 = KMAX + 1
C
C     CHECK THAT ORDER N, MATRIX ID MATNO, AND RANDOM SEED SVSEED
C     AGREE WITH THOSE IN THE HISTORY FILE.  IF NOT PROCEDURE STOPS.
C
      ITEMP = (NOLD-N)**2+(MATNO-MATOLD)**2+(SVSEED-SVSOLD)**2
C
      IF (ITEMP.EQ.0) GO TO 120
C
      WRITE(6,110)
  110 FORMAT(' PROGRAM TERMINATES'/    ' READ FROM FILE 2 CORRESPONDS TO
     1 DIFFERENT MATRIX THAN MATRIX SPECIFIED'/)
      GO TO 650
C
  120 CONTINUE
      MOLD1 = MOLD+1
C
      READ(2,130)(ALPHA(J), J=1,MOLD)
      READ(2,130)(BETA(J), J=1,MOLD1)
  130 FORMAT(4Z20)
C
      IF (KMAX.EQ.MOLD) GO TO 160
C
      READ(2,130)(V1(J), J=1,N)
      READ(2,130)(V2(J), J=1,N)
C
  140 CONTINUE
      IIX = SVSEED
C
C-----------------------------------------------------------------------
C
      CALL LANCZS(CMATV,V1,V2,ALPHA,BETA,GR,GC,G,KMAX,MOLD1,N,IIX)
C
C-----------------------------------------------------------------------
C
      KMAX1 = KMAX + 1
C
      IF (IHIS.EQ.0.AND.ISTOP.GT.0) GO TO 160
C
      WRITE(1,150) KMAX,N,SVSEED,MATNO
  150 FORMAT(2I6,I12,I8,' = KMAX,N,SVSEED,MATNO')
C
      WRITE(1,130)(ALPHA(I), I=1,KMAX)
      WRITE(1,130)(BETA(I), I=1,KMAX1)
C
      WRITE(1,130)(V1(I), I=1,N)
      WRITE(1,130)(V2(I), I=1,N)
C
      IF (ISTOP.EQ.0) GO TO 520
C
  160 CONTINUE
      BKMIN = BTOL
      WRITE(6,170)
  170 FORMAT(/' T-MATRICES (ALPHA AND BETA) ARE NOW AVAILABLE'/)
C
C-----------------------------------------------------------------------
C     SUBROUTINE TNORM CHECKS MIN|BETA|/(ESTIMATED NORM(A)) > BTOL .
C     IF THIS IS VIOLATED IB IS SET EQUAL TO THE NEGATIVE OF THE INDEX
C     OF THE MINIMAL BETA.  IF(IB < 0) THEN SUBROUTINE TNORM IS
C     CALLED FOR EACH VALUE OF MEV TO DETERMINE WHETHER OR NOT THERE
C     IS A BETA IN THE T-MATRIX SPECIFIED THAT VIOLATES THIS TEST.
C     IF THERE IS SUCH A BETA THE PROGRAM TERMINATES FOR THE USER
C     TO DECIDE WHAT TO DO.  THIS TEST CAN BE OVER-RIDDEN BY
C     SIMPLY MAKING BTOL SMALLER, BUT THEN THERE IS THE POSSIBILITY
```

```
C     THAT LOSSES IN THE LOCAL ORTHOGONALITY MAY HURT THE COMPUTATIONS.
C
C     TNORM ALSO COMPUTES TKMAX = MAX(|ALPHA(K)|,|BETA(K)|, K=1,KMAX).
C     HOWEVER, IN THE COMPLEX SYMMETRIC CASE SINCE ALL OF THE
C     EIGENVALUES OF T(1,MEV) ARE COMPUTED, TKMAX IS NOT USED TO SCALE
C     THE T-MULTIPLICITY AND SPURIOUS TOLERANCES.  THE COMPUTED
C     T-EIGENVALUE LARGEST IN MAGNITUDE IS USED INSTEAD.
C
      CALL TNORM(ALPHA,BETA,BKMIN,TKMAX,KMAX,IB)
C
C-----------------------------------------------------------------------
C
C     LOOP ON THE SIZE OF THE T-MATRIX
C
  180 CONTINUE
      MMB = MMB + 1
      MEV = NMEV(MMB)
C     IS MEV TOO LARGE ?
      IF(MEV.LE.KMAX) GO TO 200
      WRITE(6,190) MMB, MEV, KMAX
  190 FORMAT(/' TERMINATE PRIOR TO CONSIDERING THE',I6,'TH T-MATRIX'/
     1' BECAUSE THE SIZE REQUESTED',I6,' IS GREATER THAN THE MAXIMUM SIZ
     1E ALLOWED',I6/)
      GO TO 520
C
  200 MP1 = MEV + 1
      BETAM = BETA(MP1)
C
      IF (IB.GE.0) GO TO 220
C
      T0 = BTOL
C
C-----------------------------------------------------------------------
C
      CALL TNORM(ALPHA,BETA,T0,T1,MEV,IBMEV)
C
C-----------------------------------------------------------------------
C
  210 TEMP = T0/TKMAX
      IBMEV = IABS(IBMEV)
      IF (TEMP.GE.BTOL) GO TO 220
      IBMEV = -IBMEV
      GO TO 590
  220 CONTINUE
C
C-----------------------------------------------------------------------
C     SUBROUTINE COMPEV CALLS SUBROUTINE CMTQL1 TO COMPUTE THE
C     T-EIGENVALUES.  COMPEV THEN APPLIES THE T-MULTIPLICITY AND
C     SPURIOUS TESTS TO THE COMPUTED T-EIGENVALUES.  HERE INITIALIZE
C     THE TOLERANCES USED IN THE T-MULTIPLICITY AND THE SPURIOUS
C     TESTS.  THE MAX(|LAMBDA(T(1,MEV)|) WILL BE INCORPORATED
C     INSIDE THE SUBROUTINE COMPEV.  NOTE THAT THE OBSERVED ACCURACY
C     OF THE COMPUTED T-EIGENVALUES FOR THE COMPLEX SYMMETRIC CASE
C     IS APPROXIMATELY 3 DIGITS LESS THAN THAT ACHIEVED IN THE REAL
C     CASE. THUS, A FACTOR OF 500 HAS BEEN INTRODUCED.  THIS HOWEVER
C     MEANS THAT THIS TEST IS NOT AS SHARP AS IT WAS IN THE
C     REAL SYMMETRIC AND HERMITIAN CASES.  THUS, IT HAS LOWER
C     RESOLUTION AND CAN OCCASIONALLY MAKE A MISTAKE.
C
      MULTOL = 500.D0 * DFLOAT(MEV+1000) * EPSM
      SPUTOL =  MULTOL
C
C     ON RETURN FROM COMPEV
C     NDIS = NUMBER OF DISTINCT EIGENVALUES OF T(1,MEV)
C     VS = DISTINCT T-EIGENVALUES IN INCREASING ORDER OF MAGNITUDE
C     GR(K) = |VS(K)|, K = 1,NDIS,  GR(K).LE.GR(K+1)
C     MP = T-MULTIPLICITIES OF THE T-EIGENVALUES IN VS
```

```
C       MP(I) = (0,1,MI), MI>1, I=1,NDIS  MEANS:
C          (0)  VS(I) IS SPURIOUS
C          (1)  VS(I) IS SIMPLE AND GOOD
C          (MI) VS(I) IS T-MULTIPLE AND IS THEREFORE NOT ONLY GOOD BUT
C               ALSO A CONVERGED GOOD T-EIGENVALUE.
C
C
      CALL COMPEV(ALPHA,BETA,V1,V2,VS,GR,MULTOL,SPUTOL,MP,MP2,
     1MEV,NDIS,SAVTEV)
C
C-----------------------------------------------------------------------
C
      IF (NDIS.EQ.0) GO TO 630
C
C     ON EXIT FROM COMPEV MULTOL AND SPUTOL SHOULD BE SCALED
C     BY THE SIZES OF THE T-EIGENVALUES
      EVMAX = GR(NDIS)
      LOOP = NDIS
C
C-----------------------------------------------------------------------
C
      CALL LUMP(VS,V1,GR,RELTOL,SPUTOL,SCALE2,MP,MP2,LOOP)
C
C-----------------------------------------------------------------------
C
      IF (LOOP.LT.0) GO TO 650
C
      IF (NDIS.EQ.LOOP) GO TO 240
C
      WRITE(6,230) NDIS,LOOP,MEV
  230 FORMAT(/' AFTER LUMP NDIS,LOOP,MEV = ',3I6/)
C
  240 CONTINUE
      NDIS = LOOP
C
C-----------------------------------------------------------------------
C     CALCULATE MINGAPS FOR DISTINCT T(1,MEV) EIGENVALUES.
C     ON EXIT |GG(K)| = MIN(J.NE.K,|VS(K)-VS(J)|), MP2(K)=J INDEX
C     FOR MINIMUM.  GG(K)< 0 MEANS NEAREST NEIGHBOR IS SPURIOUS.
      IGAP = 0
      ITAG = 1
C
      CALL COMGAP(VS,GR,GG,MP,MP2,NDIS,IGAP,ITAG)
C
C-----------------------------------------------------------------------
C
C     SET CONVERGENCE CRITIERION
      TTOL = EPSM * EVMAX
      CONTOL = CDABS(BETAM)*1.D-10
C
  250 CONTINUE
      BETA(MP1) = BETAM
C
C-----------------------------------------------------------------------
C     THE SUBROUTINE ISOEV LABELS THOSE SIMPLE EIGENVALUES OF T(1,MEV)
C     WITH VERY SMALL GAPS BETWEEN NEIGHBORING EIGENVALUES OF T(1,MEV)
C     TO AVOID COMPUTING ERROR ESTIMATES FOR ANY SIMPLE GOOD
C     T-EIGENVALUE THAT IS TOO CLOSE TO A SPURIOUS T-EIGENVALUE.
C     MP(I) = -1 MEANS THAT THE GOOD T-EIGENVALUE IS SIMPLE AND
C     IS TOO CLOSE TO A SPURIOUS T-EIGENVALUE.
C
C     NG = NUMBER OF GOOD T-EIGENVALUES.
C     NISO = NUMBER OF ISOLATED GOOD T-EIGENVALUES.
C     GG = MINIMAL GAPS IN T(1,MEV)
C     GR(K) = |VS(K)|, K=1,NDIS
C
```

```
      CALL ISOEV(VS,GR,GG,GAPTOL,SPUTOL,SCALE1,MP,NDIS,NG,NISO)
C
C-----------------------------------------------------------------------
C
      WRITE(6,260)NG,NISO,NDIS
  260 FORMAT(/I6,' GOOD T-EIGENVALUES HAVE BEEN COMPUTED'/
     1 I6,' OF THESE ARE ISOLATED'/
     2 I6,' = NUMBER OF DISTINCT T-EIGENVALUES COMPUTED'/)
C
C     DO WE WRITE DISTINCT EIGENVALUES OF T-MATRIX TO FILE 11?
      IF (IDIST.EQ.0) GO TO 300
C
      WRITE(11,270) NDIS,NISO,MEV,N,SVSEED,MATNO
  270 FORMAT(/4I6,I12,I8,' = NDIS,NISO,MEV,N,SVSEED,MATNO'/)
C
      WRITE(11,280) (I,MP(I),VS(I),GG(I),MP2(I), I=1,NDIS)
  280 FORMAT(I4,I4,2E20.12,E12.3,I6)
C
      WRITE(11,290) NDIS, (MP(I), I=1,NDIS)
  290 FORMAT(/I6,' = NDIS, T-MULTIPLICITIES (0 MEANS  SPURIOUS)'/(20I4))
C
  300 CONTINUE
C
      IF (NISO.NE.0) GO TO 330
C
      WRITE(4,310) MEV
  310 FORMAT(/' AT MEV = ',I6,' THERE ARE NO ISOLATED T-EIGENVALUES'/
     1' SO NO ERROR ESTIMATES WERE COMPUTED/')
C
      WRITE(6,320)
  320 FORMAT(/' ALL COMPUTED GOOD T-EIGENVALUES ARE T-MULTIPLE'/
     1 ' THEREFORE THESE EIGENVALUES ARE ASSUMED TO HAVE CONVERGED')
C
      ICONV = 1
      GO TO 370
C
  330 CONTINUE
C
C-----------------------------------------------------------------------
C     SUBROUTINE INVERR COMPUTES ERROR ESTIMATES FOR ISOLATED GOOD
C     T-EIGENVALUES USING INVERSE ITERATION ON T(1,MEV). ON RETURN
C     GG(J) = MINIMUM GAP IN T(1,MEV) FOR EACH VS(J), J=1,NDIS
C     G(I) = |BETAM|*|U(MEV)| = ERROR ESTIMATE FOR ISOLATED GOOD
C              T-EIGENVALUES, WHERE I = 1, NISO AND  BETAM = BETA(MEV+1)
C              U(MEV) IS MEVTH COMPONENT OF THE UNIT EIGENVECTOR OF T
C              CORRESPONDING TO THE ITH ISOLATED GOOD T-EIGENVALUE.
C     A NEGATIVE ERROR ESTIMATE MEANS THAT FOR THAT PARTICULAR
C     T-EIGENVALUE THE INVERSE ITERATION DID NOT CONVERGE IN <= MXINIT
C     STEPS AND THAT THE CORRESPONDING ERROR ESTIMATE IS QUESTIONABLE.
C
C     ON EXIT
C     V2 CONTAINS THE ISOLATED GOOD T-EIGENVALUES
C     GR CONTAINS THE MINGAPS TO THE NEAREST DISTINCT EIGENVALUE
C        OF T(1,MEV) FOR EACH ISOLATED GOOD T-EIGENVALUE IN V2.
C     VS CONTAINS THE NDIS DISTINCT EIGENVALUES OF T(1,MEV)
C     MP CONTAINS THE CORRESPONDING CODED T-MULTIPLICITIES
C
      IT = MXINIT
C
      CALL INVERR(ALPHA,BETA,V1,V2,VS,EPSM,GR,GC,G,GG,MP,MP2,MEV,MMB,
     1NDIS,NISO,N,RHSEED,IT,IWRITE)
C
C-----------------------------------------------------------------------
C
C     SIMPLE CHECK FOR CONVERGENCE. CHECKS TO SEE IF ALL OF THE ERROR
C     ESTIMATES ARE SMALLER THAN CONTOL = CDABS(BETA(MEV+1)*1.D-10
C     IF THIS TEST IS SATISFIED, THEN CONVERGENCE FLAG, ICONV IS SET
```

```
C     TO 1.  TYPICALLY ERROR ESTIMATES ARE VERY CONSERVATIVE.
C
      WRITE(6,340) CONTOL
  340 FORMAT(/' CONVERGENCE IS TESTED USING THE CONVERGENCE TOLERANCE',
     1E13.4/)
C
      DO 350 I = 1,NISO
      IF (ABS(G(I)).GT.CONTOL) GO TO 370
  350 CONTINUE
      ICONV = 1
      MMB = NMEVS
C
      WRITE(6,360) CONTOL
  360 FORMAT(' ALL COMPUTED ERROR ESTIMATES WERE LESS THAN',E15.4/
     1 ' THEREFORE PROCEDURE TERMINATES'/)
C
  370 CONTINUE
C
C     IN REAL SYMMETRIC AND HERMITIAN LANCZOS PROGRAMS
C     AT THIS CORRESPONDING POINT THE SUBROUTINE PRTEST IS CALLED
C     TO IDENTIFY ANY T-EIGENVALUES THAT MAY HAVE BEEN MISLABELLED
C     AS SPURIOUS BECAUSE THEIR PROJECTIONS ON THE STARTING VECTOR
C     WERE TOO SMALL.  THIS CHECK WAS MADE ONLY AFTER CONVERGENCE
C     HAD OCCURRED.  HOWEVER, THE PRTEST SUBROUTINE IS BASED UPON
C     STURM SEQUENCING AND THAT IS NOT VALID FOR COMPLEX SYMMETRIC
C     MATRICES.  PERHAPS THERE IS SOME RECTANGLE ANALOG OF THE
C     PRTEST BUT WE HAVE NOT ATTEMPTED TO IDENTIFY AND INCLUDE
C     SUCH A TEST BECAUSE WE EXPECT, AS IN THE REAL SYMMETRIC AND
C     HERMITIAN CASES THAT HIDDEN EIGENVALUES WILL BE RARE.
C
C     WRITE THE GOOD T-EIGENVALUES TO FILE 3.  FIRST TRANSFER THEM
C     TO V2 AND THEIR T-MULTIPLICITIES TO THE CORRESPONDING POSITIONS
C     IN MP AND COMPUTE THE A-MINGAPS, THE MINIMAL GAPS BETWEEN THE
C     GOOD T-EIGENVALUES.  THESE GAPS WILL BE PUT IN THE ARRAY GG.
C     NOTE THAT AFTER THE SECOND CALL TO COMGAP THE ARRAY GC
C     WILL CONTAIN THE CORRESPONDING MINIMAL GAPS IN THE
C     T-MATRIX, T(1,MEV).
C
  380 CONTINUE
C
      NG = 0
      DO 390 I = 1,NDIS
      IF (MP(I).EQ.0) GO TO 390
      NG = NG+1
      MP(NG) = MP(I)
      V2(NG) = VS(I)
      GC(NG) = GG(I)
  390 CONTINUE
C
      DO 400 I = 1,NG
  400 GR(I) = CDABS(V2(I))
C
C-----------------------------------------------------------------------
C     CALCULATE MINGAPS FOR GOODEV
C     ON EXIT GG(K) = MIN(J.NE.K,|V2(K)-V2(J)|), MP2(K)=J INDEX FOR MIN
C     NG = NUMBER OF COMPUTED DISTINCT GOOD T-EIGENVALUES.
      IGAP = 0
      ITAG = 0
C
      CALL COMGAP(V2,GR,GG,MP,MP2,NG,IGAP,ITAG)
C
C-----------------------------------------------------------------------
C
C     WRITE GOOD T-EIGENVALUES OUT TO FILE 3.
C
      WRITE(6,410)MEV
```

```
  410 FORMAT(//' EIGENVALUE CALCULATION AT MEV = ',I6,'  IS COMPLETE'/)
C
      WRITE(3,420)NG,NDIS,MEV,N,SVSEED,MATNO,MULTOL,SPUTOL,IB,BTOL
  420 FORMAT(4I6,I12,I8,' = NG,NDIS,MEV,N,SVEED,MATNO'/
     1 2E15.5,I6,E13.4,' = MULTOL,SPUTOL,IB,BTOL'/
     1' EVNO',1X,'MULT',13X,'R(GOODEV)',13X,'I(GOODEV)',
     1 3X,'TMINGAP',3X,'AMINGAP',1X,'NEIGH')
C
      WRITE(3,430)(I,MP(I),V2(I),GC(I),GG(I),MP2(I), I=1,NG)
  430 FORMAT(2I5,2E22.14,2E10.3,I6)
C
C     ORDER GOODEV BY INCREASING GAP SIZE
      DO 440 I = 1,NG
      MP(I) = I
      V1(I) = V2(I)
      G(I) = GG(I)
  440 CONTINUE
C
C     WRITE(12,436)
  450 FORMAT(' MINGAPS FOR GOOD T-EIGENVALUES'/
     1 1X,'EVNUM',1X,'NEIGH',15X,'R(EV)',15X,'I(EV)',4X,'MINGAP')
C
C     WRITE(12,439) (K,MP2(K),V2(K),G(K), K = 1,NG)
  460 FORMAT(2I6,2E20.12,E10.3)
C
      DO 480 K = 2,NG
      KM1 = K-1
      DO 470 L = 1,KM1
      KK = K-L
      KP1 = KK+1
      IF (G(KP1).GE.G(KK)) GO TO 480
      Z = V1(KK)
      V1(KK) = V1(KP1)
      V1(KP1) = Z
      GTEMP = G(KK)
      G(KK) = G(KP1)
      G(KP1) = GTEMP
      ITEMP = MP(KK)
      MP(KK) = MP(KP1)
      MP(KP1) = ITEMP
  470 CONTINUE
  480 CONTINUE
C
C     WRITE(12,441)
      WRITE(3,490)
  490 FORMAT(' T-EIGENVALUES ORDERED BY INCREASING MINGAP'/
     1 1X,'GAPNUM',1X,'EVNUM',15X,'R(EV)',15X,'I(EV)',4X,'MINGAP')
C
C     WRITE(12,442) (K,MP(K),V1(K),G(K), K = 1,NG)
      WRITE(3,500) (K,MP(K),V1(K),G(K), K = 1,NG)
  500 FORMAT(I7,I6,2E20.12,E10.3)
C
  510 CONTINUE
C
C     IF CONVERGENCE FLAG ICONV.NE.1 AND NUMBER OF T-MATRICES
C     CONSIDERED TO DATE IS LESS THAN NUMBER ALLOWED, INCREMENT MEV.
C     AND LOOP BACK TO 210 TO REPEAT COMPUTATIONS.  RESTORE BETA(MEV+1).
C
      BETA(MP1) = BETAM
C
      IF (MMB.LT.NMEVS.AND.ICONV.NE.1) GO TO 180
C
C     END OF LOOP ON DIFFERENT SIZE T-MATRICES ALLOWED.
C
  520 CONTINUE
C
      IF(ISTOP.EQ.0)  WRITE(6,530)
```

```
  530 FORMAT(/' T-MATRICES (ALPHA AND BETA) ARE NOW AVAILABLE, TERMINATECSL05980
     1')                                                                 CSL05990
      IF (ISTOP.EQ.0.AND.KMAX.NE.MOLD) WRITE(1,540)                      CSL06000
      IF (IHIS.EQ.1.AND.KMAX.NE.MOLD) WRITE(1,540)                       CSL06010
  540 FORMAT(/' ABOVE ARE THE FOLLOWING VECTORS '/                       CSL06020
     1 '  ALPHA(I), I = 1,KMAX'/                                         CSL06030
     2 '  BETA(I), I = 1,KMAX+1'/                                        CSL06040
     3 ' FINAL TWO LANCZOS VECTORS OF ORDER N FOR I = KMAX,KMAX+1'/      CSL06050
     4 ' ALPHA BETA ARE IN HEX FORMAT 4Z20 '/                            CSL06060
     4 ' LANCZOS VECTORS ARE IN HEX FORMAT 4Z20 '/                       CSL06070
     5 ' ----- END OF FILE 1 NEW ALPHA, BETA HISTORY---------------'///)CSL06080
C                                                                        CSL06090
      IF (ISTOP.EQ.0) GO TO 650                                          CSL06100
C                                                                        CSL06110
      WRITE(3,550)                                                       CSL06120
  550 FORMAT(/' ABOVE ARE COMPUTED GOOD T-EIGENVALUES'/                  CSL06130
     1 ' NG = NUMBER OF GOOD T-EIGENVALUES COMPUTED'/                    CSL06140
     2 ' NDIS = NUMBER OF COMPUTED DISTINCT EIGENVALUES OF T(1,MEV)'/    CSL06150
     3 ' N = ORDER OF A,  MATNO = MATRIX IDENT'/                         CSL06160
     4 ' MULTOL = T-MULTIPLICITY TOLERANCE FOR T-EIGENVALUES'/           CSL06170
     4 ' SPUTOL = SPURIOUS TOLERANCE FOR T-EIGENVALUES'/                 CSL06180
     4 ' MULT IS THE T-MULTIPLICITY OF GOOD T-EIGENVALUE'/               CSL06190
     5 ' MULT = -1 MEANS SPURIOUS T-EIGENVALUE TOO CLOSE'/               CSL06200
     6 ' DO NOT COMPUTE ERROR ESTIMATES FOR SUCH T-EIGENVALUES'/         CSL06210
     7 ' AMINGAP = MINIMAL GAP BETWEEN THE COMPUTED A-EIGENVALUES'/      CSL06220
     9 ' TMINGAP= MINIMAL GAP W.R.T.  DISTINCT EIGENVALUES IN T(1,MEV)'/CSL06230
     2 ' ----- END OF FILE 3 GOOD T-EIGENVALUES---------------------'///CSL06240
     3 )                                                                 CSL06250
C                                                                        CSL06260
      IF (IDIST.NE.0) WRITE(11,560)                                      CSL06270
  560 FORMAT(/' ABOVE ARE THE DISTINCT EIGENVALUES OF T(1,MEV).'/        CSL06280
     2 ' THE FORMAT IS      T-MULTIPLICITY    T-EIGENVALUE    TMINGAP'/  CSL06290
     4 ' T-MULTIPLICITY = -1 MEANS THAT THE SUBROUTINE ISOEV HAS TAGGED'CSL06300
     5 /'   THIS SIMPLE T-EIGENVALUE AS HAVING A VERY CLOSE SPURIOUS'/   CSL06310
     6 '   T-EIGENVALUE SO THAT NO ERROR ESTIMATE WILL BE COMPUTED'/     CSL06320
     7 '   FOR THAT EIGENVALUE IN SUBROUTINE INVERR.'/                   CSL06330
     9 ' EACH OF THE DISTINCT T-EIGENVALUE TABLES IS FOLLOWED'/          CSL06340
     9 ' BY THE T-MULTIPLICITY PATTERN.'/                                CSL06350
     1 ' NDIS = NUMBER OF COMPUTED DISTINCT EIGENVALUES OF T(1,MEV).'/   CSL06360
     2 ' NG = NUMBER OF GOOD T-EIGENVALUES. '/                           CSL06370
     3 ' NISO = NUMBER OF ISOLATED GOOD T-EIGENVALUES. '/                CSL06380
     4 ' NISO ALSO IS THE COUNT OF +1 ENTRIES IN MULTIPLICITY PATTERN.'/CSL06390
     5 ' -----END OF FILE 11 DISTINCT T-EIGENVALUES----------------'///)CSL06400
C                                                                        CSL06410
      WRITE(4,570)                                                       CSL06420
  570 FORMAT(/' ABOVE ARE THE ERROR ESTIMATES OBTAINED FOR THE ISOLATED CSL06430
     1GOOD T-EIGENVALUES'/                                               CSL06440
     1' OBTAINED VIA INVERSE ITERATION IN THE SUBROUTINE INVERR.'/       CSL06450
     1' ALL OTHER GOOD T-EIGENVALUES HAVE CONVERGED.'/                   CSL06460
     2' ERROR ESTIMATE = CDABS(BETAM*(UM))'/                             CSL06470
     2' WHERE BETAM = BETA(MEV+1) AND UM = U(MEV).'/                     CSL06480
     3' U = UNIT EIGENVECTOR OF T WHERE T*U = EV*U AND EV = ISOLATED GOOCSL06490
     3D T-EIGENVALUE.'/                                                  CSL06500
     4' TMINGAP = GAP TO NEAREST DISTINCT EIGENVALUE OF T(1,MEV).'/      CSL06510
     6' ------ END OF FILE 4 ERRINV ------------------------------'///)CSL06520
C                                                                        CSL06530
      IF(SAVTEV.LT.0) GO TO 650                                          CSL06540
      WRITE(10,580)                                                      CSL06550
  580 FORMAT(//' ABOVE ARE THE T(1,MEV) EIGENVALUES FOLLOWED BY THE'/    CSL06560
     1 ' T(2,MEV) EIGENVALUES FOR MEV = NMEV(J), J = 1,NMEVS'/           CSL06570
     1 ' ------END OF FILE 10 T-T2EVAL----------------------'///)        CSL06580
C                                                                        CSL06590
      GO TO 650                                                          CSL06600
C                                                                        CSL06610
  590 CONTINUE                                                           CSL06620
C                                                                        CSL06630
      IBB = IABS(IBMEV)                                                  CSL06640
```

```
      TEMP = CDABS(BETA(IBB))
      IF (IBMEV.LT.0) WRITE(6,600) MEV,IBB,TEMP
  600 FORMAT(/' PROGRAM TERMINATES BECAUSE MEV REQUESTED = ',I6,' IS .GT
     1',I6/' AT WHICH AN ABNORMALLY SMALL BETA = ' ,E13.4,' OCCURRED'/)
      GO TO 650
C
  610 WRITE(6,620) SAVTEV,ISTART
  620 FORMAT(2I6,' = SAVTEV,ISTART'/' WHEN SAVTEV = -1, WE MUST HAVE IST
     1ART = 1'/)
      GO TO 650
C
  630 IF (NDIS.EQ.0.AND.ISTOP.GT.0) WRITE(6,640)
  640 FORMAT(/' INTERVALS SPECIFIED FOR BISECT DID NOT CONTAIN ANY T-EIG
     1ENVALUES'/' PROGRAM TERMINATES')
C
  650 CONTINUE
C
      STOP
C-----END OF MAIN PROGRAM FOR COMPLEX SYMMETRIC EIGENVALUE COMPUTATIONS-
      END
```

SECTION 7.4 MAIN PROGRAM, EIGENVEGTOR CALCULATIONS

```
C-----CSLEVEC (EIGENVECTORS OF COMPLEX SYMMETRIC MATRICES)--------------
C
C     CONTAINS MAIN PROGRAM FOR COMPUTING AN EIGENVECTOR CORRESPONDING
C     TO EACH OF A SET OF EIGENVALUES THAT HAVE BEEN COMPUTED
C     ACCURATELY BY THE CORRESPONDING LANCZOS EIGENVALUE PROGRAM
C     (CSLEVAL) FOR NONDEFECTIVE COMPLEX SYMMETRIC MATRICES.
C     THIS PROGRAM COULD BE MODIFIED TO COMPUTE ADDITIONAL
C     EIGENVECTORS FOR THOSE EIGENVALUES WHICH ARE MULTIPLE EIGENVALUES
C     OF THE GIVEN A-MATRIX.  THE AMOUNT OF ADDITIONAL COMPUTATION
C     REQUIRED WOULD DEPEND UPON THE GIVEN A-MATRIX AND UPON WHAT
C     PART OF THE SPECTRUM OF A IS INVOLVED.
C
C     THESE LANCZOS EIGENVECTOR COMPUTATIONS ASSUME THAT EACH
C     EIGENVALUE THAT IS BEING CONSIDERED HAS CONVERGED AS AN
C     EIGENVALUE OF THE CORRESPONDING LANCZOS TRIDIAGONAL MATRICES.
C
C     PORTABILITY:
C     THIS PROGRAM IS NOT PORTABLE DUE TO THE USE OF THE COMPLEX*16
C     VARIABLES AND CORRESPONDING COMPLEX FUNCTIONS.  MOREOVER, PFORT
C     IDENTIFIED THE FOLLOWING ADDITIONAL NONPORTABLE CONSTRUCTIONS:
C
C     1.  DATA/MACHEP/ STATEMENT
C     2.  ALL READ(5,*) STATEMENTS (FREE FORMAT)
C     3.  FORMAT(20A4) USED WITH THE EXPLANATORY HEADER, EXPLAN
C     4.  FORMAT (4Z20) USED FOR ALPHA/ BETA FILE 2.
C
C     IMPORTANT NOTE:  PROGRAM ALLOWS ENLARGEMENT OF THE ALPHA,BETA
C     ARRAYS.  IN PARTICULAR, IF ANY ONE OF THE EIGENVALUES SUPPLIED
C     IS T-SIMPLE AND NOT CLOSE TO A SPURIOUS T-EIGENVALUE, THE PROGRAM
C     REQUIRES THAT KMAX BE AT LEAST 11*MEV/8 + 12.  IF KMAX IS NOT
C     THIS LARGE, THEN THE PROGRAM WILL RESET KMAX TO THIS SIZE
C     AND EXTEND THE ALPHA, BETA HISTORY IF REQUIRED.
C     THUS, THE DIMENSIONS OF THE ALPHA AND BETA ARRAYS MUST BE
C     LARGE ENOUGH TO ALLOW FOR THIS POSSIBILITY.
C     REMEMBER THAT THE BETA ARRAY, BETA(J), IS SUCH THAT
C     J = 1,..., KMAX+1.  SO IF THE KMAX USED BY THE PROGRAM
C     IS TO BE 3000, THEN BETA MUST BE OF LENGTH AT LEAST 3001.
C
C-----------------------------------------------------------------------
      COMPLEX*16  V1(1600),V2(1600),RITVEC(10000),ZEROC,TEMPC
      COMPLEX*16  ALPHA(1600),BETA(1601),GOODEV(50),TVEC(20000)
      COMPLEX*16  EVAL,ALFA,BATA,SUMC
      DOUBLE PRECISION  GR(1600),GC(1600)
      DOUBLE PRECISION  ERTOL,SUM,TEMP,BKMIN
      DOUBLE PRECISION  MULTOL,SPUTOL,SCALEO,BTOL
      DOUBLE PRECISION  ONE,ZERO,MACHEP,EPSM
      DOUBLE PRECISION  RELTOL,ERROR,ERRMIN,TERROR,TLAST(50)
      REAL  G(1600),AMINGP(50),TMINGP(50),EXPLAN(20)
      REAL  TERR(50),ERR(50),ERRDGP(50),RNORM(50),TBETA(50)
      INTEGER  MP(50),MA(50),ML(50),MINT(50),MFIN(50),IDELTA(50)
      INTEGER  SVSEED,SVSOLD,RHSEED
      INTEGER  INTERC(1600)
      INTEGER  MBOUND,NTVCON,SVTVEC,TVSTOP,LVCONT,ERCONT,TFLAG
      DOUBLE PRECISION  DABS, DMAX1, DSQRT
      REAL ABS
      INTEGER  IABS
C-----------------------------------------------------------------------
      EXTERNAL CMATV
      DATA MACHEP/Z3410000000000000/
      EPSM = 2.D0*MACHEP
C-----------------------------------------------------------------------
C
C     ARRAYS MUST BE DIMENSIONED AS FOLLOWS:
C     1.  ALPHA:  >= KMAXN,  BETA: >= (KMAXN+1) WHERE KMAXN, THE
```

```
C              LARGEST SIZE T-MATRIX CONSIDERED BY THE PROGRAM,          CSL00650
C              IS THE LARGER OF THE SIZE OF THE ALPHA, BETA HISTORY      CSL00660
C              PROVIDED ON FILE 2 (IF ANY ) AND THE SIZE WHICH THE       CSL00670
C              PROGRAM SPECIFIES INTERNALLY, THIS LATTER IS ALWAYS       CSL00680
C              < = 11*MEV / 8  +  12, WHERE MEV IS THE SIZE              CSL00690
C              T-MATRIX THAT WAS USED IN THE CORRESPONDING EIGENVALUE CSL00700
C              COMPUTATIONS.                                             CSL00710
C     2.  V1:  >= MAX(N,KMAX)                                            CSL00720
C     3.  V2:  >= N                                                      CSL00730
C     4.  G, GR, GC:  >= MAX(N,KMAX)                                     CSL00740
C     5.  RITVEC:  >= N*NGOOD, WHERE NGOOD IS THE NUMBER OF EIGENVALUES CSL00750
C                 SUPPLIED TO THIS PROGRAM.                              CSL00760
C     6.  TVEC:  >= CUMULATIVE LENGTH OF ALL THE T-EIGENVECTORS NEEDED   CSL00770
C                   TO GENERATE THE DESIRED RITZ VECTORS.  AN EDUCATED   CSL00780
C                   GUESS AT AN APPROPRIATE LENGTH CAN BE OBTAINED       CSL00790
C                   BY RUNNING THE PROGRAM WITH THE FLAG MBOUND = 1      CSL00800
C                   AND MULTIPLYING THE RESULTING SIZE BY 5/4.           CSL00810
C     7.  INTERC:   >= KMAX                                              CSL00820
C     8.  GOODEV, AMINGP, TMINGP, TERR, ERR, ERRDGP, RNORM, TBETA,       CSL00830
C         TLAST, MP, MA, MINT, MFIN, AND IDELTA :    >= NUMBER OF        CSL00840
C         EIGENVALUES SUPPLIED.                                          CSL00850
C                                                                        CSL00860
C     OUTPUT HEADER                                                      CSL00870
      WRITE(6,10)                                                        CSL00880
   10 FORMAT(/' LANCZOS EIGENVECTOR PROCEDURE FOR COMPLEX SYMMETRIC MATRCSL00890
     1ICES'/)                                                            CSL00900
C                                                                        CSL00910
C     SET PROGRAM PARAMETERS                                             CSL00920
C     USER MUST NOT MODIFY SCALEO                                        CSL00930
      SCALEO = 5.0D0                                                     CSL00940
      ZERO = 0.0D0                                                       CSL00950
      ZEROC = DCMPLX(ZERO,ZERO)                                          CSL00960
      ONE = 1.0D0                                                        CSL00970
      MPMIN = -1000                                                      CSL00980
      MONE = -1                                                          CSL00990
C     CONVERGENCE TOLERANCE FOR T-EIGENVECTORS FOR RITZ COMPUTATIONS     CSL01000
      ERTOL = 1.D-10                                                     CSL01010
C-----------------------------------------------------------------------CSL01020
C     READ USER-SPECIFIED PARAMETERS FROM INPUT FILE 5 (FREE FORMAT)     CSL01030
C                                                                        CSL01040
C     READ USER-PROVIDED HEADER FOR RUN                                  CSL01050
      READ(5,20) EXPLAN                                                  CSL01060
      WRITE(6,20) EXPLAN                                                 CSL01070
   20 FORMAT(20A4)                                                       CSL01080
C                                                                        CSL01090
C     READ IN THE MAXIMUM PERMISSIBLE DIMENSIONS FOR THE TVEC ARRAY      CSL01100
C     (MDIMTV), FOR THE RITVEC ARRAY (MDIMRV), AND FOR THE BETA          CSL01110
C     ARRAY (MBETA).                                                     CSL01120
C                                                                        CSL01130
      READ(5,20) EXPLAN                                                  CSL01140
      READ(5,*) MDIMTV, MDIMRV, MBETA                                    CSL01150
C                                                                        CSL01160
C     READ IN RELATIVE TOLERANCE (RELTOL) USED IN DETERMINING            CSL01170
C     APPROPRIATE SIZES FOR THE T-MATRICES USED IN THE RITZ              CSL01180
C     VECTOR COMPUTATIONS                                                CSL01190
C                                                                        CSL01200
      READ(5,20) EXPLAN                                                  CSL01210
      READ(5,*) RELTOL                                                   CSL01220
C                                                                        CSL01230
C     SET FLAGS TO 0 OR 1:                                               CSL01240
C     MBOUND = 1:  PROGRAM TERMINATES AFTER COMPUTING 1ST GUESSES        CSL01250
C                  ON APPROPRIATE T-SIZES FOR USE IN THE RITZ VECTOR     CSL01260
C                  COMPUTATIONS                                          CSL01270
C     NTVCON = 0:  PROGRAM TERMINATES IF THE TVEC ARRAY IS NOT           CSL01280
C                  LARGE ENOUGH TO HOLD ALL THE T-EIGENVECTORS REQUIRED.CSL01290
C     SVTVEC = 0:  THE T-EIGENVECTORS ARE NOT WRITTEN TO FILE 11         CSL01300
C                  UNLESS TVSTOP = 1                                     CSL01310
```

```
C     SVTVEC = 1:  WRITE THE T-EIGENVECTORS TO FILE 11.
C     TVSTOP = 1:  PROGRAM TERMINATES AFTER COMPUTING THE
C                  T-EIGENVECTORS
C     LVCONT = 0:  PROGRAM TERMINATES IF THE NUMBER OF T-EIGENVECTORS
C                  COMPUTED IS NOT EQUAL TO THE NUMBER OF RITZ
C                  VECTORS REQUESTED.
C     ERCONT = 0:  MEANS FOR ANY GIVEN EIGENVALUE, A RITZ VECTOR
C                  WILL NOT BE COMPUTED FOR THAT EIGENVALUE UNLESS
C                  A T-EIGENVECTOR HAS BEEN IDENTIFIED WITH A LAST
C                  COMPONENT WHICH SATISFIES THE SPECIFIED
C                  CONVERGENCE CRITERION.
C     ERCONT = 1:  MEANS FOR ANY GIVEN EIGENVALUE, A RITZ VECTOR
C                  WILL BE COMPUTED.  IF A T-EIGENVECTOR CANNOT
C                  BE IDENTIFIED WHICH SATISFIES THE LAST
C                  COMPONENT CRITERION, THEN THE PROGRAM WILL
C                  USE THE T-VECTOR THAT CAME CLOSEST TO
C                  SATISFYING THE CRITERION
C     IWRITE = 1:  EXTENDED OUTPUT OF INTERMEDIATE COMPUTATIONS
C                  IS WRITTEN TO FILE 6
C     IREAD = 0:   ALPHA/BETA FILE IS REGENERATED.
C     IREAD = 1:   ALPHA/BETA FILE USED IN EIGENVALUE COMPUTATIONS
C                  IS READ IN AND EXTENDED IF NECESSARY.  IN BOTH
C                  CASES IREAD = 0 OR 1, THE LANCZOS VECTORS ARE
C                  ALWAYS REGENERATED FOR THE RITZ VECTOR
C                  COMPUTATIONS
C
      READ(5,20) EXPLAN
      READ(5,*) MBOUND,NTVCON,SVTVEC,IREAD
C
      READ(5,20) EXPLAN
      READ(5,*) TVSTOP,LVCONT,ERCONT,IWRITE
      IF (TVSTOP.EQ.1) SVTVEC = 1
C
C     READ IN SEED (RHSEED) FOR GENERATING RANDOM STARTING VECTOR
C     FOR INVERSE ITERATION ON THE T-MATRICES.
C
      READ(5,20) EXPLAN
      READ(5,*) RHSEED
C
C     READ IN MATNO = MATRIX/RUN IDENTIFICATION NUMBER AND
C     N = ORDER OF A-MATRIX
C
      READ(5,20) EXPLAN
      READ(5,*) MATNO,N
C
C-----------------------------------------------------------------------
C     INITIALIZE THE ARRAYS FOR THE USER-SPECIFIED MATRIX
C     AND PASS THE STORAGE LOCATIONS OF THESE ARRAYS TO THE
C     MATRIX-VECTOR MULTIPLY SUBROUTINE CMATV.
C
      CALL USPEC(N,MATNO)
C
C-----------------------------------------------------------------------
C
C     MASK UNDERFLOW AND OVERFLOW
      CALL MASK
C
C-----------------------------------------------------------------------
C     WRITE RUN PARAMETERS OUT TO FILE 6
C
      WRITE(6,30) MATNO,N
   30 FORMAT(/' MATRIX IDENTIFICATION NO. = ',I10,' ORDER OF A = ',I5)
C
      WRITE(6,40) MBOUND,NTVCON,SVTVEC,IREAD
   40 FORMAT(/3X,'MBOUND',3X,'NTVCON',3X,'SVTVEC',3X,'IREAD'/3I9,I8)
C
      WRITE(6,50) TVSTOP,LVCONT,ERCONT,IWRITE
```

```
   50 FORMAT(/3X,'TVSTOP',3X,'LVCONT',3X,'ERCONT',3X,'IWRITE'/4I9)        CSL01990
C                                                                         CSL02000
      WRITE(6,60) MDIMTV,MDIMRV,MBETA                                     CSL02010
   60 FORMAT(/3X,'MDIMTV',3X,'MDIMRV',3X,'MBETA'/2I9,I8)                 CSL02020
C                                                                         CSL02030
      WRITE(6,70) RELTOL,RHSEED                                           CSL02040
   70 FORMAT(/7X,'RELTOL',3X,'RHSEED'/E13.4,I9)                           CSL02050
C                                                                         CSL02060
C                                                                         CSL02070
C     FROM FILE 3 READ IN THE NUMBER OF EIGENVALUES (NGOOD) FOR WHICH     CSL02080
C     EIGENVECTORS ARE REQUESTED, THE ORDER (MEV) OF THE LANCZOS          CSL02090
C     TRIDIAGONAL MATRIX USED IN COMPUTING THESE EIGENVALUES, THE         CSL02100
C     ORDER (NOLD) OF THE USER-SPECIFIED MATRIX USED IN THE EIGENVALUE    CSL02110
C     COMPUTATIONS, THE SEED (SVSEED) USED FOR GENERATING THE STARTING    CSL02120
C     VECTOR THAT WAS USED IN THOSE LANCZOS EIGENVALUE COMPUTATIONS,      CSL02130
C     AND THE MATRIX/RUN IDENTIFICATION NUMBER (MATOLD) USED IN THOSE     CSL02140
C     COMPUTATIONS.  ALSO READ IN THE NUMBER (NDIS) OF DISTINCT           CSL02150
C     EIGENVALUES OF T(1,MEV) THAT WERE COMPUTED BUT THIS VALUE IS        CSL02160
C     NOT USED IN THE EIGENVECTOR COMPUTATIONS.                           CSL02170
C                                                                         CSL02180
      READ(3,80) NGOOD,NDIS,MEV,NOLD,SVSEED,MATOLD                        CSL02190
   80 FORMAT(4I6,I12,I8)                                                  CSL02200
C                                                                         CSL02210
C     READ IN THE TOLERANCES USED IN THE T-MULTIPLICITY AND SPURIOUS      CSL02220
C     TESTS DURING THE EIGENVALUE COMPUTATIONS.                           CSL02230
C     ALSO READ IN THE FLAG IB.  IF IB < 0, THEN SOME BETA(I) IN THE      CSL02240
C     T-MATRIX FILE PROVIDED ON FILE 2 FAILED THE ORTHOGONALITY           CSL02250
C     TEST IN THE TNORM SUBROUTINE.  USER SHOULD NOTE THAT THIS           CSL02260
C     PROGRAM PROCEEDS INDEPENDENTLY OF THE SIZES OF THE BETA USED.       CSL02270
C                                                                         CSL02280
      READ(3,90) MULTOL,SPUTOL,IB,BTOL                                    CSL02290
   90 FORMAT(2E15.5,I6,E13.4)                                             CSL02300
C                                                                         CSL02310
      WRITE(6,100) MULTOL,SPUTOL                                          CSL02320
  100 FORMAT(/' MULTIPLICITY TOLERANCE USED IN THE T-EIGENVALUE COMPUTATCSL02330
     1IONS WAS',E13.4/' TOLERANCE USED IN SPURIOUS CHECK',E13.4)          CSL02340
C                                                                         CSL02350
C     CONTINUE WRITE TO FILE 6 OF THE PARAMETERS FOR THIS RUN             CSL02360
C                                                                         CSL02370
      WRITE(6,110)NGOOD,NDIS,MEV,NOLD,MATOLD,SVSEED,MULTOL,SPUTOL,IB,     CSL02380
     1BTOL                                                                CSL02390
  110 FORMAT(/' EIGENVALUES SUPPLIED ARE READ IN FROM FILE 3'/' FILE 3    CSL02400
     1HEADER IS'/4X,'NG',2X,'NDIS',3X,'MEV',2X,'NOLD',2X,'MATOLD',4X,     CSL02410
     1'SVSEED'/4I6,I8,I10/7X,'MULTOL',7X,'SPUTOL',6X,'IB',9X,'BTOL'/      CSL02420
     12E13.4,I8,E13.4)                                                    CSL02430
C                                                                         CSL02440
C     IS THE ARRAY RITVEC LONG ENOUGH TO HOLD ALL OF THE DESIRED          CSL02450
C     RITZ VECTORS (APPROXIMATE EIGENVECTORS)?                            CSL02460
      NMAX = NGOOD*N                                                      CSL02470
      IF(MBOUND.EQ.1) GO TO 120                                           CSL02480
      IF(TVSTOP.NE.1.AND.NMAX.GT.MDIMRV) GO TO 1310                       CSL02490
C                                                                         CSL02500
C     CHECK THAT THE ORDER N AND THE MATRIX IDENTIFICATION NUMBER         CSL02510
C     MATNO SPECIFIED BY THE USER AGREE WITH THOSE READ IN FROM           CSL02520
C     FILE 3.                                                             CSL02530
  120 ITEMP = (NOLD-N)**2+(MATOLD-MATNO)**2                               CSL02540
      IF (ITEMP.NE.0) GO TO 1330                                          CSL02550
C                                                                         CSL02560
C     READ IN FROM FILE 3, THE T(1,MEV)-MULTIPLICITIES OF THE             CSL02570
C     EIGENVALUES WHOSE EIGENVECTORS ARE TO BE COMPUTED, THE VALUES       CSL02580
C     OF THESE EIGENVALUES AND THEIR MINIMAL GAPS AS EIGENVALUES          CSL02590
C     OF THE USER-SPECIFIED MATRIX AND AS EIGENVALUES OF THE T-MATRIX.    CSL02600
C                                                                         CSL02610
      READ(3,20) EXPLAN                                                   CSL02620
      READ(3,130) (MP(J),GOODEV(J),TMINGP(J),AMINGP(J), J=1,NGOOD)        CSL02630
  130 FORMAT(5X,I5,2E22.14,2E10.3)                                        CSL02640
C                                                                         CSL02650
```

```
      WRITE(6,140) (J,GOODEV(J),MP(J),TMINGP(J),AMINGP(J), J=1,NGOOD)
  140 FORMAT(/' EIGENVALUES READ IN, T-MULTIPLICITIES, T-GAPS AND A-GAPS
     1 '/4X,' J ',15X,' EIGENVALUE',14X,'TMULT',4X,'  TMINGAP ',4X,
     1'  AMINGAP '/(I6,2E20.12,I4,2E15.4))
C
C     READ IN ERROR ESTIMATES
      WRITE(6,170) MEV,SVSEED
C     CHECK WHETHER OR NOT THERE ARE ANY T-ISOLATED EIGENVALUES IN
C     THE EIGENVALUES PROVIDED
      DO 150 J=1,NGOOD
      IF(MP(J).EQ.1) GO TO 160
  150 CONTINUE
      GO TO 190
  160 READ(4,20) EXPLAN
      READ(4,20) EXPLAN
      READ(4,20) EXPLAN
  170 FORMAT(/' THESE EIGENVALUES WERE COMPUTED USING A T-MATRIX OF
     1ORDER ',I5/' AND SEED FOR RANDOM NUMBER GENERATOR =',I12)
      READ(4,180) NISO
  180 FORMAT(18X,I6)
      READ(4,20) EXPLAN
      READ(4,20) EXPLAN
      READ(4,20) EXPLAN
  190 DO 220 J=1,NGOOD
      ERR(J) = 0.D0
      IF(MP(J).NE.1) GO TO 220
      READ(4,200) EVAL, ERR(J)
  200 FORMAT(10X,2E20.12,E14.3)
      IF(CDABS(EVAL - GOODEV(J)).LT.1.D-10) GO TO 220
      WRITE(6,210) EVAL,GOODEV(J)
  210 FORMAT(' PROBLEM WITH READ IN OF ERROR ESTIMATES'/' EIGENVALUE REA
     1D IN',2E20.12,' DOES NOT MATCH GOODEV(J) ='/2E20.12)
      GO TO 1550
C
  220 CONTINUE
C
      WRITE(6,230) (J,GOODEV(J),ERR(J), J=1,NGOOD)
  230 FORMAT(' ERROR ESTIMATES ='/4X,' J',15X,'EIGENVALUE',20X,'ESTIMATE
     1'/(I6,2E20.12,E14.3))
C
C     READ IN THE SIZE OF THE T-MATRIX PROVIDED ON FILE 2.  READ IN
C     THE ORDER OF THE USER-SPECIFIED MATRIX , THE SEED FOR THE
C     RANDOM NUMBER GENERATOR, AND THE MATRIX/TEST IDENTIFICATION
C     NUMBER THAT WERE USED IN THE LANCZOS EIGENVALUE COMPUTATIONS.
C     IF FLAG IREAD = 0, REGENERATE HISTORY FROM SCRATCH
C     HISTORY MUST BE STORED IN MACHINE FORMAT, ((4Z20) FOR
C     IBM/3081)
C
      IF(IREAD.EQ.0)  GO TO 330
C
      READ(2,240) KMAX,NOLD,SVSOLD,MATOLD
  240 FORMAT(2I6,I12,I8)
C
      WRITE(6,250) KMAX,NOLD,SVSOLD,MATOLD
  250 FORMAT(/' READ IN HEADER FOR T-MATRICES'/' FILE 2 HEADER IS'/
     1 2X,'KMAX',2X,'NOLD',6X,'SVSOLD',2X,'MATOLD'/2I6,I12,I8/)
C
C     CHECK THAT THE ORDER, THE MATRIX/TEST IDENTIFICATION NUMBER
C     AND THE SEED FOR THE RANDOM NUMBER GENERATOR USED IN THE
C     LANCZOS COMPUTATIONS THAT GENERATED THE HISTORY FILE
C     BEING USED AGREE WITH WHAT THE USER HAS SPECIFIED.
      IF (NOLD.NE.N.OR.MATOLD.NE.MATNO.OR.SVSOLD.NE.SVSEED) GO TO 1350
C
      KMAX1 = KMAX + 1
C
C     READ IN THE T-MATRICES FROM FILE 2.  THESE ARE USED TO GENERATE
C     THE T-EIGENVECTORS THAT WILL BE USED IN THE RITZ VECTOR
```

```
C     COMPUTATIONS.  HISTORY MUST BE IN MACHINE FORMAT.
C
      READ(2,260) (ALPHA(J), J=1,KMAX)
      READ(2,260) (BETA(J), J=1,KMAX1)
  260 FORMAT(4Z20)
C
      READ(2,260) (V1(J), J=1,N)
      READ(2,260) (V2(J), J=1,N)
C
C     KMAX MAY BE ENLARGED IF THE SIZE AT WHICH THE EIGENVALUE
C     COMPUTATIONS WERE PERFORMED IS ESSENTIALLY KMAX AND
C     THERE IS AT LEAST ONE EIGENVALUE THAT IS T-SIMPLE AND
C     T-ISOLATED, IN THE SENSE THAT IF ITS CLOSEST NEIGHBOR IS TOO
C     CLOSE THAT NEIGHBOR IS A 'GOOD' T-EIGENVALUE.
      DO 270 J = 1,NGOOD
      IF(MP(J).EQ.1) GO TO 290
  270 CONTINUE
      WRITE(6,280)
  280 FORMAT(/' ALL EIGENVALUES USED ARE T-MULTIPLE OR CLOSE TO SPURIOUS
     1 T-EIGENVALUES'/' SO DO NOT CHANGE KMAX')
      IF(KMAX.LT.MEV) GO TO 1370
      GO TO 310
C
  290 KMAXN= 11*MEV/8 + 12
      IF(MBETA.LE.KMAXN) GO TO 1530
      IF(KMAX.GE.KMAXN )  GO TO 310
      WRITE(6,300) KMAX, KMAXN
  300 FORMAT(' ENLARGE KMAX FROM ',I6,' TO ',I6)
      MOLD1 = KMAX + 1
      KMAX = KMAXN
      GO TO 380
C
  310 WRITE(6,320) KMAX
  320 FORMAT(/' T-MATRICES HAVE BEEN READ IN FROM FILE 2'/' THE LARGEST
     1SIZE T-MATRIX ALLOWED IS',I6/)
C
      IF(IREAD.EQ.1) GO TO 400
C
C     REGENERATE THE ALPHA AND BETA
C
  330 MOLD1 = 1
C
C     SET KMAX
      DO 340 J = 1,NGOOD
      IF(MP(J).EQ.1) GO TO 360
  340 CONTINUE
      KMAX = MEV + 12
      WRITE(6,350) KMAX
  350 FORMAT(/' ALL EIGENVALUES FOR WHICH EIGENVECTORS ARE TO BE COMPUTE
     1D ARE EITHER T-MULTIPLE OR CLOSE TO'/' A SPURIOUS EIGENVALUE. THER
     1EFORE SET KMAX = MEV + 12 = ',I7)
      GO TO 380
C
  360 KMAXN = 11*MEV/8 + 12
      IF(MBETA.LE.KMAXN) GO TO 1530
      WRITE(6,370) KMAXN
  370 FORMAT(' SET KMAX EQUAL TO ',I6)
      KMAX = KMAXN
C
  380 WRITE(6,390) MOLD1,KMAX
  390 FORMAT(/' LANCZS SUBROUTINE GENRATES ALPHA(J), BETA(J+1), J =',
     1 I6,' TO ', I6/)
C
C-----------------------------------------------------------------------
C
      CALL LANCZS(CMATV,V1,V2,ALPHA,BETA,GR,GC,G,KMAX,MOLD1,N,SVSEED)
C
```

```
C-----------------------------------------------------------------------
C
  400 CONTINUE
C
C     SIMPLE STURM SEQUENCING IS NOT VALID FOR COMPLEX SYMMETRIC
C     MATRICES.  THUS, THE STRATEGY USED HERE FOR SELECTING
C     APPROPRIATE SIZE T-MATRICES FOR THE EIGENVECTOR COMPUTATIONS
C     MUST BE DIFFERENT FROM THAT USED IN THE REAL SYMMETRIC,
C     HERMITIAN, AND SINGULAR VALUE CASES.  AS IN THOSE CASES,
C     FOR EACH EIGENVALUE, A FIRST GUESS IS SELECTED AND THEN
C     LOOPING ON THE SIZE OF THE T-EIGENVECTOR COMPUTATIONS
C     DETERMINES APPROPRIATE SIZES FOR THE EIGENVECTOR COMPUTATIONS.
C     FIRST GUESSES AT APPROPRIATE SIZES ARE SPECIFIED BELOW.
C
      DO 430 J = 1,NGOOD
      EVAL = GOODEV(J)
C     COMPUTE A FIRST GUESS ON AN APPROPRIATE SIZE T-MATRIX EACH
C     EIGENVALUE.
      IF(MP(J).GT.1) GO TO 410
C     EIGENVALUE IS T-SIMPLE
      IF(MP(J).EQ.MONE) GO TO 420
C     EIGENVALUE IS T-SIMPLE AND T-ISOLATED
      MA(J) = (8*MEV)/9  +  1
      ML(J) = ((11*MEV)/8 + 12)
      GO TO 430
C     EIGENVALUE IS T-MULTIPLE
  410 MA(J) = (5*MEV)/(4*MP(J))  +  1
      ML(J) = (7*MEV)/(4*MP(J)) + 1
      GO TO 430
C     EIGENVALUE IS T-SIMPLE AND NOT T-ISOLATED
  420 MA(J) = (5*MEV)/8 + 1
      ML(J) = MEV
  430 CONTINUE
C
      IF (IWRITE.EQ.1) WRITE(6,440) (MA(JJ), JJ=1,NGOOD)
  440 FORMAT(/' 1ST GUESS AT APPROPRIATE SIZES FOR T-MATRICES '/
     1' ACTUAL VALUES WILL PROBABLY BE 1/4 AGAIN AS MUCH'/(13I6))
C
      WRITE(10,450) N,KMAX
  450 FORMAT(2I8,' = ORDER OF USER MATRIX AND MAX ORDER OF T(1,MEV)')
C
      WRITE(10,460)
  460 FORMAT(/' 1ST GUESS AT APPROPRIATE SIZES FOR T-MATRICES '/
     1' ACTUAL VALUES WILL PROBABLY BE 1/4 AGAIN AS MUCH'/)
      WRITE(10,470)
  470 FORMAT(5X,'J',8X,'REAL(GOODEV)',8X,'IMAG(GOODEV)',7X,'MA(J)',
     17X,'MP(J)')
C
      WRITE(10,480) (J,GOODEV(J), MA(J), MP(J), J=1,NGOOD)
  480 FORMAT(I6,2E20.12,I12,I12)
C
      IF(MBOUND.EQ.1) WRITE(10,490)
  490 FORMAT(/' GOODEV(J) IS A GOOD EIGENVALUE OF T(1,MEV)'/
     1 ' IABS(MA(J)) = APPROPRIATE SIZE T-MATRIX FOR GOODEV(J)'/
     1 ' INITIAL VALUE OF MA(J) IS CHOSEN HEURISTICALLY'/
     1 ' PROGRAM LOOPS ON SIZE OF T-MATRIX TO GET BETTER SIZE'/
     1 ' END OF SIZES OF T-MATRICES FILE 10'///)
C
C
C     TERMINATE AFTER COMPUTING 1ST GUESSES ON SIZES OF T-MATRICES
C     REQUIRED FOR THE GIVEN EIGENVALUES?
      IF(MBOUND.EQ.1) GO TO 1390
C
C
C     IS THERE ROOM FOR ALL OF THE REQUESTED T-EIGENVECTORS?
      MTOL = 0
      DO 500 J = 1,NGOOD
```

```
      MTOL = MTOL + IABS(MA(J))
  500 CONTINUE
      MTOL = (5*MTOL)/4
      IF(MTOL.GT.MDIMTV.AND.NTVCON.EQ.0) GO TO 1410
C
C-----------------------------------------------------------------------
C     GENERATE A RANDOM VECTOR TO BE USED REPEATEDLY BY
C     SUBROUTINE INVERM
C
      ILL = RHSEED
      CALL GENRAN(ILL,G,KMAX)
C
C-----------------------------------------------------------------------
C
      DO 510 I = 1,KMAX
  510 GR(I) = G(I)
C
C-----------------------------------------------------------------------
C
      CALL GENRAN(ILL,G,KMAX)
C
C-----------------------------------------------------------------------
C
      DO 520 I = 1,KMAX
  520 GC(I) = G(I)
C
C     FOR EACH EIGENVALUE LOOP ON T-EIGENVECTOR COMPUTATIONS TO
C     COMPUTE AN APPROPRIATE T-EIGENVECTOR TO USE IN THE RITZ
C     VECTOR COMPUTATIONS.
C
      MTOL = 0
      NTVEC = 0
      DO 690 J = 1,NGOOD
      ICOUNT = 0
      TFLAG = 0
      ERRMIN = 10.D0
      MABEST = MPMIN
      IF(MP(J).EQ.MPMIN) GO TO 690
      EVAL = GOODEV(J)
  530 KMAXU = IABS(MA(J))
C     SELECT A SUITABLE INCREMENT FOR THE ORDERS OF T-MATRICES
C     TO BE CONSIDERED IN DETERMINING APPROPRIATE SIZES FOR THE RITZ
C     VECTOR COMPUTATIONS
      IF(ICOUNT.GT.0) GO TO 560
C     SELECT IDELTA(J) BASED UPON THE MULTIPLICITY IN T(1,MEV)
      IF(MP(J).GT.1) GO TO 540
      IF(MP(J).LT.0) GO TO 550
C     MP(J) = 1, INITIAL MA(J) = 8*MEV/9 + 1
      IDELTA(J) = (ML(J) - IABS(MA(J)))/10  + 1
      GO TO 560
C     MULTIPLE T-EIGENVALUE: INITIAL MA(J) = 5*MEV/4*MP + 1
  540 IDELTA(J) = (ML(J) - IABS(MA(J)))/10  +  1
      GO TO 560
C     T-SIMPLE EVALUE, NEAR SPURIOUS ONE, INITIAL MA(J) = 5*MEV/8 + 1
  550 IDELTA(J) = (ML(J) - IABS(MA(J)))/10  +  1
  560 ICOUNT = ICOUNT + 1
      MTOL = MTOL+KMAXU
C
C     IS THERE ROOM IN TVEC ARRAY FOR THE NEXT T-EIGENVECTOR?
C     IF NOT, SKIP TO RITZ VECTOR COMPUTATIONS.
      IF (MTOL.GT.MDIMTV) GO TO 700
C
      IT = 3
      KINT = MTOL - KMAXU +1
C
C     RECORD THE BEGINNING AND END OF THE T-EIGENVECTOR BEING COMPUTED
      MINT(J) = KINT
```

```
      MFIN(J) = MTOL
C
C-----------------------------------------------------------------------
C     SUBROUTINE INVERM DOES INVERSE ITERATION, I.E. SOLVES
C     (T(1,KMAXU) - EVAL)*U = RHS  FOR EACH EIGENVALUE TO OBTAIN THE
C     DESIRED T-EIGENVECTOR.
C
      IF(IWRITE.EQ.1)  WRITE(6,570) J
  570 FORMAT(/I6,'TH EIGENVALUE')
C
      CALL INVERM(ALPHA,BETA,V1,TVEC(KINT),EVAL,ERROR,TERROR,EPSM,
     1    GR,GC,INTERC,KMAXU,IT,IWRITE)
C
C-----------------------------------------------------------------------
C
      TERR(J) = TERROR
      TLAST(J) = ERROR
      KMAXU1 = KMAXU + 1
      TBETA(J) = CDABS(BETA(KMAXU1))*ERROR
C
C     AFTER COMPUTING EACH OF THE T-EIGENVECTORS,
C     CHECK THE SIZE OF THE ERROR ESTIMATE, ERROR.
C     IF THIS ESTIMATE IS NOT AS SMALL AS DESIRED AND
C     |MA(J)| < ML(J), ATTEMPT TO INCREASE THE SIZE OF |MA(J)|
C     AND REPEAT THE T-EIGENVECTOR COMPUTATIONS.
C
      IF(ERROR.LT.ERTOL.OR.TFLAG.EQ.1)  GO TO 680
C
      IF(ERROR.GE.ERRMIN) GO TO 580
C     LAST COMPONENT IS LESS THAN MINIMAL TO DATE
      ERRMIN = ERROR
      MABEST = MA(J)
  580 CONTINUE
C
      IF(MA(J).GT.0)  ITEST = MA(J) + IDELTA(J)
      IF(MA(J).LT.0)  ITEST = -(IABS(MA(J)) + IDELTA(J))
      IF(IABS(ITEST).LE.ML(J).AND.ICOUNT.LE.10) GO TO 600
C     NEW MA(J) IS GREATER THAN MAXIMUM ALLOWED.
      IF(ERCONT.EQ.0.OR.MABEST.EQ.MPMIN) GO TO 620
      TFLAG = 1
      MA(J) = MABEST
      MTOL = MTOL - KMAXU
      WRITE(6,590) MA(J)
  590 FORMAT(' 10 ORDERS WERE CONSIDERED.  NONE SATISFIED THE ERROR TEST
     1'/' THEREFORE USE THE BEST ORDER OBTAINED FOR THE EIGENVECTORS'
     1,I6)
      GO TO 530
C
  600 MA(J) = ITEST
C
      MT = IABS(MA(J))
      IF(IWRITE.EQ.1) WRITE(6,610) MT
  610 FORMAT(/' CHANGE SIZE OF T-MATRIX TO',I6,' RECOMPUTE T-EIGENVECTOR
     1')
C
      MTOL = MTOL - KMAXU
C
      GO TO 530
C
C       APPROPRIATE SIZE T-MATRIX WAS NOT OBTAINED
  620 CONTINUE
      WRITE(10,630) J,EVAL,MP(J)
  630 FORMAT(/' ON 10 INCREMENTS NOT ABLE TO IDENTIFY APPROPRIATE SIZE
     1T-MATRIX FOR'/
     1' EIGENVALUE(',I4,') =',2E20.12,' T-MULTIPLICITY =',I4/)
      IF(MP(J).GT.1) WRITE(10,640)
      IF(MP(J).LT.0)  WRITE(10,650)
```

```
      IF(MP(J).EQ.1) WRITE(10,660)                                      CSL06010
  640 FORMAT(/' ORDERS TESTED RANGED FROM (5*MEV/4*MP(J)) TO APPROXIMATECSL06020
     1LY'/'  (7*MEV)/(4*MP(J)'/)                                        CSL06030
  650 FORMAT(/' ORDERS TESTED RANGED FROM (5*MEV/8)  TO MEV'/)          CSL06040
  660 FORMAT(/' ORDERS TESTED RANGED FROM 8*MEV/9 TO APPROXIMATELY 11*MECSL06050
     1V/8'/)                                                            CSL06060
      WRITE(10,670)                                                     CSL06070
  670 FORMAT(' ALLOWING LARGER ORDERS FOR THE T-MATRICES MAY RESULT IN  CSL06080
     1 SUCCESS'/' BUT PROBABLY WILL NOT.  PROBLEM IS PROBABLY DUE TO'   CSL06090
     1 /' LACK OF CONVERGENCE OF GIVEN EIGENVALUE, CHECK THE ERROR ESTIMCSL06100
     1ATE')                                                             CSL06110
      MP(J) = MPMIN                                                     CSL06120
      MTOL = MTOL - KMAXU                                               CSL06130
      GO TO 690                                                         CSL06140
  680 NTVEC = NTVEC + 1                                                 CSL06150
C                                                                       CSL06160
  690 CONTINUE                                                          CSL06170
      NGOODC = NGOOD                                                    CSL06180
      GO TO 720                                                         CSL06190
C                                                                       CSL06200
C     COME HERE IF THERE IS NOT ENOUGH ROOM FOR ALL OF T-EIGENVECTORS   CSL06210
  700 NGOODC = J-1                                                      CSL06220
      WRITE(6,710) J,MTOL,MDIMTV                                        CSL06230
  710 FORMAT(/' NOT ENOUGH ROOM IN TVEC ARRAY FOR ',I4,'TH T-EIGENVECTORCSL06240
     1'/' TVEC-DIMENSION REQUESTED = ',I6,' BUT TVEC HAS DIMENSION ',I6/CSL06250
     1)                                                                 CSL06260
      IF(NGOODC.EQ.0) GO TO 1430                                        CSL06270
      MTOL = MTOL-KMAXU                                                 CSL06280
C                                                                       CSL06290
  720 CONTINUE                                                          CSL06300
C                                                                       CSL06310
C     THE LOOP ON T-EIGENVECTOR COMPUTATIONS IS COMPLETE.               CSL06320
C     WRITE OUT THE SIZE T-MATRICES THAT WILL BE USED FOR               CSL06330
C     THE RITZ VECTOR COMPUTATIONS.                                     CSL06340
C                                                                       CSL06350
      WRITE(10,730)                                                     CSL06360
  730 FORMAT(/' SIZES OF T-MATRICES THAT WILL BE USED IN THE RITZ COMPUTCSL06370
     1ATIONS'/5X,'J',13X,'REAL(GOODEV)',13X,'IMAG(GOODEV)',1X,'MA(J)')  CSL06380
C                                                                       CSL06390
      WRITE(10,740)   (J,GOODEV(J),MA(J), J=1,NGOOD)                    CSL06400
  740 FORMAT(I6,2E25.14,I6)                                             CSL06410
      WRITE(10,490)                                                     CSL06420
C                                                                       CSL06430
      WRITE(6,750) MTOL                                                 CSL06440
  750 FORMAT(/' THE CUMULATIVE LENGTH OF THE T-EIGENVECTORS IS',I18)    CSL06450
C                                                                       CSL06460
      WRITE(6,760) NTVEC,NGOOD                                          CSL06470
  760 FORMAT(/I6,' T-EIGENVECTORS OUT OF',I6,' REQUESTED WERE COMPUTED')CSL06480
C                                                                       CSL06490
C     SAVE THE T-EIGENVECTORS ON FILE 11?                               CSL06500
      IF(TVSTOP.NE.1.AND.SVTVEC.EQ.0) GO TO 820                         CSL06510
C                                                                       CSL06520
      WRITE(11,770) NTVEC,MTOL,MATNO,SVSEED                             CSL06530
  770 FORMAT(I6,3I12,' = NTVEC,MTOL,MATNO,SVSEED')                      CSL06540
C                                                                       CSL06550
      DO 800 J=1,NGOODC                                                 CSL06560
C     IF MP(J) = MPMIN THEN NO SUITABLE T-EIGENVECTOR IS AVAILABLE      CSL06570
C     FOR THAT EIGENVALUE.                                              CSL06580
      IF(MP(J).EQ.MPMIN) WRITE(11,780) J,MA(J),GOODEV(J),MP(J)          CSL06590
  780 FORMAT(2I6,2E20.12,I6/' TH EIGVAL,T-SIZE,EVALUE,FLAG,NO EIGVEC')  CSL06600
      IF(MP(J).NE.MPMIN) WRITE(11,790) J,MA(J),GOODEV(J),MP(J)          CSL06610
  790 FORMAT(I6,I6,2E20.12,I6/' T-EIGVEC,SIZE T,EVALUE OF A,MP(J)')     CSL06620
      IF(MP(J).EQ.MPMIN) GO TO 800                                      CSL06630
      KI = MINT(J)                                                      CSL06640
      KF = MFIN(J)                                                      CSL06650
C                                                                       CSL06660
```

```
      WRITE(11,260) (TVEC(K), K=KI,KF)
C
  800 CONTINUE
C
      IF(TVSTOP.NE.1) GO TO 820
C
      WRITE(6,810) TVSTOP, NTVEC,NGOOD
  810 FORMAT(/' USER SET TVSTOP = ',I1/
     1' THEREFORE PROGRAM TERMINATES AFTER T-EIGENVECTOR COMPUTATIONS'/
     1' T-EIGENVECTORS THAT WERE COMPUTED ARE SAVED ON FILE 11'/
     1I8,' T-EIGENVECTORS WERE COMPUTED OUT OF',I7,' REQUESTED'/)
C
      GO TO 1550
C
  820 CONTINUE
C     IF NOT ABLE TO COMPUTE ALL THE REQUESTED T-EIGENVECTORS
C     CONTINUE WITH THE LANCZOS VECTOR COMPUTATIONS ANYWAY?
C
      IF(NTVEC.NE.NGOOD.AND.LVCONT.EQ.0) GO TO 1450
C
C     COMPUTE THE MAXIMUM SIZE OF THE T-MATRIX USED FOR THOSE
C     EIGENVALUES WITH GOOD ERROR ESTIMATES.
C
      KMAXU = 0
      DO 830 J = 1,NGOODC
      MT = IABS(MA(J))
      IF(MT.LT.KMAXU.OR.MP(J).EQ.MPMIN) GO TO 830
      KMAXU = MT
  830 CONTINUE
C
      IF(KMAXU.EQ.0) GO TO 1490
C
      WRITE(6,840) KMAXU
  840 FORMAT(/I6,' = LARGEST SIZE T-MATRIX TO BE USED IN THE RITZ VECTOR
     1 COMPUTATIONS')
C
C     COUNT THE NUMBER OF RITZ VECTORS NOT BEING COMPUTED
      MREJEC = 0
      DO 850 J=1,NGOODC
  850 IF(MP(J).EQ.MPMIN)  MREJEC = MREJEC + 1
      MREJET = MREJEC + (NGOOD-NGOODC)
      IF(MREJET.NE.0) WRITE(6,860) MREJET
  860 FORMAT(/' RITZ VECTORS ARE NOT COMPUTED FOR',I6,' OF THE EIGENVALU
     1ES'/)
      NACT = NGOODC - MREJEC
      WRITE(6,870) NGOOD,NTVEC,NACT
  870 FORMAT(/I6,' RITZ VECTORS WERE REQUESTED'/I6,' T-EIGENVECTORS WERE
     1COMPUTED'/I6,' RITZ VECTORS WILL BE COMPUTED'/)
C     CHECK IF THERE ARE ANY RITZ VECTORS TO COMPUTE
      IF(MREJEC.EQ.NGOODC) GO TO 1470
C
C     CONTINUE WITH THE LANCZOS VECTOR COMPUTATIONS?
      IF(LVCONT.EQ.0.AND.MREJEC.NE.0) GO TO 1450
C
C     NOW COMPUTE THE RITZ VECTORS.  REGENERATE THE
C     LANCZOS VECTORS.
C
      DO 880 I = 1,NMAX
  880 RITVEC(I) = ZEROC
C
C-----------------------------------------------------------------------
C     REGENERATE THE STARTING VECTOR. THIS MUST BE GENERATED AND
C     NORMALIZED PRECISELY THE WAY IT WAS DONE IN THE EIGENVALUE
C     COMPUTATIONS, OTHERWISE THERE WILL BE A MISMATCH BETWEEN
C     THE T-EIGENVECTORS THAT HAVE BEEN COMPUTED FROM THE T-MATRICES
C     READ IN FROM FILE 2 AND THE LANCZOS VECTORS THAT ARE
```

```
C     BEING REGENERATED.
C
      IIL = SVSEED
      CALL GENRAN(IIL,G,N)
C
C-----------------------------------------------------------------------
C
      DO 890 I = 1,N
  890 GR(I) = G(I)
C
C-----------------------------------------------------------------------
C
      CALL GENRAN(IIL,G,N)
C
C-----------------------------------------------------------------------
C
      DO 900 I = 1,N
  900 GC(I) = G(I)
C
      DO 910 I = 1,N
  910 V2(I) = DCMPLX(GR(I),GC(I))
C
C-----------------------------------------------------------------------
      CALL INPRDC(V2,V2,SUMC,N)
C-----------------------------------------------------------------------
C
      SUMC = ONE/CDSQRT(SUMC)
      DO 920 I = 1,N
      V1(I) = ZEROC
  920 V2(I) = V2(I)*SUMC
C
C     LOOP FOR GENERATING REQUIRED RITZ VECTORS (IVEC = 1,KMAXU)
C
      IVEC = 1
      BATA = ZEROC
C
      GO TO 980
C
  930 CONTINUE
C
C-----------------------------------------------------------------------
C
C    CMATV(V2,V1,BATA) CALCULATES  V1 = A*V2 - BATA*V1
      CALL CMATV(V2,V1,BATA)
      CALL INPRDC(V2,V1,ALFA,N)
C
C-----------------------------------------------------------------------
C
      DO 940 J=1,N
  940 V1(J) = V1(J)-ALFA*V2(J)
C
C-----------------------------------------------------------------------
      CALL INPRDC(V1,V1,BATA,N)
C-----------------------------------------------------------------------
C
      BATA = CDSQRT(BATA)
      SUMC = ONE/BATA
C
      TEMPC = BETA(IVEC)
      TEMP = CDABS(BATA - TEMPC)/CDABS(TEMPC)
      IF (TEMP.LT.1.0D-10)GO TO 960
C
C     IF THE BETA BEING REGENERATED DO NOT MATCH THE HISTORY FILE
C     THEN SOMETHING IS WRONG IN THE LANCZOS VECTOR GENERATION
C     AND PROGRAM TERMINATES FOR USER TO CORRECT THE PROBLEM
C     WHICH MUST BE IN THE STARTING VECTOR GENERATION OR IN
C     THE MATRIX-VECTOR MULTIPLY SUBROUTINE CMATV SUPPLIED.
```

```
C     THIS PART OF THE COMPUTATIONS MUST BE IDENTICAL TO THE
C     CORRESPONDING PART IN THE EIGENVALUE COMPUTATIONS.
C
      WRITE(6,950) IVEC,BATA,BETA(IVEC),TEMP
  950 FORMAT(/2X,'IVEC',16X,'BATA',10X,'BETA(IVEC)',14X,'RELDIF'/I6,
     13E20.12/' IN LANCZOS VECTOR REGENERATION THE ENTRIES OF THE TRIDIA
     1GONAL MATRICES BEING'/' GENERATED ARE NOT THE SAME AS THOSE IN THE
     1 MATRIX SUPPLIED ON FILE 2.'/' THEREFORE SOMETHING IS BEING INITIA
     1LIZED OR COMPUTED DIFFERENTLY FROM THE WAY'/' IT WAS COMPUTED IN T
     1HE EIGENVALUE COMPUTATIONS'/' THE PROGRAM TERMINATES FOR THE USER
     1TO DETERMINE WHAT THE PROBLEM IS'/)
      GO TO 1550
C
C
  960 CONTINUE
      DO 970 J = 1,N
      TEMPC = SUMC*V1(J)
      V1(J) = V2(J)
  970 V2(J) = TEMPC
C
  980 CONTINUE
C
      LFIN = 0
      DO 1000 J = 1,NGOODC
      LL = LFIN
      LFIN = LFIN + N
C
      IF(IABS(MA(J)).LT.IVEC.OR.MP(J).EQ.MPMIN) GO TO 1000
      II = IVEC + MINT(J) - 1
      TEMPC = TVEC(II)
C     II IS THE (IVEC)TH COMPONENT OF THE T-EIGENVECTOR CONTAINED
C     IN TVEC(MINT(J)).
C
      DO 990 K = 1,N
      LL = LL + 1
  990 RITVEC(LL) = TEMPC*V2(K) + RITVEC(LL)
C
 1000 CONTINUE
C
      IVEC = IVEC + 1
      IF (IVEC.LE.KMAXU) GO TO 930
C
C
C     RITZVECTOR GENERATION IS COMPLETE. NORMALIZE EACH RITZVECTOR.
C     NOTE THAT IF CERTAIN RITZ VECTORS WERE NOT COMPUTED THEN THE
C     CORRESPONDING PORTION OF THE RITVEC ARRAY WAS NOT UTILIZED.
C
      LFIN = 0
      DO 1050 J = 1,NGOODC
C
      KK = LFIN
      LFIN = LFIN + N
      IF(MP(J).EQ.MPMIN) GO TO 1050
C
      DO 1010 K = 1,N
      KK = KK + 1
 1010 V2(K) = RITVEC(KK)
C
C-----------------------------------------------------------------------
      CALL INPRDC(V2,V2,SUMC,N)
C-----------------------------------------------------------------------
C
      SUMC = CDSQRT(SUMC)
      RNORM(J) = CDABS(SUMC)
      TEMP = DABS(ONE-RNORM(J))
      SUMC = DCMPLX(ONE,ZERO)/SUMC
C
```

```
      KK = LFIN - N
      DO 1020 K = 1,N
      KK = KK + 1
      V2(K) = SUMC*V2(K)
 1020 RITVEC(KK) = V2(K)
C
C     COMPUTE THE 'REAL' NORM
C
C-----------------------------------------------------------------------
      CALL CINPRD(V2,V2,SUM,N)
C-----------------------------------------------------------------------
C
      IF (IWRITE.NE.0) WRITE(6,1030) J,GOODEV(J)
 1030 FORMAT(/I5,' TH EIGENVALUE CONSIDERED = ',2E20.12/)
C
      IF (IWRITE.NE.0) WRITE(6,1040) TERR(J),TBETA(J),RNORM(J),SUM
 1040 FORMAT(' NORM OF ERROR IN T-EIGENVECTOR = ',E14.3/
     1 ' CDABS(BETA(MA(J)+1)*U(MA(J))) ',E14.3/
     1 ' CDABS(EUCLIDEAN-NORM(RITVEC))  = ',E14.3/
     1 ' HERMITIAN-NORM(RITVEC)**2  = ',E14.3/)
C
      LINT = LFIN - N + 1
      EVAL = GOODEV(J)
C
C-----------------------------------------------------------------------
C
      CALL CMATV(RITVEC(LINT),V2,EVAL)
C
C-----------------------------------------------------------------------
C
C     COMPUTE ERROR IN RITZ VECTOR CONSIDERED AS A EIGENVECTOR OF A.
C     V2 = A*RITVEC - EVAL*RITVEC
C
C-----------------------------------------------------------------------
      CALL CINPRD(V2,V2,SUM,N)
C-----------------------------------------------------------------------
C
      SUM = DSQRT(SUM)
      ERR(J) = SUM
      GAP = ABS(AMINGP(J))
      ERRDGP(J) = SUM/GAP
C
 1050 CONTINUE
C
C
C     RITZVECTORS ARE NORMALIZED AND ERROR ESTIMATES ARE IN ERR ARRAY
C     AND IN ERRDGP ARRAY. STORE EVERYTHING
C
C
      WRITE(9,1060)
 1060 FORMAT(3X,'REAL(GOODEV)',3X,'IMAG(GOODEV)',1X,'MA(J)',7X,'AMINGAP'
     1 ,4X,'AERROR',2X,'AERR/GAP',4X,'TERROR')
C
      WRITE(13,1070)
 1070 FORMAT(8X,'REAL(GOODEV)',8X,'IMAG(GOODEV)',2X,'RITZNORM',3X,'AMING
     1AP',2X,'TBETA(J)',2X,'TLAST(J)')
C
      DO 1100 J=1,NGOODC
C
      IF(MP(J).EQ.MPMIN) GO TO 1100
C
      WRITE(9,1080)GOODEV(J),MA(J),AMINGP(J),ERR(J),ERRDGP(J),TERR(J)
 1080 FORMAT(2E15.8,I6,E14.6,3E10.3)
C
      WRITE(13,1090) GOODEV(J),RNORM(J),AMINGP(J),TBETA(J),TLAST(J)
 1090 FORMAT(2E20.12,4E10.3)
C
```

```
 1100 CONTINUE
C
      IF(MREJEC.EQ.0) GO TO 1180
      WRITE(9,1110)
 1110 FORMAT(/' RITZ VECTORS WERE NOT COMPUTED FOR THE FOLLOWING EIGENVA
     1LUES'/' EITHER BECAUSE THEY HAD NOT CONVERGED OR BECAUSE THE ERROR
     1 ESTIMATE'/'  WAS NOT AS SMALL AS DESIRED'/)
C
      WRITE(13,1120)
 1120 FORMAT(/' RITZ VECTORS WERE NOT COMPUTED FOR THE FOLLOWING EIGENVA
     1LUES'/' EITHER BECAUSE THEY HAD NOT CONVERGED OR BECAUSE'/' THE ER
     1ROR ESTIMATE WAS NOT AS SMALL AS DESIRED'/)
C
      DO 1170 J = 1,NGOODC
      IF(MP(J).NE.MPMIN) GO TO 1170
C     WRITE OUT MESSAGE FOR EACH EIGENVALUE FOR WHICH NO EIGENVECTOR
C     WAS COMPUTED.
C
      WRITE(9,1130)
 1130 FORMAT(6X,'GOODEV(J)',3X,'MA(J)',5X,'AMINGP(J)',6X,'TLAST(J)',3X,
     1'MP(J)')
      WRITE(9,1140) GOODEV(J),MA(J),AMINGP(J),TBETA(J),MP(J)
 1140 FORMAT(2E15.8,I8,2E14.4,I8)
C
      WRITE(13,1150)
 1150 FORMAT(6X,'REAL(GOODEV(J))',6X,'IMAG(GOODEV(J))',4X,'MA(J)',3X,
     1'MP(J)')
      WRITE(13,1160) GOODEV(J),MA(J),MP(J)
 1160 FORMAT(2E15.8,2I8)
C
 1170 CONTINUE
 1180 CONTINUE
C
      WRITE(9,1190)
 1190 FORMAT(/' ABOVE ARE ERROR ESTIMATES FOR THE A AND T EIGENVECTORS'/
     1 ' ASSOCIATED WITH THE GOODEV LISTED IN COLUMN 1'/
     1 ' AERROR = NORM(A*X - EV*X)  TERROR = NORM(T*Y - EV*Y) '/
     1 ' WHERE T = T(1,MA(J))    X = RITZ VECTOR = V*Y  V = SUCCESSIVE'/
     1 ' LANCZOS VECTORS. A MINGAP = GAP TO NEAREST A-EIGENVALUE'//)
C
      WRITE(13,1200)
 1200 FORMAT(/' ABOVE ARE ERROR ESTIMATES FOR THE A AND T EIGENVECTORS'/
     1 ' ASSOCIATED WITH THE GOODEV LISTED IN COLUMN 1'/
     1 ' AERROR = NORM(A*X-EV*X)  TERROR = NORM(T*Y-EV*Y) WHERE'
     1 /' T = T(1,MA(J))    X = RITZ VECTOR = V*Y  V = SUCCESSIVE '/
     1 ' LANCZOS VECTORS. A MINGAP = GAP TO NEAREST A-EIGENVALUE'/
     1 ' AERROR AND TERROR ARE GIVEN IN FILE 9. RNORM = NORM(X)'/
     1 ' BETA(M+1)*ABS(Y(M)) IS AN ESTIMATOR OF NORM(A*X-EV*X)'//)
C
C     NUMBER OF RITZ VECTORS COMPUTED
      NCOMPU = NGOODC - MREJEC
      WRITE(12,1210) N,NCOMPU,NGOODC,MATNO
 1210 FORMAT(3I6,I12,' SIZE A, NO.RITZVECS, NO.EVALUES,MATNO')
C
      LFIN = 0
      DO 1270 J = 1,NGOODC
      LINT = LFIN + 1
      LFIN = LFIN + N
C
      IF(MP(J).EQ.MPMIN) GO TO 1250
C     RITZ VECTOR WAS COMPUTED
      WRITE(12,1220) J, GOODEV(J), MP(J)
 1220 FORMAT(I6,4X,2E20.12,I6,' J, EIGENVAL, MP(J)')
C
      WRITE(12,1230) ERR(J),ERRDGP(J)
 1230 FORMAT(2E15.5,' = NORM(A*Z-EVAL*Z) AND  NORM(A*Z-EVAL*Z)/MINGAP')
C
```

```
      WRITE(12,1240) (RITVEC(LL), LL=LINT,LFIN)                         CSL10010
C1240 FORMAT(4Z20)                                                      CSL10020
 1240 FORMAT(2(2E20.12))                                                CSL10030
      GO TO 1270                                                        CSL10040
C     NO RITZ VECTOR WAS COMPUTED FOR THIS EIGENVALUE                   CSL10050
 1250 WRITE(12,1260) J,GOODEV(J),MP(J)                                  CSL10060
 1260 FORMAT(I6,4X,E20.12,I6,' J,EIGVALUE,NO RITZ VECTOR COMPUTED')     CSL10070
C                                                                       CSL10080
 1270 CONTINUE                                                          CSL10090
C                                                                       CSL10100
C     DID ANY T-MATRICES INCLUDE OFF-DIAGONAL ENTRIES SMALLER THAN      CSL10110
C     DESIRED, AS SPECIFIED BY BTOL?                                    CSL10120
C                                                                       CSL10130
      IF(IB.GT.0) GO TO 1300                                            CSL10140
      WRITE(6,1280) KMAXU                                               CSL10150
 1280 FORMAT(/' FOR LARGEST T-MATRIX CONSIDERED',I7,' CHECK THE SIZE OF CSL10160
     1BETAS')                                                           CSL10170
C                                                                       CSL10180
C-----------------------------------------------------------------------CSL10190
C                                                                       CSL10200
      CALL TNORM(ALPHA,BETA,BKMIN,TEMP,KMAXU,IBMT)                      CSL10210
C                                                                       CSL10220
C-----------------------------------------------------------------------CSL10230
C                                                                       CSL10240
      IF(IBMT.LT.0) WRITE (6,1290)                                      CSL10250
 1290 FORMAT(/' WARNING THE T-MATRICES FOR ONE OR MORE OF THE EIGENVALUECSL10260
     1S CONSIDERED'/' HAD AN OFF-DIAGONAL ENTRY THAT WAS SMALLER THAN THCSL10270
     1E BETA TOLERANCE THAT WAS SPECIFIED'/)                            CSL10280
 1300 CONTINUE                                                          CSL10290
C                                                                       CSL10300
       GO TO 1550                                                       CSL10310
C                                                                       CSL10320
 1310 WRITE(6,1320) NGOOD,NMAX,MDIMRV                                   CSL10330
 1320 FORMAT(/I4,' RITZ VECTORS WERE REQUESTED BUT THE REQUIRED DIMENSIOCSL10340
     1N',I6/' IS LARGER THAN THE USER-SPECIFIED DIMENSION OF RITVEC',I6 CSL10350
     1/' THEREFORE, THE EIGENVECTOR PROCEDURE TERMINATES FOR THE USER TOCSL10360
     1 INTERVENE')                                                      CSL10370
C                                                                       CSL10380
      GO TO 1550                                                        CSL10390
C                                                                       CSL10400
 1330 WRITE(6,1340) NOLD,N,MATOLD,MATNO                                 CSL10410
 1340 FORMAT(/' PARAMETERS READ FROM FILE 3 DO NOT AGREE WITH USER-SPECICSL10420
     1FIED'/' PARAMETERS, NOLD,N,MATOLD,MATNO = '/2I6,2I12/             CSL10430
     1' THEREFORE PROGRAM TERMINATES FOR USER TO RESOLVE DIFFERENCES'/) CSL10440
C                                                                       CSL10450
      GO TO 1550                                                        CSL10460
C                                                                       CSL10470
 1350 WRITE(6,1360)                                                     CSL10480
 1360 FORMAT(/' PARAMETERS IN ALPHA,BETA FILE READ IN DO NOT AGREE WITH CSL10490
     1 THOSE'/' SPECIFIED BY THE USER.  THEREFORE, THE PROGRAM TERMINATECSL10500
     1S FOR'/' THE USER TO RESOLVE THE DIFFERENCES'/)                   CSL10510
C                                                                       CSL10520
      GO TO 1550                                                        CSL10530
C                                                                       CSL10540
 1370 WRITE(6,1380) KMAX,MEV                                            CSL10550
 1380 FORMAT(/' IN ALPHA, BETA HISTORY HEADER KMAX =',I6/               CSL10560
     1' BUT EIGENVALUES WERE COMPUTED AT MEV = ',I6,' PROGRAM STOPS'/)  CSL10570
C                                                                       CSL10580
      GO TO 1550                                                        CSL10590
C                                                                       CSL10600
 1390 WRITE(6,1400)                                                     CSL10610
 1400 FORMAT(/' PROGRAM COMPUTED 1ST GUESSES AT T-MATRIX SIZES'/' READ TCSL10620
     1HEM TO FILE 10, THEN TERMINATED AS REQUESTED.')                   CSL10630
      GO TO 1550                                                        CSL10640
C                                                                       CSL10650
 1410 WRITE(6,1420) MTOL, MDIMTV                                        CSL10660
 1420 FORMAT(/' PROGRAM TERMINATES BECAUSE THE MINIMAL TVEC DIMENSION ANCSL10670
```

```
     1TICIPATED',I7/'  IS LARGER THAN THE TVEC DIMENSION',I7,' SPECIFIED
     1 BY THE USER.'/'  USER MAY RESET THE TVEC DIMENSION AND RESTART TH
     1E PROGRAM')
      GO TO 1550
C
 1430 WRITE(6,1440)
 1440 FORMAT(/' PROGRAM TERMINATES BECAUSE NO SUITABLE T-EIGENVECTORS WE
     1RE IDENTIFIED'/' FOR ANY OF THE EIGENVALUES SUPPLIED.  PROBLEM COU
     1LD BE CAUSED BY'/'  TOO SMALL A TVEC DIMENSION OR SIMPLY BE THAT T
     1-EIGENVECTORS COULD'/'  NOT BE IDENTIFIED.  USER SHOULD CHECK OUTP
     1UT'/)
      GO TO 1550
C
 1450 WRITE(6,1460) LVCONT,NTVEC,NGOOD
 1460 FORMAT(/' LVCONT FLAG =',I2,' AND NUMBER ',I5,' OF T-EIGENVECTORS
     1 COMPUTED N.E.'/' NUMBER',I5,' REQUESTED SO PROGRAM TERMINATES'/)
      GO TO 1550
 1470 WRITE(6,1480)
 1480 FORMAT(/' PROGRAM TERMINATES WITHOUT COMPUTING ANY RITZ VECTORS'/
     1' BECAUSE ALL T-EIGENVECTORS COMPUTED WERE REJECTED AS NOT SUITABL
     1E'/' FOR THE RITZ VECTOR COMPUTATIONS.  PROBABLE CAUSE IS LACK OF
     1'/' CONVERGENCE OF THE EIGENVALUES'/)
       GO TO 1550
C
 1490 WRITE(6,1500)
 1500 FORMAT(/' PROGRAM INDICATES THAT IT IS NOT POSSIBLE TO COMPUTE ANY
     1 OF THE'/' REQUESTED EIGENVECTORS. THEREFORE PROGRAM TERMINATES')
      DO 1510 J=1,NGOODC
 1510 WRITE(6,1520)  J,GOODEV(J),MP(J)
 1520 FORMAT(/4X,' J',11X,'GOODEV(J)',4X,'MP(J)'/I6,2E20.12,I9)
      GO TO 1550
C
 1530 WRITE(6,1540) MBETA,KMAXN
 1540 FORMAT(/' PROGRAM TERMINATES BECAUSE THE STORAGE ALLOTTED FOR THE
     1BETA ARRAY',I8/' IS NOT SUFFICIENT FOR THE ENLARGED KMAX =',I8,' T
     1HAT THE PROGRAM WANTS'/' USER CAN ENLARGE THE ALPHA AND BETA ARRAY
     1S AND RERUN THE PROGRAM'/)
C
 1550 CONTINUE
C
      STOP
C-----END OF MAIN PROGRAM FOR COMPLEX SYMMETRIC EIGENVECTORS------------
      END
```

SECTION 7.5 LANCZS AND SAMPLE MATRIX-VECTOR MULTIPLY SUBROUTINES

```
C-----CSLEMULT--(COMPLEX SYMMETRIC MATRICES)---------------------------
C
C     CONTAINS SUBROUTINE LANCZS USED IN THE COMPLEX SYMMETRIC
C     VERSION OF THE LANCZOS PROCEDURES PLUS SAMPLE USPEC AND
C     CMATV SUBROUTINES.
C
C     PORTABILITY:
C     THESE PROGRAMS ARE NOT PORTABLE DUE TO THE USE OF COMPLEX*16
C     VARIABLES AND CORRESPONDING FUNCTIONS.  MOREOVER, THE PFORT
C     VERIFIER IDENTIFIED THE FOLLOWING ADDITIONAL NONPORTABLE
C     CONSTRUCTIONS:
C
C     1.  ENTRIES USED TO PASS THE STORAGE LOCATIONS OF THE
C         ARRAYS AND PARAMETERS NEEDED TO SPECIFY THE GIVEN MATRIX
C         FROM THE USPEC SUBROUTINE TO THE MATRIX-VECTOR MULTIPLY
C         SUBROUTINE CMATV.
C     2.  IN THE SAMPLE USPEC SUBROUTINES PROVIDED: THE FREE FORMAT
C         READ(8,*) AND THE FORMAT (20A4).  IN THE SAMPLE CMATV:
C         THE COMPUTATION OF INDICES:  IN THE AUXILIARY SUBROUTINE
C         USED FOR COMPUTING THE KNOWN EIGENVALUES OF TEST CLASS 2
C         MATRICES, THE DATA/MACHEP DEFINITION.
C
C-----LANCZS-COMPUTE THE LANCZOS TRIDIAGONAL MATRICES------------------
C
      SUBROUTINE LANCZS(MATVEC,V1,V2,ALPHA,BETA,
     1GR,GC,G,KMAX,MOLD1,N,IIX)
C
C----------------------------------------------------------------------
      COMPLEX*16 V1(1), V2(1), BATA, ZEROC, TEMP, SUMC
      COMPLEX*16  ALPHA(1), BETA(1)
      DOUBLE PRECISION SUM, ONE, ZERO, GR(1), GC(1)
      REAL  G(1)
      EXTERNAL MATVEC
C     COMPLEX*16  CDSQRT, DCMPLX
C----------------------------------------------------------------------
C
      ZERO = 0.D0
      ONE = 1.D0
      ZEROC = DCMPLX(ZERO,ZERO)
C
      IF(MOLD1.GT.1)GO TO 50
C
C    ALPHA/BETA GENERATION STARTS AT I = 1
C    MOLD1 = 1 SET V1 = 0. AND V2 = RANDOM UNIT VECTOR
      IIL=IIX
C
C----------------------------------------------------------------------
      CALL GENRAN(IIL,G,N)
C----------------------------------------------------------------------
C
      DO 10 I = 1,N
   10 GR(I) = G(I)
C
C----------------------------------------------------------------------
      CALL GENRAN(IIL,G,N)
C----------------------------------------------------------------------
C
      DO 20 I = 1,N
   20 GC(I) = G(I)
C
      DO 30 I = 1,N
   30 V2(I) = DCMPLX(GR(I),GC(I))
C
C----------------------------------------------------------------------
```

```
      CALL INPRDC(V2,V2,SUMC,N)
C-----------------------------------------------------------------------
C
      SUMC = ONE/CDSQRT(SUMC)
      DO 40 I = 1,N
      V1(I) = ZEROC
   40 V2(I) = V2(I)*SUMC
      BETA(1) = ZEROC
C
C     ALPHA BETA GENERATION LOOP
   50 CONTINUE
C
      DO 80 I=MOLD1,KMAX
      SUMC = BETA(I)
C
C-----------------------------------------------------------------------
C     MATVEC(V2,V1,SUMC) CALCULATES  V1 = A*V2 - SUMC*V1
      CALL MATVEC(V2,V1,SUMC)
      CALL INPRDC(V2,V1,SUMC,N)
C-----------------------------------------------------------------------
C
      ALPHA(I) = SUMC
      DO 60 J=1,N
   60 V1(J) = V1(J)-SUMC*V2(J)
C
C-----------------------------------------------------------------------
      CALL INPRDC(V1,V1,SUMC,N)
C-----------------------------------------------------------------------
C
      IN = I+1
      BATA = CDSQRT(SUMC)
      BETA(IN) = BATA
      SUMC = ONE/BATA
      DO 70 J=1,N
      TEMP = SUMC*V1(J)
      V1(J) = V2(J)
   70 V2(J) = TEMP
   80 CONTINUE
C     END ALPHA, BETA GENERATION LOOP
C
C-----END OF LANCZS-----------------------------------------------------
C
      RETURN
      END
C
C-----USPEC, AND CMATV FOR COMPLEX SYMMETRIC TEST MATRICES 1------------
C
C-----START OF USPEC-(COMPLEX SYMMETRIC TEST MATRICES 1)----------------
C
C     SUBROUTINE CSPEC(N,MATNO)
      SUBROUTINE USPEC(N,MATNO)
C
C-----------------------------------------------------------------------
      DOUBLE PRECISION  C0,C1,C2,HALF,ONE,SCR,SCI,ANGLE
      COMPLEX*16 SC,TC,CL0,CL1
      REAL EXPLAN(20)
      DOUBLE PRECISION DARCOS
C     COMPLEX*16 DCMPLX
C-----------------------------------------------------------------------
      HALF = 0.5D0
      ONE  = 1.0D0
C
C     READ IN PARAMETERS TO DEFINE MATRIX
C     MATRIX IS COMPLEX DIAGONAL SIMILITARY TRANSFORM OF THE BLOCK
C     TOEPLITZ POISSON MATRICES USED TO TEST REAL SYMMETRIC MATRICES.
C     THE REAL POISSON MATRIX HAS SYMMETRIC TOEPLITZ BLOCKS ALONG THE
C     DIAGONAL.  EACH ONE OF THESE HAS THE PARAMETER C2 ALONG THE
```

```
C     DIAGONAL AND -CO ABOVE AND BELOW THE DIAGONAL.  THE OFF-DIAGONAL
C     BLOCKS ARE DIAGONAL WITH DIAGONAL ENTRIES -C1.  EACH BLOCK IS
C     KX*KX AND THERE ARE KY BLOCKS.  A HERMITIAN VERSION IS OBTAINED
C     BY APPLYING A DIAGONAL SIMILARITY TRANSFORM TO THE ABOVE
C     MATRIX WHERE THE DIAGONAL MATRIX IS SUCH THAT ITS
C     DIAGONAL ENTRIES ARE (SC)**(K-1), K=1,...,N-1.
C     THIS HERMITIAN VERSION IS TURNED INTO A COMPLEX SYMMETRIC ONE
C     IN THE MATRIX VECTOR MULTIPLY BY TREATING THE BELOW DIAGONAL
C     ENTRIES AS BEING EQUAL TO THE ABOVE DIAGONAL ENTRIES RATHER
C     THAN THEIR COMPLEX CONJUGATES.
C
      READ(8,10) EXPLAN
   10 FORMAT(20A4)
      READ(8,*) NOLD,MATOLD
      WRITE(6,20) NOLD,MATOLD
   20 FORMAT(' ORDER OF MATRIX READ FROM FILE =',I6/' MATRIX NUMBER =',
     1I8)
C
C     TEST OF PARAMETER CORRECTNESS
      ITEMP = (NOLD-N)**2 + (MATNO-MATOLD)**2
C
      IF(ITEMP.EQ.0) GO TO 40
C
      WRITE(6,30)
   30 FORMAT(' PROGRAM TERMINATES BECAUSE EITHER ORDERS OF OR LABELS FOR
     1 MATRIX DISAGREE')
      GO TO 100
C
   40 CONTINUE
C
      READ(8,10) EXPLAN
      READ(8,*)  CO,KX,KY
      READ(8,10) EXPLAN
      READ(8,*) SCR
      ANGLE = DARCOS(SCR)
      SCI = DSIN(ANGLE)
      SC = DCMPLX(SCR,SCI)
      WRITE(6,50) SC
      WRITE(9,50) SC
   50 FORMAT(' GENERATOR OF DIAGONAL TRANSFORMATION ='/2E20.12)
C
      TC = SC
      DO 60 J=2,KX
   60 TC = SC*TC
      WRITE(6,70) TC
   70 FORMAT(' TC = ',2E20.12)
C
      N = KX*KY
      C2 = ONE
      C1 = HALF-CO
      CLO = -SC*CO
      CL1 = -TC*C1
C
      WRITE(6,80) N,KX,KY,C2,CO,C1
   80 FORMAT(/5X,'N',4X,'KX',4X,'KY',7X,'DIAGONAL',3X,'X-CODIAGONAL',
     1 3X,'Y-CODIAGONAL'/3I6,3E15.8/)
C
C-----------------------------------------------------------------------
        CALL HMATVE(C2,CLO,CL1,KX,KY)
C-----------------------------------------------------------------------
C
   90 CONTINUE
      RETURN
C
C-----END OF USPEC------------------------------------------------------
  100 STOP
```

```
      END
C
C-----START OF CSMATV (FOR TEST MATRICES 1)-----------------------------
C     CALCULATE U = A*W - SUMC*U FOR COMPLEX SYMMETRIC MATRICES
C     HERE WE HAVE TAKEN A HERMITIAN VERSION OF POISSON MATRICES
C     AND TURNED IT INTO A COMPLEX SYMMETRIC TEST PROBLEM (WHOSE
C     EIGENVALUES WE DO NOT KNOW)
C
C     SUBROUTINE CSMATV(W,U,CSUM)
      SUBROUTINE CMATV(W,U,CSUM)
C
C-----------------------------------------------------------------------
      DOUBLE PRECISION  C2
      COMPLEX*16 U(1),W(1)
      COMPLEX*16 CLO,CL1,CRO,CR1,CSUM
C-----------------------------------------------------------------------
C
      N = KX*KY
      KX1 = KX-1
      KY1 = KY-1
      CRO = CLO
      CR1 = CL1
C
      KK = 1
      U(KK)=(C2*W(KK)+CRO*W(KK+1)+CR1*W(KK+KX)) - CSUM*U(KK)
      KK = KX
      U(KK)=(C2*W(KK)+CLO*W(KK-1)+CR1*W(KK+KX)) - CSUM*U(KK)
      KK = N - KX + 1
      U(KK)=(C2*W(KK)+CRO*W(KK+1)+CL1*W(KK-KX)) - CSUM*U(KK)
      KK = N
      U(KK)=(C2*W(KK)+CLO*W(KK-1)+CL1*W(KK-KX)) - CSUM*U(KK)
C
      DO 10 J = 2,KX1
      KK = J
      U(KK)=(C2*W(KK)+CLO*W(KK-1)+CRO*W(KK+1)+CR1*W(KK+KX))-CSUM*U(KK)
      KK = J+N-KX
      U(KK)=(C2*W(KK)+CLO*W(KK-1)+CRO*W(KK+1)+CL1*W(KK-KX))-CSUM*U(KK)
   10 CONTINUE
C
      DO 30 J = 2,KY1
      KK = (J-1)*KX + 1
      U(KK)=(C2*W(KK)+CRO*W(KK+1)+CL1*W(KK-KX)+CR1*W(KK+KX))-CSUM*U(KK)
      KK = J*KX
      U(KK)=(C2*W(KK)+CLO*W(KK-1)+CL1*W(KK-KX)+CR1*W(KK+KX))-CSUM*U(KK)
      DO 20 I = 2,KX1
      KK = (J-1)*KX + I
      U(KK)=(C2*W(KK)+CLO*W(KK-1)+CRO*W(KK+1)+CL1*W(KK-KX)
     1 +CR1*W(KK+KX)) - CSUM*U(KK)
   20 CONTINUE
   30 CONTINUE
C
      RETURN
C
C-----------------------------------------------------------------------
      ENTRY HMATVE(C2,CLO,CL1,KX,KY)
C-----------------------------------------------------------------------
C
C-----END OF CSMATV-----------------------------------------------------
      RETURN
      END
C
C     BELOW IS USPEC AND CMATV FOR TEST MATRICES 2.  IN THIS CASE
C     THE EIGENVALUES ARE KNOWN AND WE COMPUTE THEM TO CHECK
C     VALUES OBTAINED FROM THE LANCZOS PROGRAMS.
C
C     USES 3 SUBROUTINES BELOW, USPEC CMATV EXEVG
C
```

```
C-----START OF USPEC (TEST MATRICES 2)-----------------------------------
C
C     SUBROUTINE USPEC(N,MATNO)
      SUBROUTINE CSPEC(N,MATNO)
C
C-----------------------------------------------------------------------
      COMPLEX*16 CPAR,CC0,CC1,CC2
      DOUBLE PRECISION   C0,C1,C2,HALF,ONE
      REAL  EXPLAN(20)
C     COMPLEX*16  DCMPLX
C-----------------------------------------------------------------------
C     IVEC   = (0,-1,1) MEANS
C              (0) ONLY SET ENTRY FOR CMATV
C              (-1) CALCULATE EXACTEV AND MINGAPS AND STOP.
C              (1) CALCULATE EXACTEV AND MINGAPS AND THEN CONTINUE.
C-----------------------------------------------------------------------
      HALF = 0.5D0
      ONE  = 1.0D0
      CPAR = DCMPLX(ONE,ONE)
C-----------------------------------------------------------------------
C     READ USER-SPECIFIED PARAMETERS FROM INPUT FILE 8 (FREE FORMAT)
C
      READ(8,10) EXPLAN
      WRITE(6,10) EXPLAN
   10 FORMAT(20A4)
C
      READ(8,10) EXPLAN
      READ(8,*) KX,KY,IVEC,C0
      N = KX*KY
      C1 = HALF-C0
      C2 = ONE
      CC0 = CPAR*C0
      CC1 = CPAR*C1
      CC2 = CPAR*C2
C
      WRITE(6,20) N,KX,KY,C2,C0,C1,CPAR
   20 FORMAT(/5X,'N',4X,'KX',4X,'KY',7X,'DIAGONAL',3X,'X-CODIAGONAL',
     1 3X,'Y-CODIAGONAL'/3I6,3E15.8/7X,' COMPLEX SCALAR MULTIPLIER'/
     13X,2E15.4)
C
C-----------------------------------------------------------------------
      CALL CMATVE(CC0,CC1,CC2,KX,KY)
C-----------------------------------------------------------------------
C
      IF (IVEC.EQ.0) GO TO 30
C
C-----------------------------------------------------------------------
C     COMPUTE TRUE EIGENVALUES FOR CORRESPONDING REAL POISSON MATRIX
      CALL EXEVG(C0,C1,C2,KX,KY)
C-----------------------------------------------------------------------
C
      IF (IVEC.LT.0) STOP
C
   30 CONTINUE
C
C-----END OF USPEC-------------------------------------------------------
      RETURN
      END
C
C-----START OF CMATV (USES TEST MATRICES 2)------------------------------
C     CALCULATE U = A*W - SUM*U
C
C     SUBROUTINE CMATV(W,U,CSUM)
      SUBROUTINE CSRMAT(W,U,CSUM)
C
C-----------------------------------------------------------------------
      COMPLEX*16 U(1),W(1)
```

```
      COMPLEX*16 CCO,CC1,CC2,CLO,CL1,CRO,CR1,CSUM
C-----------------------------------------------------------------------
C
      N = KX*KY
      KX1 = KX-1
      KY1 = KY-1
      CRO = CCO
      CR1 = CC1
      CLO = CCO
      CL1 = CC1
C
      KK = 1
      U(KK)=(CC2*W(KK)+CRO*W(KK+1)+CR1*W(KK+KX)) - CSUM*U(KK)
      KK = KX
      U(KK)=(CC2*W(KK)+CLO*W(KK-1)+CR1*W(KK+KX)) - CSUM*U(KK)
      KK = N - KX + 1
      U(KK)=(CC2*W(KK)+CRO*W(KK+1)+CL1*W(KK-KX)) - CSUM*U(KK)
      KK = N
      U(KK)=(CC2*W(KK)+CLO*W(KK-1)+CL1*W(KK-KX)) - CSUM*U(KK)
C
      DO 10 J = 2,KX1
      KK = J
      U(KK)=(CC2*W(KK)+CLO*W(KK-1)+CRO*W(KK+1)+CR1*W(KK+KX))-CSUM*U(KK)
      KK = J+N-KX
      U(KK)=(CC2*W(KK)+CLO*W(KK-1)+CRO*W(KK+1)+CL1*W(KK-KX))-CSUM*U(KK)
   10 CONTINUE
C
      DO 30 J = 2,KY1
      KK = (J-1)*KX + 1
      U(KK)=(CC2*W(KK)+CRO*W(KK+1)+CL1*W(KK-KX)+CR1*W(KK+KX))-CSUM*U(KK)
      DO 20 I = 2,KX1
      KK = KK + 1
      U(KK)=(CC2*W(KK)+CLO*W(KK-1)+CRO*W(KK+1)+CL1*W(KK-KX)
     1 +CR1*W(KK+KX)) - CSUM*U(KK)
   20 CONTINUE
      KK = KK + 1
      U(KK)=(CC2*W(KK)+CLO*W(KK-1)+CL1*W(KK-KX)+CR1*W(KK+KX))-CSUM*U(KK)
   30 CONTINUE
C
      RETURN
C
C-----------------------------------------------------------------------
      ENTRY CMATVE(CCO,CC1,CC2,KX,KY)
C-----------------------------------------------------------------------
C
C-----END OF CMATV------------------------------------------------------
      RETURN
      END
C
C-----START OF EXEVG (COMPUTES EXACT EIGENVALUES FOR TEST MATRICES 2)---
C
      SUBROUTINE EXEVG(CO,C1,C2,KX,KY)
C
C-----------------------------------------------------------------------
      DOUBLE PRECISION U(2000),MACHEP
      DOUBLE PRECISION EPSM,CO,C1,C2,TO,T1,PIK,PIL,ONE,TWO,ATOLN,EE
      REAL G(2000)
      INTEGER MP(2000)
      REAL ABS
      DOUBLE PRECISION DABS, DARCOS, DFLOAT, DCOS, DMAX1
C-----------------------------------------------------------------------
      DATA MACHEP/Z3410000000000000/
      EPSM = 2.0DO*MACHEP
C-----------------------------------------------------------------------
      N = KX*KY
      ONE  = 1.0DO
      TWO  = 2.0DO
```

```
      T0 = DARCOS(-ONE)
      T1 = DFLOAT(KX+1)
      PIK = T0/T1
      T1 = DFLOAT(KY+1)
      PIL = T0/T1
C     GENERATE EXACT EIGENVALUES
      KP = 0
      DO 20 J = 1,KY
      T1 = PIL*DFLOAT(J)
      T0 = C2 - TWO*C1*DCOS(T1)
      DO 10 I = 1,KX
      KP = KP+1
      T1 = PIK*DFLOAT(I)
   10 U(KP) = T0 - TWO*C0*DCOS(T1)
   20 CONTINUE
C
C     ORDER U VECTOR BY INCREASING ALGEBRAIC SIZE
      DO 40 K = 2,N
      KM1 = K-1
      DO 30 L = 1,KM1
      JJ = K-L
      IF (U(JJ+1).GE.U(JJ)) GO TO 40
      T0 = U(JJ)
      U(JJ) = U(JJ+1)
   30 U(JJ+1) = T0
   40 CONTINUE
      ATOLN = DMAX1(DABS(U(1)),DABS(U(N)))*EPSM
C
      WRITE(9,50)
   50 FORMAT(' TRUE EIGENVALUES FOR POISSON'/)
C
      WRITE(9,60)N,KX,KY,C2,C0,C1,ATOLN
      WRITE(6,60) N,KX,KY,C2,C0,C1,ATOLN
   60 FORMAT(1X,'A-SIZE',2X,'X-DIM',2X,'Y-DIM'/3I7/
     1 5X,'A-DIAGONAL',3X,'X-CODIAGONAL',3X,'Y-CODIAGONAL',10X,'ATOLN'/
     2 4E15.8)
C
C     DETERMINE MULTIPLICITIES FOR EXACT EIGENVALUES
      I = 1
      IDEX = 1
      J = 1
      NEXACT = 0
   70 J = J+1
      IF (J.GT.N) GO TO 80
      EE = DABS(U(J)-U(I))
      IF (EE.GT.ATOLN) GO TO 80
      IDEX = IDEX+1
      GO TO 70
   80 NEXACT = NEXACT+1
      U(NEXACT) = U(I)
      MP(NEXACT) = IDEX
C     MP(K) = MULTIPLICITY OF KTH EIGENVALUE CLUSTER FOR A
      IDEX = 1
      I = J
      IF (I.GT.N) GO TO 90
      GO TO 70
   90 CONTINUE
C
C     MULTIPLICITIES HAVE BEEN DETERMINED
C     NEXACT = NUMBER OF DISTINCT A-EIGENVALUES
C
C
      WRITE(9,100)NEXACT
      WRITE(6,100)NEXACT
  100 FORMAT(I6,' = NUMBER OF TRUE A-EIGENVALUES WHICH ARE DISTINCT'/)
C
C     MINGAP CALCULATION FOR DISTINCT A-EIGENVALUES
```

```
      NM1 = NEXACT - 1
      G(NEXACT) = U(NM1)-U(NEXACT)
      G(1) = U(2)-U(1)
C
      DO 110 J = 2,NM1
      T0 = U(J)-U(J-1)
      T1 = U(J+1)-U(J)
      G(J) = T1
      IF (T0.LT.T1) G(J) = -T0
  110 CONTINUE
C
C     NEXACT DISTINCT A-EIGENVALUES ARE IN U IN ASCENDING ORDER
C     MP = MULTIPLICITIES OF THE DISTINCT EIGENVALUES OF A
C     G = TRUE MINIMUM GAP IN A FOR EACH OF THESE EIGENVALUES
C     G < 0 INDICATES THE LEFT-HAND GAP WAS MINIMAL.
C     OUTPUT MULTIPLICITIES, DISTINCT EVS, AND MINGAPS TO FILE 9
C
      WRITE(9,120)
  120 FORMAT(5X,'I',1X,'AMULT',5X,'TRUE A-EIGENVALUE(I)',
     1 3X,'A-MINGAP(I)')
C
      WRITE(9,130)(J,MP(J),U(J),G(J), J=1,NEXACT)
  130 FORMAT(2I6,E25.16,E14.3)
C
      WRITE(9,140)
  140 FORMAT(' NEXACT DISTINCT A-EIGENVALUES ARE IN ASCENDING ORDER'/
     1 ' AMULT = MULTIPLICITIES OF THE DISTINCT EIGENVALUES OF A.'/
     2 ' A-MINGAP(I) = TRUE MINIMUM GAP IN A FOR EACH EIGENVALUE.'/
     3 ' A-MINGAP(I).LT.0 INDICATES THE LEFT-HAND GAP WAS MINIMAL.'//)
C
C     WE ORDER U VECTOR BY INCREASING SIZE OF THE GAPS
C
      DO 150 K = 1,N
  150 MP(K) = K
C
      DO 170 K = 2,N
      KM1 = K-1
C
      DO 160 L = 1,KM1
      JJ = K - L
      IF (ABS(G(JJ+1)).GE.ABS(G(JJ))) GO TO 170
      EE = U(JJ)
      U(JJ) = U(JJ+1)
      U(JJ+1) = EE
      GG = G(JJ)
      G(JJ) = G(JJ+1)
      G(JJ+1) = GG
      IEE = MP(JJ)
      MP(JJ) = MP(JJ+1)
  160 MP(JJ+1) = IEE
C
  170 CONTINUE
C
      WRITE(9,180)
  180 FORMAT(5X,'K',6X,'A-MINGAP',5X,'TRUE A-EIGENVALUE(I)',2X,'A-EVNO')
C
      WRITE(9,190)(J,G(J),U(J),MP(J), J=1,NEXACT)
  190 FORMAT(I6,E14.3,E25.16,I8)
C
      WRITE(9,200)
  200 FORMAT(' NEXACT DISTINCT A-EIGENVALUES. GAPS IN ASCENDING ORDER'/
     2 ' A-MINGAP(I) = TRUE MINIMUM GAP IN A FOR EACH EIGENVALUE.'/
     3 ' A-MINGAP(I).LT.0 INDICATES THE LEFT-HAND GAP WAS MINIMAL.'/
     3 ' A-MATRIX IS BLOCK TRIDIAGONAL AND EACH DIAGONAL BLOCK IS OF ORD
     3ER NX.'/
     4 ' NX = NUMBER OF POINTS ON EACH X-LINE. THERE ARE NY DIAGONAL BLO
     4CKS.'/
```

```
     5 ' NY = NUMBER OF POINTS ON EACH Y-LINE.'/
     5 ' A-DIAGONAL    = A(K,K)'/
     6 ' X-CODIAGONAL = A(I,I+1)'/
     7 ' Y-CODIAGONAL = A(I,I+NX)'/
     8 ' ----- END OF FILE 9 EXACTEV----------------------------'//)
C
C-----END OF EXEVG------------------------------------------------------
C
      RETURN
      END
```

SECTION 7.6 OTHER SUBROUTINES USED BY THE PROGRAMS IN CHAPTER 7

```
C-----CSLESUB-(NONDEFECTIVE COMPLEX SYMMETRIC MATRICES)-----------------
C
C      NONPORTABLE CONSTRUCTIONS:
C      THESE SUBROUTINES ARE NOT PORTABLE DUE TO THE USE OF THE
C      COMPLEX*16 VARIABLES AND THE CORRESPONDING COMPLEX FUNCTIONS,
C      CDABS, DCMPLX, DREAL, DIMAG.  MOREOVER, IN SUBROUTINE
C      COMPEV THE NONPORTABLE FORMATS (4Z20) AND (20A4) ARE USED,
C      AND IN SUBROUTINE CMTQL1 THE MACHINE EPSILON IS INTRODUCED
C      VIA A NONPORTABLE DATA DEFINITION.
C
C      CONTAINS SUBROUTINES USED BY THE COMPLEX SYMMETRIC VERSION OF
C      THE LANCZOS EIGENVALUE/EIGENVECTOR CODES.
C
C      SUBROUTINES      COMPEV, CMTQL1, INVERR, TNORM, LUMP, ISOEV AND
C                       COMGAP ARE USED WITH THE LANCZOS EIGENVALUE
C                       PROGRAM CSLEVAL.  INVERM IS USED
C                       IN THE EIGENVECTOR PROGRAM CSLEVEC.  THE INNER
C                       PRODUCT SUBROUTINES CINPRD AND INPRDC ARE USED
C                       BY BOTH PROGRAMS.
C
C-----INVERSE ITERATION ON COMPLEX SYMMETRIC T(1,MEV)-------------------
C
      SUBROUTINE INVERR(ALPHA,BETA,V1,V2,VS,EPS,GR,GC,G,GG,MP,INTERC,
     1MEV,MMB,NDIS,NISO,N,IKL,IT,IWRITE)
C
C-----------------------------------------------------------------------
      COMPLEX*16  ALPHA(1),BETA(1),V1(1),V2(1),VS(1)
      COMPLEX*16  U,Z,X1,RATIO,BETAM,TEMP,ZEROC
      DOUBLE PRECISION EST,ESTR,ESTC,SUM,XU,NORM,TSUM,GSUM
      DOUBLE PRECISION  EPS,EPS3,EPS4,ZERO,ONE,GR(1),GC(1),GAP
      REAL G(1),GG(1)
      INTEGER  MP(1), INTERC(1)
      REAL  ABS
      DOUBLE PRECISION  DABS, DMIN1, DSQRT, DFLOAT, CDABS, DIMAG, DREAL
C     COMPLEX*16 DCMPLX
C-----------------------------------------------------------------------
C
C      COMPUTES ERROR ESTIMATES FOR COMPUTED ISOLATED GOOD T-EIGENVALUES
C      IN VS AND WRITES THESE EIGENVALUES AND ESTIMATES TO FILE 4.
C      BY DEFINITION A GOOD T-EIGENVALUE IS ISOLATED IF ITS CLOSEST
C      NEIGHBOR IS ALSO GOOD, OR IF ITS CLOSEST NEIGHBOR IS
C      SPURIOUS BUT THAT NEIGHBOR IS FAR ENOUGH AWAY.  SO
C      IN PARTICULAR, WE WILL COMPUTE ESTIMATES FOR ANY GOOD
C      T-EIGENVALUE THAT IS IN A CLUSTER OF GOOD T-EIGENVALUES.
C
C      USES INVERSE ITERATION ON T(1,MEV) SOLVING THE EQUATION
C      (T - X1*I)V2 = RIGHT-HAND SIDE (RANDOMLY-GENERATED)
C      FOR EACH SUCH GOOD T-EIGENVALUE X1.
C
C      PROGRAM REFACTORS T-X1*I ON EACH ITERATION OF INVERSE ITERATION.
C      TYPICALLY ONLY ONE ITERATION IS NEEDED PER T-EIGENVALUE X1.
C
C      ON ENTRY AND EXIT
C      MEV = ORDER OF T :  N = ORDER OF ORIGINAL MATRIX A
C      ALPHA, BETA CONTAIN THE NONZERO ENTRIES OF THE T-MATRIX
C      VS = COMPUTED DISTINCT EIGENVALUES OF T(1,MEV)
C      MP = T-MULTIPLICITY OF EACH T-EIGENVALUE IN VS. MP(I) = -1 MEANS
C           VS(I) IS A GOOD T-EIGENVALUE BUT THAT IT IS SITTING CLOSE TO
C           A SPURIOUS T-EIGENVALUE.  MP(I) = 0 MEANS VS(I) IS SPURIOUS.
C           ESTIMATES ARE COMPUTED ONLY FOR THOSE T-EIGENVALUES
C           WITH MP(I) = 1. FLAGGING WAS DONE IN SUBROUTINE ISOEV
C           PRIOR TO ENTERING INVERR.
C      NISO = NUMBER OF ISOLATED GOOD T-EIGENVALUES CONTAINED IN VS
C      NDIS =  NUMBER OF DISTINCT T-EIGENVALUES IN VS
```

```
C       IKL = SEED FOR RANDOM NUMBER GENERATOR                          CSL00650
C       EPS = 2. * MACHINE EPSILON                                      CSL00660
C                                                                       CSL00670
C       IN PROGRAM:                                                     CSL00680
C       ITER = MAXIMUM NUMBER OF INVERSE ITERATION STEPS ALLOWED FOR EACH CSL00690
C             X1.  ITER = IT ON ENTRY.                                  CSL00700
C       GR,GC = ARRAYS OF DIMENSION AT LEAST MEV + NISO.  USED TO STORE CSL00710
C           RANDOMLY-GENERATED RIGHT-HAND SIDE.  THIS IS NOT            CSL00720
C           REGENERATED FOR EACH X1. G IS ALSO USED TO STORE ERROR      CSL00730
C           ESTIMATES AS THEY ARE COMPUTED FOR LATER PRINTOUT.          CSL00740
C       V1,V2 = WORK SPACES USED IN THE FACTORIZATION OF T(1,MEV).      CSL00750
C       AT THE END OF THE INVERSE ITERATION COMPUTATION FOR X1, V2      CSL00760
C       CONTAINS THE UNIT EIGENVECTOR OF T(1,MEV) CORRESPONDING TO X1.  CSL00770
C       V1 AND V2 MUST BE OF DIMENSION AT LEAST MEV.                    CSL00780
C                                                                       CSL00790
C       ON EXIT                                                         CSL00800
C       GG(J) = MINIMUM GAP IN T(1,MEV) FOR EACH VS(J), J=1,NDIS        CSL00810
C       G(I) = |BETAM|*|V2(MEV)| = ERROR ESTIMATE FOR ISOLATED GOOD     CSL00820
C               T-EIGENVALUES, WHERE I = 1,NISO  AND  BETAM = BETA(MEV+1)CSL00830
C               T(1,MEV) CORRESPONDING TO ITH ISOLATED GOOD T-EIGENVALUE.CSL00840
C                                                                       CSL00850
C       IF FOR SOME X1 IT.GT.ITER THEN THE ERROR ESTIMATE IN G IS MARKED CSL00860
C       WITH A - SIGN.                                                  CSL00870
C                                                                       CSL00880
C       V2 = ISOLATED GOOD T-EIGENVALUES                                CSL00890
C       V1 = MINIMAL T-GAPS FOR THE T-EIGENVALUES IN V2.                CSL00900
C       THESE ARE CONSTRUCTED FOR WRITE-OUT PURPOSES ONLY AND NOT       CSL00910
C       NEEDED ELSEWHERE IN THE PROGRAM.                                CSL00920
C-----------------------------------------------------------------------CSL00930
C                                                                       CSL00940
C       LABEL OUTPUT FILE  4                                            CSL00950
        IF (MMB.EQ.1) WRITE(4,10)                                       CSL00960
     10 FORMAT(' INVERSE ITERATION ERROR ESTIMATES'/)                   CSL00970
C                                                                       CSL00980
C       FILE 6 (TERMINAL) OUTPUT OF ERROR ESTIMATES                     CSL00990
        IF (IWRITE.NE.0.AND.NISO.NE.0) WRITE(6,20)                      CSL01000
     20 FORMAT(/' INVERSE ITERATION ERROR ESTIMATES'/'  JISO',' JDIST',8X CSL01010
       1,'GOOD T-EIGENVALUE',4X,'BETAM*UM',5X,'TMINGAP')                CSL01020
C                                                                       CSL01030
C       INITIALIZATION AND PARAMETER SPECIFICATION                      CSL01040
        ZERO = 0.0D0                                                    CSL01050
        ONE = 1.0D0                                                     CSL01060
        ZEROC = DCMPLX(ZERO,ZERO)                                       CSL01070
        NG = 0                                                          CSL01080
        NISO = 0                                                        CSL01090
        ITER = IT                                                       CSL01100
        MP1 = MEV+1                                                     CSL01110
        MM1 = MEV-1                                                     CSL01120
        BETAM = BETA(MP1)                                               CSL01130
        BETA(MP1) = ZEROC                                               CSL01140
C                                                                       CSL01150
C       CALCULATE SCALE AND TOLERANCES                                  CSL01160
        TSUM = CDABS(ALPHA(1))                                          CSL01170
        DO 30 I = 2,MEV                                                 CSL01180
     30 TSUM = TSUM + CDABS(ALPHA(I)) + CDABS(BETA(I))                  CSL01190
C                                                                       CSL01200
        EPS3 = EPS*TSUM                                                 CSL01210
        EPS4 = DFLOAT(MEV)*EPS3                                         CSL01220
C                                                                       CSL01230
C       GENERATE SCALED RANDOM RIGHT-HAND SIDE                          CSL01240
        ILL = IKL                                                       CSL01250
C                                                                       CSL01260
C-----------------------------------------------------------------------CSL01270
        CALL GENRAN(ILL,G,MEV)                                          CSL01280
C-----------------------------------------------------------------------CSL01290
C                                                                       CSL01300
        DO 40 I = 1,MEV                                                 CSL01310
```

```
   40 GR(I) = G(I)
C
C-----------------------------------------------------------------------
      CALL GENRAN(ILL,G,MEV)
C-----------------------------------------------------------------------
C
      DO 50 I = 1,MEV
   50 GC(I) = G(I)
C
      GSUM = ZERO
      DO 60 I = 1,MEV
   60 GSUM = GSUM + DABS(GR(I)) + DABS(GC(I))
      GSUM = EPS4/GSUM
C
      DO 70 I = 1,MEV
      GR(I) = GSUM*GR(I)
   70 GC(I) = GSUM*GC(I)
C
C     LOOP ON ISOLATED GOOD T-EIGENVALUES IN VS (MP(I) = 1) TO
C     CALCULATE CORRESPONDING UNIT EIGENVECTOR OF T(1,MEV)
C
      DO 200 JEV = 1,NDIS
      IF (MP(JEV).EQ.0) GO TO 200
      NG = NG + 1
      IF (MP(JEV).NE.1) GO TO 200
      IT = 1
      NISO = NISO + 1
      X1 = VS(JEV)
C
C     INITIALIZE RIGHT HAND SIDE FOR INVERSE ITERATION
C     AND THE FLAG ON WHICH ROWS ARE INTERCHANGED
      DO 80 I = 1,MEV
      INTERC(I) = 0
   80 V2(I) = DCMPLX(GR(I),GC(I))
C
C     TRIANGULAR FACTORIZATION WITH NEAREST NEIGHBOR PIVOT
C     STRATEGY. INTERCHANGES ARE LABELLED BY SETTING INTERC = 1.
C
   90 CONTINUE
      U = ALPHA(1)-X1
      Z = BETA(2)
C
      DO 110 I = 2,MEV
      IF (CDABS(BETA(I)).GT.CDABS(U)) GO TO 100
C     NO INTERCHANGE
      V1(I-1) = Z/U
      V2(I-1) = V2(I-1)/U
      V2(I) = V2(I)-BETA(I)*V2(I-1)
      RATIO = BETA(I)/U
      U = ALPHA(I)-X1-Z*RATIO
      Z = BETA(I+1)
      GO TO 110
  100 CONTINUE
C     INTERCHANGE CASE
      RATIO = U/BETA(I)
      INTERC(I) = 1
      V1(I-1) = ALPHA(I)-X1
      U = Z-RATIO*V1(I-1)
      Z = -RATIO*BETA(I+1)
      TEMP = V2(I-1)
      V2(I-1) = V2(I)
      V2(I) = TEMP-RATIO*V2(I)
  110 CONTINUE
      IF (CDABS(U).EQ.ZERO) U = DCMPLX(EPS3,EPS3)
C
C     SMALLNESS TEST AND DEFAULT VALUE FOR LAST COMPONENT
C     PIVOT(I-1) = BETA(I) FOR INTERCHANGE CASE
```

```
C     (I-1,I+1) ELEMENT IN RIGHT FACTOR = BETA(I+1)
C     END OF FACTORIZATION AND FORWARD SUBSTITUTION
C
C     BACK SUBSTITUTION
      V2(MEV) = V2(MEV)/U
      DO 130 II = 1,MM1
      I = MEV-II
      IF (INTERC(I+1).EQ.1) GO TO 120
C     NO INTERCHANGE
      V2(I) = V2(I)-V1(I)*V2(I+1)
      GO TO 130
C     INTERCHANGE CASE
  120 CONTINUE
      V2(I) = (V2(I)-V1(I)*V2(I+1)-BETA(I+2)*V2(I+2))/BETA(I+1)
  130 CONTINUE
C
C     TESTS FOR CONVERGENCE OF INVERSE ITERATION
C     IF SUM |V2| COMPS. LE. 1 AND IT. LE. ITER DO ANOTHER INVIT STEP
C
      NORM = CDABS(V2(MEV))
      DO 140 II = 1,MM1
      I = MEV-II
  140 NORM = NORM + CDABS(V2(I))
C
      IF (NORM.GE.ONE) GO TO 160
      IT = IT+1
      IF (IT.GT.ITER) GO TO 160
      XU = EPS4/NORM
C
      DO 150 I = 1,MEV
  150 V2(I) = V2(I)*XU
C
      GO TO 90
C     ANOTHER INVERSE ITERATION STEP
C
C     INVERSE ITERATION FINISHED
C     NORMALIZE COMPUTED T-EIGENVECTOR : V2 = V2/||V2||
  160 CONTINUE
C
C-----------------------------------------------------------------------
      CALL CINPRD(V2,V2,SUM,MEV)
C-----------------------------------------------------------------------
C
      SUM = ONE/DSQRT(SUM)
C
      DO 170 II = 1,MEV
  170 V2(II) = SUM*V2(II)
C
C     SAVE ERROR ESTIMATE FOR LATER OUTPUT
      EST = CDABS(BETAM)*CDABS(V2(MEV))
      ESTR = DABS(DREAL(V2(MEV)))
      ESTC = DABS(DIMAG(V2(MEV)))
      GSUM = CDABS(BETAM)
      IF (IT.GT.ITER) EST = -EST
      G(NISO) = EST
      IF (IWRITE.EQ.0) GO TO 200
C
C     FILE 6 (TERMINAL) OUTPUT OF ERROR ESTIMATES.
      GAP = GG(JEV)
      WRITE(6,180) NISO,JEV,X1,EST,GAP
  180 FORMAT(2I6,2E20.12,2E12.3)
      WRITE(6,190) JEV, X1, EST,ESTR,ESTC
  190 FORMAT(I6,2E20.12,3E11.3)
C
  200 CONTINUE
C
C     END ERROR ESTIMATE LOOP ON ISOLATED GOOD T-EIGENVALUES.
```

```
C     GENERATE DISTINCT MINGAPS FOR T(1,MEV).  THIS IS USEFUL AS AN
C     INDICATOR OF THE GOODNESS OF THE INVERSE ITERATION ESTIMATES.
C     TRANSFER ISOLATED GOOD T-EIGENVALUES AND CORRESPONDING TMINGAPS
C     TO V2 AND V1 FOR OUTPUT PURPOSES ONLY.
C
      ISO = 0
      DO 210 J = 1,NDIS
      IF (MP(J).NE.1) GO TO 210
      ISO = ISO+1
      GR(ISO) = GG(J)
      V2(ISO) = VS(J)
  210 CONTINUE
      IF(NISO.EQ.0) GO TO 270
C
C     ERROR ESTIMATES ARE WRITTEN TO FILE 4
      WRITE(4,220)MEV,NDIS,NG,NISO,N,IKL,ITER,GSUM
  220 FORMAT(1X,'TSIZE',2X,'NDIS',1X,'NGOOD',2X,'NISO',1X,'ASIZE'/5I6/
     1 4X,'RHSEED',2X,'MXINIT',5X,'BETAM'/I10,I8,E10.3)
C
      WRITE(4,230)
  230 FORMAT(2X,'GOODEVNO',11X,'R(GOODEV)',11X,'I(GOODEV)',
     1 6X,'BETAM*UM',7X,'TMINGAP')
C
      ISPUR = 0
      I = 0
      DO 260 J = 1,NDIS
      IF(MP(J).NE.0) GO TO 240
      ISPUR = ISPUR + 1
      GO TO 260
  240 IF(MP(J).NE.1) GO TO 260
      I = I + 1
      IGOOD = J - ISPUR
      WRITE(4,250) IGOOD,V2(I),G(I),GR(I)
  250 FORMAT(I10,2E20.12,2E14.3)
  260 CONTINUE
      GO TO 290
C
  270 WRITE(4,280)
  280 FORMAT(/' THERE ARE NO ISOLATED T-EIGENVALUES SO NO ERROR ESTIMATE
     1S WERE COMPUTED')
C     RESTORE BETA(MEV+1) = BETAM
  290 BETA(MP1) = BETAM
C-----END OF INVERR------------------------------------------------------
      RETURN
      END
C-----START OF TNORM-----------------------------------------------------
C
      SUBROUTINE TNORM(ALPHA,BETA,BMIN,TMAX,MEV,IB)
C
C-----------------------------------------------------------------------
      COMPLEX*16  ALPHA(1),BETA(1)
      DOUBLE PRECISION  TMAX,BMIN,BMAX,BSIZE,BTOL,ABATA,AALFA
      DOUBLE PRECISION  DMAX1, CDABS
C     COMPLEX*16 DCMPLX
C-----------------------------------------------------------------------
C     IN REAL SYMMETRIC AND HERMITIAN VERSIONS TMAX IS USED
C     TO DETERMINE THE TOLERANCES USED IN THE T-MULTIPLICITY AND IN
C     THE SPURIOUS TESTS.  FOR THE COMPLEX SYMMETRIC CASE WE
C     HAVE TO COMPUTE ALL OF THE T-EIGENVALUES SO WE USE THEM INSTEAD
C     OF TMAX TO DETERMINE THESE TOLERANCES.  WE USE TMAX TO
C     CHECK THE RELATIVE SIZES OF THE BETA(K), K=1,...,MEV AS A
C     TEST ON THE LOCAL ORTHOGONALITY OF THE LANCZOS VECTORS.
C
C             TMAX = MAX (|ALPHA(I)|, |BETA(I)|,  I=1,MEV)
C             BMIN = MIN (|BETA(I)|, I=2,MEV)
C             BSIZE = BMIN/TMAX
C             |IB| = INDEX OF MINIMAL(BETA)
```

```
C             IB < 0 IF BMIN/TMAX < BTOL
C-----------------------------------------------------------------------
C     SPECIFY PARAMETERS
      IB = 2
      BTOL = BMIN
      BMIN = CDABS(BETA(2))
      BMAX = BMIN
      TMAX = CDABS(ALPHA(1))
C
      DO 20 I = 2,MEV
      ABATA = CDABS(BETA(I))
      IF (ABATA.GE.BMIN) GO TO 10
      IB = I
      BMIN = ABATA
   10 AALFA = CDABS(ALPHA(I))
      TMAX = DMAX1(TMAX,AALFA)
      BMAX = DMAX1(ABATA,BMAX)
   20 CONTINUE
      TMAX = DMAX1(BMAX,TMAX)
C
C     TEST OF LOCAL ORTHOGONALITY USING SCALED BETAS
      BSIZE = BMIN/TMAX
      IF (BSIZE.GE.BTOL) GO TO 40
C
C     DEFAULT.  BSIZE IS SMALLER THAN TOLERANCE BTOL SPECIFIED IN MAIN
C     PROGRAM.  PROGRAM TERMINATES FOR USER TO DECIDE WHAT TO DO
C     BECAUSE LOCAL ORTHOGONALITY OF THE LANCZOS VECTORS COULD BE
C     LOST.
C
      IB = -IB
      WRITE(6,30) MEV
   30 FORMAT(/' BETA TEST INDICATES POSSIBLE LOSS OF LOCAL ORTHOGONALITY
     1 OVER 1ST',I6,' LANCZOS VECTORS'/)
C
   40 CONTINUE
C
      WRITE(6,50) IB
   50 FORMAT(/' MINIMUM BETA RATIO OCCURS AT',I6,' TH BETA'/)
C
      WRITE(6,60) MEV,BMIN,TMAX,BSIZE
   60 FORMAT(/1X,'TSIZE',6X,'MIN BETA',5X,'TKMAX',6X,'MIN RATIO'/
     1 I6,E14.3,E10.3,E15.3/)
C
C-----END OF TNORM------------------------------------------------------
      RETURN
      END
C
C-----START OF LUMP-----------------------------------------------------
C
      SUBROUTINE LUMP(VC,V1,VA,RELTOL,SPUTOL,SCALE2,LINDEX,TFLAG,LOOP)
C
C-----------------------------------------------------------------------
      COMPLEX*16 VC(1),V1(1),ZEROC,SUMC
      DOUBLE PRECISION  VA(1),RELTOL,SPUTOL,SCALE2
      DOUBLE PRECISION  THOLD,TH1,TH2,DGAP,ZERO,ONE
      INTEGER  LINDEX(1),TFLAG(1)
      DOUBLE PRECISION  DFLOAT, DMAX1, CDABS
C     COMPLEX*16 DCMPLX
C-----------------------------------------------------------------------
C     VC(J) = JTH DISTINCT T-EIGENVALUE,  VA(J) = |VC(J)|, IN ORDER
C     OF INCREASING MAGNITUDE.
C     LINDEX(J) = T-MULTIPLICITY OF JTH DISTINCT T-EIGENVALUE
C     LOOP = NUMBER OF DISTINCT T-EIGENVALUES
C     LUMP 'COMBINES' COMPUTED 'GOOD' T-EIGENVALUES THAT ARE 'TOO CLOSE'
C     VALUE OF RELTOL IS 1.D-8.
C
C     IF IN A SET OF T-EIGENVALUES TO BE COMBINED THERE IS AN EIGENVALUE
```

```
C     WITH LINDEX=1, THEN THE VALUE OF THE COMBINED T-EIGENVALUES IS SET
C     EQUAL TO THE VALUE OF THAT EIGENVALUE.  NOTE THAT IF A SPURIOUS
C     T-EIGENVALUE IS TO BE 'COMBINED' WITH A GOOD EIGENVALUE, THEN THIS
C     IS DONE ONLY BY INCREASING THE INDEX, LINDEX, FOR THAT EIGENVALUE
C     NUMERICAL VALUES OF SPURIOUS T-EIGENVALUES ARE NEVER COMBINED WITH
C     THOSE OF GOOD T-EIGENVALUES.
C-----------------------------------------------------------------------
      ZERO = 0.0D0
      ONE = 1.D0
      ZEROC = DCMPLX(ZERO,ZERO)
      TH2 = SCALE2*SPUTOL
      DO 10 K = 1,LOOP
   10 TFLAG(K) = 0
      NLOOP = 0
      J = 0
   20 J = J+1
      IF (J.GT.LOOP) GO TO 130
      IF (TFLAG(J).EQ.1) GO TO 20
      NLOOP = NLOOP + 1
      TFLAG(J) = 1
      V1(1) = VC(J)
      ICOUNT = 1
      JN = LINDEX(J)
      TH1 = RELTOL*VA(J)
      THOLD = DMAX1(TH1,TH2)
C     THOLD = RELTOL*DMAX1(ONE,VA(J))
      IF (JN.EQ.0) GO TO 30
      INDSUM = JN
      ISPUR = 0
      SUMC = DFLOAT(JN)*VC(J)
      GO TO 40
   30 INDSUM = 1
      ISPUR = 1
      SUMC = ZEROC
   40 IF (J.EQ.LOOP) GO TO 70
      I = J
   50 I = I + 1
      IF (I.GT.LOOP) GO TO 70
      IF (TFLAG(I).EQ.1) GO TO 50
      DGAP = VA(I) - VA(J)
      IF (DGAP.GE.THOLD) GO TO 70
      DGAP = CDABS(VC(I)-VC(J))
      IF (DGAP.GE.THOLD) GO TO 50
C     LUMP VC(I) WITH VC(J)
      ICOUNT = ICOUNT + 1
      TFLAG(I) = 1
      V1(ICOUNT) = VC(I)
      IN = LINDEX(I)
      IF (IN.NE.0) GO TO 60
      ISPUR = ISPUR + 1
      INDSUM = INDSUM + 1
      GO TO 50
   60 INDSUM = INDSUM + IN
      SUMC = SUMC + DFLOAT(IN)*VC(I)
      GO TO 50
C     COMPUTE THE 'COMBINED' T-EIGENVALUE AND THE RESULTING
C     T-MULTIPLICITY
   70 CONTINUE
C
C      IF (ICOUNT.GT.1) WRITE(6,80) (K,V1(K), K = 1,ICOUNT)
   80 FORMAT(/' T-EIGENVALUES ARE LUMPED '/
     1 5X,'J',12X,'REAL(EV)',12X,'IMAG(EV)'/(I6,2E20.12))
C
       IF (ICOUNT.EQ.1) INDSUM = JN
       IDIF = INDSUM - ISPUR
       IF (IDIF.EQ.0.AND.ICOUNT.GT.1) GO TO 110
```

```
      IF (ICOUNT.EQ.1) GO TO 90
C     ICOUNT.GT.1 AND IDIF.GT.0
      SUMC = SUMC/DFLOAT(IDIF)
      VC(NLOOP) = SUMC
      VA(NLOOP) = CDABS(SUMC)
      GO TO 100
   90 VC(NLOOP) = VC(J)
      VA(NLOOP) = VA(J)
  100 LINDEX(NLOOP) = INDSUM
      GO TO 20
C     INDEX J IS FINISHED
C
C     ON RETURN VC CONTAINS THE DISTINCT T-EIGENVALUES  VA = |VC|
C     LINDEX CONTAINS THE CORRESPONDING T-MULTIPLICITIES
C
  110 WRITE(6,120) (K,V1(K), K = 1,ICOUNT)
  120 FORMAT(/' DEFAULT IN LUMP LUMPING INDICATED FOR TWO SPURIOUS EV'/
     1 ' K,V1(K) = '/(I6,2E20.12))
      NLOOP = -NLOOP
  130 CONTINUE
      LOOP = NLOOP
      RETURN
C-----END OF LUMP--------------------------------------------------------
      END
C
C-----START OF ISOEV-----------------------------------------------------
C
      SUBROUTINE ISOEV(VS,GR,GG,GAPTOL,SPUTOL,SCALE1,MP,NDIS,NG,NISO)
C
C-----------------------------------------------------------------------
      COMPLEX*16 VS(1),TO
      DOUBLE PRECISION  GR(1),SPUTOL,GAPTOL,SCALE1,TEMP,TOL,TJ,DGAP,ONE
      REAL GG(1)
      INTEGER  MP(1)
      REAL  ABS
      DOUBLE PRECISION  DMAX1, CDABS
C-----------------------------------------------------------------------
C     USE TMINGAPS TO LABEL THE ISOLATED GOOD T-EIGENVALUES
C     THAT ARE VERY CLOSE TO SPURIOUS ONES.  ERROR ESTIMATES
C     WILL NOT BE COMPUTED FOR THESE T-EIGENVALUES.
C
C     ON ENTRY AND EXIT
C     VS CONTAINS THE COMPUTED DISTINCT EIGENVALUES OF T(1,MEV)
C     GR(K) = |VS(K)|, K = 1,NDIS, GR(K).LE.GR(K+1)
C     GG(K) = MIN(J.NE.K,|VS(K)-VS(J)|)   MINGAP
C     MP CONTAINS THE CORRESPONDING T-MULTIPLICITIES
C     NDIS = NUMBER OF DISTINCT T-EIGENVALUES
C     GAPTOL = RELATIVE GAP TOLERANCE SET IN MAIN
C
C     ON EXIT
C     MP(J) IS NOT CHANGED EXCEPT THAT  MP(J)=-1, IF MP(J)=1,
C     AND A SPURIOUS T-EIGENVALUE IS TOO CLOSE.
C
C     IF MP(I)=-1 THAT SIMPLE GOOD T-EIGENVALUE WILL BE SKIPPED
C     IN THE SUBSEQUENT ERROR ESTIMATE COMPUTATIONS IN INVERR
C     THAT IS, WE COMPUTE ERROR ESTIMATES ONLY FOR THOSE GOOD
C     T-EIGENVALUES WITH MP(J)=1.
C-----------------------------------------------------------------------
      ONE = 1.0D0
      DGAP = SCALE1*SPUTOL
      NISO = 0
      NG = 0
      DO 40 J = 1,NDIS
      IF (MP(J).EQ.0) GO TO 40
      NG = NG+1
      IF (MP(J).NE.1) GO TO 40
      TJ = GR(J)
```

```
      TO = VS(J)
      TOL = DMAX1(DGAP,GAPTOL*TJ)
C     TOL = DMAX1(ONE,TJ)*GAPTOL
C     VS(J) IS NEXT SIMPLE GOOD T-EIGENVALUE
      NISO = NISO + 1
      IF (ABS(GG(J)).GT.TOL) GO TO 40
      I = J
   10 I = I-1
      IF (I.LT.1) GO TO 20
      IF (TJ-GR(I).GT.TOL) GO TO 20
      IF (MP(I).NE.0) GO TO 10
      TEMP = CDABS(TO-VS(I))
      IF (TEMP.GT.TOL) GO TO 10
      MP(J) = -MP(J)
      NISO = NISO-1
      GO TO 40
   20 I = J
   30 I = I+1
      IF (I.GT.NDIS) GO TO 40
      IF (GR(I)-TJ.GT.TOL) GO TO 40
      IF (MP(I).NE.0) GO TO 30
      TEMP = CDABS(TO-VS(I))
      IF (TEMP.GT.TOL) GO TO 30
      MP(J) = -MP(J)
      NISO = NISO-1
   40 CONTINUE
C
C-----END OF ISOEV----------------------------------------------------
      RETURN
      END
C---COMPEV---------------------------------------------------------
C
      SUBROUTINE COMPEV(ALPHA,BETA,V1,V2,VS,EVMAG,MULTOL,SPUTOL,
     1MP,T2FLAG,MEV,NDIS,SAVTEV)
C
C     USES COMPLEX SYMMETRIC VERSION OF IMTQL1, CMTQL1, TO
C     COMPUTE EIGENVALUES OF THE T-MATRIX T(1,MEV).
C
C--------------------------------------------------------------------
      COMPLEX*16 ALPHA(1),BETA(1),VS(1),V1(1),V2(1),EVAL,CTEMP
      DOUBLE PRECISION EVMAG(1)
      DOUBLE PRECISION TEMP,DGAP,TOL,DELMIN
      DOUBLE PRECISION MULTOL,SPUTOL,EVALR,EVALC
      INTEGER MP(1),T2FLAG(1),SAVTEV
      DOUBLE PRECISION CDABS, DFLOAT
C--------------------------------------------------------------------
C
      MEV1 = MEV - 1
C
      IF (SAVTEV.GE.0) GO TO 40
C
      READ(10,10) MEV
   10 FORMAT(I6)
   20 FORMAT(20A4)
      MEV1 = MEV - 1
      READ(10,30) (VS(K), K = 1,MEV)
   30 FORMAT(4Z20)
      READ(10,20) EXPLAN
      READ(10,20) EXPLAN
      READ(10,30) (V2(K), K = 1,MEV1)
      GO TO 90
C
   40 CONTINUE
C
      DO 50 J = 1,MEV
      VS(J) = ALPHA(J)
```

```
   50 V1(J) = BETA(J)
C
      WRITE(6,60) MEV
   60 FORMAT(/' COMPUTE EIGENVALUES OF T(1,',I4,') USING CMTQL1'/)
C
C-----------------------------------------------------------------------
      CALL CMTQL1(MEV,VS,V1,IERR)
C-----------------------------------------------------------------------
C
C     WRITE(6,70) IERR
   70 FORMAT(' T-EIGENVALUES VIA CMTQL1'/' IERR = ',I6/)
C
      IF (IERR.EQ.0) GO TO 90
C
      WRITE(6,80)
   80 FORMAT(' ON RETURN FROM CMTQL1 ERROR FLAG WAS NOT ZERO'/)
      GO TO 410
C
   90 CONTINUE
C
C     T-EIGENVALUES ARE IN VS IN INCREASING ORDER OF MAGNITUDE
      DO 100 J = 1,MEV
  100 EVMAG(J) = CDABS(VS(J))
C
C     THE MAGNITUDES OF THE T-EIGENVALUES ARE IN EVMAG, IN ORDER OF
C     INCREASING MAGNITUDE
C     WRITE(13,105) (EVMAG(J), J = 1,MEV)
C 105 FORMAT(' MAGNITUDES OF T-EIGENVALUES'/(4E20.12))
C
      IF(SAVTEV.NE.1) GO TO 130
      WRITE(10,110) MEV
  110 FORMAT(I6,' = ORDER OF T-MATRIX, T-EIGVALS =')
      WRITE(10,120) (VS(J), J = 1,MEV)
C 120 FORMAT(4Z20)
  120 FORMAT(4E20.12)
C
C
  130 CONTINUE
      MULTOL = MULTOL*EVMAG(MEV)
      SPUTOL = SPUTOL*EVMAG(MEV)
      TOL = 1000.0D0*SPUTOL
      WRITE(6,140) MULTOL,SPUTOL
  140 FORMAT(/' TOLERANCES USED IN T-MULTIPLICITY AND SPURIOUS TESTS ='
     1 ,2E10.3/)
C
C     T-MULTIPLICITY DETERMINATION
      J = 0
      NDIS = 0
      DO 150 I = 1,MEV
  150 T2FLAG(I) = 0
C
  160 J = J+1
      IF (J.GT.MEV) GO TO 190
      IF (T2FLAG(J).EQ.1) GO TO 160
      CTEMP = VS(J)
      EVAL = CTEMP
      TEMP = EVMAG(J)
      NDIS = NDIS + 1
      INDEX = 1
      T2FLAG(J) = 1
      I = J
  170 I = I+1
      IF (I.GT.MEV) GO TO 180
      IF (T2FLAG(I).EQ.1) GO TO 170
      DGAP = EVMAG(I)-TEMP
      IF (DGAP.GT.MULTOL) GO TO 180
      DGAP = CDABS(EVAL-VS(I))
```

```
      IF (DGAP.GT.MULTOL) GO TO 170
C     T-MULTIPLICITY INCREASES
      INDEX = INDEX + 1
      CTEMP = CTEMP + VS(I)
      T2FLAG(I) = 1
      GO TO 170
C     T-MULTIPLICITY FOR VS(NDIS) HAS BEEN DETERMINED
  180 VS(NDIS) = CTEMP/DFLOAT(INDEX)
      MP(NDIS) = INDEX
      GO TO 160
  190 CONTINUE
C     T-MULTIPLICITY CALCULATION IS COMPLETE
C
C     T(2,MEV) EIGENVALUE CALCULATION AND SPURIOUS TESTS
C
      IF (SAVTEV.LT.0) GO TO 240
C
      WRITE(6,200) MEV1
  200 FORMAT(/' COMPUTE T(2,',I4,') EIGENVALUES'/)
C
      DO 210 J = 1,MEV1
      JP1 = J+1
      V2(J) = ALPHA(JP1)
  210 V1(J) = BETA(JP1)
C
C-----------------------------------------------------------------------
      CALL CMTQL1(MEV1,V2,V1,IERR)
C-----------------------------------------------------------------------
C
C     WRITE(6,220) IERR
  220 FORMAT(' T2-HAT EIGENVALUES VIA CMTQL1'/' IERR = ',I6/)
C
      IF (IERR.EQ.0) GO TO 240
C
      WRITE(6,230)
  230 FORMAT(' ON RETURN FROM CMTQL1 ERROR FLAG WAS NOT ZERO'/)
      GO TO 410
C
  240 CONTINUE
C
      DO 250 J = 1,MEV1
  250 EVMAG(J) = CDABS(V2(J))
C
C     WRITE(13,255) (EVMAG(J), J = 1,MEV)
C 255 FORMAT(/' MAGNITUDES OF T2 EIGENVALUES'/(4E20.12))
C
      IF(SAVTEV.NE.1) GO TO 270
      WRITE(10,260) MEV1
  260 FORMAT(/I6,' = ORDER OF T2-HAT, T2EIGVALS = ')
      WRITE(10,120) (V2(J), J = 1,MEV1)
  270 CONTINUE
C
C     SPURIOUS TESTS
      DO 280 I = 1,MEV1
  280 T2FLAG(I) = 0
C
C     GO THROUGH THE EIGENVALUES OF T2-HAT.  FIND THE CLOSEST EIGENVALUE
C     OF T(1,MEV).  IF IT IS T-MULTIPLE GO ON.  IF IT IS SIMPLE DECLARE
C     SPURIOUS WHENEVER DELMIN < SPUTOL BY SETTING MP(I) = 0
      J = 0
  290 J = J+1
      IF (J.GT.MEV1) GO TO 390
C
C     WRITE(14,300) J,V2(J)
  300 FORMAT('EIGENVALUE T2-HAT =', I6,2E22.14)
C
      TEMP = EVMAG(J)
```

```
      EVAL = V2(J)
      EVALR = TEMP + SPUTOL
      EVALC = TEMP - SPUTOL
      DELMIN = 2.D0*CDABS(VS(MEV))
      IMIN = 0
C     BACKWARD SEARCH
      I = J + 1
  310 I = I - 1
      IF(I.LT.1) GO TO 320
      IF(I.GT.NDIS) I = NDIS
C
      TEMP = CDABS(VS(I))
      IF (TEMP.LT.EVALC) GO TO 320
      IF(MP(I).EQ.0) GO TO 310
      DGAP = CDABS(VS(I) - EVAL)
      IF (DGAP.GE.DELMIN) GO TO 310
      DELMIN = DGAP
      IMIN = I
C
      GO TO 310
C     FORWARD SEARCH
  320 I = J
  330 I = I + 1
      IF(I.GT.NDIS) GO TO 340
C
      TEMP = CDABS(VS(I))
      IF (TEMP.GT.EVALR) GO TO 340
      IF(MP(I).EQ.0) GO TO 330
      DGAP = CDABS(VS(I) - EVAL)
      IF (DGAP.GE.DELMIN) GO TO 330
      DELMIN = DGAP
      IMIN = I
C
      GO TO 330
C
  340 CONTINUE
      IF(IMIN.EQ.0) GO TO 370
C     WRITE(14,350) IMIN, MP(IMIN),VS(IMIN),DELMIN,J
  350 FORMAT(/I6,' TH EVALUE, MP =',I3,' EVALUE =',2E22.13/
     1' MINDEL = ',E14.3,' OCCURS FOR',I6,' TH T2-HAT EVALUE')
      IF(DELMIN.GT.SPUTOL) GO TO 290
      IF(MP(IMIN).GT.1)  GO TO 290
      MP(IMIN) = 0
C     WRITE(14,360)
  360 FORMAT(' ABOVE T-EIGENVALUE IS SPURIOUS')
      GO TO 290
  370 CONTINUE
      GO TO 290
  390 CONTINUE
C     END OF SPURIOUS TESTS
C
      DO 400 J = 1,NDIS
  400 EVMAG(J) = CDABS(VS(J))
C
      RETURN
C-----END OF COMPEV------------------------------------------------------
  410 STOP
      END
C-----CMTQL1 (EIGENVALUES OF COMPLEX SYMMETRIC TRIDIAGONAL)-------------
C
      SUBROUTINE CMTQL1(N,D,E,IERR)
C
C-----------------------------------------------------------------------
      INTEGER I,J,L,M,N,II,MML,IERR
      COMPLEX*16 D(1),E(1),B,C,F,G,P,R,S,W,CZERO,CONE
      COMPLEX*16 CDSQRT,DCMPLX
      DOUBLE PRECISION MACHEP,EPS,TEMP,T0,T1,ZERO,HALF,ONE,TWO
```

```
      DOUBLE PRECISION CDABS,DSQRT                                      CSL08000
C-----------------------------------------------------------------------CSL08010
      DATA MACHEP/Z3410000000000000/                                    CSL08020
      EPS = 100.D0*MACHEP                                               CSL08030
C-----------------------------------------------------------------------CSL08040
      ZERO = 0.0D0                                                      CSL08050
      HALF = 0.5D0                                                      CSL08060
      ONE  = 1.0D0                                                      CSL08070
      TWO  = 2.0D0                                                      CSL08080
      CZERO = DCMPLX(ZERO,ZERO)                                         CSL08090
      CONE  = DCMPLX(ONE,ZERO)                                          CSL08100
      IERR = 0                                                          CSL08110
      IF (N.EQ.1) GO TO 160                                             CSL08120
C                                                                       CSL08130
      DO 10 I = 2,N                                                     CSL08140
   10 E(I-1) = E(I)                                                     CSL08150
      E(N) = CZERO                                                      CSL08160
C                                                                       CSL08170
      DO 140 L = 1,N                                                    CSL08180
      J = 0                                                             CSL08190
C                                                                       CSL08200
C     DETERMINE FIRST NEGLIGIBLE SUBDIAGONAL ELEMENT IF ANY             CSL08210
   20 DO 30 M = L,N                                                     CSL08220
      IF (M.EQ.N) GO TO 40                                              CSL08230
      TEMP = CDABS(D(M)) + CDABS(D(M+1))                                CSL08240
      IF (CDABS(E(M)).LE.TEMP*MACHEP) GO TO 40                          CSL08250
   30 CONTINUE                                                          CSL08260
C                                                                       CSL08270
   40 P = D(L)                                                          CSL08280
C                                                                       CSL08290
      IF (M.EQ.L) GO TO 100                                             CSL08300
      IF (J.EQ.100) GO TO 150                                           CSL08310
      J = J+1                                                           CSL08320
C                                                                       CSL08330
C     FORM SHIFT AS EIGENVALUE OF (L,L+1) 2X2 CLOSEST TO D(L)           CSL08340
      G = (D(L+1) - P)*HALF                                             CSL08350
      T0 = CDABS(G)                                                     CSL08360
      T1 = CDABS(E(L))                                                  CSL08370
      IF (T0.GT.T1) GO TO 50                                            CSL08380
      W = G/E(L)                                                        CSL08390
      R = CDSQRT(CONE + W**2)                                           CSL08400
      T0 = CDABS(W + R)                                                 CSL08410
      T1 = CDABS(W - R)                                                 CSL08420
      TEMP = ONE                                                        CSL08430
      IF (T1.GT.T0) TEMP = -ONE                                         CSL08440
      G = D(M) - P + E(L)/(W + TEMP*R)                                  CSL08450
      GO TO 60                                                          CSL08460
   50 CONTINUE                                                          CSL08470
      W = E(L)/G                                                        CSL08480
      R = CDSQRT(CONE + W**2)                                           CSL08490
      T0 = CDABS(CONE + R)                                              CSL08500
      T1 = CDABS(CONE - R)                                              CSL08510
      TEMP = ONE                                                        CSL08520
      IF (T1.GT.T0) TEMP = -ONE                                         CSL08530
      G = D(M) - P + W*E(L)/(CONE + TEMP*R)                             CSL08540
   60 CONTINUE                                                          CSL08550
C                                                                       CSL08560
C     G IS SHIFTED D(M)                                                 CSL08570
C     SPECIFY PARAMETERS FOR I = M-1 CASE, I = M-1,M-2,...,L            CSL08580
C                                                                       CSL08590
      S = CONE                                                          CSL08600
      C = -CONE                                                         CSL08610
      P = CZERO                                                         CSL08620
      MML = M - L                                                       CSL08630
C                                                                       CSL08640
      DO 90 II = 1,MML                                                  CSL08650
```

```
      I = M - II
C
C     FOR I<M-1 F=T(I+2,I), B=NEW E(I), AIM OF (I,I+1) TRANSFORMATION
C     IS TO ZERO OUT F
C
      F =  S*E(I)
      B = -C*E(I)
      T0 = CDABS(G)
      T1 = CDABS(F)
      IF (T1.GT.T0) GO TO 70
C     |G| >= |F|
      W = F/G
      R = CDSQRT(CONE + W**2)
      E(I+1) = G*R
      C = CONE/R
      S = W*C
      GO TO 80
C     |F| > |G|
   70 CONTINUE
      W = G/F
      R = CDSQRT(CONE + W**2)
      E(I+1) = F*R
      S = CONE/R
      C = W*S
   80 CONTINUE
      TEMP = CDABS(W)**2 + ONE
      T0 = DSQRT(TEMP)
      T1 = CDABS(R)
      IERR = -L
      IF (T1.LE.EPS*T0) GO TO 160
      IERR = 0
C
C     C**2 + S**2 = CONE, -Q(I,I) = Q(I+1,I+1) = C, Q(I,I+1) = S
C     Q = Q-TRANSPOSE = Q-INVERSE  RR = CDSQRT(G**2 +F**2)
C     G = D(I+1) AFTER PREVIOUS TRANSFORMATION THEN G = NEW E(I)
C     NEW D(I) = D(I) - S*RR,   NEW D(I+1) = D(I+1) + S*RR
C     NEW E(I) = E(I) - C*RR,   NEW E(I+1) = RR,   P = S*RR
C
      G = D(I+1) - P
      R = (D(I) - G)*S + TWO*C*B
      P = S*R
      D(I+1) = G + P
      G = B - C*R
   90 CONTINUE
C     END OF I LOOP
C
C     UPDATE PARAMETERS FOR I = L CASE
      D(L) = D(L) - P
      E(L) = G
      E(M) = CZERO
      GO TO 20
C
C     ORDER EIGENVALUES   P = D(L)
  100 IF (L.EQ.1) GO TO 120
      DO 110 II = 2,L
      I = L+2-II
      IF (CDABS(P).GE.CDABS(D(I-1))) GO TO 130
      D(I) = D(I-1)
  110 CONTINUE
C
  120 I = 1
C
  130 D(I) = P
C
  140 CONTINUE
      GO TO 160
C
```

```
  150 IERR = L
C-----END OF CMTQL1---------------------------------------------------
  160 RETURN
      END
C
C-----COMGAP----------------------------------------------------------
C
      SUBROUTINE COMGAP(VC,VA,GG,MP,IND,M,IGAP,ITAG)
C
C---------------------------------------------------------------------
      COMPLEX*16 VC(1),Z
      DOUBLE PRECISION  VA(1),T0,T1,TU,TK
      REAL  GG(1),GTEMP
      INTEGER  MP(1),IND(1)
      REAL ABS
      DOUBLE PRECISION CDABS
C---------------------------------------------------------------------
C     IF IGAP = 0 WE DO NOT ORDER EIGENVALUES BY INCREASING GAP SIZE
C     AND WE DO NOT WRITE GAP OUTPUT TO FILE 12
C
C     VA(K) = |VC(K)|  VA(K) <= VA(K+1)
C     GG(K) = MIN |VC(K)-VC(J)|  J .NE. K.
C---------------------------------------------------------------------
      TU = VA(M) + VA(M)
      K = 0
   10 K = K+1
      IF (K.GT.M) GO TO 60
      INDEX = 0
      T1 = TU
      TK = VA(K)
      Z = VC(K)
      J = K
C     BACKWARDS
   20 J = J-1
      IF (J.LT.1) GO TO 30
      T0 = TK - VA(J)
      IF (T0.GT.T1) GO TO 30
      T0 = CDABS(Z - VC(J))
      IF (T1.LE.T0) GO TO 20
      T1 = T0
      INDEX = J
      GO TO 20
C     FORWARDS
   30 J = K
   40 J = J+1
      IF (J.GT.M) GO TO 50
      T0 = VA(J) - TK
      IF (T0.GT.T1) GO TO 50
      T0 = CDABS(Z - VC(J))
      IF (T1.LE.T0) GO TO 40
      T1 = T0
      INDEX = J
      GO TO 40
   50 IND(K) = INDEX
      GG(K) = T1
      IF(ITAG.EQ.0) GO TO 10
      IF(MP(INDEX).EQ.0)  GG(K) = -GG(K)
      GO TO 10
C
   60 CONTINUE
      IF (IGAP.EQ.0) GO TO 140
C
C     WRITE(12,70)
   70 FORMAT(' MINGAPS FOR GOOD T-EIGENVALUES'/
     1 1X,'EVNUM',1X,'NEIGH',15X,'R(EV)',15X,'I(EV)',4X,'MINGAP')
C     WRITE(12,80) (K,IND(K),VC(K),GG(K), K = 1,M)
```

```
   80 FORMAT(2I6,2E20.12,E10.3)
C
C     ORDER VC G BY INCREASING MINGAP SIZE
      DO 90 J = 1,M
      IND(J) = J
   90 CONTINUE
C
      DO 110 K = 2,M
      KM1 = K-1
      DO 100 L = 1,KM1
      KK = K-L
      KP1 = KK+1
      IF (ABS(GG(KP1)).GE.ABS(GG(KK))) GO TO 110
      Z = VC(KK)
      VC(KK) = VC(KP1)
      VC(KP1) = Z
      GTEMP = GG(KK)
      GG(KK) = GG(KP1)
      GG(KP1) = GTEMP
      ITEMP = IND(KK)
      IND(KK) = IND(KP1)
      IND(KP1) = ITEMP
  100 CONTINUE
  110 CONTINUE
C
C     WRITE(12,120)
  120 FORMAT(' T-EIGENVALUES ORDERED BY INCREASING MINGAP'/
     1 1X,'GAPNUM',1X,'EVNUM',15X,'R(EV)',15X,'I(EV)',4X,'MINGAP')
C
C     WRITE(12,130) (K,IND(K),VC(K),GG(K), K = 1,M)
  130 FORMAT(I7,I6,2E20.12,E10.3)
C
  140 CONTINUE
C-----END OF COMGAP------------------------------------------------------
      RETURN
      END
C
C-----START OF INVERM FOR TRIDIAGONAL COMPLEX SYMMETRIC MATRICES--------
C
      SUBROUTINE INVERM(ALPHA,BETA,V1,V2,X1,ERROR,ERRORV,EPS,GR,GC,
     1INTERC,MEV,IT,IWRITE)
C
C-----------------------------------------------------------------------
      COMPLEX*16  ALPHA(1),BETA(1),V1(1),V2(1)
      COMPLEX*16  X1,U,Z,TEMP,RATIO,BETAM,ZEROC
      DOUBLE PRECISION  SUM,XU,NORM,TSUM,GSUM
      DOUBLE PRECISION  EPS,EPS3,EPS4,ERROR,ERRORV,ZERO,ONE
      DOUBLE PRECISION GR(1),GC(1)
      INTEGER INTERC(1)
      DOUBLE PRECISION  DABS, DSQRT, DFLOAT, CDABS
C     COMPLEX*16 DCMPLX
C-----------------------------------------------------------------------
C
C     COMPUTES T-EIGENVECTORS FOR ISOLATED GOOD T-EIGENVALUES X1
C     USING INVERSE ITERATION ON T(1,MEV(X1)) SOLVING EQUATION
C     (T - X1*I)V2 = RIGHT-HAND SIDE (RANDOMLY-GENERATED) .
C     PROGRAM REFACTORS T- X1*I ON EACH ITERATION OF INVERSE ITERATION.
C     TYPICALLY ONLY ONE ITERATION IS NEEDED PER T-EIGENVALUE X1.
C
C     IF IWRITE = 1 THEN THERE ARE EXTENDED WRITES TO FILE 6 (TERMINAL)
C
C     ON ENTRY G CONTAINS A REAL*4 RANDOM VECTOR WHICH WAS GENERATED
C     IN MAIN PROGRAM.
C
C     ON ENTRY AND EXIT
C     MEV = ORDER OF T
C     ALPHA, BETA CONTAIN THE DIAGONAL AND OFFDIAGONAL ENTRIES OF T.
```

```
C     EPS = 2. * MACHINE EPSILON
C
C     IN PROGRAM:
C     ITER = MAXIMUM NUMBER STEPS ALLOWED FOR INVERSE ITERATION
C     ITER = IT ON ENTRY.
C     V1,V2 = WORK SPACES USED IN THE FACTORIZATION OF T(1,MEV).
C     V1 AND V2 MUST BE OF DIMENSION AT LEAST MEV.
C
C     ON EXIT
C     V2 = THE UNIT EIGENVECTOR OF T(1,MEV) CORRESPONDING TO X1.
C     ERROR = |V2(MEV)| = ERROR ESTIMATE FOR CORRESPONDING
C             RITZ VECTOR FOR X1.
C
C     ERRORV = || T*V2 - X1*V2 || = ERROR ESTIMATE ON T-EIGENVECTOR.
C     IF IT.GT.ITER THEN ERRORV = -ERRORV
C     IT = NUMBER OF ITERATIONS ACTUALLY REQUIRED
C-----------------------------------------------------------------------
C     INITIALIZATION AND PARAMETER SPECIFICATION
      ONE  = 1.0D0
      ZERO = 0.0D0
      ZEROC = DCMPLX(ZERO,ZERO)
      ITER = IT
      MP1 = MEV+1
      MM1 = MEV-1
      BETAM = BETA(MP1)
      BETA(MP1) = ZEROC
C
C     CALCULATE SCALE AND TOLERANCES
      TSUM = CDABS(ALPHA(1))
      DO 10 I = 2,MEV
   10 TSUM = TSUM + CDABS(ALPHA(I)) + CDABS(BETA(I))
C
      EPS3 = EPS*TSUM
      EPS4 = DFLOAT(MEV)*EPS3
C
C     GENERATE SCALED RANDOM RIGHT-HAND SIDE
      GSUM = ZERO
      DO 20 I = 1,MEV
   20 GSUM = GSUM + DABS(GR(I)) + DABS(GC(I))
      GSUM = EPS4/GSUM
C
C     INITIALIZE RIGHT HAND SIDE FOR INVERSE ITERATION
      DO 30 I = 1,MEV
      INTERC(I) = 0
   30 V2(I) = GSUM*DCMPLX(GR(I),GC(I))
      IT = 1
C
C     CALCULATE UNIT EIGENVECTOR OF T(1,MEV) FOR ISOLATED GOOD
C     T-EIGENVALUE X1.
C
C     TRIANGULAR FACTORIZATION WITH NEAREST NEIGHBOR PIVOT
C     STRATEGY. INTERCHANGES ARE LABELLED BY SETTING INTERC(I)=0
C
   40 CONTINUE
      U = ALPHA(1)-X1
      Z = BETA(2)
C
      DO 60 I=2,MEV
      IF (CDABS(BETA(I)).GT.CDABS(U)) GO TO 50
C     NO PIVOT INTERCHANGE
      V1(I-1) = Z/U
      V2(I-1) = V2(I-1)/U
      V2(I) = V2(I)-BETA(I)*V2(I-1)
      RATIO = BETA(I)/U
      U = ALPHA(I)-X1-Z*RATIO
      Z = BETA(I+1)
```

```
      GO TO 60
C     PIVOT INTERCHANGE
   50 CONTINUE
      RATIO = U/BETA(I)
      INTERC(I) = 1
      V1(I-1) = ALPHA(I)-X1
      U = Z-RATIO*V1(I-1)
      Z = -RATIO*BETA(I+1)
      TEMP = V2(I-1)
      V2(I-1) = V2(I)
      V2(I) = TEMP-RATIO*V2(I)
   60 CONTINUE
C
      IF (CDABS(U).EQ.ZERO) U= DCMPLX(EPS3,EPS3)
C
C     SMALLNESS TEST AND DEFAULT VALUE FOR LAST COMPONENT
C     PIVOT(I-1) = |BETA(I)| FOR INTERCHANGE CASE
C     (I-1,I+1) ELEMENT IN RIGHT FACTOR = BETA(I+1)
C     END OF FACTORIZATION AND FORWARD SUBSTITUTION
C
C     BACK SUBSTITUTION
      V2(MEV) = V2(MEV)/U
      DO 80 II = 1,MM1
      I = MEV-II
      IF (INTERC(I+1).EQ.1) GO TO 70
C     NO PIVOT INTERCHANGE
      V2(I) = V2(I)-V1(I)*V2(I+1)
      GO TO 80
C     PIVOT INTERCHANGE
   70 V2(I) = (V2(I)-V1(I)*V2(I+1)-BETA(I+2)*V2(I+2))/BETA(I+1)
   80 CONTINUE
C
C
C     TESTS FOR CONVERGENCE OF INVERSE ITERATION
C     IF SUM |V2| COMPS. LE. 1 AND IT. LE. ITER DO ANOTHER INVIT STEP
C
      NORM = CDABS(V2(MEV))
      DO 90 II = 1,MM1
      I = MEV-II
   90 NORM = NORM+CDABS(V2(I))
C
C     IS DESIRED GROWTH IN VECTOR ACHIEVED ?
C     IF NOT, DO ANOTHER INVERSE ITERATION STEP UNLESS NUMBER ALLOWED IS
C     EXCEEDED.
      IF (NORM.GE.ONE) GO TO 110
C
      IT=IT+1
      IF (IT.GT.ITER) GO TO 110
C
      XU = EPS4/NORM
      DO 100 I=1,MEV
      INTERC(I) = 0
  100 V2(I) = V2(I)*XU
C
      GO TO 40
C
C     NORMALIZE COMPUTED T-EIGENVECTOR : V2 = V2/||V2||
C
  110 CONTINUE
C
C-----------------------------------------------------------------------
      CALL CINPRD(V2,V2,SUM,MEV)
C-----------------------------------------------------------------------
C
      SUM = ONE/DSQRT(SUM)
      DO 120 II = 1,MEV
```

```
  120 V2(II) = SUM*V2(II)
C
C     SAVE ERROR ESTIMATE FOR LATER OUTPUT
      ERROR = CDABS(V2(MEV))
C
C     GENERATE ERRORV = ||T*V2 - X1*V2||.
C     LOOP IS BOTTOM UP BECAUSE LAST COMPONENTS MAY BE VERY SMALL
      V1(MEV) = ALPHA(MEV)*V2(MEV)+BETA(MEV)*V2(MEV-1)-X1*V2(MEV)
      DO 130 J = 2,MM1
      JM = MP1 - J
      V1(JM) = ALPHA(JM)*V2(JM) + BETA(JM)*V2(JM-1) + BETA(JM+1)*V2(JM+1
     1) - X1*V2(JM)
  130 CONTINUE
C
      V1(1) = ALPHA(1)*V2(1) + BETA(2)*V2(2) - X1*V2(1)
C
C-----------------------------------------------------------------------
      CALL CINPRD(V1,V1,ERRORV,MEV)
C-----------------------------------------------------------------------
C
      ERRORV = DSQRT(ERRORV)
      IF (IT.GT.ITER) ERRORV = -ERRORV
      IF (IWRITE.EQ.0) GO TO 150
C
C     FILE 6 (TERMINAL) OUTPUT OF ERROR ESTIMATES.
      WRITE(6,140) MEV,X1,ERROR,ERRORV
  140 FORMAT(1X,'TSIZE',10X,'RE(GOODEV)',10X,'IM(GOODEV)',11X,'U(M)',
     1 9X,'TERROR'/I6,2E20.12,2E15.5)
C
C     RESTORE BETA(MEV+1) = BETAM
  150 CONTINUE
      BETA(MP1) = BETAM
C-----END OF INVERM-----------------------------------------------------
      RETURN
      END
C
C-----START OF INNER PRODUCT ROUTINE------------------------------------
C
C     COMPUTES EUCLIDEAN INNER PRODUCT OF 2 COMPLEX VECTORS
C     SUMC = (V2-TRANSPOSE)*V1
C
      SUBROUTINE INPRDC(V2,V1,SUMC,N)
C
C-----------------------------------------------------------------------
      DOUBLE PRECISION  ZERO
      COMPLEX*16 V2(1),V1(1),SUMC
C-----------------------------------------------------------------------
C
      ZERO = 0.D0
      SUMC = DCMPLX(ZERO,ZERO)
      DO 10 J=1,N
   10 SUMC = SUMC + V2(J)*V1(J)
C
      RETURN
C-----END OF EUCLIDEAN INNER PRODUCT SUBROUTINE-------------------------
      END
C
C-----START OF HERMITIAN INNER PRODUCT ROUTINE--------------------------
C     COMPLEX INNER PRODUCT
C
      SUBROUTINE CINPRD(V2,V1,SUM,N)
C-----------------------------------------------------------------------
      DOUBLE PRECISION  ZERO,SUM
      COMPLEX*16 V2(1),V1(1),SUMC
C-----------------------------------------------------------------------
C     COMPUTES THE INNER PRODUCT OF THE CONJUGATE OF V2 WITH V1.
      ZERO = 0.D0
```

```
      SUMC = DCMPLX(ZERO,ZERO)
      DO 10 J=1,N
   10 SUMC = SUMC + DCONJG(V2(J))*V1(J)
      SUM = DREAL(SUMC)
C
      RETURN
C-----END OF COMPLEX INNER PRODUCT SUBROUTINE--------------------------
      END
```

SECTION 7.7 OPTIONAL PREPROCESSING PROGRAM, FILE DEFINITIONS, AND SAMPLE INPUT FILES

```
C-----LCCOMPAC-(STAND-ALONE PROGRAM)-----------------------------------
C
C     THIS PROGRAM TRANSLATES A SPARSE, COMPLEX SYMMETRIC, N X N
C     MATRIX GIVEN AS I, J, A(I,J), INTO THE SPARSE MATRIX FORMAT
C     REQUIRED BY THE SAMPLE USPEC AND CMATV PROGRAMS PROVIDED
C     FOR USE WITH THE COMPLEX SYMMETRIC LANCZOS EIGENVALUE/EIGENVECTOR
C     PROCEDURES.  PROGRAM ASSUMES THAT THE MATRIX ENTRIES ARE
C     PRESENTED TO THE PROGRAM COLUMN BY COLUMN OR ROW BY ROW.
C
C     PORTABILITY:  THIS PROGRAM IS NOT PORTABLE BECAUSE OF ITS
C                   USE OF COMPLEX*16 VARIABLES.
C
C----------------------------------------------------------------------
      COMPLEX*16  A(15000), AD(2000), ZEROC
      DOUBLE PRECISION ZERO
      INTEGER IROW(15000),ICOL(15000)
C----------------------------------------------------------------------
C    INPUT FILE 7 CONTAINS THE SPARSE COMPLEX SYMMETRIC NXN MATRIX
C    STORED AS:
C
C                 NZ,M,N,MATNO
C                 I(K)  J(K)  A(K)  K = 1,NZ
C
C     WHERE NZ IS THE TOTAL NUMBER OF NONZEROS IN THE MATRIX A,
C     N IS THE ROW AND COLUMN DIMENSION OF A,
C     AND A(K) ARE THE NONZERO ENTRIES STORED ROW BY ROW OR
C     COLUMN BY COLUMN.  PROGRAM READS THIS IN AS IROW(K) = I(K),
C     ICOL(K) = J(K), AND A(K) = A(K).
C
C     OUTPUT FILE = 8 CONTAINS THE A-MATRIX IN SPARSE FORMAT
C
C              NZS,N,NZL,MATNO
C               ICOL(K)    K = 1,NZL
C               IROW(K)    K = 1,NZS
C                 AD(K)    K = 1,N
C                  A(K)    K = 1,NZS
C
C      WHERE N IS THE ORDER OF THE INPUT MATRIX A,
C      NZ EQUALS THE NUMBER OF NONZERO ELEMENTS IN A WHICH ARE ON
C      OR BELOW THE MAIN DIAGONAL.  NZL EQUALS THE NUMBER OF THE
C      LAST COLUMN HAVING NONZEROES BELOW THE DIAGONAL IN A.
C      NZS EQUALS THE NUMBER OF NONZERO ELEMENTS BELOW THE MAIN
C      DIAGONAL.   AD(K), K=1,N, CONTAINS THE DIAGONAL ELEMENTS OF A.
C      A(K), K=1,NZS, CONTAINS THE KTH NONZERO SUB-DIAGONAL ELEMENT
C      OF THE INPUT MATRIX.  A IS STORED COLUMN BY COLUMN.
C      IROW(K), K=1,NZS, CONTAINS THE ROW INDEX OF THE  NONZERO
C      STRICTLY LOWER TRIANGULAR ELEMENT A(K).
C      ICOL(K), K=1,NZL, EQUALS THE NUMBER OF STRICTLY LOWER
C      TRIANGULAR NONZEROES IN COLUMN K OF THE INPUT MATRIX.
C
C----------------------------------------------------------------------
      ZERO = 0.D0
      ZEROC = DCMPLX(ZERO,ZERO)
C
      READ(7,10) NZ,N,MATNO,IIROW
   10 FORMAT(2I6,I8,I6)
C
      WRITE(6,20) NZ,N,MATNO,IIROW
   20 FORMAT(I10,I6,I10,' = NO. NONZERO AIJ J.GE.I, ORDER OF A, MATNO'/
     1 I6,' = IIROW IF IIROW=0 ORDERING IS BY COLS IIROW=1 BY ROWS'/)
C
      DO 30 K = 1,N
```

```
   30 AD(K) = ZEROC
C
      IF (IIROW.EQ.0) READ(7,40) (IROW(K),ICOL(K),A(K), K=1,NZ)
C
      IF (IIROW.EQ.1) READ(7,40) (ICOL(K),IROW(K),A(K), K=1,NZ)
   40 FORMAT(2I5,2E14.7)
C
      LCOUNT = 0
      K = 1
C
C     START OF A NEW COLUMN
   50 CONTINUE
      J = ICOL(K)
      ICOL(J) = 0
   60 CONTINUE
C
      IF (J.NE.IROW(K)) GO TO 70
C
C     DIAGONAL CASE
      AD(J) = A(K)
      GO TO 80
C
C     SUB-DIAGONAL NONZERO
   70 CONTINUE
      NZL = J
      LCOUNT = LCOUNT + 1
      A(LCOUNT) = A(K)
      IROW(LCOUNT) = IROW(K)
      ICOL(J) = ICOL(J) + 1
C
   80 CONTINUE
      K = K+1
C
      IF(K.GT.NZ) GO TO 90
C
      IF(ICOL(K).GT.J) GO TO 50
C
      GO TO 60
C
   90 CONTINUE
      NZS = LCOUNT
C
      WRITE(8,100) NZS,N,NZL,MATNO
      WRITE(6,100) NZS,N,NZL,MATNO
  100 FORMAT(I10,2I6,I8,' = NZS N NZL MATNO')
C
      WRITE(8,110) (ICOL(I), I=1,NZL)
      WRITE(8,110) (IROW(K), K=1,NZS)
  110 FORMAT(13I6)
C
      WRITE(8,120) (AD(K), K=1,N)
      WRITE(8,120) (A(K), K=1,NZS)
  120 FORMAT(4E19.10)
C
C-----END LCCOMPAC-------------------------------------------------------
      STOP
      END
```

-----INPUT/OUTPUT FILE DEFINITIONS FOR CSLEVAL AND CSLEVEC-----

Below is a listing of the input/output files which are accessed by the complex symmetric Lanczos eigenvalue program, CSLEVAL. Included also is a sample of the input file which CSLEVAL requires on file 5. The parameters in this file are supplied in free format. CSLEVAL computes eigenvalues of diagonalizable complex symmetric matrices.

SAMPLE DEFINITIONS OF THE INPUT/OUTPUT FILES FOR CSLEVAL

```
-----------------------------------------------------------------
*CSLEVAL EXEC LANCZOS EIGENVALUE CALCULATION COMPLEX SYMMETRIC CASE
FI 06 TERM
FILEDEF  1 DISK &1        NHISTORY  A (RECFM F LRECL 80 BLOCK 80
FILEDEF  2 DISK &1        HISTORY   A (RECFM F LRECL 80 BLOCK 80
FILEDEF  3 DISK &1        GOODEV    A (RECFM F LRECL 80 BLOCK 80
FILEDEF  4 DISK &1        ERRINV    A (RECFM F LRECL 80 BLOCK 80
FILEDEF  5 DISK CSLEVAL   INPUT     A (RECFM F LRECL 80 BLOCK 80
FILEDEF  8 DISK &1        INPUT     A (RECFM F LRECL 80 BLOCK 80
FILEDEF 10 DISK &1        T-T2EVAL  A (RECFM F LRECL 80 BLOCK 80
FILEDEF 11 DISK &1        DISTINCT  A (RECFM F LRECL 80 BLOCK 80
LOAD   CSLEVAL   CSLESUB   CSLEMULT
-----------------------------------------------------------------
```

SAMPLE INPUT FILE FOR CSLEVAL

```
-----------------------------------------------------------------
 CSLEVAL INPUT LANCZOS EIGENVALUE COMPUTATION, NO REORTHOGONALIZATION
 OF A NONDEFECTIVE COMPLEX SYMMETRIC MATRIX.
LINE 1      N     KMAX     NMEVS      MATNO
          528    2640         1     721830
LINE 2    SVSEED      RHSEED        MXINIT
        49302312     5731029             5
LINE 3    ISTART       ISTOP
               0           1
LINE 4  IHIS  IDIST  SAVTEV  IWRITE (SAVE HIST.,DISTINCT EV,TEV,WRITE
           1      0       1       1
LINE 5   RELTOL (RELATIVE TOLERANCE IN 'COMBINING' GOODEV)
    .0000000001
LINE 6   MB(1)   MB(2)   MB(3)   MB(4)    (ORDERS OF T(1,MEV) )
          1056
C      NOTE THAT WHEN READING IN PREVIOUSLY COMPUTED EIGENVALUES
C      THE VALUE OF MB(1) MUST BE EQUAL TO THE SIZE AT WHICH
C      THOSE EIGENVALUES WERE COMPUTED AND KMAX MUST BE LISTED AS
C      LARGER THAN MB(1).
-----------------------------------------------------------------
```

Below is a listing of the input/output files which are accessed by the complex symmetric Lanczos eigenvector program, CSLEVEC. Included also is a sample of the input file which CSLEVEC requires on file 5. The parameters in this file are supplied in free format. CSLEVEC computes eigenvectors for each of a user-specified subset of the eigenvalues computed by the companion program CSLEVAL.

SAMPLE DEFINITIONS OF THE INPUT/OUTPUT FILES FOR CSLEVEC

```
---------------------------------------------------------------
*CSLEVEC EXEC LANCZOS EIGENVECTOR PROGRAM COMPLEX SYMMETRIC CASE
FI 06 TERM
FILEDEF  2 DISK &1       HISTORY   A (RECFM F LRECL 80 BLOCK 80
FILEDEF  3 DISK &1       GOODEV    A (RECFM F LRECL 80 BLOCK 80
FILEDEF  4 DISK &1       ERRINV    A (RECFM F LRECL 80 BLOCK 80
FILEDEF  5 DISK CSLEVEC  INPUT     A (RECFM F LRECL 80 BLOCK 80
FILEDEF  8 DISK &1       INPUT     A (RECFM F LRECL 80 BLOCK 80
FILEDEF  9 DISK &1       ERREST    A (RECFM F LRECL 80 BLOCK 80
FILEDEF 10 DISK &1       BOUNDS    A (RECFM F LRECL 80 BLOCK 80
FILEDEF 11 DISK &1       TEIGVECS  A (RECFM F LRECL 80 BLOCK 80
FILEDEF 12 DISK &1       RITZVECS  A (RECFM F LRECL 80 BLOCK 80
FILEDEF 13 DISK &1       PAIGE     A (RECFM F LRECL 80 BLOCK 80
LOAD  CSLEVEC  CSLESUB  CSLEMULT
---------------------------------------------------------------
```

SAMPLE INPUT FILE FOR CSLEVEC

```
---------------------------------------------------------------
 CSLEVEC EIGENVECTORS COMPLEX SYMMETRIC CASE NO REORTHOGONALIZATION
LINE 1  MDIMTV    MDIMRV  MBETA (MAX.DIMENSIONS,TVEC,RITVEC AND BETA
         10000     10000   2000
LINE 2        RELTOL
        .0000000001
LINE 3  MBOUND   NTVCON SVTVEC IREAD (FLAGS
           0        1     0     1
LINE 4  TVSTOP   LVCONT ERCONT  IWRITE (FLAGS
           0        1     1       1
LINE 5    RHSEED  (RANDOM GENERATOR SEED FOR STARTING VECTOR IN INVERM)
        45329517
LINE 6  MATNO       N
          100     100
---------------------------------------------------------------
```

CHAPTER 8

REAL SYMMETRIC MATRICES, 'BLOCK' LANCZOS CODE

SECTION 8.1 INTRODUCTION

The FORTRAN codes in this chapter address the question of using an iterative 'block' Lanczos procedure to compute a 'few' extreme eigenvalues and a basis for the corresponding invariant subspace of a given real symmetric matrix A. An eigenvalue is extreme if it is one of the algebraically-smallest or the algebraically-largest eigenvalues.

For a given real symmetric matrix A, these codes compute the q algebraically-largest eigenvalues, λ_i, $1 \le i \le q$, of A and corresponding orthonormal real vectors $X_q \equiv \{x_1,...,x_q\}$ such that

$$AX_q = X_q A_q \quad \text{where } A_q \equiv X_q^T A X_q. \tag{8.1.1}$$

Typically, $A_q = \Lambda_q$, a diagonal matrix whose nonzero entries are the eigenvalues λ_i. The number q is small and specified by the user.

Real symmetric matrices are discussed in detail in Stewart [1973]. See Section 2.1 for a brief summary of the properties of real symmetric matrices which we use. The Lanczos procedure included in this chapter is not a true block Lanczos procedure. It is a hybrid Lanczos algorithm which combines ideas from the iterative block Lanczos procedures such as the one in Cullum and Donath [1974] and from the single-vector Lanczos procedure given in Chapter 2.

Several differences between the single-vector Lanczos codes in Chapter 2-5 and the iterative 'block' Lanczos codes should be stated explicitly. The single-vector Lanczos codes do not have the capability of directly computing the A-multiplicities of the computed eigenvalues. The 'block' procedures however, will determine the true A-multiplicity of a given computed eigenvalue and compute a complete invariant subspace for such an eigenvalue, as long as the number of Lanczos vectors in the first block is large enough. In order to determine A-multiplicities the single-vector codes have to do additional computation. In some cases these multiplicities and a basis for the required eigenspace can be determined without too much additional computation. This is true for example, whenever the desired eigenvalues replicate readily during the single-vector Lanczos computations.

The single-vector Lanczos procedures in Chapters 2-5 function in two stages. First the eigenvalues of the matrix being considered are computed, and then a separate program is used to compute the corresponding desired eigenvectors. The iterative 'block' Lanczos codes obtain approximations to the eigenvalues and to the eigenvectors simultaneously. Both types of codes are restartable from pre-existing computations. However, restarting has a different meaning for the two different types of codes. In the single-vector codes, restarting means computing a larger Lanczos T-matrix, starting from a pre-existing smaller one. The eigenvalue and eigenvector computations are then repeated on the larger T-matrix. In the iterative block procedures, restarting means using the current approximations to the eigenvectors (or more correctly to a basis for the desired eigenspace), to initiate another iteration of the 'block' Lanczos procedure.

The single-vector Lanczos procedures in Chapters 2-7 are iterative only in the sense that one may consider several Lanczos T-matrices of different sizes before achieving the desired convergence. However, the 'block' procedure presented here is genuinely iterative. On each iteration a block version of the Lanczos recursion is used to generate a sequence of blocks of Lanczos vectors, simultaneously generating a 'small' real symmetric Lanczos T-matrix. The eigenvalues and eigenvectors of this small Lanczos matrix are computed and mapped into approximating eigenvectors for the given matrix using the Lanczos vectors. These approximate eigenvectors then become the starting block of Lanczos vectors for the next iteration of the block Lanczos procedure. This 'block' procedure is described in detail in Section 7.5 of Chapter 7 in Volume 1.

As we said earlier, the 'block' procedure included here is a hybrid of the single-vector and of the basic iterative block Lanczos procedures. This procedure is based upon a modification of the following basic block version of the Lanczos recursion

$$Q_{j+1}B_{j+1} = AQ_j - Q_jA_j - Q_{j-1}B_j^T \equiv P_j \tag{8.1.2}$$

for $j=1,2,...,s$ where the coefficient matrices A_j and B_{j+1} are block analogs of the scalar coefficients in the single vector Lanczos recursion. In the standard block procedure,

$$A_j \equiv Q_j^T(AQ_j - Q_{j-1}B_j^T) \text{ and each } B_{j+1} \tag{8.1.3}$$

is obtained by the Gram-Schmidt orthogonalization of the columns of P_j and $s<<n$, the order of the given A-matrix. Our single-vector Lanczos procedures do not use any reorthogonalization at any point in the computations. However, in our block procedures we require near-orthogonality of the Q-blocks. This orthogonality is maintained by incorporating reorthogonalization of the blocks generated within a given iteration, with respect to certain vectors in the first Lanczos block.

The sequence of 'blocks' generated on each iteration of this hybrid procedure has the property that the first Q-block, Q_1, contains at least as many vectors as the user is trying to compute; however, the second and succeeding blocks contain exactly one vector. The corresponding Lanczos T-matrices are not block tridiagonal. Each has a border of blocks occupying the first q rows and columns and is tridiagonal below these rows and columns.

The convergence of these procedures is monitored by the subroutine DIAGOM. Convergence requires reasonable gaps between the eigenvalues requested and the eigenvalues not being approximated by the block procedure. Typically, it is the ratio of these gaps to the spread, and the distribution of the A-eigenvalues over the A-spread which controls the rate of convergence. In particular, an iterative block Lanczos procedure may have difficulty with a matrix with evenly-distributed eigenvalues. Heuristics are incorporated which allow the number of vectors used in the first Lanczos block to vary. If the convergence stagnates the procedure will terminate to allow the user to intervene and reset the program parameters if desired.

BLEVAL, the main 'block' program for these real symmetric eigenelement computations, calls the subroutine LANCZS which on each iteration then calls the subroutine LANCI1 to generate a sequence of Q-blocks for that iteration. Subroutine LANCZS then calls the subroutine DIAGOM to diagonalize the Lanczos T-matrix generated on that iteration and to compute the updated approximations to the desired eigenspace. Convergence is checked and if it has not occurred, another iteration of the block Lanczos procedure is carried out.

In this 'block' procedure there is no identification or 'spurious' test for the eigenvalues of the Lanczos T-matrix. Since near-orthogonality of the Lanczos blocks is maintained, the q algebraically-largest eigenvalues of the T-matrices are approximations to the q algebraically-largest eigenvalues of the A-matrix being used in the recursions. This statement however, is not true for the other eigenvalues of these T-matrices because the orthogonality maintained is only with respect to the eigenspace which goes with the first q eigenvalues. The accuracy of the computed eigenvalues and eigenvectors is estimated on each iteration as part of the process of computing the second block of Lanczos vectors.

All computations are in double precision real arithmetic. The user must supply a subroutine USPEC which defines and initializes the A-matrix and a subroutine BMATV which computes Ax for any specified vector x. The small T-matrix eigenelement computations use two subroutines from the EISPACK Library [1976,1977], TRED2 and IMTQL2. If the q algebraically-smallest eigenvalues are required, then the user must supply the programs with a subroutine which computes -Ax rather that Ax. The user should refer to Chapter 7 in Volume 1 for more details on iterative block Lanczos procedures.

SECTION 8.2 DOCUMENTATION FOR PROGRAMS IN CHAPTERS 8 AND 9

```
C-----DOCUMENTATION FOR-------------------------------------------------
C     BLOCK LANCZOS EIGENVALUE/EIGENVECTOR PROGRAMS FOR
C     (1) REAL SYMMETRIC MATRICES
C     (2) FACTORED INVERSES OF REAL SYMMETRIC MATRICES
C
C     AUTHORS:  JANE CULLUM, W. E. DONATH AND RALPH A. WILLOUGHBY,
C               IBM RESEARCH, YORKTOWN HEIGHTS, NY 10598.
C               PHONE: 914-945-2227
C
C
C     REAL SYMMETRIC MATRICES:
C
C     GIVEN A REAL SYMMETRIC MATRIX A THE FILES BLEVAL, BLSUB AND
C     BLMULT CAN BE USED TO COMPUTE A FEW EXTREME EIGENVALUES
C     OF A, THAT IS THE ALGEBRAICALLY-LARGEST OR THE ALGEBRAICALLY-
C     SMALLEST EIGENVALUES, AND A BASIS FOR THE CORRESPONDING
C     EIGENSPACE.
C
C     FACTORED INVERSES OF REAL SYMMETRIC MATRICES:
C
C     GIVEN A REAL SYMMETRIC MATRIX A, THE BLOCK PROCEDURE
C     CAN BE APPLIED TO AN ASSOCIATED B-MATRIX WHICH IS A
C     SCALED, SHIFTED AND PERMUTED VERSION OF A.  THAT IS,
C     B = SO*P*A*P' + SHIFT*I WHERE THE SCALE SO AND THE SHIFT
C     ARE CHOSEN BY THE USER TO PLACE THE DESIRED EIGENVALUES
C     AT THE EXTREME OF THE SPECTRUM OF B-INVERSE, AND THE
C     PERMUTATION P IS CHOSEN SO THAT THE SPARSITY OF THE A-MATRIX
C     IS PRESERVED IN THE SPARSITY OF THE FACTORIZATION OF B.
C     THE INVERSE BLOCK PROCEDURE REQUIRES A SUBROUTINE BLSOLV
C     THAT FOR A GIVEN VECTOR U, COMPUTES THE VECTOR V SUCH THAT
C     B*V = U, USING THE FACTORIZATION OF B.  THE SAMPLE BLSOLV
C     SUBROUTINE PROVIDED ASSUMES THAT THE B-MATRIX IS POSITIVE
C     DEFINITE AND THAT THE CHOLESKY FACTORS OF B ARE SUPPLIED
C     ON FILE 7.  HOWEVER, THE USER MAY REPLACE THIS SUBROUTINE
C     BY ONE THAT COMPUTES A MORE GENERAL FACTORIZATION
C     L*D*(L-TRANSPOSE) FOR AN INDEFINITE SYMMETRIC MATRIX.
C     THE BLOCK PROCEDURE USED IN THIS FASHION USES THE FILES
C     BLIEVAL, BLIMULT AND BLSUB.
C
C
C     ALGORITHM:
C     THESE PROGRAMS USE A BLOCK FORM OF LANCZOS TRIDIAGONALIZATION
C     WITH REORTHOGONALIZATION ONLY WITH RESPECT TO VECTORS
C     IN THE 1ST Q-BLOCK.  THE PROCEDURES ARE ITERATIVE, GENERATING
C     ON EACH ITERATION A SMALL SYMMETRIC LANCZOS MATRIX, T.
C     THE EIGENVALUES AND EIGENVECTORS OF THE SMALL MATRIX ARE
C     COMPUTED USING SUBROUTINES FROM THE EISPACK LIBRARY.
C     THE RELEVANT SUBSET OF THE T-EIGENVECTORS IS THEN MAPPED
C     INTO THE LARGE N-SPACE CORRESPONDING TO THE MATRIX BEING
C     USED BY THE LANCZS SUBROUTINE, CONVERGENCE IS CHECKED,
C     AND IF CONVERGENCE OF THE DESIRED EIGENVALUES AND
C     EIGENVECTORS HAS NOT YET OCCURRED, THEN THE CURRENT
C     APPROXIMATIONS TO THE DESIRED EIGENSPACE ARE USED AS
C     STARTING VECTORS FOR THE NEXT ITERATION OF BLOCK LANCZOS.
C
C     USERS SHOULD NOTE THAT TYPICALLY IN THE BLOCK LANCZOS
C     PROCEDURES, IT IS THE RATIO OF THE GAPS TO THE SPREAD THAT
C     CONTROLS THE CONVERGENCE ALONG WITH HOW THE EIGENVALUES
C     ARE DISTRIBUTED OVER THAT SPREAD.   THE BIGGER THE GAPS
C     BETWEEN THE ONES BEING COMPUTED AND THE CLOSEST ONES NOT
C     BEING COMPUTED AND THE WEAKER THE SPREAD, THE FASTER THE
C     CONVERGENCE WILL BE.  WITHOUT DECENT GAPS THIS PROCEDURE
C     WILL NOT CONVERGE.  THE PROGRAMS CONTAIN CHECKS ON
C     THE ACTUAL RATE OF CONVERGENCE WHICH WILL CAUSE THE
```

```
C     PROCEDURE TO TERMINATE IF CONVERGENCE IS NOT OCCURRING
C     SUFFICIENTLY RAPIDLY.  THE USER MAY THEN CHANGE EITHER OR
C     BOTH THE MAXIMUM SIZE T-MATRIX ALLOWED AND THE NUMBER
C     OF VECTORS IN THE FIRST Q-BLOCK AND RERUN THE PROCEDURE
C     WITH THE CURRENT APPROXIMATION TO THE DESIRED EIGENSPACE
C     AS THE STARTING BLOCK OF VECTORS.
C
C
C     THE IDEAS USED IN THESE PROGRAMS ARE DISCUSSED IN THE FOLLOWING
C     REFERENCES.
C
C     1.  JANE CULLUM AND RALPH A. WILLOUGHBY, LANCZOS ALGORITHMS
C         FOR LARGE SYMMETRIC MATRICES,  PROGRESS IN
C         SCIENTIFIC COMPUTING, EDITORS, G. GOLUB, H.O. KREISS,
C         S. ARBARBANEL, AND R. GLOWINSKI,  BIRKHAUSER BOSTON INC.,
C         CAMBRIDGE, MASSACHUSETTS, 1984.
C
C     2.  JANE CULLUM AND W.E. DONATH, A BLOCK LANCZOS ALGORITHM
C         FOR COMPUTING THE Q ALGEBRAICALLY-LARGEST EIGENVALUES AND
C         A CORRESPONDING EIGENSPACE OF LARGE, SPARSE REAL SYMMETRIC
C         MATRICES, PROCEEDINGS OF THE 1974 IEEE CONFERENCE ON
C         DECISION AND CONTROL, PHOENIX, ARIZONA, PP.505-509, NOVEMBER
C         1974.
C
C     3.  JANE CULLUM,  AN ACCELERATED 'BLOCK' LANCZOS ALGORITHM
C         FOR A FEW EXTREME EIGENVALUES OF A LARGE, SPARSE REAL
C         SYMMETRIC MATRIX.  IBM REPORT 1983.  PRESENTED AT THE
C         SPARSE MATRIX CONFERENCE, FAIRFIELD GLADE, TENNESSEE,
C         OCTOBER 1982.
C
C
C-----PORTABILITY-------------------------------------------------------
C
C     PROGRAMS WERE TESTED FOR PORTABILITY USING THE PFORT VERIFIER.
C     FOR DETAILS OF THE VERIFIER SEE FOR EXAMPLE, B. G. RYDER AND
C     A. D. HALL, "THE PFORT VERIFIER", COMPUTING SCIENCE TECHNICAL
C     REPORT 12, BELL LABORATORIES, MURRAY HILL, NEW JERSEY 07974,
C     (REVISED), JANUARY 1981.
C
C     EXCEPT FOR THE FOLLOWING CONSTRUCTIONS WHICH CAN BE EASILY
C     MODIFIED BY THE USER TO MATCH THE PARTICULAR COMPUTER BEING
C     USED, THE PROGRAM STATEMENTS ARE PORTABLE.
C
C    NONPORTABLE STATEMENTS.
C
C    IN BLEVAL, BLIEVAL (MAIN PROGRAMS)
C        1.  DATA/MACHEP STATEMENT
C        2.  ALL READ(5,*) STATEMENTS (FREE FORMAT)
C        3.  FORMAT(20A4) USED FOR THE EXPLANATORY HEADER ARRAY, EXPLAN
C        4.  FORMAT(4Z20) WHICH CAN BE USED TO WRITE LARGE VECTOR
C            FILES
C        5.  THE COMMON BLOCK: LOOPS.
C    IN BLMULT, BLIMULT
C        1.  IN BMATV, BLSOLV, AND USPEC, THE ENTRIES WHICH
C            PASS THE STORAGE LOCATIONS OF THE ARRAYS DEFINING
C            THE USER-SPECIFIED MATRIX OR FACTORIZATION.
C    IN BLSUB
C        1.  ALL STATEMENTS ARE PORTABLE EXCEPT THE ENTRY TO
C            SUBROUTINE LPERM WHICH PASSES THE PERMUTATION USED
C            TO OBTAIN THE B-MATRIX FROM SUBROUTINE USPEC.
C            SUBROUTINE LPERM IS USED ONLY IN CASE (2).
C
C
C-----MATRIX SPECIFICATION-----------------------------------------------
C
C     SUBROUTINE USPEC IS USED TO SPECIFY THE MATRIX WHICH THE BLOCK
C     LANCZOS PROCEDURE WILL USE.  IN CASE (1) THIS IS THE USER-
```

```
C       SPECIFIED A-MATRIX.   IN CASE (2) THE FACTORIZATION OF THE
C       ASSOCIATED B-MATRIX IS SPECIFIED.  SUBROUTINE USPEC HAS THE
C       CALLING SEQUENCE
C
C              CALL USPEC(N,MATNO,NNZ,AVER)
C
C       WHERE N IS THE ORDER OF THE USER-SUPPLIED MATRIX A,
C       MATNO IS AN <= 8 DIGIT INTEGER USED AS A MATRIX AND
C       TEST IDENTIFICATION NUMBER, NNZ IS THE AVERAGE NUMBER
C       OF NONZERO ENTRIES IN EACH COLUMN, AND AVER IS THE
C       AVERAGE SIZE OF THE NONZERO ENTRIES IN THE MATRIX USED
C       BY LANCZS.  NOTE THAT NNZ AND AVER ARE DEFINED AS DOUBLE
C       PRECISION SCALARS.  THE MAIN PROGRAMS ASSUME THAT THEY
C       ARE COMPUTED IN USPEC.  THE USPEC SUBROUTINE
C       DEFINES AND DIMENSIONS THE ARRAYS REQUIRED TO
C       SPECIFY THE MATRIX THAT WILL BE USED BY THE LANCZS
C       SUBROUTINE AND INITIALIZES THESE ARRAYS.  THE STORAGE
C       LOCATIONS OF THESE ARRAYS ARE THEN PASSED TO THE
C       SUBROUTINE BMATV IN CASE (1) AND TO THE SUBROUTINE BSOLV
C       IN CASE (2).  SAMPLE SUBROUTINES ARE INCLUDED FOR EACH
C       CASE.  CASE (1) ASSUMES THAT THE A-MATRIX IS STORED ON
C       FILE 8.  CASE (2) ASSUMES THAT THE FACTORIZATION OF THE
C       B-MATRIX IS STORED ON FILE 7.
C
C       IN CASE (1) :
C       BMATV IS THE SUBROUTINE USED BY THE LANCZS SUBROUTINE
C       THAT GENERATES THE LANCZOS T-MATRICES.  SUBROUTINE
C       BMATV HAS THE CALLING SEQUENCE
C
C              CALL BMATV(W,U)
C
C       WHERE U AND W ARE DOUBLE PRECISION VECTORS.  FOR A GIVEN
C       W, BMATV CALCULATES U = A*W  FOR THE USER-SPECIFIED MATRIX A.
C       A SAMPLE BMATV IS INCLUDED FOR AN ARBITRARY SPARSE,
C       SYMMETRIC A-MATRIX STORED IN THE SPARSE FORMAT SPECIFIED
C       IN THE CORRESPONDING SAMPLE USPEC SUBROUTINE.
C
C       IN CASE (2):
C       THE LANCZOS T-MATRICES ARE GENERATED USING SPARSE MATRIX
C       INVERSION, USING THE SUBROUTINE BLSOLV.  THE CALLING
C       SEQUENCE OF BLSOLV IS
C
C                     CALL BLSOLV(U,V)
C
C       WHERE U AND V ARE DOUBLE PRECISION VECTORS.  FOR A GIVEN V,
C       BLSOLV COMPUTES U = (B-INVERSE)*V USING A SPARSE
C       FACTORIZATION OF THE B-MATRIX ASSOCIATED WITH THE USER-
C       SPECIFIED A-MATRIX.
C
C       THE FOLLOWING SPARSE MATRIX FORMAT IS USED TO STORE THE
C       MATRICES IN THE SAMPLE PROGRAMS:
C       ICOL(K), K = 1,NZL, NUMBER OF SUBDIAGONAL NONZEROS IN COLUMN K.
C       IROW(K), K = 1,NZS, ROW INDEX OF ASD(K).
C       AD(K), K=1,N, CONTAINS THE DIAGONAL ELEMENTS OF THE A-MATRIX.
C       ASD(K), K=1,NZS  CONTAINS THE SUBDIAGONAL ELEMENTS OF A BY COLUMN.
C       NZS = NUMBER OF NONZERO ELEMENTS BELOW THE DIAGONAL OF A
C       NZL = INDEX OF LAST COLUMN WITH NONZERO SUBDIAGONAL ENTRIES
C       N = ORDER OF THE A-MATRIX.
C
C       IN CASE (1) THE A-MATRIX IS STORED IN THIS FORMAT ON FILE 8.
C       IN CASE (2), IN THE SAMPLE USPEC PROVIDED WHICH IS ONLY
C       FOR POSITIVE DEFINITE B-MATRICES, THE SPARSE CHOLESKY FACTOR
C       OF B, L, IS STORED ON FILE 7 IN THE ABOVE SPARSE FORMAT
C       USING ARRAYS BD AND BSD.  IN CASE (2) THE OPTIONAL AUXILIARY
C       PROGRAMS PERMUT AND LORDER ALSO REQUIRE THE A-MATRIX;
C       HOWEVER, THE BLOCK LANCZOS PROCEDURE ONLY USES THE
C       FACTORIZATION OF THE B-MATRIX.
```

```
C
C
C-----MACHEP---------------------------------------------------------
C
C
C     MACHEP IS A MACHINE DEPENDENT PARAMETER SPECIFYING THE RELATIVE
C     PRECISION OF THE FLOATING POINT ARITHMETIC USED.
C     MACHEP = 2.2 * 10**-16 FOR DOUBLE PRECISION ARITHMETIC ON
C     IBM 370-3081.
C
C     THE USER WILL HAVE TO RESET THIS PARAMETER TO
C     THE CORRESPONDING VALUE FOR THE MACHINE BEING USED.  NOTE THAT
C     IF A MACHINE WITH A MACHINE EPSILON THAT IS MUCH LARGER THAN THE
C     VALUE GIVEN HERE IS BEING USED, THEN THERE COULD BE
C     PROBLEMS WITH THE TOLERANCES.
C
C
C-----SUBROUTINES AND FUNCTIONS USER MUST SUPPLY----------------------
C
C
C     GENRAN,  FINPRO,  MASK,  USPEC,  AND
C     CASE (1)  BMATV:  CASE (2)  BLSOLV :
C
C     GENRAN = COMPUTES K PSEUDO-RANDOM NUMBERS AND STORES THEM IN
C              THE REAL ARRAY, G.  THIS SUBROUTINE IS USED TO
C              GENERATE STARTING VECTORS FOR THE BLOCK LANCZOS
C              PROCEDURE.  CALLED FROM LANCZS SUBROUTINE.
C              USER CAN SUPPLY STARTING VECTORS FOR THE BLOCK
C              PROCEDURES.  ANY ADDITIONAL VECTORS REQUIRED ARE
C              GENERATED RANDOMLY BY GENRAN.  VECTORS SUPPLIED MUST
C              BE STORED ON FILE 10.  THE NUMBER OF SUCH VECTORS TO
C              BE READ IN IS SPECIFIED BY THE PARAMETER KSET.  THE
C              EXISTING CALLING SEQUENCE IS
C
C                          CALL GENRAN(IIX,G,K).
C
C              WHERE IIX =INTEGER SEED, G = REAL ARRAY WHOSE DIMENSION
C              MUST BE >= K.  K PSEUDO-RANDOM NUMBERS ARE GENERATED
C              AND PLACED IN G.
C
C     FINPRO = DOUBLE PRECISION FUNCTION WHICH COMPUTES THE INNER
C              PRODUCT OF 2 DOUBLE PRECISION VECTORS OF DIMENSION N.
C              EXISTING CALLING SEQUENCE IS
C
C                          CALL FINPRO(N,V,J,W,K).
C
C              COMPUTES THE INNER PRODUCT OF DIMENSION N OF THE VECTORS
C              V AND W.  SUCCESSIVE COMPONENTS OF V AND OF W ARE STORED
C              AT LOCATIONS THAT ARE ,RESPECTIVELY, J AND K UNITS APART.
C
C     MASK = MASKS OVERFLOW AND UNDERFLOW.  OPTIONAL.
C            USER MUST SUPPLY OR COMMENT OUT CALL.
C
C    USPEC = DIMENSIONS AND INITIALIZES ARRAYS NEEDED TO SPECIFY
C            MATRIX USED BY LANCZS SUBROUTINE.  SEE MATRIX
C            SPECIFICATION SECTION.
C
C    BMATV = CASE (1) ONLY: COMPUTES MATRIX-VECTOR MULTIPLY FOR
C            USER-SUPPLIED A-MATRIX. SEE MATRIX SPECIFICATION SECTION.
C
C   BLSOLV = CASE (2) ONLY:  FOR GIVEN VECTOR V, COMPUTES U SUCH
C            B*U = V, GIVEN THE SPARSE FACTORIZTION OF THE B-MATRIX.
C
C
C-----PARAMETER CONTROLS---------------------------------------------
C
C
```

```
C     PARAMETER CONTROLS ARE INTRODUCED TO CONTROL VARIOUS
C     ASPECTS OF THESE PROGRAMS.
C
C     THE FLAG EFLAG SPECIFIES THE NUMBER OF COMPUTATIONAL PHASES.
C
C     EFLAG  = (0,1)  MEANS
C
C              (0) PROGRAM TERMINATES AFTER COMPLETING PHASE 1
C                  COMPUTATIONS.
C
C              (1) PROGRAM COMPLETES BOTH PHASE 1 AND PHASE 2 OF
C                  THE COMPUTATIONS.
C
C     THE FLAG OFLAG CONTROLS THE ORTHOGONALITY CHECKS BETWEEN THE
C     JTH Q-BLOCK GENERATED AND THAT VECTOR IN THE 1ST Q-BLOCK THAT
C     IS GENERATING DESCENDANTS.  FOR SAFETY, OFLAG SHOULD BE 1.
C
C     OFLAG = (0,1)  MEANS
C
C              (0) NO ORTHOGONALITY CHECKS ARE MADE ON PHASE
C                  1 PORTION OF THE COMPUTATIONS.  ORTHOGONALITY
C                  CHECKS ARE ALWAYS MADE ON PHASE 2 PORTION.
C
C              (1) PROGRAM CHECKS ORTHOGONALITY OF GENERATED
C                  Q-BLOCKS W.R.T. THAT VECTOR IN THE 1ST Q-BLOCK
C                  THAT IS GENERATING DESCENDANTS IN BOTH PHASE
C                  1 AND PHASE 2 OF THE COMPUTATIONS.
C
C     THE FLAG IWRITE DETERMINES THE AMOUNT OF OUTPUT TO FILE 6
C     DURING THE COMPUTATIONS
C
C     IWRITE = (0,1)   MEANS
C
C              (0) ABBREVIATED OUTPUT TO FILE 6.
C
C              (1) ADDITIONAL COMMENTARY ON THE COMPUTATIONS IS
C                  PRINTED TO FILE 6.
C
C     THE PROGRAM ALWAYS WRITES A LIST OF THE COMPUTED EIGENVALUES
C     AND THE BASIS FOR THE CORRESPONDING EIGENSPACE TO FILE 15,
C     ALONG WITH ESTIMATES OF THE ERRORS IN THESE COMPUTED VALUES.
C
C-----INPUT/OUTPUT FILES-----------------------------------------------
C
C     ANY INPUT DATA OTHER THAN THE A-MATRIX, THE FACTORIZATION
C     OF THE B-MATRIX OR USER-SPECIFIED STARTING VECTORS SHOULD
C     BE STORED ON FILE 5.  SEE SAMPLE INPUT/OUTPUT FROM TYPICAL RUN.
C     THE READ STATEMENTS IN THE GIVEN FORTRAN PROGRAM ASSUME THAT
C     THE DATA STORED ON FILE 5 IS IN FREE FORMAT.  USER SHOULD NOTE
C     THAT 'FREE FORMAT' IS NOT CLASSIFIED AS PORTABLE BY PFORT SO THAT
C     THE USER MAY HAVE TO MODIFY THE READ STATEMENTS FROM FILE 5 TO
C     CONFORM TO WHAT IS PERMISSIBLE ON THE COMPUTER BEING USED.
C
C     FILE 6 WAS USED AS THE INTERACTIVE TERMINAL OUTPUT FILE.
C     THIS FILE PROVIDES A RUNNING ACCOUNT OF THE PROGRESS OF THE
C     COMPUTATIONS.  THE AMOUNT OF INFORMATION PRINTED OUT IS
C     CONTROLLED BY THE PARAMETER IWRITE.
C
C DESCRIPTION OF OTHER I/O FILES
C
C FILE (K)      CONTAINS:
C
C      (7)      INPUT FILE:
C               USED IN CASE (2).  CONTAINS THE FACTORIZATION
C               OF THE B-MATRIX.
C
C      (8)      INPUT FILE:
```

```
C              USED IN CASE (1).  CONTAINS THE ARRAYS REQUIRED         BLE03330
C              TO SPECIFY THE A-MATRIX.                                BLE03340
C                                                                      BLE03350
C      (10)    INPUT FILE:                                             BLE03360
C              CONTAINS USER-SUPPLIED STARTING VECTORS, IF ANY.        BLE03370
C              TYPICALLY, THESE WOULD BE 1 OR MORE EIGENVECTOR         BLE03380
C              APPROXIMATIONS OBTAINED DURING AN EARLIER RUN.          BLE03390
C                                                                      BLE03400
C      (13)    OUTPUT FILE:                                            BLE03410
C              CONTAINS EXTRA EIGENVECTOR APPROXIMATIONS THAT          BLE03420
C              WOULD OTHERWISE BE LOST UPON ANY REDUCTION IN THE       BLE03430
C              SIZE OF THE 1ST Q-BLOCK.  IF AT ANY STAGE IN THE        BLE03440
C              BLOCK PROCEDURE, THE SIZE OF THE 1ST Q-BLOCK IS         BLE03450
C              REDUCED FROM KACT TO KACTN, THE Q-VECTORS FROM          BLE03460
C              K = KACTN+1,KACT ARE WRITTEN TO FILE 13 FOR POSSIBLE    BLE03470
C              USE AS STARTING VECTORS IN A LATER RUN OF THE           BLE03480
C              BLOCK LANCZOS PROCEDURE.                                BLE03490
C                                                                      BLE03500
C      (15)    OUTPUT FILE:                                            BLE03510
C              CONTAINS COMPUTED EIGENVALUES AND CORRESPONDING         BLE03520
C              COMPUTED EIGENSPACE AVAILABLE AT THE TIME OF            BLE03530
C              TERMINATION OF THE BLOCK LANCZOS PROCEDURE.             BLE03540
C                                                                      BLE03550
C-----PARAMETERS SET BY THE BLOCK PROGRAMS-----------------------------BLE03560
C                                                                      BLE03570
C                                                                      BLE03580
C     SPREC = TOLERANCE USED IN CHECKING ORTHOGONALITY BETWEEN         BLE03590
C           COMPUTED Q-BLOCKS AND THAT VECTOR IN THE FIRST             BLE03600
C           Q-BLOCK THAT IS GENERATING DESCENDANTS.  SEE COMMENTS      BLE03610
C           ON OFLAG.                                                  BLE03620
C                                                                      BLE03630
C-----USER-SPECIFIED PARAMETERS ---------------------------------------BLE03640
C                                                                      BLE03650
C                                                                      BLE03660
C      FOR BOTH CASES:                                                 BLE03670
C                                                                      BLE03680
C      N, MATNO = INTEGERS. SIZE OF USER-SPECIFIED MATRIX AND MATRIX   BLE03690
C                IDENTIFICATION NUMBER OF 8 OR FEWER DIGITS.           BLE03700
C                                                                      BLE03710
C      MDIMQ, MDIMTM = INTEGERS.  USER-SPECIFIED DIMENSIONS OF THE     BLE03720
C                     Q-ARRAY AND OF THE TM-ARRAY.  MDIMQ >= N*KMAX    BLE03730
C                     AND MDIMTM >= MXBLK**2.                          BLE03740
C                                                                      BLE03750
C      MAXIT,MAXIT2  = INTEGERS.  MAXIMUM NUMBER OF CALLS TO BMATV     BLE03760
C                     (CASE(1)) OR TO BLSOLV (CASE (2)) ALLOWED        BLE03770
C                     RESPECTIVELY, IN PHASE 1 AND IN PHASE 2.         BLE03780
C                                                                      BLE03790
C      RELTOL = DOUBLE PRECISION SCALAR.  RELATIVE TOLERANCE USED      BLE03800
C             TO COMPUTE CONVERGENCE CRITERION FOR PHASE 2 OF          BLE03810
C             THE BLOCK PROCEDURE.                                     BLE03820
C                                                                      BLE03830
C      SEED   = INTEGER.  SEED FOR RANDOM NUMBER GENERATOR.            BLE03840
C             USED IN GENERATION OF STARTING VECTORS FOR               BLE03850
C             THE BLOCK PROCEDURES.                                    BLE03860
C                                                                      BLE03870
C      KMAX  = INTEGER.  MXBLK = (KMAX - 1) IS MAXIMUM ALLOWED SIZE    BLE03880
C            FOR THE SMALL LANCZOS T-MATRICES.                         BLE03890
C                                                                      BLE03900
C      KM    = INTEGER.  NUMBER OF EIGENVALUES AND EIGENVECTORS        BLE03910
C            TO BE COMPUTED.                                           BLE03920
C                                                                      BLE03930
C      KACT  = INTEGER.  INITIAL NUMBER OF VECTORS IN THE 1ST Q-BLOCK. BLE03940
C            IF THERE IS ANY POSSIBILITY THAT THE KM-TH DESIRED        BLE03950
C            EIGENVALUE IS MULTIPLE, AND THE USER NEEDS TO KNOW        BLE03960
C            THIS, THEN THE USER SHOULD SET KACT > KM.  OTHERWISE,     BLE03970
C            THIS PROGRAM WILL NOT BE ABLE TO DETERMINE THAT THAT      BLE03980
C            EIGENVALUE IS MULTIPLE UNLESS THE (KM-1)-TH AND KM-TH     BLE03990
```

```
C            HAPPEN TO BE MULTIPLE.  IF IN FACT, THE KM-TH
C            EIGENVALUE IS MULTIPLE AND THE USER NEEDS A BASIS FOR
C            THE CORRESPONDING EIGENSPACE, THEN THE PROCEDURE SHOULD
C            BE RERUN WITH THE EXISTING EIGENVECTORS APPROXIMATIONS
C            AS STARTING VECTORS AND A LARGER KACT TO GUARANTEE THAT
C            A COMPLETE BASIS FOR THAT EIGENSPACE HAS BEEN OBTAINED.
C
C      KSET =  INTEGER.  NUMBER OF STARTING VECTORS SUPPLIED BY THE
C            THE USER.  THESE VECTORS SHOULD BE ON FILE 10.
C
C
C      NSTAG = INTEGER.  NUMBER OF THE ITERATION BEYOND WHICH THE
C            CHANGE IN THE KM-TH RESIDUAL OVER THE PAST 10 ITERATIONS
C            IS MONITORED AND USED AS A MEASURE OF THE RATE OF
C            CONVERGENCE OF THE BLOCK PROCEDURE.
C
C      FRACT = DOUBLE PRECISION SCALAR.  EXPECTED OR HOPED FOR
C            FRACTIONAL CHANGE IN THE KM-TH RESIDUAL OVER THE PAST
C            BLOCK LANCZOS ITERATIONS USED TO TEST FOR STAGNATION
C            OF CONVERGENCE.
C
C      NNZ   = DOUBLE PRECISION SCALAR.  AVERAGE NUMBER OF NONZERO
C            ENTRIES PER ROW IN THE MATRIX USED IN THE LANCZOS
C            PROCEDURE.
C
C
C      AVER  = DOUBLE PRECISION SCALAR.  AVERAGE SIZE OF THE NONZERO
C            ENTRIES IN THE MATRIX USED IN THE LANCZOS PROCEDURE.
C
C      CASE (2) ONLY:
C
C      SO, SHIFT = DOUBLE PRECISION SCALARS.  MATRIX USED BY LANCZS
C                SUBROUTINE IS B = SO*P*A*P' + SHIFT*I  WHERE P
C                DENOTES A PERMUTATION MATRIX SELECTED TO PRESERVE
C                THE SPARSITY OF A IN THE FACTORIZATION OF B.
C                SO AND SHIFT ARE CHOSEN BY THE USER SO THAT THE
C                DESIRED EIGENVALUES BECOME THE EXTREME EIGENVALUES
C                OF B-INVERSE.
C
C
C-----CONVERGENCE TEST-----------------------------------------------------
C
C
C      THE CONVERGENCE TEST INCORPORATED IN THIS PROGRAM IS
C      BASED UPON THE FOLLOWING FACT:  GIVEN A REAL SYMMETRIC
C      MATRIX A, A VECTOR X OF NORM 1, AND A SCALAR EVAL
C      THEN THERE EXISTS AN EIGENVALUE AEVAL OF A SUCH THAT
C      DABS(AEVAL - EVAL) .LE. NORM(A*X - EVAL*X).  WITHIN
C      EACH ITERATION OF THE BLOCK LANCZOS PROCESS THESE TYPES
C      OF NORMS ARE COMPUTED IN THE PROCESS OF COMPUTING THE
C      2ND Q-BLOCK.
C
C
C-----ARRAYS REQUIRED------------------------------------------------------
C
C
C      Q(J)      = DOUBLE PRECISON ARRAY.  ITS DIMENSION MUST BE AT
C                LEAST AS LARGE AS KMAX*N, WHERE N IS THE ORDER OF
C                THE GIVEN MATRIX, AND MXBLK = KMAX - 1 IS THE
C                MAXIMUM SIZE T-MATRIX ALLOWED ON ANY GIVEN
C                ITERATION.  THE COLUMNS OF Q HOLD THE LANCZOS
C                VECTORS GENERATED ON EACH ITERATION OF BLOCK
C                LANCZOS PLUS THERE MUST BE AN ADDITIONAL COLUMN
C                AVAILABLE FOR WORK SPACE.  THE FIRST KACT COLUMNS
C                OF Q CONTAIN THE CURRENT APPROXIMATING EIGENSPACE.
C
C      E(J)    = DOUBLE PRECISION ARRAY.  ITS DIMENSION MUST BE AT
```

```
C              LEAST MXBLK = KMAX - 1.  ON EACH ITERATION CONTAINS      BLE04670
C              THE COMPUTED EIGENVALUES OF THE LANCZOS T-MATRIX.        BLE04680
C                                                                       BLE04690
C     TM(J)   = DOUBLE PRECISION ARRAY.  ITS DIMENSION MUST BE AT       BLE04700
C              LEAST MXBLK**2 WHERE MXBLK = KMAX - 1.  CONTAINS         BLE04710
C              THE LANCZOS T-MATRIX GENERATED ON EACH ITERATION         BLE04720
C              AND THEN THE COMPUTED EIGENVECTORS OF THIS MATRIX.       BLE04730
C              EISPACK SUBROUTINES ARE USED FOR THE SMALL               BLE04740
C              EIGENELEMENT COMPUTATIONS.  EISPACK SUBROUTINE           BLE04750
C              TRED2 IS USED TO REDUCE THE GIVEN T-MATRIX TO            BLE04760
C              TRIDIAGONAL FORM.  THE EIGENELEMENT PROBLEM FOR THE      BLE04770
C              TRIDIAGONAL MATRIX IS THEN SOLVED USING THE EISPACK      BLE04780
C              SUBROUTINE IMTQL2.                                       BLE04790
C                                                                       BLE04800
C     EXPLAN(J) = REAL ARRAY.  ITS DIMENSION IS 20.  THIS ARRAY IS      BLE04810
C                USED TO ALLOW EXPLANATORY COMMENTS IN THE INPUT FILES.BLE04820
C                                                                       BLE04830
C     G(J) = REAL ARRAY.  ITS DIMENSION MUST BE >= N.  IT IS USED       BLE04840
C            FOR HOLDING THE PSEUDO-RANDOM NUMBERS USED TO GENERATE     BLE04850
C            ANY STARTING VECTORS NOT SUPPLIED BY THE USER.             BLE04860
C                                                                       BLE04870
C     RESIDL(J),  = DOUBLE PRECISION ARRAYS.  DIMENSION >= MAXIMUM      BLE04880
C     RESIDK(J),    NUMBER OF ITERATIONS ALLOWED.  MAXIMUM IS           BLE04890
C                   CURRENTLY SET TO 100. USED TO MONITOR THE           BLE04900
C                   RATE OF CONVERGENCE.                                BLE04910
C                                                                       BLE04920
C     TD(J), TOD(J), = DOUBLE PRECISION ARRAYS.  DIMENSION >= MXBLK.    BLE04930
C         SM(J)         WORK SPACES.                                    BLE04940
C                                                                       BLE04950
C     DESC(J), XLFT(J), = INTEGER ARRAYS.  DIMENSION >= MXBLK.          BLE04960
C         LEFT(J)          WORK SPACES.                                 BLE04970
C                                                                       BLE04980
C     DIR(2,J)  = 2-DIMENSIONAL INTEGER ARRAY.  COLUMN DIMENSION >=     BLE04990
C                MXBLK, ROW DIMENSION 2.  KEEPS TRACK OF NUMBER         BLE05000
C                OF VECTORS IN EACH QBLOCK.                             BLE05010
C                                                                       BLE05020
C     CASE (2) ONLY:                                                    BLE05030
C                                                                       BLE05040
C     IPR(J), IPT(J) = INTEGER ARRAYS.  EACH OF DIMENSION AT LEAST N.   BLE05050
C                     USED TO STORE THE REORDERING (IF ANY) OF          BLE05060
C                     THE GIVEN MATRIX.                                 BLE05070
C                                                                       BLE05080
C     OTHER ARRAYS                                                      BLE05090
C                                                                       BLE05100
C     THE USER IN THE SUBROUTINE USPEC MUST SPECIFY WHATEVER ARRAYS     BLE05110
C     ARE REQUIRED TO DEFINE THE MATRIX BEING USED BY LANCZS.           BLE05120
C                                                                       BLE05130
C                                                                       BLE05140
C-----SUBROUTINES INCLUDED----------------------------------------------BLE05150
C                                                                       BLE05160
C                                                                       BLE05170
C     LANCZS = CONTAINS MAJOR LOOP FOR BLOCK LANCZOS PROCEDURES.        BLE05180
C              CALLED FROM MAIN PROGRAM, CALLS SUBROUTINE LANCI1        BLE05190
C              TO GENERATE WITHIN A GIVEN ITERATION THE Q-BLOCKS        BLE05200
C              AND CORRESPONDING LANCZOS T-MATRICES.  THEN CALLS        BLE05210
C              SUBROUTINE DIAGOM TO COMPUTE THE EIGENELEMENTS           BLE05220
C              OF THE LANCZOS T-MATRIX AND TO MAP THE RELEVANT          BLE05230
C              T-EIGENVECTORS INTO RITZ VECTORS FOR THE A-MATRIX.       BLE05240
C                                                                       BLE05250
C     LANCI1 =  ON EACH ITERATION OF BLOCK LANCZOS COMPUTES             BLE05260
C               Q-SUBBLOCKS.                                            BLE05270
C                                                                       BLE05280
C     DIAGOM = CALLS EISPACK SUBROUTINES TO COMPUTE THE                 BLE05290
C              EIGENELEMENTS OF THE SMALL LANCZOS T-MATRICES            BLE05300
C              GENERATED ON EACH ITERATION OF BLOCK LANCZOS.            BLE05310
C              COMPUTES CORRESPONDING RITZ VECTORS FOR A-MATRIX.        BLE05320
C              MONITORS CONVERGENCE OF BLOCK LANCZOS PROCEDURE.         BLE05330
```

```
C                                                                       BLE05340
C     START  = GENERATES ANY REQUIRED STARTING VECTORS FOR 1ST          BLE05350
C              Q-BLOCK FOR FIRST ITERATION OF BLOCK LANCZOS.            BLE05360
C                                                                       BLE05370
C     ORTHOG = GIVEN A SET OF Q-VECTORS, Q(J), J = MA,MB,               BLE05380
C              ORTHOGONALIZES THESE VECTORS W.R.T. THE Q-VECTORS        BLE05390
C              Q(J), J = 1,MA-1.                                        BLE05400
C                                                                       BLE05410
C     LPERM  = (USED IN CASE (2) ONLY) GIVEN A MATRIX B AND A           BLE05420
C              PERMUTATION P DEFINED IN THE VECTORS IPR AND IPT,        BLE05430
C              AND A VECTOR X COMPUTE EITHER (P-TRANSPOSE)*X OR PX.     BLE05440
C                                                                       BLE05450
C     CASE (2) ONLY:                                                    BLE05460
C     FOR OPTIONAL PRELIMINARY PROCESSING:                              BLE05470
C                                                                       BLE05480
C     PERMUT  (STAND-ALONE PROGRAM):                                    BLE05490
C     USES THE NONZERO STRUCTURE OF A GIVEN MATRIX A.                   BLE05500
C     CAN BE USED TO OBTAIN A REORDERING OF A THAT WILL PRESERVE        BLE05510
C     THE SPARSENESS OF A UNDER FACTORIZATION.  PERMUT CALLS            BLE05520
C     CALLS THE SPARSPAK PACKAGE, (A. GEORGE, J. LIU, E. NG,            BLE05530
C     U. WATERLOO).  SEE THE PERMUT FORTRAN CODE FOR DETAILS.           BLE05540
C                                                                       BLE05550
C     LORDER (STAND-ALONE PROGRAM) :                                    BLE05560
C     GIVEN A MATRIX C IN SPARSE FORMAT AND A PERMUTATION P,            BLE05570
C     COMPUTES THE REORDERED MATRIX B = P*C*P' AND WRITES IT            BLE05580
C     TO FILE 9 IN SPARSE FORMAT.  SEE THE LORDER FORTRAN CODE          BLE05590
C     FOR DETAILS.                                                      BLE05600
C                                                                       BLE05610
C     LFACT (STAND-ALONE PROGRAM) :                                     BLE05620
C     GIVEN A POSITIVE DEFINITE MATRIX B IN SPARSE FORMAT,              BLE05630
C     COMPUTES THE SPARSE CHOLESKY FACTOR L OF B AND WRITES IT          BLE05640
C     TO FILE 7 IN SPARSE FORMAT.  THUS, B = L*L'.                      BLE05650
C     SEE THE LFACT FORTRAN CODE FOR DETAILS.                           BLE05660
C                                                                       BLE05670
C     LTEST (STAND-ALONE MAIN PROGRAM ) :                               BLE05680
C           (USER MUST PROVIDE 3 SUBROUTINES)                           BLE05690
C     GIVEN THE FACTORIZATION OF A SPARSE MATRIX B, COMPUTES            BLE05700
C     THE SOLUTION OF THE EQUATION B*U = B*V1 FOR A KNOWN BUT           BLE05710
C     RANDOMLY-GENERATED VECTOR V1, SOLVING WITH AND WITHOUT ITERATIVE  BLE05720
C     REFINEMENT TO OBTAIN A ROUGH CHECK ON THE NUMERICAL CONDITION     BLE05730
C     OF THE B-MATRIX.  THIS PROGRAM USES 3 USER-SUPPLIED SUBROUTINES   BLE05740
C     CMATV, CMATS AND BLSOLV.  SEE THE LTEST FORTRAN CODE FOR DETAILS. BLE05750
C                                                                       BLE05760
C                                                                       BLE05770
C-----OTHER PROGRAMS PROVIDED-------------------------------------------BLE05780
C                                                                       BLE05790
C                                                                       BLE05800
C     LECOMPAC  =  TRANSLATES A REAL SYMMETRIC MATRIX PROVIDED          BLE05810
C                  IN THE FORMAT  I, J, A(I,J) INTO THE SPARSE          BLE05820
C                  MATRIX FORMAT USED IN THE SAMPLE SUBROUTINES         BLE05830
C                  PROVIDED. IT ASSUMES THAT THE MATRIX                 BLE05840
C                  ENTRIES ARE GIVEN EITHER COLUMN BY COLUMN OR         BLE05850
C                  ROW BY ROW.  THE DATA SET CREATED IS WRITTEN TO      BLE05860
C                  FILE 8.                                              BLE05870
C                                                                       BLE05880
C                                                                       BLE05890
C-----------------------------------------------------------------------BLE05900
```

SECTION 8.3 MAIN PROGRAM, EIGENVALUE AND EIGENVECTOR CALCULATIONS

```
C-----BLEVAL (FEW EXTREME EIGENVALUES AND EIGENVECTORS)-----------------
C                    (REAL SYMMETRIC MATRICES)
C
C     CONTAINS MAIN PROGRAM FOR COMPUTING A FEW OF THE ALGEBRAICALLY-
C     LARGEST EIGENVALUES AND CORRESPONDING EIGENVECTORS OF A REAL
C     SYMMETRIC MATRIX, USING A BLOCK FORM OF LANCZOS TRIDIAGONALIZATION
C     WITH LIMITED REORTHOGONALIZATION.  PROCEDURE IS ITERATIVE.
C     PROCEDURE CAN BE USED TO COMPUTE THE ALGEBRAICALLY-SMALLEST
C     EIGENVALUES BY THE USER SUPPLYING -A*X RATHER THAN A*X, IN
C     WHICH CASE IT COMPUTES THE CORRESPONDING ALGEBRAICALLY-LARGEST
C     EIGENVALUES OF -A.  IN THIS CASE THE SIGNS OF THE COMPUTED
C     EIGENVALUES ARE CHANGED PRIOR TO WRITING TO FILE 15 SO THAT
C     ON EXIT, FILE 15 CONTAINS THE ALGEBRAICALLY-SMALLEST EIGENVALUES
C     OF A ALONG WITH THE CORRESPONDING EIGENVECTORS.
C
C     ITERATIVE 'BLOCK' LANCZOS PROCEDURE FOR WHICH ON EVERY
C     ITERATION, THE 2ND AND SUCCEEDING BLOCKS CONTAIN ONLY ONE
C     VECTOR WHICH IS SELECTED ON THE BASIS OF ITS EXPECTED INFLUENCE
C     ON THE CONVERGENCE.  Q-BLOCKS GENERATED ON A GIVEN ITERATION
C     ARE REORTHOGONALIZED ONLY W.R.T. THOSE VECTORS IN THE FIRST
C     Q-BLOCK THAT ARE NOT GENERATING DESCENDANTS ON THAT
C     ITERATION.
C
C     PFORT VERIFIER IDENTIFIED THE FOLLOWING NONPORTABLE CONSTRUCTIONS:
C     1.  DATA MACHEP DEFINITION
C     2.  FORMAT (20A4) USED FOR READING EXPLANATORY COMMENTS.
C     3.  FREE FORMAT (5,*), USED FOR PARAMETER INPUT FROM FILE 5.
C     4.  COMMON/LOOPS/ AS CONSTRUCTED IS NOT PORTABLE
C
C-----------------------------------------------------------------------
      DOUBLE PRECISION Q(44000),E(50),TM(2500),TOD(50),TD(50),EPSM,NNZ
      DOUBLE PRECISION SM(100),ERRMAX,SPREC,MACHEP,AVER,RELTOL,ERRMAN
      DOUBLE PRECISION  EVAL, RESIDL(100), RESIDK(100), RESID, FRACT
      REAL EXPLAN(20),G(2000)
      INTEGER   DIR(2,100),DESC(100),LEFT(100),XLFT(100)
      INTEGER SEED,OFLAG,EFLAG
      COMMON/LOOPS/MAXIT,ITER
      COMMON /RANDOM/SEED
      COMMON/FLAGS/EFLAG,OFLAG
      DOUBLE PRECISION  DABS, DFLOAT
C-----------------------------------------------------------------------
      EXTERNAL BMATV
      DATA MACHEP/Z3410000000000000/
C-----------------------------------------------------------------------
C
C     ARRAYS MUST DIMENSIONED AS FOLLOWS:
C
C     1.  Q:   >=  KMAX*N
C     2.  G:   >= N
C     3.  E:   >=  MXBLK
C     4.  TM:   >=  MXBLK**2
C     5.  TOD, TD, SM, DESC, LEFT, XLFT:   >= MXBLK
C     6.  DIR:  ROW DIMENSION = 2;   COLUMN DIMENSION >= MXBLK
C     7.  RESIDL, RESIDK:  >= MAXIMUM NUMBER OF ITERATIONS ALLOWED.
C         PROGRAM CURRENTLY TERMINATES IF MORE THAN 100 ITERATIONS
C         ARE REQUESTED.  USED TO MONITOR CONVERGENCE.
C     8.  EXPLAN:  DIMENSION = 20.
C
C-----------------------------------------------------------------------
C     OUTPUT HEADER
      WRITE(6,10)
   10 FORMAT(/' BLOCK LANCZOS PROCEDURE FOR REAL SYMMETRIC MATRICES'
```

```
     1 /' 2ND AND SUCCEEDING BLOCKS CONTAIN ONLY ONE VECTOR'//)
C
C     SET PROGRAM PARAMETERS
      EPSM = 2.D0*MACHEP
      SPREC = 1.D-5
      MPMIN = -1000
C
C     READ USER-SPECIFIED PARAMETERS FROM INPUT FILE 5 (FREE FORMAT)
C
C     SELECT THE AMOUNT OF INTERMEDIATE OUTPUT DESIRED (IWRITE =0,1).
C     IWRITE = 1 INCREASES THE AMOUNT OF INTERMEDIATE OUTPUT WRITTEN
C     TO FILE 6 ON EACH ITERATION OF THE BLOCK LANCZOS PROCEDURE.
      READ(5,20) EXPLAN
   20 FORMAT(20A4)
      READ(5,*)  IWRITE
C
C     READ ORDER (N) OF MATRIX AND MATRIX IDENTIFICATION NUMBER (MATNO)
      READ(5,20) EXPLAN
      READ(5,*) N,MATNO
C
C     READ USER-SPECIFIED DIMENSIONS OF Q-ARRAY (MDIMQ) AND OF THE
C     TM-ARRAY (MDIMTM).  READ MAXIMUM NUMBER (MAXIT) OF MATRIX-VECTOR
C     MULTIPLIES ALLOWED IN PHASE 1.
      READ(5,20) EXPLAN
      READ(5,*) MDIMQ, MDIMTM, MAXIT
C
C     READ FLAGS:   EFLAG = (0,1).  EFLAG = 0, MEANS PROGRAM STOPS
C     AFTER COMPLETING PHASE 1 PORTION OF BLOCK LANCZOS PROCEDURE.
C     EFLAG = 1, MEANS PROGRAM COMPLETES BOTH PHASES BEFORE
C     TERMINATING.
C     OFLAG = (0,1).  OFLAG = 0, MEANS THAT IN PHASE 1 PORTION
C     OF THE COMPUTATION, THE PROGRAM DOES NO ORTHOGONALITY CHECKS
C     ON THE Q-BLOCKS GENERATED.  OFLAG = 1 MEANS THAT IN THE
C     PHASE 1 PORTION AND IN THE PHASE 2 PORTIONS OF THE COMPUTATIONS
C     THE PROGRAM CHECKS THE ORTHOGONALITY OF THE Q-BLOCKS GENERATED
C     W.R.T. THAT VECTOR IN THE FIRST BLOCK THAT IS GENERATING
C     DESCENDANTS.  NOTE THAT IN PHASE 2, THE PROGRAM ALWAYS MAKES
C     THIS CHECK OF ORTHOGONALITY REGARDLESS OF THE VALUE OF OFLAG.
C     FOR SAFETY, OFLAG SHOULD ALWAYS BE SET TO 1, ALTHOUGH IN MANY
C     PROBLEMS THIS IS NOT NECESSARY.
      READ(5,20) EXPLAN
      READ(5,*) EFLAG,OFLAG
C
C     READ SEED USED BY SUBROUTINE GENRAN TO OBTAIN THOSE STARTING
C     VECTORS WHICH ARE GENERATED RANDOMLY.
      READ(5,20) EXPLAN
      READ(5,*) SEED
C
C     SPECIFY MAXIMUM T-SIZE ALLOWED (KMAX-1); INITIAL SIZE OF
C     STARTING BLOCK (KACT);  NUMBER OF STARTING VECTORS SUPPLIED (KSET)
C     SEE BLOCK LANCZOS HEADER FOR COMMENTS ON THE SIZE OF KACT.
      READ(5,20) EXPLAN
      READ(5,*) KMAX,KACT,KSET
C
C     SPECIFY NUMBER OF EXTREME EIGENVALUES AND EIGENVECTORS TO BE
C     COMPUTED (KM).  USER CAN SPECIFY THAT THE ALGEBRAICALLY-
C     SMALLEST EIGENVALUES ARE BEING COMPUTED BY SETTING KM < 0.
C     PROGRAM THEN ASSUMES THAT THE MATRIX-VECTOR MULTIPLY
C     SUBROUTINE WHICH THE USER HAS PROVIDED IS COMPUTING -A*X
C     INSTEAD OF A*X AND INTERNALLY IT COMPUTES THE |KM|
C     ALGEBRAICALLY-LARGEST EIGENVALUES OF -A.
      READ(5,20) EXPLAN
      READ(5,*) KM
      IF(KM.EQ.0) GO TO 490
      KML = IABS(KM)
C
C     STAGNATION OF CONVERGENCE OF THE KM-TH EIGENVALUE WILL BE
```

```
C     TESTED AFTER NSTAG ITERATIONS.  CONVERGENCE WILL BE SAID TO
C     HAVE STAGNATED IF THE RATIO OF THE SQUARE OF THE CURRENT KM-TH
C     RESIDUAL TO THE SQUARE OF THE CORRESPONDING RESIDUAL OBTAINED
C     10 ITERATIONS EARLIER IS GREATER THAN FRACT.  NSTAG SHOULD BE
C     >= 25.  IN THE TESTS FRACT WAS SET TO .01.
      READ(5,20) EXPLAN
      READ(5,*) NSTAG, FRACT
C
C     READ IN THE RELATIVE TOLERANCE (RELTOL) USED TO DETERMINE A
C     CONVERGENCE CRITERION FOR PHASE 2, AND THE MAXIMUM NUMBER (MAXIT2)
C     OF MATRIX-VECTOR MULTIPLIES ALLOWED IN PHASE 2.
      READ(5,20) EXPLAN
      IF(EFLAG.EQ.1)  READ(5,*)  RELTOL, MAXIT2
C
C     CONSISTENCY CHECKS
C     PROCEDURE REQUIRES ENOUGH ROOM IN Q-ARRAY FOR AT LEAST 2
C     BLOCKS OF SIZE KACT PLUS A WORKING VECTOR OF LENGTH N.
      MXBLK = KMAX -1
      MXBLK2 = MXBLK*MXBLK
      IF(MDIMTM.LT.MXBLK2) GO TO 470
      NKMAX = N*KMAX
      IF(MDIMQ.LT.NKMAX)   GO TO 510
      IF(KML.GT.KACT) GO TO 370
      IF(MXBLK.GT.N)  GO TO 390
      IF(2*KACT.GT.MXBLK)  GO TO 450
C
C-----------------------------------------------------------------------
C     DEFINE AND INITIALIZE THE ARRAYS FOR THE USER-SPECIFIED
C     A-MATRIX AND PASS THE STORAGE LOCATIONS OF THESE ARRAYS AND
C     OF ANY OTHER PARAMTERS NEEDED TO DEFINE THE MATRIX TO THE
C     MATRIX-VECTOR MULTIPLY SUBROUTINE BMATV.
C
      CALL USPEC(N,MATNO,NNZ,AVER)
C
C-----------------------------------------------------------------------
C     MASK OVERFLOW AND UNDERFLOW
      CALL MASK
C
C-----------------------------------------------------------------------
C     ARE THERE STARTING VECTORS TO READ IN FROM FILE 10 (KSET.NE.0) ?
      IF(KSET.EQ.0) GO TO 70
C
      READ(10,30) NOLD,KACT
   30 FORMAT(I6,I4)
      IF(NOLD.NE.N.OR.KSET.GT.KACT) GO TO 410
      DO 50 J=1,KSET
      READ(10,20) EXPLAN
      READ(10,40) EVAL,RESID
   40 FORMAT(E20.12,E13.4)
      READ(10,20) EXPLAN
      LINT= (J-1)*N + 1
      LFIN = J*N
   50 READ(10,60) (Q(JL), JL = LINT,LFIN)
   60 FORMAT(4E20.12)
C
   70 CONTINUE
C
C     WRITE TO A SUMMARY OF THE PARAMETERS FOR THIS RUN TO FILE 6
C
      MXBLK = KMAX - 1
      WRITE(6,80)  N, NNZ, AVER, MATNO
   80 FORMAT(/6X,'ORDER OF MATRIX ',5X,'AVERAGE NONZEROES PER ROW'/
     1I15,E26.4/4X,'AVERAGE SIZE OF NONZERO ENTRIES',5X,'MATRIX ID'/
     1E25.4,I21/)
C
      WRITE(6,90) MDIMQ, MDIMTM
   90 FORMAT(/18X,'USER-SPECIFIED'/2X,'MAX. DIMENSION Q-ARRAY',4X,'MAX.
```

```
     1DIMENSION TM-ARRAY'/I16,I26/)                                     BLE01970
C                                                                       BLE01980
      WRITE(6,100)  OFLAG, EFLAG                                        BLE01990
  100 FORMAT(/4X,'OFLAG',4X,'EFLAG'/I8,I9/)                             BLE02000
C                                                                       BLE02010
      IF(EFLAG.EQ.1)  WRITE(6,110) MAXIT,RELTOL,MAXIT2                  BLE02020
  110 FORMAT(/4X,' MAXIT ',8X,'  RELTOL  ',6X,' MAXIT2 '/I10,E20.6,I12/)BLE02030
      IF(EFLAG.EQ.0)  WRITE(6,120) MAXIT                                BLE02040
  120 FORMAT(/4X,' MAXIT '/I10/)                                        BLE02050
C                                                                       BLE02060
      WRITE(6,130)  SEED                                                BLE02070
  130 FORMAT(/' SEED FOR RANDOM NUMBER GENERATOR'/I24/)                 BLE02080
C                                                                       BLE02090
      IF(KM.GT.0)  WRITE(6,140) KML                                     BLE02100
  140 FORMAT(/' COMPUTE THE',I3,' ALGEBRAICALLY-LARGEST EIGENVALUES AND BLE02110
     1CORRESPONDING VECTORS'/)                                          BLE02120
      IF(KM.LT.0)  WRITE(6,150) KML                                     BLE02130
  150 FORMAT(/' COMPUTE THE',I3,' ALGEBRAICALLY-SMALLEST EIGENVALUES ANDBLE02140
     1 CORRESPONDING VECTORS'/' PROGRAM ASSUMES THAT USER IS PROVIDING -BLE02150
     1A*X INSTEAD OF A*X'/' AND COMPUTES THE ALGEBRAICALLY-LARGEST EIGENBLE02160
     1VALUES OF -A.'/' HOWEVER ON EXIT, FILE 15 CONTAINS THE ALGEBRAICALBLE02170
     1LY-SMALLEST EIGENVALUES OF'/' THE ORIGINAL A-MATRIX AND CORRESPONDBLE02180
     1ING EIGENVECTORS.'/)                                              BLE02190
      IF(KM.LT.0) KM = - KM                                             BLE02200
C                                                                       BLE02210
C     COMPUTE PHASE 1 CONVERGENCE TOLERANCE                             BLE02220
      IF(AVER.GE.1.)                                                    BLE02230
     1ERRMAX = 2.D0*DFLOAT(N+1000)*NNZ*AVER*MACHEP                      BLE02240
      IF(AVER.LT.1.)                                                    BLE02250
     1ERRMAX = 2.D0*DFLOAT(N+1000)*NNZ*AVER**2*MACHEP                   BLE02260
C                                                                       BLE02270
      WRITE(6,160) KACT,MXBLK,KSET                                      BLE02280
  160 FORMAT(/' ON INITIAL ITERATIONS, THE FIRST BLOCK CONTAINS ',I3,' VBLE02290
     1ECTORS'/' HOWEVER THE SIZE OF THE FIRST BLOCK MAY CHANGE AS THE ITBLE02300
     1ERATIONS PROCEED'/' THE MAXIMUM SIZE T-MATRIX THAT CAN BE GENERATEBLE02310
     1D IS ',I4/' THE USER SUPPLIED ',I3,' STARTING VECTORS'/)          BLE02320
C                                                                       BLE02330
      WRITE(6,170)                                                      BLE02340
  170 FORMAT(/' ITERATIVE PROCEDURE'/' PROCEDURE MONITORS THE SIZES OF TBLE02350
     1HE NORM(GRADIENTS)**2 ON EACH'/' ITERATION.  CONVERGENCE IS SAID  BLE02360
     1TO HAVE OCCURRED WHEN ALL'/' RELEVANT (NORMS)**2 ARE LESS THAN ERRBLE02370
     1MAX',E10.3/' TYPICALLY, PHASE 1 ERRMAX YIELDS SOMEWHAT LESS THAN'/BLE02380
     1' SINGLE PRECISION ACCURACY.  PHASE 2 REFINES THE VECTORS OBTAINEDBLE02390
     1'/' ON PHASE 1, ACCORDING TO THE ACCURACY SPECIFIED BY THE USER'/)BLE02400
C                                                                       BLE02410
      WRITE(6,180) ERRMAX                                               BLE02420
  180 FORMAT(//' PHASE 1 CONVERGENCE CRITERION, ERRMAX '/E22.3/)        BLE02430
C                                                                       BLE02440
C-----------------------------------------------------------------------BLE02450
C     PASS STORAGE LOCATIONS OF VARIOUS ARRAYS TO LANCZS AND LANCI1     BLE02460
C     SUBROUTINES                                                       BLE02470
C                                                                       BLE02480
      CALL LANZP(DIR,DESC,SM,TM,TOD,TD,G,XLFT,LEFT,SPREC)               BLE02490
      CALL LANCP1(DIR,DESC,TM,SM,XLFT,LEFT)                             BLE02500
C                                                                       BLE02510
C-----------------------------------------------------------------------BLE02520
C                                                                       BLE02530
C     ENTER PHASE 1 OF BLOCK LANCZOS PROCEDURE.  BLOCK PROCEDURE        BLE02540
C     HAS 2 POSSIBLE PHASES.  USER SPECIFIES PHASE 1 ONLY OR PHASE 1    BLE02550
C     AND PHASE 2 BY SETTING EFLAG = 0 OR 1, RESPECTIVELY. PHASE 1      BLE02560
C     COMPUTES VECTORS THAT MAY BE SOMEWHAT LESS ACCURATE THAN SINGLE   BLE02570
C     PRECISION.  PHASE 2 TAKES THE VECTORS OBTAINED IN PHASE 1         BLE02580
C     AND ATTEMPTS TO REFINE THEM.  THE USER SPECIFIES THE DEGREE       BLE02590
C     OF REFINEMENT DESIRED BY SETTING THE VALUES OF RELTOL AND MAXIT2. BLE02600
C     BOTH PHASES SHOULD BE USED.                                       BLE02610
      IPHASE = 1                                                        BLE02620
      NITER = 0                                                         BLE02630
```

```
  190 ITER = 0
      RESIDL(1) = FRACT
      RESIDL(2) = NSTAG
C
C-----------------------------------------------------------------------
C     CALL INITIATES THE BLOCK LANCZOS PROCEDURE.
C     ON RETURN EIGENVALUE APPROXIMATIONS ARE IN E(I), I=1,KACT
C     IN ALGEBRAICALLY DECREASING ORDER.   EIGENVECTOR APPROXIMATIONS
C     ARE IN FIRST N*KACT LOCATIONS IN THE Q-ARRAY.
C
      CALL LANCZS(BMATV,KML,KSET,KACT,MXBLK,N,Q,E,RESIDL,RESIDK,ERRMAX,
     1 IPHASE,NITER,IWRITE)
C
C-----------------------------------------------------------------------
C
      IF(IPHASE.EQ.MPMIN) WRITE(15,200) N,KACT
  200 FORMAT(2I10,' PHASE 2 TERMINATED '/' PROGRAM INDICATES ACCURACY SP
     1ECIFIED BY USER IS NOT ACHIEVABLE'/)
C
      ITERA = IABS(ITER)
      IF(IWRITE.NE.MPMIN.AND.ITER.GT.0)  WRITE(6,210) IPHASE,ITERA
  210 FORMAT(/1X,'PHASE COMPLETED',5X,' NUMBER MATRIX-VECTOR MULTIPLIES
     1USED'/I10,I30)
C
      IF(IWRITE.EQ.MPMIN.OR.ITER.LT.0)  WRITE(6,220) IPHASE,ITERA
  220 FORMAT(/1X,'PHASE TERMINATED',5X,' NUMBER MATRIX-VECTOR MULTIPLIES
     1 USED'/I10,I30)
C
      IF(ITER.GT.0.AND.IWRITE.NE.MPMIN)  GO TO 250
C
      IF(ITER.LT.0)  WRITE(6,230)
  230 FORMAT(//' SMALL EIGENVALUE SUBROUTINE DEFAULTED'/' BLOCK LANCZOS
     1 PROCEDURE STOPS AFTER SAVING CURRENT EIGENVECTOR APPROXIMATIONS'/
     1/)
C
      WRITE(15,240)
      WRITE(6,240)
  240 FORMAT(//' BLOCK LANCZOS PROCEDURE TERMINATES WITHOUT CONVERGENCE
     1'/' USER SHOULD EXAMINE OUTPUT TO DETERMINE REASONS FOR TERMINATIO
     1N'//)
C
C     WRITE EIGENVALUE AND EIGENVECTOR APPROXIMATIONS CONTAINED IN
C     THE FIRST Q-BLOCK TO FILE 15
C
  250 IF(IPHASE.EQ.1)  WRITE(15,260) N,KACT,SEED
  260 FORMAT(I6,I4,I12,' PHASE 1, ORDER A-MATRIX, SIZE OF Q(1), SEED')
      IF(IPHASE.EQ.2)  WRITE(15,270) N,KACT,SEED
  270 FORMAT(I6,I4,I12,' PHASE 2, ORDER A-MATRIX, SIZE OF Q(1), SEED')
C
      JJ=KACT
      LINT = -N+1
      LFIN = 0
      DO 290 J=1,KACT
      LINT = LINT + N
      LFIN = LFIN + N
      JJ=JJ+1
C
C     NOTE THAT RESIDUAL PRINTED OUT CORRESPONDS TO VALUE OBTAINED
C     PRIOR TO FINAL PROJECTION Q(1)-TRANSPOSE*AQ(1) DONE BEFORE
C     TERMINATION
C
      IF(KM.LT.0)   E(J) = -E(J)
      WRITE(15,280) E(J), SM(JJ)
  280 FORMAT(/E20.12,E13.4,'= EIGENVALUE, NORM(ERROR)**2,EIGENVECTOR='/)
  290 WRITE(15,300) (Q(L), L=LINT,LFIN)
      WRITE(15,310)
  300 FORMAT(4E20.12)
```

```
  310 FORMAT(/' ABOVE ARE COMPUTED APPROXIMATE EIGENVECTORS'/)              BLE03310
C                                                                           BLE03320
      IF(ITER.GT.MAXIT) WRITE(15,320) ITER,MAXIT                            BLE03330
  320 FORMAT(//' PROCEDURE TERMINATED BECAUSE NUMBER OF MATRIX-VECTOR MUBLE03340
     1LTIPLIES ',I6/' EXCEEDED MAXIMUM NUMBER ',I6,' ALLOWED'//)            BLE03350
C                                                                           BLE03360
      IF(ITER.LT.0) WRITE(15,330)                                           BLE03370
  330 FORMAT(//' USER BEWARE.  EIGENELEMENT COMPUTATIONS DEFAULTED BECAUBLE03380
     1SE'/'   EISPACK SUBROUTINE DEFAULTED.   EIGENVALUE AND EIGENVECTORBLE03390
     1 APPROXIMATIONS'/'   ABOVE WERE THOSE AVAILABLE AT THE TIME OF DEFBLE03400
     1AULT'/' SOMETHING IS SERIOUSLY WRONG.'//)                             BLE03410
C                                                                           BLE03420
C     CHECK FOR TERMINATION AFTER PHASE 1                                   BLE03430
C     ITER < 0 MEANS EISPACK SUBROUTINE DEFAULTED                           BLE03440
C     IPHASE = MPMIN MEANS THAT PHASE 2 TERMINATED DUE TO ORTHOGONALITY BLE03450
C     IWRITE = MPMIN MEANS THAT CONVERGENCE APPEARS TO HAVE STAGNATED   BLE03460
C     ITER > MAXIT MEANS MAXIMUM NUMBER OF MATRIX-VECTOR MULTIPLIES     BLE03470
C           ALLOWED BY USER WAS EXCEEDED                                    BLE03480
      IF(ITER.LT.0.OR.ITER.GT.MAXIT) GO TO 530                              BLE03490
      IF(IPHASE.EQ.MPMIN.OR.IWRITE.EQ.MPMIN) GO TO 530                      BLE03500
      IF(EFLAG.NE.1.OR.IPHASE.EQ.2) GO TO 530                               BLE03510
C                                                                           BLE03520
C     ENTER 2ND PHASE OF COMPUTATION TO ATTEMPT TO OBTAIN MORE              BLE03530
C     ACCURATE EIGENVECTOR APPROXIMATIONS.                                  BLE03540
C     USER CONTROLS THE SIZE OF THE ERROR TOLERANCE BY SPECIFYING           BLE03550
C     THE PARAMETER RELTOL.                                                 BLE03560
C                                                                           BLE03570
      IPHASE = 2                                                            BLE03580
      MAXIT = MAXIT2                                                        BLE03590
      KSET = KACT                                                           BLE03600
C                                                                           BLE03610
C     ERROR TOLERANCE USES THE CONVERGED EIGENVALUE LARGEST IN              BLE03620
C     MAGNITUDE.                                                            BLE03630
      TD(1) = DABS(E(1))                                                    BLE03640
      IF(KML.EQ.1) GO TO 350                                                BLE03650
      DO 340 J = 2,KML                                                      BLE03660
  340 IF(DABS(E(J)).GT.TD(1))    TD(1) = DABS(E(J))                         BLE03670
  350 TD(1) = DMAX1(TD(1),1.D0)                                             BLE03680
      ERRMAN = RELTOL**2 * TD(1)**2                                         BLE03690
      IF(ERRMAN.GE.ERRMAX) GO TO 430                                        BLE03700
      ERRMAX = ERRMAN                                                       BLE03710
C                                                                           BLE03720
      WRITE(6,360) ERRMAX, MAXIT2                                           BLE03730
  360 FORMAT(//' ENTER PHASE 2 OF COMPUTATION'/' CONVERGENCE CRITERION IBLE03740
     1S REDUCED TO ',E13.4/' NO MORE THAN ',I5,' MATRIX VECTOR MULTIPLIEBLE03750
     1S WILL BE ALLOWED.'/' PROGRAM WILL TERMINATE IF BLOCK ORTHGONALITYBLE03760
     1 PROBLEMS MATERIALIZE'/)                                              BLE03770
C                                                                           BLE03780
      GO TO 190                                                             BLE03790
C                                                                           BLE03800
C     INCONSISTENCIES IN THE DATA                                           BLE03810
C                                                                           BLE03820
  370 WRITE(6,380) KM,KACT                                                  BLE03830
  380 FORMAT(/' PROGRAM TERMINATES BECAUSE THE NUMBER OF EIGENELEMENTS  BLE03840
     1REQUESTED, KM =',I3/'  IS LARGER THAN THE SIZE OF THE FIRST Q BLOCBLE03850
     1K, KACT =',I3,' SPECIFIED'/'  USER MUST RESET KM OR KACT'/)           BLE03860
       GO TO 530                                                            BLE03870
C                                                                           BLE03880
  390 WRITE(6,400) KMAX,N                                                   BLE03890
  400 FORMAT(/' PROGRAM TERMINATES BECAUSE KMAX = ',I5,' IS TOO LARGE FOBLE03900
     1R THE SIZE, N = ',I5,', OF THE GIVEN MATRIX'/'  USER MUST DECREASEBLE03910
     1THE SIZE OF KMAX.'/)                                                  BLE03920
      GO TO 530                                                             BLE03930
C                                                                           BLE03940
  410 WRITE(6,420) NOLD,N,KACT,KSET                                         BLE03950
  420 FORMAT(/' PROGRAM TERMINATES BECAUSE FAULT OCCURRED IN READING IN BLE03960
     1THE EIGENVECTOR APPROXIMATIONS'/' EITHER THE SIZE MATRIX SPECIFIEDBLE03970
```

```
     1ON THE EIGENVECTOR FILE' ,I6/'  DID NOT MATCH THE SIZE SPECIFIED 'BLE03980
     1,I5,' IN THE PROGRAM OR THE NUMBER'/'     OF VECTORS IN FILE 10 = 'BLE03990
     1,I4,' IS LESS THAN THE NUMBER ',I3/'   USER SAID WERE THERE'/)    BLE04000
      GO TO 530                                                         BLE04010
C                                                                       BLE04020
  430 WRITE(6,440)  ERRMAN, ERRMAX                                      BLE04030
  440 FORMAT(/' COMPUTED PHASE 2 CONVERGENCE CRITERION ',E13.4/'  IS LARBLE04040
     1GER THAN PHASE 1 CRITERION ',E13.4/'  SO PROGRAM TERMINATES'/)    BLE04050
      GO TO 530                                                         BLE04060
C                                                                       BLE04070
  450 WRITE(6,460)  KACT,MXBLK                                          BLE04080
  460 FORMAT(/' PROGRAM TERMINATES BECAUSE THERE IS NOT ENOUGH ROOM TO  BLE04090
     1GENERATE 2 BLOCKS',' BECAUSE KACT = ',I3,' AND MXBLK = ', I4/)    BLE04100
      GO TO 530                                                         BLE04110
C                                                                       BLE04120
C                                                                       BLE04130
  470 WRITE(6,480)  MDIMTM, MXBLK                                       BLE04140
  480 FORMAT(/' PROGRAM TERMINATES BECAUSE THE DIMENSION ',I6,' OF THE TBLE04150
     1M ARRAY'/'    IS TOO SMALL FOR THE LARGEST T-MATRIX ALLOWED ',I4) BLE04160
      GO TO 530                                                         BLE04170
C                                                                       BLE04180
  490 WRITE(6,500)                                                      BLE04190
  500 FORMAT(/' USER SPECIFIED NUMBER OF EIGENVALUES OF INTEREST AS 0'/'BLE04200
     1 PROGRAM TERMINATES FOR USER TO RESET KM TO DESIRED NONZERO VALUE'BLE04210
     1/)                                                                BLE04220
      GO TO 530                                                         BLE04230
C                                                                       BLE04240
  510 WRITE(6,520)  MDIMQ, KMAX,N                                       BLE04250
  520 FORMAT(/' PROGRAM TERMINATES BECAUSE THE DIMENSION ',I6,' OF THE QBLE04260
     1-ARRAY'/'    IS TOO SMALL TO HOLD ',I5, ' VECTORS OF LENGTH ',I4) BLE04270
      GO TO 530                                                         BLE04280
C                                                                       BLE04290
  530 CONTINUE                                                          BLE04300
C                                                                       BLE04310
      STOP                                                              BLE04320
C-----END OF MAIN PROGRAM FOR BLOCK LANCZOS PROCEDURE-------------------BLE04330
      END                                                               BLE04340
```

SECTION 8.4 SAMPLE MATRIX-VECTOR MULTIPLY SUBROUTINES

```
C-----BLMULT------------------------------------------------------------BLM00010
C                                                                       BLM00020
C     CONTAINS SAMPLE USPEC AND BMATV SUBROUTINES FOR USE WITH          BLM00030
C     THE BLOCK LANCZOS PROCEDURE FOR REAL SYMMETRIC MATRICES.          BLM00040
C     PROGRAMS ARE USED WITH BLEVAL AND BLSUB FILES.                    BLM00050
C                                                                       BLM00060
C     NONPORTABLE CONSTRUCTIONS:                                        BLM00070
C     1.  THE ENTRY MECHANISM USED TO PASS THE STORAGE                  BLM00080
C         LOCATIONS OF THE USER-SPECIFIED MATRIX FROM THE               BLM00090
C         SUBROUTINE USPEC TO THE MATRIX-VECTOR SUBROUTINE              BLM00100
C         BMATV.                                                        BLM00110
C     2.  IN THE SAMPLE USPEC AND BMATV SUBROUTINES FOR DIAGONAL        BLM00120
C         TEST MATRICES: FREE FORMAT (8,*) AND THE FORMAT (20A4).       BLM00130
C                                                                       BLM00140
C-----USPEC (GENERAL SYMMETRIC SPARSE MATRICES)-------------------------BLM00150
C                                                                       BLM00160
C     SUBROUTINE USPEC(N,MATNO,NNZ,AVER)                                BLM00170
      SUBROUTINE GUSPEC(N,MATNO,NNZ,AVER)                               BLM00180
C                                                                       BLM00190
C-----------------------------------------------------------------------BLM00200
      DOUBLE PRECISION  ASD(10000),AD(5010),AVER,NNZ                    BLM00210
      INTEGER  IROW(10000),ICOL(5010)                                   BLM00220
C-----------------------------------------------------------------------BLM00230
C    USPEC DIMENSIONS AND INITIALIZES THE ARRAYS NEEDED TO DEFINE       BLM00240
C    THE USER-SPECIFIED MATRIX AND THEN PASSES THE STORAGE LOCATIONS    BLM00250
C    OF THESE ARRAYS TO THE MULTIPLY SUBROUTINE BMATV.                  BLM00260
C                                                                       BLM00270
C    MATRIX IS STORED IN FOLLOWING SPARSE MATRIX FORMAT:                BLM00280
C    N = ORDER OF A-MATRIX,                                             BLM00290
C    NZS = NUMBER OF NONZERO SUBDIAGONAL ENTRIES,                       BLM00300
C    NZL = INDEX OF LAST COLUMN CONTAINING NONZERO SUBDIAGONAL ENTRIES, BLM00310
C    ICOL(J), J=1,NZL IS THE NUMBER OF NONZERO SUBDIAGONAL ELEMENTS     BLM00320
C            IN COLUMN J.                                               BLM00330
C    IROW(K), K = 1,NZS IS THE CORRESPONDING ROW INDEX FOR ASD(K).      BLM00340
C    AD(I), I=1,N CONTAINS DIAGONAL ENTRIES (INCLUDING ANY 0            BLM00350
C          DIAGONAL ENTRIES).                                           BLM00360
C    ASD(K), K=1,NZS CONTAINS NONZERO SUBDIAGONAL ENTRIES, BY COLUMN    BLM00370
C    FOR J > NZL THERE ARE NO NONZERO SUBDIAGONAL ELEMENTS IN COLUMN J. BLM00380
C    ICOL(J) = 0 IS ALLOWED                                             BLM00390
C                                                                       BLM00400
C-----------------------------------------------------------------------BLM00410
C     ARRAYS THAT DEFINE THE MATRIX ARE READ IN FROM FILE 8             BLM00420
C                                                                       BLM00430
      READ(8,10) NZS,NOLD,NZL,MATOLD                                    BLM00440
   10 FORMAT(I10,2I6,I8)                                                BLM00450
C                                                                       BLM00460
      WRITE(6,20) NZS,NOLD,NZL,MATOLD                                   BLM00470
   20 FORMAT(I10,2I6,I8,' = NZS,NOLD,NZL,MATOLD'/)                      BLM00480
C                                                                       BLM00490
C     TEST OF PARAMETER CORRECTNESS                                     BLM00500
      ITEMP = (NOLD-N)**2 + (MATNO-MATOLD)**2                           BLM00510
C                                                                       BLM00520
      IF(ITEMP.EQ.0) GO TO 40                                           BLM00530
C                                                                       BLM00540
      WRITE(6,30) NOLD,N,MATOLD,MATNO                                   BLM00550
   30 FORMAT(/' PROGRAM TERMINATES BECAUSE EITHER THE SIZE ',I4,' OF THEBLM00560
     1 MATRIX'/' READ FROM FILE 8 DIFFERS FROM THE SIZE ',I4,' SPECIFIEDBLM00570
     1 BY'/' THE USER OR THE MATNO ',I8,' READ IN DIFFERS FROM THE MATNOBLM00580
     1 '/ I8,' SPECIFIED BY THE USER'/)                                 BLM00590
      GO TO 100                                                         BLM00600
C                                                                       BLM00610
   40 CONTINUE                                                          BLM00620
C                                                                       BLM00630
```

```
C     NUMBER OF NONZERO SUBDIAGONAL ENTRIES IN EACH COLUMN IS READ
C     THEN THE CORRESPONDING ROW INDEX FOR EACH SUCH ENTRY IS READ
      READ(8,50) (ICOL(K), K=1,NZL)
      READ(8,50) (IROW(K), K=1,NZS)
   50 FORMAT(13I6)
C
C     DIAGONAL IS READ FIRST, THEN NONZERO BELOW DIAGONAL ENTRIES
      READ(8,60) (AD(K), K=1,N)
      READ(8,60) (ASD(K), K=1,NZS)
   60 FORMAT(4E19.10)
C
C     COMPUTE NNZ, THE AVERAGE NUMBER OF NONZEROS PER COLUMN, AND
C     AVER, THE AVERAGE SIZE OF NONZERO ENTRIES.
      ITCOL = 0
      AVER = 0.D0
      DO 70 K = 1,N
      IF(DABS(AD(K)).EQ.0.D0)  GO TO 70
      ITCOL = ITCOL + 1
      AVER = AVER + DABS(AD(K))
   70 CONTINUE
      NTCOL = ITCOL
      DO 80 K = 1,N
   80 ITCOL = ITCOL + 2*ICOL(K)
      NNZ = DFLOAT(ITCOL)/DFLOAT(N)
      DO 90 K = 1,NZS
   90 AVER = AVER + DABS(ASD(K))
      AVER = AVER/DFLOAT(NZS + NTCOL)
C
C-----------------------------------------------------------------------
C     PASS STORAGE LOCATIONS OF ARRAYS THAT DEFINE THE MATRIX TO
C     THE MATRIX-VECTOR MULTIPLY SUBROUTINE BMATV
C
      CALL BMATVE(ASD,AD,ICOL,IROW,N,NZL)
C-----------------------------------------------------------------------
C
      RETURN
  100 STOP
C-----END OF USPEC------------------------------------------------------
      END
C
C-----MATRIX-VECTOR MULTIPLY FOR REAL SPARSE SYMMETRIC MATRICES---------
C
C     SUBROUTINE BMATV(W,U)
      SUBROUTINE GBMATV(W,U)
C
C-----------------------------------------------------------------------
      DOUBLE PRECISION   U(1),W(1),ASD(1),AD(1)
      INTEGER  IROW(1),ICOL(1)
      COMMON/LOOPS/MAXIT,ITER
C-----------------------------------------------------------------------
C    SPARSE MATRIX-VECTOR MULTIPLY FOR LANCZS  U = A*W
C    SEE USPEC SUBROUTINE FOR DESCRIPTION OF THE ARRAYS THAT DEFINE
C    THE A-MATRIX
C-----------------------------------------------------------------------
C     INCREMENT THE A*W COUNTER
      ITER = ITER + 1
C     COMPUTE THE DIAGONAL TERMS
      DO 10 I = 1,N
   10 U(I) = AD(I)*W(I)
C
C     COMPUTE BY COLUMN
      LLAST = 0
      DO 30 J = 1,NZL
C
      IF (ICOL(J).EQ.0) GO TO 30
      LFIRST = LLAST + 1
```

```
      LLAST = LLAST + ICOL(J)
C
      DO 20 L = LFIRST,LLAST
      I = IROW(L)
C
      U(I) = U(I) + ASD(L)*W(J)
      U(J) = U(J) + ASD(L)*W(I)
C
   20 CONTINUE
C
   30 CONTINUE
C
      RETURN
C
C-----------------------------------------------------------------------
C    STORAGE LOCATIONS OF ARRAYS ARE PASSED TO BMATV FROM USPEC
C
      ENTRY BMATVE(ASD,AD,ICOL,IROW,N,NZL)
C-----------------------------------------------------------------------
C
      RETURN
C-----END OF BMATV------------------------------------------------------
      END
C
C-----MATRIX-VECTOR MULTIPLY FOR DIAGONAL TEST MATRICES-----------------
C     BMATV COMPUTES  U = (DIAGONAL MATRIX) * W
C
      SUBROUTINE BMATV(W,U)
C     SUBROUTINE DBMATV(W,U)
C
C-----------------------------------------------------------------------
      DOUBLE PRECISION  W(1),U(1),D(1)
      COMMON/LOOPS/MAXIT,ITER
C-----------------------------------------------------------------------
C      INCREMENT THE LOOP COUNTER
      ITER = ITER + 1
C
      DO 10 I=1,N
   10 U(I)= D(I)*W(I)
C  10 U(I)= -D(I)*W(I)
C
      RETURN
C
C-----------------------------------------------------------------------
      ENTRY MVDIAE(D,N)
C-----------------------------------------------------------------------
C
      RETURN
C-----END OF DIAGONAL TEST MATRIX  MULTIPLY-----------------------------
      END
C
C-----START OF USPEC FOR DIAGONAL TEST MATRIX---------------------------
C
      SUBROUTINE USPEC(N,MATNO,NNZ,AVER)
C     SUBROUTINE DUSPEC(N,MATNO,NNZ,AVER)
C
C-----------------------------------------------------------------------
      DOUBLE PRECISION  D(1000),SPACE,SHIFT,AVER,NNZ
      DOUBLE PRECISION  DABS, DFLOAT
      REAL  EXPLAN(20)
C-----------------------------------------------------------------------
C
      READ(8,10) EXPLAN
   10 FORMAT(20A4)
      READ(8,*) NOLD,NUNIF,SPACE,D(1),SHIFT
      NNUNIF = NOLD - NUNIF
      WRITE(6,20) NOLD,SPACE,NNUNIF,D(1),SHIFT
```

```
   20 FORMAT(/' DIAGONAL TEST MATRIX, SIZE = ',I4/' MOST ENTRIES ARE ',
     1E10.3,' UNITS APART.',I3,' ENTRIES'/' ARE IRREGULARLY SPACED. FIRS
     1T ENTRY IS ',E10.3,' SHIFT = ',E10.3/)
C
      IF(N.NE.NOLD) GO TO 100
C     COMPUTE THE UNIFORM PORTION OF THE SPECTRUM
      DO 30 J=2,NUNIF
   30 D(J) = D(1) - DFLOAT(J-1)*SPACE
      NUNIF1=NUNIF + 1
      READ(8,10)  EXPLAN
      DO 40 J=NUNIF1,N
   40 READ(8,*) D(J)
C
      IF(SHIFT.EQ.0.) GO TO 60
      DO 50 J=1,N
   50 D(J) = D(J) + SHIFT
C
C     PRINT OUT THE EIGENVALUES OF INTEREST
   60 WRITE(6,70) (D(I), I=1,10 )
      NB = NUNIF - 2
      WRITE(6,80) (D(I), I = NB,N)
   70 FORMAT(/' BLOCK LANCZOS TEST, 1ST 10 ENTRIES OF DIAGONAL TEST MATR
     1IX = '/(3E22.14))
   80 FORMAT(/' MIDDLE UNIFORM PORTION OF MATRIX IS NOT PRINTED OUT'/
     1' END OF UNIFORM PLUS NONUNIFORM SECTION = '/(3E25.16))
C
C     DIAGONAL GENERATION COMPLETE
C     COMPUTE NNZ AND AVER
      NNZ = 1.D0
      AVER = 0.D0
      DO 90 K = 1,N
   90 AVER = AVER + DABS(D(K))
      AVER = AVER/DFLOAT(N)
C
C-----------------------------------------------------------------------
C    CALL ENTRY TO MATRIX-VECTOR MULTIPLY SUBROUTINE  TO PASS
C    STORAGE LOCATION OF D-ARRAY AND ORDER OF A-MATRIX.
C
       CALL MVDIAE(D,N)
C-----------------------------------------------------------------------
C
      RETURN
  100 WRITE(6,110) NOLD,N
  110 FORMAT(' PROGRAM TERMINATES BECAUSE NOLD = ',I5,'DOES NOT EQUAL N
     1 =',I5)
C-----END OF USPEC SUBROUTINE FOR 'DIAGONAL' TEST MATRICES--------------
      STOP
      END
```

SECTION 8.5 OTHER SUBROUTINES USED BY THE PROGRAMS IN CHAPTERS 8 AND 9

```
C-----BLSUB-------------------------------------------------------------
C
C     PFORT VERIFIER IDENTIFIED THE FOLLOWING NONPORTABLE
C     CONSTRUCTIONS:
C     1. ENTRY MECHANISMS USED TO PASS THE STORAGE LOCATIONS OF
C        SEVERAL ARRAYS FROM THE MAIN PROGRAM TO THE SUBROUTINES
C        LANCZS AND LANCI1.
C     2. COMMON BLOCK: LOOPS: USED IN LANCZS AND LANCI1.
C
C     SUBROUTINES:    LANCZS, LANCI1, ORTHOG, START, AND DIAGOM
C                     ARE USED WITH THE BLOCK LANCZOS PROGRAMS
C                     BLEVAL AND BLIEVAL.  LPERM IS USED WITH BLIEVAL.
C
C
C-----LANCZS FOR BLOCK LANCZOS PROCEDURE--------------------------------
C
C     ON EACH ITERATION CALLS LANCI1 SUBROUTINE TO GENERATE
C     THE Q-SUBBLOCKS AND THEN CALLS DIAGOM SUBROUTINE TO
C     DIAGONALIZE THE SMALL SYMMETRIC MATRIX WHICH IS THE PROJECTION
C     OF THE MATRIX BEING USED BY LANCZS ONTO THE SUBSPACE SPANNED
C     BY THESE Q-BLOCKS.
C
      SUBROUTINE LANCZS(MATVEC,KML,KSET,KACT,MXBLK,N,Q,E,RESIDL,
     1 RESIDK,ERRMAX,IPHASE,NITER,IWRITE)
C
C-----------------------------------------------------------------------
      DOUBLE PRECISION  E(1),Q(1),ERRMAX,SPREC,RESN,FRACT,RKM,SUM
      DOUBLE PRECISION  TM(1),SM(1),TD(1),TOD(1),RESIDL(1),RESIDK(1)
      REAL  G(1)
      INTEGER  EFLAG,OFLAG,DIR(2,1),DESC(1),LEFT(1),XLFT(1)
      DOUBLE PRECISION FINPRO
      COMMON /LOOPS/MAXIT,ITER
      COMMON/FLAGS/EFLAG,OFLAG
      EXTERNAL MATVEC
C-----------------------------------------------------------------------
      KM = KML
      MMT = MXBLK*MXBLK
      MPMIN = -1000
      IKACT = KACT + 10
      FRACT = RESIDL(1)
      NSTAG = RESIDL(2)
      IORTHO = 0
C
C     CONSTRUCT STARTING VECTORS
      IF(KSET.EQ.0) GO TO 10
C-----------------------------------------------------------------------
      CALL ORTHOG(1,KSET,N,Q)
   10 CALL START (KSET+1,KACT,N,Q,G,ERRMAX)
C-----------------------------------------------------------------------
   20 CONTINUE
C     INITIALIZE THE LANCZOS T-MATRIX.
      DO 30 J=1,MMT
   30 TM(J)=0.D0
C
C     INITIALIZE THE Q-BLOCK DIRECTORY
      DIR(1,1)=1
      DIR(2,1)=KACT
C
C     ORTHOGONALIZE THE STARTING VECTORS
      IF(NITER.EQ.0) GO TO 40
C-----------------------------------------------------------------------
      CALL ORTHOG(1,KACT,N,Q)
C-----------------------------------------------------------------------
```

```
   40 CONTINUE
C
C     GENERATE THE QSUBBLOCKS USED ON ITERATION NITER AND STORE IN
C     THE Q-ARRAY
C
      DO 90 I=1,MXBLK
C
C-----------------------------------------------------------------------
      CALL LANCI1(MATVEC,MXBLK,NITER,I,N,Q,KACT,KML,ERRMAX,RESN,RKM,
     1 IND,KACTN,IWRITE)
C-----------------------------------------------------------------------
C
C     HAS CONVERGENCE OCCURRED?
      II = I+1
      IF (I.EQ.1.AND.DIR(2,I).EQ.DIR(2,II)) GO TO 140
C
C     WAS THERE ROOM FOR ANOTHER Q-BLOCK?
      IF (DIR(2,II).LT.DIR(1,II))  GO TO 100
C
C     IF OFLAG = 1 OR IPHASE = 2, CHECK THE ORTHOGONALITY OF
C     THE Q-SUBBLOCKS GENERATED WITH RESPECT TO THAT VECTOR
C     IN THE 1ST Q-BLOCK WHICH IS GENERATING DESCENDANTS.
C     IN PHASE 2 LOSSES IN ORTHOGONALITY ARE USED TO
C     DETERMINE WHEN THE LIMITS ON THE ACHIEVABLE ACCURACY HAVE
C     BEEN REACHED.
C
      IF(OFLAG.EQ.0.AND.IPHASE.EQ.1) GO TO 90
C
      L1=DIR(1,II)
      LL1 = (L1-1)*N + 1
      IND1 = (IND-1)*N + 1
C-----------------------------------------------------------------------
      SUM = FINPRO(N,Q(IND1),1,Q(LL1),1)
C-----------------------------------------------------------------------
C
      IF(DABS(SUM).LT.SPREC)GO  TO 80
C
      IF(IWRITE.EQ.1) WRITE(6,50)  IND,L1,SUM,I
   50 FORMAT(/' INNER PRODUCT OF VECTORS ',I3,' AND ',I3,' = ',E13.3/
     1' THIS VIOLATES ORTHOGONALITY TEST.  TERMINATE BLOCK GENERATION'
     1/' WITH ',I3,'TH BLOCK '/)
C
C     ORTHOGONALITY TEST VIOLATED, TERMINATE BLOCK GENERATION
C     FOR THIS ITERATION.  IN PHASE 2 KEEP TRACK OF NUMBER OF
C     SUCH VIOLATIONS THAT LIMIT THE NUMBER OF BLOCKS TO < 10.
C     TERMINATE AFTER 3 SUCH VIOLATIONS IN PHASE 2.
      IF(IPHASE.NE.1.AND.I.LT.IKACT)  IORTHO = IORTHO + 1
      IF(IORTHO.LT.3.AND.II.NE.2)  GO TO 70
      WRITE(6,60)
   60 FORMAT(/' THE ORTHOGONALITY TEST HAS FAILED THREE TIMES'/
     1' TERMINATE THE BLOCK PROCEDURE'/)
      IPHASE = -1000
C     BEFORE TERMINATING WRITE THE CURRENT EIGENVECTOR/EIGENVALUE
C     APPROXIMATIONS TO FILE 15
      GO TO 160
C
C     TERMINATE THE Q-BLOCK GENERATION ON THIS ITERATION
   70 DIR(2,II)=DIR(2,I)
      GO TO 100
C
   80 CONTINUE
C
C     END OF ORTHOGONALITY TESTS
C
   90 CONTINUE
C
```

```
C     END OF RECURSIVE Q-BLOCK GENERATION
C
  100 CONTINUE
      MM = DIR(2,II)
      IF(IWRITE.EQ.1)  WRITE (6,110)  MM,I
  110 FORMAT(' T-MATRIX IS OF ORDER ',I3, ' NUMBER OF BLOCKS = ',I3)
C
C-----------------------------------------------------------------------
C     DIAGONALIZE THE PROJECTION MATRIX TM.  ON RETURN THE
C     UPDATED APPROXIMATIONS TO THE DESIRED EIGENVECTORS ARE IN THE
C     FIRST KACT COLUMNS OF THE Q-ARRAY.
C     UPDATED EIGENVALUE APPROXIMATIONS ARE IN E.
      TD(1) = RKM
      TD(2) = FRACT
      IERR = NSTAG
C
      CALL DIAGOM(MXBLK,MM,TM,KACT,N,Q,E,RESIDL,RESIDK,
     1 RESN,IND,KACTN,KM,TD,TOD,NITER,IERR,IWRITE)
C-----------------------------------------------------------------------
C
C     INCREMENT COUNTER FOR NUMBER OF BLOCK LANCZOS ITERATIONS
      NITER = NITER + 1
C     IWRITE = MPMIN MEANS BLOCK LANCZOS PROCEDURE TERMINATED ABNORMALLY
      IF(IWRITE.EQ.MPMIN)  GO TO 160
C     IERR .NE. 0 MEANS EISPACK SUBROUTINE DEFAULTED
      IF(IERR.EQ.0) GO TO 130
      WRITE(6,120)
  120 FORMAT(/' EISPACK SIGNALS TROUBLE IN SMALL IMTQL2 EIGENVALUE SUBRO
     1UTINE,'/' SO BLOCK LANCZOS PROGRAM TERMINATES'/)
      ITER = -ITER
C
      RETURN
C
  130 IF (ITER.GE.MAXIT) GO TO 160
C
C     UPDATED APPROXIMATIONS WERE OBTAINED WITHOUT EXCEEDING
C     MAXIMUM NUMBER OF MATRIX-VECTOR MULTIPLIES SET BY THE USER.
C     CONTINUE BLOCK LANCZOS LOOP ITERATIONS
C
      GO TO 20
C
  140 WRITE(6,150)
  150 FORMAT(//' BLOCK LANCZOS PROCEDURE CONVERGED'//)
C
C     BLOCK LANCZOS PROCEDURE HAS CONVERGED.
C     ATTEMPT TO IMPROVE THE APPROXIMATE EIGENVECTORS BY DIAGONALIZING
C     THE SMALL PROJECTION MATRIX OBTAINED BY USING ONLY THE
C     FIRST BLOCK IN Q-ARRAY.
C
  160 KACT2 = KACT*MXBLK
      DO 170 KK = 1,KACT2
  170 TM(KK) = 0.D0
C-----------------------------------------------------------------------
      CALL ORTHOG(1,KACT,N,Q)
C-----------------------------------------------------------------------
      KK0 = 1-N
      KACTP1 = (KACT)*N + 1
      JJ0 = -MXBLK-1
      DO 190 K=1,KACT
      JJ0 = JJ0 + MXBLK + 1
      KK0 = KK0 + N
C-----------------------------------------------------------------------
      CALL MATVEC(Q(KK0),Q(KACTP1))
C-----------------------------------------------------------------------
      LL0 = (K-2)*N + 1
      JJ = JJ0
      DO 180 L=K,KACT
```

```
      LLO = LLO + N
      JJ=JJ+1
C-----------------------------------------------------------------------
      TM(JJ) = FINPRO(N,Q(LLO),1,Q(KACTP1),1)
C-----------------------------------------------------------------------
  180 CONTINUE
C
  190 CONTINUE
C
C-----------------------------------------------------------------------
C     USE EISPACK SUBROUTINE TRED2 TO TRIDIAGONALIZE TM-MATRIX
C     TM = (1ST Q-BLOCK)-TRANSPOSE*A*(1ST Q-BLOCK).
C     ON RETURN DIAGONAL ELEMENTS COMPUTED ARE IN TD, OFF-DIAGONAL
C     ELEMENTS ARE IN TOD, TRANSFORMATIONS USED ARE IN TM.
C     THEN USE EISPACK SUBROUTINE IMTQL2 TO DIAGONALIZE THE T-MATRIX.
C     ON RETURN.  EIGENVALUES ARE IN TD IN ASCENDING ORDER.
C     CORRESPONDING EIGENVECTORS ARE IN TM.
C
      CALL TRED2(MXBLK,KACT,TM,TD,TOD,TM)
      CALL IMTQL2(MXBLK,KACT,TD,TOD,TM,IERR)
C-----------------------------------------------------------------------
C
      IF(IERR.EQ.0)  GO TO 200
      WRITE(6,120)
      ITER = -ITER
C
      RETURN
C
C     COMPUTE SUCCESSIVELY THE JTH-COMPONENTS OF THE RITZ VECTORS.
C     REORDER THE EIGENVALUES (AND EIGENVECTORS) SO THAT THEY
C     ARE IN ALGEBRAICALLY DECREASING ORDER.
C
  200 DO 220 J=1,N
      JJO = - MXBLK
      JLO = -N + J
      DO 210 K=1,KACT
      TOD(K)=0.D0
      JJO = JJO + MXBLK
      JJ= JJO
      JL = JLO
      DO 210 L=1,KACT
      JJ=JJ+1
      JL = JL + N
  210 TOD(K)=TOD(K)+TM(JJ)*Q(JL)
      JK = JLO
      DO 220 K=1,KACT
      JK = JK + N
      KACTK = KACT - K + 1
      Q(JK)=TOD(KACTK)
  220 CONTINUE
      DO 230 K=1,KACT
      KACTK = KACT - K + 1
  230 E(K)=TD(KACTK)
C
C     HAS CONVERGENCE OCCURRED?
      IF(I.EQ.1.AND.DIR(2,I).EQ.DIR(2,I+1))  GO TO 250
C
C     CONVERGENCE HAS NOT OCCURRED, PROCEDURE TERMINATED FOR SOME
C     OTHER REASON
      WRITE(6,240)
  240 FORMAT(//' BLOCK LANCZOS PROCEDURE TERMINATES WITHOUT CONVERGENCE'
     1/' AFTER WRITING THE CURRENT EIGENVALUE AND EIGENVECTOR APPROXIMAT
     1IONS'/' TO FILE 15'/)
C
      RETURN
C
  250 IF(IPHASE.EQ.1) WRITE(6,260) (E(K), K=1,KACT)
```

```
      IF(IPHASE.EQ.2) WRITE(6,270) (E(K), K=1,KACT)
  260 FORMAT(/' AT END OF PHASE 1, COMPUTED EIGENVALUES ='/(4E20.12))
  270 FORMAT(/' AT END OF PHASE 2, COMPUTED EIGENVALUES ='/(4E20.12))
C
      RETURN
C
C-----------------------------------------------------------------------
C     ENTRY RECEIVES STORAGE LOCATIONS OF SEVERAL OF THE ARRAYS
C     USED BY THE LANCZS SUBROUTINE.  THIS ALLOWS USER TO SPECIFY
C     THE DIMENSIONS OF THESE ARRAYS IN THE MAIN PROGRAM.
C
      ENTRY LANZP(DIR,DESC,SM,TM,TOD,TD,G,XLFT,LEFT,SPREC)
C-----------------------------------------------------------------------
C
C-----END OF LANCZS-----------------------------------------------------
      RETURN
      END
C
C-----START OF LANCI1---------------------------------------------------
C     GENERATES THE Q-SUBBLOCKS ON EACH ITERATION OF THE BLOCK LANCZOS
C     PROCEDURE.
C
      SUBROUTINE LANCI1(MATVEC,MXBLK,NITER,I,N,Q,KACT,KML,ERRMAX,
     1RESN,RKM,IND,KACTN,IWRITE)
C
C-----------------------------------------------------------------------
      DOUBLE PRECISION  Q(1),TM(1),S,SM(1),T,ERRMAX,SUM,RESN,RKM
      INTEGER  DIR(2,1),DESC(1),LEFT(1),XLFT(1)
      DOUBLE PRECISION FINPRO, DSQRT
      EXTERNAL MATVEC
C-----------------------------------------------------------------------
C     SIZE OF FIRST BLOCK CAN CHANGE.
      IF(I.EQ.1) KACTN = KACT
C
C     XLFT(I+2) IS CUMULATIVE TOTAL OF VECTORS IN 1ST QBLOCK NOT
C     GENERATING DESCENDANTS.
C
      IF(I.GT.1) GO TO 10
      XLFT(1) = 0
      XLFT(2) = 0
   10 XLFT(I+2) = XLFT(I+1)
C
C     INITIALIZE THE DIRECTORY FOR NEXT QBLOCK Q(I+1)
C
      I2=DIR(2,I)
      I1=DIR(1,I)
      DIR(1,I+1)=I2+1
      DIR(2,I+1)=I2
C
C     IS THERE ROOM FOR ANOTHER QBLOCK?
C
      MS = I2-I1+1
      IF (MS+I2.LE.MXBLK) GO TO 70
C
C     NOT ENOUGH ROOM TO GENERATE ANOTHER BLOCK
C     COMPLETE THE TM-MATRIX. NOTE THAT THE TM-MATRIX IS
C     DIMENSIONED AS (MXBLK,1) AND THE EISPACK SUBROUTINES
C     REQUIRE THE LOWER TRIANGULAR PART OF THIS MATRIX.
C
      I3=I2+1
      JI30 = (I3-1)*N
      JI31 = JI30 + 1
      JK1 = (I1-2)*N + 1
      DO 60 K=I1,I2
      JK1 = JK1 + N
C-----------------------------------------------------------------------
```

```
      CALL MATVEC(Q(JK1),Q(JI31))                                       BLS03300
C-----------------------------------------------------------------------BLS03310
C     COMPUTE LAST DIAGONAL BLOCK IN TM-MATRIX FOR THIS ITERATION       BLS03320
C                                                                       BLS03330
      JL1 = (K-2)*N + 1                                                 BLS03340
      KK = (K-1)*MXBLK +  K - 1                                         BLS03350
   20 DO 30 L=K,I2                                                      BLS03360
      KK = KK + 1                                                       BLS03370
      JL1 = JL1 + N                                                     BLS03380
C-----------------------------------------------------------------------BLS03390
      TM(KK) = FINPRO(N,Q(JL1),1,Q(JI31),1)                             BLS03400
C-----------------------------------------------------------------------BLS03410
   30 CONTINUE                                                          BLS03420
C                                                                       BLS03430
C     COMPUTE ASSOCIATED CORRECTION TERMS IN TM-MATRIX.                 BLS03440
      IF(XLFT(I).EQ.0) GO TO 50                                         BLS03450
      LUP = XLFT(I)                                                     BLS03460
      DO 40 JJ = 1,LUP                                                  BLS03470
      L= LEFT(JJ)                                                       BLS03480
      JL1 = (L-1)*N + 1                                                 BLS03490
C-----------------------------------------------------------------------BLS03500
      SUM = FINPRO(N,Q(JI31),1,Q(JL1),1)                                BLS03510
C-----------------------------------------------------------------------BLS03520
      KK = (L-1)*MXBLK + K                                              BLS03530
      TM(KK) = SUM + TM(KK)                                             BLS03540
   40 CONTINUE                                                          BLS03550
C                                                                       BLS03560
   50 CONTINUE                                                          BLS03570
C                                                                       BLS03580
   60 CONTINUE                                                          BLS03590
C                                                                       BLS03600
      RETURN                                                            BLS03610
C                                                                       BLS03620
C     ON EVERY BLOCK PASS THROUGH HERE TO GENERATE THE ITH-BLOCK        BLS03630
C     DIAGONAL ENTRY A(I) OF THE TM-MATRIX, EXCEPT THE LAST DIAGONAL    BLS03640
C     BLOCK WHICH IS GENERATED ABOVE                                    BLS03650
C                                                                       BLS03660
   70 CONTINUE                                                          BLS03670
C     COMPUTE (A-MATRIX)*(ITH-Q-BLOCK)                                  BLS03680
      KA=I2                                                             BLS03690
      DO 80 K=I1,I2                                                     BLS03700
      KA=KA+1                                                           BLS03710
      JKA1 = (KA-1)*N + 1                                               BLS03720
      JK1 = (K-1)*N + 1                                                 BLS03730
C-----------------------------------------------------------------------BLS03740
      CALL MATVEC(Q(JK1),Q(JKA1))                                       BLS03750
C-----------------------------------------------------------------------BLS03760
      DESC(K)=KA                                                        BLS03770
   80 DESC(KA)=K                                                        BLS03780
C                                                                       BLS03790
C     COMPUTE (A-MATRIX)*(ITH-Q-BLOCK) - ((I-1)TH-Q-BLOCK)*B(I)-TRANS   BLS03800
C     WHERE B(I) DENOTES THE ITH SUBDIAGONAL BLOCK                      BLS03810
C                                                                       BLS03820
      IF(I.EQ.1) GO TO 110                                              BLS03830
      J1 = DIR(1,I-1)                                                   BLS03840
      J2 = DIR(2,I-1)                                                   BLS03850
      DO 100 K=I1,I2                                                    BLS03860
      KD=DESC(K)                                                        BLS03870
      JKD0 = (KD-1)*N                                                   BLS03880
      KK = (J1-2)*MXBLK + K                                             BLS03890
      DO 90 L=J1,J2                                                     BLS03900
      JL = (L-1)*N                                                      BLS03910
      KK = KK + MXBLK                                                   BLS03920
      S=TM(KK)                                                          BLS03930
      JKD = JKD0                                                        BLS03940
      DO 90 J=1,N                                                       BLS03950
      JKD = JKD + 1                                                     BLS03960
```

```
      JL = JL + 1                                                       BLS03970
   90 Q(JKD) = Q(JKD) - S*Q(JL)                                         BLS03980
  100 CONTINUE                                                          BLS03990
      LINT = (KD-1)*N + 1                                               BLS04000
      LFIN = KD*N                                                       BLS04010
C                                                                       BLS04020
C     COMPUTE A(I)                                                      BLS04030
C                                                                       BLS04040
  110 DO 130 K=I1,I2                                                    BLS04050
      KKMX = (K-1)*MXBLK                                                BLS04060
      KD=DESC(K)                                                        BLS04070
      JKD1 = (KD-1)*N+ 1                                                BLS04080
      JL1 = (K-2)*N + 1                                                 BLS04090
      DO 120 L=K,I2                                                     BLS04100
      JL1 = JL1 + N                                                     BLS04110
      KK = KKMX + L                                                     BLS04120
C-----------------------------------------------------------------------BLS04130
      TM(KK) = FINPRO(N,Q(JL1),1,Q(JKD1),1)                             BLS04140
C-----------------------------------------------------------------------BLS04150
  120 CONTINUE                                                          BLS04160
  130 CONTINUE                                                          BLS04170
C                                                                       BLS04180
C     COMPUTE P(I) = P(I) - (ITH-Q-BLOCK)*A(I)                          BLS04190
C                                                                       BLS04200
      DO 170 K=I1,I2                                                    BLS04210
      KKMX = (K-1)*MXBLK                                                BLS04220
      KD=DESC(K)                                                        BLS04230
      JKD0 = (KD-1)*N                                                   BLS04240
      JL = (I1-1)*N                                                     BLS04250
      DO 140 L=I1,I2                                                    BLS04260
      KK = KKMX + L                                                     BLS04270
      IF(L.LT.K) KK=(L-1)*MXBLK + K                                     BLS04280
      S=TM(KK)                                                          BLS04290
      JKD = JKD0                                                        BLS04300
      DO 140 J=1,N                                                      BLS04310
      JL = JL + 1                                                       BLS04320
      JKD = JKD + 1                                                     BLS04330
  140 Q(JKD) = Q(JKD) - S*Q(JL)                                         BLS04340
C                                                                       BLS04350
C     REORTHOGONALIZE THE BLOCK P(I) WITH RESPECT TO ALL VECTORS        BLS04360
C     IN THE 1ST QBLOCK THAT ARE NOT CURRENTLY GENERATING ANY           BLS04370
C     DESCENDANTS.  NOTE THAT 2ND Q-BLOCK IS REORTHOGONALIZED           BLS04380
C     ELSEWHERE.                                                        BLS04390
      IF(XLFT(I).EQ.0) GO TO 170                                        BLS04400
      LUP = XLFT(I)                                                     BLS04410
      DO 160 JJ = 1,LUP                                                 BLS04420
      L= LEFT(JJ)                                                       BLS04430
      JL0 = (L-1)*N                                                     BLS04440
      LLMX = (L-1)*MXBLK                                                BLS04450
      JL1 = JL0 + 1                                                     BLS04460
      JKD1 = JKD0 + 1                                                   BLS04470
C-----------------------------------------------------------------------BLS04480
      SUM = FINPRO(N,Q(JL1),1,Q(JKD1),1)                                BLS04490
C-----------------------------------------------------------------------BLS04500
      JKD = JKD0                                                        BLS04510
      JL = JL0                                                          BLS04520
      DO 150 J=1,N                                                      BLS04530
      JKD = JKD + 1                                                     BLS04540
      JL = JL + 1                                                       BLS04550
  150 Q(JKD) = Q(JKD) - SUM* Q(JL)                                      BLS04560
      KK = LLMX + K                                                     BLS04570
      TM(KK) = SUM + TM(KK)                                             BLS04580
C                                                                       BLS04590
  160 CONTINUE                                                          BLS04600
  170 CONTINUE                                                          BLS04610
C                                                                       BLS04620
C                                                                       BLS04630
```

```
C   GENERATE B(I+1)
C
      K1=DESC(I1)
      K2=DESC(I2)
      IFLAG=0
C
C     COMPUTE NORMS
C
  180 CONTINUE
      JK1 = (K1-2)*N + 1
      DO 190 K=K1,K2
      JK1 = JK1 + N
C-----------------------------------------------------------------------
      SM(K) = FINPRO(N,Q(JK1),1,Q(JK1),1)
C-----------------------------------------------------------------------
  190 CONTINUE
C
        IF(I.EQ.1.AND.K1.EQ.I2+1)  WRITE(6,200) NITER,
     1 (K,SM(K), K =K1,K2)
  200 FORMAT(//' ON ITERATION', I4,' NORM(GRADIENTS)**2 OF 1ST BLOCK = '
     1/5(I4,E12.3))
C
C     TEST FOR CONVERGENCE OF BLOCK LANCZOS
C
      IF(I.GT.1.OR.K1.GT.I2+1)  GO TO 250
C
C     TEST THE FIRST KM OF THE EIGENVALUES FOR CONVERGENCE
      K2L = K1 + KML - 1
      RKM = SM(K2L)
      DO 210 K = K1,K2L
      IF(SM(K).GT.ERRMAX )  GO TO 220
  210 CONTINUE
      GO TO 430
C
C     CAN WE REDUCE KACT?  IF A SMALL RESIDUAL (GRADIENT) IS IDENTIFIED,
C     SIZE OF 1ST BLOCK MAY BE REDUCED.
  220 IF(KML.EQ.KACT) GO TO 250
      DO 230 K = K2L,K2
      IF(SM(K).GT.ERRMAX) GO TO 230
      KSAV = K
      KACTN = KSAV - KACT
      GO TO 240
C
  230 CONTINUE
      GO TO 250
C
  240 K2 = KSAV
C
C     GENERATE THE TRANSPOSE OF B(I)
C
  250 CONTINUE
C
C     DETERMINE THE MAXIMAL NORM
      K=K1
      S=SM(K)
      DO 260 L=K1,K2
      IF (SM(L).LT.S) GOTO 260
      K=L
      S=SM(L)
  260 CONTINUE
C     FOR 2ND QBLOCK, SAVE INDEX AND SIZE OF MAXIMAL NORM
      IF(I.GT.1) GO TO 270
      IND = K - KACT
      RESN = SM(K)
C
  270 IF(S.LE.ERRMAX)GO TO 340
C
```

```
      IF(IFLAG.EQ.1) GO TO 340
C
      S=DSQRT(S)
      JK0 = (K-1)*N
      JK = JK0
      DO 280 J=1,N
      JK = JK + 1
  280 Q(JK)=Q(JK)/S
      JL0 = (K1-2)*N
      DO 310 L=K1,K2
      JL0 = JL0 + N
      LL=(DESC(L) - 1)*MXBLK + K1
      IF (L.NE.K) GOTO 290
      TM(LL)=S
      GO TO 310
  290 JK1 = JK0 + 1
      JL1 = JL0 + 1
C-----------------------------------------------------------------------
      T = FINPRO(N,Q(JK1),1,Q(JL1),1)
C-----------------------------------------------------------------------
      TM(LL)=T
      JK = JK0
      JL = JL0
      DO 300 J=1,N
      JK = JK + 1
      JL = JL + 1
  300 Q(JL) = Q(JL) - T*Q(JK)
  310 CONTINUE
      IF (K.EQ.K1) GOTO 330
C
      JK1 = (K1-1)*N
      JK = JK0
      DO 320 J=1,N
      JK = JK + 1
      JK1 = JK1 + 1
      T=Q(JK1)
      Q(JK1)=Q(JK)
  320 Q(JK)=T
      MA=DESC(K)
      MB=DESC(K1)
      DESC(K1)=MA
      DESC(K)=MB
      DESC(MA)=K1
      DESC(MB)=K
  330 CONTINUE
C
      DIR(2,I+1)=K1
C
      IFLAG=1
C
      K1=K1+1
      IF(I.EQ.1) GO TO 340
      IF (K1.LE.K2) GO TO 180
C     RETURN TO LANCZS
C
      RETURN
C
C    IMPLICIT VECTOR DEFLATION
C
  340 CONTINUE
      J= XLFT(I+2)
      IF(K1.GT.K2) GO TO 360
      DO 350 L= K1,K2
      J = J+1
  350 LEFT(J) = DESC(L)
  360 XLFT(I+2) = J
C
```

```
C     FORCE REORTHGONALIZATION OF 2ND AND 3RD QBLOCKS W.R.T. THOSE
C     VECTORS IN 1ST QBLOCK THAT ARE NOT GENERATING DESCENDANTS
C     ON THIS ITERATION.
      IF(I.GT.1)  GO TO 370
      XLFT(1) = XLFT(3)
      XLFT(2) = XLFT(3)
  370 IJJ = I + 2
      IJJJ= XLFT(IJJ)
C
      IF(IJJJ.EQ.0) GO TO 390
      IF(IWRITE.EQ.1)  WRITE(6,380) (LEFT(IJ),IJ= 1,IJJJ)
  380 FORMAT(' VECTORS NOT GENERATING DESCENDANTS ARE '/(10I6))
C
  390 IF(I.EQ.1.AND.KML.GT.1) GO TO 400
C
      RETURN
C
C     REORTHOGONALIZE 2ND QBLOCK W.R.T VECTORS IN 1ST BLOCK NOT
C     GENERATING DESCENDANTS
  400 IF(XLFT(I).EQ.0) RETURN
      LUP = XLFT(I)
      KD = DIR(2,I+1)
      JKD0 = (KD-1)*N
      DO 420 JJ = 1,LUP
      L = LEFT(JJ)
      JL0 = (L-1)*N
      JL1 = JL0 + 1
      JKD1 = JKD0 + 1
C-----------------------------------------------------------------------
      SUM = FINPRO(N,Q(JKD1),1,Q(JL1),1)
C-----------------------------------------------------------------------
      JL = JL0
      JKD = JKD0
      DO 410 J=1,N
      JL = JL + 1
      JKD = JKD + 1
  410 Q(JKD) = Q(JKD) - SUM *Q(JL)
  420 CONTINUE
C
      RETURN
C
C     EXIT IF CONVERGENCE OF DESIRED EIGENVECTORS IS CONFIRMED.
C
  430 CONTINUE
      DO 440 L=K1,K2
      M=DESC(L)
  440 DESC(M)=0
      DIR(2,2)=DIR(2,1)
C
      WRITE(6,450) ERRMAX
  450 FORMAT(/' CONVERGENCE OBSERVED, ALL RESIDUALS**2 .LT. ERRMAX = ',
     1 E20.12)
C
      RETURN
C
C-----------------------------------------------------------------------
C     ALLOWS PASSAGE OF LOCATIONS OF SOME OF THE ARRAYS USED BY LANCI1
C     SO THAT THESE ARRAYS CAN BE DIMENSIONED IN THE MAIN PROGRAM
C
      ENTRY LANCP1(DIR,DESC,TM,SM,XLFT,LEFT)
C-----------------------------------------------------------------------
C
      RETURN
C-----END OF LANCI1-----------------------------------------------------
      END
C
C-----ORTHOG------------------------------------------------------------
```

```
C     ORTHOGONALIZE COLUMNS M = MA,MB OF Q-ARRAY W.R.T COLUMNS M = 1,MB
C
      SUBROUTINE ORTHOG(MA,MB,N,Q)
C
C-----------------------------------------------------------------------
      DOUBLE PRECISION  Q(1), S
      DOUBLE PRECISION FINPRO, DSQRT
C-----------------------------------------------------------------------
C     MAIN LOOP
      DO 50 M = MA,MB
      MMO = (M-1)*N
      LLO = -N
      DO 40 L = 1,M
      LLO = LLO + N
      LL = LLO + 1
      MM = MMO + 1
C-----------------------------------------------------------------------
      S = FINPRO(N,Q(LL),1,Q(MM),1)
C-----------------------------------------------------------------------
C
      IF (M.EQ.L) GO TO 20
C
      MM = MMO
      LL = LLO
      DO 10 I=1,N
      LL = LL + 1
      MM = MM + 1
   10 Q(MM) = Q(MM) - S*Q(LL)
      GO TO 40
C
   20 S = DSQRT(S)
      MM = MMO
      DO 30 I=1,N
      MM = MM + 1
   30 Q(MM) = Q(MM)/S
C
   40 CONTINUE
   50 CONTINUE
C
      RETURN
C-----END OF ORTHOG-----------------------------------------------------
      END
C
C-----START-------------------------------------------------------------
C     GENERATES PSEUDO-RANDOM STARTING VECTORS.
C
      SUBROUTINE START(KA,KB,N,Q,G,ERRMAX)
C
C-----------------------------------------------------------------------
      DOUBLE PRECISION  Q(1), ERRMAX, S
      REAL  G(1)
      COMMON/RANDOM/IIX
      DOUBLE PRECISION FINPRO, DSQRT
C-----------------------------------------------------------------------
      IF(KA.GT.KB) RETURN
C
      IIL = IIX
      DO 110 K = KA,KB
      KKO = (K-1)*N
C
C-----------------------------------------------------------------------
      CALL GENRAN(IIL,G,N)
C-----------------------------------------------------------------------
C
      KK = KKO
      DO 10 I = 1,N
      KK = KK + 1
```

```
   10 Q(KK) = G(I)
      LLO = -N
   20 DO 70 L=1,K
      LLO = LLO + N
      LL = LLO + 1
      KK = KKO + 1
C-----------------------------------------------------------------------
      S = FINPRO(N,Q(LL),1,Q(KK),1)
C-----------------------------------------------------------------------
C
      IF (K.EQ.L) GO TO 40
C
      LL = LLO
      KK = KKO
      DO 30 I=1,N
      LL = LL + 1
      KK = KK + 1
   30 Q(KK) = Q(KK) - S*Q(LL)
      GO TO 70
C
   40 S = DSQRT(S)
      IF(S.LE.ERRMAX) GO TO 80
      KK = KKO
      DO 50 I=1,N
      KK = KK + 1
   50 Q(KK) = Q(KK)/S
C
      WRITE(6,60) K
   60 FORMAT(I6,' TH STARTING VECTOR IS GENERATED RANDOMLY')
C
   70 CONTINUE
      GO TO 110
C
C-----------------------------------------------------------------------
   80 CALL GENRAN(IIX,G,N)
C-----------------------------------------------------------------------
C
      WRITE(6,90) K
   90 FORMAT(/I6,' TH RANDOM VECTOR REJECTED, GENERATE ANOTHER'/)
C
      KK = KKO
      DO 100 I = 1,N
      KK = KK + 1
  100 Q(KK) = G(I)
      GO TO 20
C
  110 CONTINUE
      RETURN
C-----END OF START------------------------------------------------------
      END
C
C-----START OF DIAGOM---------------------------------------------------
C     DIAGOM CALLS THE EISPACK SUBROUTINES TRED2 AND IMTQL2 TO
C     DIAGONALIZE THE SMALL SYMMETRIC MATRICES GENERATED AT EACH
C     ITERATION OF BLOCK LANCZOS.
C
      SUBROUTINE DIAGOM(MXBLK,MM,TM,KACT,N,Q,E,RESID,RESK,RESN,IND,
     1 KACTN,KM,TD,TOD,NITER,IERR,IWRITE)
C
C-----------------------------------------------------------------------
      DOUBLE PRECISION  TM(MXBLK,1),Q(1),E(1),TD(1),TOD(1),RESID(1)
      DOUBLE PRECISION  RESK(1),RESN,RATIO,FRACT,RKM,EMAX,SPREAD,EGAP
      DOUBLE PRECISION  DABS,DFLOAT,DMAX1
C-----------------------------------------------------------------------
      IF(NITER.GE.100) GO TO 270
      RKM = TD(1)
      FRACT = TD(2)
```

```
      NSTAG = IERR
      KWANT = KACT
C
C     STORE KM-TH RESIDUALS**2 FOR CHECK ON STAGNATION OF CONVERGENCE
      NITER1 = NITER + 1
      RESK(NITER1) = RKM
      IF(NITER.LE.NSTAG) GO TO 10
C     TEST FOR STAGNATION
      NITERM = NITER - 10
      RATIO = RKM / RESK(NITERM)
      IF(RATIO.GT.FRACT) GO TO 250
C
   10 CONTINUE
C
C     TEST GAPS TO DETERMINE IF SIZE OF 1ST Q-BLOCK CAN BE REDUCED
      IF(NITER.EQ.0) GO TO 40
      IF(KM.EQ.KACT.OR.NITER.LT.10) GO TO 30
      KACT1 = KACT - 1
      DO 20 K = KM,KACT1
      RATIO =  DABS(E(K+1) - E(K))
      IF(RATIO.LT.25*EGAP) GO TO 20
      KACT = K
      GO TO 40
   20 CONTINUE
C
C     IF KACT.NE.KACTN, THEN SUBROUTINE LANCI1 IDENTIFIED A VERY
C     SMALL RESIDUAL FOR SOME E(J), J>= KM.
   30 IF(KACT.EQ.KACTN)  GO TO 50
      RATIO = DABS(E(KACTN+1) - E(KACTN))
      IF(RATIO.LE.EGAP)  GO TO 50
      KACT = KACTN
   40 ICOUNT = 1
      INDEXP = IND
      RESID(1) = RESN
      GO TO 80
C
   50 CONTINUE
      IF(IND.NE.INDEXP) GO TO 70
C     INDEX OF VECTOR OF MAXIMUM NORM IS SAME AS ON PREVIOUS ITERATION
      ICOUNT = ICOUNT + 1
      IF(ICOUNT.LE.5) GO TO 60
      ITEST = ICOUNT - 4
      RATIO = RESID(ITEST)/RESN
      IF(DABS(RATIO).GT.10.D0) GO TO 60
C
C     CONVERGENCE STAGNATED, ADD NEXT RITZ VECTOR IN THE CHAIN
C     TO THE 1ST Q-BLOCK AND RESET THE FLAGS THAT KEEP TRACK OF
C     CONVERGENCE.
      INDEXP = IND
      ICOUNT = 0
      KACT = KACT + 1
      KWANT = KACT
C     CHECK THAT THERE IS ENOUGH ROOM TO ENLARGE THE 1ST QBLOCK
      IF(2*KACT.GT.MXBLK) GO TO 230
      GO TO 80
C
   60 RESID(ICOUNT) = RESN
      INDEXP = IND
      GO TO 80
C
   70 ICOUNT = 1
      RESID(1) = RESN
      INDEXP = IND
C
C-----------------------------------------------------------------------
C     USE EISPACK SUBROUTINES TO DIAGONALIZE THE SMALL TM-MATRIX.
C
```

```
   80 CALL TRED2(MXBLK,MM,TM,TD,TOD,TM)
      CALL IMTQL2(MXBLK,MM,TD,TOD,TM,IERR)
C-----------------------------------------------------------------------
      IF(IERR.EQ.0) GO TO 90
      RETURN
C
C     SELECT RELEVANT EIGENVALUES AND EIGENVECTORS OF THE T-MATRIX.
   90 CONTINUE
C
C     IMTQL2 RETURNS EIGENVALUES (AND CORRESPONDING EIGENVECTORS) IN
C     ALGEBRAICALLY-ASCENDING ORDER.  REARRANGE TO DESCENDING ORDER.
C
      DO 100 L=1,MM
      MML = MM-L+1
  100 E(L) = TD(MML)
C
  110 WRITE(6,120) KACT, (E(J), J=1,KACT)
  120 FORMAT(' COMPUTED',I4,' ALGEBRAICALLY-LARGEST EIGENVALUES'/(4E20.1
     12))
C
C      COMPUTE ESTIMATE MAXIMUM EIGENVALUE AND OF SPREAD
      IF(NITER.GT.1) GO TO 140
      EMAX = DMAX1(DABS(E(1)),DABS(E(MM)))
      SPREAD = DABS(E(1) - E(MM))
      EGAP = SPREAD/DFLOAT(N)
      IF(NITER.EQ.1)  WRITE(6,130) EMAX,SPREAD,EGAP
  130 FORMAT(/4X,'ESTIMATED NORM OF MATRIX',4X,'ESTIMATED SPREAD',6X,'SP
     1READ*(SIZE)*(-1)'/E28.4,E20.4,E24.3)
  140 CONTINUE
C
C      COMPUTE RITZ VECTORS
      DO 180 I=1,N
      DO 150 KK=1,KWANT
      TOD(KK)=0.D0
      K = MM - KK + 1
      IL = - N + I
      DO 150 L = 1,MM
      IL = IL + N
  150 TOD(KK) = TOD(KK) + TM(L,K)*Q(IL)
      IKK = -N + I
  160 DO 170 KK=1,KACT
      IKK = IKK + N
  170 Q(IKK)=TOD(KK)
  180 CONTINUE
C
C      ON FILE 13 SAVE ANY EXTRA VECTORS NO LONGER NEEDED IN 1ST Q-BLOCK
      IF(KWANT.EQ.KACT) GO TO 290
      K1 = KACT + 1
      K2 = KWANT
      DUMMY = 100.
      DO 190 K = K1,K2
      LINT = (K-1)*N + 1
      LFIN = K*N
      WRITE(13,210) E(K),DUMMY,K
      WRITE(13,220) (Q(L), L=LINT,LFIN)
  190 CONTINUE
      KDELTA = KWANT - KACT
      WRITE(13,200) KDELTA
  200 FORMAT(/' ABOVE ARE ',I3,' VECTORS STRIPPED FROM A 1ST Q-BLOCK'/
     1' DURING A BLOCK LANZCOS RUN WHICH COULD BE USED AS STARTING VECTO
     1RS'/' IN A LATER RUN IF THE USER DECIDES THAT THESE EIGENVALUES SH
     1OULD'/' BE COMPUTED AFTER ALL.  FORMAT USED IN THE SAME AS WAS USE
     1D'/' IN THE CORRESPONDING BLSTARTV FILE'/)
  210 FORMAT(/E20.12,E13.4,I6,' = EVAL,DUMMY,EVAL NUMBER,EVEC='/)
  220 FORMAT(4E20.12)
```

```
      GO TO 290
C
C     DEFAULT, SIZE OF 1ST Q-BLOCK TOO LARGE FOR MXBLK
  230 IWRITE = -1000
      WRITE(6,240) KACT,MXBLK
      WRITE(15,240) KACT,MXBLK
  240 FORMAT(//' BLOCK LANCZOS PROCEDURE TRIED TO INCREASE THE SIZE OF 1
     1ST QBLOCK'/' TO ',I3,' BUT THIS IS NOT FEASIBLE BECAUSE TWICE THIS
     1 SIZE'/' IS G.T. MXBLK WHICH EQUALS ',I4/' USER CAN RERUN PROGRAM
     1WITH LARGER MXBLK'/)
      GO TO 290
C
C     DEFAULT, CONVERGENCE RATE IS TOO SLOW
  250 IWRITE = -1000
      WRITE(6,260) NITER,RATIO,FRACT
      WRITE(15,260) NITER,RATIO,FRACT
  260 FORMAT(//' ON ITERATION ',I3,' CONVERGENCE APPEARS TO BE STAGNATED
     1'/' RATIO OF SQUARE OF CURRENT KM-TH RESIDUAL TO CORRESPONDING SQU
     1ARE'/' 10 ITERATIONS EARLIER IS ',E10.3,' COMPARED TO '/
     1' USER-SPECIFIED RATIO ',E10.3,'. THEREFORE, PROGRAM TERMINATES'/'
     1 USER SHOULD LOOK AT THE OUTPUT. IF CONVERGENCE HAS STAGNATED, USE
     1R'/' CAN EITHER INCREASE KACT OR KMAX OR RESET THE STAGNATION PARA
     1METERS'/' NSTAG AND FRACT, AND RESTART THE BLOCK PROCEDURE USING T
     1HE'/' CURRENT EIGENVECTOR APPROXIMATIONS AS STARTING VECTORS'/)
      GO TO 290
C
  270 IWRITE = -1000
      WRITE(6,280)
      WRITE(15,280)
  280 FORMAT(//' SOMETHING IS SERIOUSLY WRONG.  NUMBER OF ITERATIONS IS
     1EXCESSIVE'/' PROGRAM TERMINATES FOR USER TO DECIDE WHAT TO DO'/'
     1ALTERNATIVES INCLUDE INCREASING KACT OR KMAX OR BOTH, AND RESTARTI
     1NG'/' USING THE CURRENT APPROXIMATIONS AS STARTING VECTORS'//)
C
  290 CONTINUE
      RETURN
C-----END OF DIAGOM------------------------------------------------------
      END
C-----LPERM PERMUTES VECTORS--------------------------------------------
C
      SUBROUTINE LPERM(W,U,IPERM)
C
C-----------------------------------------------------------------------
      DOUBLE PRECISION  U(1),W(1)
      INTEGER  IPR(1),IPT(1)
C-----------------------------------------------------------------------
C     SUBROUTINE HAS 2 BRANCHES:  IPERM = 1,  CALCULATES
C     U = P*W  WHERE  P IS THE PERMUTATION REPRESENTED BY IPR
C     LET J = IPR(K) THEN U(K) = W(J), K = 1,N. WE SET W(K)=U(K), K=1,N
C     IPERM = 2, USING THE PERMUTATION IPT (P-TRANSPOSE) U = P'*W, W=U
C     LET J = IPT(K) THEN U(K) = W(J), K=1,N. WE SET W(K) = U(K), K=1,N
C-----------------------------------------------------------------------
C
      IF(IPERM.EQ.2) GO TO 10
C     IPERM = 1
      DO 20 K = 1,N
      J = IPR(K)
   20 U(K) = W(J)
      DO 30 K = 1,N
   30 W(K) = U(K)
      GO TO 60
C     IPERM = 2
   10 DO 40  K = 1,N
      J = IPT(K)
   40 U(K) = W(J)
      DO 50 K = 1,N
   50 W(K) = U(K)
```

```
   60 CONTINUE                                                          BLS09980
C                                                                       BLS09990
      RETURN                                                            BLS10000
C                                                                       BLS10010
C-----------------------------------------------------------------------BLS10020
      ENTRY LPERME(IPR,IPT,N)                                           BLS10030
C-----------------------------------------------------------------------BLS10040
C                                                                       BLS10050
C-----END OF LPERM---------------------------------------------         BLS10060
      RETURN                                                            BLS10070
      END                                                               BLS10080
```

SECTION 8.6 FILE DEFINITIONS AND SAMPLE INPUT FILE

Below is a listing of the input/output files which are accessed by the real symmetric block Lanczos eigenvalue/eigenvector program, BLEVAL. BLEVAL computes a few extreme eigenvalues and corresponding eigenvectors of a real symmetric matrix. Also below is a sample of the input file which BLEVAL requires on file 5. The parameters in this file are supplied in free format.

SAMPLE DEFINITIONS OF THE INPUT/OUTPUT FILES FOR BLEVAL

```
-----------------------------------------------------------------
*BLEVAL EXEC
FI 06 TERM
FILEDEF  5 DISK BLEVAL    INPUT     A (RECFM F LRECL 80 BLOCK 80
FILEDEF  8 DISK &1        INPUT     A (RECFM F LRECL 80 BLOCK 80
FILEDEF 10 DISK &1        BLSTARTV  A (RECFM F LRECL 80 BLOCK 80
FILEDEF 13 DISK &1        BLEXTRAV  A (RECFM F LRECL 80 BLOCK 80
FILEDEF 15 DISK &1        BLEIGVEC  A (RECFM F LRECL 80 BLOCK 80
*IMTQL2 AND TRED2 ARE 2 EISPACK LIBRARY SUBROUTINES
LOAD  BLEVAL  BLSUB  BLMULT  IMTQL2  TRED2
-----------------------------------------------------------------
```

SAMPLE INPUT FILE FOR BLEVAL

```
-----------------------------------------------------------------
 LINE 1 IWRITE (SPECIFY MESSAGE LEVEL TO FILE 6: 1 MEANS DETAILED
             1
 LINE 2     N       MATNO  (SIZE OF A-MATRIX, MATRIX IDENT. NUMBER
          528        528
 LINE 3 MDIMQ           MDIMTM        MAXIT (DIMS. Q, TM, MAX A*X'S
        40000            2500          1000
 LINE 4 EFLAG   OFLAG (EFLAG=(0,1) 1=2PHASES. OFLAG: 1=ORTHOG CHECK
             1       1
 LINE 5 SEED       (STARTING VECTOR SEED, RANDOM NUMBER GENERATOR
     3482736
 LINE 6 KMAX  KACT  KSET (MAX T SIZE +1,SIZE 1ST BLOCK,VECS SUPPLIED
          21     4     0
 LINE 7 KM   (NUMB. EVS FOR ALG-LARGEST, -(NUMB. EVS) FOR ALG-SMALLEST
         4
 LINE 8 NSTAG  FRACT (NO. ITNS BEFORE TEST CONVERGENCE, TEST FRACTION
          25     .01
 LINE 9 RELTOL     MAXIT2 (PHASE 2, CONVERGE. TOL. ,  NO. A*X'S ALLOWED
     .00000001       1000
-----------------------------------------------------------------
```

CHAPTER 9

FACTORED INVERSES, REAL SYMMETRIC BLOCK LANCZOS

SECTION 9.1 INTRODUCTION

The FORTRAN codes in this chapter address the question of using an iterative block Lanczos procedure to compute a 'few' eigenvalues and a basis for the corresponding eigenspace of a real symmetric matrix A by computing a few extreme eigenvalues and a corresponding basis for the inverse of a real symmetric matrix B obtained from A by scaling, shifting and permuting A. For a given real symmetric matrix A, the codes consider the inverse of a matrix B where

$$B \equiv PCP^T \text{ with } C \equiv S0*A + SHIFT*I, \tag{9.1.1}$$

the scale S0 and the shift SHIFT are specified by the user, and the permutation matrix P is chosen so that for a sparse matrix A (or C), the resulting factorization of the associated B matrix is also sparse. An eigenvalue is extreme if it is one of the algebraically-smallest or the algebraically-largest eigenvalues in the eigenvalue spectrum.

Specifically, for a given real symmetric matrix A and associated B-matrix as defined in Eqn(9.1.1), the codes in this chapter compute the q algebraically-largest eigenvalues, λ_i, $1 \le i \le q$, of B^{-1} and corresponding orthonormal real vectors $X_q \equiv \{x_1,...,x_q\}$ such that

$$B^{-1}X_q = X_q A_q, \text{ where } A_q \equiv X_q^T A X_q. \tag{9.1.2}$$

Typically, $A_q = \Lambda_q$, a diagonal matrix whose nonzero entries are the eigenvalues λ_i. The number q is small and specified by the user.

Real symmetric matrices and factorizations of real symmetric matrices are discussed in Stewart [1973]. See also Bunch and Kaufman [1977] and George and Liu [1981]. Chapter 2, Section 2.1 contains a brief summary of the properties of real symmetric matrices which we use in these codes.

The Lanczos code contained in this chapter is a simple modification of the hybrid 'block' Lanczos procedure given in Chapter 8 to handle the factored inverse of the B-matrix given in Eqn(9.1.1). Therefore please see Chapter 8, Section 8.1, for comments about this procedure and for comments regarding the differences between iterative block Lanczos procedures and single-vector Lanczos procedures.

BLIEVAL is the main 'block' program for the factored inverse version of the 'block' Lanczos codes in Chapter 8. BLIEVAL uses the same subroutines as the real symmetric codes in Chapter 8, with the exception of the user-supplied subroutines. The user must supply a subroutine USPEC which defines and initializes the matrix which is to be used by the LANCZS and LANCI1 subroutines. In the factored inverse case, USPEC specifies the factorization of the particular B-matrix being used. These Lanczos programs do not require the A-matrix. However, the user must supply the scalars S0 and SHIFT, and the permutation P (if any). The user must also supply a subroutine BLSOLV which solves the system of equations $Bu = x$ for any given vector x.

The sample USPEC and BLSOLV subroutines provided assume that the B-matrix being used is positive definite and that the Cholesky factors of B

$$B = LL^T \text{ where L is a lower triangular matrix,} \tag{9.1.3}$$

are used for the matrix-vector multiply, $B^{-1}x$, for any given vector x. However, the user may replace these subroutines by subroutines which define and use a more general factorization. These Lanczos codes only require that the BLSOLV subroutine solves the system $Bu = x$, rapidly and accurately.

All computations are in double precision real arithmetic. On each iteration, the accuracy of the computed eigenvectors is checked in the process of computing the second block of Lanczos vectors on that iteration. Note that the eigenvectors of B^{-1} are simple permutations of the eigenvectors of A. These permutations are undone prior to the termination of the block procedure. The corresponding eigenvalues of A are obtained from those of B^{-1} by a simple scalar transformation which is included in the codes. The eigenelement computations for the small Lanczos matrices use two subroutines from the EISPACK Library [1976,1977], TRED2 and IMTQL2.

Several optional preprocessing programs are provided, PERMUT, LORDER, LFACT, and LTEST. Listings for these programs are given in Chapter 4. PERMUT calls the SPARSPAK Library [1979] to attempt to identify a reordering or permutation P of a given matrix A for which sparseness is preserved under factorization of the permuted matrix. LORDER takes a given matrix C and permutation P and computes the sparse matrix format for the permuted matrix, $B \equiv PCP^T$. LFACT computes the Cholesky factors of a given positive definite matrix. LTEST performs a very crude check on the numerical condition of the matrix supplied to it, by solving a system of equations with and without iterative refinement, LINPACK [1979].

The usefulness of this code for computing a few interior eigenvalues of a given real symmetric matrix is dubious. For such an application one would have to select a shift SHIFT that places the desired eigenvalues of the A-matrix on the extreme of the spectrum of the associated matrix B^{-1} and is chosen so that the B-matrix is well-conditioned numerically. This is not a trivial task. The user should refer to Chapter 7 of Volume 1 of this book for more details on iterative block Lanczos procedures.

SECTION 9.2 MAIN PROGRAM, EIGENVALUE AND EIGENVECTOR COMPUTATIONS

```
C-----BLIEVAL (FEW EXTREME EIGENVALUES AND EIGENVECTORS)----------------
C            (USING FACTORED INVERSE OF A REAL SYMMETRIC MATRIX)
C
C      CONTAINS MAIN PROGRAM FOR COMPUTING A FEW EIGENVALUES
C      AND CORRESPONDING EIGENVECTORS OF A REAL SYMMETRIC MATRIX
C      BY COMPUTING A FEW OF THE ALGEBRAICALLY-LARGEST OR
C      ALGEBRAICALLY-SMALLEST EIGENVALUES OF THE INVERSE OF A SCALED,
C      SHIFTED, AND PERMUTED VERSION B OF THE ORIGINAL A-MATRIX
C      USING A BLOCK FORM OF LANCZOS TRIDIAGONALIZATION WITH LIMITED
C      REORTHOGONALIZATION.  THIS BLOCK PROCEDURE IS ITERATIVE AND
C      REQUIRES A SUBROUTINE BLSOLV THAT FOR ANY GIVEN VECTOR W
C      COMPUTES U SUCH THAT B*U = W.   THE SAMPLE BLSOLV SUBROUTINES
C      PROVIDED FOR SPARSE MATRICES ARE ONLY FOR THE CASE THAT B IS
C      POSITIVE DEFINITE AND USE THE CHOLESKY FACTORS OF B.  HOWEVER,
C      THE USER COULD REPLACE THESE BY A SUBROUTINE WHICH COMPUTES
C      FOR AN INDEFINITE MATRIX THE FACTORIZATION L*D*(L-TRANSPOSE).
C
C      THIS BLOCK PROCEDURE COMPUTES THE ALGEBRAICALLY-LARGEST
C      EIGENVALUES OF THE INVERSE OF THE B-MATRIX, UNLESS THE USER
C      SUPPLIES -(B-INVERSE)*X RATHER THAN (B-INVERSE)*X, IN WHICH
C      CASE IT COMPUTES THE CORRESPONDING ALGEBRAICALLY-SMALLEST
C      EIGENVALUES OF (B-INVERSE) BY COMPUTING THE ALGEBRAICALLY-
C      LARGEST EIGENVALUES OF -(B-INVERSE).  IN THIS CASE THE SIGNS
C      OF THE COMPUTED EIGENVALUES ARE CHANGED PRIOR TO WRITING TO
C      FILE 15 SO THAT ON EXIT, FILE 15 CONTAINS THE ALGEBRAICALLY-
C      SMALLEST EIGENVALUES OF B-INVERSE ALONG WITH THE CORRESPONDING
C      EIGENVALUES OF THE ORIGINAL A-MATRIX AND CORRESPONDING
C      EIGENVECTORS.  THE MATRIX B = SO*P*A*P' + SHIFT*I WHERE THE
C      SCALE SO AND SHIFT ARE READ IN THIS PROGRAM, AND THE
C      PERMUTATION P IS DEFINED IN THE CORRESPONDING USPEC SUBROUTINE.
C      THE PROGRAM ASSUMES THAT THE FACTORIZATION READ IN USPEC
C      CORRESPONDS TO THE SO, SHIFT AND PERMUTATION READ IN.  THE SO
C      AND SHIFT ARE CHOSEN SO THAT THE DESIRED EIGENVALUES ARE AT
C      THE EXTREME OF THE SPECTRUM OF B-INVERSE.
C
C      THIS IS AN ITERATIVE 'BLOCK' LANCZOS PROCEDURE FOR WHICH ON
C      EVERY ITERATION, THE 2ND AND SUCCEEDING BLOCKS CONTAIN ONLY ONE
C      VECTOR WHICH IS SELECTED ON THE BASIS OF ITS EXPECTED INFLUENCE
C      ON THE CONVERGENCE.   Q-BLOCKS GENERATED ON A GIVEN ITERATION
C      ARE REORTHOGONALIZED ONLY W.R.T. THOSE VECTORS IN THE FIRST
C      Q-BLOCK WHICH ARE NOT ALLOWED TO GENERATE DESCENDANTS ON
C      THAT ITERATION.
C
C      PFORT VERIFIER IDENTIFIED THE FOLLOWING NONPORTABLE CONSTRUCTIONS:
C      1.  DATA MACHEP DEFINITION
C      2.  FORMAT (20A4) USED FOR READING EXPLANATORY COMMENTS.
C      3.  FREE FORMAT (5,*), USED FOR PARAMETER INPUT FROM FILE 5.
C      4.  COMMON/LOOPS/ AS CONSTRUCTED IS NOT PORTABLE
C
C-----------------------------------------------------------------------
      DOUBLE PRECISION Q(44000),E(50),TM(2500),TOD(50),TD(50),EPSM,NNZ
      DOUBLE PRECISION SM(100),ERRMAX,SPREC,MACHEP,AVER,RELTOL,ERRMAN
      DOUBLE PRECISION  EVAL, RESIDL(100), RESIDK(100), RESID, FRACT
      DOUBLE PRECISION SO,SHIFT
      REAL EXPLAN(20),G(2000)
      INTEGER   DIR(2,100),DESC(100),LEFT(100),XLFT(100)
      INTEGER SEED,OFLAG,EFLAG
      COMMON/LOOPS/MAXIT,ITER
      COMMON /RANDOM/SEED
      COMMON/FLAGS/EFLAG,OFLAG
      DOUBLE PRECISION  DABS, DFLOAT
C-----------------------------------------------------------------------
      EXTERNAL BLSOLV
```

```
      DATA MACHEP/Z3410000000000000/
C-----------------------------------------------------------------------
C
C     ARRAYS MUST DIMENSIONED AS FOLLOWS:
C
C     1.  Q:    >=  KMAX*N
C     2.  G:    >= N
C     3.  E:    >=  MXBLK
C     4.  TM:     >=  MXBLK**2
C     5.  TOD, TD, SM, DESC, LEFT, XLFT:    >= MXBLK
C     6.  DIR:  ROW DIMENSION = 2;   COLUMN DIMENSION >= MXBLK
C     7.  RESIDL, RESIDK:  >= MAXIMUM NUMBER OF ITERATIONS ALLOWED.
C         PROGRAM CURRENTLY TERMINATES IF MORE THAN 100 ITERATIONS
C         ARE REQUESTED.  USED TO MONITOR CONVERGENCE. SEE SUBROUTINE
C         DIAGOM.
C     8.  EXPLAN:  DIMENSION = 20.
C
C-----------------------------------------------------------------------
C     OUTPUT HEADER
      WRITE(6,10)
   10 FORMAT(/' BLOCK LANCZOS PROCEDURE, USES FACTORED INVERSE OF A USER
     1-SPECIFIED MATRIX'/' 2ND AND SUCCEEDING BLOCKS GENERATED ON EACH B
     1LOCK ITERATION '/' CONTAIN ONLY ONE VECTOR'//)
C
C     SET PROGRAM PARAMETERS
      EPSM = 2.D0*MACHEP
      SPREC = 1.D-5
      MPMIN = -1000
C
C     READ USER-SPECIFIED PARAMETERS FROM INPUT FILE 5 (FREE FORMAT)
C
C     SELECT THE AMOUNT OF INTERMEDIATE OUTPUT DESIRED (IWRITE =0,1).
C     IWRITE = 1 INCREASES THE AMOUNT OF INTERMEDIATE OUTPUT WRITTEN
C     TO FILE 6 ON EACH ITERATION OF THE BLOCK LANCZOS PROCEDURE.
      READ(5,20) EXPLAN
   20 FORMAT(20A4)
      READ(5,*)  IWRITE
C
C     READ ORDER (N) OF MATRIX AND MATRIX IDENTIFICATION NUMBER (MATNO)
C     READ SCALE (S0) AND SHIFT (SHIFT) APPLIED TO MATRIX AND
C     FLAG JPERM.  JPERM = (0,1):  JPERM = 1 MEANS MATRIX HAS BEEN
C     PERMUTED
      READ(5,20) EXPLAN
      READ(5,*) N,MATNO,S0,SHIFT,JPERM
C
C     READ USER-SPECIFIED DIMENSIONS OF Q-ARRAY (MDIMQ) AND OF THE
C     TM-ARRAY (MDIMTM).  READ MAXIMUM NUMBER (MAXIT) OF CALLS TO THE
C     BLSOLV SUBROUTINE ALLOWED IN PHASE 1.
      READ(5,20) EXPLAN
      READ(5,*) MDIMQ, MDIMTM, MAXIT
C
C     READ FLAGS:   EFLAG = (0,1).  EFLAG = 0, MEANS PROGRAM STOPS
C     AFTER COMPLETING PHASE 1 PORTION OF BLOCK LANCZOS PROCEDURE.
C     EFLAG = 1, MEANS PROGRAM COMPLETES BOTH PHASES BEFORE
C     TERMINATING.
C     OFLAG = (0,1).  OFLAG = 0, MEANS THAT IN PHASE 1 PORTION
C     OF THE COMPUTATION, THE PROGRAM DOES NO ORTHOGONALITY CHECKS
C     ON THE Q-BLOCKS GENERATED.  OFLAG = 1 MEANS THAT IN THE
C     PHASE 1 PORTION AND IN THE PHASE 2 PORTIONS OF THE COMPUTATIONS
C     THE PROGRAM CHECKS THE ORTHOGONALITY OF THE Q-BLOCKS GENERATED
C     W.R.T. THAT VECTOR IN THE FIRST BLOCK THAT IS GENERATING
C     DESCENDANTS.  NOTE THAT IN PHASE 2, THE PROGRAM ALWAYS MAKES
C     THIS CHECK OF ORTHGONALITY REGARDLESS OF THE VALUE OF OFLAG.
C     FOR SAFETY, OFLAG SHOULD ALWAYS BE SET TO 1, ALTHOUGH FOR MANY
C     PROBLEMS THIS IS NOT NECESSARY.
      READ(5,20) EXPLAN
```

```
      READ(5,*) EFLAG,OFLAG
C
C     READ SEED USED BY SUBROUTINE GENRAN TO OBTAIN THOSE STARTING
C     VECTORS WHICH ARE GENERATED RANDOMLY.
      READ(5,20) EXPLAN
      READ(5,*) SEED
C
C     SPECIFY MAXIMUM T-SIZE ALLOWED (KMAX-1); INITIAL SIZE OF
C     STARTING BLOCK (KACT);  NUMBER OF STARTING VECTORS SUPPLIED (KSET)
C     SEE BLOCK LANCZOS HEADER FOR COMMENTS REGARDING THE SIZE OF KACT.
      READ(5,20) EXPLAN
      READ(5,*) KMAX,KACT,KSET
C
C     SPECIFY NUMBER (KM) OF EXTREME EIGENVALUES AND EIGENVECTORS
C     OF B-INVERSE TO BE COMPUTED.  THE BLOCK PROCEDURE WORKS WITH THE
C     INVERSE OF THE MATRIX B = SO*P*A*P' + SHIFT*I, USING A
C     FACTORIZATION OF B.  TO INDICATE THAT THE ALGEBRAICALLY-
C     SMALLEST EIGENVALUES OF B-INVERSE ARE BEING COMPUTED SET KM < 0.
C     IF KM < 0, THE PROGRAM ASSUMES THAT BLSOLV SUBROUTINE WHICH
C     THE USER HAS PROVIDED IS COMPUTING -(B-INVERSE)*X
C     INSTEAD OF (B-INVERSE)*X AND INTERNALLY IT COMPUTES THE |KM|
C     ALGEBRAICALLY-LARGEST EIGENVALUES OF -(B-INVERSE).
      READ(5,20) EXPLAN
      READ(5,*) KM
      IF(KM.EQ.0) GO TO 540
      KML = IABS(KM)
C
C     STAGNATION OF CONVERGENCE OF THE KM-TH EIGENVALUE WILL BE
C     TESTED AFTER NSTAG ITERATIONS.  CONVERGENCE WILL BE SAID TO
C     HAVE STAGNATED IF THE RATIO OF THE SQUARE OF THE CURRENT KM-TH
C     RESIDUAL TO THE SQUARE OF THE CORRESPONDING RESIDUAL OBTAINED
C     10 ITERATIONS EARLIER IS GREATER THAN FRACT.  NSTAG SHOULD BE
C     >= 25.  FRACT WAS SET EQUAL TO .01 IN THE TESTS.
      READ(5,20) EXPLAN
      READ(5,*) NSTAG, FRACT
C
C     READ IN THE RELATIVE TOLERANCE (RELTOL) USED TO DETERMINE A
C     CONVERGENCE CRITERION FOR PHASE 2, AND THE MAXIMUM NUMBER (MAXIT2)
C     OF CALLS TO SUBROUTINE BLSOLV ALLOWED IN PHASE 2.
      READ(5,20) EXPLAN
      IF(EFLAG.EQ.1)  READ(5,*)  RELTOL, MAXIT2
C
C     CONSISTENCY CHECKS
C     PROCEDURE REQUIRES ENOUGH ROOM IN THE Q-ARRAY FOR AT LEAST 2
C     BLOCKS OF SIZE KACT PLUS A WORKING VECTOR OF LENGTH N.
      MXBLK = KMAX -1
      MXBLK2 = MXBLK*MXBLK
      IF(MDIMTM.LT.MXBLK2) GO TO 520
      NKMAX = N*KMAX
      IF(MDIMQ.LT.NKMAX)   GO TO 560
      IF(KML.GT.KACT) GO TO 420
      IF(MXBLK.GT.N)  GO TO 440
      IF(2*KACT.GT.MXBLK)  GO TO 500
C
C-----------------------------------------------------------------------
C     DEFINE AND INITIALIZE THE ARRAYS NEEDED TO DEFINE THE
C     FACTORIZATION OF THE B-MATRIX.  PASS THE STORAGE LOCATIONS
C     OF THESE ARRAYS TO THE SUBROUTINE BLSOLV.
C
      CALL USPEC(N,MATNO,NNZ,AVER)
C
C-----------------------------------------------------------------------
C     MASK OVERFLOW AND UNDERFLOW
      CALL MASK
C
C-----------------------------------------------------------------------
C     ARE THERE STARTING VECTORS TO READ IN FROM FILE 10 (KSET.NE.0) ?
```

```
      IF(KSET.EQ.0) GO TO 70
C
      READ(10,30) NOLD,KACT
   30 FORMAT(I6,I4)
      IF(NOLD.NE.N.OR.KSET.GT.KACT) GO TO 460
      DO 50 J=1,KSET
      READ(10,20) EXPLAN
      READ(10,40) EVAL,RESID
   40 FORMAT(E20.12,E13.4)
      READ(10,20) EXPLAN
      LINT= (J-1)*N + 1
      LFIN = J*N
   50 READ(10,60) (Q(JL), JL = LINT,LFIN)
   60 FORMAT(4E20.12)
C
   70 CONTINUE
C
C     WRITE TO A SUMMARY OF THE PARAMETERS FOR THIS RUN TO FILE 6
C
      MXBLK = KMAX - 1
      WRITE(6,80)  N, NNZ, AVER, MATNO
   80 FORMAT(/4X,'ORDER OF B-MATRIX ',5X,'AVERAGE NUMBER NONZEROES PER R
     1OW IN FACTOR'/
     1I15,E47.4/3X,'CRUDE ESTIMATE OF SIZE NONZERO ENTRIES',5X,'MATRIX I
     1D'/E31.4,I21/)
C
      WRITE(6,90)  SO, SHIFT
   90 FORMAT(/4X,'SCALE USED ON A-MATRIX',5X,'SHIFT USED ON A-MATRIX'/
     1E26.4,E27.4/)
C
      WRITE(6,100) MDIMQ, MDIMTM
  100 FORMAT(/18X,'USER-SPECIFIED'/2X,'MAX. DIMENSION Q-ARRAY',4X,'MAX.
     1DIMENSION TM-ARRAY'/I16,I26/)
C
      WRITE(6,110)  OFLAG, EFLAG
  110 FORMAT(/4X,'OFLAG',4X,'EFLAG'/I8,I9/)
C
      IF(OFLAG.EQ.1)  WRITE(6,120)  SPREC
  120 FORMAT(/4X,'ORTHOGONALITY TEST TOLERANCE'/E25.2)
C
      IF(EFLAG.EQ.1)  WRITE(6,130) MAXIT,RELTOL,MAXIT2
  130 FORMAT(/4X,' MAXIT ',8X,'  RELTOL  ',6X,' MAXIT2 '/I10,E20.6,I12/)
      IF(EFLAG.EQ.0)  WRITE(6,140) MAXIT
  140 FORMAT(/4X,' MAXIT '/I10/)
C
      WRITE(6,150)  SEED
  150 FORMAT(/' SEED FOR RANDOM NUMBER GENERATOR'/I24/)
C
      IF(KM.GT.0)  WRITE(6,160) KML
  160 FORMAT(/' COMPUTE THE',I3,' ALGEBRAICALLY-LARGEST EIGENVALUES AND
     1CORRESPONDING VECTORS'/' OF THE INVERSE OF B =  (SO*P*A*P-TRANS +
     1HIFT*I)'/)
      IF(KM.LT.0)  WRITE(6,170) KML
  170 FORMAT(/' COMPUTE THE',I3,' ALGEBRAICALLY-SMALLEST EIGENVALUES AND
     1 CORRESPONDING VECTORS'/' OF THE INVERSE OF THE MATRIX B = (SO*P*A
     1*P-TRANS + SHIFT*I).'/' PROGRAM ASSUMES THAT USER IS PROVIDING -(B
     1-INVERSE)*X INSTEAD OF (B-INVERSE)*X'/' AND COMPUTES THE ALGEBRAIC
     1ALLY-LARGEST EIGENVALUES OF -(B-INVERSE).'/' HOWEVER ON EXIT, FILE
     1 15 CONTAINS THE ALGEBRAICALLY-SMALLEST EIGENVALUES'/' OF B-INVERS
     1E, THE CORRESPONDING EIGENVALUES OF THE ORIGINAL A-MATRIX'/' AND T
     1HE CORRESPONDING EIGENVECTORS OF A.'/)
C
C     NOTE THAT THE ESTIMATE FOR AVER IN THE INVERSE CASE IS VERY CRUDE
C     COMPUTE PHASE 1 CONVERGENCE TOLERANCE
      IF(AVER.GE.1.)
     1ERRMAX = 2.D0*DFLOAT(N+1000)*NNZ*AVER*MACHEP
      IF(AVER.LT.1.)
```

```
     1ERRMAX = 2.D0*DFLOAT(N+1000)*NNZ*AVER**2*MACHEP                   BLI02640
C                                                                       BLI02650
      WRITE(6,180) KACT,MXBLK,KSET                                      BLI02660
  180 FORMAT(/' ON INITIAL ITERATIONS, THE FIRST BLOCK CONTAINS ',I3,' VBLI02670
     1ECTORS'/' HOWEVER THE SIZE OF THE FIRST BLOCK MAY CHANGE AS THE ITBLI02680
     1ERATIONS PROCEED'/' THE MAXIMUM SIZE T-MATRIX THAT CAN BE GENERATEBLI02690
     1D IS ',I4/' THE USER SUPPLIED ',I3,' STARTING VECTORS'/)          BLI02700
C                                                                       BLI02710
      WRITE(6,190)                                                      BLI02720
  190 FORMAT(/' ITERATIVE PROCEDURE'/' PROCEDURE MONITORS THE SIZES OF TBLI02730
     1HE NORM(GRADIENTS)**2 ON EACH'/' ITERATION.  CONVERGENCE IS SAID  BLI02740
     1TO HAVE OCCURRED WHEN ALL'/' RELEVANT (NORMS)**2 ARE LESS THAN ERRBLI02750
     1MAX',E10.3/' PHASE 1 ERRMAX MAY YIELD SOMEWHAT LESS THAN SINGLE PRBLI02760
     1ECISION ACCURACY.'/'  PHASE 2 REFINES THE VECTORS OBTAINED ON PHASBLI02770
     1E 1, ACCORDING TO'/' THE ACCURACY SPECIFIED BY THE USER'/)        BLI02780
C                                                                       BLI02790
      WRITE(6,200) ERRMAX                                               BLI02800
  200 FORMAT(//' PHASE 1 CONVERGENCE CRITERION, ERRMAX '/E22.3/)        BLI02810
C                                                                       BLI02820
C-----------------------------------------------------------------------BLI02830
C     PASS STORAGE LOCATIONS OF VARIOUS ARRAYS TO LANCZS AND LANCI1     BLI02840
C     SUBROUTINES                                                       BLI02850
C                                                                       BLI02860
      CALL LANZP(DIR,DESC,SM,TM,TOD,TD,G,XLFT,LEFT,SPREC)               BLI02870
      CALL LANCP1(DIR,DESC,TM,SM,XLFT,LEFT)                             BLI02880
C                                                                       BLI02890
C-----------------------------------------------------------------------BLI02900
C                                                                       BLI02910
C     ENTER PHASE 1 OF BLOCK LANCZOS PROCEDURE.  BLOCK PROCEDURE        BLI02920
C     HAS 2 POSSIBLE PHASES.  USER SPECIFIES PHASE 1 ONLY OR PHASE 1    BLI02930
C     AND PHASE 2 BY SETTING EFLAG = 0 OR 1, RESPECTIVELY. PHASE 1      BLI02940
C     COMPUTES VECTORS THAT ARE USUALLY ACCURATE TO SINGLE PRECISION.   BLI02950
C     PHASE 2 TAKES THE VECTORS OBTAINED IN PHASE 1 AND REFINES THEM.   BLI02960
C     THE USER SPECIFIES THE DEGREE OF REFINEMENT DESIRED BY SELECTING  BLI02970
C     THE VALUE OF RELTOL AND MAXIT2.  BOTH PHASES SHOULD BE USED.      BLI02980
      IPHASE = 1                                                        BLI02990
      NITER = 0                                                         BLI03000
  210 ITER = 0                                                          BLI03010
      RESIDL(1) = FRACT                                                 BLI03020
      RESIDL(2) = NSTAG                                                 BLI03030
C                                                                       BLI03040
C-----------------------------------------------------------------------BLI03050
C     CALL INITIATES THE BLOCK LANCZOS PROCEDURE.                       BLI03060
C     ON RETURN EIGENVALUE APPROXIMATIONS ARE IN E(I),                  BLI03070
C     I = 1,KACT, IN ALGEBRAICALLY DECREASING ORDER.   CORRESPONDING    BLI03080
C     EIGENVECTOR APPROXIMATIONS ARE IN FIRST N*KACT LOCATIONS IN       BLI03090
C     THE Q-ARRAY.                                                      BLI03100
C                                                                       BLI03110
      CALL LANCZS(BLSOLV,KML,KSET,KACT,MXBLK,N,Q,E,RESIDL,RESIDK,ERRMAX,BLI03120
     1 IPHASE,NITER,IWRITE)                                             BLI03130
C                                                                       BLI03140
C-----------------------------------------------------------------------BLI03150
C                                                                       BLI03160
      IF(IPHASE.EQ.MPMIN) WRITE(15,220) N,KACT                          BLI03170
  220 FORMAT(2I10,' PHASE 2 TERMINATED '/' PROGRAM INDICATES ACCURACY SPBLI03180
     1ECIFIED BY USER IS NOT ACHIEVABLE'/)                              BLI03190
C                                                                       BLI03200
      ITERA = IABS(ITER)                                                BLI03210
      IF(IWRITE.NE.MPMIN.AND.ITER.GT.0)  WRITE(6,230) IPHASE,ITERA      BLI03220
  230 FORMAT(/1X,'PHASE COMPLETED',5X,' NUMBER CALLS TO BLSOLV SUBROUTINBLI03230
     1E USED'/I10,I32)                                                  BLI03240
C                                                                       BLI03250
      IF(IWRITE.EQ.MPMIN.OR.ITER.LT.0)  WRITE(6,240) IPHASE,ITERA       BLI03260
  240 FORMAT(/1X,'PHASE TERMINATED',5X,' NUMBER CALLS TO BLSOLV SUBROUTIBLI03270
     1NE USED'/I10,I32)                                                 BLI03280
C                                                                       BLI03290
```

```
      IF(ITER.GT.0.AND.IWRITE.NE.MPMIN)  GO TO 270
C
      IF(ITER.LT.0)  WRITE(6,250)
  250 FORMAT(//' SMALL EIGENVALUE SUBROUTINE DEFAULTED'/' BLOCK LANCZOS
     1 PROCEDURE STOPS AFTER SAVING CURRENT EIGENVECTOR APPROXIMATIONS'/
     1/)
C
      WRITE(15,260)
      WRITE(6,260)
  260 FORMAT(//' BLOCK LANCZOS PROCEDURE TERMINATES WITHOUT CONVERGENCE
     1'/' USER SHOULD EXAMINE OUTPUT TO DETERMINE REASONS FOR TERMINATIO
     1N'//)
C
C     WRITE EIGENVALUE AND EIGENVECTOR APPROXIMATIONS CONTAINED IN
C     THE FIRST Q-BLOCK TO FILE 15
C
  270 IF(IPHASE.EQ.1)  WRITE(15,280) N,KACT,SEED
  280 FORMAT(I6,I4,I12,' PHASE 1, ORDER A-MATRIX, SIZE OF Q(1), SEED')
      IF(IPHASE.EQ.2)  WRITE(15,290) N,KACT,SEED
  290 FORMAT(I6,I4,I12,' PHASE 2, ORDER A-MATRIX, SIZE OF Q(1), SEED')
C
C     PERMUTE THE EIGENVECTORS IF NECESSARY
      IF(JPERM.EQ.0) GO TO 310
      LINT = -N + 1
      KACT1 = KACT*N + 1
      DO 300 J = 1,KACT
      LINT = LINT + N
C-----------------------------------------------------------------------
      IPERM = 2
      CALL LPERM(Q(LINT),Q(KACT1),IPERM)
C-----------------------------------------------------------------------
  300 CONTINUE
C
C     COMPUTE THE EIGENVALUES OF THE A-MATRIX
  310 DO 320 J = 1,KACT
      IF(KM.LT.0)    E(J) = -E(J)
      TD(J) = 1.D0/E(J)
  320 TD(J) = (TD(J) - SHIFT)/S0
C
C     NOTE THAT RESIDUAL PRINTED OUT CORRESPONDS TO VALUE OBTAINED
C     PRIOR TO FINAL PROJECTION Q(1)-TRANSPOSE*AQ(1) DONE BEFORE
C     TERMINATION
      JJ=KACT
      LINT = -N + 1
      LFIN = 0
      DO 340 J=1,KACT
      LINT = LINT + N
      LFIN = LFIN + N
      JJ=JJ+1
C
C     NOTE THAT RESIDUAL PRINTED OUT CORRESPONDS TO VALUE OBTAINED
C     PRIOR TO FINAL PROJECTION Q(1)-TRANSPOSE*(B-INVERS)*Q(1) DONE
C     BEFORE TERMINATION
C
      WRITE(15,330) E(J), SM(JJ),TD(J)
  330 FORMAT(/E20.12,E13.4,E20.12,'BI-EVAL,ER**2,A-EVAL,A-EVEC'/)
  340 WRITE(15,350) (Q(L), L=LINT,LFIN)
      WRITE(15,360)
  350 FORMAT(4E20.12)
  360 FORMAT(/' ABOVE ARE COMPUTED APPROXIMATE EIGENVECTORS'/)
C
      IF(ITER.GT.MAXIT) WRITE(15,370) ITER,MAXIT
  370 FORMAT(//' PROCEDURE TERMINATED BECAUSE NUMBER OF CALLS TO BLSOLV
     1 SUBROUTINE',I6/' EXCEEDED MAXIMUM NUMBER ',I6,' ALLOWED'//)
C
      IF(ITER.LT.0) WRITE(15,380)
  380 FORMAT(//' USER BEWARE.  EIGENELEMENT COMPUTATIONS DEFAULTED BECAU
```

```
     1SE'/'   EISPACK SUBROUTINE DEFAULTED.   EIGENVALUE AND EIGENVECTOR
     1 APPROXIMATIONS'/'   ABOVE WERE THOSE AVAILABLE AT THE TIME OF DEF
     1AULT'/' SOMETHING IS SERIOUSLY WRONG.'//)
C
C      CHECK FOR TERMINATION AFTER PHASE 1
C      ITER < 0 MEANS EISPACK SUBROUTINE DEFAULTED
C      IPHASE = MPMIN MEANS THAT PHASE 2 TERMINATED DUE TO ORTHOGONALITY
C      IWRITE = MPMIN MEANS THAT CONVERGENCE APPEARS TO HAVE STAGNATED
C      ITER > MAXIT MEANS MAXIMUM NUMBER OF CALLS TO BLSOLV
C           ALLOWED BY USER WAS EXCEEDED
      IF(ITER.LT.0.OR.ITER.GT.MAXIT) GO TO 580
      IF(IPHASE.EQ.MPMIN.OR.IWRITE.EQ.MPMIN) GO TO 580
      IF(EFLAG.NE.1.OR.IPHASE.EQ.2) GO TO 580
C
C      ENTER 2ND PHASE OF COMPUTATION TO ATTEMPT TO OBTAIN MORE
C      ACCURATE EIGENVECTOR APPROXIMATIONS.
C      USER CONTROLS THE SIZE OF THE ERROR TOLERANCE BY SPECIFYING
C      THE PARAMETER RELTOL.
C
      IPHASE = 2
      MAXIT = MAXIT2
      KSET = KACT
C
C      ERROR TOLERANCE USES THE CONVERGED EIGENVALUE LARGEST IN
C      MAGNITUDE.
      TD(1) = DABS(E(1))
      IF(KML.EQ.1) GO TO 400
      DO 390 J = 2,KML
  390 IF(DABS(E(J)).GT.TD(1))    TD(1) = DABS(E(J))
  400 TD(1) = DMAX1(TD(1),1.D0)
      ERRMAN = RELTOL**2 * TD(1)**2
      IF(ERRMAN.GE.ERRMAX) GO TO 480
      ERRMAX = ERRMAN
C
      WRITE(6,410) ERRMAX, MAXIT2
  410 FORMAT(//' ENTER PHASE 2 OF COMPUTATION'/' CONVERGENCE CRITERION I
     1S REDUCED TO ',E13.4/' NO MORE THAN ',I5,' CALLS TO SUBROUTINE BLS
     1OLV WILL BE ALLOWED.'/' PROGRAM WILL TERMINATE IF BLOCK ORTHGONALI
     1TY PROBLEMS MATERIALIZE'/)
C
      GO TO 210
C
C      INCONSISTENCIES IN THE DATA
C
  420 WRITE(6,430) KM,KACT
  430 FORMAT(/' PROGRAM TERMINATES BECAUSE THE NUMBER OF EIGENELEMENTS
     1REQUESTED, KM =',I3/'  IS LARGER THAN THE SIZE OF THE FIRST Q BLOC
     1K, KACT =',I3,' SPECIFIED'/'  USER MUST RESET KM OR KACT'/)
      GO TO 580
C
  440 WRITE(6,450) KMAX,N
  450 FORMAT(/' PROGRAM TERMINATES BECAUSE KMAX = ',I5,' IS TOO LARGE FO
     1R THE SIZE, N = ',I5,', OF THE GIVEN MATRIX'/'  USER MUST DECREASE
     1THE SIZE OF KMAX.'/)
      GO TO 580
C
  460 WRITE(6,470) NOLD,N,KACT,KSET
  470 FORMAT(/' PROGRAM TERMINATES BECAUSE FAULT OCCURRED IN READING IN
     1THE EIGENVECTOR APPROXIMATIONS'/' EITHER THE SIZE MATRIX SPECIFIED
     1ON THE EIGENVECTOR FILE' ,I6/'  DID NOT MATCH THE SIZE SPECIFIED '
     1,I5,' IN THE PROGRAM OR THE NUMBER'/'    OF VECTORS IN FILE 10 = '
     1,I4,' IS LESS THAN THE NUMBER ',I3/'   USER SAID WERE THERE'/)
      GO TO 580
C
  480 WRITE(6,490)  ERRMAN, ERRMAX
  490 FORMAT(/' COMPUTED PHASE 2 CONVERGENCE CRITERION ',E13.4/'  IS LAR
     1GER THAN PHASE 1 CRITERION ',E13.4/'  SO PROGRAM TERMINATES'/)
```

```
      GO TO 580
C
  500 WRITE(6,510)  KACT,MXBLK
  510 FORMAT(/' PROGRAM TERMINATES BECAUSE THERE IS NOT ENOUGH ROOM TO
     1GENERATE 2 BLOCKS',' BECAUSE KACT = ',I3,' AND MXBLK = ', I4/)
      GO TO 580
C
C
  520 WRITE(6,530)  MDIMTM, MXBLK
  530 FORMAT(/' PROGRAM TERMINATES BECAUSE THE DIMENSION ',I6,' OF THE T
     1M ARRAY'/'   IS TOO SMALL FOR THE LARGEST T-MATRIX ALLOWED ',I4)
      GO TO 580
C
  540 WRITE(6,550)
  550 FORMAT(/' USER SPECIFIED NUMBER OF EIGENVALUES OF INTEREST AS 0'/'
     1 PROGRAM TERMINATES FOR USER TO RESET KM TO DESIRED NONZERO VALUE'
     1/)
      GO TO 580
C
  560 WRITE(6,570)  MDIMQ, KMAX,N
  570 FORMAT(/' PROGRAM TERMINATES BECAUSE THE DIMENSION ',I6,' OF THE Q
     1-ARRAY'/'   IS TOO SMALL TO HOLD ',I5, ' VECTORS OF LENGTH ',I4)
      GO TO 580
C
  580 CONTINUE
C
      STOP
C-----END OF MAIN PROGRAM FOR INVERSE BLOCK LANCZOS PROCEDURE-----------
      END
```

SECTION 9.3 SAMPLE MATRIX-VECTOR MULTIPLY SUBROUTINES

```
C---BLIMULT-(INVERSES OF REAL SYMMETRIC MATRICES)-----------------------
C
C     CONTAINS SUBROUTINES  LANCZS AND SAMPLE USPEC AND BLSOLV
C     USED BY THE VERSION OF THE BLOCK LANCZOS ALGORITHMS FOR
C     FACTORED INVERSES OF REAL SYMMETRIC MATRICES, BLIVAL.
C     NOTE THAT SAMPLE BLSOLV FOR SPARSE MATRICES ASSUMES THAT
C     B-MATRIX IS POSITIVE DEFINITE AND USES CHOLESKY FACTORS.
C     HOWEVER, THE USER CAN DIRECTLY REPLACE THAT SUBROUTINE BY
C     A SUBROUTINE FOR INDEFINITE MATRICES THAT COMPUTES THE
C     GENERALIZED FACTORIZATION L*D*(L-TRANSPOSE).
C
C     NONPORTABLE CONSTRUCTIONS:
C     1.  THE ENTRY MECHANISM USED TO PASS THE STORAGE LOCATIONS
C         OF THE FACTORIZATION OF THE MATRIX THAT WILL BE USED
C         BY THE LANCZS SUBROUTINE TO THE SUBROUTINE BLSOLV.
C     2.  IN THE SAMPLE USPEC AND BLSOLV SUBROUTINES PROVIDED:
C         THE FREE FORMAT (7,*), THE FORMAT (20A4) USED FOR
C         READING EXPLANATORY COMMENTS IN THE MATRIX SPECIFICATION
C         FILES, AND THE HEX FORMAT (4Z20) USED IN THE USPECS.
C     3.  THE COMMON BLOCK: LOOPS
C
C-----USPEC FOR FACTORED INVERSES OF REAL SYMMETRIC MATRICES------------
C
      SUBROUTINE CUSPEC(N,MATNO,NNZ,AVER)
C     SUBROUTINE USPEC(N,MATNO,NNZ,AVER)
C
C-----------------------------------------------------------------------
      DOUBLE PRECISION  BD(2200),BSD(10000),NNZ,AVER
      INTEGER  KCOL(2200),KROW(10000),IPR(2200),IPT(2200)
C-----------------------------------------------------------------------
C     THIS SAMPLE SUBROUTINE ASSUMES THAT B IS POSITIVE DEFINITE
C     USER COULD REPLACE BY SIMILAR SUBROUTINE FOR GENERAL FACTORIZATION
C     DIMENSIONS ARRAYS NEEDED TO DEFINE CHOLESKY FACTOR OF B-MATRIX,
C     READS CHOLESKY FACTOR FROM FILE 7, AND THEN PASSES STORAGE
C     LOCATIONS OF THESE ARRAYS TO THE B-MATRIX SOLVE SUBROUTINE BLSOLV.
C
C     HERE WE HAVE B = P*C*P' = L*L'  WHERE  C = SO*A + SHIFT*I.
C     P IS A PERMUTATION MATRIX DEFINED BY THE VECTOR MAPS IPR AND IPT.
C     THE ITH ROW OF B CORRESPONDS TO THE JTH ROW OF C (A) WHERE
C     J = IPR(I) AND I = IPT(J).
C
C     THE B-CHOLESKY FACTOR IS STORED IN THE FOLLOWING SPARSE FORMAT:
C     N = ORDER OF THE B-MATRIX.
C     NZT = NUMBER OF NONZERO SUBDIAGONAL ENTRIES IN THE CHOLESKY
C           FACTOR, L.
C     KCOL(J), J=1,N IS THE NUMBER OF NONZERO SUBDIAGONAL ELEMENTS IN
C              COLUMN J OF L.
C     KROW(K), K=1,NZT IS THE ROW INDEX FOR CORRESPONDING ENTRY BSD(K).
C     BD(J), J = 1,N CONTAINS THE DIAGONAL ENTRIES OF L.
C     BSD(K), K =1,NZT CONTAINS THE NONZERO SUBDIAGONAL ENTRIES OF L
C     JPERM = (0,1):  1 MEANS CHOLEKSY FACTOR CORRESPONDS TO
C                     PERMUTED C.  0 MEANS NO PERMUTATION WAS USED.
C-----------------------------------------------------------------------
C     READ CHOLESKY FACTOR FROM FILE 7.  MUST BE STORED
C     IN SPARSE MATRIX FORMAT.
      READ(7,10) NZT,NOLD,NZL,MATOLD,JPERM
   10 FORMAT(I10,2I6,I8,I6)
C
      WRITE(6,20) NZT,NZL,N,NOLD,MATOLD,JPERM
   20 FORMAT(' HEADER, CHOLESKY FACTOR FILE'/
     1 3X,'NZT',3X,'NZL',5X,'N',2X,'NOLD',2X,'MATOLD',1X,'JPERM'/
     1 4I6,I8,I6/)
C
```

```
      IF (N.NE.NOLD.OR.MATNO.NE.MATOLD) GO TO 100
C
      READ(7,30) (KCOL(K), K = 1,NZL)
      READ(7,30) (KROW(K), K = 1,NZT)
   30 FORMAT(13I6)
      READ(7,40) (BD(K), K = 1,N)
      READ(7,40) (BSD(K), K = 1,NZT)
   40 FORMAT(4Z20)
C  20 FORMAT(3E25.16)
C
C     DOES CHOLESKY FACTOR CORRESPOND TO PERMUTED B?
      IF(JPERM.EQ.0)  GO TO 60
      READ(7,30) (IPR(K), K = 1,N)
C
      DO 50 K = 1,N
      J = IPR(K)
   50 IPT(J) = K
C-----------------------------------------------------------------------
      CALL LPERME(IPR,IPT,N)
C-----------------------------------------------------------------------
   60 CONTINUE
C
C     COMPUTE NNZ, THE AVERAGE NUMBER OF NONZEROS PER COLUMN, AND
C     AVER, THE AVERAGE SIZE OF NONZERO ENTRIES IN THE FACTORS
C     OF THE B-MATRIX.  FROM THIS, ESTIMATE (TOO CRUDELY) THE
C     AVERAGE FOR B-INVERSE AS AVER = 1/AVER.
      ITCOL = 0
      AVER = 0.D0
      DO 70 K = 1,N
      IF(DABS(BD(K)).EQ.0.D0)  GO TO 70
      ITCOL = ITCOL + 1
      AVER = AVER + DABS(BD(K))
   70 CONTINUE
      NTCOL = ITCOL
      DO 80 K = 1,N
   80 ITCOL = ITCOL + 2*KCOL(K)
      NNZ = DFLOAT(ITCOL)/DFLOAT(N)
      DO 90 K = 1,NZS
   90 AVER = AVER + DABS(BSD(K))
      AVER = AVER/DFLOAT(NZS + NTCOL)
      AVER = 1.D0/AVER
C
C-----------------------------------------------------------------------
C     PASS STORAGE LOCATIONS OF FACTORS TO INVERSION SUBROUTINE BLSOLV
      CALL BSOLVE(BSD,BD,KCOL,KROW,N,NZT,NZL)
C-----------------------------------------------------------------------
C
      GO TO 120
C
  100 CONTINUE
C     DEFAULT EXIT
      WRITE(6,110)
  110 FORMAT(/' TERMINATE.  PARAMETERS IN CHOLESKY FACTOR FILE'/
     1' DO NOT AGREE WITH THOSE SPECIFIED BY THE USER'/)
      STOP
C
  120 CONTINUE
C-----END OF USPEC------------------------------------------------------
      RETURN
      END
C
C-----BLSOLV-(FACTORED INVERSES OF REAL SYMMETRIC MATRICES)-------------
C
C     SUBROUTINE BLSOLV(V,U)
      SUBROUTINE CBSOLV(V,U)
C
C-----------------------------------------------------------------------
```

```
      DOUBLE PRECISION  BD(1),BSD(1),U(1),V(1),TEMP,ZERO,ONE
      INTEGER  KCOL(1),KROW(1)
      COMMON/LOOPS/MAXIT,ITER
C-----------------------------------------------------------------------
      ITER = ITER + 1
      ZERO = 0.0D0
      ONE  = 1.0D0
C     SOLVE B*U = V FOR U WHERE  B = L*L'
C     SET U = V. FIRST SOLVE L*U = U FOR U, THEN SOLVE L'*U = U FOR U
      KL = 0
      DO 10 K = 1,N
   10 U(K) = V(K)
      DO 30 K = 1,N
      TEMP = U(K)/BD(K)
      U(K) = TEMP
      IF (KCOL(K).EQ.0.OR.K.EQ.N) GO TO 30
      KF = KL + 1
      KL = KL + KCOL(K)
      DO 20 KK = KF,KL
      KR = KROW(KK)
   20 U(KR) = U(KR) - TEMP*BSD(KK)
   30 CONTINUE
      NP1 = N+1
      KF = NZT + 1
      DO 50 K = 1,N
      L = NP1 - K
      TEMP = U(L)
      IF (KCOL(L).EQ.0.OR.L.EQ.N) GO TO 50
      KL = KF - 1
      KF = KF - KCOL(L)
      DO 40 LL = KF,KL
      LR = KROW(LL)
   40 TEMP = TEMP - BSD(LL)*U(LR)
   50 U(L) = TEMP/BD(L)
   60 CONTINUE
C
      RETURN
C
C-----------------------------------------------------------------------
      ENTRY BSOLVE(BSD,BD,KCOL,KROW,N,NZT,NZL)
C-----------------------------------------------------------------------
C
      RETURN
C-----END OF BLSOLV-----------------------------------------------------
      END
C
C-----SUBROUTINES FOR DIAGONAL TEST MATRICES----------------------------
C     BLSOLV AND USPEC SUBROUTINES FOR DIAGONAL TEST MATRICES
C
C-------BLSOLV DIAGONAL TEST MATRIX-------------------------------------
C
C     SUBROUTINE DBSOLV(V,U)
      SUBROUTINE BLSOLV(V,U)
C
C-----------------------------------------------------------------------
      DOUBLE PRECISION  V(1),U(1),D(1)
      COMMON/LOOPS/MAXIT,ITER
C-----------------------------------------------------------------------
      ITER = ITER + 1
   10 DO 20 I=1,N
   20 U(I)= V(I)/D(I)
C  20 U(I)= -V(I)/D(I)
C
   30 CONTINUE
      RETURN
C
```

```
C-----------------------------------------------------------------------
C     BELOW ENTRY IS FOR A DIAGONAL TEST MATRIX
      ENTRY DSOLVE(D,N)
C-----------------------------------------------------------------------
      RETURN
C-----END OF 'DIAGONAL' TEST MATRIX  BLSOLV-----------------------------
      END
C
C-----START OF USPEC FOR DIAGONAL TEST MATRIX---------------------------
C
      SUBROUTINE USPEC(N,MATNO,NNZ,AVER)
C     SUBROUTINE DUSPEC(N,MATNO,NNZ,AVER)
C
C-----------------------------------------------------------------------
      DOUBLE PRECISION  D(1000),DI(1000),SHIFT,SPACE,NNZ,AVER
      DOUBLE PRECISION  DABS, DFLOAT
      REAL  EXPLAN(20)
C-----------------------------------------------------------------------
C
      READ(7,10) EXPLAN
   10 FORMAT(20A4)
      READ(7,*) NOLD,NUNIF,SPACE,D(1),SHIFT
      NNUNIF = NOLD - NUNIF
      WRITE(6,20) NOLD,SPACE,NNUNIF,D(1),SHIFT
   20 FORMAT(/' DIAGONAL TEST MATRIX, SIZE = ',I4/' IS THE INVERSE OF MA
     1TRIX WITH MOST ENTRIES',E10.3/' UNITS APART AND WITH ',I3,' ENTRIE
     1S IRREGULARLY SPACED'/' FIRST ENTRY WAS ',E13.4,' SHIFT = ',E10.3
     1/)
C
      IF(N.NE.NOLD) GO TO 120
C     COMPUTE THE UNIFORM PORTION OF THE SPECTRUM
      DO 30 J=2,NUNIF
   30 D(J) = D(1) - DFLOAT(J-1)*SPACE
      NUNIF1=NUNIF + 1
      READ(7,10)  EXPLAN
      DO 40 J=NUNIF1,N
   40 READ(7,*) D(J)
      NB = NUNIF - 2
C
      IF(SHIFT.EQ.0.) GO TO 60
      DO 50 J=1,N
   50 D(J) = D(J) + SHIFT
C
C     COMPUTE EIGENVALUES OF INVERSE FOR PRINTOUT ONLY
   60 DO 70 J = 1,N
   70 DI(J) = 1.D0/D(J)
      WRITE(6,80) (J,DI(J), J=1,N )
   80 FORMAT(/' INVERSE BLOCK LANCZOS TEST, LANCZS USES INVERSE OF GIVEN
     1MATRIX'/' ENTRIES OF INVERSE OF DIAGONAL TEST MATRIX = '/(I4,E20.1
     12,I4,E20.12,I4,E20.12))
C
C     DIAGONAL GENERATION COMPLETE
C
C     COMPUTE NNZ AND AVER
      NNZ = 1.D0
      AVER = 0.D0
      DO 90 K = 1,N
   90 AVER = AVER + DABS(DI(K))
      AVER = AVER/DFLOAT(N)
      AVER = 1.D0/AVER
C
C     COMPUTE THE GAPS
      N1 = N-1
      DO 100 K = 1,N1
  100 DI(K) = DI(K+1) - DI(K)
      WRITE(6,110) (K,DI(K), K=1,N1)
  110 FORMAT(/' GAPS BETWEEN EIGENVALUES'/(I4,E13.4,I4,E13.4,I4,E13.4,I4
```

```
     1,E13.4))
C
C-----------------------------------------------------------------------
C      PASS STORAGE LOCATIONS OF D AND N TO DSOLV SUBROUTINE
       CALL DSOLVE(D,N)
C-----------------------------------------------------------------------
C
       RETURN
   120 WRITE(6,130) NOLD,N
   130 FORMAT(' PROGRAM TERMINATES BECAUSE NOLD = ',I5,'DOES NOT EQUAL N
     1 =',I5)
C-----END OF USPEC SUBROUTINE FOR 'DIAGONAL' TEST MATRICES--------------
       STOP
       END
```

SECTION 9.4 FILE DEFINITIONS AND SAMPLE INPUT FILE

Below is a listing of the input/output files which are accessed by the real symmetric block Lanczos eigenvalue/eigenvector program, BLIEVAL. BLIEVAL calculates a few extreme eigenvalues and corresponding eigenvectors of the inverse of a real symmetric matrix (scaled and shifted). Also below is a sample of the input file which BLIEVAL requires on file 5. The parameters in this file are supplied in free format.

```
SAMPLE DEFINITIONS OF THE INPUT/OUTPUT FILES FOR BLIEVAL
-----------------------------------------------------------------
*BLIEVAL EXEC
FI 06 TERM
FILEDEF  5 DISK BLIEVAL   INPUT     A (RECFM F LRECL 80 BLOCK 80
FILEDEF  8 DISK &1        INPUT     A (RECFM F LRECL 80 BLOCK 80
FILEDEF 10 DISK &1        BLSTARTV  A (RECFM F LRECL 80 BLOCK 80
FILEDEF 13 DISK &1        BLEXTRAV  A (RECFM F LRECL 80 BLOCK 80
FILEDEF 15 DISK &1        BLEIGVEC  A (RECFM F LRECL 80 BLOCK 80
*IMTQL2 AND TRED2 ARE 2 EISPACK LIBRARY SUBROUTINES
LOAD  BLIEVAL  BLSUB  BLIMULT  IMTQL2  TRED2
------------------------------------------------------------------
```

```
SAMPLE INPUT FILE FOR BLIEVAL
------------------------------------------------------------------
 LINE 1 IWRITE (SPECIFY MESSAGE LEVEL TO FILE 6: 1 MEANS DETAILED
            1
 LINE 2 N     MATNO    SO    SHIFT   JPERM (SIZE,ID,SCALE,SHIFT,PERM?
      1250   1250     1.       0.       0
 LINE 3 MDIMQ          MDIMTM       MAXIT (DIMS. Q, TM, MAX A*X'S
       40000             2500        1000
 LINE 4 EFLAG  OFLAG ( EFLAG=(0,1) 1=2PHASES. OFLAG: 1=ORTHOG CHECK
            1      1
 LINE 5 SEED      (STARTING VECTOR SEED, RANDOM NUMBER GENERATOR
    3482736
 LINE 6 KMAX  KACT  KSET (MAX T SIZE +1,SIZE 1ST BLOCK,VECTORS SUPPLIED
         31     3     0
 LINE 7 KM (NUMB. EVS FOR ALG-LARGEST, -(NUMB. EVS) FOR ALG-SMALLEST
         3
 LINE 8 NSTAG  FRACT (NO. ITNS BEFORE TEST CONVERGENCE, TEST FRACTION
          25    .05
 LINE 9 RELTOL           MAXIT2 (PHASE 2,CONVERGE.TOL.,NO. A*X'S ALLOWED
     .00000001             1000
----------------------------------------------------------------------
```

REFERENCES[a]

Björck, Å (1967). Solving linear least squares problems by Gram-Schmidt orthogonalization. *BIT* **7**, 1-21.

Bunch, J. R., and Kaufman, L. (1977). Some stable methods for calculating inertia and solving symmetric linear systems. *Math. Comp.* **31**, 163-179.

Cullum, J., and Donath, W. E. (1974). A block Lanczos algorithm for computing the q algebraically largest eigenvalues and a corresponding eigenspace for large, sparse symmetric matrices. *Proc. 1974 IEEE Conference on Decision and Control,* 505-509.

Cullum, J., and Willoughby, R. A. (1984). Lanczos Algorithms, Large Symmetric Matrices: Volume 1, Theory. Progress in Scientific Computing Series, Birkhauser, Boston.

EISPACK Guide (1976). Matrix Eigensystem Routines - EISPACK Guide. Smith, B. T., Boyle, J. M., Garbow, B. S., Ikebe, Y., Klema, V. C., and Moler, C. B. *Lecture Notes in Computer Science* **6**, Second Edition, Springer, New York.

EISPACK Guide (1977). Matrix Eigensystem Routines - EISPACK Guide Extension. Garbow, B. S., Boyle, J. M., Dongarra, J. J., and Moler, C. B. *Lecture Notes in Computer Science* **51**, Second Edition, Springer, New York.

George, A., and Liu, J. W. H. (1981). Computer Solution of Large Sparse Positive Definite Systems. Prentice-Hall, Englewood Cliffs, NJ.

Golub, G. H., and Kahan, W. (1965). Calculating the singular values and pseudoinverse of a matrix. *SIAM J. Numer. Anal.* **2**, 205-224.

Golub, G. H., Luk, F. T., and Overton, M. L. (1981). A block Lanczos method for computing the singular values and corresponding singular vectors of a matrix. *ACM TOMS* **7**, 149-169.

Lanczos, C. (1961). Linear Differential Operators. Van Nostrand, New York.

LINPACK User's Guide. (1979). Dongarra, J. J., Bunch, J. R., Moler, C. B., and Stewart, G. W. SIAM Press, Philadelphia.

Moro, G., and Freed, J. H. (1981). Calculation of ESR spectra and related Fokker-Planck forms by the use of the Lanczos algorithm. *J. Chem. Phys.* **74**, 3757-3773.

Paige, C. C. (1971). The computation of eigenvalues and eigenvectors of very large sparse matrices. *Ph.D. Thesis,* U. London.

Paige, C. C. (1972). Computational variants of the Lanczos method for the eigenproblem. *J. Inst. Math. Appl.* **10**, 373-381.

Paige, C. C. (1976). Error analysis of the Lanczos algorithms for tridiagonalizing a symmetric matrix. *J. Inst. Math. Appl.* **18**, 341-349.

Paige, C. C. (1980). Accuracy and effectiveness of the Lanczos algorithm for the symmetric eigenproblem. *Linear Algebra Appl.* **34**, 235-258.

PFORT Verifier (1981). Ryder, B. G., and Hall, A. D. Bell Laboratory Technical Report 12.

SPARSPAK User Guide (1979). George, A., Liu, J. W. H., and Ng, E. CS Dept. Tech. Report, U. Waterloo.

Stewart, G. W. (1973). Introduction to Matrix Computations. Academic Press, New York.

Wilkinson, J. H. (1965). The Algebraic Eigenvalue Problem. Oxford University Press, New York.

[a] Extended set of references is contained in Volume 1.

Index

REAL SYMMETRIC MATRICES

SINGLE VECTOR CODES

HERMITIAN MATRICES

SINGLE-VECTOR CODES

FACTORED INVERSES, REAL SYMMETRIC MATRICES

SINGLE-VECTOR CODES

REAL, SYMMETRIC GENERALIZED PROBLEMS

SINGLE-VECTOR CODES

REAL RECTANGULAR MATRICES

SINGLE-VECTOR CODES

NONDEFECTIVE, COMPLEX SYMMETRIC MATRICES

SINGLE-VECTOR CODES

REAL SYMMETRIC MATRICES

BLOCK CODES

FACTORED INVERSES, REAL SYMMETRIC MATRICES

BLOCK CODES